The Defocusing Nonlinear Schrödinger Equation

The Defocusing Nonlinear Schrödinger Equation

From Dark Solitons to Vortices and Vortex Rings

Panayotis G. Kevrekidis
University of Massachusetts
Amherst, Massachusetts

Dimitri J. Frantzeskakis
University of Athens
Athens, Greece

Ricardo Carretero-González
San Diego State University
San Diego, California

siam.
Society for Industrial and Applied Mathematics
Philadelphia

10 9 8 7 6 5 4 3 2 1

Publisher	David Marshall
Acquisitions Editor	Elizabeth Greenspan
Developmental Editor	Gina Rinelli
Managing Editor	Kelly Thomas
Production Editor	Louis R. Primus
Copy Editor	Matthew Bernard
Production Manager	Donna Witzleben
Production Coordinator	Cally Shrader
Compositor	Techsetters, Inc.
Graphic Designer	Lois Sellers

Library of Congress Cataloging-in-Publication Data
Kevrekidis, Panayotis G.
The defocusing nonlinear Schrödinger equation : from dark solitons to vortices and vortex rings / Panayotis G. Kevrekidis, University of Massachusetts, Amherst, Massachusetts, Dimitri J. Frantzeskakis, University of Athens, Athens, Greece, Ricardo Carretero-González, San Diego State University, San Diego, California.
pages cm. – (Other titles in applied mathematics ; 143)
Includes bibliographical references and index.
ISBN 978-1-611973-93-8
1. Gross-Pitaevskii equations. 2. Schrödinger equation. I. Frantzeskakis, Dimitri J. II. Carretero-González, Ricardo. III. Title.
QC174.26.W28K4785 2015
530.1201'515353–dc23
2015016537

To Maria, Despina, and Athena;
Eleni, Sofia, and Apostolos-Ioannis;
Alexandra and Adrián,

for their love and support, and the joy
they bring to our lives,

and to our parents,
for all that they have done for us.

Contents

Preface

The phenomenon of Bose–Einstein condensation is a phase transition originally predicted by Bose and Einstein in 1924. In particular, it was shown that below a critical temperature T_c, a finite fraction of particles of a boson gas condenses into the same quantum state, known as the Bose–Einstein condensate (BEC). Although Bose–Einstein condensation was known to be a fundamental phenomenon, connected, e.g., to superfluidity and superconductivity, BECs were only experimentally realized 70 years later. This major achievement took place in 1995 and has already been recognized through the 2001 Nobel Prize in Physics. This first unambiguous manifestation of a macroscopic quantum state in a many-body system sparked an explosion of activity, as reflected by the publication of several thousand papers related to BECs since then. Nowadays there exist more than a hundred experimental BEC groups around the world, while an enormous amount of theoretical work has followed and driven the experimental efforts, with an impressive impact on many branches of physics.

From a theoretical standpoint, and for experimentally relevant conditions, the static and dynamical properties of a BEC can be described by means of an effective mean-field equation known as the Gross–Pitaevskii equation (GPE). This model is of the form

$$i\hbar\frac{\partial\Psi}{\partial t} = -\frac{\hbar^2}{2m}\Delta\Psi + g|\Psi|^2\Psi + V_{\text{ext}}(\mathbf{r})\Psi, \tag{1}$$

where $\Psi = \Psi(\mathbf{r}, t)$ is the BEC wavefunction (the atomic density is proportional to $|\Psi|^2$), Δ is the Laplacian, m is the atomic mass, and the prefactor g is proportional to the atomic scattering length. Notice that $g > 0$ (e.g., for ^{87}Rb and ^{23}Na atoms) or $g < 0$ (e.g., for ^{7}Li and ^{85}Rb atoms) corresponds, respectively, to repulsive or attractive interatomic interactions; accordingly, the nonlinearity in the GPE model is defocusing (repulsive) or focusing (attractive).

The GPE is a variant of the famous nonlinear Schrödinger (NLS) equation incorporating an external confining potential, V_{ext}. The NLS equation is known to be a universal model describing the evolution of complex field envelopes in nonlinear dispersive media. As such, the NLS equation is a key model appearing in a variety of physical contexts, ranging from optics to fluid dynamics and plasma physics, while it has also attracted much interest from a mathematical viewpoint. The relevance and importance of the NLS model is not limited to the case of conservative systems; in fact, the NLS equation is directly connected to dissipative universal models, such as the complex Ginzburg–Landau equation, which have been studied extensively in the context of pattern formation. In the case of BECs, the nonlinearity in the GPE (NLS) model is introduced by the interatomic interactions through an effective mean-field.

Importantly, the mean-field approach, and the study of the GPE, allows the prediction and description of important and experimentally relevant nonlinear effects and nonlinear waves, such as solitons and vortices. These so-called matter-wave solitons and vortices can be viewed as fundamental nonlinear excitations of BECs, and as such have attracted considerable attention. Importantly, they have also been observed in many elegant experiments using various relevant techniques. These include, among others, phase engineering in order to create vortices or dark matter-wave solitons in them, the stirring (or rotation) of the condensates creating vortices and vortex lattices, and the change of scattering length (from repulsive to attractive via Feshbach resonances) to produce bright matter-wave solitons and soliton trains. As far as vortices and vortex lattices are concerned, it should be noted that their description and connection to phenomena as rich and profound as superconductivity and superfluidity were one of the themes of 2003 Nobel Prize in Physics.

In an earlier project entitled "Emergent Nonlinear Phenomena in Bose–Einstein Condensates," we asked a number of experts to contribute their perspectives on a series of select topics at the interface of nonlinear science and atomic physics, within the exciting juncture of BECs. Yet, this was already 2007, and the level of developments in the field during the last 6–7 years has been remarkable. New results have enabled observations of nonlinear phenomena pertaining to the existence, evolution, and mutual interactions of dark solitons, vortices, and vortex rings. For instance, in quasi-1D geometries, studies of dark solitons and their collisions have emerged, and counterflow experiments have revealed spontaneous realizations of symbiotic solitons such as dark-bright and dark-dark ones. Quasi-2D experiments have showcased, through novel minimally destructive imaging techniques, the formation and direct observation of clusters containing a few vortices, and have offered a profound understanding of vortices, their motions induced by inhomogeneities of the background density, and their mutual interactions. Finally, the state-of-the-art experimental techniques are on the verge of being able to extract 3D in situ images that will provide long-standing feedback upon current frontiers, such as vortex lines and vortex rings, their intrinsic motions, and mutual interactions with direct implications to quantum turbulence, among many others.

While "The *Focusing* Nonlinear Schrödinger Equation" has enjoyed its share of attention through, among others, a volume on this very subject, we felt, in light of these recent developments, that "The *Defocusing* Nonlinear Schrödinger Equation" has received considerably less attention in the form of a coherent volume that would collect many of these recent results. The aim of the present book is to bridge this gap, by presenting a wide array of findings on the realm of BECs and on the NLS-type models that arise therein.

In addition to an overarching introduction to the relevant class of models, the book contains a chapter focused on 1D and quasi-1D settings, which is centered around dark-soliton excitations. There is also a chapter on 2D and quasi-2D case examples which, in turn, chiefly evolves around vortices and their dynamics. Finally, we also give a 3D formulation of the relevant system, with a particular aim towards understanding the dynamics of vortex lines and vortex rings. While the presentation is naturally rather biased towards numerous themes of our particular interest and recent work, we hope that the collection of the relevant background, theoretical developments, numerical computations, and experimental results will be of rather wide interest to atomic physicists, nonlinear scientists, and applied mathematicians. Furthermore, we hope that this book will be equally valuable to beginners on the field, as well as to experienced researchers in the area.

Panayotis Kevrekidis	Dimitri Frantzeskakis	Ricardo Carretero-González
Amherst, Massachusetts	Athens, Greece	San Diego, California

February 2015

Acknowledgments

Most of the material included in this manuscript would not have come to fruition without the invaluable contribution of our collaborators (including many of our past and present research students). We are very grateful to all of our collaborators for the time they have lent us towards the fascinating study of nonlinear waves and especially nonlinear Schrödinger-type problems. Thanking each individual would be unwieldy (there are several hundred); however, it is clear in our minds that the research journey we have taken over the last 15 years would not have been possible without the continuous exchange of ideas with all the collaborators, co-authors, and colleagues with whom we have had the pleasure to interact, as well as our students and postdocs due to their timeless effort and energy invested in this program.

We would also like to acknowledge the financial support that has been generously given to us towards our research efforts. In particular we would like to thank the following organizations for their support over the last decade or so: the National Science Foundation (NSF), The University of Athens, the Computational Science Research Center (CSRC) at San Diego State University (SDSU), the Los Alamos National Laboratory (LANL), the Department of Energy (DoE), the Binational Science Foundation, the Alexander von Humboldt Foundation, Germany, the European Research Council (ERC), and the US Air Force Office of Scientific Research (US-AFOSR).

Chapter 1

Introduction

1.1 ▪ Introduction

Nonlinearity is at the heart of the most interesting phenomena in science. It is responsible for the formation and evolution of complex patterns that drive the dynamics of spatio-temporal systems. Much of the relevant pattern formation phenomena arise through the emergence of localized structures with finite spatial support or with sufficiently fast spatial decay. These structures often behave like individual robust entities, or quasiparticles, and hence are dubbed *solitons* or solitary waves. Solitons not only have nontrivial existence and stability properties, but they have complex mutual interactions and/or interactions with inhomogeneities in the medium (created by, for example, external potentials, impurities, or the presence of other nonhomogeneous solutions) and nucleation or merging properties.

Within the realm of modulated amplitude waves, there are two main types of such solitons depending on the nature of the effective nonlinearity (attractive or repulsive) of the underlying medium. In the case of an attractive (or focusing) medium, the nonlinearity leads for the "lumping" of mass/density together to form a coherent structure called a *bright soliton*.[1] While in low-dimensional settings bright solitons are typically robust, the generalization of these solitons to higher dimensions (2D or 3D) often yields to blow-up, where the self-attraction cannot be arrested by the internal structure of the soliton and the solution increases without bound (sometimes even in finite time). On the other hand, in a repulsive (or defocusing) medium, nonlinearity is responsible for the creation of localized troughs or density "dips," the so-called *dark solitons*.[1] These dark solitons typically correspond, in one dimension, to solutions where the field under consideration has a phase jump when going across the density notch. This type of solution constitutes a *topologically* charged structure since it supports a phase jump that is topologically embedded on its domain. The most striking consequence of this topological feature is that these structures are typically structurally stable since no smooth deformation is able to remove the topological phase jump.

This book is devoted to the study of the complex patterns, and especially the nonlinear coherent structures, that arise in the lowest-order nonlinear model for envelope waves: the *nonlinear Schrödinger (NLS) equation*. The NLS, being the *normal form* for envelope waves, represents a *universal* model at the root of an extremely wide range of

[1] The term "bright" ("dark") is coined from the context of optics, where a pertinent optical soliton appears as a light intensity hump (dip) against a zero (finite) light background.

physical and other natural phenomena and applications. Furthermore, due to its rich and complex phenomenology, the NLS is also a paradigm for nonlinear spatio-temporal dynamics and is at the forefront of intense and challenging mathematical research. We start our narrative with a general, self-contained introduction to the basic one-dimensional structure supported by the repulsive, or defocusing, NLS equation: the *dark soliton*. This will also serve as a prelude for the presentation we offer on the higher-dimensional (2D and 3D) extensions of the model, where vortices (2D), vortex lines, and vortex rings (3D) are the basic topologically charged solutions.

As we mentioned above, a dark soliton is a nonlinear solitary wave having the form of a density dip with a phase jump across its density minimum. This localized nonlinear wave exists on top of a stable continuous (or extended finite-width) background. Dark solitons are the most fundamental nonlinear excitations of the universal NLS model with a defocusing nonlinearity, and, as such, they have been studied in diverse branches of physics. Importantly, apart from a vast amount of literature devoted to relevant theoretical works, there exist many experimental results on dark solitons, including the observation of optical dark solitons, either as temporal pulses in optical fibers [1, 2], or as spatial structures in bulk media and waveguides [3, 4], the excitation of a nonpropagating kink in a parametrically driven shallow liquid [5], dark soliton standing waves in a discrete mechanical system [6], high-frequency dark solitons in thin magnetic films [7], dissipative dark solitons in a complex plasma [8], dark solitons in water wave tanks [9], and so on.

Theoretical studies on dark solitons started as early as in 1971 [10] in the context of Bose–Einstein condensates (BECs). In particular, in Ref. [10], exact soliton solutions of the Gross–Pitaevskii equation (GPE) (which is a variant of the NLS model incorporating an external trapping potential) [11] were found and were connected, in the small-amplitude limit, with the solitons of the Korteweg–de Vries (KdV) equation. Later, and shortly after the demonstration of integrability of the focusing NLS equation [12], the defocusing NLS equation was also shown [13] to be completely integrable by means of the inverse scattering transform (IST) [14]; this way, single as well as multiple dark soliton solutions of arbitrary amplitudes were found analytically. The IST approach allowed for an understanding of the formation of dark solitons [15, 16, 17, 18, 19, 20], the interaction and collisions between dark solitons [13, 21] (see also Refs. [22, 23, 24, 25, 26] and [27] for relevant theoretical and experimental studies, respectively), and paved the way for the development of perturbation methods for investigating their dynamics in the presence of perturbations [26, 28, 29, 30, 31, 32, 33]. From a physical standpoint, dark solitons were mainly studied in the field of nonlinear optics—from which the term "dark" was coined. The first theoretical work in this context, namely the prediction of dark solitons in nonlinear optical fibers at the normal dispersion regime [34], was subsequently followed by extensive studies of optical dark solitons [35, 36].

A new era for dark solitons started shortly after the experimental realization of atomic BECs [37, 38, 39]: this achievement was awarded the Nobel Prize in Physics of 2001 [40, 41], and has been recognized as one of the most fundamental recent developments in quantum and atomic physics over the last decades (see, e.g., the books [42, 43] for reviews). In an effort to understand the properties of this exciting state of matter, there has been much interest in the macroscopic nonlinear excitations of BECs (see reviews in Refs. [44, 45]). In that regard, the so-called matter-wave dark solitons were among the first purely nonlinear states that were experimentally observed in BECs [46, 47, 48, 49, 50].

The interest on matter-wave dark solitons is not surprising for a number of reasons. First of all, for BECs confined in parabolic traps, these structures are the nonlinear analogues of the excited states of a "prototype" quantum system [51, 52], namely the quantum harmonic oscillator [53]. On the other hand, the topological nature of matter-wave dark solitons (due to the phase jump at their density minimum) renders them a "degenerate," one-dimensional (1D) analogue of *vortices*, which are of paramount importance in diverse branches of physics [54]. Additionally, and perhaps more importantly, matter-wave dark solitons are—similarly to vortices [55, 56, 57]—quite fundamental structures arising spontaneously upon crossing the BEC phase transition [58, 59], with properties which may be used as diagnostic tools probing the rich physics of a purely quantum system (BEC) at the mesoscale [60]. As concerns applications, it has been proposed that the dark soliton position can be used to monitor the phase acquired in an atomic matter-wave interferometer in the nonlinear regime [61, 62] (see also relevant experiments of Refs. [63, 64] devoted to atom-chip interferometry of BECs).

The early matter-wave dark soliton experiments, as well as previous works on dark solitons mainly in optics, inspired many theoretical efforts towards a better understanding of the stability as well as the static and dynamical properties of matter-wave dark solitons. Thus, it is probably not surprising that a new series of experimental results from various groups have appeared [65, 66, 67, 68, 69, 70, 71, 72, 73], while still other experiments—not directly related to dark solitons—reported observation of these structures [63, 64, 74]. These very recent experimental results were obtained with an unprecedented control over the condensate as compared to the earlier soliton experiments. Thus, these "new age" experiments were able not only to experimentally verify various theoretical predictions, but also to open new exciting possibilities (see, e.g., the recent review of Ref. [75]). Given this emerging interest, and how new experiments in BEC physics inspire novel ideas—both in theory and in experiments—new exciting results are expected to appear.

The present book aims to start by providing an overview of the theoretical and experimental progress on the study of dark solitons. Subsequently, it will focus on higher-dimensional topological structures, including single and multiple vortices [54, 76, 77] as well as vortex rings [78, 79]. We will highlight the close connection between such coherent structures that appear in defocusing NLS systems, with the physical motivation originating from the atomic BECs context. In many cases, we will also discuss similarities between atomic solitons and vortex structures with their corresponding optical siblings (see, e.g., Ref. [80] for a relevant work on dark solitons). Our presentation will cover the basic theory emphasizing, in particular, the connection between the theoretical results and experimental observations; this way, in most cases, the theoretical discussion will be immediately followed by a presentation of pertinent experimental results. Our theoretical approach will chiefly be based on the mean-field theory: as will be shown, the latter can be used as a foundation for building an understanding of most effects and experimental findings related to matter-wave dark solitons; consequently, thermal and quantum effects—which may be particularly relevant and important in certain cases—will only be briefly covered.

The organization of the book is as follows. First, in this chapter, we present a basic introduction which is devoted to an overarching presentation of our setting. In particular, we start with the mean-field description of BECs, present the GPE, and discuss its connection with the respective full quantum many-body problem. Next, we present the ground state of the condensate and discuss how its small-amplitude excitations can be studied by means of the Bogoliubov–de Gennes (BdG) equations. Lower-dimensional versions of the GPE model, pertinent to highly anisotropic trapping potentials, are also discussed; this way, depending on the shape of the trap, we start from purely three-dimensional (3D) BECs and introduce elongated (alias "cigar-shaped") BECs, quasi-one-dimensional (1D) BECs and quasi-two-dimensional (2D) (alias "disk-shaped") ones, as well as discuss cases relevant to the dimensionality crossover regimes. The topics of strongly interacting Bose gases, and their relevant mean-field description, are also briefly covered.

Based on the above basic material we then proceed with the presentation of results for the 1D, 2D, and 3D settings, in Chapters 2, 3, and 4, respectively. As a natural partition of the corresponding presentations, Chapter 2 is devoted to dark solitons, Chapter 3 to vortices, and Chapter 4 to vortex structures in 3D, such as vortex lines and vortex rings. We now briefly summarize the different topics that are covered in these three chapters.

Chapter 2: The one-dimensional case. The presentation of Chapter 2 starts with the description of the theoretical basis for the study of dark solitons. Specifically, first we present the completely integrable 1D NLS equation, its basic properties, and the dark soliton solutions. Relevant mathematical tools, such as inverse scattering transform, the renormalization of the integrals of motion of dark solitons, and the small-amplitude approximation (leading to the connection of shallow dark solitons to KdV solitons) are discussed. Furthermore, the generation of matter-wave dark solitons by means of various methods is also presented. We also provide the multiple-dark soliton solutions of the NLS equation, and discuss their interactions and collisions.

Next, we study matter-wave dark solitons in quasi-1D Bose gases. Particularly, we first discuss the adiabatic dynamics of dark solitons in the presence of a harmonic trap or an optical lattice, by means of the Hamiltonian and Lagrangian approaches of the perturbation theory, the Landau dynamics, and the small-amplitude approximation approaches. Furthermore, we present a connection between the stability, statics, and dynamics of dark solitons, relying on a study of the Bogoliubov spectrum of single- and multiple-dark solitons, and discuss the role of the pertinent anomalous modes. Nonadiabatic effects, namely emission of radiation of solitons in the form of sound waves, as some rigorous results concerning the persistence and stability of matter-wave dark solitons, are also covered.

As a preliminary discussion of the material that will be presented in the next chapters, we also present results on "lower-dimensional" multidimensional dark solitons. In particular, we briefly discuss the transverse (alias "snaking") instability of rectilinear dark solitons in higher dimensions, and the concomitant soliton decay into vortex pairs or vortex rings. The theme of matter-wave dark solitons of radial symmetry, namely ring dark solitons and spherical shell solitons, is also covered. Furthermore, we present results concerning the stability of dark solitons in cigar-shaped (3D) BECs, and in BECs in the dimensionality crossover regime from 3D to 1D; in the latter experimentally relevant setting, both single- and multiple-dark soliton statics and dynamics are analyzed.

In addition, and based on the previous analysis, we discuss various experimentally relevant settings and parameter regimes for multicomponent (pseudospinor and spinor) matter-wave dark solitons in atomic BECs. In this context, first we study the homogeneous system and present dark-bright, as well as regular and beating dark-dark solitons, stemming from SU(2) rotations of the dark-bright ones. Then, following the lines of presentation of previous sections, we discuss multiple vector solitons and their interactions, their dynamics in the trap, as well as the role of dimensionality of the system.

Next, we analyze some "special" 1D settings: in particular, we explore how the breakdown of BEC superfluidity is connected to the generation of matter-wave dark solitons, present results on the spontaneous symmetry breaking phenomenon in double-well potentials, and analyze some settings relevant to the so-called nonlinearity management. Finally, we conclude Chapter 2 by discussing dissipative effects, and particularly the statics and dynamics of dark solitons in single- and multicomponent systems at finite temperatures.

Chapter 3: The two-dimensional case. Chapter 3 pertains to the existence, stability, dynamics, and interactions of coherent structures in 2D, mainly focusing on vortex structures which are the prototypical excited (nonlinear) states in 2D. The chapter begins with a general overview of the basic nonlinear solutions that arise in the 2D defocusing case, namely the radial ground states and their excitations continued from the linear limit, including vortices and other radially symmetric structures. Special emphasis will be placed on the stability and dynamical properties of single charge vortices embedded within the stereotypical harmonic trapping potential that is generically employed in the physical experiments. The spiraling dynamics of vortices due to finite temperature effects will also be covered.

We then present the emergence and stability for the spatial 2D extension of a dark soliton, namely the dark soliton stripe. Our presentation is based on a detailed bifurcation analysis, starting from the linear limit of a small number of atoms, of the dark soliton stripe. Importantly, the dark soliton stripe is connected, through a cascade of symmetry-breaking bifurcations, with vortex-carrying states with increasing number of vortices (vortices with alternating charges arranged in a line where the original dark soliton stripe resided). Other, more complicated vortex arrangements will also be identified as bifurcating from multiple dark-soliton lines and/or crosses and from ring dark soliton stripes as well.

After covering the emergence of vortex structures in the vicinity of the linear limit, we turn to the high density BEC limit where the vortices can be considered as quasiparticles that retain their shape as they evolve interacting with the inhomogeneous background (induced by the presence of an external trapping potential) and interacting with other vortices. The approach we take here is to consider each vortex as a point vortex with effective equations of motion (ordinary differential equations, ODEs) depending only on their positions. Then, using dynamical system tools, we explore the multiple static and dynamical scenarios of this particle model for the case of a small number of vortices (up to 5 or so). Importantly, we emphasize the connection and comparison of the steady states, bifurcations, and dynamics of these configurations, bearing a small number of vortices, with recent experimental results based on state-of-the-art techniques to image small vortex clusters in situ in a nondestructive manner. An alternative approach for the description of steady state vortex configurations based in generating functions and classical polynomials is also presented. Furthermore, we describe a coarse graining technique to treat the large vortex number case by means of following the continuous limit description for the vortex density. This coarse graining approach allows us to start with the microscopic equations of motion (ODEs on the vortex-vortex and vortex-trap interactions) and gradually move towards the steady state density limit of many vortices (typically arranged in vortex lattice configurations).

An attractive aspect of BECs is the broad control that one has on the external potential (not only from the model perspective but also experimentally as well!). A salient feature of this control is the possibility of rendering the trapping potential anisotropic. This potential anisotropy is responsible for the destabilization of some vortex configurations that are stable in the isotropic case or the stabilization of otherwise unstable configurations. The approach we take here in order to address such features is a dynamical systems one, based on bifurcations of the effective particle equations. Another interesting case that emerges from the manipulation of the external potential in 2D is the few-well potential case. Using Galerkin projections we study the features of a few-mode reduction leading to symmetry breaking bifurcations due to the presence of nonlinearity in the system. Furthermore, we describe the combined effects of a harmonic external trap with a periodic optical lattice on the stability and dynamics of single vortices.

In the 2D chapter we also describe the nucleation of vortices by means of different techniques relevant to experiments such as (i) controlling the snaking instabilities of dark soliton stripes, (ii) nonlinear interference of colliding BECs fragments, and (iii) the breakdown of superfluidity from localized defects dragged through the condensate. We also study the possibility of controlling the position and motion of vortices using localized defects as tweezers.

The topic of 2D quantum turbulence in BEC and its connection with the dynamics, formation, and annihilation of vortices is also touched upon. Then, we describe the realm of multicomponent settings where nonlinear structures in one component interact with other structures in the other component(s). Salient examples in this case include vortex-bright and vortex-vortex solitary wave complexes. Finally, we briefly touch upon the fascinating, and very active, topic of exciton polariton condensates. These condensates are characterized by a departure from the Hamiltonian (GPE) model by including loss and gain mechanisms that are inherent to the corresponding experiments.

Chapter 4: The three-dimensional case. Finally, in Chapter 4 we cover the coherent structures that emerge in the defocusing 3D GPE system. We start with the description of vortex lines (or vortex solitons) that are the generalization of 2D vortices by homogeneously extending the latter about their perpendicular direction. We briefly describe the stability and vibrational (Kelvin) modes of vortex lines and their bent counterpart.

We then describe the realm of vortex rings that are vortex lines that are looped back onto themselves. We start by giving an account of the possible mechanisms to generate vortex rings. Next, we describe the dynamics of single vortex rings that have an intrinsic velocity (depending on their charge and radius). Interestingly, the velocity of vortex rings can be dramatically changed by the presence of internal excitations of the vortex ring. Equipped with the dynamics of single vortex rings, we proceed to study their mutual interactions and scattering collisions. Provided the vortex rings are kept well apart, their dynamics can be described using classical tools for fluid dynamics. In particular, we explore the possibility of using the Biot–Savart law for interacting vortex filaments. Using this dynamical reduction we describe the interaction of coaxial vortex rings and compare it favorably with the original GPE model. The dynamics is typically governed by periodic leapfrogging that can be generalized to multiple vortex rings (although the stability is greatly affected). Finally, within the realm of vortex ring interactions, we (numerically) study the scattering of vortex rings. We note that for a wide range of parameters the scattering of vortex rings can be practically considered as an elastic collision where the vortex rings strongly interact (even merge) and then separate following a very simple scattering angle formula.

In Chapter 4 we also cover some effects of the dimensionality on the vortex rings. In particular, elongated (cigar-shaped) condensates support dark solitons that, while evolving in the longitudinal direction, oscillate between effective 1D structures (dark solitons) and 3D vortex rings. We also briefly touch upon multicomponent BECs where the existence of two speeds of sound induces different nucleation regimes when impurities are dragged along the condensate.

Finally, motivated by the intensive numerical simulations needed in the 3D case, we devote a section to describe an efficient numerical method to integrate NLS/GPE-type models. Although this section is placed in the chapter pertaining to 3D structures (where it is more relevant, given the computational expense), we also describe the codes corresponding to the 1D and 2D cases. The numerical method that we put forward is based on the graphical processor units (GPU) concept where it is possible to use a high-end graphics card to parallelize the bulk of the computations involved in the numerical integration of the GPE. For this purpose, we describe a two-step explicit high-order scheme for the

Laplacian, together with a novel modulus squared Dirichlet boundary condition and the numerical stability analysis for the ensuing temporal integrator.

1.2 ▪ Mean-Field Description of Bose–Einstein Condensates

Bose–Einstein condensation of dilute atomic gases is an unambiguous manifestation of a macroscopic quantum state in a many-body system. As such, this phenomenon has triggered an enormous amount of experimental and theoretical work [42, 43]. Importantly, this field is intimately connected with branches of physics such as superfluidity, superconductivity, lasers, coherent optics, nonlinear optics, and physics of nonlinear waves. Many of the common elements between BEC physics and the above areas, and in particular optics, rely on the existence of macroscopic coherence in the many-body state of the system. From a theoretical standpoint, this can be understood by the fact that many effects related to BEC physics can be described by a *mean-field* model, namely the Gross–Pitaevskii equation (GPE) [11]. The latter is a partial differential equation (PDE) of the NLS type, which plays a key role—among other fields—in nonlinear optics [36]. Thus, BEC physics is closely connected to nonlinear optics (and the physics of nonlinear waves), with vortices and solitons being perhaps the most prominent examples of common nonlinear structures arising in these areas [44, 45].

Below we will briefly discuss the theoretical background for the description of BECs. We emphasize, in particular, lowest-order mean-field theory, as this can be used as a basis to understand the nonlinear dynamics of matter-wave dark solitons. Interesting effects naturally arise beyond the GPE mean-field, both due to thermal and quantum fluctuations. Such effects become particularly relevant in extremely tightly confining geometries, or when the Bose–Einstein condensation transition region is approached.

1.2.1 ▪ The Gross–Pitaevskii Equation

In order to describe theoretically the statics and dynamics of BECs a quantum many-body approach is required [42, 43] (see also Ref. [81] for a review on the many-body aspects of BECs). Particularly, a sufficiently dilute ultracold atomic gas, composed of N interacting bosons of mass m confined by an external potential $V_{\text{ext}}(\mathbf{r})$, can be described by the many-body Hamiltonian; the latter can be expressed, in second quantization form, through the boson annihilation and creation field operators $\hat{\Psi}(\mathbf{r},t)$ and $\hat{\Psi}^{\dagger}(\mathbf{r},t)$ (which create and annihilate a particle at the position $\mathbf{r}$), namely,

$$\hat{H} = \int d\mathbf{r}\hat{\Psi}^{\dagger}(\mathbf{r},t)\hat{H}_0\hat{\Psi}(\mathbf{r},t) + \frac{1}{2}\int d\mathbf{r}d\mathbf{r}'\hat{\Psi}^{\dagger}(\mathbf{r},t)\hat{\Psi}^{\dagger}(\mathbf{r}',t)V_{\text{tb}}(\mathbf{r}-\mathbf{r}')\hat{\Psi}(\mathbf{r}',t)\hat{\Psi}(\mathbf{r},t),$$

where $\hat{H}_0 = -(\hbar^2/2m)\nabla^2 + V_{\text{ext}}(\mathbf{r})$ is the single-particle Hamiltonian operator and $V_{\text{tb}}(\mathbf{r}-\mathbf{r}')$ is the two-body interatomic potential. Apparently, the underlying full many-body problem is very difficult to treat analytically (or even numerically) as N increases and, thus, for convenience, a mean-field approach can be adopted. The mean-field approach is based on the separation of the condensate contribution from the boson field operator as follows [82]:

$$\hat{\Psi}(\mathbf{r},t) = \langle\hat{\Psi}(\mathbf{r},t)\rangle + \hat{\Psi}'(\mathbf{r},t) = \Psi(\mathbf{r},t) + \hat{\Psi}'(\mathbf{r},t). \tag{1.1}$$

In the above expression, the expectation value of the field operator, $\langle\hat{\Psi}(\mathbf{r},t)\rangle \equiv \Psi(\mathbf{r},t)$, is known as the *macroscopic wavefunction of the condensate*, while $\hat{\Psi}'(\mathbf{r},t)$ describes the noncondensate part, which accounts for quantum and thermal fluctuations. Considering the case of a dilute ultracold gas with binary collisions at low energy, characterized by the s-wave scattering length a, the interatomic potential can be replaced by an effective delta-function interaction potential, $V_{\text{tb}}(\mathbf{r}'-\mathbf{r}) = g\delta(\mathbf{r}'-\mathbf{r})$ [42, 43], with the coupling constant g given by $g = 4\pi\hbar^2 a/m$. Under these assumptions, a nontrivial zeroth-order theory for the BEC wavefunction can be obtained by means of the Heisenberg evolution equation $i\hbar(\partial\hat{\Psi}/\partial t) = [\hat{\Psi}, \hat{H}]$, upon replacing the field operator $\hat{\Psi}$ with the classical field Ψ, i.e., ignoring the quantum and thermal fluctuations described by $\hat{\Psi}'(\mathbf{r}', t)$. Such a consideration leads to the GPE [11], which has the form

$$i\hbar\frac{\partial}{\partial t}\Psi(\mathbf{r},t) = \left[-\frac{\hbar^2}{2m}\nabla^2 + V_{\text{ext}}(\mathbf{r}) + g|\Psi(\mathbf{r},t)|^2\right]\Psi(\mathbf{r},t). \tag{1.2}$$

In the above equation, $\Psi(\mathbf{r}, t)$ is normalized to the number of atoms N, namely,

$$N = \int |\Psi(\mathbf{r},t)|^2 d\mathbf{r}, \tag{1.3}$$

and the nonlinearity (which is obviously introduced by the interatomic interactions) is characterized by the s-wave scattering length a, which is $a > 0$ or $a < 0$ for repulsive or attractive interatomic interactions, respectively. Notice that Eq. (1.2) can be written in canonical form, $i\hbar\partial_t\Psi = \delta E/\delta\Psi^*$ (with $*$ denoting complex conjugate), where the dynamically conserved energy functional E is given by

$$E = \int\left[\frac{\hbar^2}{2m}|\nabla\Psi|^2 + V_{\text{ext}}|\Psi|^2 + \frac{1}{2}g|\Psi|^4\right]d\mathbf{r}, \tag{1.4}$$

with the three terms in the right-hand side representing, respectively, the kinetic energy, the potential energy, and the interaction energy.

A time-independent version of the GPE can be obtained upon expressing the BEC wavefunction as $\Psi(\mathbf{r}, t) = \Psi_0(\mathbf{r})\exp(-i\mu t/\hbar)$, where $\mu = \partial E/\partial N$ is the chemical potential. This way, Eq. (1.2) yields the following equation for the stationary state Ψ_0:

$$\left[-\frac{\hbar^2}{2m}\nabla^2 + V_{\text{ext}}(\mathbf{r}) + g|\Psi_0|^2(\mathbf{r})\right]\Psi_0(\mathbf{r}) = \mu\Psi_0(\mathbf{r}). \tag{1.5}$$

1.2.2 ▪ The Mean-Field Approach versus the Many-Body Quantum Mechanical Problem

Although the GPE has been known since the early 1960s [11], it was almost 40 years later that it was shown that this model can be derived rigorously from a self-consistent treatment of the respective many-body quantum mechanical problem [83]. In particular, in Ref. [83]—which dealt with the stationary GPE (1.5)—it was proved that the GPE energy functional describes correctly the energy and the particle density of a trapped gas of N bosons to the leading order in the small parameter $\bar{n}|a|^3$, where $\bar{n}$ is the average density of the gas.[2] The above results were proved in the limit where the number of particles $N \to \infty$ and the scattering length $a \to 0$, such that the product Na stays constant. Importantly, although Ref. [83] referred to the full three-dimensional (3D) Bose gas, extensions of this work for lower-dimensional settings were also reported (see the review [84] and references therein).

[2] The condition $\bar{n}|a|^3 \ll 1$, which is also required for the derivation of the GPE (1.2) as per the analysis of the previous section, implies that the Bose-gas is "dilute" or "weakly-interacting"; typically, in BEC experiments, $\bar{n}|a|^3 < 10^{-3}$ [43].

As mentioned above, these results (as well as the ones in Ref. [84]) were proved for stationary solutions of the GPE, and, in particular, for the ground state solution. Shortly after these first efforts, the time-dependent GPE (1.2) was also analyzed within a similar asymptotic analysis in Ref. [85]. In this work, it was proved that the limit points of the k-particle density matrices of $\Psi_{N,t}$ (which is the solution of the N-particle Schrödinger equation) satisfy asymptotically the GPE (and the associated hierarchy of equations) with a coupling constant given by $\int v(x)dx$, where $v(x)$ describes the interaction potential.

The above works paved the way for the rigorous proof of the fact that the time-dependent Gross–Pitaevskii equations describes the dynamics of a trapped BEC. This proof was given in Refs. [86, 87, 88, 89], and the essence of the arguments is outlined in Ref. [86]. For this reason, and since in this latter work the next crucial step on the rigorous validation of the GPE was established, below we will briefly present the main results presented in Ref. [86].

As in the work of Lieb, Seiringer, and Yngvason [83], the starting point of the analysis of Ref. [86] is the effective Hamiltonian of N identical bosons, which can be expressed (in units so that $\hbar = 2m = 1$) as follows:

$$\hat{H}_N = \sum_{j=1}^{N} \left[-\Delta_{\mathbf{r}_j} + V_{\text{ext}}(\mathbf{r}_j) \right] + \sum_{i<j} V_N(\mathbf{r}_i - \mathbf{r}_j), \tag{1.6}$$

where $V_{\text{ext}}(\mathbf{r}) \geq 0$ with $V_{\text{ext}}(\mathbf{r}) \to \infty$ as $|\mathbf{r}| \to \infty$ is a confining potential. For the interaction potential V_N, a nonnegative potential V is fixed and rescaled by defining $V_N(\mathbf{r}) = N^2 V(N\mathbf{r})$. This scaling is chosen so that the scattering length of V_N is of the order $1/N$. The Hamiltonian (1.6) therefore describes a system of N bosons confined by the external potential V_{ext} into a volume of order 1, interacting through a potential with effective range of the order $1/N$. Since the typical distance between neighboring particles is of order $N^{-1/3}$, and thus, much bigger than the effective range of the interaction, Eq. (1.6) describes a very dilute system.

In the work of Ref. [83], and for repulsive interactions, it was proved that the ground state energy per particle in the GPE limit ($V \to \infty$, $a_0 = Na =$ fixed) is a solution of the constrained minimization problem

$$\min \left\{ \mathscr{E}_{\text{GP}} \;:\; ||\phi||_{L^2(\mathbb{R}^3)} = 1 \right\}, \tag{1.7}$$

where

$$\mathscr{E}_{\text{GP}} = \int_{\mathbb{R}^3} \left(|\nabla \phi|^2 + V_{\text{ext}} |\phi|^2 + 4\pi a_0 |\phi|^4 \right) d^3\mathbf{r}, \tag{1.8}$$

is the GPE functional. The extension of this variational result to the case of rotating Bose systems was proved in Ref. [91]. The fact that the ground state of the trapped Bose gas exhibits a complete Bose–Einstein condensation was proved in Ref. [90]. In particular, in this work it was proved that the one-particle density matrix of the ground state converges to the 1D projection $|\phi_{\text{GP}}\rangle\langle\phi_{\text{GP}}|$. The study of the correlation properties of the ground state starts from the work of Dyson [92], proving the upper bound on the ground state energy obtained from Bogoliubov theory, for the homogeneous dilute Bose gas. Almost 40 years later, Lieb and Yngvason justified the lower bound in Ref. [93], by showing that low energy states necessarily contain a characteristic short-scale pair correlation. The trapping potential introduces a second length scale, but it leaves the short-scale structure intact, as also proved in Ref. [83].

The strategy of the proof in Refs. [86, 87] is based on the study of the so-called Bogoliubov-Born-Green-Kirkwood-Yvon hierarchy

$$
\begin{aligned}
i\partial_t \gamma_{N,t}^{(k)} = \sum_{j=1}^{k} \left[-\Delta_{\mathbf{r}_j}, \gamma_{N,t}^{(k)}\right] + \sum_{j=1}^{k} \left[V_N(\mathbf{r}_i - \mathbf{r}_j), \gamma_{N,t}^{(k)}\right] \\
+ (N-k) \sum_{j=1}^{k} \mathrm{Tr}_{k+1} \left[V_N(\mathbf{r}_j - \mathbf{r}_{k+1}), \gamma_{N,t}^{(k+1)}\right]
\end{aligned}
\tag{1.9}
$$

for $k = 1, 2, \ldots, N$. In Eq. (1.9), Tr_{k+1} stands for the partial trace over the $(k+1)$th variable. The key observation is that the sequence of families $\{\gamma_{N,t}^{(k)}\}_{k=1}^{N}$ which are the solutions of (1.9), converge as $N \to \infty$, to limit points $\{\gamma_{\infty,t}^{(k)}\}_{k \geq 1}$ which are solutions of the infinite hierarchy of equations

$$
i\partial_t \gamma_{\infty,t}^{(k)} = \sum_{j=1}^{k} \left[-\Delta_{\mathbf{r}_j}, \gamma_{\infty,t}^{(k)}\right] + B^{(k)} \gamma_{\infty,t}^{(k+1)}, \tag{1.10}
$$

where

$$
B^{(k)} \gamma_{\infty,t}^{(k+1)} = 8\pi a_0 \sum_{j=1}^{k} \mathrm{Tr}_{k+1} \left[\delta(\mathbf{r}_j - \mathbf{r}_{k+1}), \gamma_{\infty,t}^{(k+1)}\right] \tag{1.11}
$$

is the collision operator. Furthermore, the product ansatz $\gamma_{\infty,t}^{(k)} = |\phi\rangle\langle\phi|$ is a solution of Eq. (1.10) if and only if ϕ is a solution of the GPE

$$
i\partial_t \phi = -\Delta\phi + 8\pi a_0 |\phi|^2 \phi, \tag{1.12}
$$

with initial data $\phi_{t=0} = \phi_{\mathrm{GP}}$—the ground state minimizer of Eqs. (1.7) and (1.8). The key steps in proving the convergence of Eq. (1.9) to Eq. (1.10) consist of finding the appropriate topology in which this convergence holds; in other words, to recover the appropriate normed space of density matrices to study the convergence as well as the uniqueness of the limit solution of Eq. (1.10). For this highly nontrivial step, a Poincaré-type inequality of mixed Sobolev norms has a crucial role, as well as for the derivation of the necessary energy estimates on Eq. (1.9).

These rigorous results, as well as a large number of experimental results related to the physics of BECs, indicate that (under certain conditions) the GPE is a good starting point for analyzing the statics and dynamics of BECs.

1.2.3 ▪ Ground State and Excitations of the Condensate

Let us now consider a condensate confined in a harmonic external potential, namely,

$$
V_{\mathrm{ext}}(\mathbf{r}) = \frac{1}{2} m(\omega_x^2 x^2 + \omega_y^2 y^2 + \omega_z^2 z^2), \tag{1.13}
$$

where ω_x, ω_y, and ω_z are the (generally different) trap frequencies along the three directions. In this setting, and in the case of repulsive interatomic interactions ($a > 0$) and sufficiently large number of atoms N, Eq. (1.5) can be used to determine analytically the *ground state* of the system. In particular, in the asymptotic limit of $Na/a_{\mathrm{ho}} \gg 1$ (where

$a_{\rm ho} = \sqrt{\hbar/(m\omega_{\rm ho})}$ is the harmonic oscillator length associated with the geometrical average $\omega_{\rm ho} = (\omega_x \omega_y \omega_z)^{1/3}$ of the trap frequencies), it is expected that the atoms are pushed towards the rims of the condensate, resulting in slow spatial variations of the density profile $n(\mathbf{r}) \equiv |\Psi_0(\mathbf{r})|^2$. Thus, the latter can be obtained as an algebraic solution stemming from Eq. (1.5) when neglecting the kinetic energy term—the so-called *Thomas–Fermi* (TF) limit [42, 43, 44]:

$$n(\mathbf{r}) = g^{-1}[\mu - V_{\rm ext}(\mathbf{r})], \tag{1.14}$$

in the region where $\mu > V_{\rm ext}(\mathbf{r})$, and $n = 0$ outside, and the value of μ being determined by the normalization condition [cf. Eq. (1.3)]. Notice that the TF approximation becomes increasingly accurate for larger values of μ.

Small-amplitude excitations of the BEC can be studied upon linearizing Eq. (1.5) around the ground state. Particularly, we consider small perturbations of this state, i.e.,

$$\Psi(\mathbf{r},t) = e^{-i\mu t/\hbar}\left[\Psi_0(\mathbf{r}) + \sum_j \left(u_j(\mathbf{r})e^{-i\omega_j t} + \upsilon_j^*(\mathbf{r})e^{i\omega_j t}\right)\right], \tag{1.15}$$

where u_j and υ_j are the components of the linear response of the BEC to the external perturbations that oscillate at frequencies $\pm\omega_j$ [these are (generally complex) eigenfrequencies]. Substituting Eq. (1.15) into Eq. (1.5), and keeping only linear terms in u_j and υ_j, we obtain the so-called Bogoliubov–de Gennes (BdG) equations:

$$\begin{aligned}
\left[\hat{H}_0 - \mu + 2g\,|\Psi_0|^2(\mathbf{r})\right]u_j(\mathbf{r}) + g\,\Psi_0^2(\mathbf{r})\upsilon_j(\mathbf{r}) &= \hbar\,\omega_j\,u_j(\mathbf{r}),\\
\left[\hat{H}_0 - \mu + 2g\,|\Psi_0|^2(\mathbf{r})\right]\upsilon_j(\mathbf{r}) + g\,\Psi_0^{*2}(\mathbf{r})u_j(\mathbf{r}) &= -\hbar\,\omega_j\,\upsilon_j(\mathbf{r}),
\end{aligned} \tag{1.16}$$

where $\hat{H}_0 \equiv -(\hbar^2/2m)\nabla^2 + V_{\rm ext}(\mathbf{r})$ is the single-particle Hamiltonian. Importantly, these equations can also be used, apart from the ground state, for any other stationary state (including, e.g., solitons) with the function Ψ_0 being chosen accordingly. In such a general context, the BdG equations provide the eigenfrequencies $\omega \equiv \omega_r + i\omega_i$ and the amplitudes u_j and υ_j of the normal modes of the system. Note that due to the Hamiltonian nature of the system, if ω is an eigenfrequency of the Bogoliubov spectrum, so are $-\omega$, ω^* and $-\omega^*$. In the case of stable configurations with $\omega_i = 0$, the solution of BdG equations with frequency ω represents the same physical oscillation as the solution with frequency $-\omega$ [43].

In the case of a homogeneous gas [$V_{\rm ext}(\mathbf{r}) = 0$] characterized by a constant density $n_0 = |\Psi_0|^2$, the amplitudes u_j and υ_j in the BdG equations are plane waves, $\sim \exp(i\mathbf{k}\cdot\mathbf{r})$, of wave vector $\mathbf{k}$. Then, Eqs. (1.16) lead to the dispersion relation

$$(\hbar\omega)^2 = \left(\frac{\hbar^2\mathbf{k}^2}{2m}\right)\left(\frac{\hbar^2\mathbf{k}^2}{2m} + 2gn_0\right). \tag{1.17}$$

In the case of repulsive interatomic interactions ($g > 0$), Eq. (1.17) indicates that small-amplitude harmonic excitations of the stationary state

$$\Psi = \sqrt{n_0}\exp(-i\mu t/\hbar), \tag{1.18}$$

with $\mu = n_0$, are always stable since $\omega_i = 0$ for every $\mathbf{k}$. Thus, this state is not subject to the modulational instability (see, e.g., Ref. [94] and references therein). This fact is important, as the wavefunction of Eq. (1.18) can serve as a stable background (alias "pedestal"), on top of which strongly nonlinear localized excitations may be formed; such excitations may be, e.g., matter-wave dark solitons (or vortices and vortex rings in 2D and 3D, respectively) which are of particular interest herein. Notice that the above mentioned small-amplitude harmonic excitations are in fact sound waves, characterized by the phonon dispersion relation $\omega = |\mathbf{k}|c_s$ [see Eq. (1.17) for small momenta $\hbar\mathbf{k}$], where

$$c_s = \sqrt{gn_0/m} \tag{1.19}$$

is the speed of sound. We should note in passing that in the case of attractive interatomic interactions ($g < 0$), the speed of sound becomes imaginary, which indicates that long wavelength perturbations grow or decay exponentially in time. Thus, the stationary state of Eq. (1.18) is subject to the modulational instability, which is responsible for the formation of matter-wave bright solitons [95, 96, 97] in attractive BECs (see also the reviews [44, 45, 94, 98] and references therein).

1.2.4 ▪ Lower-Dimensional Condensates and Relevant Mean-Field Models

Let us consider again a condensate confined in the harmonic trap of Eq. (1.13). In this case, the trap frequencies set characteristic length scales for the spatial size of the condensate through the harmonic oscillator lengths $a_j \equiv (\hbar/m\omega_j)^{1/2}$ ($j \in \{x,y,z\}$). Another important length scale, introduced by the effective mean-field nonlinearity, is the so-called healing length defined as $\xi = (8\pi n_0 a)^{-1/2}$ (with n_0 being the maximum condensate density). The healing length, being the scale over which the BEC wavefunction "heals" over defects, sets the spatial scale corresponding to the width of nonlinear excitations, such as matter-wave dark solitons and vortices.

Based on the above, as well as the form of the ground state [cf. Eq. (1.14)], it is clear that the shape of the BEC is controlled by the relative values of the trap frequencies. For example, if $\omega_x = \omega_y \equiv \omega_\perp \approx \omega_z$ (i.e., for an isotropic trap), the BEC is almost spherical, while for $\omega_z < \omega_\perp$ (i.e., for an anisotropic trap) the BEC is "cigar shaped." It is clear that such a cigar-shaped BEC

(a) may be a purely 3D object,

(b) acquires an almost 1D character (for strongly anisotropic traps with $\omega_z \ll \omega_\perp$ and $\mu \ll \hbar\omega_\perp$), or

(c) is in the so-called dimensionality crossover regime from 3D to 1D.

These regimes can be described by the dimensionless parameter [99]

$$d = N\Omega\frac{a}{a_\perp}, \tag{1.20}$$

where $\Omega = \omega_z/\omega_\perp$ is the so-called "aspect ratio" of the trap. Particularly, if the dimensionality parameter is $d \gg 1$, the BEC locally retains its original 3D character (although it may have an elongated, quasi-1D shape) and its ground state can be described by the TF approximation in all directions. On the other hand, if $d \ll 1$, excited states along the transverse direction are not energetically accessible and the BEC is effectively 1D. This regime is extremely useful for an analytical study of matter-wave dark solitons. Finally, if $d \approx 1$, the BEC is in the crossover regime between 1D and 3D, which is particularly relevant as recent matter-wave dark soliton experiments have been conducted in this regime [70, 72].

Let us now discuss in more detail lower-dimensional mean-field models describing cigar-shaped BECs. First, we consider the quasi-1D regime ($d \ll 1$) characterized by an extremely tight transverse confinement. In this case, following Refs. [51, 100, 101], the BEC wavefunction is separated into transverse and longitudinal components, namely $\Psi(\mathbf{r},t)=\Phi(r;t)\psi(z,t)$. Then, the transverse component $\Phi(r;t)$ is described by the Gaussian ground state of the transverse harmonic oscillator (and, thus, the transverse width of the condensate is set by the transverse harmonic oscillator length $a_\perp$), while the longitudinal wavefunction $\psi(z,t)$ obeys the following effectively 1D GPE:

$$i\hbar\partial_t\psi(z,t)=\left[-\frac{\hbar^2}{2m}\partial_z^2+V(z)+g_{1D}|\psi(z,t)|^2\right]\psi(z,t), \tag{1.21}$$

where the effective 1D coupling constant is given by $g_{1D}=g/2\pi a_\perp^2=2a\hbar\omega_\perp$ and $V(z)=(1/2)m\omega_z^2z^2$. Notice that in the case under consideration, if the additional condition $[(N/\sqrt{\Omega})(a/a_\perp)]^{1/3}\gg 1$ is fulfilled, then the longitudinal condensate density $n(z,t)\equiv|\psi(z,t)|^2$ can be described by the TF approximation—see Eq. (1.14) with μ now being the 1D chemical potential (and $g\to g_{1D}$) [99]. Following the terminology of Ref. [70], this regime will hereafter be referred to as the TF-1D regime.

Next, let us consider the effect of the deviation from one-dimensionality on the longitudinal condensate dynamics. In this case, the wavefunction can be factorized as before, but with the transverse component Φ assumed to depend also on the longitudinal variable z (and time t) [102, 103, 104, 105]. Physically speaking, it is expected that the transverse direction will no longer be occupied by the ground state, but Φ would still be approximated by a Gaussian function with a width $w=w(z,t)$ that can be treated as a variational parameter [103, 104, 105]. This way, it is possible to employ different variational approaches and derive the following NLS equation for the longitudinal wavefunction,

$$i\hbar\frac{\partial\psi}{\partial t}=\left[-\frac{\hbar^2}{2m}\frac{\partial^2}{\partial z^2}+V(z)+f(n)\right]\psi. \tag{1.22}$$

The nonlinearity function $f(n)$ in Eq. (1.22) depends on the longitudinal density $n(z,t)$ and may take different forms. Particularly, in Ref. [103] (where variational equations related to the minimization of the action functional were used) $f(n)$ is found to be

$$f(n)=\frac{g}{2\pi a_\perp^2}\frac{n}{\sqrt{1+2an}}+\frac{\hbar\omega_\perp}{2}\left(\frac{1}{\sqrt{1+2an}}+\sqrt{1+2an}\right), \tag{1.23}$$

and the respective NLS equation is known as the nonpolynomial Schrödinger equation (NPSE). On the other hand, in Refs. [104, 105] (where variational equations related to the minimization of the transverse chemical potential were used) the result for $f(n)$ is

$$f(n)=\hbar\omega_\perp\sqrt{1+4an}, \tag{1.24}$$

and the respective NLS equation will be called the Gerbier–Muñoz Mateo–Delgado equation (GMDE). Since, as explained above, the derivation of the mean-field models with the nonlinearity functions in Eqs. (1.23) and (1.24) is based on different approaches, these nonlinearity functions are quite different. Nevertheless, they can be "reconciled" in the weakly interacting limit of $an\ll 1$: in this case, the width of the transverse wavefunction becomes $w=a_\perp$ and Eq. (1.22)—with either the nonlinearity function of Eq. (1.23) or that of Eq. (1.24)—is reduced to the 1D GPE model of Eq. (1.21).

The above effective 1D models predict accurately ground state properties of quasi-1D condensates, such as the chemical potential, the axial density profile, the speed of sound,

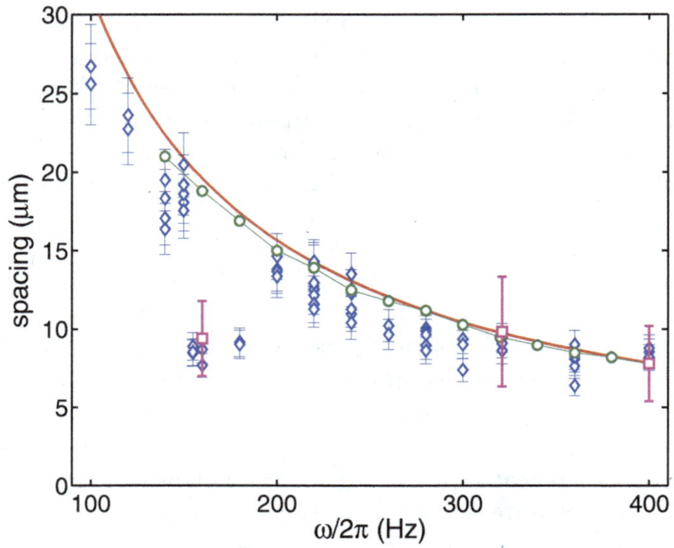

Figure 1.1. *Average spacing of adjacent maxima of the longitudinal Faraday patterns in a periodically excited BEC as a function of the transverse driving frequency. Blue diamonds depict the experimental data of Ref. [107]. Green circles correspond to spacings extracted from the 1D NPSE numerics while magenta squares correspond to the spacings extracted from the full 3D numerics. The red solid line shows the results from the NPSE from an analytical modulational stability analysis. Adapted with permission from Ref. [108].*

collective oscillations, and others. Importantly, these models are particularly useful in the dimensionality crossover regime, where they describe the axial dynamics of cigar-shaped BECs in a very good approximation to the 3D GPE (see, e.g., theoretical work related to matter-wave dark solitons in Ref. [106] and relevant experimental results in Ref. [70]). We will further elaborate on the application of these models below (see Sec. 2.3.5, where statics and dynamics of dark solitons will be discussed).

Nevertheless, it would be relevant to present, at this point, an example not connected with solitons but illustrating the usefulness of the NPSE in accurately describing the effective 1D dynamics of a 3D setup from the model GPE and the experimental perspective. This example concerns the work presented in Ref. [108] where the NPSE was used (1.22)–(1.23) to describe the Faraday patterns observed in the experiment of Ref. [107]. In this experiment, a quasi-1D BEC was periodically excited by modulating the transverse trapping strength and thus forming higher density stripes along the longitudinal extent of the BEC in a manner akin to the original Faraday fluid experiment. In Fig. 1.1 we depict the main results of Ref. [108] that present a comparison of the spacing of the Faraday stripes as a function of the periodic forcing of the transverse confinement. Specifically, note the very good agreement between the NPSE results (green circles), the full 3D GPE numerics (magenta squares), and the actual experiments (blue diamonds).

Extremely weak deviations from one-dimensionality can also be treated by means of a rather simple noncubic nonlinearity that can be obtained by Taylor expanding $f(n)$, namely,

$$f(n) = g_1 n - g_2 n^2, \tag{1.25}$$

where $g_1 = g_{1D}$ and g_2 depends on the form of $f(n)$. In this case, Eq. (1.22) becomes a cubic-quintic NLS (cqNLS) equation. This model was derived self-consistently in

Ref. [102], where dynamics of matter-wave dark solitons in elongated BECs was considered; there, the coefficient g_2 was found to be equal to $g_2 = 24\ln(4/3)a^2\hbar\omega_\perp$.

Here, it is worth mentioning that the quintic term in the cqNLS equation may have a different physical interpretation, namely to describe *three-body interactions*, regardless of the dimensionality of the system. In this case, the coefficients g_{1D} and g_2 in Eq. (1.25) are generally complex, with the imaginary parts describing *inelastic* two- and three-body collisions, respectively [109]. As concerns the rate of the three-body collision process, it is given by $(dn/dt) = -Ln^3$ [42], where L is the loss rate (which is of order of 10^{-27}–10^{-30} cm^{6}s^{-1} for various species of alkali atoms [110]). Accordingly, the decrease of the density is accounted for by the term $-i(L/2)|\psi|^4\psi$ in the time-dependent GPE, i.e., to the quintic term in the cqNLS equation.

It is also relevant to note that the NLS Eq. (1.22) has also been used as a mean-field model describing strongly interacting 1D Bose gases and, particularly, the so-called Tonks-Girardeau gas of impenetrable bosons [111] (see also Refs. [112, 113] for relevant experimental observations). In this case, the function $f(n)$ takes the form [114]

$$f(n) = \frac{\pi^2\hbar^2}{2m}n^2, \tag{1.26}$$

and, thus, Eq. (1.22) becomes a quintic NLS equation. Although the applicability of this equation has been criticized (as in certain regimes it fails to predict correctly the coherence properties of the strongly interacting 1D Bose gases [115]), the corresponding hydrodynamic equations for the density n and the phase φ arising from the quintic NLS equation under the Madelung transformation $\psi = \sqrt{n}\exp(i\varphi)$ are well documented in the context of the local density approximation [116]. Additionally, it should be noticed that the time-independent version of the quintic NLS equation has been rigorously derived from the many-body Schrödinger equation [117].

We finally mention that another lower-dimensional version of the fully 3D GPE can be derived for "disk-shaped" (alias "pancake") condensates confined in strongly anisotropic traps with $\omega_\perp \ll \omega_z$ and $\mu \ll \hbar\omega_z$. In such a case, a procedure similar to the one used for the derivation of Eq. (1.21) leads to the following $(2+1)$-dimensional NLS equation:

$$i\hbar\partial_t\psi(x,y,t) = \left[-\frac{\hbar^2}{2m}\nabla_\perp^2 + V(r) + g_{2D}|\psi(x,y,t)|^2\right]\psi(x,y,t), \tag{1.27}$$

where $r^2 = x^2 + y^2$, $\nabla_\perp^2 = \partial_x^2 + \partial_y^2$, the effectively 2D coupling constant is given by $g_{2D} = g/\sqrt{2\pi}a_z = 2\sqrt{2\pi}aa_z\hbar\omega_z$, while the potential is given by $V(r) = (1/2)m\omega_\perp^2 r^2$. It should also be noticed that other effective 2D mean-field models (involving systems of coupled 2D equations [103] or 2D GPEs with generalized nonlinearities [103, 105]) have also been proposed for the study of the transverse dynamics of disk-shaped BECs.

The above models will be used below to investigate the static and dynamical properties of matter-wave dark solitons, and higher-dimensional analogues thereof, arising in the respective settings.

Chapter 2

The One-Dimensional Case

2.1 ▪ General Background for the Study of Dark Solitons

2.1.1 ▪ NLS Equation and Dark Soliton Solutions

We start by considering the case of a quasi-1D condensate described by Eq. (1.21). These can be expressed in the following dimensionless form:

$$i\partial_t \psi(z,t) = \left[-\frac{1}{2}\partial_z^2 + V(z) + |\psi(z,t)|^2\right]\psi(z,t), \tag{2.1}$$

where the density $|\psi|^2$, length, time, and energy are respectively measured in units of $2a$, $a_\perp$, $\omega_\perp^{-1}$, and $\hbar\omega_\perp$, while the external potential $V(z)$ is given by

$$V(z) = \frac{1}{2}\Omega^2 z^2. \tag{2.2}$$

In the case under consideration, the normalized trap strength (aspect ratio) is $\Omega \ll 1$ and, thus, as a first step in our analysis, the potential $V(z)$ is ignored.[3] In such a case, the condensate is homogeneous and can be described by the *completely integrable* defocusing NLS equation [13] (see also the review [35]):

$$i\partial_t \psi(z,t) = \left[-\frac{1}{2}\partial_z^2 + |\psi(z,t)|^2\right]\psi(z,t). \tag{2.3}$$

This equation possesses an infinite number of conserved quantities (integrals of motion); the lowest-order ones are the number of particles N, the momentum P, and the energy E, respectively given by

$$N = \int_{-\infty}^{+\infty} |\psi|^2 dz, \tag{2.4}$$

$$P = \frac{i}{2}\int_{-\infty}^{+\infty} (\psi\partial_z\psi^* - \psi^*\partial_z\psi)\,dz, \tag{2.5}$$

$$E = \frac{1}{2}\int_{-\infty}^{+\infty} (|\partial_z\psi|^2 + |\psi|^4)\,dz. \tag{2.6}$$

[3] Note that in the limit of $z \to \pm\infty$ this approximation always breaks down.

It is also noted that the NLS Eq. (2.3) can be obtained by the Euler–Lagrange equation $\delta\mathscr{L}/\delta\psi^* = \partial_t(\partial_{\partial_t\psi^*}\mathscr{L}) + \partial_z(\partial_{\partial_z\psi^*}\mathscr{L})\partial_{\psi^*}\mathscr{L} = 0$, where the Lagrangian density $\mathscr{L}$ is given by (see Ref. [118] and references therein)

$$\mathscr{L} = \frac{i}{2}(\psi\partial_t\psi^* - \psi^*\partial_t\psi) - \frac{1}{2}\left(|\partial_z\psi|^2 + |\psi|^4\right). \tag{2.7}$$

The simplest nontrivial solution of Eq. (2.3) is a plane wave solution of wavenumber k and frequency ω, namely,

$$\psi = \sqrt{n_0}\exp[i(kz - \omega t + \theta_0)], \qquad \omega = \frac{1}{2}k^2 - \mu, \tag{2.8}$$

where the constant BEC density n_0 sets the chemical potential, i.e., $n_0 = \mu$, and θ_0 is an arbitrary constant phase. This solution, which is reduced to the stationary state of Eq. (1.18) for $k = 0$, is also modulationally stable as can be confirmed by a simple stability analysis (see, e.g., Refs. [35, 94]). For small densities, $n_0 \ll 1$, the above plane wave satisfies the linear Schrödinger equation, $i\partial_t\psi + \frac{1}{2}\partial_z^2\psi = 0$, and the pertinent linear wave solutions of the NLS equation are characterized by the dispersion relation $\omega = \frac{1}{2}k^2$. Notice that if the system is characterized by a length L, then the integrals of motion for the stationary solution in Eq. (2.8) take the values

$$N = 2n_0 L, \qquad P = kn_0 L, \qquad E = \frac{1}{2}(k^2 - n_0)n_0 L. \tag{2.9}$$

The NLS equation admits nontrivial solutions, in the form of dark solitons, which can be regarded as strongly nonlinear excitations of the plane wave solution (2.8). In the most general case of a moving background [$k \neq 0$ in Eq. (2.8)], a single dark soliton solution may be expressed as [13]

$$\psi(z,t) = \sqrt{n_0}(B\tanh\zeta + iA)\exp[i(kz - \omega t + \theta_0)], \tag{2.10}$$

where $\zeta \equiv \sqrt{n_0}B[z - z_0(t)]$; here, $z_0(t) = vt + z_0$ is the soliton center, z_0 is an arbitrary real constant representing the initial location of the dark soliton, v is the relative velocity between the soliton and the background given by $v = A\sqrt{n_0} + k$, the frequency ω is provided by the dispersion relation of the background plane wave, $\omega = (1/2)k^2 + n_0$ [cf. Eq. (2.8)],[4] and, finally, the parameters A and B are connected through the equation $A^2 + B^2 = 1$. In some cases it is convenient to use one parameter instead of two and, thus, one may introduce

$$A = \sin\phi, \qquad B = \cos\phi, \tag{2.11}$$

where ϕ is the so-called soliton phase angle ($|\phi| < \pi/2$). Notice that although the asymptotics of the dark soliton solution (2.10) coincide with the ones of Eq. (2.8), the plane waves at $z \to \pm\infty$ have different phases; as a result, there exists a nontrivial phase jump $\Delta\phi$ across the dark soliton, given by

$$\Delta\phi = 2\left[\tan^{-1}\left(\frac{B}{A}\right) - \frac{\pi}{2}\right] = -2\tan^{-1}\left(\frac{A}{B}\right). \tag{2.12}$$

Note that, hereafter, we will consider the simpler case where the background of matter-wave dark solitons is at rest, i.e., $k = 0$; then the frequency ω actually plays the role of the

[4] Here, this dispersion relation implies that $\omega > k^2$, and thus the allowable region in the (k,ω) plane for dark solitons is located *above* the parabola $\omega = \frac{1}{2}k^2$ corresponding to the linear waves.

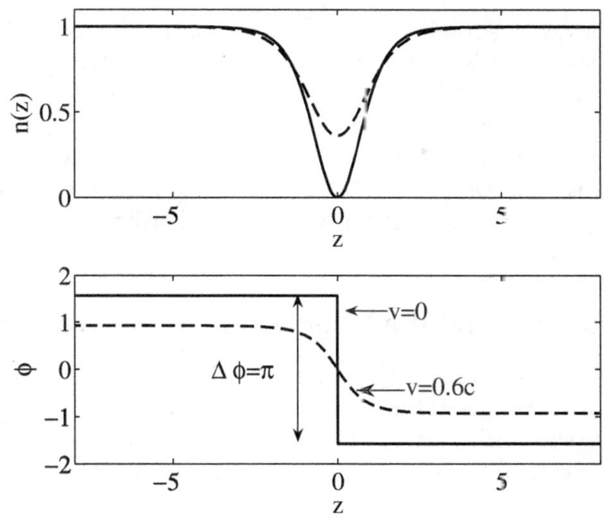

Figure 2.1. *Examples of the density (top panel) and phase (bottom panel) of a black (solid line) and a gray (dashed line) soliton on top of a background with density $n_0 = 1$. The black soliton's parameters are $A = 0$ and $B = 1$, i.e., $v = 0$, $n_{\min} = 0$, and $\Delta\phi = \pi$. The gray soliton's parameters are $A = 0.6$ and $B = 0.8$, i.e., $v = 0.6c_s$ (here $c_s = \sqrt{n_0} = 1$), $n_{\min} = n_0(1-B^2) = n_0A^2 = 0.36$, and $\Delta\phi = 0.31\pi$. Adapted with permission from Ref. [75].*

normalized chemical potential, namely $\omega = \mu = n_0$, which is determined by the number of atoms of the condensate.

The soliton phase angle describes also the *darkness* of the soliton's density,

$$|\psi|^2 = n_0(1-\cos^2\phi\,\mathrm{sech}^2\zeta), \tag{2.13}$$

with minimum density $n_{\min} = n_0(1-\cos^2\phi)$. This way, the cases $\phi = 0$ and $0 < \phi < \pi/2$ correspond to the so-called *black* and *gray* solitons, respectively. The amplitude and velocity of the dark soliton are given (for $k = 0$) by $\sqrt{n_0}\cos\phi$ and $\sqrt{n_0}\sin\phi$, respectively; thus, the black soliton

$$\psi = \sqrt{n_0}\tanh(\sqrt{n_0}z)\exp(-i\mu t), \tag{2.14}$$

is characterized by a zero velocity, $v = 0$ (and, thus, it is also called *stationary kink*), while the gray soliton moves with a finite velocity $v \neq 0$. Examples of the forms of a black and a gray soliton are illustrated in Fig. 2.1.

In the limiting case of a very shallow (small-amplitude) dark soliton with $\cos\phi \ll 1$, the soliton velocity is close to the speed of sound which, in our units, is given by

$$c_s = \sqrt{n_0}. \tag{2.15}$$

The speed of sound is, therefore, the maximum possible velocity of a dark soliton which, generally, travels with a velocity less than the speed of sound. We finally note that the dark soliton solution (2.10) has two independent parameters (for $k = 0$), one for the background, n_0, and one for the soliton, ϕ, while there is also a freedom (translational invariance) in selecting the initial location of the dark soliton z_0.[5]

[5] Recall that the underlying model, namely the completely integrable NLS equation, has infinitely many symmetries, including translational and Galilean invariance.

In the case of a condensate confined in a harmonic trap [cf. Eq. (1.13)], the background of the dark soliton is, in fact, of *finite extent*, being the ground state of the BEC [which may be approximated, for large μ, by the Thomas–Fermi cloud, cf. Eq. (1.14)]. For example, in the quasi-1D setting of the 1D GPE (2.1) with the harmonic potential in Eq. (2.2), the "composite" wavefunction (describing both the background and the soliton) can be approximated as $\psi(z,t) = \Phi(z)\exp(-i\mu t)\psi_{\rm ds}(z,t)$, where $\Phi(z)$ is the TF background and $\psi_{\rm ds}(z,t)$ is the dark soliton wavefunction of Eq. (2.10), which satisfies the 1D GPE for $V(z)=0$.

2.1.2 ▪ Dark Solitons and the Inverse Scattering Transform

The single dark soliton solution of the NLS Eq. (2.3) presented in the previous section, as well as multiple dark soliton solutions (see Sec. 2.1.6 below), can be derived by means of the inverse scattering transform (IST) [13]. A basic step of this approach is the solution of the Zakharov–Shabat (ZS) eigenvalue problem, with eigenvalue λ, for the auxiliary two-component eigenfunction $U=(u_1,u_2)^T$, namely,

$$LU=\begin{pmatrix} i\partial_z & \psi(z,0) \\ \psi^*(z,0) & -i\partial_z \end{pmatrix}\begin{pmatrix} u_1 \\ u_2 \end{pmatrix}=\lambda U, \tag{2.16}$$

with the boundary conditions at time $t=0$ given by $\psi(z\to+\infty,0)=\sqrt{n_0}$ and $\psi(z\to-\infty,0)=\sqrt{n_0}\exp(i\theta)$. Here, $\sqrt{n_0}$ is the amplitude of the background wavefunction and θ is a constant phase. Since the operator L is self-adjoint, the ZS eigenvalue problem possesses real discrete eigenvalues λ_j, with magnitudes $|\lambda_j|<\sqrt{n_0}$. Importantly, each real discrete eigenvalue $\lambda_j=\sqrt{n_0}\sin\phi_j$ corresponds to a dark soliton of depth $\sqrt{n_0}\cos\phi_j$ and velocity $\sqrt{n_0}\sin\phi_j$. To make a connection to the dark soliton solutions of the NLS equation presented in the previous section, we note that the dark soliton of Eq. (2.10) corresponds to a *single* eigenvalue $\lambda=\sqrt{n_0}\sin\phi$.

Although the system of ZS equations (2.16) is linear, its general solution for arbitrary initial condition is not available. Thus, various methods have been developed for the determination of the spectrum of the ZS problem, such as the so-called quasiclassical method [15, 16] (see also Ref. [20]), the variational approach [119], as well as other techniques that can be applied to the case of dark soliton trains [120, 121]. In any case, the *generation* of single- as well as multiple-dark solitons (see Sec. 2.1.6 below) can be studied in the framework of the IST method, and many useful results can be obtained. In that regard, first we note that a pair of dark solitons—corresponding to a discrete eigenvalue pair in the associated scattering problem—can always be generated by an *arbitrarily* small dip on a background of constant density [15] (see also Ref. [16]). This means that the generation of dark solitons is a *thresholdless* process, contrary to the case of bright solitons which are created when the number of atoms exceeds a certain threshold [122]. In another example, as dark solitons are characterized by a phase jump across them, we may assume that they can be generated by an antisymmetric initial wavefunction profile of the form

$$\psi(z,0)=\sqrt{n_0}\tanh(\alpha z), \tag{2.17}$$

characterized by a background density n_0 and a width α^{-1} (the ratio $\sqrt{n_0}/\alpha$ is assumed to be arbitrary). In such a case, the ZS eigenvalue problem (2.16) can be solved exactly [17, 18, 19] and the resulting eigenvalues of the discrete spectrum are given by $\lambda_1=0$ and $\lambda_{2j}=-\lambda_{2j+1}=\sqrt{n_0-\mu_j^2}$, where positive μ_j are defined as $\mu_j=\sqrt{n_0}-j\alpha$, $j=1,2,\ldots,N_0$, and N_0 is the largest integer such that $N_0<\sqrt{n_0}/\alpha$. These results show that for arbitrary $\sqrt{n_0}/\alpha$ the initial wavefunction profile of Eq. (2.17) will always produce a black soliton [cf. Eq. (2.14)] at $z = 0$ (corresponding to the first, zero eigenvalue) and additional N_0 pairs of symmetric gray solitons (corresponding to the even number of the secondary, nonzero eigenvalues), propagating to the left and to the right of the primary black soliton. Apparently, the total number of eigenvalues and, thus, the total number of solitons, is $2N_0 + 1$ and depends on the ratio $\sqrt{n_0}/\alpha$. Apart from the above example, dark soliton generation was systematically studied in Ref. [16] for a variety of initial conditions (such as box-like dark pulses, phase steps, and others). Notice that, generally, initial wavefunction profiles with odd symmetry will produce an odd number of dark solitons, while profiles with an even symmetry (as, e.g., in the study of Ref. [15]) produce pairs of dark solitons; this theoretical prediction was also confirmed in experiments with optical dark solitons [123]. Furthermore, the initial phase change across the wavefunction plays a key role in dark soliton formation, while the number of dark solitons that are formed can be changed by small variations of the phase.

2.1.3 ▪ Integrals of Motion and Properties of Dark Solitons

Let us now proceed by considering the integrals of motion for dark solitons. Taking into regard that Eqs. (2.4)–(2.6) refer to both the background and the soliton, one may follow Refs. [28, 29, 124, 125], and *renormalize* the integrals of motion so as to extract the contribution of the background [see Eqs. (2.9)]. This way, the *renormalized integrals of motion* become finite and, when calculated for the dark soliton solution (2.10), provide the following results (for $k = 0$). The number of particles $N_{\rm ds}$ of the dark soliton reads

$$N_{\rm ds} = \int_{-\infty}^{+\infty} (n_0 - |\psi|^2) dz = 2\sqrt{n_0} B. \tag{2.18}$$

The momentum $P_{\rm ds}$ of the dark soliton is given by

$$\begin{aligned} P_{\rm ds} &= \frac{i}{2}\int_{-\infty}^{+\infty} (\psi \partial_z \psi^* - \psi^* \partial_z \psi) dz - n_0 \Delta\phi \\ &= \frac{i}{2}\int_{-\infty}^{+\infty} (\psi \partial_z \psi^* - \psi^* \partial_z \psi)\left(1 - \frac{n_0}{|\psi|^2}\right) dz \\ &= -2v(c_s^2 - v^2)^{1/2} + 2c_s^2 \tan^{-1}\left[\frac{(c_s^2 - v^2)^{1/2}}{v}\right], \end{aligned} \tag{2.19}$$

where $\Delta\phi$ is given by Eq. (2.12) and $c_s = \sqrt{n_0}$ is the speed of sound. Furthermore, the energy $E_{\rm ds}$ of the dark soliton is given by

$$E_{\rm ds} = \frac{1}{2}\int_{-\infty}^{+\infty} \left[|\partial_z \psi|^2 + (|\psi|^2 - n_0)^2\right] dz = \frac{4}{3}(c_s^2 - v^2)^{3/2}, \tag{2.20}$$

while the renormalized Lagrangian density takes the form [26]

$$\mathcal{L}_{\rm ds} = \frac{i}{2}(\psi \partial_t \psi^* - \psi^* \partial_t \psi)\left(1 - \frac{n_0}{|\psi|^2}\right) - \frac{1}{2}\left[|\partial_z \psi|^2 + (|\psi|^2 - n_0)^2\right]. \tag{2.21}$$

The renormalized integrals of motion can now be used for a better understanding of basic features of dark solitons. To be more specific, one may differentiate the expressions

(2.19) and (2.20) over the soliton velocity $v \equiv A\sqrt{n_0}$ to obtain the result

$$\frac{\partial E_{\rm ds}}{\partial P_{\rm ds}} = v, \tag{2.22}$$

which shows that the dark soliton effectively behaves like a *classical particle*, obeying a standard equation of classical mechanics. Furthermore, it is also possible to associate an *effective mass* to the dark soliton, according to the equation $m_{\rm ds} = \partial P_{\rm ds}/\partial v$. This way, using Eq. (2.19), it can readily be found that

$$m_{\rm ds} = -4\sqrt{n_0}B, \tag{2.23}$$

which shows the dark soliton is characterized by a *negative* effective mass. The same result, but for almost black solitons ($B \approx 1$) with sufficiently small soliton velocities ($v^2 \ll c_s^2$), can also be obtained using Eq. (2.20) [126]: in this case, the energy of the dark soliton can be approximated as $E_{\rm ds} \approx (4/3)c_s^3 - 2c_s v^2$ or, equivalently,

$$E_{\rm ds} = E_0 + \frac{1}{2} m_{\rm ds} v^2, \tag{2.24}$$

where $E_0 \equiv \frac{4}{3}c_s^3$, and the soliton's effective mass is $m_{\rm ds} = -4\sqrt{n_0}$.

2.1.4 ▪ The Small-Amplitude Approximation

As mentioned above, the case of $B^2 = \cos^2\phi \ll 1$ corresponds to a small-amplitude (shallow) dark soliton, which travels with a speed v close to the speed of sound, i.e., $v \approx c_s$. In this case, it is possible to apply the *reductive perturbation method* [127] and show that, in the small-amplitude limit, the NLS dark soliton can be described by an effective KdV equation (see, e.g., Ref. [128] for various applications of the KdV model). The basic idea of this so-called *small-amplitude approximation* can be understood in terms of the similarity between the KdV soliton and the shallow dark soliton's density profile; indeed, the KdV equation for a field $u(z,t)$ expressed as

$$\partial_t u + 6u\partial_z u + \partial_z^3 u = 0 \tag{2.25}$$

possesses a single-soliton solution (see, e.g., Ref. [14]),

$$u(z,t) = 2\varkappa^2 \mathrm{sech}^2[\varkappa(z - 4\varkappa^2 t)] \tag{2.26}$$

(with $\varkappa$ being an arbitrary constant), which shares the same functional form with the density profile of the shallow dark soliton of the NLS equation [see Eqs. (2.26) and (2.13)]. The reduction of the cubic NLS equation to the KdV equation was first presented in Ref. [10] and later the formal connection between several integrable evolution equations was investigated in detail [129]. Importantly, such a connection is still possible even in cases of strongly perturbed NLS models, a fact that triggered various studies on dark soliton dynamics in the presence of perturbations (see, e.g., Refs. [130, 131, 132] for studies in the context of optics, as well as the recent review [133] and references therein). Generally, the advantage of the small-amplitude approximation is that it may predict approximate analytical dark soliton solutions in models where exact analytical dark soliton solutions are not available, or can only be found in an implicit form [130].

Let us now consider a rather general case, and discuss small-amplitude dark solitons of the generalized NLS Eq. (1.22); in the absence of the potential [$V(z) = 0$], this equation is expressed in dimensionless form as

$$i\partial_t\psi = -\frac{1}{2}\partial_z^2\psi + f(n)\psi, \tag{2.27}$$

where the nonlinearity f is a function of the density $n \equiv |\psi|^2$ and the units are the same to the ones used for Eq. (2.1). Then, we use the Madelung transformation $\psi(z,t) = \sqrt{n(z,t)}\exp[i\varphi(z,t)]$, with φ representing the BEC phase, to express Eq. (2.27) in the hydrodynamic form

$$\partial_t\varphi + f(n) + \frac{1}{2}(\partial_z\varphi)^2 - \frac{1}{2}n^{-1/2}\partial_z^2 n^{1/2} = 0, \tag{2.28}$$

$$\partial_t n + \partial_z(n\partial_z\varphi) = 0. \tag{2.29}$$

The simplest solution of Eqs. (2.28) and (2.29) is $n = n_0 \equiv |\psi_0|^2$ and $\phi = -\mu t = -f_0 t$, where $f_0 \equiv f(n_0) = f(|\psi_0|^2)$. Note that in the model of Eq. (1.23) one has $f_0 = \frac{2+3n_0}{2\sqrt{1+n_0}}$, for the model of Eq. (1.24), $f_0 = \sqrt{1+2n_0}$, and so on. Next, assuming slow spatial and temporal variations, we define the slow variables

$$Z = \epsilon^{1/2}(z - ct), \quad T = \epsilon^{3/2}t, \tag{2.30}$$

where ϵ is a formal small parameter ($0 < \epsilon \ll 1$) connected with the soliton amplitude. Additionally, we introduce asymptotic expansions for the density and phase:

$$n = n_0 + \epsilon n_1(Z,T) + \epsilon^2 n_2(Z,T) + \cdots, \tag{2.31}$$

$$\varphi = -f_0 t + \epsilon^{1/2}\varphi_1(Z,T) + \epsilon^{3/2}\varphi_2(Z,T) + \cdots. \tag{2.32}$$

Then, substituting Eqs. (2.31) and (2.32) into Eqs. (2.28) and (2.29), and Taylor expanding the nonlinearity function $f(n)$ as $f(n) = f_0 + \epsilon f_0' n_1 + \epsilon^2[(1/2)f_0'' n_1^2 + f_0' n_2] + \mathcal{O}(\epsilon^3)$ (where $f_0^{(\ell)} \equiv \frac{d^\ell f}{dn^\ell}|_{n=n_0}$), we obtain a hierarchy of equations. In particular, Eqs. (2.28) and (2.29) lead, respectively, at the orders $\mathcal{O}(\epsilon)$ and $\mathcal{O}(\epsilon^{3/2})$, to the following linear system:

$$-c\partial_Z\varphi_1 + f_0' n_1 = 0, \quad n_0\partial_Z^2\varphi_1 - c\partial_Z n_1 = 0. \tag{2.33}$$

The compatibility condition of the above equations is the algebraic equation $c^2 = f_0' n_0$, which shows that the velocity c in Eq. (2.30) is equal to the speed of sound, $c \equiv c_s$. Additionally, Eqs. (2.33) connect the phase φ_1 and the density n_1 through the equation

$$\partial_Z\varphi_1 = \frac{c_s}{n_0}n_1. \tag{2.34}$$

To the next orders, viz. $\mathcal{O}(\epsilon^2)$ and $\mathcal{O}(\epsilon^{5/2})$, Eqs. (2.28) and (2.29), respectively, yield

$$\partial_T\varphi_1 - c_s\partial_Z\varphi_2 + f_0' n_2 + \frac{1}{2}f_0'' n_1^2 + \frac{1}{2}(\partial_Z\varphi_1)^2 - \frac{1}{4}n_0^{-1}\partial_Z^2 n_1 = 0, \tag{2.35}$$

$$\partial_T n_1 - c_s\partial_Z n_2 + \partial_Z(n_1\partial_Z\varphi_1) + n_0\partial_Z^2\varphi_2 = 0. \tag{2.36}$$

The compatibility conditions of Eqs. (2.35) and (2.36) are the algebraic equation $c_s^2 = f_0' n_0$, along with a KdV equation [see Eq. (2.25)] for the unknown density n_1:

$$2c_s\partial_T n_1 + (3f_0' + n_0 f_0'')n_1\partial_Z n_1 - \frac{1}{4}\partial_Z^3 n_1 = 0. \tag{2.37}$$

Thus, the density n_1 of the shallow dark soliton can be expressed as a KdV soliton [see Eq. (2.26)]. In terms of the original time and space variables, n_1 is expressed as follows:

$$n_1(z,t) = -\frac{3\varkappa^2}{2(3f_0' + n_0 f_0'')}\operatorname{sech}^2\left[\epsilon^{1/2}\varkappa(z - vt)\right], \tag{2.38}$$

where $\varkappa$ is (as before) an arbitrary parameter [assumed to be of order $\mathcal{O}(1)$], while v is the soliton velocity; the latter is given by

$$v = c_s - \epsilon\frac{\varkappa^2}{2c_s}, \tag{2.39}$$

and, clearly, $v \lesssim c_s$. Equation (2.38) describes a small-amplitude *dip* [of order $\mathcal{O}(\epsilon)$—see Eq. (2.31)] on the background density of the condensate, with a phase φ_1 that can be found using Eq. (2.34); in terms of the variables z and t, the result is

$$\varphi_1(z,t) = -\frac{3\varkappa c_s}{2n_0(3f_0' + n_0 f_0'')}\tanh\left[\epsilon^{1/2}\varkappa(z - vt)\right]. \tag{2.40}$$

The above expression shows that the density dip is accompanied by a tanh-shaped phase jump. Thus, the wavefunction characterized by the density n_1 in Eq. (2.38) and the phase φ_1 in Eq. (2.40) is an approximate *shallow dark soliton* solution of the GPE (2.27), obeying the effective KdV Eq. (2.37).

Notice that the above analysis applies for $f(n) = n$ (i.e., for the cubic NLS model), as well as for all forms of the nonlinearity function in Eqs. (1.23)–(1.26). Furthermore, variants of the reductive perturbation method have also been applied for the study of matter-wave dark solitons in higher-dimensional settings [134, 135], multicomponent condensates [136, 137] (see also Sec. 2.4) and combinations thereof [138].

2.1.5 ▪ On the Generation of Dark Solitons

Matter-wave dark solitons can be created in experiments by means of various methods, namely phase imprinting, density engineering, quantum-state engineering (which is a combination of phase imprinting and density engineering methods), the matter-wave interference methods, and by dragging an obstacle sufficiently fast through a condensate. In connection to Sec. 2.1.2—and following the historical evolution of the subject—here we will discuss the first four methods (phase-imprinting, density-engineering, quantum-state engineering, and matter-wave interference methods), while the remaining fifth method will be presented in Sec. 2.5.1 below.

The Phase-Imprinting Method

The earlier results of Sec. 2.1.2, as well as more recent theoretical studies in the BEC context [139, 140] (see also Ref. [141]), paved the way for the generation of matter-wave dark solitons by means of the phase-imprinting method. This technique was used in the earlier [46, 47, 50]—but also in recent [68, 69]—matter-wave dark soliton experiments. The phase-imprinting method involves a manipulation of the BEC phase, without changing the BEC density, which can be implemented experimentally by illuminating part of the condensate by a short off-resonance laser beam (i.e., a laser beam with a frequency far from the relevant atomic resonant frequency—see details in the review [142]). This procedure can be described in the framework of Eq. (1.21), by considering a time-dependent potential of the form $V(z;t) \propto \phi(z)p(t)$, where $p(t)$ is the laser pulse envelope and $\phi(z)$ is the imprinted phase, given by [143]

$$\phi(z) = \frac{\Delta\phi}{2}\left[1+\tanh\left(\frac{z-z_*}{bW}\right)\right], \tag{2.41}$$

where $\Delta\phi$ is the phase gradient, while the width W of the potential edge sets the steepness of the phase gradient at z_*. Note that since experimentally relevant values correspond to a 10–20% absorption width of the phase step, an empirical factor $b = 0.45$ is also introduced in Eq. (2.41) [143].

From a theoretical standpoint, phase imprinting can be studied (in the absence of the trapping potential) in the framework of the IST method, upon considering an initial wavefunction of the form $\psi(z,0) = \exp[i\phi(z)]$; see details in Ref. [140]. Here, the imprinted phase $\phi(z)$ is assumed to increase from left to right and approach constants as $z \to \pm\infty$ [as, e.g., in Eq. (2.41)]. The pertinent ZS eigenvalue problem can be solved by mapping Eqs. (2.16) to a damped driven pendulum problem. This way, a formula for the number of both the even and the odd number of generated dark solitons, traveling in both directions, can be derived analytically [140].

In some experiments (see, e.g., Ref. [46]), the generation of the "dominant" dark soliton is followed by the generation of a secondary wave packet traveling in the opposite direction with a velocity near the speed of sound. This effect can also be understood in the framework of IST: small perturbations of the dark soliton produce shallow "satellite" dark solitons moving with velocities $v \lesssim c_s$ [19].

The Density-Engineering Method

The density-engineering method involves a direct manipulation of the BEC density, without changing the BEC phase, such that local reductions of the density are created which eventually evolve into dark solitons. This technique was used in the Harvard experiments [48, 66], where a compressed pulse of slow light was used to create a defect on the condensate density. This defect induced the formation of shock waves that shed dark solitons (or other higher-dimensional topological structures, such as vortex rings [66]). Notice that the use of a compressed pulse of slow light is not really necessary or beneficial in order to create dark solitons by means of the density engineering method: in fact, a local reduction of the BEC density can also be created by modifying the (harmonic) trapping potential with an additional barrier potential, which may be induced by an optical dipole potential or a far-detuned laser beam; this barrier can then be switched off nonadiabatically (while the harmonic trap is kept on), creating the desired local reduction of the density [143]. This technique was employed in a recent experiment [74], where such a dipole beam was used in different setups to induce merging and splitting rubidium condensates; depending on the parameters, this process leads to the formation of dark soliton trains, or a high-density bulge and dispersive shock waves.

As in the case of phase imprinting, the density-engineering technique can be studied by means of the IST method (in the absence of the trapping potential). In fact, earlier works [15, 16] (see also Ref. [121]) have already addressed the problem of dark soliton generation induced by initial change of the density: for example, in the case of a box-like initial condition, namely $\psi(z,0) = \sqrt{n_0}$ for $|z| > z_0$ and $\psi(z,0) = \sqrt{n_1}$ for $|z| < z_0$ (with $n_1 < n_0$), the ZS spectral problem admits an explicit solution, as it can be solved exactly on the intervals $|z| < z_0$ and $|z| > z_0$. In particular, it can be shown that there appear two discrete eigenvalues $\lambda_{1,2} = \pm 2\sqrt{n_0}\left[1 - 2z_0^2(\sqrt{n_0} - \sqrt{n_1})^2\right]$ (for $\sqrt{n_0} - \sqrt{n_1} \ll \sqrt{n_0}$) and, thus, two small-amplitude dark solitons are generated.

The Quantum-State Engineering Method

A combination of the phase-imprinting and density-engineering methods is also possible, leading to the so-called quantum-state engineering technique [143, 144]. This method, which involves manipulation of both the BEC density and phase, has been used in experiments at JILA [49] and Hamburg [68] with a two-component ^{87}Rb BEC (see Sec. 2.4 below): in one component, a so-called filled dark soliton was created, with the hole in this component being filled by the other component. Depending on the trap geometry, the created dark soliton was found to be either unstable or stable. Particularly, in the JILA experiment [49], the dark soliton evolved in a quasispherical trap (after the filling from the other component was selectively removed) and, due to the onset of the so-called snaking instability, the soliton was found to decay into vortex rings (see Sec. 2.3.2 below). On the other hand, in the Hamburg experiment [68], the filled dark soliton in one component was allowed to evolve (in the presence of the other component) in an elongated cigar-shaped trap; this way, a so-called dark-bright soliton pair was created (see Sec. 2.4), which was found to be stable, performing slow oscillations in the trap as predicted in theory [145].

Notice that a similar two-component engineering technique was also used for the creation of vortices [146], while earlier experimental results from the JILA group could be interpreted as a formation of a stack of filled dark solitons in a single BEC [147].

The Matter-Wave Interference Method

Matter-wave interference experiments (see, e.g., the seminal work of Ref. [148]) are known to demonstrate, apart from self-interference, the interference between two BECs confined in a trap, divided into separate parts by means of a barrier potential induced by a laser beam. In particular, the BECs are left to expand and overlap forming interference fringes, similar to the ones known in optics. Much interest has been drawn to a better understanding of this fundamental phenomenon, especially as concerns the coherence properties of the interfering BECs (see, e.g., Ref. [149]). On the other hand, since most of the relevant experimental findings can be quantitatively reproduced in the framework of the GPE mean-field theory [150], we will proceed by adopting this approach in order to discuss the connection between dark solitons and matter-wave interference.

An interesting variation of the interference process, which is naturally attributed to the inherent nonlinearity of BECs due to interatomic interactions, is that, under certain conditions, the collision of two initially separated condensates can lead to the creation of dark solitons. This "nonlinear interference" effect was first observed in simulations [151], and was subsequently analyzed theoretically [152]. Other studies, basically relying on the self-interference of BECs, have also been proposed as well [153, 154, 155, 156, 157]. Importantly, relevant experiments employing this so-called matter-wave interference method have already been reported, demonstrating the generation of vortices [158] (see also theoretical work in Refs. [156, 159, 160]) and dark solitons [70, 71, 72, 74] (see also the experiment of Ref. [63]).

To get a deeper insight into the physics of the matter-wave interference process, let us follow the arguments of Ref. [152] and consider the interference between two separated quasi-1D BECs colliding in the presence of a harmonic trap. There exist two different regimes characterizing this process, namely a linear and a nonlinear one, depending on the competition between the kinetic and the interaction energies. In the linear regime, the kinetic energy of the condensates exceeds the nonlinear interaction energy of the atoms. In this case, and at any time t, the total wavefunction of the system can be well approximated by a linear superposition of the wavefunctions that each individual condensate would have at t. The two initially *well-separated* BECs interfere at the trap center, produce a linear interference pattern, and then separate again regaining their initial shape. The fringe spacing l of the interference pattern is determined by the k vector that each individual condensate would have if performing a dipole oscillation alone in the trap and, at the time of maximum overlap, $l = \pi/D(\hbar/2m\omega_z)$ (here D is the initial distance between the condensates). It is clear that the higher the kinetic energy is, the higher the number of fringes and the smaller the fringe spacing is. Approximating the individual wavefunctions in the TF limit, it can be found that the kinetic energy (estimated from the curvature of a $\cos^2$ interference pattern) exceeds the peak nonlinear energy (at the center of the fringes) when the initial distance between the two BECs exceeds critical distance, namely $D > D_c$, where D_c is given by [152]

$$D_c = \pi\left(12\pi\frac{N\hbar a}{m\omega_z}\right)^{1/3}. \tag{2.42}$$

Here, N is the number of atoms, a the s-wave scattering length, ω_z the longitudinal trap frequency, and m the atomic mass. If the above condition is not fulfilled, namely $D < D_c$, then the system enters in the nonlinear regime. In the latter, the interference pattern consists of stable fringes with a phase jump of order of π across them, which can naturally be identified as genuine dark solitons. Notice that in the nonlinear regime, the initially individual condensates instead of reforming as separate objects, they form a combined condensate undergoing a quadrupole oscillation.

The results of Heidelberg experiments [70, 72] can be compared directly to the above theoretical predictions. In these experiments, dark solitons were created by releasing a ^{87}Rb BEC from a double-well trap into a harmonic trap in the dimensionality crossover regime from 1D to 3D. For the parameters used, the initial distance of the individual condensates was approximately five times smaller than the critical distance and the trap frequencies were ramped, with ramping times chosen so as to minimize the excitation of the quadrupole mode. It was shown that, in accordance to the observations of Ref. [63], the number of created solitons is even for a zero phase-difference between the two initially separated condensates, while it is odd for a phase difference close to π. If the phase difference is exactly equal to π, a standing (black) dark soliton in the middle of the trap is always created. Notice that the total number of the created solitons depends on the momentum of the merging condensates, which may be controlled by varying, e.g., the distance between the condensate fragments, the number of atoms, or the aspect ratio of the trap [74, 70, 72]. More details on these experiments can be found below in Sec. 2.3.5.

2.1.6 ▪ Multiple Dark Solitons and Dark Soliton Interactions

The Two-Soliton State and Dark Soliton Collisions

Apart from the single dark soliton solution, the NLS Eq. (2.3) possesses exact analytical *multiple* dark soliton solutions, which can be found by means of the IST [13, 21] (see also Refs. [22, 23]). Such solutions describe the elastic collision between dark solitons as, in the asymptotic limit of $t \to \pm\infty$, the multiple-soliton solution can be expressed as a linear superposition of individual single-soliton solutions, which remain unaffected by the collision apart from a collision-induced phase shift. To be more specific, let us consider the two-soliton wavefunction $\psi = \psi(z,t)$, which can be asymptotically expressed as

$$\psi \to \psi(z-\sqrt{n_0}A_1 t, A_1, z_1^+) + \psi(z-\sqrt{n_0}A_2 t, A_2, z_2^-), \quad t \to +\infty, \tag{2.43}$$

$$\psi \to \psi(z-\sqrt{n_0}A_1 t, A_1, z_1^-) + \psi(z-\sqrt{n_0}A_2 t, A_2, z_2^-), \quad t \to -\infty, \tag{2.44}$$

where $z^{\pm}_{1,2}$ denote the position of each individual soliton (in the above expressions, the parameters A_j and B_j ($j = 1,2$), with $A_j^2 + B_j^2 = 1$, characterize the velocity and depth of the soliton j). Apparently the shape and the parameters of each soliton are preserved, while the phase shift of each soliton is given by

$$\Delta z_1 \equiv z_1^+ - z_1^- = +\frac{1}{2B_1} \ln\left[\frac{(A_1 - A_2)^2 + (B_1 + B_2)^2}{(A_1 - A_2)^2 + (B_1 - B_2)^2}\right], \tag{2.45}$$

$$\Delta z_2 \equiv z_2^+ - z_2^- = -\frac{1}{2B_2} \ln\left[\frac{(A_1 - A_2)^2 + (B_1 + B_2)^2}{(A_1 - A_2)^2 + (B_1 - B_2)^2}\right]. \tag{2.46}$$

Note that if the soliton velocities are equal, i.e., $A_1 = -A_2 = A$ (hence, $B_1 = B_2 = B$), then the phase shift is equal for both solitons and is given by $\Delta z = (2B)^{-1} \ln(1 + B^2/A^2)$.

Equations (2.45) and (2.46) show that the spatial shift of each soliton trajectory is in the same direction as the velocity of each individual soliton and, thus, the dark solitons always *repel* each other. Here it should be mentioned, however, that this important result (as well as the collision dynamics near the collision point) can be better understood upon studying the explicit form of the two-soliton wavefunction rather than its asymptotic limit considered above. To do so, we consider again the case of a two-soliton solution, assuming for simplicity that the two solitons are moving with equal velocities (i.e., $A_1 = -A_2 = A$). In such a case, the two-soliton wavefunction is given by [22, 23]

$$\psi(z,t) = \frac{F(z,t)}{G(z,t)} \exp(-i\mu t), \tag{2.47}$$

where

$$\begin{aligned} F(z,t) = {} & 2(n_0 - 2n_{\min}) \cosh(2n_0 ABt) \\ & - 2n_0 A \cosh(2\sqrt{n_0} Bz) + i \sinh(2n_0 ABt), \end{aligned} \tag{2.48}$$

$$G(z,t) = 2\sqrt{n_0} \cosh(2n_0 ABt) + 2\sqrt{n_{\min}} \cosh(2\sqrt{n_0} Bz), \tag{2.49}$$

while $n_{\min} = n_0(1 - B^2) = n_0 A^2$ is the minimum density (i.e., the density at the center of each soliton). The density profile of the two-soliton solution in Eq. (2.47) is sketched in the top panel of Fig. 2.2.

To study analytically the interaction and collision between dark solitons, we follow the approach of Ref. [72] and find, at first, the trajectory of the soliton coordinate z_0 as a function of time; this can be done as follows. First, let us recall that the dark soliton coordinate z_0 is the location of the minimum density (see Fig. 2.2). Then, using the auxiliary equation $\partial_z |\psi|^2 = 0$, where the density $|\psi|^2$ is determined by Eq. (2.47), the following result is obtained:

$$\cosh(2\sqrt{n_0} B z_0) = \sqrt{\frac{n_0}{n_{\min}}} \cosh(2n_0 ABt) - 2\sqrt{\frac{n_{\min}}{n_0}} \frac{1}{\cosh(2n_0 ABt)}. \tag{2.50}$$

Then, Eq. (2.50) determines the distance $2z_0^*$ between the two solitons at the point of their closest proximity, i.e., the collision point corresponding to $t = 0$:

$$z_0^* = \frac{1}{2\sqrt{n_0 - n_{\min}}} \cosh^{-1}\left(\sqrt{\frac{n_0}{n_{\min}}} - 2\sqrt{\frac{n_{\min}}{n_0}}\right). \tag{2.51}$$

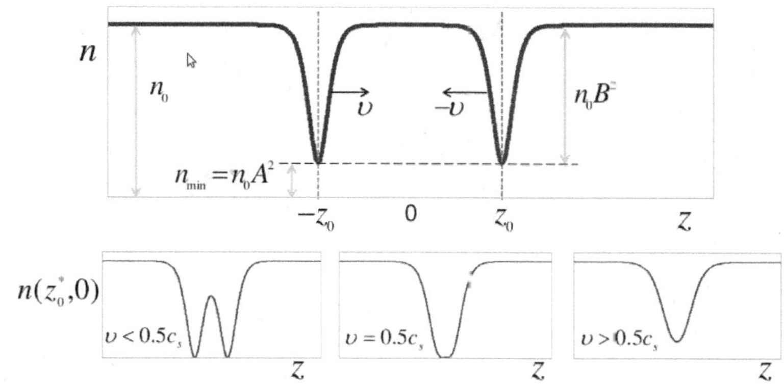

Figure 2.2. *The top panel shows the density profile of the two-soliton solution in Eq. (2.47). The bottom panels show the density profile of two dark solitons at their collision point corresponding to $z = z_0^*$ and $t = 0$. The density of low-speed solitons, $v < v_c$, is characterized by two distinguishable minima (bottom left panel), while in the case of high-speed solitons, $v > v_c$, the density exhibits a single minimum (bottom right panel); in the critical case, $v = v_c$, the density has a flat single minimum (bottom middle panel). Adapted with permission from Ref. [75].*

This equation [which holds for $n_{\min}/n_0 = A^2 \leq 1/4$, otherwise Eq. (2.51) provides a complex (unphysical) value for z_0^*] shows that $z_0^* = 0$ for $n_{\min}/n_0 = A^2 = 1/4$. Thus, it is clear that there exists a critical value of the soliton velocity, namely $v_c = \sqrt{n_0}/2 \equiv c_s/2$, which defines two types of dark solitons, exhibiting different behavior during their collision: "low-speed" solitons, with $v < v_c$, which are *reflected* by each other, and "high-speed" solitons, with $v > v_c$, which are *transmitted* through each other. In fact, as shown in Figs. 2.2 and 2.3, the density profile of the low-speed (high-speed) two-soliton state exhibits two separate minima (a single nonzero minimum) at the collision point, namely $n(z_0^*,0) = 0$ $[n(z_0^*,0) \neq 0]$. In the case of solitons moving with the critical velocity, $v = v_c = c_s/2$, the two-soliton density exhibits a "flat" single zero minimum at the collision point (see bottom-middle panel of Fig. 2.2). In other words, low-speed solitons are, in fact, *well-separated* solitons, which can always be characterized by two individual density minima—even at the collision point—while high-speed solitons completely overlap at the collision point. Notice that the case of gray soliton collision can effectively be described—in the small-amplitude approximation—by the collision dynamics of the KdV equation [161].

The Repulsive Interaction between Slow Dark Solitons

Let us now investigate in more detail the case of well-separated solitons, which are always reflected by each other, with their interaction resembling the one of hard-sphere-like particles. In particular, we consider the limiting case of extremely slow solitons, i.e., $n_0/n_{\min} = A^2 \ll 1/4$, for which the soliton separation is large for every time (i.e., $z_0^* \gg 0$); in this case, the second term in the right-hand side of Eq. (2.50) is much smaller than the first one and can be ignored. This way, the soliton coordinate is expressed as

$$z_0 = \frac{1}{2\sqrt{n_0}B} \cosh^{-1}\left[A^{-1}\cosh(2n_0ABt)\right]. \tag{2.52}$$

The above equation yields the soliton velocities

$$\frac{dz_0}{dt} = \frac{\sqrt{n_0}\sinh(2n_0ABt)}{\sqrt{A^{-1}\cosh^2(2n_0ABt)-1}}, \tag{2.53}$$

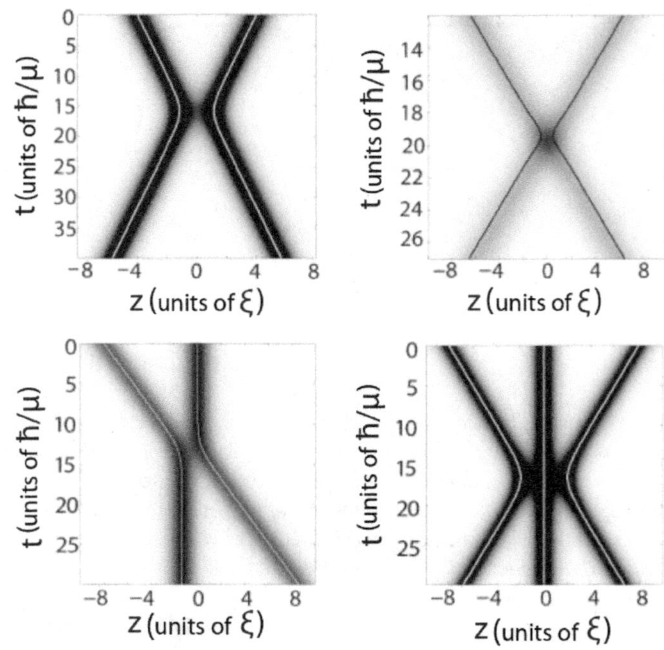

Figure 2.3. *Top panels: Soliton trajectories for symmetric two-soliton collisions for different initial velocities: $A = 0.2$ (left) and $A = 0.8$ (right). In the former (latter) case the soliton are reflected by (transmitted through) each other. Bottom panels: Soliton trajectories for an asymmetric two-soliton collision for initial velocities $A_1 = 0.5$ and $A_2 = 0$ (left), and for a three-soliton collision for initial velocities $A_1 = -A_3 = 0.4$ and $A_2 = 0$ (right). In all panels, shown are density plots of the wavefunction carrying the solitons as obtained by direct numerical integration of the homogeneous NLS equation. The solid lines correspond to the respective analytical predictions based on the effective interaction potential. In all cases the normalized chemical potential is $\mu = 1$. Adapted with permission from Ref. [72].*

which, in the limit $t \to 0$, become $dz_0/dt = 0$. Thus, as the dark solitons approach each other, their depth (velocity) is increased (decreased), and they become black at the collision point ($t = 0$), while remaining at some distance away from each other. Afterwards, the dark solitons are reflected by each other and continue their motion in opposite directions, with their velocities approaching the asymptotic values $dz_0/dt = \pm\sqrt{n_0}A$ for $t \to \pm\infty$ [see Eq. (2.53)], i.e., the velocity values of each individual soliton.

Next, differentiating Eq. (2.52) twice with respect to time, and using Eq. (2.50) (without the second term which is negligible for well-separated solitons), one may derive a Newton-type equation of motion for the soliton coordinate in the form $d^2z_0/dt^2 = -\partial V_{\text{int}}(z_0)/\partial z_0$, where the interaction potential $V_{\text{int}}(z_0)$ is given by

$$V_{\text{int}}(z_0) = \frac{1}{2}\frac{n_0B^2}{\sinh^2(2\sqrt{n_0}Bz_0)}. \tag{2.54}$$

It is clear that V_{int} is a *repulsive* potential, indicating that the dark solitons *repel each other*. If the separation between the dark solitons is sufficiently large (i.e., $2z_0 \gg 1$) then the hyperbolic sinh function in Eq. (2.54) can be approximated by its exponential asymptote, and the potential in Eq. (2.54) can be simplified as

$$V_{\text{int}}(z_0) \approx 2n_0B^2 \exp(-4\sqrt{n_0}Bz_0). \tag{2.55}$$

The latter expression can also be derived by means of a Lagrangian approach [26]. Importantly, although the above result refers to a symmetric two-soliton collision, the results of Ref. [72] show that it is possible to use the repulsive potential (2.54) in the cases of *non symmetric* collisions—using an "average depth" of the two solitons—and *multiple* dark solitons—with each soliton interacting with its neighbors; see, as examples the bottom panels of Fig. 2.3 and relevant discussion in Sec. 2.3.5.

Experiments on Multiple Dark Solitons

Multiple dark solitons were first created in a ^{23}Na BEC in the NIST experiment [47] by the phase-imprinting method (see Sec. 2.1.5), while the interaction and collision between two dark solitons in a ^{87}Rb BEC was first studied in the Hannover experiment of Ref. [50]. Nevertheless, in this early experiment the outcome of the collision was not sufficiently clear due to the presence of dissipation caused by the interaction of the condensate with the thermal cloud. In the more recent Hamburg experiment [69], the phase-imprinting method was also used to create two dark solitons in a ^{87}Rb BEC with slightly different depths. These solitons propagated to opposite sides of the condensate, reflected near the edges of the BEC, and subsequently underwent a single collision near the center of the trap. In addition, in the recent Heidelberg experiment [70] two dark solitons were created in a ^{87}Rb BEC by the so-called *interference method* (see Sec. 2.1.5 below). The solitons observed in this experiment, which were "well-separated" ones, propagated in opposite directions, were reflected and then underwent multiple genuine elastic collisions, from which the solitons emerged essentially unscathed. Notice that the experimentally observed dynamics of the oscillating and interacting dark-soliton pair of Ref. [70], as well as the one of multiple dark solitons in another Heidelberg experiment [72], was in a very good agreement with theoretical predictions based on the effective particle-like picture for dark solitons (see Sec. 2.2.2 and Sec. 2.3.5 below) and the interaction potential of Eq. (2.54).

Stationary Dark Solitons in the Trap

At this point, it is relevant to briefly discuss the case where multiple dark solitons are considered in a trapped condensate. In this case, both the single dark soliton and all other multiple dark soliton states can be obtained in a stationary form from the noninteracting (linear) limit of Eq. (2.1), i.e., in the absence of the nonlinear term. In this case, Eq. (2.1) is reduced to a linear Schrödinger equation for a confined single-particle state. For the harmonic potential of Eq. (1.13), this Schrödinger equation describes the quantum harmonic oscillator, characterized by discrete energies and corresponding localized eigenmodes in the form of Hermite–Gauss polynomials [53]. As shown in Refs. [51, 52], all these eigenmodes exist also in the fully nonlinear problem, and describe an analytical continuation of the above mentioned linear modes to a set of nonlinear stationary states. Additionally, analytical and numerical results of the work [162] suggest that in the case of a harmonic trapping potential there are no solutions of the 1D GPE (2.1) without a linear counterpart. This actually means that interatomic interactions (i.e., the effective mean-field nonlinearity in the GPE model) transforms all higher-order stationary modes into a sequence of stationary dark solitons confined in the harmonic trap [51, 52]; note that as concerns its structure, this chain of, say n, stationary dark solitons shares the same spatial profile with the linear eigenmode of quantum number n. From a physical point of view, multiple stationary dark soliton states exist due to the fact that the repulsion between dark solitons is counterbalanced by the restoring force induced by the trapping potential.

Having explored the context of multiple dark solitons in a homogeneous BEC, in the absence of the external potential, we now proceed to the case of an inhomogeneous BEC, in the presence of an external trapping potential.

2.2 ▪ Matter-Wave Dark Solitons in Quasi-1D Bose Gases

2.2.1 ▪ General Comments

We consider again the quasi-1D setup of Eq. (2.1), but now incorporating the external potential $V(z)$. In this setting, the dynamics of matter-wave dark solitons can be studied analytically by means of various perturbation methods, assuming that the trapping potential $V(z)$ is smooth and slowly varying on the soliton scale. This means that in the case, e.g., of the conventional harmonic trap [cf. Eq. (2.2)], the normalized trap strength is taken to be $\Omega \sim \epsilon$, where $\epsilon \ll 1$ is a formal small (perturbation) parameter. In such a case, Eq. (2.1) can be expressed as a perturbed NLS equation, namely,

$$i\partial_t \psi + \frac{1}{2}\partial_z^2 \psi - |\psi|^2 \psi = R(\psi) \equiv V(z)\psi. \tag{2.56}$$

Then, according to the perturbation theory for solitons [163], one may assume that a perturbed soliton solution of Eq. (2.56) can be expressed in the following general form:

$$\psi(z,t) = \psi_{\mathrm{ds}}(z,t) + \epsilon \psi_r(z,t). \tag{2.57}$$

Here, $\psi_{\mathrm{ds}}(z,t)$ has the functional form of the dark soliton solution (2.10), but with the soliton parameters depending on time, and ψ_r is the radiation—in the form of sound waves—emitted by the soliton. Generally, the latter term is strong only for sufficiently strong perturbations (see, e.g., Refs. [164, 165], as well as Ref. [80] and discussion in Sec. 2.2.4). Thus, the simplest possible approximation for a study of matter-wave dark solitons in a trap corresponds to the so-called *adiabatic approximation* of the perturbation theory for solitons [163], namely $\psi(z,t) \approx \psi_{\mathrm{ds}}(z,t)$. In any case, the study of matter-wave dark solitons in a trap should take into regard that the trap changes the boundary conditions for the wavefunction, and BEC density, namely $n \to 0$ [instead of $n \to n_0$ in the homogeneous case—see, e.g., Eq. (2.13)] as $z \to \pm\infty$. From a physical viewpoint, and based on the particle-like nature of dark solitons (see Sec. 2.1.3), one should expect that dark solitons could be reflected from the trapping potential; such a mechanism should then result in an oscillatory motion of dark solitons in the parabolic trap.

There exist many theoretical works devoted to the oscillations of individual dark solitons in trapped BECs. The earlier works on this subject reported that solitons oscillate in a condensate confined in a harmonic trap of strength Ω, and provided estimates for the oscillation frequency. In particular, in Ref. [151] soliton oscillations were observed in simulations and a soliton's equation of motion was presented without derivation; in the same work, it was stated that the solitons oscillate with frequency Ω (rather than the correct asymptotic result which is $\Omega/\sqrt{2}$—see below). The same result was derived in Ref. [166], considering the dipole mode of the condensate supporting the dark soliton. Other works [101, 167, 168] also considered oscillations of dark solitons in trapped BECs. An analytical description of the dark-soliton motion, and the correct asymptotic result, in the large μ limit, for the soliton oscillation frequency, $\Omega/\sqrt{2}$, were first presented in Ref. [169] by means of a multiple-time-scale boundary-layer theory (this approach is commonly used for vortices [54]). The same result was obtained in Refs. [126, 170] by solving the Bogoliubov–de Gennes (BdG) equations (for almost black solitons performing small-amplitude oscillations around the trap center—see Sec. 2.2.3 below), using a time-independent version of the boundary-layer theory. Furthermore, in Ref. [126] a kinetic-equation approach was used to describe dissipative dynamics of the dark soliton due to the interaction of the BEC with the thermal cloud.

Matter-wave dark soliton dynamics in trapped BECs was also analyzed in other works by means of different techniques that were originally developed for optical dark solitons [35]. In particular, in Ref. [171] the problem was analyzed by means of the adiabatic perturbation theory for dark solitons devised in Ref. [29], in Ref. [172] by means of the small-amplitude approximation (see Sec. 2.1.4), while in Ref. [173] by means of the perturbation theory of Ref. [30]. Later, in Refs. [174, 175] the so-called "Landau dynamics" approach was developed, based on the use of the renormalized soliton energy [cf. Eq. (2.20)], along with a local density approximation. Models relevant to the dynamics of matter-wave dark solitons in 1D strongly interacting Bose gases, were also considered and analyzed by means of the small-amplitude approximation [176, 177] (see also work for dark solitons in this setting in Refs. [115, 178, 179, 180, 181]). In other works, a Lagrangian approach for matter-wave dark solitons was presented [182] (see also Ref. [183]), and an asymptotic multiscale perturbation method was used to describe dark soliton oscillations and the inhomogeneity-induced emission of radiation [184]. More recently, the motion of dark solitons was rigorously analyzed in Ref. [185] (where a wider class of traps was considered), while in Ref. [186] the same problem was studied in the framework of a generalized NLS model.

Finally, as far as experiments are concerned, the oscillations of dark solitons were observed in the Hamburg [68, 69] and Heidelberg [70, 72] experiments. In these works, the experimentally determined soliton oscillation frequencies were found to deviate from the theoretically predicted value $\Omega/\sqrt{2}$. This deviation was explained in Refs. [68, 69] by the anharmonicity of the trap, while in Refs. [70, 72] by the dimensionality of the system and the soliton interactions (see details in Sec. 2.3.5 below).

2.2.2 ▪ Adiabatic Dynamics of Matter-Wave Dark Solitons

The Perturbed NLS Equation

The adiabatic dynamics of dark matter-wave solitons may be studied analytically by means of the Hamiltonian [28, 29] or the Lagrangian [26] approach of the perturbation theory for dark solitons, which were originally developed for the case of a constant background. These approaches were later modified (see Ref. [171] for the Hamiltonian approach and Refs. [182, 183] for the Lagrangian approach) to take into regard that, in the context of BECs, the background is inhomogeneous due to the presence of the external potential. The basic steps of these perturbation methods are (a) determine the background wavefunction carrying the dark soliton, (b) derive from Eq. (2.56) a perturbed NLS equation for the dark soliton wavefunction, and (c) determine the evolution of the dark soliton parameters by means of the renormalized Hamiltonian [cf. Eq. (2.20)] or the renormalized Lagrangian [cf. Eq. (2.21)] of the dark soliton. Here, we will present the first two steps of the above approach and, in the following two subsections, we will describe the adiabatic soliton dynamics in the framework of the Hamiltonian and Lagrangian approaches.

We consider again Eq. (2.56) and seek the background wavefunction in the form

$$\psi(z,t) = \Phi(z)\exp(-i\mu t + i\theta_{\rm o}), \tag{2.58}$$

where μ is the normalized chemical potential, $\theta_{\rm o}$ is an arbitrary phase, while the unknown real function $\Phi(z)$ satisfies the following equation:

$$\mu\Phi + \frac{1}{2}\frac{d^2\Phi}{dz^2} - \Phi^3 = V(z)\Phi. \tag{2.59}$$

Then, we seek for a dark soliton solution of Eq. (2.56) on top of the inhomogeneous background satisfying Eq. (2.59), namely, $\psi = \Phi(z)\exp(-i\mu t + i\theta_{\rm o})\psi_{\rm ds}(z,t)$, where the unknown wavefunction $\psi_{\rm ds}(z,t)$ represents a dark soliton. This way, employing Eq. (2.59), the following evolution equation for the dark soliton wavefunction is readily obtained:

$$i\partial_t \psi_{\rm ds} + \frac{1}{2}\partial_z^2 \psi_{\rm ds} - \Phi^2(|\psi_{\rm ds}|^2 - 1)\psi_{\rm ds} = -\frac{d}{dz}\ln(\Phi)\partial_z \psi_{\rm ds}. \tag{2.60}$$

It is clear that if the trapping potential $V(z)$ is smooth and slowly varying on the soliton scale, then the right-hand side, and also part of the nonlinear terms of Eq. (2.60), can be treated as a perturbation. To obtain this perturbation in an explicit form, we use the TF approximation to express the background wavefunction as $\Phi(z) = \sqrt{1 - V(z)}$ [see Eq. (1.14) for $g = 1$ and $\mu = 1$][6] and approximate the logarithmic derivative of Φ as

$$-\frac{d}{dz}\ln\Phi \approx \frac{1}{2}\frac{dV}{dz}\left(1 + V + V^2\right). \tag{2.61}$$

This way, Eq. (2.60) leads to the following perturbed NLS equation:

$$i\frac{\partial \psi_{\rm ds}}{\partial t} + \frac{1}{2}\frac{\partial^2 \psi_{\rm ds}}{\partial z^2} - (|\psi_{\rm ds}|^2 - 1)\psi_{\rm ds} = Q(\psi_{\rm ds}), \tag{2.62}$$

where the perturbation $Q(\psi_{\rm ds})$ is given by

$$Q(\psi_{\rm ds}) = \left(1 - |\psi_{\rm ds}|^2\right)\psi_{\rm ds} V + \frac{1}{2}\partial_z \psi_{\rm ds}\frac{dV}{dz}(1 + V + V^2). \tag{2.63}$$

Hamiltonian Approach of the Perturbation Theory

First we note that in the absence of the perturbation (2.63), Eq. (2.62) has a dark soliton solution of the form

$$\psi_{\rm ds}(z,t) = \cos\phi \tanh\zeta + i\sin\phi, \tag{2.64}$$

where $\zeta \equiv \cos\phi\,[x - (\sin\phi)t]$ [see Eq. (2.10)]. Then, considering an adiabatic evolution of the dark soliton, we assume that in the presence of the perturbation the dark soliton parameters become slowly varying unknown functions of t [28, 29, 171]. Thus, the soliton phase angle becomes $\phi \to \phi(t)$ and, as a result, the soliton coordinate becomes $\zeta \to \zeta(t) = \cos\phi(t)[z - z_0(t)]$. In the latter expression, the dark soliton center $z_0(t)$ is connected to the soliton phase angle through the following equation:

$$\frac{dz_0(t)}{dt} = \sin\phi(t). \tag{2.65}$$

The evolution of the soliton phase angle can be found by means of the evolution of the renormalized soliton energy. In particular, employing Eq. (2.20) (for $\mu = 1$), it is readily found that $dE_{\rm ds}/dt = -4\cos^2\phi\sin\phi(d\phi/dt)$. On the other hand, using Eq. (2.62) and its complex conjugate, it can be found that the evolution of the renormalized soliton energy is given by $dE_{\rm ds}/dt = -\int_{-\infty}^{+\infty}\left[Q(\psi_{\rm ds})\partial_t\psi_{\rm ds}^* + Q^*(\psi_{\rm ds})\partial_t\psi_{\rm ds}\right]dz$. Then, the above expressions for $dE_{\rm ds}/dt$ yield the evolution of ϕ, namely [29],

$$\frac{d\phi}{dt} = \frac{1}{2\cos^2\phi\sin\phi}\mathrm{Re}\left[\int_{-\infty}^{+\infty} Q(\psi_{\rm ds})\partial_t\psi_{\rm ds}^* dz\right]. \tag{2.66}$$

[6] It can easily be shown that the main result of the analysis [cf. Eq. (2.68)] can be generalized for every value of μ such that the system is in the TF-1D regime.

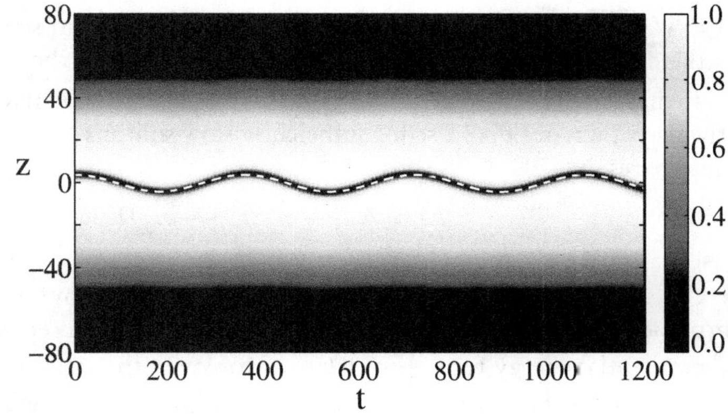

Figure 2.4. *Contour plot showing the evolution of the density of a harmonically confined BEC, as obtained by direct numerical integration of the GPE (2.56). The initial condition is* $\psi = [\mu - (1/2)\Omega^2 z^2]^{1/2}\tanh(z - z_0(0))$, *i.e., a TF cloud, characterized by a chemical potential* μ, *and carrying a dark soliton initially placed at* $z = z_0(0)$. *The parameter values are* $\mu = 1$, $\Omega = 0.025$, *and* $z_0(0) = 4$. *The dark soliton oscillates with frequency* $\omega_{osc} = \Omega/\sqrt{2} \approx 0.018$. *The dotted line across the soliton trajectory corresponds to the analytical prediction of Eq. (2.68). Adapted with permission from Ref. [75].*

Next, we Taylor expand the potential $V(z)$ around the soliton center z_0, and assume that the dark soliton is moving in the vicinity of the trap center, i.e., $\mu \equiv 1 \gg V$, which means that the last two terms in the right-hand side of Eq. (2.63) can be neglected. This way, one may further simplify the expression for the perturbation in Eq. (2.63) which, when inserted into Eq. (2.66), yields the following result:

$$\frac{d\phi}{dt} = -\frac{1}{2}\cos\phi \frac{\partial V}{\partial z_0}. \tag{2.67}$$

To this end, combining Eq. (2.67) with Eq. (2.65), we obtain the following equation of motion for nearly stationary (black) solitons with $\cos\phi \approx 1$,

$$\frac{d^2 z_0}{dt^2} = -\frac{1}{2}\frac{\partial V}{\partial z_0}. \tag{2.68}$$

The above result indicates that the dark soliton center can be regarded as a Newtonian particle: Eq. (2.68) has the form of a Newtonian equation of motion of a classical particle, of an effective mass $M_{\rm eff} = 2$, in the presence of the external potential V. In the case of the harmonic potential [cf. Eq. (2.2)], Eq. (2.68) becomes the equation of motion of the classical linear harmonic oscillator, $d^2 z_0/dt^2 = -(1/2)\Omega^2 z_0$, and shows that the dark soliton oscillates with frequency

$$\omega_{osc} = \frac{\Omega}{\sqrt{2}}, \tag{2.69}$$

or, in physical units, with $\omega_z/\sqrt{2}$. An example of an oscillating matter-wave dark soliton is shown in Fig. 2.4.

At this point, it is relevant to follow the considerations of Ref. [126] (see also Refs. [169, 170]) and estimate the energy E_{ds} of this almost dark soliton in the trap. Taking into regard that in the case of a homogeneous BEC this energy is given by Eq. (2.24), one may use a local density approximation and use in Eq. (2.20) the local speed of sound,

$c(z)=\sqrt{n_0(z)}$ [82] (here, $n_0(z)$ is the density of the ground state of the BEC), rather than the constant value $c_s=\sqrt{n_0}$ [cf. Eq. (2.15)]. Then, in the TF limit, the density is expressed as $n_0(z)=\mu-\frac{1}{2}\Omega^2z^2=c_s^2-\frac{1}{2}\Omega^2z^2$ and, thus, one may follow the lines used for the derivation of Eq. (2.24) (for sufficiently slow solitons and weak trap strengths) and obtain the result

$$E_{\rm ds}=E_0+\frac{1}{2}m_{\rm ds}v^2+\frac{1}{4}m_{\rm ds}\Omega^2z^2, \tag{2.70}$$

where $E_0\equiv\frac{4}{3}c_s^3$ and $m_{\rm ds}=-4\sqrt{n_0}$ as in Eq. (2.24). The above equation shows that the incorporation of the harmonic trap results in a decrease of the energy of the dark soliton by the potential energy term $\frac{1}{4}|m_{\rm ds}|\Omega^2z^2$. Moreover, the ratio of the soliton mass over this potential energy is given by $(\Omega^2z^2/4)^{-1}$, which is exactly two times the ratio of the atomic mass (which is equal to $m=1$ in our units) over the external potential, namely $(\Omega^2z^2/2)^{-1}$. This is another interpretation of the result that the effective mass of the dark soliton center is $M_{\rm eff}=2$.

Lagrangian Approach for Matter-Wave Dark Solitons

The perturbed NLS Eq. (2.62), with the perturbation of Eq. (2.63), can also be treated by means of a variational approach as discussed in the beginning of Sec. 2.2.2. First, we assume that the solution of Eq. (2.62) is expressed as [see Eqs. (2.10) and (2.11)]

$$\psi_{\rm ds}(z,t)=B\tanh\zeta+iA. \tag{2.71}$$

Here, A and B are unknown slowly varying functions of time (with $A^2+B^2=1$) representing, respectively, the velocity and amplitude of the dark soliton (which become time dependent due to the presence of the perturbation), while $\zeta\equiv B(t)[z-z_0(t)]$, where $z_0(t)$ is the dark soliton center. Note that in the unperturbed case $dz_0/dt\equiv A$, but in the perturbed case under consideration this simple relationship may not be valid (see below). Next, the evolution of the unknown soliton parameters $\alpha_j(t)$ [which is a generic name for $z_0(t)$ and $A(t)$] are obtained via the Euler–Lagrange equations [26, 182]

$$\frac{\partial L_{\rm ds}}{\partial\alpha_j}-\frac{d}{dt}\left(\frac{\partial L_{\rm ds}}{\partial\dot{\alpha}_j}\right)=2{\rm Re}\left\{\int_{-\infty}^{+\infty}Q^*(\psi_{\rm ds})\frac{\partial\psi_{\rm ds}}{\partial\alpha_j}dz\right\}, \tag{2.72}$$

where $\dot{\alpha}_j\equiv d\alpha_j/dt$ and $L_{\rm ds}=\int_{-\infty}^{+\infty}dz\mathcal{L}_{\rm ds}\{\psi_{\rm ds}\}$ represents the averaged Lagrangian of the dark soliton of the unperturbed NLS equation [namely for $Q(\psi_{\rm ds})=0$], with the Lagrangian density $\mathcal{L}_{\rm ds}$ being given by Eq. (2.21) (for $n_0=1$). The averaged Lagrangian can readily be obtained by substituting the ansatz (2.71) into Eq. (2.21):

$$L_{\rm ds}=2\frac{dz_0}{dt}\left[-AB+\tan^{-1}\left(\frac{B}{A}\right)\right]-\frac{4}{3}B^3. \tag{2.73}$$

Therefore, substituting Eqs. (2.73) and (2.63) into Eq. (2.72), it is straightforward to derive evolution equations for the soliton parameters. For completeness, we will follow Ref. [182] and present the final result taking also into account the last two terms in the right-hand side of Eq. (2.63)—which were omitted in the previous subsection—so as to describe the motion of shallower solitons as well. This way, and employing a Taylor expansion of the potential around the soliton center (as in the previous subsection), we obtain the following evolution equations for $z_0(t)$ and $A(t)$:

$$\frac{1}{A}\frac{dz_0}{dt} = \left[1 - \frac{1}{2}V(z_0)\right] - \left(\frac{15-\pi^2}{36B^2}\right)\left(\frac{\partial V}{\partial z_0}\right)^2 [1-2V(z_0)], \tag{2.74}$$

$$\frac{1}{B^2}\frac{dA}{dt} = -\frac{1}{2}\frac{\partial V}{\partial z_0} - \frac{V(z_0)}{3}\frac{\partial V}{\partial z_0} - \frac{\partial V}{\partial z_0}\left[\frac{V^2(z_0)}{3} + \frac{6-\pi^2}{36}\left(\frac{\partial V}{\partial z_0}\right)^2\right]. \tag{2.75}$$

Equations (2.74) and (2.75) describe the dark soliton dynamics in the trap, in both cases of nearly black solitons ($A \approx 0$ or $B \approx 1$) and gray ones (with arbitrary A or B). In the former case, and neglecting the higher-order corrections arising from the inclusion of the last two terms in the right-hand side of Eq. (2.63), the result of Eq. (2.68) is recovered: nearly black solitons oscillate near the trap center with the characteristic frequency given in Eq. (2.69). On the other hand, numerical simulations in Ref. [182] have shown that the full system of Eqs. (2.74)–(2.75) predicts that shallow solitons oscillate in the trap with the *same* characteristic oscillation frequency. Therefore, there is a clear indication that the oscillation frequency of Eq. (2.69) does *not* depend on the dark soliton amplitude. This result is proved by means of the Landau dynamics approach that will be discussed below.

Landau Dynamics of Dark Solitons

The oscillations of dark solitons of arbitrary amplitudes in a trap can also be studied by means of the so-called *Landau dynamics* approach devised in Refs. [169, 175]. This approach, which further highlights the particle-like nature of the matter-wave dark solitons, relies on a clear physical picture: when a dark soliton moves in a weakly inhomogeneous background, its local energy stays constant. Hence, one may employ the local density approximation, and rewrite the energy conservation law of Eq. (2.20) as $c^2(z_0) - v^2 = (3E_{\mathrm{ds}}/4)^{2/3}$, where $c^2(z_0)$ is the local speed of sound evaluated at the dark soliton center z_0. Then, in the TF limit, one has $c^2(z_0) = c_s^2 - \frac{1}{2}\Omega^2 z_0^2$ (as before), and taking into regard that the soliton velocity is $v = dz_0/dt$, the following equation for the energy of the dark soliton is readily obtained:

$$\frac{1}{2}M_{\mathrm{eff}}\left(\frac{dz_0}{dt}\right)^2 + \frac{1}{2}\Omega^2 z_0^2 = \tilde{E}_{\mathrm{ds}}, \tag{2.76}$$

where $\tilde{E}_{\mathrm{ds}} = c_s^2 - (3E_{\mathrm{ds}}/4)^{2/3}$ and the effective mass of the dark soliton center is again found to be $M_{\mathrm{eff}} = 2$. It is readily observed that Eq. (2.76) can be reduced to Eq. (2.68) and, thus, it leads to the oscillation frequency of Eq. (2.69). Nevertheless, the result obtained in the framework of the Landau dynamics approach is more general, as it actually refers to dark solitons of arbitrary amplitudes. Moreover, the same approach can be used also in the case of more general models, including, e.g., the cases of nonharmonic traps and/or more general nonlinearities, such as the physically relevant ones described by Eqs. (1.23)–(1.26) [175]. Nevertheless, it should be noted that in such more general cases the problem can be treated analytically for almost black solitons ($v \ll c$), performing small-amplitude oscillations. In this case, the conservation law $E_{\mathrm{ds}}(c(z_0), v) = E_0$ can be Taylor expanded around $z_0 = 0$ and $v = 0$, leading to expressions for the soliton's effective soliton mass and oscillation frequency [175].

The Small-Amplitude Approximation

Next, we discuss the adiabatic dynamics of *small-amplitude* dark solitons in trapped 1D Bose gases. In this case, one may formally reduce the more general GPE model of Eq. (2.27) (including the potential term $V\psi$) to a KdV equation with variable coefficients—see Ref. [45] for details and Ref. [187] for applications of this KdV model. The main result of such an analysis is that the density and the phase of the approximate shallow dark soliton solution of Eq. (2.27) have, to the leading order of approximation, the functional form of their counterparts in Eqs. (2.38) and (2.40), but with the soliton parameter $\varkappa$ depending on a slow variable, say Z (see earlier work for a calculation of $\varkappa(Z)$ in Refs. [188, 189]). This way, approximate analytical shallow soliton solutions have been found in various works [161, 172, 176, 177] for different forms of the nonlinearity function $f(n)$. Nevertheless, there are some subtle issues concerning the validity of this approximation, as discussed in Refs. [173, 175, 184], which is, strictly speaking, valid *away from the turning points* (where the soliton velocity vanishes). On the other hand, numerical results (see, e.g., Ref. [177]), illustrate that the range of validity of the above results is, in fact, wider than what may be expected based on the limitations of this approach.

The small-amplitude approximation, along with a local density approximation, has also been used to estimate the shallow soliton's oscillation frequency: the shallow soliton's velocity v, which in the homogeneous problem was found to be close to the speed of sound, namely $v \approx c_s = \sqrt{f_0' n_0}$ [see Eq. (2.39)], can be approximated in the inhomogeneous system as follows:

$$\frac{dZ}{dt} \approx c_s(Z) = \sqrt{f_0' n_0(Z)}. \tag{2.77}$$

In appropriate cases, Eq. (2.77) can be used for the derivation of physically relevant results. For example, following Ref. [175], we assume that $f(n) = n^\alpha$, where $\alpha = 1$ for weakly interacting BECs, or $\alpha = 2$ for strongly interacting Tonks gases [114]. Then, in the TF limit, $n_0(Z) = [\mu - V(Z)]^{1/\alpha}$, and Eq. (2.77) is reduced to the form $dZ/\sqrt{\mu - V(Z)} = \sqrt{\alpha}dt$. The latter is integrated and yields [for $V(Z) = (1/2)\Omega^2 Z^2$] the soliton trajectory

$$Z = L\sin[(\Omega\sqrt{\alpha/2})t], \tag{2.78}$$

where $L = \sqrt{2\mu}/\Omega$ is the length of the TF cloud. Equation (2.78) predicts that the (gray) shallow dark soliton will perform oscillations approximately in the entire spatial region occupied by the gas, with an oscillation frequency which takes the following values (in physical units): for $\alpha = 1$, i.e., for quasi-1D BECs described by the cubic GPE, $\omega_{\text{osc}} = \omega_z/\sqrt{2}$, while for $\alpha = 2$, i.e., for the Tonks-Girardeau gas described by a quintic NLS equation, $\omega_{\text{osc}} = \omega_z$. Note that the latter result was first obtained via a many-body calculation [190], and later was derived by means of the KdV approximation [176].

2.2.3 ▪ Bogoliubov–de Gennes Analysis for Dark Solitons

The Single-Soliton State

The above result, namely the fact that almost black solitons perform oscillations around the trap center with the frequency given in Eq. (2.69), can also be derived by means of a BdG analysis as was first demonstrated in Ref. [170] (see also results in Refs. [191, 192, 193]). Such an analysis can be done in the TF limit for the stationary dark soliton state, namely the black soliton ψ_0 [see Eq. (2.14)] located at the trap center, i.e., $z_0 = 0$, which is actually the first excited state of the condensate. Then, following the discussion in Sec. 1.2.3, the excitation spectrum can be found as follows: using the ansatz $\psi(z,t) = [\psi_0(z) + u(z)e^{-i\omega t} + \upsilon^*(z)e^{i\omega t}]e^{-i\mu t}$ [where $\omega = \omega_r + i\omega_i$ is a (generally complex)

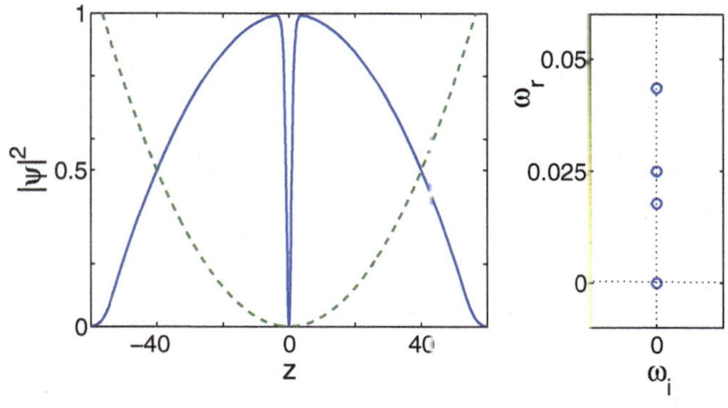

Figure 2.5. *Left panel: The density of a condensate carrying a stationary (solid line) soliton located at z = 0. The normalized chemical potential is $\mu = 1$. The dashed line shows the trapping potential with normalized trap strength $\Omega = 0.025$. Right panel: The lowest characteristic eigenfrequencies of the Bogoliubov excitation spectrum: The eigenfrequency located at the origin corresponds to the Goldstone mode, the one at $\Omega = 0.025$ to the Kohn mode, and the one at $\sqrt{3}\Omega$ to the quadrupole mode. Finally, there exists an anomalous mode with frequency approximately equal to $\omega_A = \Omega/\sqrt{2}$. Adapted with permission from Ref. [75].*

eigenfrequency and (u, v) are perturbation eigenmodes], we derive from Eq. (2.1) the following BdG equations:

$$\left[\hat{H} - \mu + 2|\psi_0|^2\right]u + \psi_0^2 v = \omega u, \tag{2.79}$$

$$\left[\hat{H} - \mu + 2|\psi_0|^2\right]v + \psi_0^{*2} u = -\omega v, \tag{2.80}$$

where $\hat{H} = -(1/2)\partial_z^2 + (1/2)\Omega^2 z^2$ is the single-particle operator. A typical example showing the initial configuration, i.e., the condensate and the stationary dark soliton (which can be found, e.g., by a Newton–Raphson method), as well as a part of the corresponding spectral plane (ω_r, ω_i), are shown in Fig. 2.5.

The BdG analysis reveals that all the eigenfrequencies of the spectrum are real, which indicates that the stationary dark soliton is dynamically stable. The four smallest magnitude eigenfrequency pairs[7] and their corresponding eigenmodes have the following physical significance (see, e.g., Ref. [43]). First, there exists a zero eigenfrequency, located at the origin of the spectral plane (ω_r, ω_i), which reflects the phase invariance of the 1D GPE. The respective eigenfunction is the so-called Goldstone mode and does not result in any physical excitation (oscillation) of the system. The solutions with eigenfrequencies $\omega_r = \pm\Omega$ correspond to the so-called *dipole mode* (or *Kohn mode*), which is relevant to the motion of the center of mass of the system; note that as the system is harmonically confined, the center of mass oscillates with the frequency of the harmonic trap [194]. The solutions with eigenfrequencies having the next larger magnitude correspond to the *quadrupole mode*, with the location of the eigenfrequencies at $\omega_r = \pm\sqrt{3}\Omega$, being particular to the one-dimensionality of the system [99]. Note that the excitation of the quadrupole mode (induced, e.g., by a time-modulation of the trap strength) results in a breathing behavior of the condensate, with its width oscillating with the above-mentioned frequency.

Of particular interest are the solutions with eigenfrequencies $\omega_r = \omega_A \equiv \Omega/\sqrt{2}$, which correspond to the so-called *anomalous mode*. This mode appears in the Bogoliubov

[7] Recall that due to the Hamiltonian nature of the system, the eigenfrequencies $\pm\omega_r$ correspond to the same physical oscillation.

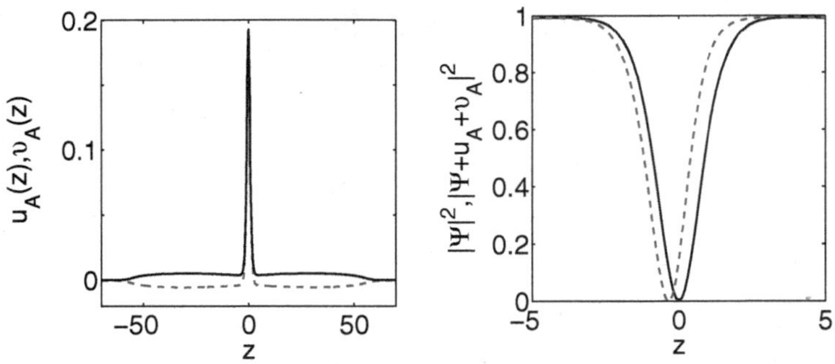

Figure 2.6. *Left panel: The eigenfunctions u_A (solid line) and υ_A (dashed line) of the anomalous mode. Right panel: The solid line shows the density of the stationary dark soliton in a region around the trap center. The dashed line shows the density of the dark soliton when excited by the anomalous mode. The parameter values are the same to the ones of Fig. 2.5. Adapted with permission from Ref. [75].*

analysis only when topological excitations of the condensate are involved, namely dark solitons or in higher dimensions vortices [76]. A characteristic property of the anomalous mode is that the integral of the norm × energy product, $\int(|u|^2-|\upsilon|^2)\omega\, dz$ (in our units), is negative rather than positive as is the case for all the positive frequency modes associated with the ground state of the system [43]. The mathematically equivalent statement is that the anomalous mode possesses a topological property of the so-called *negative Krein signature* [195], namely $K \equiv \text{sgn}\{\int(|u|^2-|\upsilon|^2)\omega\, dz\} < 0$ (for positive eigenfrequencies ω). Practically, this means that the anomalous mode becomes structurally unstable upon collision with other eigenvalues, or when dissipation is present [196]. In our case, finite temperature automatically implies the presence of dissipation which, as discussed in more detail in Sec. 2.6 below, may be described—in the simplest possible approach—by including a phenomenological temperature-induced damping term in the GPE model.

In order to better clarify the above, and discuss in more detail the stability of the excitation corresponding to the anomalous mode, namely of the dark soliton, we note the following. At temperatures $T \to 0$ (as is the case under consideration), the negative energy of the dark soliton does not imply any instability (e.g., a decay process) and, thus, the soliton is dynamically stable. Nevertheless, at finite temperatures, i.e., in the presence of a thermal cloud, the above mentioned properties of the anomalous mode indicate that the soliton will become unstable: in this case, the presence of the temperature-dependent damping results in the decay of the soliton (see discussion in Refs. [126, 170] as well as in Sec. 2.6 below). From a physical point of view, the decay mechanism resembles the one of the low-energy excitations of trapped BECs [197], and originates from the scattering of thermal particles on the dark soliton. Thus, according to these arguments, matter-wave dark solitons can be regarded as thermodynamically unstable excitations as, in the presence of the temperature-induced dissipation, the system will be driven towards configurations with lower energy; in other words, the dark solitons will decay to the ground state. This scenario is also often referred to as *energetic instability* [198].

As mentioned above, the eigenfrequency of the anomalous mode is equal to the oscillation frequency of a dark soliton around the center of the harmonic trap in the TF limit. On the other hand, the eigenfunctions u_A and υ_A of the anomalous mode, shown in Fig. 2.6, are localized within the notch of the dark soliton [170, 193], and their sum can be approximated as $u_A + \upsilon_A \propto \text{sech}^2(\sqrt{n_0}z)$. Notice that in the case of a uniform condensate (i.e., in the absence of the trap), there exists a translational mode with zero frequency which has the same functional form, namely $\partial_z \psi_0$, due to the translational

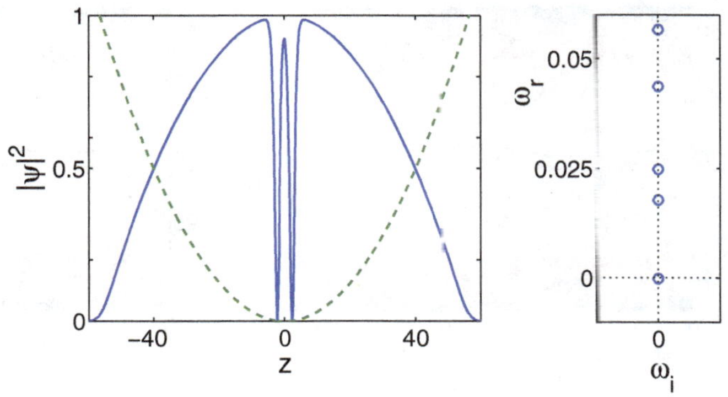

Figure 2.7. *Left panel: The density of a condensate carrying a stationary two-dark-soliton state (parameter values are* $\mu = 1$ *and* $\Omega = 0.025$*, as in Fig. 2.5). The solitons are located at* $z \approx \pm 2.3$*. Right panel: The lowest characteristic eigenfrequencies of the Bogoliubov excitation spectrum: The eigenfrequency located at the origin corresponds to the Goldstone mode, the one at* $\Omega = 0.025$ *to the Kohn mode, and the one at* $\sqrt{3}\Omega = 0.043$ *to the quadrupole mode. Finally, there exist two anomalous modes with eigenfrequencies* $\omega_1 = 0.0179$ *and* $\omega_2 = 0.0566$*. Adapted with permission from Ref. [75].*

invariance of the dark soliton solution. When the trap is present, however, this symmetry is broken, which suggests that the anomalous mode can be regarded as the "ghost" of the broken translational invariance of the dark soliton solution. We finally mention that the direct connection of the anomalous mode to the oscillation of the dark soliton can be better explained by the fact that an excitation of the stationary black soliton ψ_0 by the anomalous mode results in a displacement of the soliton from the trap center (see right panel in Fig. 2.6). In other words, the GPE (2.1) with the initial condition $\psi(z;t=0) = \psi_0(z) + \delta[u_A(z) + v_A^*(z)]$, for small δ, will naturally lead to dark soliton oscillations studied in Sec. 2.2.2.

The Multisoliton State

The BdG analysis can also be performed in the more general case of multisoliton states [199], which may be found in a stationary form (as explained in Sec. 2.1.6). In this case, starting from the noninteracting limit, it can be found that the Bogoliubov spectrum of the nth excited state consists of one zero eigenvalue (corresponding to the Goldstone mode), n double eigenvalues (accounted for by the presence of the harmonic trap), and infinitely many simple ones. Then, in the nonlinear regime, one of the eigenvalues of each double pair becomes an anomalous mode of the system (characterized by a negative valued integral of the norm × energy product) and, thus, the number of anomalous modes in the excitation spectrum equals to the number of dark solitons [199]. This is in agreement with the fact that the number of eigenvalues with negative Krein signature equals to the number of the nodes of the stationary state [200]. Notice that in the framework of the 1D GPE (2.1)—and in particular in the TF-1D regime—the first anomalous mode coincides with the oscillation frequency $\omega_{\text{osc}} = \Omega/\sqrt{2}$ of the single dark soliton. An example of a condensate with a stationary two-dark-soliton state, as well as the pertinent spectral plane, are shown in Fig. 2.7.

The physical significance of the n anomalous modes has been discussed in Refs. [72, 199]: for example, in the case of a two-dark-soliton state, the smallest of the anomalous modes corresponds to an in-phase oscillation (i.e., when the two dark solitons oscillate together without changing their relative spatial separation), the largest anomalous mode

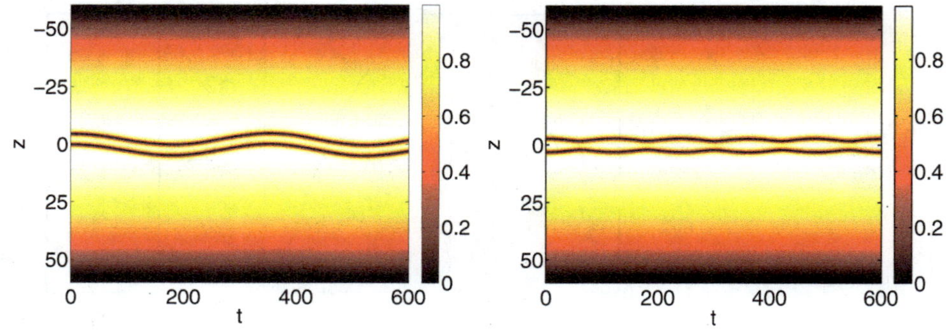

Figure 2.8. *Spatio-temporal contour plot of the condensate density, for parameter values* $\mu = 1$ *and* $\Omega = 0.025$ *(as in Fig. 2.7). Left panel: The dark solitons, initially placed at* $z_1 = 0$ *and* $z_2 = 5$*, oscillate in phase with a frequency* $\omega_{\rm osc} = 0.018 \approx \omega_1 = 0.0179 \approx \Omega/\sqrt{2} = 0.0176$*. Right panel: The dark solitons, initially placed at* $z = \pm 3$*, oscillate out of phase with a frequency* $\omega_{\rm osc} = 0.057 \approx \omega_2 = 0.0566$*. Here,* $\omega_{1,2}$ *are the eigenfrequencies of the first and second anomalous modes, respectively. Adapted with permission from Ref. [75].*

corresponds to an out-of-phase oscillation (i.e., when the two dark solitons move to opposite directions with the same velocity and undergo head-on collision), and so on. An example of the in-phase and the out-of-phase oscillations of the two-dark-soliton state shown in the left panel of Fig. 2.7 (when the solitons are properly displaced from their equilibrium positions) is illustrated in Fig. 2.8. Here it should be noticed that since the starting point of the above considerations is the noninteracting limit, a similar analysis can also be performed in the case of other mean-field models with noncubic nonlinearity, as the ones describing cigar-shaped BECs in the dimensionality crossover regime from 3D to 1D (see Sec. 1.2.4 and Sec. 2.3.5 below).

Finally, as concerns the stability of nonlinear modes (see relevant investigations in Refs. [72, 201, 52, 202]), all higher-order nonlinear modes are unstable near the noninteracting limit [72, 202] (but can be stabilized by using anharmonic traps [202]); nevertheless, the instability ceases to exist sufficiently deep inside the nonlinear regime (i.e., for sufficiently large BECs, with large N) [72].

2.2.4 ▪ Radiation Effects: Sound Emission by the Soliton

As indicated in Sec. 2.2.1, the dark soliton experiences a background density gradient in the presence of the perturbation $R(\psi)$ in Eq. (2.56) and, thus, it continuously emits energy in the form of sound waves. Here, we will study this effect in more detail, considering the case of $R(\psi) = V(z)\psi$ with $V(z)$ being a harmonic potential. A particular feature of this setup is that the emitted sound energy remains confined within the spatial region of the trap and, hence, the soliton continuously re-interacts with the emitted sound waves [164]; in fact, this process is such that, on average, the dark soliton re-absorbs the radiation it emits. Thus, in the case of harmonic traps, an investigation of the inhomogeneity-induced sound emission, as well as an estimation of the rate of emission of energy, is relevant for short timescales, i.e., $0 \le t \le T_{\rm osc} \equiv 2\pi/\omega_{\rm osc}$ [where $\omega_{\rm osc}$ is given by Eq. (2.69), with $\Omega \ll 1$]. On the other hand, if dark solitons evolve in the presence of nonharmonic potentials (such as localized barriers [80, 165, 171, 183, 203], disordered potentials [204, 205], anharmonic traps [206], or other "properly designed" potentials—see below), the problem may be easier—at least in terms of a numerical investigation. In fact, as is explained below, it is possible to consider suitable setups that either damp off the emitted sound density or cause the emitted sound to dephase.

Various such setups were proposed and analyzed in the past; the most prominent example is, perhaps, a tight inner "dimple" trap, confining a dark soliton, located within a much weaker outer harmonic potential (such a configuration can be realized by focusing an off-resonant laser beam within a harmonic trap) [164]. In this case, if the depth of the dimple trap is sufficiently small, the sound waves can escape (to the outer trap), while the soliton can remain confined in this region. In this limit, sound energy is damped off for short enough timescales, until it bounces off the weaker outer trap and thus becomes forced to re-interact with the soliton. An alternative setup considered in Refs. [207, 208] consists of a harmonic trap perturbed by an optical lattice potential (see Sec. 2.2.5). In this case, the optical lattice can confine a soliton within a single or a few lattice sites, with the sound (again for short enough times) escaping to neighboring sites. Although in this case the sound re-interacts with the soliton on faster timescales than in the case of the dimple trap mentioned above, the presence of the lattice dephases the emitted sound waves, and hence accelerates the soliton decay.

The radiation-induced dissipation of matter-wave dark solitons in harmonic traps was studied analytically in Ref. [184] by means of an asymptotic multiscale expansion method. Particularly, assuming that $R(\psi)=\epsilon^2 z^2\psi$ (with ϵ being a formal small parameter defined by the aspect ratio Ω), the following results were obtained. In the limit $\epsilon\to 0$, the dark soliton evolves adiabatically so that the dark soliton center $z_0(t)=vt+z_{\mathrm{o}}\to s(T)/\epsilon$, i.e., it becomes a function of the slow timescale $T=\epsilon t$, while the soliton velocity is given by $v(T)=\dot{s}\equiv ds/dT$. The adiabatic dynamics is followed by generation of sound waves, which can be taken into regard as per Eq. (2.57). In fact, in the decomposition of the wavefunction ψ into an inner and an outer asymptotic scale, the leading-order radiative effects are taken into account when the phase θ_{o} [see Eq. (2.58)] depends also on $T=\epsilon t$, i.e., $\theta_{\mathrm{o}}\equiv\theta(T)$, and the first-order corrections to the dark soliton (2.10) grow linearly in z. Neglecting reflections from the trap, the extended dynamical equation for the position $s(T)$ of the dark soliton (2.10) takes the form

$$\ddot{s}+s=\frac{\epsilon\dot{s}}{2\sqrt{(1-s^2)^3}\sqrt{1-s^2-\dot{s}^2}}+\mathrm{O}(\epsilon^2). \tag{2.81}$$

The left-hand side of Eq. (2.81) represents the leading-order adiabatic dynamics of the dark soliton [see also Eq. (2.68)] oscillating on the ground state of the trap, namely a harmonic oscillator with the obvious solution $s(T)=s_0\cos(T+\delta_0)$ (with s_0 and δ_0 being arbitrary parameters representing the initial position and phase of the soliton). As long as $s^2+\dot{s}^2<1$, the adiabatic dynamics approximation remains valid for large values of the position $s(0)$ and speed $\dot{s}(0)$ of the dark soliton. In other words, dark solitons oscillate in the trap with a *uniform* oscillation frequency for solitons of *all* amplitudes and velocities, in agreement with the prediction of the Landau dynamics approach (see Sec. 2.2.2). Obviously, in the limiting case of $s^2+\dot{s}^2\to 1$ (i.e., for extremely shallow dark solitons) Eq. (2.81) is not applicable.

Next, letting $E=\frac{1}{2}\left(\dot{s}^2+s^2\right)$ be the energy of the harmonic oscillator, one may employ Eq. (2.81) to calculate the rate of change of E due to the leading-order radiative effects appearing in the right-hand side of Eq. (2.81). The result is

$$\dot{E}=\frac{\epsilon\dot{s}^2}{2\sqrt{(1-s^2)^3}\sqrt{1-s^2-\dot{s}^2}}+\mathrm{O}(\epsilon^2)>0, \tag{2.82}$$

and shows that due to the energy pumping (2.82), the amplitude of the harmonic oscillator increases in time. In the limit $s^2+\dot{s}^2 \to 0$, Eqs. (2.81) and (2.82) can be simplified. First, the energy of the dark soliton oscillations accelerates by the squared law $\dot{E}=\epsilon\dot{s}^2/2$, which was confirmed in numerical simulations in the setup of Ref. [164]. Second, the nonlinear equation (2.81) is linearized as follows:

$$\ddot{s}+s-\frac{\epsilon}{2}\dot{s}=O(\epsilon^2,s^3), \tag{2.83}$$

Equation (2.83) includes an *antidamping* term accounting for the emission of radiation, indicating that the center point $(0,0)$ becomes an unstable spiral point on the plane $(s,\dot{s})$. Apparently, the leading-order solution reads $s(T)=s_0e^{\epsilon T/4}\cos(T+\delta_0)$; thus, the amplitude of oscillations of a dark soliton increases while its own amplitude decreases.

The above results of the asymptotic analysis were confirmed by the numerical simulations of Ref. [184], but also by numerical findings reported in other works: the radiation-induced effects were also observed for dark solitons oscillating between two Gaussian humps [165], or for dark solitons that are parametrically driven by a pair of two periodically modulated Gaussian barriers, oscillating in antiphase at a frequency close to the soliton frequency [209]. It is worth noticing that the mechanism proposed in Ref. [209] pumps energy into the dark soliton, which may compensate the inhomogeneity-induced emission of radiation, as well as the damping due to the presence of the thermal cloud [126] (see Sec. 2.6 below).

2.2.5 ▪ Dark Solitons in Optical Lattices

General Considerations

Bose–Einstein condensates loaded into periodic optical potentials, namely optical lattices (OLs), have attracted much attention as they demonstrate rich physical properties and nonlinear dynamics (see, e.g., Refs. [44, 94, 210, 211, 212, 213] for reviews). Optical lattices are generated by a pair of interfering laser beams forming a standing wave which induces a periodic potential; thus, for a BEC confined in an optical lattice, the trapping potential in the GPE model can be regarded as a superposition of a harmonic (magnetic or optical) trap and a periodic potential. Particularly, in a quasi-1D setting (the generalization to higher-dimensional cases is straightforward), cf. Eq. (2.56), the trap takes the following dimensionless form (see, e.g., Ref. [45]):

$$V(z)=\frac{1}{2}\Omega^2z^2+V_0\cos^2(kz), \tag{2.84}$$

where Ω and V_0 denote, respectively, the harmonic trap and OL strengths, $L\equiv\pi/k=(\lambda/2)\sin(\phi/2)$ is the periodicity of the lattice, with λ being the common wavelength of the two interfering laser beams, and ϕ the angle between them.

In some cases (as, e.g., in the experiments of Refs. [214, 215]), the harmonic potential is very weak as compared to the optical lattice and, thus, it can be ignored. Then the stationary states of the pertinent GPE—including solely the OL potential—can be found in the form of infinitely extended waves (Bloch waves), with the periodicity of the OL. Particularly, in the linear limit where the Floquet–Bloch theory is relevant, the stationary states can be found upon employing the usual ansatz, $\psi(z,t)=F(z)\exp(-i\mu t)$ (here μ is the dimensionless chemical potential), where the function $F(z)=u_{k,\alpha}(z)\exp(ikz)$ and $u_{k,\alpha}(z)=u_{k,\alpha}(z+nL)$ with $n\in\mathbb{Z}$ [216]. If the Floquet exponent (also called "quasi-momentum") k is real then the wavefunction ψ has the form of an infinitely extended wave, known as a *Bloch wave*. Such waves exist in *bands* (labeled by the index α), while they do not exist in *gaps*, which are spectral regions characterized by $\mathrm{Im}(k) \neq 0$. Importantly, the concept of Bloch waves can also be extended in the nonlinear case, giving rise to nonlinear Bloch waves (see, e.g., Chapter 6 in Ref. [44] and references therein). In the case of weak nonlinearity, the nonlinear band-gap spectrum and the nonlinear Bloch waves are similar to the ones in the linear case [217]. However, deeper in the nonlinear regime (or, physically speaking, when the local BEC density grows), the chemical potential of the nonlinear Bloch wave is increased for repulsive BECs and, thus, the nonlinearity effectively "shifts" the edges of the linear band.

In the absence of the harmonic potential, and if the OL is very deep compared to the chemical potential, the strongly spatially localized wavefunctions at the lattice sites are approximated by Wannier functions in the so-called superfluid regime (see, e.g., Ref. [218]) and the *tight-binding approximation* can be applied. then, the continuous GPE is reduced to the discrete NLS (DNLS) equation (see, e.g., Refs. [94, 45, 210] and Chapter 13 in Ref. [44], as well as Refs. [219, 220] for reviews for the DNLS model).

Let us now focus on the case of weak optical lattices, which is more relevant for studies of solitons. In fact, apart from dark solitons, which can naturally exist in the presence of the OL (see discussion below), matter waves can also feature a nonlinear localization in the gaps of the linear spectrum, and form fundamental nonlinear structures, namely *gap solitons*—cf. the experiment reported in Ref. [214]. The underlying mechanism can effectively be described by means of the so-called Bloch-wave envelope approximation near the band edge, first formulated in the context of optics [221], and then used in the BEC context as well (see, e.g., the review [210] and references therein). In particular, a multiscale asymptotic method was used to show that the BEC wavefunction can effectively be described as $\psi(z,t) = U(z,t)u_{k,0}(z)\exp(ikz)$, where $u_{k,0}(z)\exp(ikz)$ represents the Bloch state in the lowest band $\alpha = 0$ (at the corresponding central quasimomentum k), while the envelope function $U(z,t)$ is governed by the following dimensionless NLS equation:

$$i\frac{\partial U}{\partial t} = -\frac{1}{2m_{\text{eff}}}\frac{\partial^2 U}{\partial z^2} + g'_{1D}|U|^2U, \tag{2.85}$$

where g'_{1D} is a renormalized (due to the presence of the OL) effectively 1D coupling constant, and m_{eff} is the effective mass. Importantly, the latter is proportional to the inverse effective diffraction coefficient $\partial^2\mu/\partial k^2$ whose sign may change, as it is actually determined by the curvature of the band structure of the linear Bloch waves. Obviously, the NLS Eq. (2.85) directly highlights the above-mentioned nonlinear localization and soliton formation, which occurs for $m_{\text{eff}} \times g'_{1D} < 0$. Note that the above method can be extended in higher-dimensional (2D and 3D) settings [222], in which the effective mass becomes a tensor. In such a case, and for repulsive BECs, all components of the tensor have to be negative for the formation of multidimensional localized gap nonlinear structures [223].

As mentioned above, dark soliton states can exist in the OL setting. In particular, dark solitons have been studied both in combined harmonic and OL potentials [207, 208, 224, 225] and in optical lattices, in the absence of the parabolic trap. In the latter case, various studies have been performed in the frameworks of the continuous GPE, as well as its tight-binding approximation counterpart [226, 227, 228, 229, 230]. Matter-wave dark solitons have also been studied in double-periodic optical superlattices [230, 231]; such potentials are of the form

$$V(z) = V_1\cos(k_1 z) + V_2\cos(k_2 z), \tag{2.86}$$

where k_1 and $k_2 > k_1$ are the primary and secondary lattice wavenumbers, and V_1 and V_2 are the sublattice amplitudes [232]. Finally, we note in passing that there exists a vast amount of work concerning dark solitons in periodic media arising in various contexts, such as nonlinear optics [233, 234, 235, 236, 237, 238], solid-state physics [239], and the theory of nonlinear waves [240].

Below, we briefly discuss results of various studies devoted to the stability and dynamics of dark solitons in the presence of the OL potential.

Dark Solitons in Combined Harmonic and OL Potentials

The stability of matter-wave dark solitons in the combined harmonic and OL potential was first studied in Ref. [224] by means of a BdG analysis that was performed in the framework of both the continuous quasi-1D GPE and its DNLS counterpart. It was found that in the discrete model stationary dark solitons located at the minimum of the harmonic trap are, generally, subject to a weak oscillatory instability, which manifests itself as a shift of the soliton from its initial location, accompanied by quasiperiodic oscillations. On the other hand, in the continuous GPE model, dark solitons may be stable, with the (in)stability determined by the period and amplitude of the OL. In any case, the dark solitons are robust and if the oscillatory instability is present, it sets in at large times.

The dynamics of dark solitons in the combined harmonic and OL potential can be studied upon distinguishing physically relevant cases, depending on the competition of the characteristic spatial scales of the problem [208]. Particularly, assuming that the harmonic trap varies slowly on the soliton scale, i.e., $w = 1/\cos\phi \sim \xi \ll \Omega^{-1}$, where w is the soliton width for chemical potential $\mu = 1$, ϕ is the soliton phase angle, and ξ the healing length, the following three cases can readily be identified:

(a) the case of a long-period OL, with $L \gg \xi$,

(b) the case of a short-period OL, with $L \ll \xi$, and

(c) the intermediate case, with $L \sim \xi$.

Then, if the OL strength is sufficiently small, the soliton dynamics in cases (a) and (b) can be treated in the framework of the adiabatic approximation (see Sec 2.2.2). Particularly, as shown in Ref. [208], case (a) can be studied by means of the Hamiltonian approach of the perturbation theory, and case (b) by means of a multiscale expansion method (treating k^{-1} as a small parameter); this way, it can be shown that in both cases the dark soliton behaves as an effective classical particle, performing harmonic oscillations in the presence of the trap of Eq. (2.84). The oscillation frequency, which is different from its characteristic value $\Omega/\sqrt{2}$, is modified by the presence of the lattice according to the equations

$$\omega_{\rm osc} = \sqrt{\frac{1}{2}\Omega^2 - V_0 k^2}, \qquad \omega_{\rm osc} = \frac{\Omega}{\sqrt{2}}\left(1 - \frac{7}{256}\frac{V_0^2}{k^4}\right), \tag{2.87}$$

for cases (a) and (b), respectively. As concerns the more interesting case (c) (see Refs. [207, 208]), it can be shown that if the dark soliton is initially placed quite close to the bottom of a well of the OL potential, it remains there for a rather long time; eventually, however, it escapes due to the radiation-loss mechanism, and then performs large-amplitude oscillations in the condensate. Furthermore, if the harmonic trap is weak enough, the soliton eventually decays. In fact, as discussed in Ref. [207], the OL causes a dynamical instability (because the dark soliton has to "traverse" the potential humps caused by the lattice) resulting in a faster decay of the soliton than if it was evolving in the presence of the harmonic trap only: the presence of the lattice dephases the sound waves emitted by the soliton, hence reducing the effectiveness of the soliton to get stabilized by reabsorbing the sound waves (see discussion in Sec. 2.2.4). Nevertheless, according to the observations of Ref. [208] that the soliton may remain stationary for a relatively long time, in Ref. [225] (see also Ref. [208]) it was proposed that a time-dependent OLs may either (i) capture a moving dark soliton or (ii) capture and drag a stationary soliton, bringing it to a preselected final destination. Notice that the transfer mechanism is robust as long as adiabaticity of the process is ensured (i.e., for sufficiently small speeds of the moving OL).

Dark Solitons in Optical Lattices and Superlattices

As mentioned above, dark solitons in OLs and superlattices have also been studied, in the absence of the harmonic trap, in the frameworks of the continuous and discrete NLS equations. Particularly, in Ref. [226], a DNLS model was derived in the tight-binding approximation—i.e., for a single-isolated band in the Floquet–Bloch spectrum—which was used to study matter-wave dark solitons. Later, in Ref. [227], a continuous coupled-mode model was used to study the existence and stability of so-called dark lattice solitons, while a more general analysis was presented in Ref. [228]; in that work, a continuous GPE model with periodic potential was shown to support stable stationary dark solitons for both attractive and repulsive interatomic interactions, which were found numerically [228] (see also relevant results in [229]).

In a more recent work [230], where both regular optical lattices and superlattices were considered, it was shown that each type of nonlinear Bloch wave can serve as a stable background supporting dark solitons. This way, different families of dark solitons, originating within the bands of the Floquet–Bloch spectrum, were found and their dynamical properties were analyzed. In particular, considering the continuous analogue of the Peierls–Nabarro potential (see, e.g., Ref. [241]) in discrete lattices, it was shown that the mobility and interaction properties of the dark solitons can be effectively controlled by changing the structure of the optical superlattice; moreover, following the ideas of Ref. [225], time-dependent superlattices were also shown to control the static and dynamical properties of matter-wave solitons [231].

Here we should point out that all the above mentioned studies on the dynamics of matter-wave dark solitons in optical lattices were carried out in the framework of the GPE mean-field theory. Nevertheless, it is worth emphasizing that the GPE is inadequate for dealing with several important aspects of ultracold bosons in optical lattices, such as the superfluid–to–Mott-insulator phase transition (see, e.g., Refs. [242, 243]) or, more generally, strong correlation effects (see, e.g., the review [213]). Thus, more recently, studies on the quantum dynamics of dark solitons have started to appear. In that regard, it is relevant to mention that matter-wave dark solitons were studied in the context of the Bose–Hubbard model [244, 245], and it was found that dark soliton collisions become inelastic, in strong contrast to the predictions of mean-field theory. A conclusion of the above works [244, 245] is that the lifetime and collision properties of matter-wave dark solitons in optical lattices may provide a clear signature of quantum effects. Additionally, in another recent work [246], dark solitons were studied in the vicinity of the superfluid–to–Mott-insulator transition; particularly, in this work [246], antisymmetric eigenstates corresponding to standing solitons, as well as propagating solitons created by phase imprinting, were presented and the soliton characteristics were found to depend on quantum fluctuations.

From the viewpoint of experiments, trains of stationary dark solitons were observed in a ^{87}Rb condensate confined in a 3D harmonic trap and a 1D OL [65]. The underlying mechanism for the formation of such structures were multiple Bragg reflections caused by displacing the harmonic trap and, thus, setting the BEC into motion. Due to the dimensionality of the system, the solitons were found to be subject to the snaking instability, giving rise to the subsequent formation of vortex rings (see Sec. 2.3.2), similarly to the observations of the pertinent JILA experiment (but without the OL) [49].

2.2.6 ▪ Persistence and Stability of Dark Solitons

As was highlighted in this section, there exist many alternative approaches for the study of the statics and dynamics of matter-wave dark solitons in the quasi-1D setting. Nevertheless, rigorous results concerning the persistence and stability of dark solitons in a generalized GPE-like model [cf. Eq. (1.22)] were obtained only recently [185, 247]. Particularly, in Ref. [247], the existence and stability of a black soliton of Eq. (2.27) were studied in the absence of the potential term, while in Ref. [185] a more general model, incorporating the potential term, was considered. More specifically, the model used in Ref. [185] was of the following form:

$$i\partial_t\psi = -\frac{1}{2}\partial_z^2\psi + f(n)\psi + \varepsilon V(z)\psi, \tag{2.88}$$

where ε is a formal small parameter setting the strength of the potential. The results obtained in Ref. [185] for bounded and exponentially decaying potentials (as, e.g., ones corresponding to red-detuned laser beams—see, e.g., the experiment of Ref. [248]) can be summarized as follows.

Let us consider that, in the absence of the potential, Eq. (2.88) admits a black soliton solution of the form $\psi(z,t) = q(z-s)\exp[-if(n_0)t+i\theta]$ (here, s is the soliton center and θ an arbitrary constant phase), with boundary conditions $q \to \pm\sqrt{n_0}$ as $z \to \pm\infty$. Then the above solution persists in the presence of the perturbation induced by the potential term in Eq. (2.88) provided that the function $M(s)$ defined through

$$M'(s) = \frac{dM}{ds} = \int_{-\infty}^{+\infty} V'(z)[n_0 - q^2(z-s)]dz \tag{2.89}$$

possesses at least one single root, say s_0. Then, the stability of the dark soliton solution depends on the sign of the first derivative of the function in Eq. (2.89), evaluated at s_0: an instability occurs, with one imaginary eigenfrequency pair for $\varepsilon M''(s_0) < 0$, and with exactly one complex eigenfrequency quartet for $\varepsilon M''(s_0) > 0$. In fact, this instability is dictated by the translational eigenvalue, which bifurcates from the origin as soon as the perturbation is present. For $\varepsilon M''(s_0) < 0$, the relevant eigenfrequency pair moves along the imaginary axis, leading to an instability associated with exponential growth of a perturbation along the relevant eigendirection. On the other hand, for $\varepsilon M''(s_0) > 0$, although the eigenfrequency should move along the real axis, it can not do so because the latter is filled with continuous spectrum; thus, since the translation mode and the eigenmodes of the continuous spectrum have opposite Krein signature, the collision of the eigenfrequency of the translational mode with the continuous spectrum results in a complex eigenvalue quartet, signaling the presence of an oscillatory instability. The relevant eigenfrequencies can be determined by a quadratic characteristic equation which, in the case of the cubic GPE model (2.88) with $f(n) = n$ and $n_0 = 1$, takes the form [185]

$$\lambda^2 + \frac{\varepsilon}{4}M''(s_0)\left(1 - \frac{\lambda}{2}\right) = O(\varepsilon^2), \tag{2.90}$$

and the eigenvalues λ are related to the eigenfrequencies ω through $\lambda^2 = -\omega^2$. For sufficiently small $\varepsilon > 0$, this equation has only one real root $\lambda(\varepsilon) > 0$ for $M''(s_0) < 0$ and two complex-conjugate roots, with $\mathrm{Re}\{\lambda(\varepsilon)\} > 0$ for $M''(s_0) > 0$.

It is interesting to observe that if the characteristic equation (2.90) is formally applied to the cubic GPE model (2.88) with $f(n)=n$, $n_0=1$ and $V(z)=z^2$, one obtains $M''(s_0)=2\int_{-\infty}^{+\infty}\text{sech}^2(z)dz=4$ and, thus, Eq. (2.90) takes the form $\lambda^2-(\varepsilon/2)\lambda+\varepsilon=O(\varepsilon^2)$. Using appropriate rescalings, it can easily be shown that the latter characteristic equation can be derived from Eq. (2.83) of Sec. 2.2.4. Although the validity of the radiative boundary conditions for $V(z)=z^2$ cannot be verified by the analysis of Ref. [185], the above observation leads to the following conjecture [185]: in the most general GPE model [cf. Eq. (2.88)], the two complex-conjugate eigenvalues with positive real part for $M''(s_0)>0$ result from the following Newton's particle equation of motion for the soliton center $s(t)$:

$$\mu_0\ddot{s}-\varepsilon\lambda_0 M''(s)\dot{s}=-\varepsilon M'(s), \tag{2.91}$$

where $M(s)$ is the effective potential implied by Eq. (2.89), while the constants μ_0 and λ_0 represent, respectively, the soliton's mass and antidamping —as per the discussion of Sec. 2.2.4.

The validity of Eq. (2.91), as well as the other theoretical predictions presented in this section, were tested against numerical simulations in Ref. [185] for small decaying potentials, and the agreement between the analytical and numerical results was found to be very good. Notice that although the above results of Ref. [185] can only be rigorously applied to the case of small, bounded, and exponentially decaying potentials, the basic qualitative features may formally persist for other types of external potentials as well. A pertinent example is the work of Ref. [249], where the persistence and stability of matter-wave black solitons were studied in a condensate characterized by a periodic, piecewise-constant scattering length as shown in Ref. [249].[8] A formal application of the predictions of Ref. [185] concerning the persistence and stability of dark solitons in this setting, was found to be in very good agreement with relevant numerical findings. Nevertheless, an analysis similar to the one presented in Ref. [185], but for other types of potentials (such as confining and periodic ones) is still missing.

2.3 ▪ "Low-Dimensional" Multidimensional Dark Solitons

2.3.1 ▪ Rectilinear Solitons

Quasi-1D matter-wave dark solitons may naturally exist in higher-dimensional settings. For example, in the experimentally relevant case of a cigar-shaped trap, the actual dimensionality of the BEC density is 3D rather than 1D, despite the fact that the BEC can be treated as an effectively 1D object using the NLS Eq. (1.22) with the generalized nonlinearities of Eqs. (1.23)–(1.25)—see Sec. 1.2.4. The density of such a cigar-shaped BEC with a dark soliton on top of it is illustrated in the left panel of Fig. 2.9. Furthermore, quasi-1D dark solitons may also exist in disk-shaped BECs (see Sec. 1.2.4), which are described by the (2+1)-dimensional GPE (1.27). The latter, can be expressed in the following dimensionless form:

$$i\partial_t\psi=\left[-\frac{1}{2}\nabla^2+V(r)+|\psi|^2\right]\psi, \tag{2.92}$$

where $\nabla^2=\partial_x^2+\partial_y^2$, and the density $|\psi|^2$, length, time, and energy are respectively measured in units of $2\sqrt{2\pi}aa_z$, a_z, ω_z^{-1}, and $\hbar\omega_z$, while the potential $V(r)$ is

$$V(r)=\frac{1}{2}\Omega^2 r^2, \tag{2.93}$$

[8] BECs with spatially varying nonlinearity coefficient g, so-called collisionally inhomogeneous condensates [250], have attracted much attention, as they provide a variety of interesting phenomena—cf. Sec. 2.5.3 below.

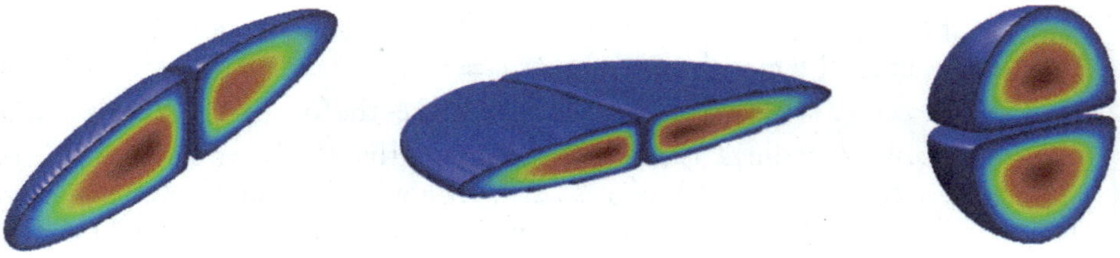

Figure 2.9. *Examples of the 3D densities of condensates, confined in various types of anisotropic harmonic traps and carrying quasi-1D dark solitons. Shown are (from left to right) the longitudinal cuts of a cigar-shaped BEC, a disk-shaped BEC, and a spherical BEC. Adapted with permission from Ref. [75].*

with the aspect ratio being $\Omega = \omega_\perp/\omega_z \ll 1$. In this case, the soliton of Eq. (2.10), with the variable z being replaced by x, is an exact analytical solution of Eq. (2.92) for $V(z) = 0$. This "rectilinear" soliton has the form of a dark "stripe" on top of a 2D TF cloud, and the BEC wavefunction can be expressed (similarly to the 1D case) as $\psi = \psi_{\rm TF}(r)\exp(-i\mu t)\psi_{\rm ds}(x,t)$. It is also natural to consider the full $(3+1)$-dimensional version of Eq. (2.92), with $\nabla^2 = \partial_x^2 + \partial_y^2 + \partial_z^2$, where quasi-1D dark soliton stripes exist as well. In this case, the potential and the 3D TF cloud are modified according to the relative values of the confining frequencies in the three directions. Examples of the densities of a disk-shaped BEC and a spherical BEC carrying a rectilinear dark soliton are shown, respectively, in the middle and right panels of Fig. 2.9.

Apart from the quasi-1D dark solitons, purely 2D dark solitons without vorticity have also been predicted to occur in theory (but they have not been observed so far in experiments). Such dark soliton solutions of the GPE (2.92), which have been derived in the framework of the small-amplitude approximation (see Sec. 2.1.4), may have the form of lumps satisfying an effective Kadomtsev–Petviashvili (KP) equation [134] or dromions satisfying an effective Davey–Stewartson system [135]; quasi-1D and 2D dark solitons of the dromion type have also been predicted to occur in disk-shaped multicomponent condensates [138].

We also note that a more "exotic" dark soliton configuration in the 2D setting was presented in Ref. [251] (see also Ref. [252]). This configuration, which refers to a two-component BEC (see Sec. 2.4), consists of a "cross" formed by the intersection of two rectilinear domain walls, with the wavefunctions of the same species filling each pair of opposite quadrants having a π phase difference. This way, a dark soliton configuration is formed, which was found to be stable for long times—and even in the presence of rotation of the trap—in a large parametric region (see Sec. 3.11.4 below for more detales on these solutions).

2.3.2 ▪ Snaking Instability: Preliminary Discussion

An important issue arising in higher-dimensional settings is the stability of dark solitons which, for simplicity, will be studied at first in the $(2+1)$-dimensional geometry (relevant to disk-shaped BECs) and in the absence of the potential $V(r)$. The stability of the 1D dark soliton stripe (lying, say, along the x direction) in such a 2D setting was first studied in Ref. [253] (see also Refs. [254, 255]). In this work, it was shown that the soliton is prone to transverse modulational instability, i.e., it is unstable against long-wavelength transverse periodic perturbations $\sim \cos(Qy)$, where Q is the wavenumber of the perturbation. In particular, the instability band is defined by $Q < Q_{\rm cr}$, where the critical value of the perturbation wavenumber is given by (for $\mu = 1$)

$$Q_{\rm cr}^2 \equiv \cos^2\phi - 2 + 2\sqrt{\cos^4\phi + \sin^2\phi}. \tag{2.94}$$

The above expression is a result of a linear stability analysis, which indicates that the amplitude of the rectilinear dark soliton will grow exponentially in the transverse direction. Nonlinear regimes of this instability were also studied analytically by means of asymptotic expansion techniques (see, e.g., Ref. [256] and references therein). This instability was extensively studied in the context of nonlinear optics, both theoretically [35] and experimentally [257, 258], and was found to be responsible for a possible decay of a plane dark soliton into a chain of vortices of opposite topological charges (vortex-antivortex pairs). Particularly, when the transverse modulational instability sets in, a plane black soliton undergoes a transverse "snake" deformation (hence the name "snaking instability") [35, 256], causing the nodal plane to decay into vortex pairs. On the other hand, unstable gray solitons may not decay into vortices, but rather perform long lived oscillations accompanied by emission of radiation in the form of sound waves.

In the following chapters, we will further elaborate on the formation and subsequent dynamics of the above mentioned vortex structures in the higher-dimensional settings (2D and 3D). Nevertheless, as a preliminary discussion of the structures that will be studied in more detail below, we will introduce the special class of dark solitons in higher-dimensional settings, namely ones exhibiting a radial symmetry.

2.3.3 ▪ Dark Solitons of Radial Symmetry

Radially symmetric dark solitons can be realized by wrapping the nodal plane around on itself. Such structures were first introduced in the context of nonlinear optics [259], with the motivation being that these dark solitons may not be prone to the snaking instability: indeed, if a dark stripe is bent so as to form a dark ring of length $L < 2\pi/Q_{\rm cr}$ (in the $(2+1)$-dimensional geometry), then the snaking instability will be suppressed. Such ring dark solitons (RDSs) were studied in nonlinear optics both theoretically [260, 261] and experimentally [262, 263, 264], while later were also predicted to occur in BECs [265]. In this context, and in the 2D setting (i.e., in a disk-shaped BEC), the RDS has the form of an annular "trough." On the other hand, in a 3D setting (i.e., in a spherical BEC), the radially symmetric dark soliton is called spherical dark soliton [182], or spherical shell soliton (SSS)—according to the nomenclature of Ref. [266]—and has the form of a nodal spherical "shell." Examples of the 3D densities of a disk-shaped and a spherical BEC, carrying a RDS and a SSS, are shown in Fig. 2.10.

As was originally proposed in Ref. [265], RDSs may be generated in BECs by means of phase-engineering techniques (i.e., by a proper phase-imprinting method—see Sec. 2.1.5), similar to the ones used for the generation of optical RDSs [262, 263, 264]. Another technique that has been proposed for the generation of RDSs in BEC is the matter-wave interference method (see Sec. 2.1.5): if the condensate is initially trapped in a narrow cylindrical box-like potential, and then is allowed to coherently expand in the presence of a wider cylindrical impenetrable hard-wall potential, it is reflected from the boundary, and the self-interference pattern has the form of a sequence of nonstationary concentric RDSs [153, 154, 155] (see also relevant work in Ref. [156]).

From a mathematical standpoint, matter-wave dark solitons of radial symmetry can be considered as quasi-1D objects and, accordingly, be analyzed by means of a quasi-1D GPE. In particular, either the RDS or SSS can be described by Eq. (2.92), with the Laplacian

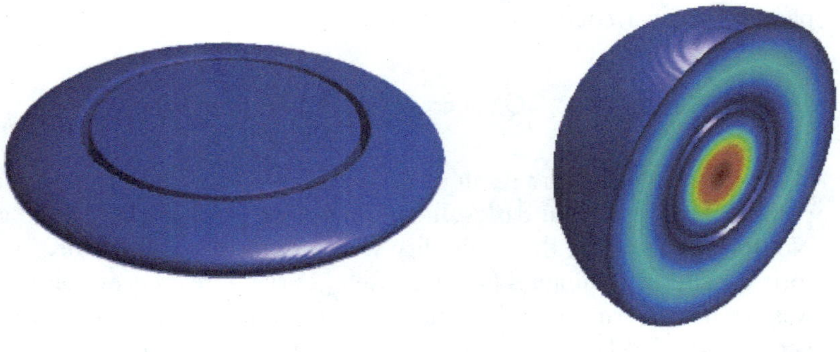

Figure 2.10. *Examples of the 3D densities of condensates, confined in a disk-shaped (left panel) and a spherical (right panel) traps, and carrying a ring dark soliton and a spherical shell soliton, respectively. Adapted with permission from Ref. [75].*

being of the form

$$\nabla^2 = \partial_r^2 + \frac{(D-1)}{r}\partial_r, \tag{2.95}$$

with $D = 1,2,3$. In this setup, the simplest case of $D = 1$ reduces Eq. (2.92) to the 1D GPE (2.1) describing a quasi-1D BEC (here, $r \equiv z$). The higher-dimensional setups correspond to the cases of $D = 2$ and $D = 3$: in the former case, Eq. (2.92) describes a disk-shaped BEC in the (x,y) plane (with r given by $r = \sqrt{x^2+y^2}$), while in the latter case Eq. (2.92) describes a spherical BEC (with r given by $r = \sqrt{x^2+y^2+z^2}$).

In such a quasi-1D setup, the Hamiltonian or the Lagrangian perturbation theory for dark solitons (see Sec. 2.2.2) may also be applied for the study of the dynamics of the RDS and SSS. In particular, Eq. (2.92), with the Laplacian of Eq. (2.95), can be treated as a perturbed 1D NLS equation [similar to Eq. (2.56)] provided that both the potential term and the term $\sim r^{-1}$ can be considered as small perturbations. The latter assumption is physically relevant for radially symmetric solitons of large radius r_0. Then, approximating the functional form of the RDS or SSS as [cf. Eq. (2.64)],

$$\psi_s(r,t) = \cos\phi(t)\tanh\zeta + i\sin\phi(t), \quad \zeta = \cos\phi(t)[r - r_0(t)], \tag{2.96}$$

where $r_0(t)$ is the soliton radius, it can be found [265] (see also Ref. [182]) that r_0 is governed by the following Newtonian equation of motion:

$$\frac{d^2 r_0}{dt^2} = -\frac{1}{2}\frac{\partial V_{\text{eff}}}{\partial r_0}. \tag{2.97}$$

Here, the effective potential is given by

$$V_{\text{eff}}(r_0) = V(r_0) - \ln r_0^{2(D-1)/3}, \tag{2.98}$$

and $V(r_0) = (1/2)\Omega^2 r_0^2$ is the trapping potential evaluated at the soliton radius r_0. In the 1D limit of $D = 1$, the last term in the right-hand side of Eq. (2.98) vanishes and Eq. (2.97) is reduced to Eq. (2.68). In the higher-dimensional cases of $D = 2$ or $D = 3$, the equation of the soliton motion (2.97) is clearly nonlinear (even for a nearly black RDS or SSS) due to the presence of the repulsive curvature-induced logarithmic potential.

Equation (2.97) predicts the existence of both the oscillating (gray) and stationary (black) RDS or SSS: the former perform oscillations on top of the TF cloud, changing their radii between a minimum and a maximum value, while the latter correspond to the minimum of the effective potential Eq. (2.98) (such stationary states do not exist in the case of a uniform ground state—as in the context of nonlinear optics [259]). As was shown in Ref. [265] (see also Ref. [182]), such a particle-like approach can describe quite effectively the above generic scenarios and the RDS dynamics up to a certain time (before the development of instabilities). Furthermore, in Refs. [267, 268] it was shown that the dynamics of small-amplitude RDSs, as well as the collisions between them, can be described in the framework of an effective cylindrical KdV equation [128] (see also Refs. [259, 260, 261] for similar findings in optics). On the other hand, numerical simulations of Ref. [265] revealed that RDSs are generally unstable, as they either decay to radiation (the small-amplitude ones) or are subject to the snaking instability (the moderate- and large-amplitude ones). Interestingly, as will be shown in the next chapter, the snaking instability of the RDS results in the formation of vortex-antivortex pairs in multiples of four, which are initially set along a ring, forming a so-called vortex necklace. Eventually, this pattern relaxes to a set of four pairs located on a ring, which oscillate in the radial direction between the same limits which confined the oscillations of the original RDS; simultaneously, the pairs perform oscillatory motion along the ring [265].

Matter-wave dark solitons of radial symmetry were also analyzed by means of other approaches. For example, in Ref. [266] (see also Chapter 7 in Ref. [44]) RDSs and SSSs were considered as nonlinear generalizations of Bessel functions, namely solutions of the equation

$$q'' + \frac{1}{r}q' - \frac{S^2}{r^2}q + 2\mu q - 2q^3 = 0, \tag{2.99}$$

resulting from Eq. (2.92) [with $V(r) = 0$] when the ansatz $\psi = q(r)\exp(-i\mu t + iS\phi)$ is introduced (in the latter expression, S is the topological charge of a central vortex). In this setting (i.e., in the absence of the trap), the solutions of Eq. (2.99) include, apart from the ground state, singly and multiply charged vortices, as well as infinitely many RDSs; the nodes of the latter correspond to the nodes of the nonlinear Bessel function governed by Eq. (2.99). On the other hand, if an external harmonic potential is incorporated in Eq. (2.99), then it is possible to find infinite branches of nonlinear bound states, with each branch stemming from the respective mode of the underlying linear problem (the radially symmetric 2D quantum harmonic oscillator) [269]. We will further elaborate on this analysis in the next chapter.

A stability analysis performed in Refs. [266, 269] also revealed that the radially symmetric dark solitons are generically unstable (see Chapter 3 for details) but, in agreement to the findings of Ref. [265], their lifetimes may be considerably long. Additionally, as shown in the work [270], the lifetime of RDSs may be extended employing the so-called nonlinearity management (alias Feshbach resonance management (FRM) [271]) technique, which is based on the use of external fields to periodically modulate in time the s-wave scattering length (see Sec. 2.5.3 below); indeed, as shown in Ref. [270], a deep RDS becomes shallower in the presence of a modulated nonlinearity coefficient, resulting to a delay in the onset of the snaking instability and an increase of the soliton lifetime. In any case, the theoretical investigations indicate that RDSs and SSSs have a good chance to be experimentally observed. In fact, structures reminiscent of stationary SSSs have already been observed as transients in the Harvard experiment of Ref. [66] (see also Ref. [272] were SSSs are predicted to occur as a result of collisions of vortex rings).

2.3.4 ▪ Dark Soliton Stability in Cigar-Shaped Condensates

The transverse (in)stability of dark solitons confined in a purely 3D setting, namely in a cylindrical trap of the form $V(z,r) = (1/2)m[\omega_z^2 z^2 + \omega_\perp^2(x^2+y^2)]$, was first studied in

Ref. [170]. In that work, it was shown that dynamical stability of a black soliton (stationary kink), say ψ_0, with a nodal plane perpendicular to the axis of the cylindrical trap (see left panel of Fig. 2.9), requires a strong radial confinement [as in the $(2+1)$-dimensional case]. Particularly, it was shown that the instability can be suppressed if the transverse (radial) condensate component is *not* in the TF regime, which is guaranteed as long as $\hbar\omega_\perp > \mu$ (where μ is the 3D chemical potential). This criterion can physically be understood as follows: if the condensate is confined in a highly anisotropic (cigar-shaped) trap, then the energy of the lowest possible radial excitation, $\hbar\omega_\perp$, must exceed the kink-related kinetic energy, $K_0 = -(1/2)\psi_0\partial_z^2\psi_0 \sim \mu$, so that the latter cannot be transferred to the BEC's unstable transverse modes by the interatomic interaction. Moreover, a systematic study in Ref. [170] revealed a criterion of dynamical stability for the black soliton in terms of the ratio $\omega_\perp/\omega_z$, namely,

$$\gamma \equiv \frac{\mu}{\hbar\omega_\perp} < \gamma_c. \tag{2.100}$$

The critical value γ_c was calculated for various values of $\omega_\perp/\omega_z$ and it was found that a characteristic value of γ_c, pertinent to the limiting case of $\omega_\perp \gg \omega_z$, is $\gamma_c \approx 2.4$.

In a more quantitative picture, a detailed study of the BdG equations in Ref. [170] (see also relevant work in Refs. [191, 192]) revealed the emergence of complex eigenvalues in the excitation spectrum and their connection to oscillatory dynamical instabilities, including the snaking instability. Additionally, in Ref. [273] it was found that the emergence of real eigenvalues is directly connected to bifurcations of the rectilinear black solitons to other stationary states that may exist in BECs confined in cylindrical traps. In particular, an investigation of the dependence of the excitation energy on the dimensionality parameter d [cf. Eq. (1.20)] led to the following results: for sufficiently low excitation energies, the black soliton may bifurcate to a *solitonic vortex*[9] or to an axisymmetric vortex ring (see also Chapter 7 in Ref. [44] and references therein), with the corresponding bifurcation points occurring at a low and a higher value of d. It was also found that the emergence of the first (second) real eigenvalue in the Bogoliubov spectrum coincides with the above mentioned bifurcation points. Therefore, according to the above results, it can be concluded that the emergence of the complex eigenvalues in the excitation spectrum (a) denotes the onset of dynamical instabilities of black solitons and (b) indicates the excitation of lower energy topological states. Since these states are energetically preferable, the onset of the dynamical instability will result in decay of the "high-energy" black soliton to these "low-energy" states carrying vorticity.

On the other hand, the stability of moving (gray) solitons was analyzed in Ref. [102]. According to this work, and following the arguments of Ref. [170], a criterion for *not* being in the radial TF regime (which is required for dynamical stability of the gray soliton) is $w \lesssim R \sim \xi$, where R is the radial size of the BEC and w is the soliton width (ξ is the healing length). Recalling that the soliton width w and velocity v depend on the soliton phase angle as $w \sim 1/\cos\phi$ and $v \sim \sin\phi$, it is clear that as the soliton is moving towards the boundaries of the BEC, its width (velocity) is increased (decreased). Thus, the instability border $R \sim w$ for the gray soliton is reached for larger values of the parameter γ [see Eq. (2.100)] than the ones pertaining to the black soliton. In the case of strongly anisotropic traps, the critical value of the chemical potential required for the dynamical stability of the gray soliton is proportional to the soliton velocity. In other words, the stability domain of gray solitons is wider than the one of black solitons. In fact, the shallower the soliton is, the more stable it becomes, similarly to the case of homogeneous systems;

[9] We will elaborate further on these structures in Chapter 3.

see Eq. (2.94), which indicates that the instability band vanishes for shallow solitons with $\cos\phi \to 0$.

The numerical simulations of Ref. [102] have also revealed that for $\gamma > 10$ a phase-imprinted dark soliton—with a π-phase jump—always decays in a cigar-shaped BEC (on a time scale of order of $\omega_\perp^{-1}$), while for $\gamma \lesssim 5$ it transforms into a dark soliton characterized by a flat notch region and r-independent velocity. Here, it is relevant to mention that in the Technion experiment [71], an interesting nonlinear excitation that evolves periodically between a dark soliton and a vortex ring was observed in a ^{87}Rb BEC, for $\gamma = 4.95$ (see also relevant theoretical work in Refs. [192, 274]). We will further describe these structures in Sec. 4.3.5 below.

2.3.5 ▪ Dimensionality Crossover Regime from 3D to 1D

The Single-Soliton State

As discussed in Sec. 1.2.4, if the dimensionality parameter [cf. Eq. (1.20)] takes values $d \approx 1$, then a cigar-shaped BEC is in the so-called dimensionality crossover regime from 3D to 1D. However, in such an experimentally relevant regime,[10] exact analytical dark soliton solutions (of arbitrary amplitudes) of the pertinent effectively 1D mean-field models (see Sec. 1.2.4) are not available. As a result, the analytical techniques presented in Sec. 2.2.2 cannot be applied for the study of matter-wave dark soliton dynamics in this regime. Nevertheless, the results of Secs. 2.2.2 and 2.3.4 indicate that the evolution of dark solitons should be similar to the one pertaining to the TF-1D regime, while the solitons would not be prone to the snaking instability, as in the case of the purely higher-dimensional setups.

The statics and dynamics of matter-wave dark solitons in the crossover regime between 1D and 3D were studied in Ref. [106]. Particularly, numerical simulations of the 3D GPE revealed that matter-wave dark solitons are indeed dynamically stable, and perform harmonic oscillations in the harmonic trap. In the same work, the frequency of oscillation of a dark soliton inside the harmonic trap was compared from three different models: the original 3D GPE and the reduced models corresponding to the 1D GPE (2.1) and the effectively 1D NPSE [cf. Eqs. (1.22) and (1.23)]. Note that the latter model can be expressed in dimensionless form as

$$i\partial_t \psi = \left[-\frac{1}{2}\partial_z^2 + \frac{1}{2}\Omega^2 z^2 + \frac{3|\psi|^2+2}{2(1+|\psi|^2)^{1/2}}\right]\psi, \tag{2.101}$$

where units are the same to the ones used for Eq. (2.1).

Figure 2.11 depicts this comparison as the effective trapping frequency along the transverse direction is varied. This, in turn, corresponds to varying the effective dimensionality of the system (1D to 3D from left to right in the figure). As it is clear from the figure, for very tight traps in the transverse direction (i.e., in the 1D regime, cf. left of the leftmost vertical dashed line), all the models yield approximately the same oscillation frequency. However, as the BEC becomes more three-dimensional, the 1D GPE clearly fails to reproduce the right oscillation frequency. On the other hand, the NPSE is able to accurately predict the oscillation frequency all the way until the BEC becomes truly three-dimensional (see right-most vertical dashed line). Thus, summarizing, the soliton oscillation frequency is a decreasing function of the dimensionality parameter d, taking values ranging from $\omega_{\rm osc} = \Omega$ (corresponding to the non-interacting limit of $d \to 0$), to

[10] Note that the Heidelberg experiments [70, 72] were conducted in this regime.

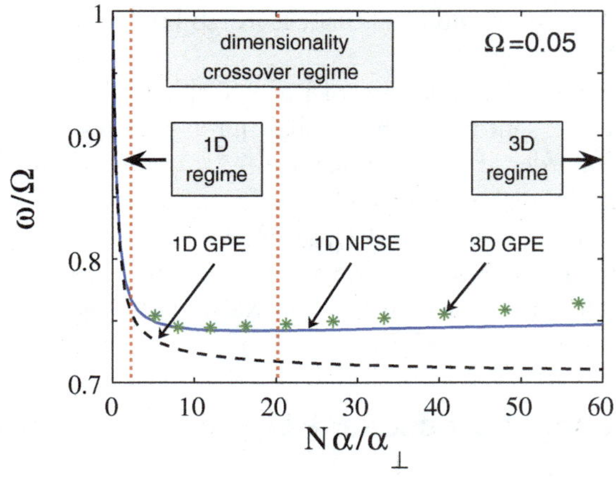

Figure 2.11. *Normalized frequency ω/Ω as a function of the dimensionality parameter $d = N\alpha/\alpha_{\perp}$. Increasing d from zero corresponds to changing the effective dimensionality of the system from 1D to 3D. Solid (dashed) line corresponds to the anomalous mode frequency obtained by the BdG analysis of the NPSE (1D GPE), and stars denote the oscillation frequency of the dark soliton obtained by the 3D GPE. The vertical dotted lines depict the different dimensionality regimes. Adapted with permission from Ref. [106].*

$\omega_{\rm osc} = \Omega/\sqrt{2}$ [corresponding to the TF-1D regime of $d \ll 1$—cf. Eq. (2.69)], with Ω being the normalized strength of the harmonic trap.

Importantly, as shown in Fig. 2.11, the oscillation frequency $\omega_{\rm osc}$ resulting from the 3D GPE is equal to the eigenfrequency ω_A of the anomalous mode of the effectively 1D NPSE model (2.101). This finding leads, in turn, to the following conclusion: an equation of motion for the center z_0 of a dark soliton in a condensate in the dimensionality crossover regime can be expressed as

$$\frac{d^2 z_0}{dt^2} = -\frac{\partial V_{\rm eff}}{\partial z_0}, \quad V_{\rm eff} = \frac{1}{2}\omega_{\rm osc}^2 z_0^2, \quad \omega_{\rm osc} \equiv \omega_A. \tag{2.102}$$

Hence, the oscillation frequency is *up-shifted* from its TF-1D value and, as a result, the effective trapping potential felt by the dark soliton during its motion [see Eq. (2.68)] will effectively become steeper. In that regard, it is relevant to mention that substantial shifts of the oscillation frequency (which may be of order of 10%) have been predicted in Ref. [106] and later confirmed in the Heidelberg experiment of Ref. [70] (see discussion in Sec. 2.3.5). It should also be noticed that the soliton oscillation frequency is also a decreasing function of the soliton amplitude (or, in other words, of the oscillation amplitude), contrary to the result corresponding to the TF-1D regime [70, 72].[11]

Results similar to the ones obtained in the framework of the NPSE model [106] can also be obtained in the framework of the generalized NLS Eq. (1.22) with the nonlinearity function of Eq. (1.24) (see, e.g., the analysis of Ref. [72]). Furthermore, we should mention that the oscillations of dark solitons were also analyzed in the framework of a GPE model with generalized nonlinearities, and specific results in the physically relevant case of a cubic-quintic nonlinearity (modeling two- and three-body interactions—see Sec. 1.2.4) were presented [186]. The same model was also studied in Ref. [275], where various stationary states—including dark solitons—were found and analyzed in detail.

[11] Recall that the results of Sec. 2.2.2 indicate that the soliton oscillation frequency does not depend on the soliton amplitude in the TF-1D regime.

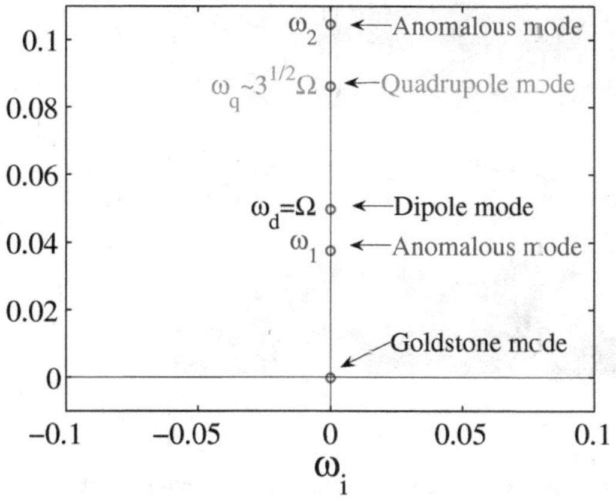

Figure 2.12. *The lowest characteristic eigenfrequencies of the Bogoliubov excitation spectrum for a condensate, in the dimensionality crossover regime from 3D to 1D, carrying two dark solitons. The model used is the NPSE (2.101), with parameters* $\Omega = 0.05$*, and* $\mu = 1.86$*; the dimensionality parameter is* $d = N\Omega\alpha/\alpha_\perp = 0.82$*. The lowest characteristic eigenfrequencies of the Bogoliubov excitation spectrum: shown are the one located at the origin, the one at* $\Omega = 0.05$*, the one at* $\sqrt{3}\Omega = 0.087$ *(corresponding to the Goldstone mode, the Kohn mode, and the quadrupole mode, respectively), as well as the two anomalous modes, one with* $\omega_1 = 0.756\Omega = 0.0378$ *and one with* $\omega_2 = 2.094\Omega = 0.1047$*. Adapted with permission from Ref. [75].*

The Multiple-Soliton State

Apart from the single-dark-soliton state, the case of multiple-dark solitons can also be analyzed in the dimensionality crossover regime using, as a guideline, the methodology exposed in Secs. 2.1.6, 2.2.2, and 2.2.3. In particular, in the case of well-separated and symmetrically interacting dark solitons in a harmonic trap, one may follow the analysis of Refs. [70, 72] and analyze this problem by adopting a simple physical picture: each soliton in the multisoliton state follows the evolution of the single-soliton state, i.e., it oscillates in the trap with an oscillation frequency $\omega_{\rm osc}$ equal to the eigenfrequency ω_1 of the *first* anomalous mode (determined by a BdG analysis of the pertinent 1D mean-field models of Sec. 1.2.4), and interacts with the neighboring solitons via the effective repulsive potential of Eq. (2.54).

In order to further elaborate on the above, let us consider—as an example—the two-dark-soliton state of the NPSE model of Eq. (2.101). This state (which, in the linear limit, corresponds to the second excited state of the quantum harmonic oscillator) can be obtained as a nonlinear stationary state of the system by means of, e.g., a Newton–Raphson fixed point method. The pertinent configuration, has the form of two dark solitons, placed at $z_1 = -z_2 = 2.185$ with a fixed relative distance $\delta z_0 \approx 4.37$. The corresponding Bogoliubov excitation spectrum, namely the spectral plane (ω_r, ω_i) of the eigenfrequencies $\omega \equiv \omega_r + i\omega_i$, is shown in Fig. 2.12; as it can clearly be observed, among the lowest eigenfrequencies (such as the ones at $\omega = 0$, $\omega_d = \Omega$, and $\omega_q \approx \sqrt{3}\Omega$, corresponding to the Goldstone, dipole, and quadrupole modes, respectively), there exist two anomalous modes with eigenfrequencies $\omega_1 = 0.756\Omega = 0.0378$ and $\omega_2 = 2.094\Omega = 0.1047$.

According to the analysis of Sec. 2.2.3, *small* displacements of the dark solitons from their equilibrium points lead to the in-phase and out-of-phase oscillatory motion of the dark soliton pair (see Fig. 2.13), with the respective oscillation frequencies being equal to the eigenfrequencies ω_1 and ω_2 of the two anomalous modes. Importantly, the value of

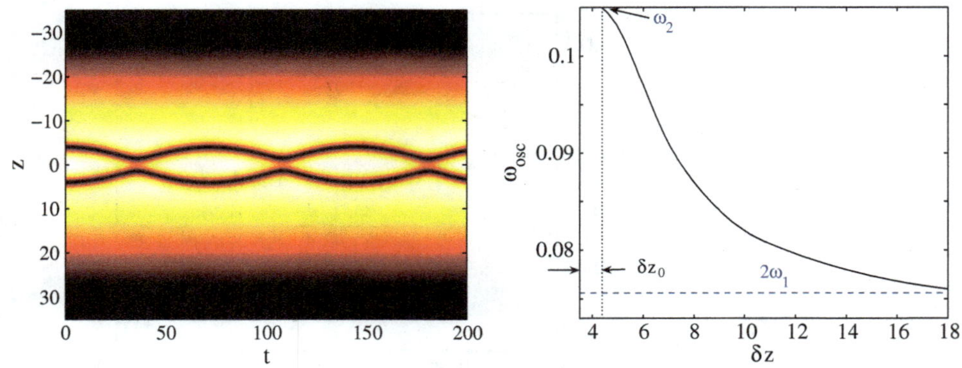

Figure 2.13. *Left panel: Spatio-temporal contour plot of the density of a cigar-shaped BEC confined in a trap of strength $\Omega = 0.05$. The condensate is in the dimensionality crossover regime from 3D to 1D, and the model used is the NPSE Eq. (2.101). The dark solitons, initially placed at $z = \pm 4$ (i.e., $\delta z = 8 > \delta z_0 = 4.37$), oscillate out-of-phase with a frequency $\omega_{\rm osc} = 1.74\Omega < \omega_2 = 2.094\Omega$. Right panel: The oscillation frequency of the dark solitons, $\omega_{\rm osc}$, as a function of their initial relative distance δz. For $\delta z \to \delta z_0$ (corresponding to the stationary state) we obtain $\omega_{\rm osc} \to \omega_2$, while for $\delta z \gg \delta z_0$ we obtain $\omega_{\rm osc} \to 2\omega_1$. Adapted with permission from Ref. [75].*

the eigenfrequency of the first anomalous mode, $\omega_1 = 0.0378$, is quite close to the oscillation frequency of a *single* dark soliton, $\omega_{\rm osc} = 0.0375$, in the same setup (i.e., with the same parameter values), with the percentage difference being $\approx 1.6\%$. This generic example suggests that, generally, the dynamics of the two-dark-soliton state can be described by an effective Lagrangian for the two solitons, namely $L_{\rm eff} = T - V$; here, T and V are the kinetic and potential energies, respectively, depending on the soliton centers, z_i $(i = 1, 2)$, and soliton velocities, $\dot{z}_i \equiv dz_i/dt$, as follows:

$$T \equiv \sum_{i=1}^{2} \frac{1}{2}(\dot{z}_i)^2, \quad V \equiv \sum_{i=1}^{2} \frac{1}{2}\omega_{\rm osc}^2 z_i^2 + V_{\rm int}(z_2 - z_1), \tag{2.103}$$

where $V_{\rm int}(z_2 - z_1)$ (with $z_2 - z_1 \equiv 2z_0$) is the repulsive potential of Eq. (2.54). Then, the evolution of the soliton centers can readily be determined by the Euler–Lagrange equations $d(\partial_{\dot{z}_i} L_{\rm eff})/dt - \partial_{z_i} L_{\rm eff} = 0$. The latter, may be simplified upon using the approximate form of the repulsive potential [cf. Eq. (2.55)], thus leading to the following equations of motion:

$$\ddot{z}_1 = -\omega_{\rm osc}^2 z_1 - 8n_0^{3/2} \exp[-2\sqrt{n_0}(z_2 - z_1)], \tag{2.104}$$

$$\ddot{z}_2 = -\omega_{\rm osc}^2 z_2 + 8n_0^{3/2} \exp[-2\sqrt{n_0}(z_2 - z_1)], \tag{2.105}$$

where we have assumed well-separated, almost black solitons [i.e., in Eq. (2.55) we have set $B \approx 1$]. Apparently, the above analysis can readily be generalized for multiple solitons [72], with each one interacting with its neighbors. Importantly, if $\omega_{\rm osc}$ in the system of Eqs. (2.104) and (2.105) was considered to be unknown, then it would be possible to be directly obtained in the form of the characteristic frequencies of the normal modes of this system (see details in Ref. [72]); these characteristic frequencies coincide to the ones determined via the BdG analysis. This result justifies *a posteriori* the considered decomposition of the principal physical mechanisms (oscillations and interactions of solitons) characterizing the system.

We should also note that apart from the case of small-amplitude oscillations of two well-separated, almost black solitons, the more general case of the dynamics of n-interacting dark solitons (which may also perform large amplitude oscillations—see below) is possible using the full set of the above mentioned Euler–Lagrange equations; the latter lead to the following n coupled equations of motion [72]:

$$\ddot{z}_i - \sum_{k=1}^{n} \left(\frac{\partial^2 V}{\partial z_k \partial \dot{z}_i} \dot{z}_k + \frac{\partial^2 V}{\partial \dot{z}_k \partial \dot{z}_i} \ddot{z}_k \right) + \frac{\partial V}{\partial z_i} = 0, \tag{2.106}$$

where $V \equiv \sum_{i=1}^{n} V_i$ is the potential energy, with

$$V_i = \sum_{j=1, j\neq i}^{n} \frac{n_C B_{ij}^2}{2\sinh^2[\sqrt{n_C} B_{ij}(z_i - z_j)]}$$

being the interaction potential felt by the ith soliton due to the presence of the other solitons, while $z_{ij} = (1/2)(z_i - z_j)$ and $B_{ij} = (1/2)(B_i + B_j)$ denote, respectively, the relative coordinate and the average depth for solitons i and j.

Let us now return to the above example of the two-dark-soliton state of Eq. (2.101) and consider again the out-of-phase oscillation (we will use the parameter values of Fig. 2.12). In this case, if the initial soliton separation is significantly larger than $\delta z_0 = 4.37$—or, in other words, if the displacements of solitons around their equilibrium positions are *not* small—then $\omega_{\rm osc}$ differs from (in fact, it is quite smaller than) ω_2; considering, e.g., that $\delta z = 8$ (corresponding to initial locations of the soliton centers $z = \pm 4$—see Fig. 2.13), the out-of-phase oscillation of the two solitons is characterized by a frequency $\omega_{\rm osc} = 1.74\Omega < \omega_2 = 2.094\Omega$. A qualitative explanation for this difference is the following. As discussed above, the evolution of two dark solitons can be effectively described by the equations of motion (2.104) and (2.105), in the presence of the repulsive potential which depends exponentially on their relative distance. If the relative distance between the two solitons is not significantly different than the one pertaining to the corresponding stationary state, i.e., $\delta z \approx \delta z_0$, the effective repulsive force is strong and throughout their motion, and, as a result, their motion is strongly affected by their coupling. On the other hand, if their initial separation becomes larger (as, e.g., in the case of the example under consideration, with $\delta z = 8$), the repulsive force becomes exponentially small as they spend time far away from each other, and, as a result, the motion of each individual soliton is not significantly affected by the presence of the other.

In the right panel of Fig. 2.13 we show the oscillation frequency of the dark soliton—obtained by the direct integration of the NPSE model of Eq. (2.101)—as a function of the initial relative distance between the two solitons. It is clear that the oscillation frequency, which takes values in the interval $\delta z \geq \delta z_0$ (the value corresponding to the stationary state), exhibits two different asymptotic regimes: when the initial soliton separation is small enough, $\delta z \to \delta z_0$ (i.e., for strong coupling between the solitons), the oscillation frequency tends to the eigenfrequency ω_2 of the largest anomalous mode; on the other hand, when the initial soliton separation is large enough, $\delta z \gg \delta z_0$ (i.e., when the solitons are actually decoupled), the oscillation frequency tends to $2\omega_1$. The latter value can be explained by the fact that the period of oscillation of each individual soliton is the half of the one that would correspond to a single-soliton oscillation since the solitons only have to travel on their own half of the background density as they bounce off of each other.

Connection to Experiments: The Two-Soliton State

The above analysis and results can directly be compared to experimental observations reported in the Heidelberg experiment of Ref. [70]. In this work, an oscillating and interacting dark soliton pair, in the dimensionality crossover regime between 1D and 3D, was experimentally observed in a BEC of ^{87}Rb in the $|F=2, m_F=2\rangle$ state containing about $N=1500$–1700 atoms. The thermal phase fluctuations in this experiment were negligible for the measured temperature of $T \approx 10$ nK (the critical temperature for condensation was $T_c \approx 110$ nK).

In this experiment, solitons performing large amplitude oscillations were formed by the method of matter-wave interference (see Sec. 2.1.5 and Ref. [70] for details). The formation of dark solitons was confirmed by 3D GPE simulations, as shown in Fig. 2.14. Including the optical and time resolution, the experimentally observed density profile evolution is well reproduced. A dominant pair of solitons oscillates close to the center

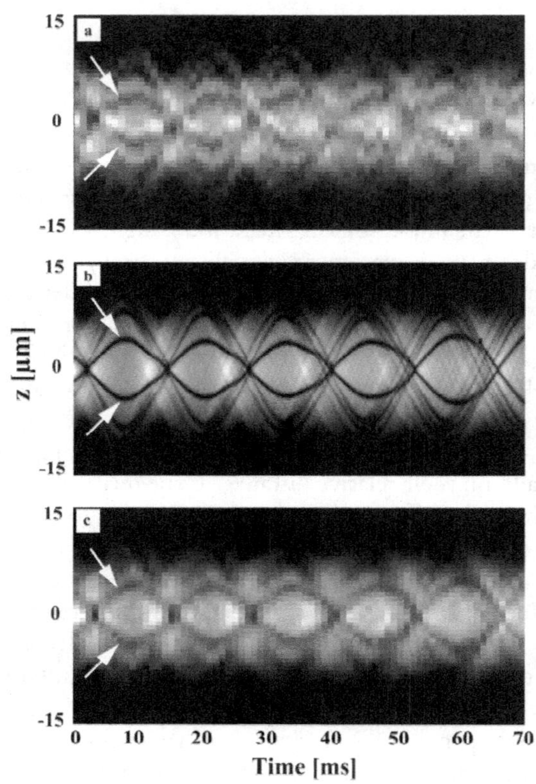

Figure 2.14. *Observation of the time evolution of dark solitons in a harmonic trap in the Heidelberg experiment [70]. The dominant soliton pair is indicated by arrows.* (a) *Experimental observation of the dynamics of the longitudinal atomic density. Each longitudinal density profile (vertical lines), corresponding to a given evolution time, is deduced from typically 10 experimental realizations. The obtained absorption images of the condensate at each time step are averaged and integrated over their transverse direction. The number of atoms in the shown case is* $N = 1700$ *and the trapping frequencies are* $(\omega_z, \omega_\perp) = 2\pi \times (53, 890)$ *Hz.* (b) *Result of the numerical integration of the 3D GPE taking into account the full preparation process of the solitons.* (c) *Same as* (b), *taking into account the finite spatial (1 μm) as well as temporal resolution (1 ms) of the experiment. The loss of contrast due to the convolution process explains the experimentally observed fading out of the solitons with time. Adapted with permission from Ref. [70].*

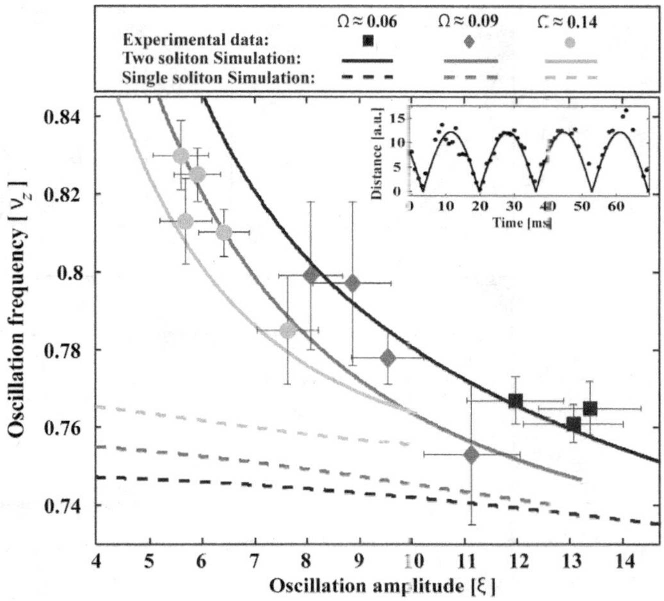

Figure 2.15. *Comparison between experimentally obtained soliton oscillation frequencies and NPSE simulation for one and two solitons. Each frequency point is deduced from the temporal evolution of the soliton distance as shown in the inset. Different symbols correspond to different aspect ratios* Ω *of the trap. For each aspect ratio the oscillation amplitude is varied as explained in the text. NPSE simulations are represented by solid lines for the two soliton case, and by dashed lines for the respective single-soliton oscillations. The error bars on the measured frequencies account for statistical errors on the measured soliton and trap frequencies and systematic errors on the atom number used to calculate the healing length. Adapted with permission from Ref. [70].*

of the cloud and additional pairs of solitons with much lower contrast can also be distinguished. Below, we focus on the dynamics of the dominant central pair and show that its oscillation frequency is well described within a two-soliton approximation.

First, the dependence of the oscillation frequency on the distance between the two solitons can be found and compared with experimental data. In particular, the oscillation frequency of the dominant soliton pair for different trap parameters and different inter-soliton distances was investigated experimentally. A typical data set consisted of 50 time steps and 10 pictures per time step. The data did not allow to distinguish if the two solitons do or do not cross at the collision point. Hence, for each time step, the soliton distance, which is well defined, was measured and its time evolution was reconstructed, from which an oscillation frequency can be extracted as shown in the inset of Fig. 2.15. The obtained frequency is divided by 2 in order to compare it to the oscillation frequency expected for a single trapped soliton. The shot-to-shot reproducibility of the soliton dynamics up to 100 ms allows the observation of up to 7 oscillation periods and hence the deduction of the frequency with high accuracy. The typical statistical experimental error in the frequency measurement was $\pm 1.5\%$. Figure 2.15 shows the results of the frequency measurements and their comparison with numerical simulations for the motion of two trapped solitons using the NPSE model [103]; obviously, the result is in accordance with the one shown in the right panel of Fig. 2.13. Notice that for evolution times longer than 100 ms, solitons can still be observed in single shots of the experiment, but the ability to average several pictures is lost, which appears as decreasing contrast of the solitons. The reason for this

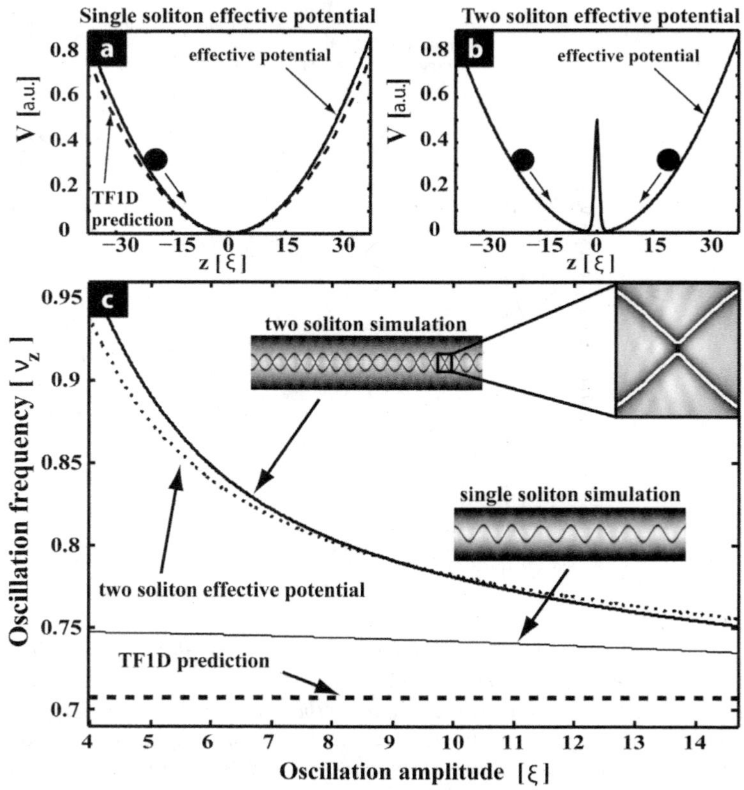

Figure 2.16. *The oscillation dynamics of dark solitons in a trapped BEC is well captured in an effective particle picture.* (a) *For one soliton, the particle moves in a harmonic trap.* (b) *For two solitons, an additional barrier due to the repulsive interaction appears.* (c) *For the single- and two-soliton case, the dependence of the oscillation frequencies (in units of ν_z) on the oscillation amplitude from the trap center is shown for one experimental parameter set with $\Omega = \omega_z/\omega_\perp \approx 0.06$. The dashed line shows the TF-1D prediction ($\nu_z/\sqrt{2}$). The thin solid line indicates the up-shift of the single-soliton frequency mainly due to the dimensionality. For the case of two solitons, the thick solid line also includes the up-shift due to the intersoliton interaction deduced from the NPSE. The dotted line represents the result of the simple effective particle model stemming from Eq. (2.103). Density profile evolutions obtained from the NPSE are shown in the insets. A detail of the collision between the two solitons is also shown, demonstrating that they do not cross each other. The white lines correspond to the trajectories of the density minima. Adapted with permission from Ref. [70].*

might be slightly different starting conditions of the experiment from shot-to-shot and a possible heating during the collisions of the solitons.

Good estimates for the single-soliton oscillation frequency (denoted by ν_{1_s} here) and the soliton interaction strength can be calculated as follows. The single-soliton frequency ν_{1_s} is obtained by numerical integration of the NPSE describing a single soliton. Because the experimental parameters $\Omega \approx 0.06 - 0.14$ and $d = N\Omega a_s/a_\perp \approx 1.2 - 1.8$ are both in the crossover regime and far from the Thomas–Fermi limit, substantial corrections to the asymptotic value $\nu_z/\sqrt{2}$ ($\omega_z = 2\pi\nu_z$) are expected, as per the discussion above. Indeed, the oscillation frequency of a single dark soliton is up-shifted by a few percent from the asymptotic value as found by the BdG analysis of the NPSE. The simulations reveal that for the above parameters, and for $\Omega = \omega_z/\omega_\perp \approx 0.06$, this up-shift is 5% [see Fig. 2.16(c)]. Note that the frequencies obtained by 1D GPE simulations are approximately 2% higher than the asymptotic limit, because the Thomas–Fermi limit is not reached for these experimental parameters as discussed in Ref. [173]. The effect of dimensionality of the system, i.e. the role of the transverse degrees of freedom which is captured only by the NPSE or the 3D GPE, accounts for the remaining 3%. Figure 2.16(c) shows the comparison between the asymptotic limit $\nu_z/\sqrt{2}$ and the single-soliton NPSE simulation for one specific trap. The simulation results for the three different parameter sets used in the experiment are shown in Fig. 2.15. As expected, the single-soliton frequency increases with the aspect ratio.

As is also shown in Fig. 2.16(c), the repulsive interaction between the solitons results in an up-shift of the oscillation frequency, compared to the single-soliton case, that strongly depends on the oscillation amplitude. The presented model accurately reproduces the up-shift if μ is set to be the chemical potential of the condensate obtained from the 3D GPE model. In the experimentally accessible parameter range, the agreement of the model with NPSE simulations is better than 5%. This allows one to clearly identify the significant role of the repulsive interactions and shows that the effective repulsive potential in Eq. (2.103) obtained in the 1D homogeneous case is a good approximation to the complex situation of the Heidelberg experiment.

Thus, summarizing, and in accordance to the theoretical analysis of this section, considerable up-shifts—up to 16%—of the soliton oscillation frequency from the value of $\Omega/\sqrt{2}$ are observed in the study of Ref. [70], and are quantitatively attributed to the dimensionality of the system and the soliton interactions.

Connection to Experiments: Multiple Solitons

As mentioned above, the two oscillating and interacting dark solitons were created in the Heidelberg experiment by means of the matter-wave interference method. Such an experimental procedure merging two BECs, which have been initially separated by an optical double-well potential, in a harmonic trap, can also be used for the generation of more than two solitons, or even for the generation of a single soliton, as mentioned in Sec. 2.1.5. Indeed, it is possible to vary the number of solitons by increasing the kinetic energy during the collision process [152], and by doing so, to lower the distance between the created solitons, and to create additional solitons further away from the trap center. The following possibilities can be used:

- lowering the number of atoms,
- lowering the ramping-down time of the optical lattice,[12] or
- increasing the aspect ratio $\Omega = \omega_z/\omega_\perp$ of the trap.

Even though it is additionally possible in the simulations to vary the soliton distance by the height of the optical lattice, this possibility is not relevant for the experiment, because this possibility is used to adjust the tunneling coupling between the two initial wells providing the initial phase coherence of the condensate.

The experimental creation of four solitons plus two additional weak ones with high oscillation amplitude is shown in Fig. 2.17. The illustrated oscillation dynamics includes the creation process of the solitons with a time resolution of 1 ms.

The method also offers the possibility of creating an odd number of solitons. Indeed, changing the symmetry of the double-well potential results in an energy difference, ΔE,

[12] In the Heidelberg experiments [70, 72], the double-well potential is created by the superposition of a crossed optical dipole trap and a 1D optical lattice Removing the optical lattice leads to the merger of the two initially separated BECs in the harmonic trap.

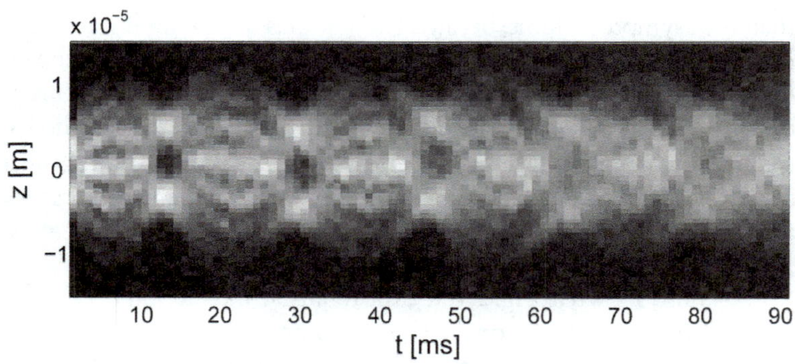

Figure 2.17. *Observation of the oscillation of four dark solitons including the creation process. The evolution is averaged over 10 experimental runs. Adapted with permission from Ref. [72].*

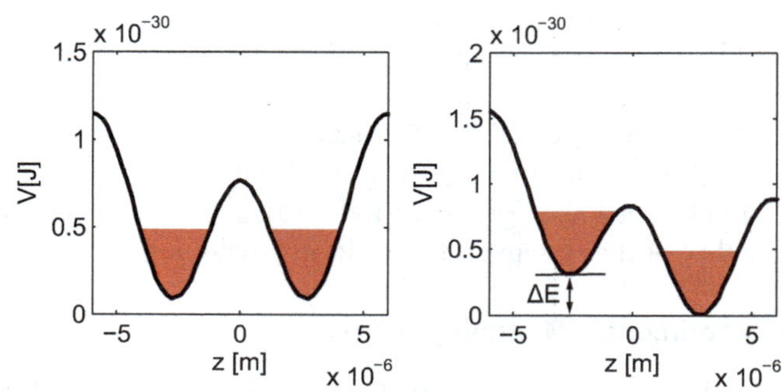

Figure 2.18. *Left: Symmetric double-well potential. Right: A shift in the symmetry of the double-well potential also leads to a relative shift of the energy levels of the two wells.*

between the levels of the two wells, as shown in Fig. 2.18. This can be realized by shifting the second beam of the dipole trap with respect to the optical lattice. Holding an asymmetry for a certain time accumulates a phase difference between the BECs in the two wells. In a simple approximation, this phase difference is given by $\Delta\phi \approx \Delta E/\hbar \cdot t$. By changing the asymmetry and the time of phase accumulation before releasing the two condensates from the double well, arbitrary phase differences can be achieved. After the collision of the two BECs, an initial phase difference will affect the resulting soliton pattern. A small phase difference leads to an asymmetric evolution pattern of the solitons, while a phase difference close to π leads to the creation of an additional soliton between the other solitons yielding an odd number of solitons. Especially interesting is the case were the phase difference is exactly π. Then one soliton is standing between the oscillating other ones in the center of the trap. In this case the interaction potential is increased compared to the two-soliton case as the resting soliton is really black, meaning $B = 1$.

Figure 2.19 shows the experimental realization of three dark solitons in a harmonic trap created by the above discussed method. The trap frequencies used in the experiment are $\nu_z = 36.1 \pm 0.25$ Hz (longitudinal frequency), $\nu_\perp = 407.5 \pm 40.8$ Hz (transverse frequency). The mean number of atoms in the BEC is $N = 1570 \pm 146$. In this measurement the optical lattice is ramped down on a timescale of $\tau_{\mathrm{OL}} = 2$ ms. Since the oscillation amplitude of the two outer solitons, $A_{\mathrm{osc}} = (21 \pm 0.6)\xi$, is very high in the shown experiment, their oscillation frequency ν_d compared to the trap frequency ν_z is quite low: $\nu_d/\nu_z = 0.775 \pm 0.006$.

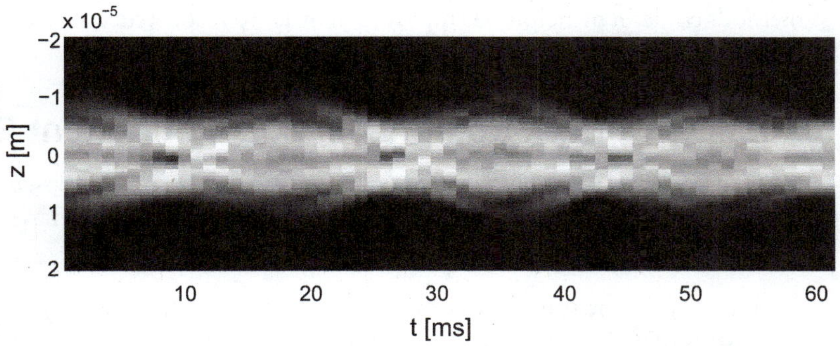

Figure 2.19. *Experimental observation of three dark solitons in a harmonic trap averaged over 16 runs. The soliton in the center of the trap is at rest (i.e., it is a black soliton), whereas the two outer ones are undergoing an oscillation. The time evolution plot was obtained by integrating the images over the transverse axis, meaning that each vertical line shows the longitudinal density of the BEC at a certain point in time. The time resolution is* 1 *ms. Adapted with permission from Ref. [72].*

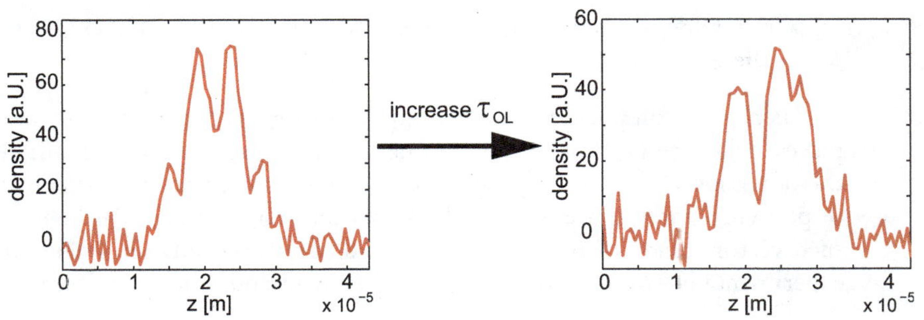

Figure 2.20. *Single shots of the longitudinal density of the BEC. Increasing the ramping-down time of the intensity of the optical lattice from $\tau_{\rm OL} = 2$ ms to $\tau_{\rm OL} = 5$ ms in principle offers the possibility of creating a single soliton. However, the experimental creation of the single-soliton state could not be achieved in a reproducible way. Left: Single shot of the three soliton state shown in Fig. 2.19. Note that the stationary soliton in the middle appears with finite density on its minimum in the experimental images. This is probably due to the limited optical resolution. Right: Single shot of a single dark soliton state. Adapted with permission from Ref. [72].*

In principle this method should offer the possibility of creating one standing soliton. It should be possible to achieve this by increasing the ramping-down time of the optical lattice $\tau_{\rm OL}$ further, which decreases the kinetic energy of the collision process. In the lowest collisional state only one interference fringe is produced which converts into a single dark soliton. However, due to technical problems[13] the experiments in Heidelberg [70, 72] neither created a standing single-soliton state nor a moving single-soliton state in a reproducible way. The created single solitons (see Fig. 2.20) were fluctuating in position from shot to shot, probably due to technical fluctuations of the initial phase difference between the two colliding condensates. Therefore, a time evolution plot like the one of Fig. 2.19 could not be obtained for the single-soliton case. Nevertheless, phase imprinting

[13] The long ramping-down time $\tau_{\rm OL}$ necessary to create the single-soliton state results in a shallow slope of the intensity of the optical lattice. This intensity slope cannot be locked by the feedback loop of the position control of the optical lattice. As a result the phase difference between the two initially separated condensates is fluctuating. This leads to a shot-to-shot fluctuation of the position of the created solitons.

methods could in principle be employed in order to reconstruct the oscillations of a single soliton [69].

2.4 ▪ Matter-Wave Dark Solitons in Multicomponent BECs

Multicomponent ultracold atomic gases and BECs may be composed by two or more atomic gases, which have been observed in experiments, and may have the form of mixtures of

- two different spin states of the same atom species, so-called *pseudospinor* condensates [276, 277, 278] (see also Ref. [279]);
- different Zeeman sublevels of the same hyperfine level, so-called *spinor* condensates [280] (see also the reviews [281, 282] and references therein);
- different atom species (so called *heteronuclear* mixtures) [283];
- degenerate boson-fermion clouds [284];
- purely degenerate fermion clouds [285] (see also Ref. [42, 43, 44] for reviews and references therein).

Such multicomponent systems support various types of matter-wave soliton complexes, with the type of soliton in one species being the same or different to that in the other species. Obviously, in the case of repulsive interactions under consideration, of particular interest are vector solitons with the one component being a dark soliton; such vector solitons have mainly been studied theoretically—but also, in some works, experimentally—in the context of two-component and spinor condensates.

2.4.1 ▪ General Overview

Model, Ground States, and Phase Separation

Generally speaking, a mixture of $\mathcal{N}$ purely bosonic components can be described in the framework of mean-field theory by a system of $\mathcal{N}$ coupled GPEs. In the case where different components belong to the same atom species, this GPE system of equations can be expressed in the following dimensionless form:

$$i\frac{\partial \psi_n}{\partial t} = -\frac{1}{2}\nabla^2\psi_n + V_n(\mathbf{r})\psi_n + \sum_{k=1}^{\mathcal{N}}\left[g_{nk}|\psi_k|^2\psi_n - \varkappa_{nk}\psi_k + \Delta_{nk}\psi_n\right]. \tag{2.107}$$

Here, ψ_n is the wavefunction of the nth component ($n = 1,\ldots,\mathcal{N}$), $V_n(\mathbf{r})$ is the trapping potential confining the nth component, Δ_{nk} is the chemical potential difference between components n and k, the nonlinearity coefficients $g_{nk} = g_{kn}$ characterize interatomic collisions, while the linear coupling coefficients $\varkappa_{nk} = \varkappa_{kn}$ account for spin state interconversion, usually induced by a spin-flipping resonant electromagnetic wave (see, e.g., Ref. [286]). It should be mentioned that in some works (see, e.g., Refs. [287, 288, 289]) fermionic mixtures are also described in the framework of the mean-field theory, with the self-interacting nonlinear terms being replaced by $g_{nn}|\psi_n|^{4/3}\psi_n$. Notice that the GPEs (2.107) are characterized by conservation of both the energy E and the total number of atoms, $N \equiv \sum_{k=1}^{\mathcal{N}} N_k = \sum_{k=1}^{\mathcal{N}} \int |\psi_k|^2 d\mathbf{r}$; furthermore, in the absence of linear interconversions ($\varkappa_{nk} = 0$), the number of atoms of each component N_k is separately conserved.

Let us consider the case of two bosonic species ($\mathcal{N} = 2$), and assume that the system is homogeneous ($V_n = 0$). If, additionally, there is no spin-state interconversion ($\varkappa_{nk} = 0$) and chemical potential difference [$(\Delta\mu_{nk}) = 0$], then the binary mixture is immiscible provided that the following immiscibility condition holds [290] (see also Ref. [291]):

$$\Delta \equiv (g_{12}^2 - g_{11}g_{22})/g_{11}^2 > 0, \tag{2.108}$$

where Δ is the so-called miscibility parameter. In practice, this parameter may take values of the order of 10^{-3} or even less; for instance, $\Delta \approx 9 \times 10^{-4}$ or $\Delta \approx 0.036$ for a mixture of two spin states of a ^{87}Rb BEC [276, 277] or a ^{23}Na BEC [278], respectively. The condition (2.108) indicates that if the mutual repulsion between species is stronger than the repulsion between atoms of the same species then the two species do not mix. In such a case, the two species tend to separate by filling two different spatial regions, thus forming, e.g., a "ball" and "shell" configuration (see, e.g., Ref. [277] for relevant experimental results). This way, the ground state of the system—i.e., the state minimizing the energy—may take the form of domain-wall solutions of the GPEs (2.107) [292, 293, 294, 295, 296, 297]. In accordance to the experimental observations, these solutions represent configurations of the following form: in the Thomas–Fermi limit (where kinetic energy is negligible), one species occupies the region around the trap center, and it is separated by two domain walls from side domains occupied by the other species; on the other hand, kinetic energy favors a configuration where a single domain wall at the trap center separates two domains occupied by the different species [296].

The dynamics of phase separation of two-component BECs has been studied in various works both theoretically—see, e.g., Refs. [298, 299, 300, 301, 302, 303] and also Refs. [304, 305, 306] for proposed applications—and experimentally [279, 307, 308]. Importantly, magnetic-field Feshbach resonances can be used to controllably change the interspecies [307] or the intraspecies [308] scattering length, and thus controllably change the (im)miscibility between the two species [308]. Alternatively, the (im)miscibility properties of two spin states of a BEC can be controlled by the presence of linear coupling [set by parameters $\varkappa_{nk}$ in Eq. (2.107)], as predicted theoretically [298, 309, 310] and demonstrated experimentally [311]. More recently, in the context of artificial gauge potentials and spin-orbit coupling in cold atoms [312], it has been shown that the (im)miscibility characteristics of spinor gases can be changed using Raman coupling—see, e.g., Refs. [313, 314] and [315, 316] for experimental and theoretical results, respectively.

Vector Solitons in Binary BECs

Apart from domain walls, a trapped two-component quasi-1D BEC supports vector solitons, with the one component being a dark soliton; in such a case, typically, the other component may be

- a bright soliton, so that the vector soliton is a dark-bright soliton ("DB soliton") [145, 299, 317, 318, 319, 320, 321, 322, 323, 324, 325];
- a dark soliton, so that the vector soliton is a dark-dark soliton ("DD soliton") [136, 138, 326, 327, 299, 328, 329, 330, 331, 332, 317];
- an antidark soliton, so that the vector soliton is a dark-antidark soliton [331, 326, 299].

Here we should note that an antidark soliton is actually a dark soliton with reverse-sign amplitude, i.e., it has the form of a hump (instead of a dip) on top of the background density (see, e.g., Refs. [131, 132]).

As indicated above, all of these types of vector solitons have been studied in many theoretical works. Importantly, DB solitons have also recently been observed in experiments using either the phase-imprinting method [68] or the counterflow between two miscible condensates [333, 334, 335, 336, 337]. These experimental studies have chiefly involved the dynamics of a single DB soliton in a trap [68, 334], the generation of multiple DB solitons [333], the study of their interactions [335], as well as the creation of a particular type of dark-dark solitons stemming from SU(2)-rotated DB solitons; as we will discuss below in more detail, these vector solitons have the form of "beating dark-dark" solitons [336, 337]. On the other hand, vector solitons of the dark-antidark type have been predicted to exist in theory, either in a stationary form [331] or as dynamical entities resulting from instabilities [326, 299].

Below we will focus on the most generic vector matter-wave solitons, namely the DB and DD ones, presenting results corresponding to physically relevant setups, which include the standard time-independent harmonic trap. Notice that in the absence of the trap, vector solitons of the above mentioned types have been extensively studied in the context of nonlinear optics: there, multicomponent solitons occur when fields of one frequency, or one polarization, become coupled to fields of other frequencies, or other polarizations (see, e.g., the review [35] and references therein). Mathematically speaking, the existence (and stability) of multicomponent optical solitons (and also matter-wave solitons in the miscible case) can be understood by the fact that the relevant coupled NLS equations rely on the so-called Manakov system [338]: the latter has the form of a vector NLS equation, namely,

$$i\partial_t \mathbf{u} = -\frac{1}{2}\partial_z^2 \mathbf{u} \pm |\mathbf{u}|^2 \mathbf{u}, \tag{2.109}$$

where $\mathbf{u}(z,t) = (u_1(z,t), u_2(z,t), \ldots, u_n(z,t))$ is an n-component vector. This system is known to be completely integrable [339, 340, 341] (in fact, it can be integrated by extending the IST method that has been used to integrate the scalar NLS equation [12, 13]) and admits such vector n-soliton solutions [342, 343, 344].

2.4.2 ▪ Vector Solitons in the Homogeneous Case

We start our considerations with the simpler possible case, corresponding to a two-component elongated (along the x direction) repulsive BEC, composed of two different hyperfine states of the same alkali isotope, e.g., of ^{87}Rb. Such states can be $|1,-1\rangle$ and $|2,1\rangle$, as in the case of the experiment of Ref. [279], or $|1,-1\rangle$ and $|2,-2\rangle$, as in the case of the experiments of Refs. [333, 334, 335, 336, 337]. In the first case the scattering lengths take the values $a_{11} = 100.4a_0$, $a_{12} = 97.66a_0$, and $a_{22} = 95.00a_0$, while in the second case the respective values are $a_{11} = 100.4a_0$, $a_{12} = 98.98a_0$, and $a_{22} = 98.98a_0$ (where a_0 is the Bohr radius). In either case, it is clear that the scattering lengths take approximately the same values, say $a_{ij} \approx a$. Furthermore, assuming that the trap is highly anisotropic (i.e., if the longitudinal and transverse trapping frequencies are such that $\omega_x \ll \omega_\perp$), the system of Eqs. (2.107) can be reduced to the form

$$i\partial_t \psi_1 = -\frac{1}{2}\partial_x^2 \psi_1 + V(x)\psi_1 + (|\psi_1|^2 + |\psi_2|^2 - \mu)\psi_1, \tag{2.110}$$

$$i\partial_t \psi_2 = -\frac{1}{2}\partial_x^2 \psi_2 + V(x)\psi_2 + (|\psi_1|^2 + |\psi_2|^2 - \mu - \Delta)\psi_2, \tag{2.111}$$

where Δ is the difference between chemical potentials of the two components, while the potential in Eqs. (2.110) and (2.111) is now given by $V(x) = (1/2)\Omega^2 x^2$, where $\Omega =$

$\omega_x/\omega_\perp$ is a natural small parameter of the system. Note that in our analysis below we will assume that the chemical potentials characterizing each component are equal, and take $\Delta = 0$.

Our presentation of the vector soliton solutions will be as follows. Since, in the absence of the potential (and $\Delta = 0$), the system of Eqs. (2.110) and (2.111) is the completely integrable Manakov system, we will start from an exact vector soliton solution of Eqs. (2.110) and (2.111), i.e., the DB soliton, and then derive other types of vector solitons, such as the regular and beating DD solitons, by means of SU(2) rotations.

Considering the boundary conditions $|\psi_1|^2 \to \mu$ and $|\psi_2|^2 \to 0$ as $|x| \to \infty$, Eqs. (2.110) and (2.111) possess an exact analytical single DB soliton solution of the following form (see, e.g., Ref. [145]):

$$\psi_1(x,t) = \sqrt{\mu}(\cos\phi \tanh\xi + i\sin\phi), \tag{2.112}$$

$$\psi_2(x,t) = \eta \operatorname{sech}\xi \exp[ikx + i\theta(t)], \tag{2.113}$$

where $\xi = D(x - x_0(t))$, ϕ is the dark soliton's phase angle, $\cos\phi$ and η represent the amplitudes of the dark and bright solitons, and D and $x_0(t)$ are associated with the inverse width and the center position of the DB soliton. Furthermore, $k = D\tan\phi$ and $\theta(t)$ are the (constant) wavenumber and phase of the bright soliton, respectively. The above parameters of the DB soliton are connected through the following equations:

$$D^2 = \mu\cos^2\phi - \eta^2, \tag{2.114}$$

$$\dot{x}_0 = k = D\tan\phi, \tag{2.115}$$

$$\dot{\theta} = \frac{1}{2}(D^2 - k^2), \tag{2.116}$$

with $\dot{x}_0$ and $\dot{\theta}$ denoting the DB soliton velocity and angular frequency, respectively (overdots denote time derivatives). Thus, the DB soliton (2.112)–(2.113) is characterized by three free parameters [seven parameters μ, ϕ, η, k, D, $\dot{x}_0$, $\dot{\theta}$ and four constraints (2.114)–(2.116)]. Notice that the amplitude η of the bright soliton, the chemical potential μ of the dark soliton, as well as the (inverse) width parameter D of the DB soliton are connected to the number of atoms N_b of the bright soliton by means of the following equation:

$$N_b \equiv \int_{-\infty}^{+\infty} |\psi_2|^2\, dx = \frac{2\sqrt{\mu}\eta^2}{D}. \tag{2.117}$$

A typical example of the densities of the dark and bright solitons composing the vector DB soliton is shown in Fig. 2.21. Note that the DB solitons are frequently called *symbiotic solitons*; this is due to the fact that the bright soliton component (which does not exist in the system with repulsive interactions, but is only supported by attractive interactions [98]) can be sustained in the system due to the nonlinear coupling with the dark soliton component.

It is now particularly useful to observe that the system of Eqs. (2.110) and (2.111) is invariant under SU(2) rotations [344]. To be more specific, let us first recall that a general matrix element of SU(2) takes the form

$$U = \begin{pmatrix} \alpha & -\beta^* \\ \beta & \alpha^* \end{pmatrix},$$

where α and β are complex constants such that $|\alpha|^2 + |\beta|^2 = 1$. Then, it can be shown

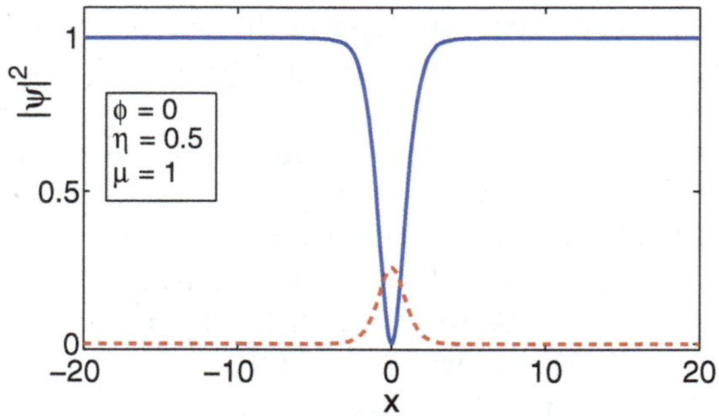

Figure 2.21. *Example of the density (at $t=0$ and for $\phi=0$ and $\mu=1$) of a DB soliton for $\eta=0.5$; solid and dashed lines correspond to the individual densities $|\psi_1|^2$ (dark soliton) and $|\psi_2|^2$ (bright soliton), respectively.*

that if $(\psi_1,\psi_2)^T$ are solutions of Eqs. (2.110) and (2.111), then

$$\begin{pmatrix} \psi_1' \\ \psi_2' \end{pmatrix} \equiv U \begin{pmatrix} \psi_1 \\ \psi_2 \end{pmatrix} = \begin{pmatrix} \alpha\psi_1 - \beta^*\psi_2 \\ \beta\psi_1 + \alpha^*\psi_2 \end{pmatrix}$$

are also solutions of Eqs. (2.110) and (2.111). This suggests that one can now start from the DB soliton solution and construct SU(2) rotated solutions, in the following form:

$$\psi_1(x,t) = \alpha\sqrt{\mu}\{\cos\phi\tanh\xi + i\sin\phi\} - \beta^*\eta \operatorname{sech}\xi \exp\{ikx + i\theta(t)\}, \quad (2.118)$$

$$\psi_2(x,t) = \beta\sqrt{\mu}\{\cos\phi\tanh\xi + i\sin\phi\} + \alpha^*\eta \operatorname{sech}\xi \exp\{ikx + i\theta(t)\}. \quad (2.119)$$

With the additional four parameters $\alpha, \beta \in \mathbb{C}$ and the constraint $|\alpha|^2 + |\beta|^2 = 1$, the solution (2.118)–(2.119) is characterized by six free parameters. Introducing a new parameter c, the velocity of the background fluid, still another solution can be constructed from Eqs. (2.118)–(2.119) via a Galilean boost:

$$\psi_{1,2}(x,t) \to \psi_{1,2}(x-ct,t)\exp[i(cx - c^2t/2)].$$

Thus, in the most general case, this soliton solution is characterized by seven free parameters. One natural set of parameters can be found from the far-field, $|x| \to \infty$ behavior consisting of two densities, an overall fluid velocity, and four phases.

Due to Galilean invariance and phase invariance, $\psi_j'(x,t) = \psi_j(x,t)\exp(i\varphi_j)$, we will assume, without loss of generality, that the background is at rest ($c=0$) and focus, more specifically, on the case of the SO(2) rotated DB soliton. In this case, the corresponding orthogonal matrix is given by

$$U = \begin{pmatrix} \cos(\delta) & -\sin(\delta) \\ \sin(\delta) & \cos(\delta) \end{pmatrix}, \quad (2.120)$$

where δ is an arbitrary angle. This way, the relevant SO(2) rotated soliton solution takes the form

$$\psi_1 = \cos(\delta)\sqrt{\mu}\{\cos\phi\tanh(\xi) + i\sin\phi\} - \sin(\delta)\eta\operatorname{sech}(\xi)\exp\{ikx + i\theta\}, \quad (2.121)$$

$$\psi_2 = \sin(\delta)\sqrt{\mu}\{\cos\phi\tanh(\xi) + i\sin\phi\} + \cos(\delta)\eta\operatorname{sech}(\xi)\exp\{ikx + i\theta\}. \quad (2.122)$$

Note that, for $\delta = 0$, the solution (2.121) and (2.122) is reduced to the DB soliton (2.112) and (2.113), while for $\eta = 0$ it becomes a "regular" DD soliton, meaning one where two dark solitons are colocated. This becomes evident by considering the asymptotics of the vector soliton, which are $|\psi_1|^2 \to \mu\cos^2(\delta)$ and $|\psi_2|^2 \to \mu\sin^2(\delta)$ as $|x| \to \infty$. When both parameters δ and η are nonzero, then each soliton is composed by a succession of a dip and a hump on top of constant background densities, respectively corresponding to the dark and bright components of the expressions (2.121) and (2.122). Importantly, the asymptotics in this general case remain unchanged and, thus, the solution (2.121) and (2.122) is, in fact, a DD soliton solution, characterized by four free parameters.

Let us now consider the densities of the individual dark solitons, which read

$$\begin{aligned} n_1 \equiv |\psi_1|^2 &= \mu\cos^2(\delta) - (\mu\cos^2(\delta)\cos^2\phi - \eta^2\sin^2(\delta))\operatorname{sech}^2\xi - \sqrt{\mu}\eta\sin(2\delta) \\ &\quad\times\{\sin\phi\sin[kx+\theta(t)] + \cos\phi\cos[kx+\theta(t)]\tanh\xi\}\operatorname{sech}\xi, \end{aligned} \tag{2.123}$$

$$\begin{aligned} n_2 \equiv |\psi_2|^2 &= \mu\sin^2(\delta) - (\mu\sin^2(\delta)\cos^2\phi - \eta^2\cos^2(\delta))\operatorname{sech}^2\xi + \sqrt{\mu}\eta\sin(2\delta) \\ &\quad\times\{\sin\phi\sin[kx+\theta(t)] + \cos\phi\cos[kx+\theta(t)]\tanh\xi\}\operatorname{sech}\xi, \end{aligned} \tag{2.124}$$

while the total density n_{tot} of the DD soliton is given by

$$n_{\text{tot}} = n_1 + n_2 = \mu - D^2\operatorname{sech}^2\xi. \tag{2.125}$$

Notice that the total density of the DD soliton is time independent and has the form of a dark soliton density of depth D^2 on top of a background density μ. The above density is, in fact, identical to the density of the DB soliton; this is due to the fact that under SO(2) rotation the total density, as well as all other conserved quantities of the system, remain unchanged. This point is particularly important when considering the motion of these solitons in a trap (see below).

Examples of the densities of the individual solitons, as well as the total densities of the above DD solitons, are shown in Fig. 2.22. In the more general case corresponding to $\eta \neq 0$ (e.g., for $\eta = 0.8$—cf. top right panel), the density of each soliton is characterized by a dip and a hump as mentioned above. On the other hand, for $\eta = 0$, the vector soliton is a regular DD soliton, with the background densities being generally unequal (cf. bottom left panel, for $\delta = \pi/8$); for $\delta = \pi/4$ (cf. bottom right panel), the background densities are equal. In all cases, the total densities follow the well-known pattern of the density of a single dark soliton (see thin lines).

It is now useful to consider the individual dark soliton densities, n_1 and n_2, across the trajectory of the DD soliton, i.e., for $\xi = 0$; in such a case, $x = x_0(t) = kt$ and the densities read

$$\begin{aligned} n_1(\xi = 0) &= \mu\cos^2(\delta)\sin^2\phi + \eta^2\sin^2(\delta) \\ &\quad - \sqrt{\mu}\eta\sin(2\delta)\sin\phi\sin\left[\frac{1}{2}(k^2+D^2)t\right], \end{aligned} \tag{2.126}$$

$$\begin{aligned} n_2(\xi = 0) &= \mu\sin^2(\delta)\sin^2\phi + \eta^2\cos^2(\delta) \\ &\quad + \sqrt{\mu}\eta\sin(2\delta)\sin\phi\sin\left[\frac{1}{2}(k^2+D^2)t\right]. \end{aligned} \tag{2.127}$$

It is clear that if $\eta = 0$, i.e., in the case of regular DD solitons, the densities $n_{1,2}(\xi = 0)$ are time independent. However, in the general case $\eta \neq 0$, the densities $n_{1,2}(\xi = 0)$ are periodic functions of time and, thus, in this case, the solutions (2.121) and (2.122) will be called beating DD solitons [336, 337]. The relevant angular frequency, which constitutes

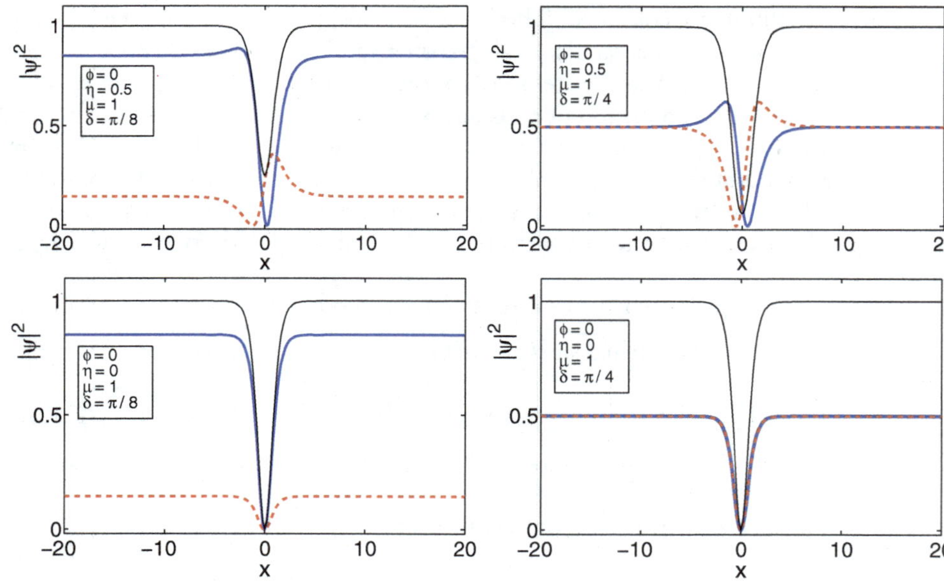

Figure 2.22. *Examples of the densities (at $t = 0$ and for $\phi = 0$ and $\mu = 1$) of beating (top panels) and regular (bottom panels) DD solitons: top left and top right panels show the densities of beating DD solitons with $\eta = 0.5$, with the individual dark solitons having, respectively, unequal ($\delta = \pi/8$) or equal ($\delta = \pi/4$) densities; bottom left and bottom right panels show the densities of regular DD solitons with $\eta = 0$, with the individual dark solitons having, respectively, unequal ($\delta = \pi/8$) or equal ($\delta = \pi/4$) densities. Thick solid and thick dashed lines correspond, respectively, to the individual densities $|\psi_1|^2$ and $|\psi_2|^2$, while the thin line corresponds to the total density $|\psi_1|^2 + |\psi_2|^2$.*

the internal beating frequency of the DD soliton, is given by

$$\omega_0 = \frac{1}{2}(k^2 + D^2) = \frac{1}{2}(\mu - \eta^2 \sec^2 \phi), \tag{2.128}$$

where we have also used Eq. (2.114). The frequency ω_0 is bounded by two limiting values. First, in the case $\eta \to 0$, the beating DD soliton becomes a regular DD soliton, characterized by a width $D = \sqrt{\mu}\cos\phi$ and a velocity $k = \sqrt{\mu}\sin\phi$; in this case, $\omega_0 \to (1/2)\mu$. Second, in the limiting case $D \to 0$, the beating DD soliton is reduced to a plane wave; in this case, $\omega_0 \to (1/2)k^2$. In other words, the intrinsic oscillation frequency assumes values in the range

$$\frac{1}{2}k^2 < \omega_0 < \frac{1}{2}\mu. \tag{2.129}$$

Before proceeding further, it is interesting to present results of recent experiments, utilizing the counterflow of two miscible BECs [333, 334, 335, 336, 337], where the observation of various types of vector solitons was reported. A characteristic example (see Ref. [337]) is shown in Fig. 2.23. The BEC in this figure is comprised of two different hyperfine states of ^{87}Rb, and the solitons are generated by subjecting the BEC to intercomponent counterflow (see Ref. [333] for details of this technique). In each panel, the atom clouds of the two components are vertically offset for imaging only, while all the dynamics leading to the soliton formation occurs in overlapped clouds.

There are clear examples of DB solitons (marked "DB" in the figure), consisting of a dark soliton in one component coupled to a bright soliton in the second component.

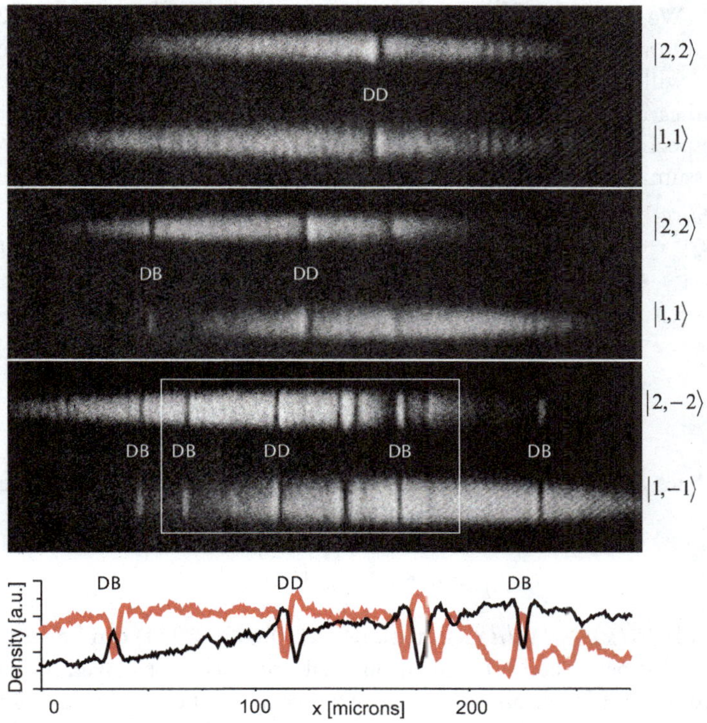

Figure 2.23. *Prototypical experimental images of dark-bright (DB) and dark-dark (DD) solitons in a two-component BEC. The two components are vertically offset for separate imaging. All dynamics occur with vertically overlapped components before the imaging procedure. Clear examples of DB and DD solitons are observed and marked. In the fourth panel, the red (thick) line shows a radially integrated cross section of the upper component in the boxed region of the third panel, while the black (thin) line shows the cross section of the lower component. The $|F, m_F\rangle$ hyperfine states used for these images are given to the right of each component. Adapted with permission from Ref. [337].*

The experimental result of Fig. 2.23 demonstrates not only the generation of several DB solitons but also the formation of solitons of the DD type (and even coexistence of solitons of different types, in a single BEC). It is interesting to compare the density profiles of the DD solitons (bottom panel of Fig. 2.23) with the ones of the SO(2) rotated DB solitons (i.e., the beating DD solitons) shown, e.g., in the top right panel of Fig. 2.22: a remarkable agreement is observed. The above results provide the motivation for the detailed study of the stability and dynamics of vector solitons in the trap (see next sections).

2.4.3 ▪ Dark-Bright Solitons in the Trap

Adiabatic Perturbation Theory for Dark-Bright Solitons in the Trap

In this section we will present the basic formulation of the adiabatic perturbation theory for DB solitons. In fact, our aim is to extend the considerations of Sec. 2.2.2, which referred to single-component condensates, to the case of multicomponent systems. As in Sec. 2.2.2, the prototypical perturbation that we will consider refers to the presence of a shallow parabolic trap. Nevertheless, as we will discuss in more detail below, the same methodology can be applied to cases where other physically relevant perturbations come into play; these include, e.g., the presence of thermal effects [325], localized defects [324], and others.

We start our analysis by considering the full problem of Eqs. (2.110) and (2.111), i.e., assuming that $\Delta \neq 0$ and the potential $V(x) = (1/2)\Omega^2 x^2$ is present. For convenience, we will slightly change the notation for the wavefunctions using $\psi_1 = u_d$ and $\psi_2 = u_b$, indicating that the component 1 (2) is supposed to support a dark (bright) soliton; the respective chemical potentials of the two components are now $\mu_d = \mu$ and $\mu_b = \mu + \Delta$. Assuming that the dark soliton is on top of a TF cloud in component u_d, with density $|u_{d,\mathrm{TF}}|^2 = \max\{\mu - V(x), 0\}$, we substitute the density $|u_d|^2$ in Eqs. (2.110) and (2.111) by $|u_d|^2 \to |u_{d,\mathrm{TF}}|^2 |u_d|^2$. Furthermore, introducing the transformations $t \to \mu t$, $x \to \sqrt{\mu} x$, $|u_b|^2 \to \mu^{-1}|u_b|^2$, we cast Eqs. (2.110) and (2.111) into the following form:

$$i\partial_t u_d + \frac{1}{2}\partial_x^2 u_d - \left(|u_d|^2 + |u_b|^2 - 1\right) u_d = R_d, \tag{2.130}$$

$$i\partial_t u_b + \frac{1}{2}\partial_x^2 u_b - \left(|u_b|^2 + |u_d|^2 - \tilde{\mu}\right) u_b = R_b, \tag{2.131}$$

where $\tilde{\mu} = 1 + \Delta/\mu$ and the functional perturbations R_d and R_b are given by

$$R_d \equiv (2\mu^2)^{-1}[2(1 - |u_d|^2)V(x)u_d + V'(x)\partial_x u_d], \tag{2.132}$$

$$R_b \equiv \mu^{-2}[(1 - |u_d|^2)V(x)u_b]. \tag{2.133}$$

while $V'(x) \equiv dV/dx$. Equations (2.130) and (2.131) can be viewed as a system of two coupled perturbed NLS equations, with perturbations given by Eqs. (2.132) and (2.133). Notice that in the absence of the perturbations, i.e., in the case of the homogeneous system $[V(x) = 0]$ subject to the boundary conditions $|u_d|^2 \to 1$ and $|u_b|^2 \to 0$ as $|x| \to \infty$, the NLS Eqs. (2.130) and (2.131) possess an exact analytical one-DB-soliton solution of the form of Eqs. (2.112) and (2.113), with parameters D and x_0 as per Eqs. (2.114) and (2.115) and for $\theta(t) = (1/2)(D^2 - k^2)t + (\Delta/\mu)t$.

To study the effect of perturbations on the dynamics of the DB soliton, we can adopt the effective particle approach of Sec. 2.2.2 and employ the Hamiltonian approach of the perturbation theory for matter-wave solitons. We start by considering the Hamiltonian (total energy) of the system of Eqs. (2.130) and (2.131) for $R_b = R_d = 0$, namely,

$$E = \frac{1}{2}\int_{-\infty}^{+\infty} |\partial_x u_d|^2 + |\partial_x u_b|^2 + (|u_d|^2 + |u_b|^2 - 1)^2 - 2(\tilde{\mu} - 1)|u_b|^2\, dx. \tag{2.134}$$

The energy of the system, when calculated for the DB soliton solution of Eqs. (2.112) and (2.113), takes the following form:

$$E = \frac{4}{3}D^3 + \chi\left(\frac{1}{2}D^2 \sec^2\phi - \frac{\Delta}{\mu}\right) \quad \text{with} \quad \chi \equiv \frac{N_b}{\sqrt{\mu}}. \tag{2.135}$$

We now consider an adiabatic evolution of the DB soliton and, particularly, we assume that, in the presence of the perturbations of Eqs. (2.132) and (2.133), the DB soliton parameters become slowly varying unknown functions of time t. Thus, the DB soliton parameters become $\phi \to \phi(t)$ and $D \to D(t)$ and, as a result, Eqs. (2.114) and (2.115) read

$$D^2(t) = \cos^2\phi(t) - \frac{\chi}{2}D(t), \tag{2.136}$$

$$\dot{x}_0(t) = D(t)\tan\phi(t), \tag{2.137}$$

where we have used Eq. (2.117). The evolution of the parameters $\phi(t)$, $D(t)$ and $x_0(t)$ can be found by means of the evolution of the DB soliton energy. In particular, employing

Eq. (2.135), it is readily found that

$$\frac{dE}{dt} = 4\dot{D}D^2 + \chi D \sec^2\phi(\dot{D} + D\dot{\phi}\tan\phi). \tag{2.138}$$

On the other hand, using Eqs. (2.130) and (2.131) and their complex conjugates, it can be found that the evolution of the DB soliton energy, due to the presence of the perturbations, is given by

$$\frac{dE}{dt} = -2\mathrm{Re}\left\{\int_{-\infty}^{+\infty} (R_d^* \partial_t u_d + R_b^* \partial_t u_b) dx\right\}, \tag{2.139}$$

where the asterisk denotes complex conjugation. Substituting R_d and R_b into Eq. (2.139) and evaluating the integrals, we finally obtain from Eqs. (2.138) and (2.139) the following result:

$$\begin{aligned} &4\dot{D}D^2 + \chi D \sec^2\phi(\dot{D} + D\dot{\phi}\tan\phi) \\ &= \frac{1}{\mu^2}\left(2\cos^3\phi\sin\phi - \chi D\sin\phi\cos\phi\right) V'(x_0). \end{aligned} \tag{2.140}$$

Equation (2.140), together with Eqs. (2.136) and (2.137), constitute a system of equations for the unknown soliton parameters $\phi(t)$, $D(t)$ and $x_0(t)$. We focus on the case of a DB soliton near the center of the trap, with an almost "black" dark soliton component (i.e., $x_0 \approx 0$ and $\cos\phi \approx 1$), for which the above system has a fixed point:

$$x_{0,\mathrm{eq}} = 0, \qquad \phi_{\mathrm{eq}} = 0, \qquad D_{\mathrm{eq}} = \sqrt{1 + \left(\frac{\chi}{4}\right)^2} - \frac{\chi}{4}. \tag{2.141}$$

Considering now small perturbations around the fixed points, i.e., $x_0 \to 0 + x_0$, $\phi \to 0 + \phi$ and $D \to D_{\mathrm{eq}} + D_1$, we linearize Eqs. (2.136) and (2.137) and Eq. (2.140) with respect to x_0, ϕ, and D_1, and obtain the following results:

$$D_1 = -\tilde{D}\phi^2, \qquad \tilde{D} \equiv \left(2D_{\mathrm{eq}} + \chi/2\right)^{-1}, \tag{2.142}$$

$$\dot{\phi} = \frac{-2 + \chi D_{\mathrm{eq}}}{D_{\mathrm{eq}}[-8D_{\mathrm{eq}}\tilde{D} - \chi(2\tilde{D} - D_{\mathrm{eq}})]} V'(x_0), \tag{2.143}$$

$$\dot{x}_0 = D_{\mathrm{eq}}\phi. \tag{2.144}$$

Here we should mention again that the above system of equations was derived for the potential-induced perturbations given in Eqs. (2.132) and (2.133). As we will see below, solution of the system of Eqs. (2.142)–(2.144) will describe the oscillatory dynamics of DB solitons in the trap.

Oscillations and Spectra of Dark-Bright Solitons in the Trap

It is now possible to derive from the system of Eqs. (2.142)–(2.144) an equation of motion for the DB soliton center, near its equilibrium position. Indeed, differentiating Eq. (2.144) with respect to time once, and using Eqs. (2.143) and (2.144), we obtain after some straightforward algebraic manipulations the following equation for x_0:

$$\ddot{x}_0 = -\left(\frac{1}{2} - \frac{\chi}{\chi_0}\right) V'(x_0), \tag{2.145}$$

where it is recalled that $\chi = N_b/\sqrt{\mu}$ [cf. Eq. (2.135)], while $\chi_0 \equiv 8\sqrt{1 + (\chi/4)^2}$.

Here, we should note that the result described by Eqs. (2.145) was first derived in Ref. [145]. In this work, it was assumed that each component is confined in a different trapping potential, namely V_d and V_b for the dark and bright solitons, respectively, which are slowly varying on the DB soliton scale D^{-1}, similarly to our analysis above. Then, for sufficiently slow solitons, a multiple-time-scale boundary-layer theory—similar to the one employed in Ref. [169]—was used to derive the following equation of motion for the soliton center:

$$\ddot{x}_0 = -\frac{1}{2}V_d'(x_0) - \frac{\chi}{8\sqrt{1+(\chi/4)^2 - V_d(x_0)}}\left[V_d'(x_0) - 2V_b'(x_0)\right]. \tag{2.146}$$

It is clear that for solitons confined in the same trap, and moving sufficiently close to the trap center (i.e., $V_b = V_d \ll \mu \equiv 1$), Eq. (2.146) coincides with Eq. (2.145).

Let us now focus on the physical significance of the result of Eq. (2.145). First, we should stress that, similarly to the case of dark solitons, the DB soliton center can be considered as a classical particle obeying a Newtonian equation of motion in the presence of the external potential V. In the case of the harmonic potential $V(x) = (1/2)\Omega^2 x^2$, Eq. (2.145) becomes the equation of motion of the classical linear harmonic oscillator, $\ddot{x}_0 + \omega_{\mathrm{osc}}^2 x_0 = 0$, where the oscillation frequency ω_{osc} of the DB soliton is given by

$$\omega_{\mathrm{osc}}^2 = \Omega^2\left(\frac{1}{2} - \frac{\chi}{\chi_0}\right). \tag{2.147}$$

Notice that in the limit $\chi \to 0$, i.e., $N_b \to 0$, Eq. (2.145) is reduced to Eq. (2.68) (recall that the latter predicts dark soliton oscillations with a frequency $\Omega/\sqrt{2}$—cf. Eq. (2.147) for $\chi = 0$), while the motion of the DB soliton becomes more sensitive to the presence of the bright component as N_b is increased. To be more specific, the oscillation frequency is *down-shifted* as compared to the characteristic value of $\Omega/\sqrt{2}$, i.e., the DB soliton pair executes slower oscillations, as the bright component is enhanced.

Apart from the analytical approximations used above to describe the oscillations of DB solitons in the trap, it is now particularly relevant to also consider the excitation spectra of stationary DB solitons [i.e., ones with the dark soliton component being a stationary kink with $\phi = 0$—cf. Eqs. (2.112)–(2.116)]. Such states, in the form $\psi_1(x,t) = u(x)$ and $\psi_2(x,t) = v(x)$, can be obtained numerically in the framework of Eqs. (2.110) and (2.111), using a fixed-point algorithm and an initial guess relying on the DB soliton solutions (2.112) and (2.113) of the respective homogeneous problem. Then, the stability of these states are determined by means of the BdG analysis, as in the case of scalar dark solitons (see Sec. 2.2.3). Particularly, we introduce the following ansatz into Eqs. (2.110) and (2.111):

$$\psi_1(x,t) = u(x) + \varepsilon\left[a(x)e^{\lambda t} + b^*(x)e^{\lambda^* t}\right], \tag{2.148}$$

$$\psi_2(x,t) = v(x) + \varepsilon\left[c(x)e^{\lambda t} + d^*(x)e^{\lambda^* t}\right], \tag{2.149}$$

and linearize the resulting equations (keeping only terms of order of the small parameter ε). Then, the ensuing eigenvalue problem for eigenvalues $\lambda = \lambda_r + i\lambda_i$ and eigenmodes $\{a(x), b(x), c(x), d(x)\}$ is numerically solved. In the case of a single DB soliton under consideration, the excitation spectrum can be well understood in both cases, corresponding to the absence and the presence of the harmonic trap, using the following arguments.

First, in the absence of the trap, the system of Eqs. (2.110) and (2.111) features not only a U(1) (phase) invariance in each of the components, but also a translational invariance;

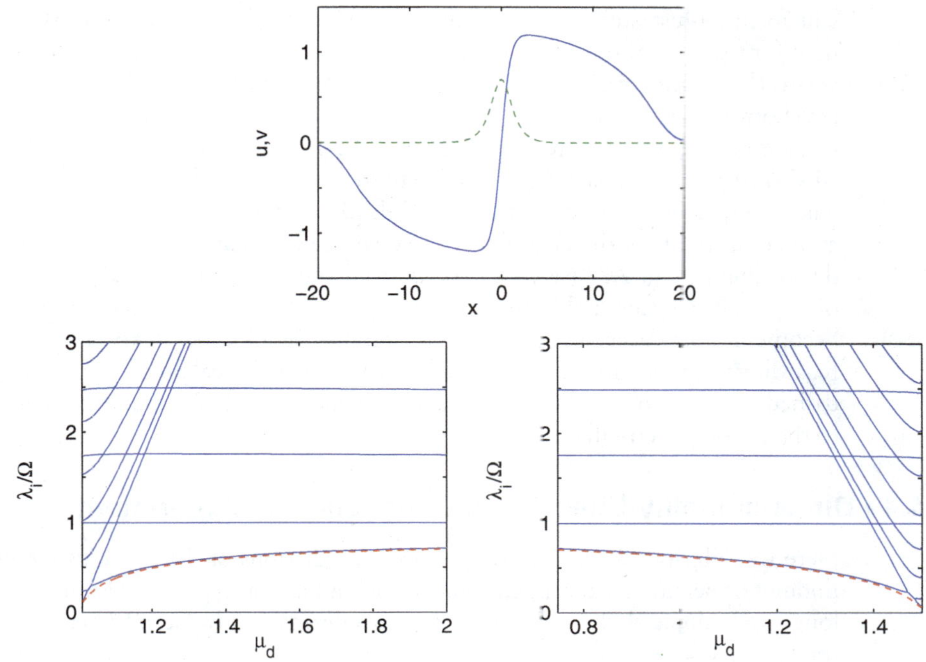

Figure 2.24. *The top panel depicts the stationary solution for a single DB-soliton for $\mu_d = 3/2$, $\mu_b = 1$, and $\Omega = 0.1$. The bright (dark) components are shown by the dashed green (solid blue) lines. The bottom left (right) panel shows the normalized imaginary part λ_i/Ω of the eigenvalues for the single DB soliton as a function of μ_d (μ_b) for $\mu_d = 1$ ($\mu_b = 3/2$). The (red) dashed line depicts the analytical prediction for the DB-soliton oscillation frequency [cf. Eq. (2.147)], providing an excellent approximation to the anomalous mode eigenfrequency. Adapted with permission from Ref. [335].*

thus, the system has three pairs of eigenvalues (each associated with one of the above symmetries) at the origin of the spectral plane (λ_r, λ_i). In this case, the phonon band (associated with the continuous spectrum of the problem) covers the entire imaginary axis of the spectral plane.

Second, in the presence of the trap, the single DB soliton "lives," as explained above, on the background of the confined ground state $\{\psi_1, \psi_2\} = \{\psi_{\rm GS}, 0\}$, which, can be approximated (for large values of μ_d) by $\{\psi_1, \psi_2\} \approx \{\psi_{\rm TF}, 0\}$. It is well known [45] (see also Sec. 2.2.3) that the harmonic potential introduces a discrete (point) BdG spectrum for this spatially confined ground state. In addition to that, the translational invariance of the unconfined system is broken and, due to the presence of the DB soliton, a single eigenvalue $\lambda^{(\rm AM)}$ emerges. The respective (negative energy) eigenmode is the anomalous mode (AM), while the associated eigenvalue $\lambda^{(\rm AM)}$ is directly connected with the oscillation frequency of the DB soliton in the harmonic trap, similarly to the case of a dark soliton in one-component BECs (see Ref. [126] and discussion in Sec. 2.2.3). In fact, the imaginary part of the eigenvalue $\lambda^{(\rm AM)}$ reads $\lambda_i^{(\rm AM)} = \omega_{\rm osc}$, where $\omega_{\rm osc}$ is the oscillation frequency of the single DB soliton, given by Eq. (2.147).

The above results are illustrated in Fig. 2.24, where a typical example of a stationary single DB-soliton state in the trap has been numerically computed and is depicted in the top panel; additionally, the eigenvalues λ_i characterizing the numerically obtained excitation (BdG) spectra of such stationary states are shown as functions of the chemical potentials $\mu_d = \mu$ and $\mu_b = \mu + \Delta$ in the bottom left and right panels of the figure, respectively.

The former illustrates the dependence on the chemical potential μ_d (of the dark component for fixed μ_b of the bright), while the latter showcases the corresponding frequencies versus the bright component chemical potential μ_b (for fixed μ_d). As observed in these two bottom panels, there exist two types of spectral lines, namely "slowly varying" ones (analogous to ones that are present in the spectrum of a dark soliton in one-component BECs [70]) and "fast varying" ones due to the presence of the bright-soliton component. The latter, as was pointed out also in Ref. [334] may, in fact, collide with the internal anomalous mode of the DB soliton and give rise to instability quartets which are barely discernible in Fig. 2.24 (see, e.g., the bottom right panel for $\mu_b > 1.4$ where a merger of eigenvalues occurs); this effect is much more pronounced when deviations from 1D become important, as will be explained in more detail in the next section. Importantly, regarding the anomalous mode, it is found that the analytical prediction of Eq. (2.147) (red dashed line) is remarkably good at capturing the anomalous mode eigenvalue pertaining to the DB-soliton oscillation.

2.4.4 ▪ Dimensionality Effects and Connection to Experiments

Here we will discuss the connection of the predictions of the previous section with the findings of pertinent experiments performed in Hamburg [68] and Pullman [334]: there, long-lived, single dark-bright matter-wave solitons—with the number of atoms N_b of the bright soliton being $\approx 10\%$ of the total number of atoms—were observed in a two-component quasi-1D ^{87}Rb BEC.

In particular, in the experiment reported in Ref. [68], a dark soliton was created in the spin state in the $|1,0\rangle$ of a ^{87}Rb BEC by means of the phase-imprinting method, and the density dip was filled by atoms, forming the bright soliton, in the spin state $|2,0\rangle$ of the condensate; note that in this experiment atoms were confined in an optical dipole trap with trapping frequencies $\omega_{x,y,z} = 2\pi \times \{5.9, 85, 133\}$ Hz (in our notation). On the other hand, in the experiment of Ref. [334], individual DB solitons were created during the counterflow of two BEC components, as per the same method first introduced in the experiment of Ref. [333]). Particularly, the dark and bright component were formed by ^{87}Rb atoms in the $|1,-1\rangle$ and $|2,-2\rangle$ states, respectively; the atoms, in this experiment too, were held in an elongated optical dipole trap with frequencies $\omega_{x,y,z} = 2\pi \times \{1.3, 163, 116\}$ Hz.

From the above description, it turns out that a direct comparison of the findings of these experiments with the above analytical results for the quasi-1D setting could only be done on a qualitative basis. For this reason, we follow the analysis of Ref. [334] and discuss the effect of dimensionality on the dynamics of DB solitons in more detail. For the sake of physical clarity, we start by considering the dimensional version of Eqs. (2.107) for the macroscopic wavefunctions of a two-component BEC:

$$i\hbar\frac{\partial \psi_k}{\partial t} = \left(-\frac{\hbar^2}{2m}\nabla^2 + U(\mathbf{r}) + g_{kk}|\psi_k|^2 + g_{12}|\psi_{3-k}|^2\right)\psi_k, \tag{2.150}$$

where the wavefunctions $\psi_k(\mathbf{r})$ ($k = 1,2$) for the dark and bright soliton components are normalized to the atom numbers N_d and N_b, $g_{ij} = 4\pi\hbar^2 a_{ij}/m$ are the effective nonlinear coefficients due to the s-wave scattering for $i,j = 1,2$, and $U(\mathbf{r})$ is the confining potential. For sufficiently anisotropic traps, we may decompose the transverse and axial dynamics of the system; this way, following the procedure described in Sec. 1.2.4, we factorize the wavefunction as $\psi_k(\mathbf{r},t) = \phi_k(\mathbf{r}_\perp; x) f_k(x,t)$, we substitute in Eq. (2.150), multiply by $\phi_k^*(\mathbf{r}_\perp; x)$, and integrate over the transverse directions. This leads to the following effective

1D model:

$$\left[i\hbar\frac{\partial}{\partial t}+\frac{\hbar^2}{2m}\frac{\partial^2}{\partial x^2}-V(x)\right]f_k=\mu_{\perp k}[f_k]f_k, \tag{2.151}$$

where $V(x)=(1/2)m\omega_x^2x^2$ is the axial potential and the transverse chemical potential $\mu_{\perp k}=\mu_{\perp k}[f_k(x,t)]$ is a functional of f_k:

$$\begin{aligned}\mu_{\perp k}[f_k]=\int d^2\mathbf{r}_\perp\phi_k^*\Big(&-\frac{\hbar^2}{2m}\nabla_\perp{}^2+\frac{1}{2}m\omega_\perp^2r_\perp^2\\&+g_{kk}|\phi_k|^2|f_k|^2+g_{12}|\phi_{3-k}|^2|f_{3-k}|^2\Big)\phi_k.\end{aligned} \tag{2.152}$$

Here, $(\omega_x,\omega_\perp)$ are the trap frequencies along the longitudinal (axial) and transverse directions, and we have assumed that the derivatives of ϕ_k do not depend on the axial variable x. For an effectively 1D system, we assume that the transverse wavefunction remains in its Gaussian ground state, namely: $\phi_k=\frac{1}{\sqrt{\pi}\sigma_k}\exp\left(-\frac{r_\perp^2}{2\sigma_k^2}\right)$. To account for axial effects, we allow the width σ_k to be a variational parameter, $\sigma_k=\sigma_k[f_k(x,t)]$; this yields the following equation:

$$\mu_{\perp k}=\frac{\hbar^2}{2m}\sigma_k^{-2}+\frac{m}{2}\omega_\perp^2\sigma_k^2+\frac{g_{kk}|f_k|^2}{2\pi}\sigma_k^{-2}+\frac{g_{12}|f_{3-k}|^2}{\pi(\sigma_1^2+\sigma_2^2)}.$$

As was explained in Sec. 1.2.4, there are two different approaches to determine σ_k: (a) one can minimize the chemical potential $\mu_{\perp k}$ with respect to σ_k for given $f_k(x,t)$ [104, 105]; (b) one can use the Euler–Lagrange equations from the Lagrangian associated to Eq. (2.151), minimizing the total energy [103]. These two approaches lead to the following expression for σ_k:

$$\sigma_k^4=\frac{\hbar^2}{m^2\omega_\perp^2}+\frac{g_{kk}|f_k|^2}{A\pi m\omega_\perp^2}+\frac{2g_{12}|f_{3-k}|^2}{\pi m\omega_\perp^2(\sigma_1^2+\sigma_2^2)^2}\sigma_k^4, \tag{2.153}$$

where the parameter $A=1$ ($A=2$) corresponds to the GMDE (NPSE) system—see Sec. 1.2.4—for the two-component condensate. Notice that Eqs. (2.151) and (2.153) constitute a set of coupled nonlinear equations, which have to be solved consistently in order to obtain $f_k(x,t)$ and $\sigma_k[f_k(x,t)]$.

It is now relevant to use the above effective 1D models to derive the oscillation period of DB solitons, and compare the results with the ones obtained in the framework of the full 3D GPEs. This is done in Fig. 2.25, where we have considered a cylindrical trap with frequencies $\omega_\perp=2\pi\times 133$ Hz and $\omega_x=2\pi\times 5.9$ Hz, similar to the ones used in the experiment of Ref. [68]. The top panel illustrates the dependence of the period of the DB soliton on the number of atoms (N_d,N_b) of the two components. Clearly, variation of the number of atoms, especially in the bright component by a factor of 2, may lead to a significant (approximately two fold) variation of the DB soliton frequency. The agreement between 1D and 3D generally becomes worse when N_b is increased and, to a lesser extent, when N_d is increased. Notice that the NPSE model yields generally more accurate predictions than the GMDE one.

The bottom panels of Fig. 2.25 show the DB soliton spectrum. When the anomalous mode collides with a mode of positive energy, the DB soliton becomes dynamically unstable, i.e., the amplitude of its oscillation increases. Such collisions are present in the DB soliton spectrum, highlighting the critical difference in comparison with the case of dark solitons in single-component BECs. Furthermore, as observed in the right-most bottom panel of Fig. 2.25 (and also corroborated by 3D GPE simulations), this oscillatory

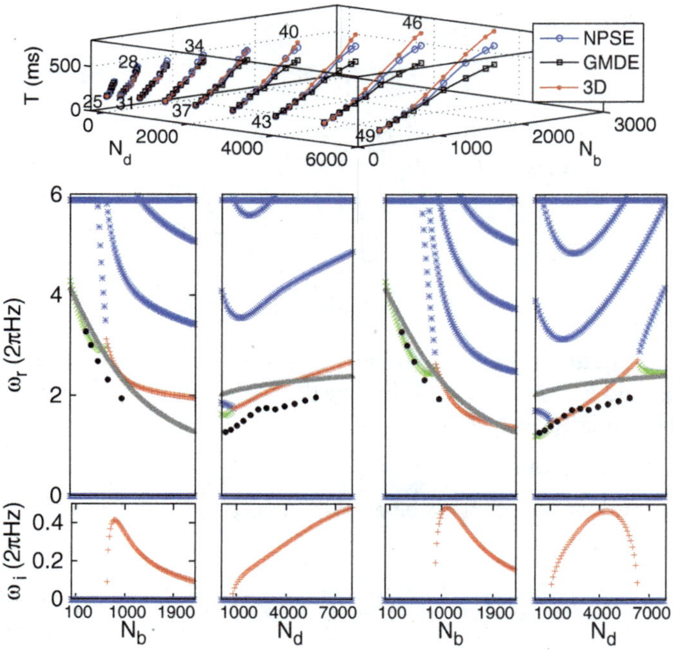

Figure 2.25. *Top: DB soliton oscillation period vs. N_d and N_b for fixed μ_d (chemical potential for the dark soliton) indicated by the numbers in the panel. Circles (blue) and squares (black) depict the results from the BdG analysis of the GMDE and NPSE, respectively; small (red) dots represent results of the 3D GPE. Bottom: BdG analysis of GMDE (left) and NPSE (right) [real ω_r and imaginary ω_i parts of the eigenfrequencies]: anomalous mode (green), dynamically unstable mode (red), analytical 1D result of Ref. [145] (gray), and 3D GPE results (black points). Adapted with permission from Ref. [334].*

instability disappears for sufficiently large N_d. Additionally, from Fig. 2.25, it is inferred that for progressively larger atomic populations, a departure from the quasi-1D behavior emerges and a larger (smaller) oscillation frequency is obtained, if N_d (N_b) is increased.

Let us now compare the above findings with the ones of the experiment of Ref. [68] in more detail. First, we should note that, in this experiment, the created DB soliton was observed to perform slow oscillations, with a frequency 0.24Ω, which is quite smaller than the frequency of the corresponding single dark soliton in the same setting (which is $\Omega/\sqrt{2} \approx 0.707\Omega$). Qualitatively speaking, this clearly underscores the fact that the "filled" dark soliton is heavier than its "bare" counterpart, as predicted in the framework of both the 1D and effective 1D models. On a quantitative basis, an observation of the top panel of Fig. 2.25 shows that the atom numbers used in Ref. [68] (up to $N_b \sim 8000$ and $N_d \sim 92000$) are out of the realm of validity of the effective 1D models. Thus, in this case, only the 3D GPE is relevant. Fully 3D simulations performed in Ref. [334] (for trap frequencies and atom numbers comparable to those used in Ref. [68]) have shown that the oscillation period of a DB soliton is approximately 827 ms, which is comparable with the period observed in Ref. [68] (slightly larger than 1 s). Possible sources for the discrepancy between the numerical and experimental oscillation periods include

(i) high sensitivity of the period on the N_d to N_b ratio,

(ii) sensitivity of the dynamics to uncertainties in the measured scattering lengths, and, perhaps more importantly,

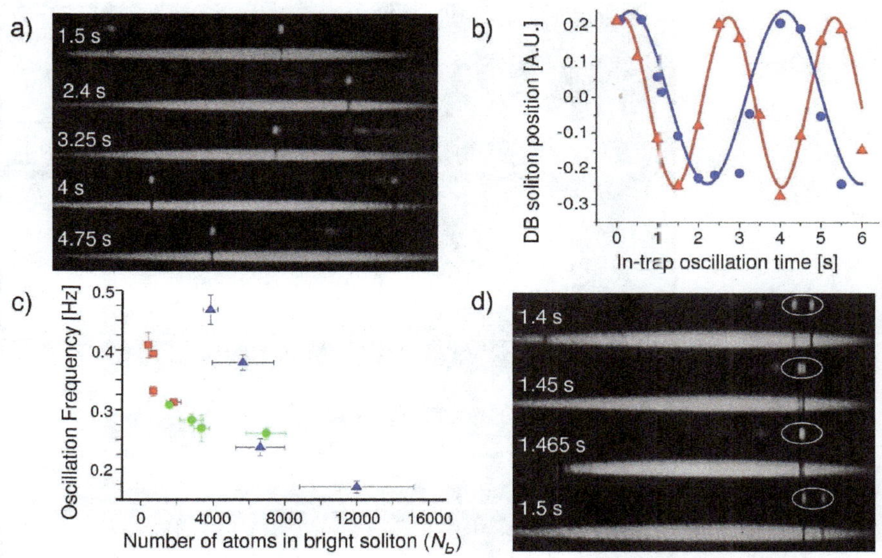

Figure 2.26. (a) *Experimental images showing the oscillation of the DB soliton at the in-trap evolution times indicated.* (b) *In-trap DB soliton oscillation. Red triangles and blue squares correspond to, respectively,* $(N_b, N_d) \approx (680, 27\,000)$ *and* $(N_b, N_d) \approx (9\,000, 650\,000)$*. Solid lines correspond to fitted harmonic oscillations with frequencies* 0.39 *Hz and* 0.27 *Hz, respectively.* (c) *Oscillation frequency vs.* N_b *for different number of atoms in the dark component; red squares:* $N_d \approx 30\,000$*; green triangles:* $N_d \approx 200\,000$*; and blue circles:* $N_d \approx 430\,000$*.* (d) *Experimental expansion images showing the oscillation and collision (in circled regions) of two DB solitons. Images in* (a) *and* (d) *are taken after 7 ms and 8 ms of free expansion for the bright and dark component, respectively. The components are vertically overlapped prior to expansion. Adapted with permission from Ref. [334].*

(iii) the numerical simulations confirm that the oscillations are not harmonic and tend to have increasing periods for higher oscillating amplitudes.

We conclude this section by discussing the findings of the experiment of Ref. [334], which can be summarized in Fig. 2.26. First, panel (a) depicts the experimental images demonstrating the oscillation of the DB soliton in the trap. Panel (b) shows pertinent soliton trajectories for two experimental sets of (N_b, N_d) and their comparison with fitted harmonic oscillations of frequencies 0.39 Hz and 0.27 Hz, respectively. In panel (c), the oscillation frequency of the DB soliton is shown as a function of the number of atoms N_b of the bright component, for three different number of atoms N_d in the dark component. Here, we should point out that the lack of exact cylindrical symmetry, as well as the large atom number in the experiment Ref. [334], preclude a direct comparison with the analytical results presented above. However, this experiment too, clearly shows the anticipated decrease of oscillation frequency with increasing number of atoms in the bright and decreasing number of atoms in the dark component. Finally, in panel (d) experimental results of the collision between two DB solitons are shown, indicating their near-elastic nature. This latter result motivate us to study multiple DB solitons and interactions between DB solitons in more detail (see next section).

2.4.5 ▪ Multiple Dark-Bright Solitons

As was discussed above (and also shown in Figs. 2.23 and 2.25), multiple DB solitons can be created experimentally by means of counterflow between two miscible conden-

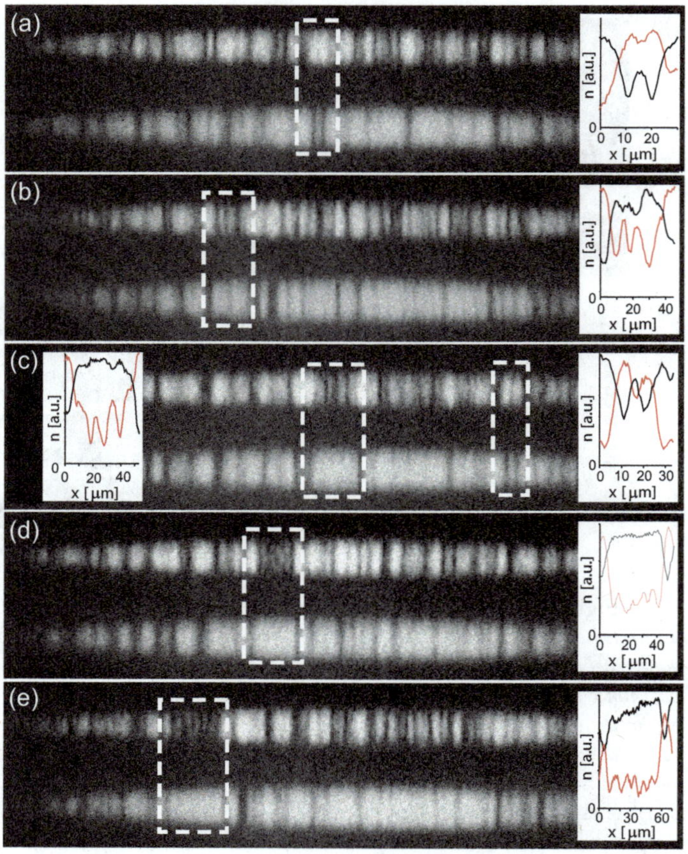

Figure 2.27. *Experimental images indicating DB-soliton clustering in a two-component BEC. The upper cloud in each image (and red curve in inset) shows atoms in the $|2,-2\rangle$ state, while the lower cloud (black curve) shows atoms in the $|1,-1\rangle$ state. Prior to imaging, the two components are overlapped in trap for 5 s. Insets show integrated cross sections of the boxed region. Adapted with permission from Ref. [335].*

sates. Perhaps the most characteristic example demonstrating the emergence of multiple DB solitons in a real experiment is the one shown in Fig. 2.27 [335]: there, DB-soliton clustering occurring during the counterflow of two rubidium BEC species, namely the hyperfine states $|1,-1\rangle$ and $|2,-2\rangle$, confined in an elongated optical dipole trap with frequencies of $2\pi\times\{1.5, 140, 178\}$ Hz is shown. It can readily be observed that frequently formed large gaps in one component (which supports the dark-solitons) are filled by bright solitons in the other component. Interestingly, these gaps are structured by small, periodic density bumps, indicating that these regions are composed of merged solitons. Some of these features are marked by the boxed regions in Fig. 2.27, with corresponding cross sections shown as insets. Clusters of two and three merged solitons [see Figs. 2.27(a)–(c)] are clearly observed, while there exist indications of clusters composed of four- to five-solitons—see Figs. 2.27(d), (e).

Interactions Between Dark-Bright Solitons

We start our analysis by considering the simplest case corresponding to the absence of the trap, i.e., $\Omega = 0$ in Eqs. (2.130) and (2.131). In this case, and in the Manakov limit, exact multiple DB soliton solutions are available (see, e.g., Ref. [319]). However, since the analytical form of these states is quite complicated, an analytical treatment of interactions between DB solitons becomes a particularly complicated problem, whose solution does not provide a clear physical picture (see, e.g., Ref. [321]). Instead, one may employ a variational approach (in the lines of the analysis of Ref. [26] for the interaction between dark solitons in a single-component system) which, as we will see, sheds light in the physical aspects of the problem, providing a better understanding for the interaction process and the formation of bound states composed by multiple DB solitons (also called "DB soliton molecules" [323]).

Thus, following the analysis of Ref. [335], we consider, in the framework of Eqs. (2.130) and (2.131) for $\Omega = 0$, a pair of equal-amplitude single DB solitons, traveling in opposite directions. This state can be approximated by the following ansatz:

$$u_d(x,t) = \left(\cos\phi \tanh X_- + i \sin\phi\right)\left(\cos\phi \tanh X_+ - i \sin\phi\right), \tag{2.154}$$

$$u_b(x,t) = \eta \,\mathrm{sech} X_- \, e^{i[kx+\theta)+(\tilde{\mu}-1)t]} + \eta \,\mathrm{sech} X_+ \, e^{i[-kx+\theta)+(\tilde{\mu}-1)t]} \, e^{i\Delta\theta}, \tag{2.155}$$

where $\tilde{\mu} = \mu_b/\mu_d$ is the ratio of the chemical potentials. The solitons' coordinates are given by $X_\pm = D(x \pm x_0(t))$ with $2x_0$ being the relative distance between the two solitons, while $\Delta\theta$ is the relative phase between the two bright solitons (assumed to be constant). Notice that $\Delta\theta = 0$ and $\Delta\theta = \pi$ correspond to *in-phase* and *out-of-phase* bright solitons, respectively. We should also mention that the ansatz (2.154) is a symmetric form of two dark solitons on the common background which (provided that the separation distance $2x_0$ is sufficiently large) interact weakly with each other; recall that the same considerations were used in Sec. 2.1.6, for the study of intersoliton interactions in a dark soliton pair in a single-component system. Similarly, the ansatz (2.155) is a superposition of two bright solitons of equal amplitudes, placed at the locations of their respective dark solitons siblings; such a form of the bright soliton pair is commonly used for the study of interactions between bright solitons (see, e.g., Chapter 3.2.2 of Ref. [36]). We finally note that in the case of two well-separated DB solitons, the parameters N_b, η, μ_d (chemical potential of the dark solitons component), and D are connected through the equation $N_b \approx 4\eta^2\sqrt{\mu_d}/D$. Namely, N_b is approximately twice as large compared to the case of one DB soliton—see Eq. (2.117).

Since the ansatz of Eqs. (2.154) and (2.155) does not correspond to an exact solution of Eqs. (2.130) and (2.131), we will now employ the adiabatic perturbation theory for DB solitons of Sec. 2.4.3. In particular, we will again assume that the DB soliton parameters are slowly varying functions of time and determine their evolution by means of the evolution of the DB soliton energy. To do this, first we substitute the ansatz (2.154) and (2.155) into Eq. (2.134) and perform the integrations under the assumption that the soliton velocity is sufficiently small, such that $\cos(kx) \approx 1$ [and $\sin(kx) \approx 0$]. Then, we further simplify the result assuming that the solitons are well separated, i.e., their relative distance is $x_0 \gg 1$. This way, we find that the energy of the system takes the form

$$E = 2E_1 + E_{\mathrm{DD}} + E_{\mathrm{B3}} + 2E_{\mathrm{DB}}, \tag{2.156}$$

where E_1 is the energy of a single DB soliton [cf. Eq. (2.135)], while the remaining terms account for the interaction between the two DB solitons. In particular, E_{DD}, E_{BB}, and E_{DB} denote, respectively, the interaction energy between the two dark solitons, the two bright ones, and the interaction energy between the dark soliton of one component and the bright soliton in the other component. Approximate expressions for the above interaction energies are given by

$$E_{DD} = 16\cos^2\phi\left[\frac{1}{3}D\cos^2\phi + D + 2(\cos^2\phi - D^2)x_0 - \frac{3+4\cos^2\phi}{3D}\cos^2\phi\right]e^{-4Dx_0}, \tag{2.157}$$

$$E_{BB} = \chi\left[2D\left(D(1-Dx_0) - k^2x_0\right) + D\chi\right]\cos\Delta\theta e^{-2Dx_0} + \chi\left[\chi D(2Dx_0-1)\left(1+2\cos^2\Delta\theta\right)\right]e^{-4Dx_0}, \tag{2.158}$$

$$E_{DB} = -4\chi\cos^2\phi\cos\Delta\theta e^{-2Dx_0} + \chi\cos^2\phi\left[\frac{16}{3}\cos^2\phi - 16Dx_0 + 8\right]e^{-4Dx_0}, \tag{2.159}$$

where terms of order $O(e^{-6Dx_0})$ and higher have been neglected (it has been checked that their contribution does not alter our main results). We can now use the above expressions and find the evolution of the soliton parameters from the energy conservation, $dE/dt = 0$. In fact, focusing on the case of low velocity, almost black solitons [with $\dot{D}(t) \approx 0$ and $\cos\phi(t) \approx 1$], the energy conservation leads to the following nonlinear evolution equation for the DB soliton center:

$$\ddot{x}_0 = F_{int}, \tag{2.160}$$

$$F_{int} \equiv F_{DD} + F_{BB} + 2F_{DB}. \tag{2.161}$$

In the above equations, F_{int} is the interaction force between the two DB solitons (depending on the soliton coordinate x_0), which contains the following three distinct contributions: the interaction forces F_{DD} and F_{BB} between the two dark and the two bright solitons, respectively, as well as the interaction force F_{DB} of the dark soliton of one DB soliton pair with the bright soliton of the other pair. The functional form of the above forces is

$$F_{DD} = \frac{1}{\chi_o}\left[\frac{1}{3}(544 - 352D_0^2) + 128D_0\left(D_0^2 - 1\right)x_0\right]e^{-4D_0x_0}, \tag{2.162}$$

$$F_{BB} = \frac{\chi}{\chi_o}\left[-6D_0 + 4D_0^2x_0 - 2\chi\right]D_0^2\cos\Delta\theta e^{-2D_0x_0} + \frac{\chi^2}{\chi_o}\left[\left(1+2\cos^2\Delta\theta\right)(6-8D_0x_0)\right]D_0^2e^{-4D_0x_0}, \tag{2.163}$$

$$F_{DB} = \frac{\chi}{\chi_o}\left[(8D_0\cos\Delta\theta)e^{-2D_0x_0} + \left(64D_0x_0 - \frac{208}{3}\right)D_0e^{-4D_0x_0}\right], \tag{2.164}$$

where $D(t) \approx D_0$ since we are assuming that $\dot{D}(t) \approx 0$.

The main result of the above analysis, namely the equation of motion for the two-DB-soliton state [cf. Eq. (2.160)] provides a clear physical picture for the interaction between the two DB solitons. To better understand this result, it is now useful to investigate the role of each force term, and to also consider some relevant limiting cases.

First, we note that (to the leading order of approximation) the interaction force between the bright soliton components has a longer range than the interaction forces between the dark soliton components, since $F_{BB} \propto \exp(-2D_0 x_0)$ while $F_{DD} \propto \exp(-4D_0 x_0)$ [cf. Eqs. (2.162) and (2.163)] and, as a result, it is decaying more slowly for large soliton separation x_0 than the one between the two dark solitons; additionally, the interaction between dark and bright solitons is also, to leading order, $F_{DB} \propto \exp(-2D_0 x_0)$. This is in accordance with earlier predictions, where the same dependence of the force over the soliton separation was found (see, e.g., Refs. [345] and [18, 26, 72] for bright and dark solitons, respectively).

Let us now consider the role of the bright-soliton component. In its absence, i.e., for $\chi = 0$ [cf. Eq. (2.114)], it is clear that $F_{BB} = F_{DB} = 0$ and Eq. (2.160) describes the interaction between two dark (almost black) solitons; in this case, taking into regard that $D_0 = 1$, it can readily be found that the pertinent (repulsive) interaction potential is $\propto 2\exp(-4x_0)$, which coincides with the result of Ref. [26] obtained by means of a variational approach (see also a relevant discussion in Ref. [72]). On the other hand, when bright solitons are present (i.e., for $\chi \neq 0$), the principal nature of the bright-bright-soliton interaction—and also of part of the dark-bright-soliton interaction—depends on the relative phase $\Delta\theta$ between the two bright solitons through the factor $\cos\Delta\theta$ [see Eqs. (2.163) and (2.164)]. In particular, if $\Delta\theta = 0$ (in-phase case) the interaction is repulsive, while if $\Delta\theta = \pi$ (out-of-phase case), the interaction is attractive; this result stems from the fact that the coefficients of the terms $\propto \cos\Delta\theta$ in Eqs. (2.163) and (2.164) are positive definite since the parameter $D_0 \geq 1$ for every $\chi > 0$.

According to the above, it is clear that the competition between repulsive (for dark solitons) and attractive (for out-of-phase bright solitons) forces leads to the emergence of fixed points in the equation of motion (2.160); note that the interaction between dark and bright solitons does not alter qualitatively (although it obviously affects quantitatively) this interplay between repulsive terms $\propto \exp(-4D_0 x_0)$ and attractive ones $\propto \exp(-2D_0 x_0)$ (if $\Delta\theta = \pi$). Thus, even in the absence of the trap (which exerts extra potential forces to the solitons), the above analysis predicts that there exists a *stationary* DB-soliton "molecule" composed of two DB solitons. Note that stationary two-DB solitons were also found numerically and experimentally in Ref. [346] in the context of nonlinear optics, but their existence details and stability properties were not considered.

The fixed (equilibrium) points x_{eq} of Eq. (2.160), corresponding to equilibrium distances between the constituent DB solitons forming the stationary molecule, can be determined as solutions of the transcendental equation resulting from Eq. (2.160) for $\ddot{x}_0 = 0$ in the out-of-phase case ($\Delta\theta = \pi$). Once the fixed points x_{eq} are found, their stability can be studied by introducing the ansatz $x_0(t) = x_{eq} + \delta(t)$ into Eq. (2.160), and linearizing with respect to the small-amplitude perturbation $\delta(t)$; this way, we derive the following equation:

$$\ddot{\delta} + \omega_0^2 \delta = 0, \tag{2.165}$$

where the oscillation frequency ω_0 is given by

$$\omega_0^2 = -\left.\frac{\partial F_{\text{int}}}{\partial x_0}\right|_{x_0 = x_{eq}}. \tag{2.166}$$

Notice that the oscillation frequency ω_0 represents the internal out-of-phase motion of the two DB solitons; in the considered case where the trap is absent, the in-phase motion of the solitons is associated with the neutral translation mode due to the translational invariance of the system (the respective in-phase Goldstone mode has a vanishing frequency).

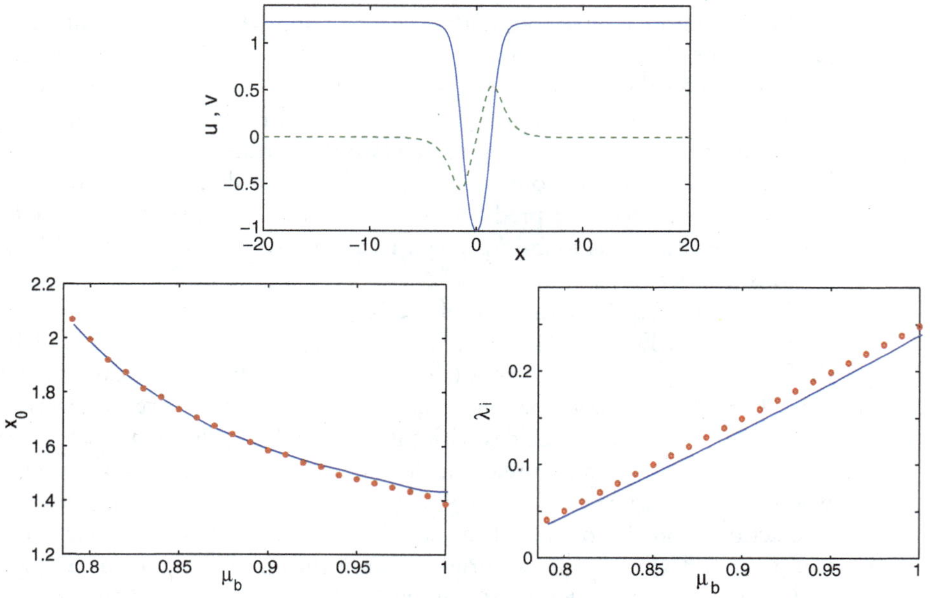

Figure 2.28. *Top panel: A stationary DB-soliton pair: the solid blue line denotes the two-dark-soliton state (recall that each dark soliton is associated with a zero crossing), while the green dashed line denotes the respective two-bright-soliton state. The chemical potentials are $\mu_d = 3/2$ and $\mu_b = 1$. Bottom left panel: the equilibrium center of mass x_0 as a function of the chemical potential μ_b (for $\mu_d = 3/2$). The solid blue line depicts the analytical prediction of Eq. (2.160), while red circles depict the numerically obtained soliton center x_0. Bottom right panel: the oscillation frequency for the out-of-phase motion of the DB-soliton pair as a function of μ_b (for $\mu_d = 3/2$). The solid blue line depicts the analytical result for ω_0 [cf. Eq. (2.166)], while red circles depict the corresponding numerically obtained imaginary eigenvalue λ_i (for the out-of-phase soliton motion) of the excitation spectrum. Adapted with permission from Ref. [335].*

The numerical simulations confirm the existence of the stationary two-DB-soliton state (in the out-of-phase case), with an example of such a state shown in the top panel of Fig. 2.28 (for chemical potentials $\mu_d = 3\mu_b/2 = 3/2$). It is also possible to determine the dependence of the equilibrium soliton positions and the effective frequency ω_0 [cf. Eq. (2.166)] on the chemical potential μ_b of the bright soliton component. The respective analytical and numerical results are shown in the bottom left and right panels of Fig. 2.28. To obtain these numerical results, a (least squares) fitting algorithm was used to accurately identify the amplitude η, inverse width D, and equilibrium center of mass x_0 of the bright component. The numerical findings for x_0 and ω_0 (the latter is numerically obtained via a BdG analysis, as the imaginary eigenvalue λ_i pertaining to the out-of-phase motion of the stationary two-DB-soliton state) are directly compared with the semi-analytical results of Eqs. (2.160) and (2.166), respectively. Obviously, there is a very good quantitative agreement between the analytical and numerical results (see bottom panels of Fig. 2.28). Notice that despite the motion of this eigenvalue through the continuous spectrum, no instability is observed in the parametric window shown in Fig. 2.28.

Multiple Dark-Bright Solitons in the Trap

We now consider the case of multiple DB solitons in the presence of the harmonic trap. In this case, each of the multiple DB solitons is subject to two forces: (a) the restoring force of the trap, F_{tr} [in the case of a single DB soliton, this force induces an in-trap oscillation with a frequency ω_{osc}—see Eq. (2.147)], and (b) the pairwise interaction force F_{int} [cf. Eq. (2.161)] from other DB solitons. Thus, taking into regard that $F_{\text{tr}} = -\omega_{\text{osc}}^2 x_0$, the effective equation of motion for the center x_0 of a two-DB-soliton state can be written as follows:

$$\ddot{x}_0 = F_{\text{tr}} + F_{\text{int}}. \tag{2.167}$$

One can thus straightforwardly generalize the above equation for $\mathcal{N}$-interacting DB-soliton states, similarly to the case of multiple dark solitons in one-component BECs (see discussion in Sec. 2.3.5 and Refs. [70, 72, 347]).

Importantly, the presence of the trap-induced restoring force F_{tr} allows for the existence of stationary DB-soliton molecules not only for out-of-phase bright solitons (as in the homogeneous case, whereby such a state could be stationary even without the trap as found above), but also for in-phase bright solitons. In the latter case, the repulsion between both the dark- and the bright-soliton component(s), is balanced by the restoring force F_{tr}. In the case of two DB solitons placed at $x = \pm x_0$, the equilibrium points, x_{eq}, can readily be found (as before) as solutions of the transcendental equation resulting from Eq. (2.167) for $\ddot{x}_0 = 0$, in both the in- and out-of-phase cases. To study the stability of these equilibrium points in the framework of Eq. (2.167), we may again use the ansatz $x_0(t) = x_{\text{eq}} + \delta(t)$, and obtain a linear equation for the small-amplitude perturbation $\delta(t)$, similar to that of Eq. (2.165), namely, $\ddot{\delta} + \omega_1^2 \delta = 0$, where the frequency ω_1 is given by

$$\omega_1^2 = \omega_{\text{osc}}^2 + \omega_0^2, \tag{2.168}$$

where ω_0 is given by Eq. (2.166). Similarly to the case of dark solitons in one-component BECs [72] (see also Ref. [75]), by construction, this mode captures the out-of-phase motion of the DB-soliton pair. Furthermore, by symmetry, the in-phase oscillation of the DB-soliton pair in the trap will occur with the frequency

$$\omega_2 = \omega_{\text{osc}}. \tag{2.169}$$

These two characteristic frequencies (ω_1, ω_2) coincide with the eigenfrequencies of the two anomalous modes of the BdG spectrum of the trapped DB-soliton pair.

The above features of the multiple-DB-soliton states can be tested numerically. First, we consider the two-DB-soliton state in the trap; for this case, results are shown in Figs. 2.29 and 2.30, both for the in-phase and the out-of-phase states. In particular, the top left and right panels of Fig. 2.29 show examples of two stable, in-phase and out-of-phase stationary DB-soliton pairs, respectively (both for $\mu_d = 3/2$ and $\mu_b = 1$). The bottom panels of the figure show the equilibrium positions of the soliton centers. In the in-phase case (bottom left panel), it is observed that larger chemical potential (number of atoms) in the second component leads to stronger repulsion and, hence, larger distance from the trap center. In the out-of-phase case (bottom right panel), we observe a similar effect but in the reverse direction (due to the attraction of the out-of-phase bright-soliton components) for smaller values of the chemical potential. In both cases a good agreement is observed between the numerically observed equilibrium separations and the theoretically predicted ones from Eq. (2.167).

Next, to study the validity of Eq. (2.168)—pertinent to small-amplitude oscillations around the fixed points—we show in Fig. 2.30 the eigenvalues λ of the excitation spectrum as functions of μ_b, both for the in-phase (left column) and for the out-of-phase (right column) cases. The imaginary and real parts, λ_i and λ_r, of the respective eigenvalues, normalized over the trap strength Ω, are respectively shown in the top and bottom panels of Fig. 2.30. In the top panels, it is straightforward to compare the analytical result of Eq. (2.168) with the BdG result, namely the second anomalous mode of the

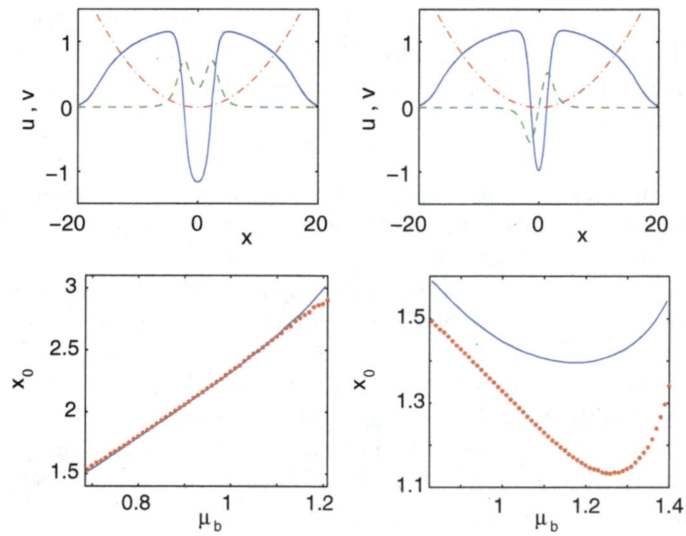

Figure 2.29. *The left and right columns correspond, respectively, to an in-phase and an out-of-phase DB-soliton pair in a harmonic trap with* $\Omega = 0.1$. *The top row of panels depicts the profiles of the DB-soliton pairs (solid blue lines and dashed green lines corresponding, respectively, to the dark and bright components) and the trapping potential (dashed-dotted red line). The bottom row of panels depicts the numerical (red markers) and the analytical (solid blue line) results for the equilibrium distance between the solitons as a function of* μ_b; *the theoretical prediction is based on Eq. (2.167) Adapted with permission from Ref. [335].*

spectrum, corresponding to the out-of-phase oscillations of the DB-soliton pair. Once again, good agreement is observed between the two; the differences may be partially attributed to the "interaction" (i.e., collisions) of these modes with other modes of the BdG spectrum. Notice that there exist narrow instability windows, arising due to the crossing of the anomalous mode(s) of the DB-soliton pair with eigenmodes of the background of the two-component system. These instabilities arise in the form of Hamiltonian–Hopf bifurcations [348] through the emergence of quartets of complex eigenvalues resulting from the collision of two pairs. Clearly, the growth rates of the pertinent oscillatory instabilities are fairly small (i.e., the instabilities are weak) in both the in- and out-of-phase cases; it should be mentioned, however, that in the latter case, the formation of the quartets appears to be occurring in very narrow intervals.

The above considerations can also be generalized for the cases of three or more DB solitons, although the analytical calculations become increasingly more tedious. As an example, in Fig. 2.31 we show two different three-DB-soliton configurations: the first column in the figure corresponds to the in-phase three-DB-soliton state, while the second column corresponds to the out-of-phase variant thereof. Notice that in the case under consideration, there exist narrow parametric intervals of dynamical instability, which are narrower for the out-of-phase case (as in the case of the two-DB-soliton states).

2.4.6 ▪ Dark-Dark and Beating Dark-Dark Solitons

In this section, we consider the case of dark-dark (DD) solitons. The dynamics and interactions of these structures in a two-component quasi-1D BEC were first studied in Refs. [328, 329] by means of a variational approach. In the case of equal chemical potentials, the latter is based on the use of the following ansatz [pertinent to the stationary

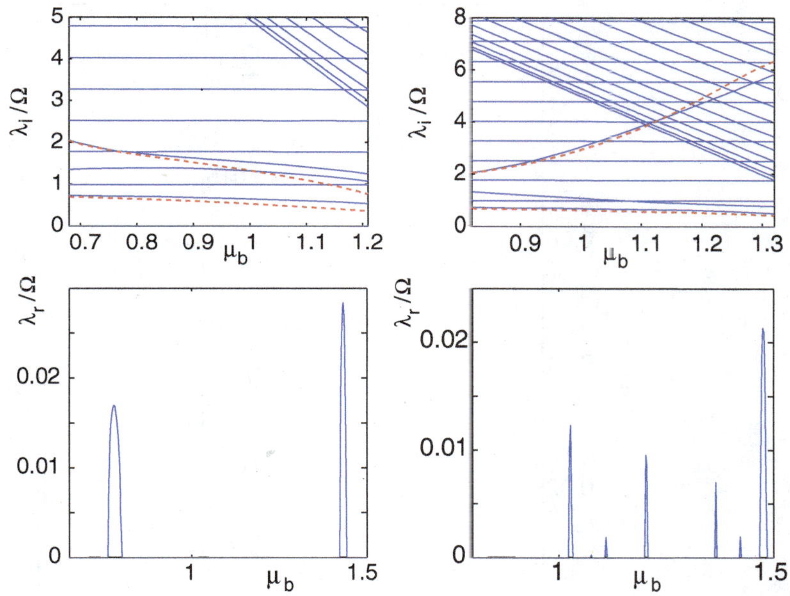

Figure 2.30. *The left and right columns of panels correspond, respectively, to an in-phase and an out-of-phase DB-soliton pair in a harmonic trap with $\Omega = 0.1$. Shown are the imaginary (top row of panels) and the real (bottom row of panels) parts of the eigenvalues as functions of μ_b for $\mu_d = 3/2$. In the top panels, the theoretical predictions for the eigenfrequencies of the anomalous modes of the system, pertaining to the in-phase (ω_2) and out-of-phase (ω_1) oscillations of the DB solitons [see Eqs. (2.168) and (2.169)] are depicted by red dashed lines. Notice that collisions of modes (eigenvalue crossings) observed in the top panels indicate the emergence of instability windows observed in the bottom panels. The instabilities are of the Hamiltonian–Hopf type and result in the emergence of eigenvalue quartets. Adapted with permission from Ref. [335].*

version of Eqs. (2.110) and (2.111)] for the soliton wavefunctions:

$$\psi_1(x,t) = B\tanh[B(x - x_0(t))] + iA, \tag{2.170}$$

$$\psi_2(x,t) = B\tanh[B(x + x_0(t))] \mp iA, \tag{2.171}$$

where $2x_0(t)$ denotes the relative distance between the two solitons, $A = \dot{x}_0$, $A^2 + B^2 = 1$, and the $\mp$ signs correspond, respectively, to a kink-antikink state (where the solitons' phase fronts are facing each other) and a kink-kink state (where the solitons' phase fronts are in the same direction). Both miscible and immiscible cases where studied in Refs. [328, 329] and the main results of the analysis can be summarized as follows. In the marginal case ($g_{11} = g_{22} = g_{12}$), and for the kink-antikink state, the trajectories in the $(x_0, \dot{x}_0)$ phase-plane are either periodic surrounding the center $(0,0)$ (indicating the formation of a bound state, i.e., a "soliton molecule"), or free [indicating acceleration (deceleration) of the approaching (outgoing) solitons]; contrarily, in the case of the kink-kink state, where solitons move in the same direction, the solitons form a bound state which can never be broken. On the other hand, in the immiscible case (i.e., when domain-walls are present), it was shown that if a dark soliton exceeds a critical velocity then it can be transferred from one component to the other at the domain wall; however, for lower velocities, multiple reflections within the domain were observed. In such a case, the soliton is accelerated after each reflection and eventually escapes from the domain.

Here, it is interesting to note that, for sufficiently small distances $2x_0(t)$, the kink-antikink state studied in Refs. [328, 329] bears resemblance to the beating DD soliton

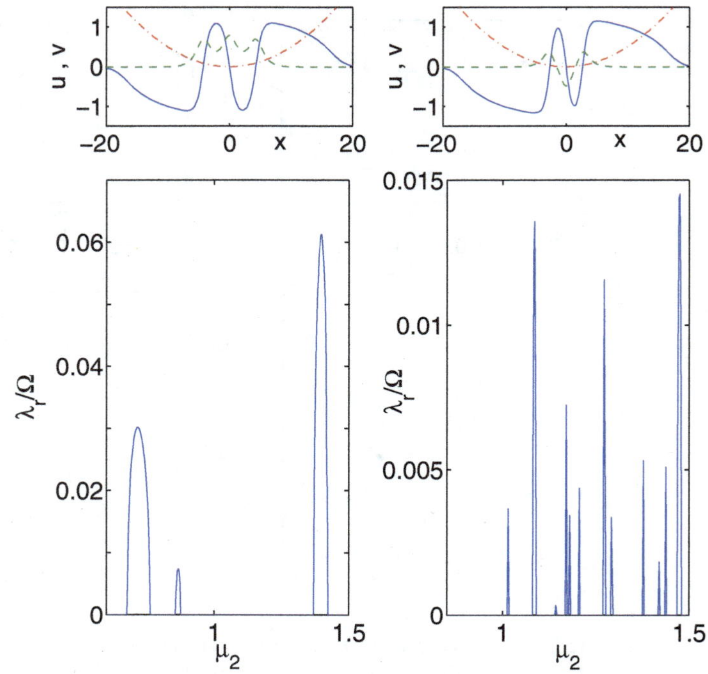

Figure 2.31. *The left and right columns of panels correspond, respectively, to in-phase and out-of-phase three-DB-soliton configurations. The top panels depict the respective stationary states, for* $\mu_d = 3/2$, $\mu_b = 1$, *and* $\Omega = 0.1$*; solid blue lines depict the dark-soliton components, dashed green lines the bright ones, while the dashed-dotted red line shows the harmonic trap. The bottom panels show the growth rates for these parameter values for the case of the in- and out-of-phase three-DB-solition configurations. Adapted with permission from Ref. [335].*

introduced in Sec. 2.4.2 [cf. Eqs. (2.121) and (2.122)]; in fact, the density of the kink-antikink features a time-periodic density profile, analogous to the one of the beating DD soliton (see, e.g., Fig. 2 of Ref. [328]). It is thus relevant to consider in more detail the stability, dynamics, oscillations, and interactions of these beating DD soliton structures.

Stability of Beating Dark-Dark Solitons

We start with the homogeneous case, and consider beating DD solitons as periodic solutions of Eqs. (2.110) and (2.111). In order to compute the soliton profile and determine its stability, we consider, at first, the solution of Eqs. (2.110) and (2.111), with $g_{11} = g_{22} = g_{12} = 1$, as a Fourier series expansion of period ω_0, namely,

$$u_1(x,t) = \sum_{k=-\infty}^{\infty} z_k(x) e^{ik\omega_0 t}, \qquad u_2(x,t) = \sum_{k=-\infty}^{\infty} y_k(x) e^{ik\omega_0 t}, \tag{2.172}$$

with $\{z_k\}, \{y_k\} \in \mathbb{R}$. Then, the dynamical equations are reduced to the following coupled equations:

$$[\mu - k\omega_0 - V(x)] z_k + \frac{1}{2}\partial_x^2 z_k = \sum_p \sum_q (z_p z_q^* + y_p y_q^*) z_{k-p+q}, \tag{2.173}$$

$$[\mu - k\omega_0 - V(x)] y_k + \frac{1}{2}\partial_x^2 y_k = \sum_p \sum_q (z_p z_q^* + y_p y_q^*) y_{k-p+q}, \tag{2.174}$$

where we have used the notation $z_k \equiv z_k(x)$, $y_k \equiv y_k(x)$. In the absence of the trap $[V(x)=0]$, it is straightforward to observe that

$$z_0(x)=\sqrt{\frac{\mu}{2}}\tanh(\sqrt{2\omega_0}x)=y_0(x), \tag{2.175}$$

$$z_1(x)=-\sqrt{\frac{\mu}{2}-\omega_0}\,\mathrm{sech}(\sqrt{2\omega_0}x)=-y_1(x), \tag{2.176}$$

$$z_j(x)=y_j(x)=0, \qquad |j|>1 \quad \text{or} \quad j=-1, \tag{2.177}$$

is actually the solution (2.121) and (2.122) for $\chi=\pi/4$, $\phi=k=0$, and $\omega_0=D^2/2$. Notice that it is also possible to numerically find a DD soliton solution in the system with the trap: this can be done upon multiplying the dark component $\{z_k\}$ of the previous solution by the Thomas–Fermi cloud $u_1^{(\mathrm{TF})}=\sqrt{\max\{\mu-V(x),0\}}$, and using this as an initial guess in a fixed-point algorithm for the system of Eqs. (2.173) and (2.174). It is worth remarking that solutions in the trap exist for $\mu>2\omega_0$, as predicted in Sec. 2.4.2.

Once a periodic solution is found, its (linear) spectral stability can be analyzed by means of Floquet analysis. To this end, the time evolution of small perturbations $\xi_1(x,t)$, $\xi_2(x,t)$ to a periodic solution $\{u_{1,0}(x,t),u_{2,0}(x,t)\}$ must be traced. For the double indices of $u_{i,j}$, i represents the component index, i.e., $i=1,2$ is the first and second component of the DD soliton solution respectively, and the index $j=0$ denotes that this is the (numerically) exact periodic solution for u_1 and u_2, around which we linearize in our Floquet analysis. The perturbations are introduced in the dynamical equations (2.110) and (2.111) as

$$u_1(x,t)=\left[u_{1,0}(x,t)+\delta\xi_1(x,t)\right], \qquad u_2(x,t)=\left[u_{2,0}(x,t)+\delta\xi_2(x,t)\right], \tag{2.178}$$

where $\delta \ll 1$, and the resulting equation at order $\mathcal{O}(\delta)$ reads:

$$i\partial_t\xi_1=\left[-\frac{1}{2}\partial_x^2+2|u_{1,0}|^2+|u_{2,0}|^2-\mu+V\right]\xi_1+u_{1,0}^2\xi_1^*+u_{2,0}^*u_{1,0}\xi_2+u_{2,0}u_{1,0}\xi_2^*,$$

$$i\partial_t\xi_2=\left[-\frac{1}{2}\partial_x^2+|u_{1,0}|^2+2|u_{2,0}|^2-\mu+V\right]\xi_2+u_{2,0}^2\xi_2^*+u_{1,0}^*u_{2,0}\xi_1+u_{1,0}u_{2,0}\xi_1^*.$$

Then, in the framework of Floquet analysis, the stability properties of periodic orbits are found by diagonalizing the monodromy matrix $\mathcal{M}$, which is defined as

$$\begin{bmatrix} \mathrm{Re}(\xi_1(x,T)) \\ \mathrm{Im}(\xi_1(x,T)) \\ \mathrm{Re}(\xi_2(x,T)) \\ \mathrm{Im}(\xi_2(x,T)) \end{bmatrix} = \mathcal{M} \begin{bmatrix} \mathrm{Re}(\xi_1(x,0)) \\ \mathrm{Im}(\xi_1(x,0)) \\ \mathrm{Re}(\xi_2(x,0)) \\ \mathrm{Im}(\xi_2(x,0)) \end{bmatrix}, \tag{2.179}$$

where $T=2\pi/\omega_0$ is the period of the oscillation. As the system is symplectic and Hamiltonian, the linear stability of the solutions requires that the monodromy eigenvalues, λ (also called Floquet multipliers) must lie on the unit circle (see, e.g., Refs. [349, 350] for details). As shown in Ref. [337], this is indeed the case: DD soliton solutions corresponding to the Manakov limit are indeed stable both in the absence and in the presence of the trap.

On the other hand, we should mention that substantial variations of the scattering lengths values from the Manakov limit can have a significant impact on the robustness of the beating DD solitons. For example, following Ref. [337], scattering lengths with ratios $g_{11}:g_{12}:g_{22}=g:1:1$ were constructed and it was assumed that $g=1.6$—see Fig. 2.32 in a setting where the trap is absent. In this case, it is observed that the intrinsic beating

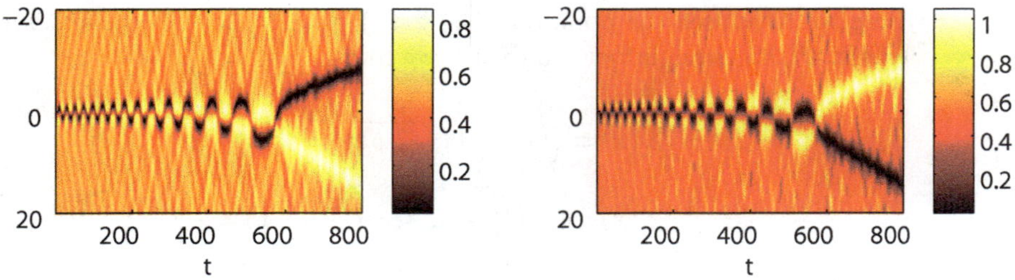

Figure 2.32. *The dynamics of the densities of a DD soliton in the absence of a trap but for* $g_{11} = 1.6$, $g_{12} = g_{22} = 1$*; left and right panels show the densities of the first and second dark soliton components, respectively. It is observed that the coherent structure executes a growing oscillation which eventually results in its splitting into a pair of dark-antidark solitons (i.e., a dark soliton in the one component coupled to a lump in the other). The parameters used here are* $\eta = 0.6$, $\chi = \pi/4$, $\theta = k = 0$, *and* $\mu = 1$. *Adapted with permission from Ref. [337].*

oscillation is growing and, as a result, the soliton is eventually split into two fragments, which clearly resemble dark-antidark solitons [331] (see also Ref. [326]). In particular, each of the components acquires a dark soliton coupled to an antidark soliton, i.e., a density hump (instead of a dip) on top of a finite background, in the second component.

Oscillations of Beating Dark-Dark Solitons in the Trap

In the presence of the trap, the dynamics of the center of the beating DD soliton is still described by the dynamics of the original (unrotated) DB soliton center x_0. The latter was shown to perform harmonic oscillations in the trap, with a frequency given by Eq. (2.147). This fact is verified in Fig. 2.33, where we show a DD soliton oscillating in a parabolic trap of strength $\Omega = 0.05$, for parameter values $\mu = 1$, $\eta = 0.6$, $\chi = \pi/4$, $\theta = k = 0$, and initial soliton position $x_0(t = 0) = 2.5$. Note that for these runs, the initial profile of the beating DD soliton in the trap is approximated by the numerically found (in-trap) ground state—i.e., the Thomas–Fermi cloud—multiplied by the beating DD solution (without a trap) of Eqs. (2.121) and (2.122). Then the evolution of the densities of the oscillating solitons is found by numerically integrating the corresponding GPE equations. The bottom left panel shows the center of mass of the beating DD soliton in the trap. Using Fourier analysis, we can infer the numerical frequency of in-trap oscillation, which can, in turn, be compared to the analytical one; cf. Eq. (2.147). As shown in the bottom right panel of the figure, there is very good agreement between the two.

Here we should note that, as shown in Ref. [337], the above phenomenology persists for small deviations from the Manakov limit, i.e., for cases where the nonlinearity coefficients are not equal. However, for large deviations from this limit, the solitons feature a strong radiation emission when oscillating in the trap. This affects the oscillation frequency, and induces growing intrinsic oscillations that eventually result in the breakup and formation of a single dark-antidark soliton, similar to the one shown in Fig. 2.32.

Interactions Between Beating Dark-Dark Solitons

We now consider the interactions of two beating DD solitons. Following the considerations of Sec. 2.4.5, first we start from the untrapped case; then, using as an initial ansatz a two-DB-soliton state [cf. Eqs. (2.154) and (2.155)], we employ a SO(2) rotation [cf. Eq. (2.120)] and obtain a two-beating-DD-soliton ansatz in the form

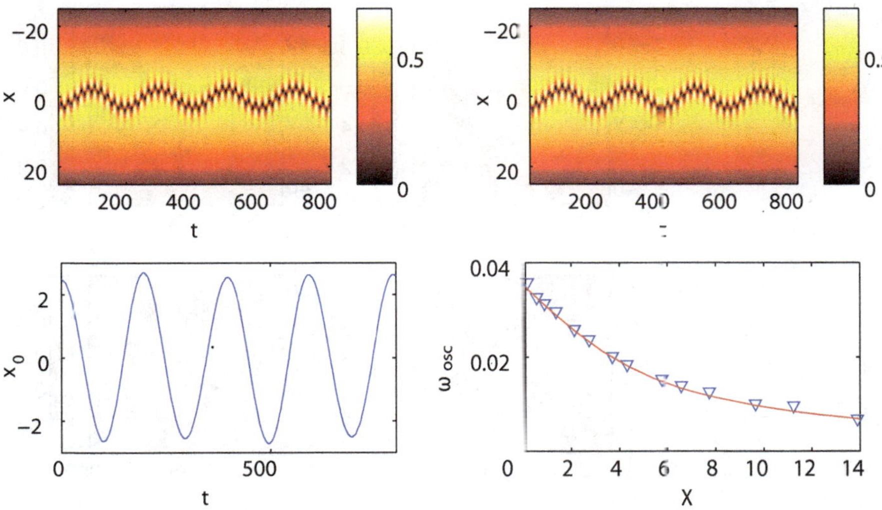

Figure 2.33. *Motion of a beating dark-dark soliton in a trap of strength $\Omega = 0.05$. Other parameters are $\mu = 1$, $\eta = 0.6$, $\chi = \pi/4$, $\theta = k = 0$, and initial soliton position $x_0(t = 0) = 2.5$. The top two panels are contour plots of the density, where the oscillation of the soliton components are shown. The bottom left panel shows the center of mass of the beating DD soliton in the upper panels; the value of the oscillation frequency, as found analytically via Eq. (2.147) is 0.03123, while the numerical value, calculated by Fourier transform, is 0.03238. The lower right panel yields the comparison between the analytically calculated frequencies (line) versus the numerical obtained ones (triangles), as functions of parameter χ. Adapted with permission from Ref. [337].*

$$u_1 = \cos(\chi)\left(\cos\phi\tanh\xi_- + i\sin\phi\right)\left(\cos\phi\tanh\xi_+ - i\sin\phi\right) - \sin(\chi)\left(\eta\,\mathrm{sech}\,\xi_- e^{i(kx+\theta(t))} + e^{i\Delta\theta}\eta\,\mathrm{sech}\,\xi_+ e^{i(-kx+\theta(t))}\right), \tag{2.180}$$

$$u_2 = \sin(\chi)\left(\cos\phi\tanh\xi_- + i\sin\phi\right)\left(\cos\phi\tanh\xi_+ - i\sin\phi\right) + \cos(\chi)\left(\eta\,\mathrm{sech}\,\xi_- e^{i(kx+\theta(t))} + e^{i\Delta\theta}\eta\,\mathrm{sech}\,\xi_+ e^{i(-kx+\theta(t))}\right), \tag{2.181}$$

where, as above, $\Delta\theta$ is the relative phase between the two bright solitons [recall that $\Delta\theta = \pi$ and $\Delta\theta = 0$ correspond to the out-of-phase (OOP) and the in-phase (IP) cases, respectively]. Using initial conditions corresponding to the above ansatz, systematic numerical simulations in Ref. [337] led to the following results.

First, in the Manakov limit (as well as for small deviations from this limit), and in the IP case, the repulsion between the beating DD solitons results in a strong separation of the two waves (which still perform their internal beating); on the other hand, in the OOP case, the competition between the repulsion of the dark components and the attraction between the bright components of the progenitor DB solitons leads to a nearly stationary configuration for a lengthy evolution interval. However, the repulsive interaction prevails and the solitons eventually separate.

Second, when deviations from the Manakov limit become significant, as e.g., in the case $g_{11} : g_{12} : g_{22} = 1.6 : 1 : 1$, the dynamics become more complicated, as shown in Fig. 2.34. It is observed that in the IP case, the two beating DD solitons initially separate and move away from each other, then they are reflected from the domain boundary and a new collision occurs (note that the reflection from the domain boundary is a byproduct of the no-flux boundary conditions used in the simulations). After this collision, one

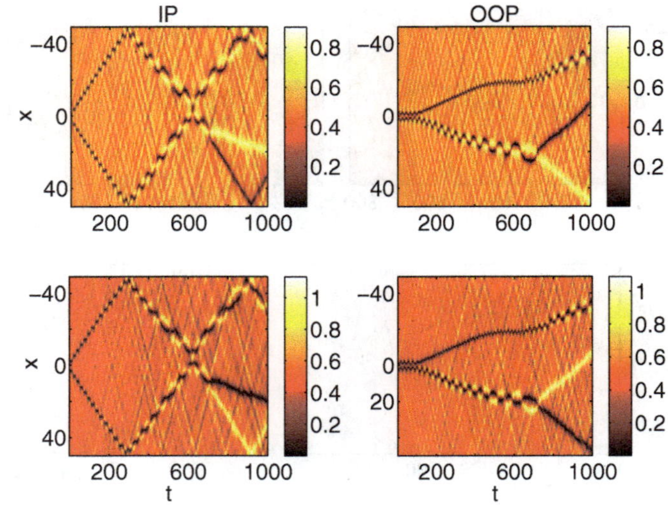

Figure 2.34. *Space-time contour plots of two DD soliton densities in phase (left) and out of phase (right) for* $g = 1.6$ *in the set* $g_{11} : g_{12} : g_{22} = 1.6 : 1 : 1$. *Here* $\chi = \pi/4$, $\eta = 0.5$, $x_0 = 1.5$, *and* $D = 1.2$. *Adapted with permission from Ref. [337].*

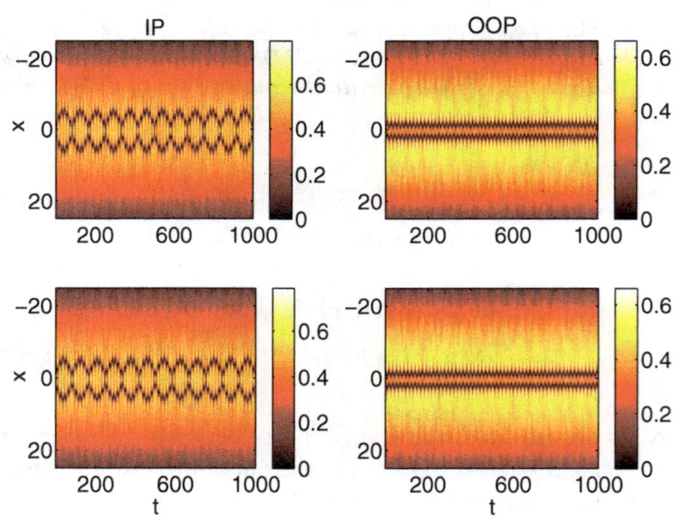

Figure 2.35. *Space-time contour plots of two beating DD soliton densities in phase (left) and out of phase (right) in the case of equal* g_{ij}*'s within a harmonic trap with trap frequency* $\Omega = 0.05$. *Here* $\chi = \pi/4$, $\eta = 0.5$, $x_0 = 1.5$, $D = 1.2$, *and* $\mu = 1$. *Adapted with permission from Ref. [337].*

of the two beating DD solitons is decomposed into a dark-antidark soliton pair, with each of these solitons moving with different velocities. For the OOP case, the separation arises much faster than for the Manakov limit and, interestingly, results in an asymmetric evolution with one of the DD solitons again breaking up in a pair of dark-antidark solitons (as in Fig. 2.32). Notice that, as in the IP case, the other soliton is not broken up in a similar way during the horizon of the simulation although it is likely that such an event will also occur for that wave.

Next, we consider two beating DD solitons in the parabolic trap $V(x) = \frac{1}{2}\Omega^2 x^2$ (with $\Omega = 0.05$), in the case of unit coefficients and chemical potential $\mu = 1$. From Fig. 2.35, we infer that the two beating DD solitons are now trapped and oscillate around an equilibrium position. Notice that in the IP case the solitons perform out-of-phase oscillations and undergo quasi-elastic collisions, while in the OOP case the weak residual repulsion is counter balanced by the presence of the trap, and we observe that the two beating DD solitons remain in a close distance to each other.

Finally, we note that in the case of two beating DD solitons, but with $g_{11} : g_{12} : g_{22} = 1.6 : 1 : 1$ (as before), within the same trap it can be found [337] that—despite the presence of the trap—it is not possible to sustain a robust set of oscillations and interactions between the beating DD solitons. This is especially true in the OOP case, where the oscillatory growth of the beating eventually results in the breakup of the DD soliton states into dark-antidark ones (which generally appear more robust for such higher values of g).

2.4.7 ▪ Spinor Condensates

The Model—Ferromagnetic and Polar Spinor Condensates

The spin degree of freedom of spinor BECs gives rise to important new phenomena (including, among others, the formation of spin domains [280], spin textures [351], and vortices [352], as well as spin oscillations [353]), which are not present in other types of BECs. Generally, a spinor condensate formed by atoms with spin F can be described in the framework of mean-field theory by a $(2F+1)$-component macroscopic wavefunction; accordingly, a spinor $F=1$ condensate is characterized by a vector order parameter, namely $\mathbf{\Psi}(\mathbf{r},t) = [\Psi_{-1}(\mathbf{r},t), \Psi_0(\mathbf{r},t), \Psi_{+1}(\mathbf{r},t)]^T$, with the components corresponding to the three values of the vertical spin component, $m_F = -1, 0, +1$.

Let us focus for simplicity on the quasi-1D case, assuming that the condensate is kept in a highly anisotropic trap, with the longitudinal and transverse trapping frequencies chosen so that $\omega_x \ll \omega_\perp$. Then, following the considerations of Sec. 1.2.4, we may assume approximately separable wavefunctions, $\Psi_{0,\pm1} \approx \psi_{0,\pm1}(x)\psi_\perp(y,z)$, where $\psi_\perp(y,z)$ is the ground state of the respective harmonic oscillator. Then, averaging of the underlying system of the coupled 3D GPEs in the transverse plane (y,z) [100] leads to the following system of coupled 1D equations for the longitudinal components of the wavefunctions (see also Refs. [354, 355, 356, 357, 358, 359]):

$$i\hbar\partial_t\psi_{\pm1} = \mathcal{H}_0\psi_{\pm1} + c_2^{(1\mathrm{D})}(|\psi_{\pm1}|^2 + |\psi_0|^2 - |\psi_{\mp1}|^2)\psi_{\pm1} - c_2^{(1\mathrm{D})}\psi_0^2\psi_{\mp1}^*, \tag{2.182}$$

$$i\hbar\partial_t\psi_0 = \mathcal{H}_0\psi_0 + c_2^{(1\mathrm{D})}(|\psi_{-1}|^2 + |\psi_{+1}|^2)\psi_0 + 2c_2^{(1\mathrm{D})}\psi_{-1}\psi_0^*\psi_{+1}, \tag{2.183}$$

where the asterisk denotes complex conjugation,

$$\mathcal{H}_0 \equiv -(\hbar^2/2m)\partial_x^2 + V(x) + c_0^{(1\mathrm{D})}n_{\mathrm{tot}} \tag{2.184}$$

is the spin-independent part of the system, where $V(x) = (1/2)m\omega_x^2x^2$ (m is the atomic mass), and

$$n_{\mathrm{tot}} = n_{-1} + n_0 + n_{+1} \equiv |\psi_{-1}|^2 + |\psi_0|^2 + |\psi_{+1}|^2 \tag{2.185}$$

is the total density. The nonlinearity coefficients have an effectively 1D form, namely $c_0^{(1\mathrm{D})} = c_0/2\pi a_\perp^2$ and $c_2^{(1\mathrm{D})} = c_2/2\pi a_\perp^2$, where $a_\perp = \sqrt{\hbar/m\omega_\perp}$ is the transverse harmonic oscillator length, which defines the size of the transverse ground state. Coupling constants c_0 and c_2 which account, respectively, for the spin-independent and spin-dependent interactions between identical spin-1 bosons, are given by

$$c_0 = \frac{4\pi\hbar^2(a_0 + 2a_2)}{3m}, \qquad c_2 = \frac{4\pi\hbar^2(a_2 - a_0)}{3m}, \tag{2.186}$$

where a_0 and a_2 are the s-wave scattering lengths in the symmetric channels with total spin of the colliding atoms $F = 0$ and $F = 2$, respectively. Note that the $F = 1$ spinor condensate may be either *ferromagnetic* (such as the ^{87}Rb), characterized by $c_2 < 0$, or *polar*—alias *antiferromagnetic*—(such as the ^{23}Na), with $c_2 > 0$ [360, 361]; this characterization is explained in more detail below.

Measuring time, length and density in units of $\hbar/c_0^{(1\mathrm{D})}n_0$, $\hbar/\sqrt{mc_0^{(1\mathrm{D})}n_0}$ and n_0, respectively (where n_0 is the peak density), we cast Eqs. (2.182) and (2.183) in the dimensionless form

$$i\partial_t\psi_{\pm1} = H_0\psi_{\pm1} + \delta\left[(|\psi_{\pm1}|^2 + |\psi_0|^2 - |\psi_{\mp1}|^2)\psi_{\pm1} + \psi_0^2\psi_{\mp1}^*\right], \tag{2.187}$$

$$i\partial_t\psi_0 = H_0\psi_0 + \delta\left[(|\psi_{-1}|^2 + |\psi_{+1}|^2)\psi_0 + 2\psi_{-1}\psi_0^*\psi_{+1}\right], \tag{2.188}$$

where $H_0 \equiv -(1/2)\partial_x^2 + V(x) + n_{\mathrm{tot}}$, the potential is $V(x) = (1/2)\Omega^2x^2$, the normalized trap's strength is

$$\Omega = \frac{3}{2(a_0 + 2a_2)n_0}\left(\frac{\omega_x}{\omega_\perp}\right), \tag{2.189}$$

and we define

$$\delta \equiv \frac{c_2^{(1D)}}{c_0^{(1D)}} = \frac{a_2 - a_0}{a_0 + 2a_2}. \tag{2.190}$$

Obviously, the parameter δ represents the ratio of the strengths of the spin-dependent and spin-independent interatomic interactions. According to what was said above, $\delta < 0$ and $\delta > 0$ correspond, respectively, to ferromagnetic and polar spinor BECs. Notice that in the relevant cases of ^{87}Rb and ^{23}Na atoms with $F = 1$, this parameter takes values $\delta = -4.66\times10^{-3}$ [362] and $\delta = +3.14\times10^{-2}$ [363], respectively; i.e., in either case, it is a small parameter in Eqs. (2.187) and (2.188). Nevertheless, the above values may in principle be modified by employing the so-called confinement-induced Feshbach resonance [364].

The system of Eqs. (2.187) and (2.188) possesses two integrals of motion: the total number of atoms $N = \int_{-\infty}^{+\infty} n_{\mathrm{tot}}\,dx$ and energy $E = \int_{-\infty}^{+\infty} \mathcal{E}\,dx$, with the energy density $\mathcal{E}$ being given by

$$\mathcal{E} = \sum_{j=0,\pm1}\left(\frac{1}{2}|\partial_x\psi_j|^2 + V(x)n_j + \frac{1}{2}n_j^2\right) + \sum_{j\neq k,k=0,\pm1} n_jn_k + \frac{\delta}{2}\left|\vec{f}\right|^2, \tag{2.191}$$

where the spin-dependent part of the energy, $(\delta/2)|\vec{f}|^2$, is proportional to the square of the local spin expectation vector; the latter is given by $\vec{f} = \vec{\psi}^\dagger\vec{S}\vec{\psi}$, where $\vec{\psi} \equiv (\psi_{+1}, \psi_0, \psi_{-1})^T$ is the three-component wavefunction, and $\vec{S} = (S_x, S_y, S_z)^T$ is the vector of spin operators [281], so that the local spin expectation vector yields $\vec{f} = (f_x, f_y, f_z)^T$, where

$$f_x = \sqrt{2}\,\mathrm{Re}\left[\psi_0(\psi_{+1}^* + \psi_{-1}^*)\right], \tag{2.192}$$

$$f_y = -\sqrt{2}\,\mathrm{Im}\left[\psi_0^*(\psi_{+1} - \psi_{-1})\right], \tag{2.193}$$

$$f_z = |\psi_{+1}|^2 - |\psi_{-1}|^2, \tag{2.194}$$

with the total mean-field spin given by $\left|\vec{f}(x;t)\right|^2 = \sqrt{f_x^2 + f_y^2 + f_z^2}$.

Stationary states of a spinor $F = 1$ condensate can be found in the form

$$\psi_j = \sqrt{n_j(x)}\exp\left(i\theta_j - i\mu_jt\right),\ j = -1, 0, +1, \tag{2.195}$$

where n_j and θ_j are densities and phases of the components, and μ_j are their chemical potentials. Note that such states are also called *spin-polarized states* of the system, characterized by a constant population of each spin component. Substituting Eq. (2.195) into Eqs. (2.187) and (2.188), it can readily be found that the conditions for the existence of spin-polarized states are

$$2\mu_0 = \mu_{-1} + \mu_{+1}, \tag{2.196}$$

$$\Delta\theta \equiv 2\theta_0 - \theta_{+1} - \theta_{-1} = 0 \quad \text{or} \quad \pi. \tag{2.197}$$

Then, the spatial profiles of the densities $n_j(x)$ are governed by the following coupled time-independent equations:

$$\mu_{\pm 1}\sqrt{n_{\pm 1}} = H_0\sqrt{n_{\pm 1}} + \delta\left[(n_{\pm 1} + n_0 - n_{\mp 1})\sqrt{n_{\pm 1}} \pm n_0\sqrt{n_{\pm 1}}\right], \tag{2.198}$$

$$\mu_0\sqrt{n_0} = H_0\sqrt{n_0} + \delta\left[(n_{-1} + n_{+1})\sqrt{n_0} \pm 2\sqrt{n_{-1}n_0n_{+1}}\right]. \tag{2.199}$$

The plus sign in front of the last term in Eqs. (2.198) and (2.199) corresponds to $\Delta\theta = 0$, leading to stationary states with all three components being *in phase*. On the other hand, the minus sign corresponds to $\Delta\theta = \pi$, which means that one of the components ($m_F = +1$ or $m_F = -1$) of the spinor stationary state is *out of phase* with the others. We note that both types of stationary states always exist in the system, regardless of the sign of δ. Nevertheless, the in-phase stationary states minimize the spin-dependent part $(\delta/2)|\vec{f}|^2$ of the total energy [cf. Eq. (2.191)] for $\delta < 0$, while the out-of-phase stationary states minimize the spin-dependent energy for $\delta > 0$ [280, 360] (see also [281]). Thus, the spin-dependent part of the energy determines the *ground state* of the spinor condensate: *ferromagnetic* for $\delta < 0$ and *polar* for $\delta > 0$.

We finally note that once a stationary state, $\Psi_s(x) = [\tilde{\psi}_{-1}(x), \tilde{\psi}_0(x), \tilde{\psi}_{+1}(x)]^T$, is found, it is possible to analyze its stability in the framework of the BdG analysis: one can linearize around the unperturbed spin-polarized state, assuming a perturbed solution of the form

$$\psi_j(x,t) = \left[\tilde{\psi}_j(x) + \epsilon\left(u_j(x)e^{i\omega t} + v_j^*(x)e^{i\omega^* t}\right)\right]e^{-i\mu t}, \tag{2.200}$$

where u_j and v_j represent infinitesimal perturbations with eigenfrequencies ω, and it was assumed, for simplicity, that the chemical potentials for all components are equal: $\mu \equiv \mu_{-1} = \mu_0 = \mu_{+1}$. Then, the solution of the ensuing BdG equations for ω and associated eigenfunctions u_j, v_j^* provides complete information about the stability of the underlying stationary state, Ψ_s.

Single-Mode Approximation and Modulational Instability

A quite convenient method to analyze the dynamics of spinor BECs is the so-called single-mode approximation (SMA) [365, 366]. According to this approach, it is assumed that all spin sub-states share the same profile defined by the wavefunction $\Phi(x,t)$. In particular, the three hyperfine components are presented in the form

$$\psi_{\pm 1,0}(x,t) = \lambda_{\pm 1,0}\Phi(x,t),$$

where the (generally complex) constants $\lambda_{\pm 1,0}$ represent the relative population of each spinor component in a steady state, with $|\lambda_{-1}|^2 + |\lambda_0|^2 + |\lambda_{+1}|^2 = 1$. This way, substituting the above ansatz into Eqs. (2.187) and (2.188) we obtain the following set of equations:

$$i\partial_t\Phi+\frac{1}{2}\partial_x^2\Phi-\left[1+\delta\left(|\lambda_{-1}|^2+|\lambda_0|^2-|\lambda_{+1}|^2+\frac{\lambda_0^2\lambda_{+1}^*}{\lambda_{-1}}\right)\right]|\Phi|^2\Phi=V(x)\Phi,$$

$$i\partial_t\Phi+\frac{1}{2}\partial_x^2\Phi-\left[1+\delta\left(-|\lambda_{-1}|^2+|\lambda_0|^2+|\lambda_{+1}|^2+\frac{\lambda_0^2\lambda_{-1}^*}{\lambda_{+1}}\right)\right]|\Phi|^2\Phi=V(x)\Phi,$$

$$i\partial_t\Phi+\frac{1}{2}\partial_x^2\Phi-\left[1+\delta\left(|\lambda_{-1}|^2+|\lambda_{+1}|^2+\frac{2\lambda_{-1}\lambda_0^*\lambda_{+1}}{q_0}\right)\right]|\Phi|^2\Phi=V(x)\Phi.$$

The above nonlinear evolution equations are then reduced to the scalar GPE

$$i\partial_t\Phi+\frac{1}{2}\partial_x^2\Phi-p|\Phi|^2\Phi=V(x)\Phi, \tag{2.201}$$

if the following consistency conditions hold:

$$\lambda_0^2=2\lambda_{-1}\lambda_{+1} \quad \text{and} \quad p\equiv 1+\delta \tag{2.202}$$

or

$$\lambda_0\lambda_{+1}^*+\lambda_0^*\lambda_{-1}=0 \quad \text{and} \quad p\equiv 1. \tag{2.203}$$

Notice that in the homogeneous case $[V(x)=0]$, the corresponding scalar NLS Eq. (2.201) is completely integrable by means of the IST method [13]; this is an integrable limit of the GPEs (2.187) and (2.188) (other integrable limits, again in the homogeneous case, will be discussed below).

The same analysis can be done by using an ansatz for the densities, namely $n_j(x)=q_j n(x)$, but with the populations $q_j\in\mathbb{R}$ of each spin component, now subject to the normalization condition $q_{-1}+q_0+q_{+1}=1$. Then, in the case of equal chemical potentials, $\mu\equiv\mu_{-1}=\mu_0=\mu_{+1}$, Eqs. (2.187) and (2.188) lead to the GPE model

$$\left(-\mu-\frac{1}{2}\partial_x^2+\frac{1}{2}\Omega^2x^2+pn\right)\sqrt{n}=0 \tag{2.204}$$

for the density n if the following consistency conditions hold:

$$q_0=2\sqrt{q_{-1}q_{+1}}, \quad p\equiv 1+\delta \quad \text{for} \quad \Delta\theta=0, \tag{2.205}$$

or

$$q_{-1}=q_{+1}, \quad p\equiv 1 \quad \text{for} \quad \Delta\theta=\pi. \tag{2.206}$$

It is clear that various families of solutions of the GPEs (2.187) and (2.188) may arise in the framework of SMA. In the homogeneous system, the simplest class of solutions are plane waves. Apparently, such solutions are modulationally stable (in the defocusing setting under consideration) for $\delta=0$, but they may become unstable in the general case of finite δ. Since below we are going to study excited states (such as various types of polarized states, domain walls, and solitons) on top of such plane wave solutions, it is relevant to study, at this point, the modulational instability (MI) of plane waves in the spinor $F = 1$ condensate. This problem was studied in Ref. [367] in the most general case of three plane waves; here, for simplicity (and also for reasons that will become clear below), we will follow the analysis of Ref. [137], and analyze the MI of particular plane wave solutions, having at least one component equal to zero. This class of exact plane wave solutions (in the spatially homogeneous system with $\Omega = 0$) can be obtained in the framework of SMA in the special cases of (a) $q_0 = 0$, $q_{-1} = q_{+1} = 1/2$ and (b) $q_0 = 1$, $q_{-1} = q_{+1} = 0$; the respective plane waves have the form

$$\psi_{-1} = \psi_{+1} = \sqrt{\frac{\mu}{2}} \exp(-i\mu t), \quad \psi_0 = 0; \tag{2.207}$$

$$\psi_{-1} = \psi_{+1} = 0, \quad \psi_0 = \sqrt{\mu} \exp(-i\mu t). \tag{2.208}$$

Let us now study the stability of solution (2.207) [the analysis of solution (2.208) follows the same lines and leads to the same result] by introducing the ansatz

$$\psi_{-1} = \psi_{+1} = \sqrt{\frac{\mu}{2} + \epsilon \tilde{n}(x,t)} \exp[-i\mu t + i\epsilon \tilde{\phi}(x,t)],$$
$$\psi_0 = \epsilon \tilde{\Phi}_0(x,t) \exp(-i\mu t), \tag{2.209}$$

where the real functions $\tilde{n}(x,t)$ and $\tilde{\phi}(x,t)$ represent density and phase perturbations of the fields $\psi_{\pm 1}$, the complex function $\tilde{\Phi}_0$ represents perturbation of the zero solution for the field ψ_0, and ϵ is a formal small parameter. Substituting the above ansatz into Eqs. (2.187) and (2.188), we derive at order $O(\epsilon)$ the following system:

$$i\left(\partial_t \tilde{n} + \frac{\mu}{2}\partial_x^2 \tilde{\phi}\right) - \mu\left(\partial_t \tilde{\phi} + 2\tilde{n} - \frac{\mu}{4}\partial_x^2 \tilde{n}\right) = 0, \tag{2.210}$$

$$i\partial_t \tilde{\Phi}_0 + \frac{1}{2}\partial_x^2 \tilde{\Phi}_0 - \delta \cdot \mu\left(\tilde{\Phi}_0 + \tilde{\Phi}_0^*\right) = 0. \tag{2.211}$$

Combining real and imaginary parts of Eq. (2.210), we arrive at a dispersive wave equation

$$\partial_t^2 \tilde{n} - \mu \partial_x^2 \tilde{n} + (\mu^2/8)\partial_x^4 \tilde{n} = 0, \tag{2.212}$$

which gives rise to a stable dispersion relation between wavenumber k and frequency ω (the absence of complex roots for ω at real k implies the *modulational stability* of the underlying plane wave):

$$\omega^2 = \mu k^2 \left(1 + \mu k^2/8\right). \tag{2.213}$$

It follows from Eq. (2.213) that, in the long-wave limit ($k \to 0$), small-amplitude waves can propagate on top of plane wave solution (2.207) with the *speed of sound*,

$$c = \sqrt{\mu}. \tag{2.214}$$

A similar analysis for Eq. (2.211), which is decoupled from Eq. (2.210), leads to the dispersion relation

$$\omega^2 = k^2 \left(\delta \cdot \mu + k^2/4\right). \tag{2.215}$$

It is clear that if $\delta > 0$ (which corresponds to the *polar state*, as mentioned above) then Eq. (2.215) has no complex roots for ω, hence the trivial solution to Eq. (2.188), $\psi_0 = 0$,

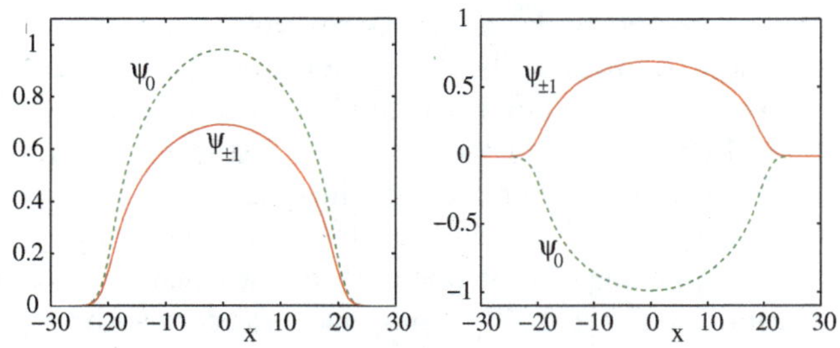

Figure 2.36. *Examples of two stable spin-polarized states for $\Delta\theta = 0$ (left panel) and $\Delta\theta = \pi$ (right panel), both obtained for $\Omega = 0.1$ and $\mu = 2$. Wavefunctions $\psi_{\pm 1}$ are identical and are depicted by the solid lines, while the wavefunction ψ_0 is depicted by the dashed line. In the left panel, $q_{-1} = q_{+1} = 0.5$ and $q_0 = 1$, while in the right panel $q_{-1} = q_{+1} = 0.25$ and $q_0 = 0.5$. Adapted with permission from Ref. [368].*

is modulationally stable. However, $\delta < 0$ (corresponding to the *ferromagnetic state*) gives rise to modulational instability of the solution $\psi_0 = 0$ against the perturbations whose wavenumbers belong to the instability band, $k \leq 2\sqrt{|\delta|\mu}$. Notice that these results comply with those reported in Ref. [367].

Below, we focus on modulationally stable cases, which pertain to polar states with $\delta > 0$.

Polarized States in the Trap

Let us now consider a polar spinor $F = 1$ condensate confined in a sufficiently weak trap. Then, employing the SMA, one can solve approximately Eq. (2.204) by means of the TF approximation: neglecting the kinetic-energy term ($\sim \partial_x^2 \sqrt{n}$) in Eq. (2.204), we find density profiles of the three spin components,

$$n_j = \frac{q_j}{p} \max\left\{\mu - \frac{1}{2}\Omega^2 x^2, 0\right\}. \tag{2.216}$$

Obviously, all three components of the TF solution have the same spatial width (the TF radius), $R_{\mathrm{TF}} = \sqrt{2\mu}/\Omega$.

Spin-polarized solutions of Eqs. (2.187) and (2.188), with profiles close to those produced by the TF approximation of Eq. (2.216), can be found numerically. In particular, in Ref. [368], such states were found by means of a fixed-point algorithm (the Newton-Raphson method), using as an initial guess, three identical profiles of the form $\psi_j(x) = \sqrt{n(x)}\exp(i\theta_j)$, with $n(x) = \max\{\mu - (1/2)\Omega^2 x^2, 0\}$ and $\Delta\theta = 0$ or π. Then, keeping the trap's strength Ω fixed and varying the chemical potential μ, the numerical solution converged to stable spin-polarized states, which were indeed close to the approximate one given by Eq. (2.216). Two typical examples are shown in the top panels of Fig. 2.36 for both cases, $\Delta\theta = 0$ (left panel) and $\Delta\theta = \pi$ (right panel), with $\mu = 2$. The stability of the TF states was examined too (by means of both the linear-stability analysis and direct simulations), and it was found that these states are always stable [368].

Apart from the smooth spin-polarized states of the TF type, there exist other ones, which feature holes in some of the components, or in all of them. An example of such states is shown in Fig. 2.37 for $\Delta\theta = 0$. As seen in the two top left panels of this figure, the ψ_0 component has a pulse-like shape concentrated about the trap's center, while the

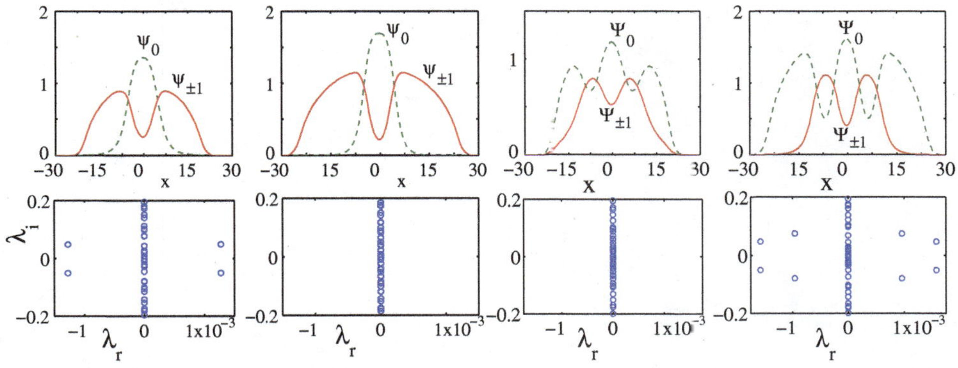

Figure 2.37. *Top panels: Spin-polarized states with a hole in the $\psi_{\pm1}$ components and a pulse-like shape of the ψ_0 field (first for $\mu = 2$, and second for $\mu = 3$, from left to right), and with one hole in each of the $\psi_{\pm1}$ components and two holes in the ψ_0 component (third for $\mu = 2$, and fourth for $\mu = 3$, from left to right); solid and dashed lines depict components $\psi_{\pm1}$ and ψ_0. The panels in the bottom row display spectral planes (λ_r, λ_i) of the (in)stability eigenvalues for the corresponding states shown in the top panels. Note that the instability of the first state (pertaining to $\mu = 2$) is of the oscillatory type, being accounted for by a quartet of eigenvalues with nonzero real parts; on the other hand, the instability of the fourth state (pertaining to $\mu = 3$) is driven by two quartets in the eigenvalue spectrum (see also text). Adapted with permission from Ref. [368].*

$\psi_{\pm1}$ components feature a large hole at the same spot.[14] The shape of this structure can explained by the fact that the interaction between components is repulsive, hence a peak (hole) in ψ_0 ($\psi_{\pm1}$) induces a hole (peak) in $\psi_{\pm1}$ (ψ_0).

A BdG analysis shows that the state with $\mu = 3$ (second state from left to right) is stable, as all the eigenvalues are imaginary, while the one with $\mu = 2$ (left column of panels) is unstable. Further analysis demonstrates that all such unstable states are destabilized by a *Hamiltonian–Hopf bifurcation*, which gives rise to a quartet of eigenvalues with nonzero real parts. As concerns the unstable state shown in Fig. 2.37, the instability interval is $1.81 \leq \mu \leq 2.15$, with the instability growth rate, $\max(\lambda_r) \approx 1.3 \times 10^{-3}$, corresponding to $\mu \approx 2$. A direct numerical integration of Eqs. (2.187) and (2.188) shows that, indeed, the oscillatory instability of this unstable state sets in at a very large time, $t \approx 4000$ (as a consequence of the extremely small growth rate). As a result of the instability, the pulse in the ψ_0 component broadens and its amplitude accordingly decreases, while the hole in the $\psi_{\pm1}$ components becomes shallower. Nevertheless, eventually, the system settles down to a steady state, which is qualitatively similar to the initial one [368].

Apart from the states considered above, it is also possible to find spin-polarized ones which feature, e.g., one hole in each of the $\psi_{\pm1}$ components, and two holes in ψ_0.[15] Examples of such a state are shown in the two right columns of panels in Fig. 2.37, for $\mu = 2$ (third from left to right) and $\mu = 3$ (fourth from left to right). As is observed, one may consider components ψ_{-1} and ψ_{+1} as built of two overlapping pulses, which induce two holes in the ψ_0 component due to the repulsive intercomponent interactions.

[14] A similar state with the field profiles in $\psi_{\pm1}$ and ψ_0 exchanged, i.e., the ψ_0 component featuring the hole, and $\psi_{\pm1}$ ones concentrated in narrow pulses, was also found. Moreover, such states were found too with either ψ_0 or $\psi_{\pm1}$ having the opposite sign (i.e., for $\Delta\theta = \pi$).

[15] Similar states with one hole in ψ_0 and two holes in each of the $\psi_{\pm1}$ components, as well their counterparts corresponding to $\Delta\theta = \pi$, can also been found; the (in)stability of these additional states is similar to that reported in Fig. 2.37.

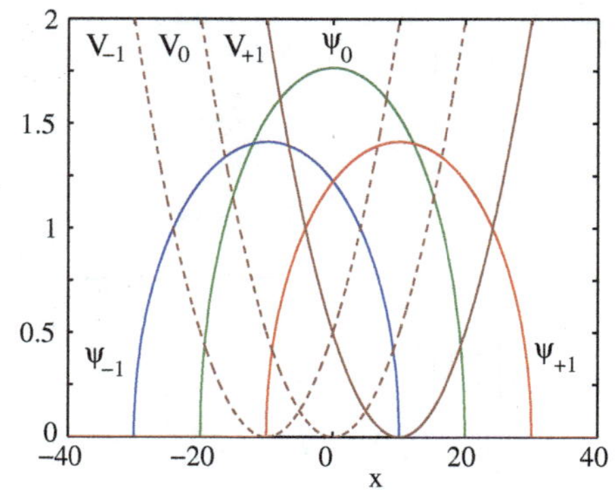

Figure 2.38. *Initialization of the system when three spatially separated traps, V_{-1}, V_0 (shown by dashed parabolas), and V_{+1} (the solid parabola), with equal strengths ($\Omega = 0.1$) and centers placed at $x = -10$, 0, and $+10$, hold the TF states of the ψ_{-1}, ψ_0, and ψ_{+1} components, respectively. Then, two traps (V_{-1} and V_0), centered at $x = -10$ and $x = 0$, are turned off, and a stationary solution, supported solely by the trap (V_{+1}) centered at $x = 10$, is looked for by means of the fixed-point algorithm, using the configuration with the three mutually shifted components as an input. Adapted with permission from Ref. [368].*

Results of the stability analysis for the above mentioned states show that the one for $\mu = 2$ is stable, while the one for $\mu = 3$ is unstable. In this case, two quartets of eigenvalues with nonzero real parts are found in the spectral plane (see bottom right panel of Fig. 2.37). These lead to instability in the interval $2.58 \leq \mu \leq 3.22$, with the respective instability growth rate being $(\lambda_r)_{\max} \approx 1.8 \times 10^{-3}$ for $\mu \approx 2.9$ (i.e., of the same order of magnitude as in the previous case). The development of the instability was studied in Ref. [368], in direct simulations, starting with initial conditions in the form of a perturbed solution pertaining to $\mu = 3$. It was shown that, in this case, the instability results in a strong oscillatory deformation of the three components (after $t \approx 5500$), contrary to the establishment of the new stationary pulse-hole state observed in the case of the configuration in the top right panel of Fig. 2.36. It was also noted that the resulting states are asymmetric with respect to $x \to -x$ transformations, indicating the possibility of asymmetric states in the system such as domain walls (see also below).

Domain Walls in the Trap

In the previous section, we focused on spin-polarized states in which all the three spin components were spatially overlapping, since they were confined by the same potential trap. However, it is also possible to use three different traps, each confining a different component, to initially separate them, and then allow the system to evolve in the presence of only one of these traps (i.e., turning off the other two). This way, one can also find spin-polarized states having the form of domain-wall (DW) structures.

First, we describe the initialization of the system. We assume that the three TF-shaped components are initially loaded into three different traps, $V_j(x)$, of the same strength, Ω, centered at different positions:

$$V_j(x) = \frac{1}{2}\Omega^2 (x - j\Delta x)^2, \qquad j = -1, 0, +1. \tag{2.217}$$

We choose $\Delta x = \Omega^{-1}$ (i.e., $\Delta x = 10$ for $\Omega = 0.1$), which induces the initially separated TF configurations; see Fig. 2.38.

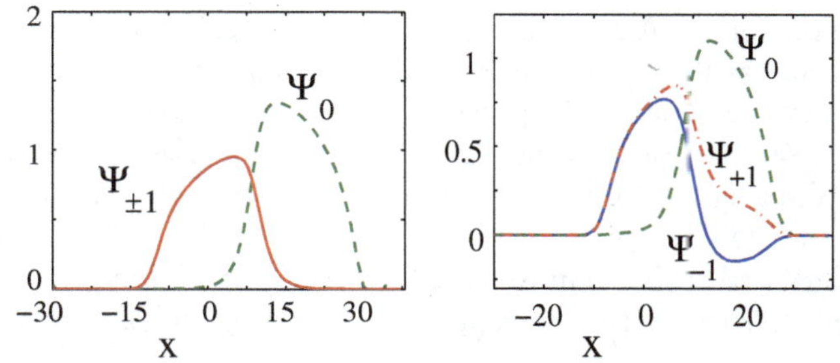

Figure 2.39. *Left panel: the wavefunctions of the components (ψ_{+1} and ψ_{-1} are identical) in a stationary state found from the initial configuration prepared as shown in Fig. 2.38. The resulting spin-polarized state has the form of a domain-wall structure between the ψ_0 and $\psi_{\pm 1}$ components. The parameters are $\Omega = 0.1$ and $\mu = 2$. The right panel shows the wavefunctions of the domain-wall state found at $\mu = 1.43$. Adapted with permission from Ref. [368].*

After preparing this state, we turn off the traps $V_{-1}(x)$ and $V_0(x)$, keeping only the rightmost one, $V_{+1}(x)$, which now acts on all the three components. The so-defined initial configuration is fed, as an initial seed, into the fixed-point algorithm, to find a spin-polarized state generated by it. Other possibilities, such as turning off potentials $V_{\pm 1}$ and keeping V_0, arranging the three components in a different way, etc., eventually lead to retrieving the spin-polarized states presented in the previous section, while the approach outlined above [keeping $V_{+1}(x)$ and switching $V_{-1}(x)$ and $V_0(x)$ off] generates new DW patterns, which are displayed in Fig. 2.39, and could not be obtained otherwise (in fact, the asymmetry of the procedure is instrumental in generating the new states).

The most interesting spin-polarized DW states found following this procedure correspond to values of the chemical potential $\mu \geq 1.43$ (or norm $N \geq 5400$), for $\Omega = 0.1$; for smaller μ, we typically found structures of the TF type. Two examples, one for $\mu = 2$, and another exactly corresponding to $\mu = 1.43$, are shown in Fig. 2.39.

In the former case ($\mu = 2$), the ψ_0 component (which has the larger norm) is centered to the right of the midpoint of the remaining trap ($x_{+1} = 10$), while the identical $\psi_{\pm 1}$ components are pushed to the left, due to the repulsion from ψ_0, with a DW created between $\psi_{\pm 1}$ and ψ_0.

In the state found at the above-mentioned special value, $\mu = 1.43$, which is shown in the right panel of Fig. 2.39, the shape of the ψ_0 component is similar to that displayed in the left panel for $\mu = 2$, while the $\psi_{\pm 1}$ components are not identical, in contrast to the previous example. In the present case, the ψ_0 and $\psi_{\pm 1}$ components overlap over a wider spatial region, and ψ_{-1} changes its sign at $x \approx 13$, featuring a structure resembling the waveform of a dark soliton embedded in a bright one [369].

The stability of the DW states was also investigated in the framework of the BdG equations [368]. It was concluded that there are no unstable eigenvalues, i.e., those with a nonzero with a real part, in the interval $1.43 \leq \mu \leq 5$, or, equivalently, $5400 \leq N \leq 35000$ for $\Omega = 0.1$. Thus, the DWs are stable in this region.

2.4.8 ▪ Dark Solitons in Spinor Condensates

Integrable Cases, Solitons, and Nonlinear Structures: Overview

Having discussed various states in a polar spinor BEC, we now proceed with the presentation of soliton solutions in this setting. First of all, it is interesting to discuss limiting cases of Eqs. (2.187) and (2.188) that lead to completely integrable systems (in the absence of the potential). A scalar integrable limit, namely one corresponding to the scalar NLS equation, has already been presented in the framework of SMA. On the other hand (and again in the absence of the potential), there exist two other limiting cases of Eqs. (2.187) and (2.188) that lead to completely integrable vectorial models. The first one corresponds to the limiting case of $\delta = 0$, for which the system of Eqs. (2.187) and (2.188) is reduced to the completely integrable Manakov system [cf. Eq. (2.109)]. The second one, which was identified and analyzed in Ref. [354], corresponds to the case $\delta = 1$ (i.e., for interatomic and anti-ferromagnetic interactions of equal magnitude). In this case, the resulting matrix NLS equation, expressed in the form

$$i\partial_t Q + \partial_x^2 Q - 2QQ^\dagger Q = 0, \qquad Q \equiv \begin{pmatrix} \psi_{+1} & \psi_0 \\ \psi_0 & \psi_{-1} \end{pmatrix}, \tag{2.218}$$

with nonvanishing boundary conditions, is completely integrable by means of the IST method [370] and admits exact analytical vector $\mathcal{N}$-dark soliton solutions (i.e., single- and multiple-vector dark solitons of the dark-dark-dark type in terms of the $m_F = -1, 0, +1$ spinor components) [355] (see also Ref. [366]). The one-dark-soliton state of this system can be classified as (a) ferromagnetic (i.e., with nonzero total spin), which has domain-wall shaped wavefunctions, and (b) polar (i.e., with zero total spin), characterized by the familiar hole soliton profile. In this setting, it was found that the collisions of two solitons give rise to interesting spin-dependent phenomena, such as spin mixing or spin transfer [355].

Generally, we should point out that the analysis of GPE systems modeling spinor BECs has become a subject of increasing attention. Thus, apart from the above-mentioned works referring to integrable cases, there exist works devoted to solitons and other nonlinear structures in spinor BECs. In particular, in the defocusing case under consideration [corresponding to $c_2 > 0$; cf. Eqs. (2.182) and (2.183)], apart from the homogeneous case studied in Ref. [355], dark solitons were also studied in systems with various types of traps; this way, gap (bright) solitons [359, 371] and dark solitons [372] have also been predicted to occur in spinor BECs in optical lattices, while weak dark solitons—obeying an effective inhomogeneous KdV equation—were found in harmonically confined spinor BECs [373]. Topological states, such as dark solitons and vortices [374], monopoles [375], polar-core spin vortices [376], and topological Wigner crystals of half-solitons [377], as well as periodic solutions [378, 379] were studied as well. We finally note in passing that bright solitons have also been investigated in the focusing case (i.e., for $c_2 < 0$), both for $F = 1$ [354, 355, 356, 358, 380, 381, 382, 383, 384] and $F = 2$ [385, 384] spinor BECs.

Coming back to the fundamental model presented above, namely the GPEs (2.187) and (2.188), it is interesting to note that mixed DB solitons of the dark-dark-bright (DDB) or bright-bright-dark type (BDD)—again in terms of the $m_F = -1, 0, +1$ spinor components—were also predicted to occur in antiferromagnetic spinor $F = 1$ BECs [137] (see also relevant work in Ref. [386]). Below we will focus on this case, as we aim to connect these results with pertinent ones presented in the case of binary BECs, and also present interesting connections with other integrable cases appearing in the framework of the small-amplitude approximation.

Dark-Bright Solitons in Spinor $F = 1$ Condensate

Let us consider again Eqs. (2.187) and (2.188) and recall that δ is a naturally occurring small parameter in the problem. Based on this observation, we can use δ as a formal small parameter in a perturbation scheme that we will present below. In particular, aiming to find solutions of Eqs. (2.187) and (2.188) close to the plane wave given by Eq. (2.207), we start the analysis by introducing the following ansatz:

$$\psi_{\pm 1} = \sqrt{n(x,t)}\exp[-i\mu t + i\phi(x,t)], \quad \psi_0 = \Phi_0(x,t)\exp(-i\mu t), \tag{2.219}$$

where $n(x,t)$ and $\phi(x,t)$ are the real density and phase of fields $\psi_{\pm 1}$, while Φ_0 is, generally, a complex function. Substituting Eq. (2.219) into Eqs. (2.187) and (2.188), we derive the following system:

$$\begin{aligned} &\frac{i}{2}[\partial_t n + \partial_x(n\partial_x\phi)] - n\left[\partial_t\phi + 2n - \mu + (1+\delta)|\Phi_0|^2\right] \\ &-n\left[\frac{1}{2}(\partial_x\phi)^2 - \frac{1}{2\sqrt{n}}\partial_x^2\sqrt{n} + \delta\Phi_0^2 e^{-2i\phi}\right] = 0, \end{aligned} \tag{2.220}$$

$$i\partial_t\Phi_0 + \frac{1}{2}\partial_x^2\Phi_0 - (2n - \mu + |\Phi_0|^2)\Phi_0 - 2\delta n\left(\Phi_0 + \Phi_0^* e^{-2i\phi}\right) = 0. \tag{2.221}$$

Notice that the plane wave (2.207) corresponds to an obvious solution of Eqs. (2.220) and (2.221) with $n = \mu/2$, $\phi = 0$, $\Phi_0 = 0$.

We now adopt the small-amplitude approximation, and consider solutions for small deviations from the plane wave state. Then, using δ as a small parameter, we introduce the stretched variables

$$X \equiv \sqrt{\delta}(x - \sqrt{\mu}t), \quad T \equiv \delta t, \tag{2.222}$$

and seek for solutions to Eqs. (2.220) and (2.221) as

$$n = (\mu/2) + \delta \cdot \rho, \quad \phi = \sqrt{\delta}\alpha, \quad \Phi_0 = \delta^{3/4} q, \tag{2.223}$$

$$q \equiv q_1\cos(kx - \omega t) + iq_2\sin(kx - \omega t), \tag{2.224}$$

where $\rho = \rho(X,T)$, $\alpha = \alpha(X,T)$, $q_{1,2} = q_{1,2}(X,T)$, while k and ω are unknown wavenumber and frequency. Substituting Eqs. (2.223) in Eq. (2.220), at order $O(\delta)$, we derive a relation between density ρ and phase α,

$$\sqrt{\mu}\partial_X\alpha = 2\rho. \tag{2.225}$$

On the other hand, at the next order, $O(\delta^{3/2})$, the resulting equation is complex:

$$-(i\mu/4)(2\partial_X\rho - \sqrt{\mu}\partial_X^2\alpha) + \partial_T\alpha + |q|^2 = 0. \tag{2.226}$$

The imaginary part of the expression on the left side of Eq. (2.226) vanishes due to the validity of Eq. (2.225), while the real part leads to equation $\partial_T\alpha + |q|^2 = 0$. The condition for its compatibility with Eq. (2.225) is

$$\partial_T\rho = -(\sqrt{\mu}/2)\partial_X(|q|^2). \tag{2.227}$$

We now proceed to Eq. (2.221), which, to leading order in δ, i.e., at $O(\delta^{3/4})$, yields the following system:

$$\omega q_1 - (k^2/2)q_2 = 0, \qquad -[(k^2/2) + 2\mu]q_1 + \omega q_2 = 0. \tag{2.228}$$

Nontrivial solutions to Eqs. (2.228) are possible when the following dispersion relation for ω and k holds:

$$\omega^2 = k^2(\mu + k^2/4). \tag{2.229}$$

Note that the difference between Eq. (2.229) and the dispersion relation in Eq. (2.215) arises from the fact that the terms proportional to δ in Eq. (2.29) appear to the leading order in the linear regime [see Eq. (2.211)], but to first order in the nonlinear regime. Next, to order $O(\delta^{5/4})$, Eq. (2.29) leads to a system

$$-\sqrt{\mu}\partial_X q_1 + k\partial_X q_2 = 0, \qquad -k\partial_X q_1 + \sqrt{\mu}\partial_X q_2 = 0, \tag{2.230}$$

which has nontrivial solutions if $k^2 = \mu$. In combination with Eq. (2.229), the latter relation selects the frequency, $\omega = 5\mu^2/4$. Finally, at order $O(\delta^{7/4})$, Eq. (2.221) leads to the equation

$$i\partial_T q + \frac{1}{2}\partial_X^2 q - 2\rho q = 0. \tag{2.231}$$

Equations (2.227) and (2.231), which are the basic result of the above analysis, constitute the *Yajima-Oikawa* (YO) system. This model describes the interaction of low-frequency and high-frequency waves, and was originally proposed in the context of plasma physics, as a special case of the so-called Zakharov equations [387] (see also Ref. [128]): in this context, the YO system applies to Langmuir (high-frequency) waves, which form a wave packet (soliton) moving at velocities close to the speed of sound, and are thus strongly coupled to the ion-acoustic (low-frequency) waves [388].

As shown in Ref. [388], the YO system is integrable by means of the inverse-scattering transform, and gives rise to soliton solutions (see also Ref. [386]). The solitons have a $-\mathrm{sech}^2$ shape for field ρ, and a sech shape for q, which correspond to a density dip for components $\psi_{\pm 1}$ and a bright soliton for ψ_0, as per Eqs. (2.223). According to Eq. (2.225), the phase profile of the $\psi_{\pm 1}$ components, in the form of tanh, is associated to the density dip, hence the patterns in these components, generated by the exact solution of the YO system, are dark solitons. The full form of the approximate (asymptotic) DDB soliton solution to Eqs. (2.187) and (2.188), into which the YO soliton is mapped by Eqs. (2.219), (2.222), and (2.223), is

$$\begin{aligned}\psi_{\pm 1}(x,t) &= \sqrt{(\mu/2) - 2\delta\eta^2\,\mathrm{sech}^2(2\sqrt{\delta}\eta Z)}\\ &\quad\times \exp\left[-i\mu t - 2i\eta\sqrt{\delta/\mu}\tanh(2\sqrt{\delta}\eta Z)\right],\end{aligned} \tag{2.232}$$

$$\begin{aligned}\psi_0(x,t) &= 2^{3/2}\delta^{3/4}\eta\mu^{-1/4}\sqrt{\xi}\,\mathrm{sech}(2\sqrt{\delta}\eta Z)\\ &\quad\times \exp\left[-i\mu t + i\sqrt{\mu}x - 2i\sqrt{\delta}\xi Z + 2i\delta(\eta^2 - \xi^2)t\right],\end{aligned} \tag{2.233}$$

where $Z \equiv x - (\sqrt{\mu} - 2\sqrt{\delta}\xi)t$, while η, ξ are arbitrary parameters of order $O(1)$.

A similar analysis can be performed to derive asymptotic soliton solutions of the BBD type. In that case, starting from the plane wave solution (2.208), we seek for solutions of Eqs. (2.187) and (2.188) in the form of

$$\psi_{\pm 1} = \Phi_0(x,t)\exp(-i\mu t), \quad \psi_0 = \sqrt{n(x,t)}\exp[-i\mu t + i\phi(x,t)]. \tag{2.234}$$

Next, following the same analytical approach that led above to the DDB soliton, we again end up with the YO system, in a form similar to Eqs. (2.227) and (2.231), namely,

$$\partial_T \rho = -2\sqrt{\mu}\partial_X |q|^2, \tag{2.235}$$

$$i\partial_T q + \frac{1}{2}\partial_X^2 q - \rho q = 0. \tag{2.236}$$

Eventually, the approximate k BBD soliton solution to Eqs. (2.187) and (2.188), generated by the YO soliton, is

$$\psi_{\pm 1}(x,t) = 2\delta^{3/4}\eta\mu^{-1/2}\sqrt{\xi}\,\mathrm{sech}(2\sqrt{\delta}\eta Z) \times \exp\left[-i\mu t + i\sqrt{\mu}x - 2i\sqrt{\delta}\xi Z + 2i\delta(\eta^2 - \xi^2)t\right], \tag{2.237}$$

$$\psi_0(x,t) = \sqrt{(\mu/2) - 4\delta\eta^2\,\mathrm{sech}^2(2\sqrt{\delta}\eta Z)} \times \exp\left[-i\mu t - 2i\eta\sqrt{\delta/\mu}\tanh(2\sqrt{\delta}\eta Z)\right] \tag{2.238}$$

It is worthwhile to note that both types of solutions, namely DDB and BBD solitons, are *genuinely traveling* ones, i.e., they do not exist with zero speed.

Numerical simulations in Ref. [137] demonstrated that, for small amplitudes, such DB solitons feature genuine soliton behavior, i.e., they propagate undistorted and undergo quasi-elastic collisions. To show this, one may consider a collision scenario, taking initial conditions in the form of a superposition of two different pulses:

$$\psi_{\pm 1}(x,0) = \sqrt{\frac{\mu}{2} - \frac{\nu}{2}\left[\mathrm{sech}^2(x_+) + \mathrm{sech}^2(x_-)\right]} \times \exp\left(-i\sqrt{\frac{\nu}{\mu}}\tanh(x_+) + i\sqrt{\frac{\nu}{\mu}}\tanh(x_-)\right), \tag{2.239}$$

$$\psi_0(x,0) = \nu^{3/4}\sqrt{\frac{\xi}{\eta\sqrt{\mu}}}\left[\mathrm{sech}(x_+)\exp\left(+i\sqrt{\frac{\mu}{\nu}}x_+ - i(\xi/\eta)x_+\right) + \mathrm{sech}(x_-)\exp\left(-i\sqrt{\frac{\mu}{\nu}}x_- + i(\xi/\eta)x_-\right)\right], \tag{2.240}$$

where $x_\pm = \sqrt{\nu}(x \pm x_0)$ and x_0 are initial positions of centers of the two pulses. As seen in Eqs. (2.239) and (2.240), the soliton components are lent opposite initial momenta and, as a result, they propagate in opposite directions, as shown in the top panel of Fig. 2.40. We stress that even though a small amount of radiation is emitted in the course of the evolution (see four bottom panels of Fig. 2.40), the two dark solitons in the $\psi_{\pm 1}$ fields, coupled to their bright counterparts in the ψ_0 component, propagate practically undistorted, and around $t = 19$ they undergo a *quasi-elastic* collision; moreover, it is clearly observed that the solitons remain unscathed after the collision. This result is consistent with the asymptotic calculations performed above, indicating that the small-amplitude limit (for small δ) of the nonintegrable system of Eqs. (2.187) and (2.188) behaves like the integrable YO system.

Apart from small-amplitude DB solitons, in Ref. [137], DB solitons of moderate and large amplitudes (and also solitons corresponding to large values of parameter δ) were studied too; it was shown that such solitons can exist as long-lived objects, as well. Fur-

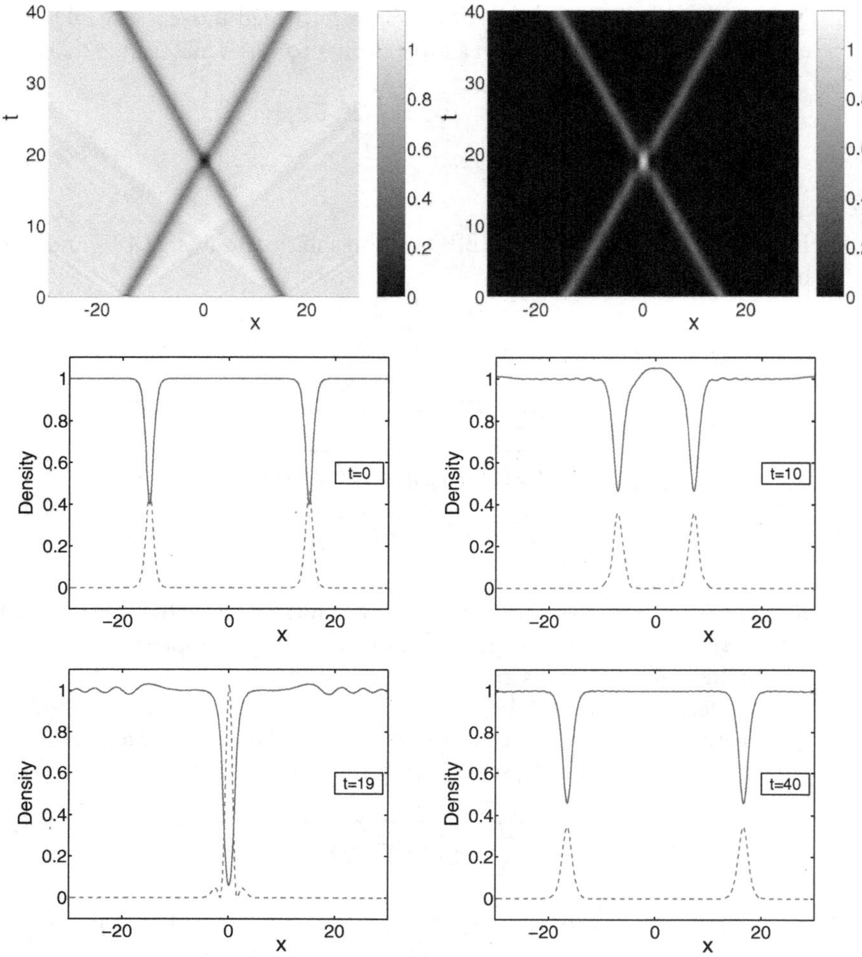

Figure 2.40. *The top two panels show contour plots of the densities of the $\psi_{\pm 1}$ (left) and ψ_0 (right) components of the spinor condensate with $\delta = 0.0314$ in the homogeneous system ($V(x) = 0$). The $\psi_{\pm 1}$ components contain a pair of dark pulses, initially placed at $x_0 = \pm 15$, that, together with the bright components in the ψ_0 field coupled to them, undergo a quasi-elastic collision at $t \approx 19$, and propagate unscathed afterwards. The parameters are $\mu = 2$, $\xi = 1.54$, $\eta = 3.091$, and $\nu = 1.2$. Four bottom panels show snapshots of the densities at $t = 0$, $t = 10$ (before the collision), $t = 19$ (when the collision occurs), and $t = 40$ (after the collision). Adapted with permission from Ref. [137].*

thermore, the dynamics of solitons in the presence of the trap was also investigated in detail. The main findings of this investigation can be summarized as follows: for sufficiently small number of atoms of the bright soliton, the bright components are guided by the dark ones, and the vector soliton performs harmonic oscillations; the oscillation frequency is different for small- and moderate-amplitude solitons, and it is respectively given by

$$\omega_{\rm osc} = \frac{\Omega}{\sqrt{2}}(1-\alpha_0\sqrt{\delta})-\epsilon_0, \qquad \omega_{\rm osc} = \Omega_{\rm osc}(1-\alpha_1\delta^2)+\epsilon_1. \tag{2.241}$$

In the above expression, $\Omega_{\rm osc}$ is given by Eq. (2.147), while the constants $\alpha_{0,1}$ and $\epsilon_{0,1}$ (with $\epsilon_{0,1} \ll \alpha_{0,1}$) depend on the normalized number of atoms of the bright component.

The above results show that the characteristic oscillation frequency of the dark soliton ($\Omega/\sqrt{2}$), as well as the oscillation frequency of the DB soliton (see Ref. [145] and Sec. 2.4.3 above), is modified by the spin-dependent interactions that are present in a spinor $F = 1$ BEC.

2.5 ▪ Some Distinct One-Dimensional Settings

2.5.1 ▪ Superfluidity and Dark Solitons

A flow past an obstacle is known to be one of the most fundamental contexts for studying superfluidity. Particularly, according to the Landau criterion for superfluidity [389], a superfluid flow past an obstacle is stable (unstable) for group velocities smaller (larger) than the speed of sound. Actually, breakdown of superfluidity is caused by the opening of channels for emission of excitations in the fluid, whose formation manifests itself as an effective dissipation. In the BEC context, early experiments from the MIT group [248, 390] demonstrated the onset of dissipation induced by the motion of an obstacle (in the form of a strongly repulsive dipole beam). From a theoretical standpoint, the problem can be studied by using a NLS equation (or a GPE) that includes a localized external potential, or "impurity," of the form $V_{\mathrm{imp}}(\mathbf{r}-v t)$ [with $V_{\mathrm{imp}}(\mathbf{r}) \to 0$ as $|\mathbf{r}| \to \infty$] accounting for the presence of the obstacle moving with velocity v; this potential may be naturally superimposed to the usual trapping potential confining the condensate (see more details below). Note that in relevant earlier two-dimensional studies [391], where the NLS equation as a model of superflow was used, vortex formation induced by the superfluid flow around an obstacle was predicted.

One-Dimensional Condensate Flow Past an Obstacle: Critical Velocity

Let us consider the simpler lower-dimensional setting, namely the 1D flow of a repulsive NLS fluid in the presence of an obstacle (see, e.g., Refs. [392, 393, 394]). To better explain the basic ideas and phenomenology, we start with the simpler case of an untrapped condensate, and follow the analysis of Ref. [392]. There, the following model was considered:

$$i\partial_t \psi = -\partial_x^2 \psi - \psi - |\psi|^2 \psi + V_{\mathrm{imp}}(x-vt)\psi, \tag{2.242}$$

where the external localized repulsive potential V_{imp}, moving with velocity $v > 0$, represents the motion of an impurity in the NLS fluid; note that the boundary condition $|\psi|^2 = 1$ is imposed as $x \to \pm\infty$. In the frame of the moving impurity, the above model reads

$$i(\partial_t \psi - v\partial_x \psi) = -\partial_x^2 \psi - \psi - |\psi|^2 \psi + V_{\mathrm{imp}}(x)\psi. \tag{2.243}$$

Time-independent solutions of Eq. (2.243), corresponding to steady flows, can be sought in the form $\psi(x,t) = R(x)\exp[i\phi(x)]$, which leads to the following system:

$$v\partial_x R = 2\partial_x R \partial_x \phi + R\partial_x^2 \phi, \tag{2.244}$$

$$-vR\partial_x \phi = \partial_x^2 R - R(\partial_x \phi)^2 + R - R^3 - V_{\mathrm{imp}}(x)R. \tag{2.245}$$

Requiring that $R = 1$ and $\partial_x \phi = 0$ as $x \to \pm\infty$, Eq. (2.244) can readily be integrated, leading to the result $\partial_x \phi = (v/2)(1-R^{-2})$; substituting this expression into Eq. (2.245), the following equation for R is obtained:

$$\partial_x^2 R = \frac{v^2}{4}\left(R + \frac{1}{R^3}\right) - R + R^3 + V_{\mathrm{imp}}(x)R. \tag{2.246}$$

Then, employing again the boundary conditions, and assuming that $R(x) = 1 + r(x)$ for large $|x|$, a linearization of Eq. (2.246) with respect to r leads to the result

$$\partial_x^2 r = (2 - v^2) r. \tag{2.247}$$

The above simple equation implies that the stability of the flow is guaranteed only if the relevant eigenvalues λ (determined by the equation $\lambda^2 = 2 - v^2$) are real. Thus, the flow is stable (unstable) for $v < c_s$ ($v > c_s$), where $c_s \equiv \sqrt{2}$ is the speed of sound in the units of Eq. (2.242).

Here, we should mention that if $v > c_s$, the modes of Eq. (2.247) are oscillating for large $|x|$, which prevents the existence of a localized solution for a generic potential $V_{\mathrm{imp}}(x)$, due to the boundary conditions associated with Eq. (2.246) (recall that $R = 1$ as $x \to \pm\infty$). On the other hand, if $v < c_s$, for large $|x|$ there exists one bounded and one unbounded mode. In fact, as was shown in Ref. [392], a steady dissipationless motion is possible in this regime only below an obstacle-dependent [i.e., $V_{\mathrm{imp}}(x)$-dependent] critical velocity v_c, and the flow profile can be determined analytically in several physically relevant cases. The steady flow disappears at the critical velocity, by merging with an unstable solution in a saddle-node bifurcation. This unstable solution represents the transition state for emission of dark solitons, which are repeatedly generated for $v > v_c$. In fact, the onset of dissipation corresponds to nonstationary flow with a wake asymptotically extending upstream to infinity, and downstream periodic emission of dark solitons [393].

Coming back to the critical velocity, it should be pointed out that in the presence of nonlinearity it turns out to be smaller than the speed of sound (i.e., $v_c < c_s$) [392, 393], which indicates a correction to the classical Landau criterion. For instance, as was shown in Ref. [392] in the framework of Eq. (2.243), for a sufficiently weak impurity potential $V_{\mathrm{imp}}(x)$, the critical velocity is approximately given by

$$v_c \approx \sqrt{2} - \frac{3}{2\sqrt{2}} \left(\int_{-\infty}^{+\infty} V_{\mathrm{imp}}(x)\, dx \right)^{2/3}. \tag{2.248}$$

Note that in the case of a delta-like obstacle of strength $V_0 \ll 1$, namely for $V_{\mathrm{imp}}(x) = V_0 \delta(x)$, Eq. (2.248) implies that $v_c = \sqrt{2} - (3/2\sqrt{2}) V_0^{2/3}$ [392].

Generally, the fact that $v_c < c_s$ may be explained by the fact that, in the region of the potential, the local fluid velocity can reach values higher than the local sound velocity (note that critical velocity values smaller than the speed of sound were also observed in experiments [248, 390]).

A Trapped Quasi-1D Condensate Past an Obstacle

The results presented above (which are mostly referring to *homogeneous* superfluids) paved the way for a better understanding of the BEC flow past an obstacle, and inspired further investigations [395, 396, 397]. Here, we will consider an interesting question, namely how phenomenology changes in the physically relevant situation where a quasi-1D condensate is confined in a parabolic trap. This problem was studied in Ref. [397], where the differences from the case of homogeneous BECs were highlighted. In particular, in this work, the following GPE was considered:

$$i \partial_t u = -\partial_x^2 u + |u|^2 u + \left[V_{\mathrm{tr}}(x) + V_{\mathrm{imp}}(x,t) \right] u, \tag{2.249}$$

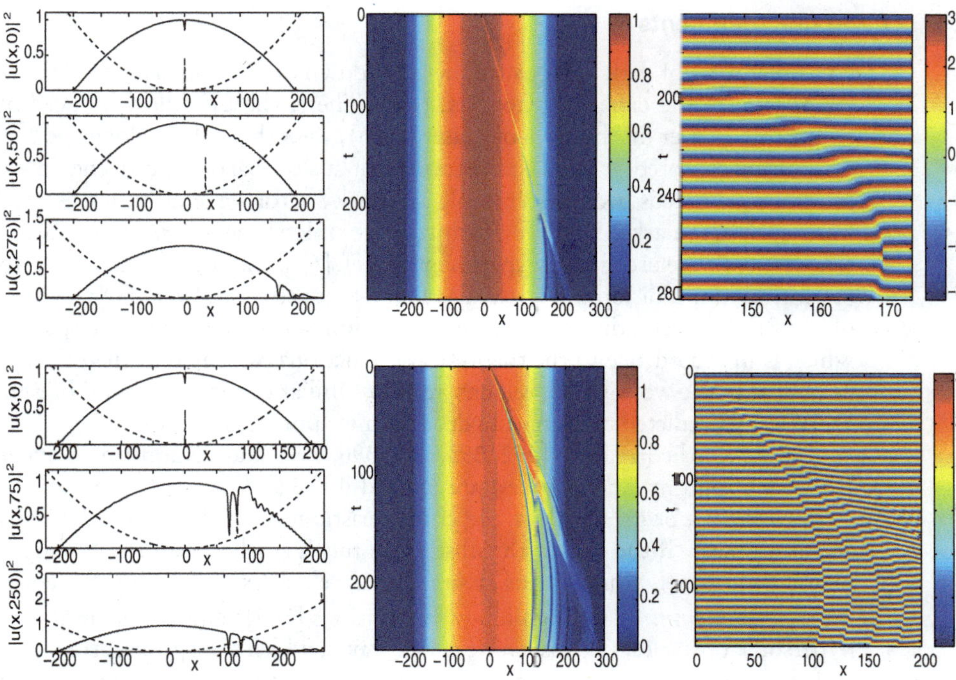

Figure 2.41. *Two cases of the motion of the obstacle, characterized by an amplitude $V_0 = 0.5$ and width $\epsilon^2 = 0.09$, through a trapped condensate with a TF radius $R_{\mathrm{TF}} = 200$ (the chemical potential is $\mu = 1$). Top and bottom panels correspond, respectively, to velocities $v = 0.75$ and $v = 1.1$. In each case, the left panels show the spatial profiles of the BEC density (squared absolute value of the wavefunction) for three different times (the potential is shown by the dashed line). The middle panel shows contour plots of the density and its spatio-temporal evolution. The right panel displays the spatio-temporal evolution of the phase, revealing the presence of phase dislocations—and thus demonstrating the formation of gray solitons. Adapted with permission from Ref. [397].*

where $V_{\mathrm{tr}}(x)$ is a harmonic trapping potential of strength Ω, and $V_{\mathrm{imp}}(x)$ accounts for an impurity moving across the condensate. These potentials are given by

$$V_{\mathrm{tr}}(x) = \frac{1}{2}\Omega^2 x^2, \qquad V_{\mathrm{imp}}(x,t) = V_0 \exp\left(-\frac{(x-vt)^2}{\epsilon^2}\right), \tag{2.250}$$

where the parameters V_0, ϵ and $v > 0$ describe the amplitude, width, and velocity of the impurity, respectively. Actually, if this moving obstacle emulates a laser-beam spot, the Gaussian shape adopted above is the most natural one.

The numerical simulations of Ref. [397] have shown that in the trapped BEC, emission of coherently structured radiation *always* occurs, i.e., for *any* speed of the body moving across the condensate. This is due to the fact that any given velocity turns out to be supersonic in a part of the system where the density is small enough. A typical example of the numerical experiments is depicted in Fig. 2.41, where two examples for different impurity velocities are presented. As it is clear from the figure, the fastest impurity (bottom panels) is able to emit radiation, in the form of dark solitons (as evidenced by the phase dislocations depicted in the right panels), earlier (i.e., already for higher densities). To be more precise, the emitted dark solitons are gray ones, since they have an intrinsic velocity (after completely detaching from the impurity, they slide to the left).

Connection to Experiments

The nucleation of dark solitons via the breakdown of BEC superfluidity was studied systematically in the experiment of Ref. [67]. There, the BEC flow induced by a broad, penetrable barrier (in the form of a laser beam) swept through an elongated ^{87}Rb condensate was investigated. It was demonstrated that at slow barrier speeds the flow is stable, at intermediate speeds becomes unstable and dark soliton generation is observed, while at faster speeds, remarkably, soliton formation completely ceases. Both repulsive and attractive barriers were used in the experiment of Ref. [67], and were found to lead to dark soliton formation; additionally, it was also found that the critical velocity for the breakdown of the BEC superfluidity and soliton generation was smaller than the speed of sound—which is in accordance to the theoretical predictions presented in the previous sections. Note that in the work [398], velocity regimes similar to the ones found in Ref. [67] were analytically predicted by using a hydrodynamic approach.

As shown theoretically [392, 393, 395, 396, 397] and demonstrated experimentally [67], dark solitons (and vortices) are formed if the size of the "hypersonic" obstacle is of the order of, or greater than, the characteristic healing length of the condensate. On the other hand, if the size of the obstacle is much smaller than the healing length, the main loss channel, which opens at supersonic velocities of the obstacle, corresponds to the *Čerenkov emission* of Bogoliubov's excitations [399]. Notice that in the case of large hypersonic obstacles, two dispersive shock waves, which start propagating from the front and the rear parts of the obstacle, are formed. Far from the obstacle, the shock front gradually transforms into a linear "ship wave" located outside the Mach cone [400, 401, 402, 403], whereas the rear zone of the shock is converted into a "fan" of oblique dark solitons located inside the Mach cone [404, 405, 406] (see also relevant experimental results in Refs. [400, 407]).

An important result reported in Ref. [408] is that although such dark solitons are unstable in higher-dimensional settings with respect to transverse perturbations (see Sec. 2.3.2), the instability becomes *convective*—rather than being absolute—for sufficiently large flow velocities and, thus, dark solitons are effectively stable in the region around the obstacle.

Binary Superfluid Flow Past an Obstacle

It is natural to extend the above ideas in the case of multicomponent BECs. The relevant problem, namely the flow of a multicomponent superfluid past an obstacle, was studied in the literature, where both cases of a two-component [326, 409, 327] and a spinor $F=1$ condensate [410] were analyzed. Below, we briefly describe the main findings of these works, as well as results from pertinent studies in the same context.

We start with the simpler case of a quasi-1D two-component BEC, in the miscible regime, whose dynamics is described by a system of GPEs expressed in the following dimensionless form:

$$i\partial_t \psi_j = \left(-\frac{1}{2}\partial_x^2 + \sum_{k=1}^{2} g_{jk}|\psi_k|^2 + V\right)\psi_j. \tag{2.251}$$

Following Ref. [326], we assume that the external potential V is a combination of a repulsive potential (describing, e.g., a blue-detuned laser beam) V_{imp}, and a harmonic trap V_{tr}, i.e.,

$$V = V_{\mathrm{tr}} + V_{\mathrm{imp}}, \quad V_{\mathrm{tr}} = \frac{1}{2}\Omega^2 x^2, \quad V_{\mathrm{imp}} = V_0 \exp\left(-\frac{(x-vt)^2}{2\epsilon^2}\right), \tag{2.252}$$

where A, ϵ and v are, respectively, the strength, width and velocity of the laser impurity, and Ω the harmonic trap strength. The nonlinearity coefficients are chosen to be $g_{11}:g_{12}:g_{22}=1.5:1:1.03$, having in mind that two of them have the ratios that they typically do for ^{87}Rb, while the third is tuned to a different value (so as to ensure miscibility since the more realistic value of $g_{11}=0.97$ would lead to immiscibility). Below we use the following parameter values: $\mu_1=1.2$, $\mu_2=1$, $\epsilon=0.5$, and $\Omega=0.02$ when a harmonic trap is present.

The uniform solutions of Eqs. (2.251) will then satisfy

$$|\psi_1^{(0)}|^2=\frac{\mu_1 g_{22}-\mu_2 g_{12}}{\Delta}, \qquad |\psi_2^{(0)}|^2=\frac{\mu_2 g_{11}-\mu_1 g_{12}}{\Delta}, \tag{2.253}$$

where $\Delta=g_{11}g_{22}-g_{12}^2$. Expressing Eqs. (2.251) in the traveling-wave frame (i.e., $x\to x-vt$) and linearizing around these uniform states according to $\psi_j=\psi_j^{(0)}+\psi_j^{(1)}$ we obtain the equations for the small amplitudes $\psi_j^{(1)}$,

$$\frac{1}{4}\frac{d^2}{dx^2}\psi_j^{(1)}=\left(c_j^2-v^2\right)\psi_j^{(1)}+g_{jk}\psi_j^{(0)}\psi_k^{(0)}\psi_k^{(1)}, \tag{2.254}$$

with $j,k\in\{1,2\}$, $k\neq j$, and $c_j^2=g_{jj}(\psi_j^{(0)})^2$. We now write this system as one of four first-order equations (for $\Psi_j^{(1)}=\dot{\psi}_j^{(1)}\equiv(d/dx)\psi_j^{(1)}$ and $\psi_j^{(1)}$), namely,

$$\begin{pmatrix}\dot{\Psi}_1^{(1)}\\ \dot{\psi}_1^{(1)}\\ \dot{\Psi}_2^{(1)}\\ \dot{\psi}_2^{(1)}\end{pmatrix}=\begin{pmatrix}0 & c_1^2-v^2 & 0 & b\\ 1 & 0 & 0 & 0\\ 0 & b & 0 & c_2^2-v^2\\ 0 & 0 & 1 & 0\end{pmatrix}\begin{pmatrix}\Psi_1^{(1)}\\ \psi_1^{(1)}\\ \Psi_2^{(1)}\\ \psi_2^{(1)}\end{pmatrix}, \tag{2.255}$$

where $b=g_{12}\psi_1^{(0)}\psi_2^{(0)}$. Then, the 4×4 matrix has eigenvalues

$$\lambda^2=\tilde{c}^2-v^2\pm\sqrt{\tilde{c}^4-\Delta(\psi_1^{(0)}\psi_2^{(0)})^2},$$

with $\tilde{c}^2=(c_1^2+c_2^2)/2$. For stability, we need the eigenvalues to be real, hence $\lambda^2>0$ implies that

$$v<c_1<c_2,$$

where the critical velocities, corresponding to the two distinct speeds of sound, are given by

$$c_{1,2}=\tilde{c}^2\mp\sqrt{\tilde{c}^4-\Delta\left(\psi_1^{(0)}\psi_2^{(0)}\right)^2}, \tag{2.256}$$

where the smaller (larger) speed, c_1 (c_2), corresponds to the $-$ (+) sign. These critical velocities correspond to the speeds of sound of the two components. For the parameters mentioned above, $c_1=0.34393$ and $c_2=1.04796$. We expect that superfluidity will break down when the speed v of the defect moving across the condensate overcomes these speeds; in fact, it should be recalled that the actual critical point should be expected to be lower than the above Landau prediction (see previous subsections and Refs. [392, 393]).

In Fig. 2.42, we test numerically the theoretical prediction for the existence of two critical velocities for the dynamical evolution in the two components. Particularly, the top panel of the figure shows a relevant "bifurcation diagram," where the dependence of the

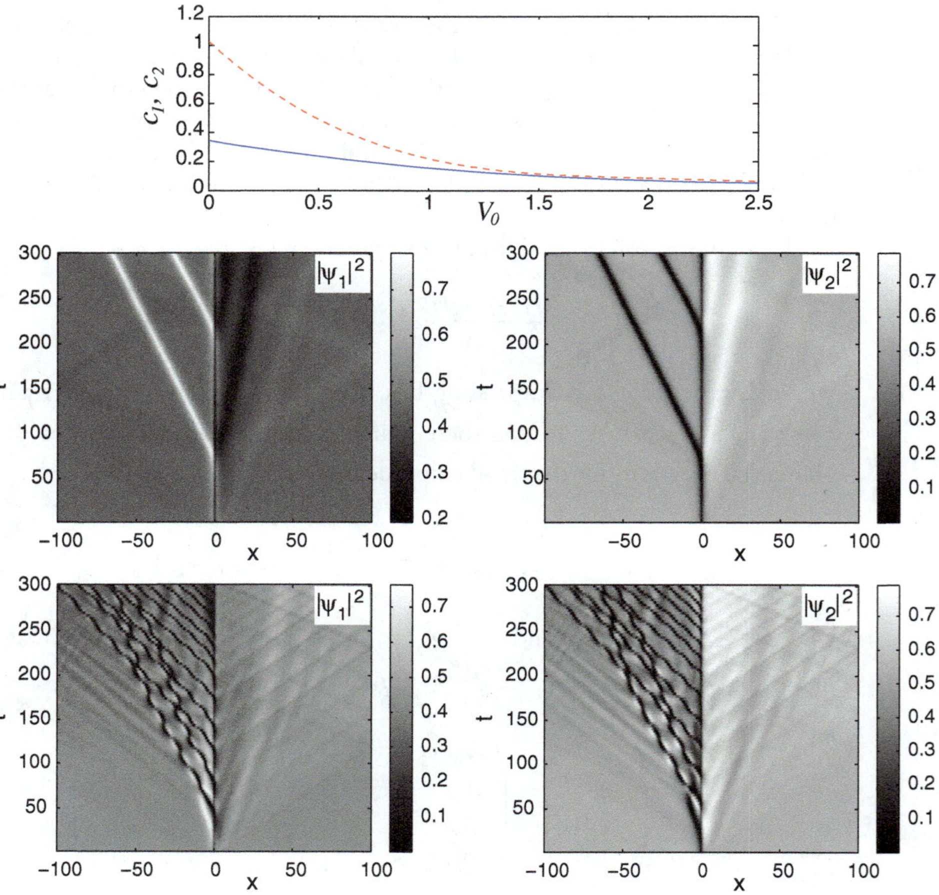

Figure 2.42. *Top panel: Critical velocities for the first (dashed line) and second (solid line) components as a function of the impurity strength* V_0. *Middle row of panels: Space-time evolution of the density for the two components in moving coordinate frame with velocity* $v = 0.2$ *(the speed of the impurity); clearly the impurity induces the radiation of dark-antidark pairs. Bottom row of panels: Similar to the middle row of panels but with* $v = 0.3$; *dark-dark solitons are emitted. Impurity parameters:* $V_0 = 0.9$ *and* $\epsilon = 0.5$. *Adapted with permission from Ref. [326].*

critical velocities on the "strength" V_0 of the impurity is numerically evaluated. Note that as the strength of the impurity tends to zero ($V_0 \to 0$) one recovers the numerical values for $c_{1,2}$ stated above. It is important to recall, as mentioned above, that c_1 corresponds to the component with the *smallest* chemical potential (and not necessarily to the component with the smallest background uniform state $\psi_j^{(0)}$). In the presented case $\mu_1 = 1.2 > \mu_2 = 1$ and thus, as the velocity of the impurity is increased, dark solitons are first nucleated in the *second* component—despite the fact that $\psi_1^{(0)} < \psi_2^{(0)}$. The critical velocities are computed by finding (i) the speed c_1 above which, apparently, one component is supercritical emitting *dark solitons*, while the other is subcritical emitting *antidark solitons* (i.e., bright solitons on a finite background) that "accompany" the dark ones; and (ii) the speed c_2 above which *both* components nucleate dark solitons (see also bottom panels of the figure for a space-time evolution of the density contour plot for the two components for such a supercritical—in both components—velocity v). It is noteworthy that such structures had been predicted in a steady form in Ref. [331], but, to our knowledge, this is the first demonstration of their dynamical nucleation. A further observation is worth making about the case of $v > c_2$. Note that, especially for early times, the impurity (placed at $x = 0$ in the computations of Fig. 2.42 performed in the co-moving frame) initially emits structures that appear to be beating dark-dark solitons of the type explored in Sec. 2.4.6. This is again the first manifestation of such structures. It is worth mentioning that the phenomenology observed in Fig. 2.42 in the case where the harmonic trap was absent, also persists in the presence the trap, as shown in numerical simulations of Ref. [326].

Concluding this section, we can summarize the above results, as well as pertinent ones for the higher-dimensional setting (see Ref. [326]) as follows. In the case of a two-component BEC, the existence of two different speeds of sound provides the possibility for three dynamical regimes: when both components are subcritical, nucleation of coherent structures does not occur; when both components are supercritical they form dark-dark solitons in 1D and vortices or rotating vortex dipoles in 2D; in the intermediate regime, the nucleation of a dark–anti-dark soliton in 1D or a vortex-lump configuration in 2D is observed. Furthermore, as shown in Ref. [327], dark solitons can be convectively stabilized in the 2D setting at sufficiently high values of the obstacle velocity, similarly to the case of one-component BECs [408].

Spinor $F = 1$ Condensate Flow Past an Obstacle

Next, the motion of a localized defect through a quasi-1D spinor $F = 1$ condensate is considered. Following the analysis of Ref. [410], this problem can be analyzed in the framework of Eqs. (2.187) and (2.188), but with the external potential $V(x)$ in the Hamiltonian H_0 being given by Eq. (2.252). Then, following the lines of the analysis presented above for the two-component case, and adopting the single-mode approximation, it is again possible to derive the critical velocities, corresponding to distinct speeds of sound. Interestingly enough, despite the three-component nature of the spinor $F = 1$ system, the analysis of the small-amplitude excitation problem performed in Ref. [410] revealed that the nature of the nonlinearity is such that there appear *not three*, but merely *two* critical velocities in the system. In the case of in-phase ($\Delta\theta = 0$) stationary uniform states of the system the two critical velocities are given by:

$$c_1 = \sqrt{\delta}\, c_2, \quad c_2 = \sqrt{\mu}. \tag{2.257}$$

It is clear that the first critical velocity is characteristic for the spinor $F = 1$ condensate under consideration (as it depends on the normalized spin-dependent interaction strength δ), while the second one is the standard speed of sound appearing in the one-component GPE [392]. Note that these speeds of sound were already implicit in the seminal work of Ref. [360] [see, e.g., Eqs. (8)–(11) therein and the accompanying discussion, as well as Sec. 2.4.8]. It is worthwhile to recall that as $\delta > 0$ for antiferromagnetic bosonic spin-1 atoms (e.g., ^{23}Na), while it is $\delta < 0$ for *ferromagnetic* ones (e.g., ^{87}Rb), then only c_2 is relevant (i.e., will exist) in the case of, e.g., the polar ^{23}Na spinor condensate.

Numerical simulations of Ref. [410] tested the dynamics for different values of defect speeds in comparison to the above mentioned two critical speeds. Surprisingly, in all simulations (and both in 1D and in higher dimensions), the physical irrelevance of the first critical speed c_1 was definitively confirmed: this speed is *not* activated and no emission of nonlinear wave excitations emerges when this threshold is crossed—see middle column of panels in Fig. 2.43. From a physical perspective, a potential explanation of this feature can stem from the fact that this speed is associated with spin excitations [360], rather than density excitations (from the latter one would normally anticipate the emergence/emission

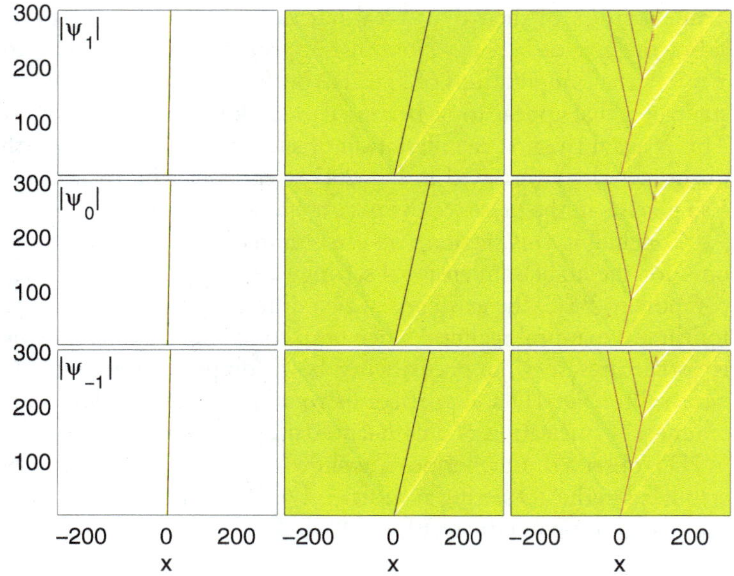

Figure 2.43. *Time evolution of the three hyperfine components (top, middle, and bottom rows, respectively) for different defect velocities v; in particular: $v = 0.025 < c_1$ (left column), $c_1 < v = 0.325 \lesssim c_2$ (middle column), and $v = 0.335 > c_2$ (right column). Here, the initial condition corresponds to the in-phase stationary uniform state, while $c_1 = 0.058$ and $c_2 = 0.329$ are the two critical velocities and the defect strength is $V_0 = 0.9$. While the results in the left and middle columns show a steady flow (apart from an oscillatory structure that is detached at $t = 0$), the evolution shown in the panels of the right column is characterized by the emission of dark solitons, even from the early stages of the process. Notice that the analytically predicted first (lower) critical velocity would fall between the velocities of the results depicted in the left and middle columns, but no significant change is observed in the dynamics between these two cases. Adapted with permission from Ref. [410].*

of nonlinear excitations, as is the case in the two-component system studied in the previous section). On the other hand, when the defect speed exceeds the second critical one, c_2, then emission of coherent structures arises independently of dimension; the resulting waveforms are dark solitons in the 1D setting (see right columns of Fig. 2.43), vortex-antivortex pairs in two dimensions, and spinor vortex rings in the fully 3D case (see more details on the higher-dimensional settings in Sec. 4.4.1).

2.5.2 ▪ Double-Well Potentials

An important element in the studies of BEC is the variety of external potentials that may be used to confine the ultracold atomic gases. The most typical forms of such trapping potentials that we have considered so far are harmonic-oscillator traps and optical lattices. A similar situation is relevant also in the context of nonlinear optics, where the basic model, namely the NLS equation, may incorporate harmonic or periodic potentials describing graded-index waveguides and periodic waveguiding arrays, respectively [36]. A combination of harmonic and periodic traps may be used to create a *double-well potential*, which has drawn a great deal of attention. BECs confined in double-well potentials were studied experimentally in Refs. [411] and [412], and many theoretical works addressed this setting [413, 414, 415, 416, 417, 418, 419, 420, 421].

A fundamental effect that was studied, in diverse forms, in these works is the spontaneous-symmetry-breaking (SSB) phenomenon, which can be qualitatively described as follows. It is well known that, in quantum mechanics, the double-well gives rise to alternating symmetric and antisymmetric bound states. However, the nonlinearity breaks the symmetry, giving rise to *asymmetric* bound states, i.e., to the SSB effect. Apart from BECs, studies on double-well potentials and the SSB effect were performed also in nonlinear optics: there, twin-core self-guided laser beams in Kerr media [422], optically induced dual-core waveguiding structures in photorefractive crystals [423], and trapped light beams in an annular core of an optical fiber [424] among others, also lead to manifestations of phenomenology associated with double-well potentials. Notice that in addition to the studies of the SSB in static double-well potentials, analysis was extended to diverse situations with the double-well or the BEC itself subject to "management" (i.e., the depth of the well, or height of the barrier between the wells, or the nonlinearity coefficient of the GPE) were made periodically varying functions of time [425]. Recent works on this subject can be found in the review [426].

In most of the above mentioned works, the analysis was based on variants of the so-called two-mode approximation, which replaces the underlying GPE by a system of two coupled ODEs. Below, following Ref. [420], we will present the basic ideas and results for a GPE model with a double-well potential.

Model and Analytical Considerations

We start by presenting the model, namely the following normalized GPE:

$$i\partial_t u = -\frac{1}{2}\partial_x^2 u + s|u|^2 u + V(x)u - \mu u. \tag{2.258}$$

Here, for the sake of completeness, we consider both cases of attractive or repulsive interatomic interactions, corresponding to $s = \pm 1$, respectively. Additionally, μ is the chemical potential and $V(x)$ is the double-well potential, which is assumed to be composed of a parabolic trap (of strength Ω) and a sech^2-shaped barrier (of strength V_0, width w and location x_0), namely,

$$V(x) = \frac{1}{2}\Omega^2 x^2 + V_0 \mathrm{sech}^2\left(\frac{x - x_0}{w}\right), \tag{2.259}$$

with the choice $x_0 = 0$ ($x_0 \neq 0$) corresponding to a symmetric (asymmetric) double well. Below, we will focus on the case of the symmetric double-well potential; details on the analysis of the asymmetric case (which follows the lines of the symmetric one) can be found in Ref. [420].

The spectrum of the underlying linear Schrödinger equation ($s = 0$) consists of a ground state, $u_0(x)$, and excited states, $u_l(x)$ ($l \geq 1$). In the weakly nonlinear regime of the nonlinear problem, using a Galerkin-type approach, we expand $u(x,t)$ as

$$u(x,t) = c_0(t)u_0(x) + c_1(t)u_1(x) + \cdots, \tag{2.260}$$

and truncate the expansion, keeping solely the first two modes; here $c_{0,1}(t)$ are unknown time-dependent complex prefactors. As mentioned above, such an approximation (involving the truncation of higher-order modes and the spatio-temporal factorization of the wavefunction), is expected to be quite useful for a weakly nonlinear analysis, i.e., for a sufficiently small L^2 norm (i.e., number of atoms) of the solution. Our aim is to identify nonlinear states that stem from the linear ones, as well as their bifurcations.

Substituting Eq. (2.260) into Eq. (2.258), and projecting the result to the corresponding eigenmodes, we obtain the following ODEs for the projections to u_0 and u_1 respectively:

$$\begin{aligned} i\dot{c}_0 &= (\omega_0-\mu)c_0 - sA_0|c_0|^2c_0 - sB(2|c_1|^2c_0 + c_1^2c_0^*) \\ &\quad -s\Gamma_1|c_1|^2c_1 - s\Gamma_0(2|c_0|^2c_1 + c_0^2c_1^*), \end{aligned} \tag{2.261}$$

$$\begin{aligned} i\dot{c}_1 &= (\omega_1-\mu)c_1 - sA_1|c_1|^2c_1 - sB(2|c_0|^2c_1 + c_0^2c_1^*) \\ &\quad -s\Gamma_0|c_0|^2c_0 - s\Gamma_1(2|c_1|^2c_0 + c_1^2c_0^*). \end{aligned} \tag{2.262}$$

In Eqs. (2.261) and (2.262), dots denote time derivatives, asterisks denote complex conjugates, $\omega_{0,1}$ are the eigenvalues corresponding to the eigenstates $u_{0,1}$, while $A_0 = \int u_0^4 dx$, $A_1 = \int u_1^4 dx$, $B = \int u_0^2 u_1^2 dx$, $\Gamma_0 = \int u_0 u_1^3 dx$ and $\Gamma_1 = \int u_1 u_0^3 dx$ are constants. Recall that u_0 and u_1 are real (due to the Hermitian nature of the underlying linear Schrödinger problem) and are also orthonormal. Notice also that in the symmetric case ($x_0 = 0$), due to the parity of the eigenfunctions, the constants $\Gamma_0 = \Gamma_1 = 0$.

We now use amplitude-phase (action-angle) variables, $c_j = \rho_j e^{i\phi_j}$, $j = 0,1$ (ρ_j and ϕ_j are assumed to be real), to derive from the ODEs (2.261) and (2.262) a set of four equations. Introducing the function $\varphi \equiv \phi_1 - \phi_0$, we find that the equations for ρ_0 and ϕ_0 are

$$\dot{\rho}_0 = s[(\Gamma_0\rho_0^2 + \Gamma_1\rho_1^2)\sin(\varphi) + B\rho_1^2\rho_0\sin(2\varphi)], \tag{2.263}$$

$$\begin{aligned} \dot{\phi}_0 &= (\mu-\omega_0) - sA_0\rho_0^2 - 2sB\rho_1^2 - sB\rho_1^2\cos(2\varphi) \\ &\quad -s\left(\frac{\Gamma_1\rho_1^3}{\rho_0} + 3\rho_0\rho_1\Gamma_0\right)\cos(\varphi), \end{aligned} \tag{2.264}$$

while the equations for ρ_1 and ϕ_1 are found by interchanging indices 1 and 0 in the above equations. Next, taking into regard the conservation of the number of atoms $N = \int |u|^2 dx$, we obtain the equation $\rho_0^2 + \rho_1^2 = N$. Finally, subtracting Eq. (2.264) for $\dot{\phi}_0$ and the corresponding one for $\dot{\phi}_1$, we obtain

$$\begin{aligned} \dot{\varphi} &= -\Delta\omega + s(A_0\rho_0^2 - A_1\rho_1^2) - sB(2+\cos(2\varphi))(\rho_0^2 - \rho_1^2) \\ &\quad -s\frac{\cos(\varphi)}{\rho_0\rho_1}\left[\Gamma_0\rho_0^2(\rho_0^2 - 3\rho_1^2) + \Gamma_1\rho_1^2(3\rho_0^2 - \rho_1^2)\right], \end{aligned} \tag{2.265}$$

where $\Delta\omega = \omega_1 - \omega_0$. Equations (2.263) and (2.265) constitute a dynamical system, which, in principle, can be investigated in detail using phase-space analysis (such an approach was presented in Refs. [413, 414] for similar systems that were derived using different expansion of the field u). Here, we will focus on the fixed points of the system [corresponding to the nonlinear eigenstates of Eq. (2.258)], and analyze their stability and bifurcations. Below we will analyze the cases $s = \pm 1$ and $x_0 = 0$, using—for our numerical computations—the values $V_0 = 1$, $\Omega = 0.1$, and $w = 0.5$.

Bifurcations and Stability Analysis

We consider a symmetric double-well potential ($x_0 = 0$), the form of which, as well as the eigenvalues, the ground state u_0 and the first excited state u_1 of the corresponding linear problem ($s = 0$) are shown in Fig. 2.44. In this case, the parameters involved in Eqs. (2.263) and (2.265) are found to be $A_0 = 0.09078$, $A_1 = 0.09502$, $B = 0.08964$, $\omega_0 = 0.13282$, and $\omega_1 = 0.15571$. Recall that because of the even and odd parity of the linear states u_0 and u_1 we have $\Gamma_0 = \Gamma_1 = 0$.

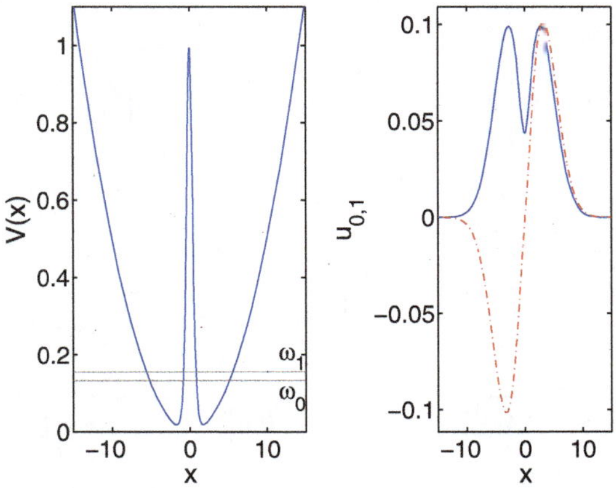

Figure 2.44. *The left panel shows a symmetric ($x_0 = 0$) double-well potential $V(x)$, with parameters $\Omega = 0.1$, $V_0 = 1$, and $w = 0.5$; the horizontal solid lines (ω_0, ω_1) indicate the eigenvalues corresponding to the eigenstates u_0 and u_1 of the linear Schrödinger equation ($s = 0$). The right panel shows the wavefunctions of the ground state u_0 (solid line) and the first excited state u_1 (dashed-dotted line). Adapted with permission from Ref. [420].*

Since we are interested in real solutions of Eq. (2.258), we will confine our considerations to steady states with $\varphi = 0 \pmod{\pi}$ [note that, as observed from (2.260), one can realize that the function u is real if $\phi_{0,1} = k\pi$, with k integer]. This way, Eq. (2.263) is automatically satisfied, while Eq. (2.265) becomes

$$-\Delta\omega - s(A_0\rho_0^2 - A_1\rho_1^2) + 3sB(\rho_0^2 - \rho_1^2) = 0. \tag{2.266}$$

Then, utilizing the definition of the number of atoms $N = \rho_0^2 + \rho_1^2$, it is readily found that Eq. (2.266) leads to

$$\rho_0^2 = \frac{\Delta\omega + sN(A_1 - 3B)}{s(A_0 + A_1 - 6B)}, \tag{2.267}$$

while, solving Eq. (2.266) with respect to ρ_1^2, we obtain

$$\rho_1^2 = \frac{\Delta\omega + sN(3B - A_0)}{s(6B - A_1 - A_0)}. \tag{2.268}$$

First, we consider the case of attractive nonlinearity, i.e., $s = -1$. As the denominator on the right-hand side of Eq. (2.267) is $6B - A_0 - A_1 > 0$, the numerator has to be nonnegative, i.e., $N \geq \Delta\omega/(A_1 - 3B)$. Since the inequality in the latter expression is always true [i.e., for all N, as $\Delta\omega/(A_1 - 3B) < 0$], it is concluded that no solution bifurcates from the antisymmetric branch with $\rho_0 = 0$. However, Eq. (2.268) implies that $N \geq \Delta\omega/(3B - A_0)$. As the latter requirement can now be satisfied, it is clear that asymmetric solutions, with $\rho_1 \neq 0$, can bifurcate from the symmetric ones with $\rho_0^2 = N$ and $\rho_1^2 = 0$. These solutions account for the symmetry breaking due to the inclusion of the odd eigenfunction $u_1(x)$ in Eq. (2.260). To this end, it is readily found that the critical value of the norm at which the bifurcation occurs is given by

$$N_c = \frac{\Delta\omega}{3B - A_0}. \tag{2.269}$$

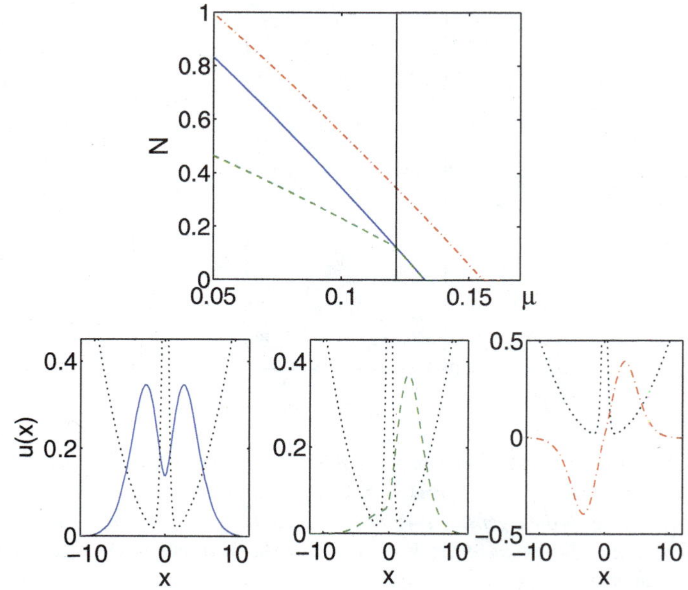

Figure 2.45. *Top panel: The norm of the solutions of Eq. (2.258) for attractive nonlinearity ($s=-1$) as a function of μ for a symmetric ($x_0=0$) double well. The potential parameters are $\Omega=0.1$, $V_0=1$, and $w=0.5$. The blue solid line denotes the symmetric solution, the red dashed-dotted line denotes the antisymmetric one, while the green dashed line denotes the asymmetric solution that is generated from the bifurcation at $\mu_c\approx 0.122$. The analytical threshold for the bifurcation $\mu_c=\omega_0-A_0N=0.12115$ is depicted by the thin vertical black line. Bottom panels: The profiles of the wavefunctions corresponding to the symmetric (left), asymmetric (middle), and antisymmetric (right) branches for $\mu=0.05$. The black dashed line shows the double-well potential. Adapted with permission from Ref. [420].*

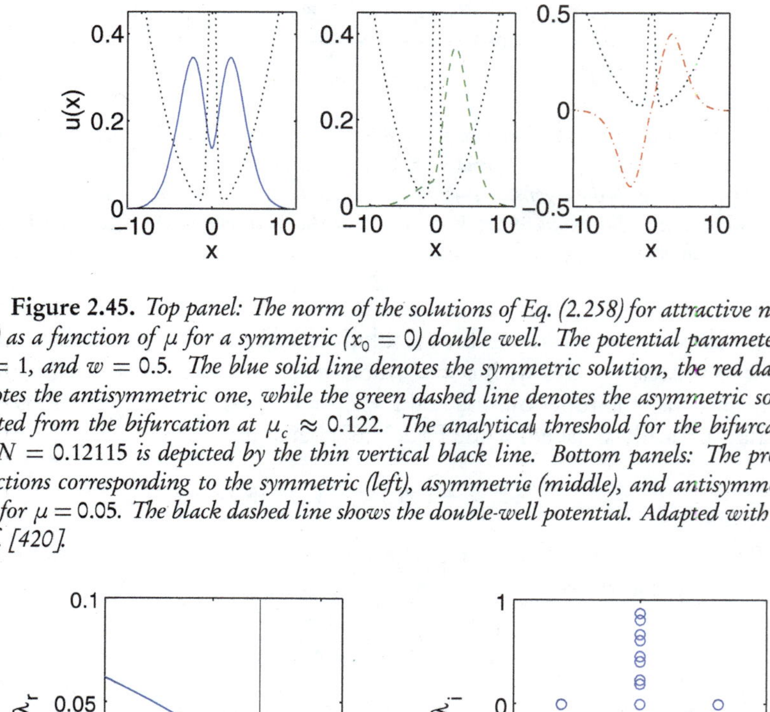

Figure 2.46. *Left panel: The real part λ_r of the most unstable eigenvalue of the symmetric solution as a function of μ; note that the symmetry-breaking bifurcation destabilizes the symmetric solution for $\mu<\mu_c=0.122$. Right panel: The result of the linear stability analysis around the symmetric solution for $\mu=0.05<\mu_c$ in the complex plane (λ_r,λ_i). The existence of an eigenvalue with positive real part implies instability of the solution. Adapted with permission from Ref. [420].*

One can also determine, using Eq. (2.261) and (2.262), the critical chemical potential (propagation constant) at which the symmetry breaking is expected, $\mu_c=\omega_0-A_0N$.

The above analytical predictions are in excellent agreement with the numerical results, which, in the case under consideration, are summarized in Fig. 2.45 (bifurcation diagram) and Fig. 2.46 (stability diagram). To construct the bifurcation diagram (Fig. 2.45), we find numerically the different branches of solutions (including the bifurcating ones) of the time-independent version of Eq. (2.258) (with $\partial_t u = 0$). This is done upon employing a fixed point algorithm (Newton–Raphson) and using the continuation of the solutions for the parameter μ. In the top panel of Fig. 2.45 the norm $N = \int |u|^2 dx$ of the solutions is shown as a function of the chemical potential μ. The red dashed-dotted branch denotes the antisymmetric solution of Eq. (2.258); as expected, continuation of this branch to the linear limit ($N \to 0$) shows that it starts (bifurcates) from $\mu = \omega_1$, namely the eigenvalue of the first excited state. The blue solid and green dashed lines correspond, respectively, to the symmetric and asymmetric solutions; the latter bifurcates from the former through symmetry breaking, and the bifurcation point is found to be $\mu_c = 0.122(\pm 0.001)$, in excellent agreement with the analytically predicted value ($\mu_c = 0.12115$). Note that continuation of the symmetric branch to the linear limit ends with the eigenvalue ω_0, corresponding to the ground state of the linear problem. The bottom panels of Fig. 2.45 illustrate the profiles of the symmetric (left), asymmetric (middle), and antisymmetric (right) branches for $\mu = 0.05$. It is clearly seen that the symmetric and antisymmetric wavefunctions follow the form of the corresponding linear ones. As far as the asymmetric branch is concerned, we show the one bearing the larger part of the density in the right well (notice that two such asymmetric states arise, being mirror images of each other with respect to the $x = 0$ axis).

The results of the linear stability analysis are shown in Fig. 2.46. The left panel, displaying the real part λ_r of the most unstable eigenvalue of the symmetric solution as a function of μ, shows that the symmetry-breaking bifurcation destabilizes the symmetric solution (for $\mu < \mu_c = 0.122$). This occurs through the appearance of a pair of real eigenvalues, as shown in the right panel of Fig. 2.46 (for $\mu = 0.05$) in the complex plane (λ_r, λ_i) of the stability eigenvalues. As the asymmetric solution emerging at the bifurcation point is stable, it is concluded that the bifurcation considered in this case is a (supercritical) pitchfork.

We now consider the case of the repulsive nonlinearity ($s = +1$). In this case, as the denominator on the right-hand side of Eq. (2.267) is $A_0 + A_1 - 6B < 0$, the numerator has to be nonpositive, i.e., $N \geq \Delta\omega/(3B - A_1)$. The equality in the latter expression can be satisfied, and, thus, the asymmetric solution can bifurcate from the antisymmetric branch with $\rho_0 = 0$. This way, the bifurcation does not originate from the symmetric branch but rather from the antisymmetric one, with $(\rho_0, \rho_1) = (0, \sqrt{N})$, giving again rise to symmetry breaking. The critical value of the norm N at which the bifurcation occurs is given by

$$N_c = \frac{\Delta\omega}{3B - A_1}, \tag{2.270}$$

while the critical value of the chemical potential is $\mu_c = \omega_0 + 3BN$. The numerical results are summarized in Fig. 2.47 (bifurcation diagram) and Fig. 2.48 (stability diagram). As expected from the above analysis, the asymmetric solution of Eq. (2.258) emerges from the antisymmetric branch (see the top panel of Fig. 2.47) through a pitchfork bifurcation. Additionally, Fig. 2.48 clearly shows that the symmetry-breaking bifurcation destabilizes the antisymmetric solution (see also Ref. [427] for a similar analysis and results). Moreover, the numerically found critical value of μ where the bifurcation takes place is $\mu_c = 0.168(\pm 0.001)$, once again in excellent agreement with the analytical result ($\mu_c = 0.16822$). Finally, similarly to the previous case (pertaining to the attractive nonlinearity), the bottom panels of Fig. 2.47 illustrate the wavefunction profiles for $\mu = 0.22$, while the right panel of Fig. 2.48 is a typical example of the linear stability analysis around the unstable antisymmetric solution (for the same value of the chemical potential).

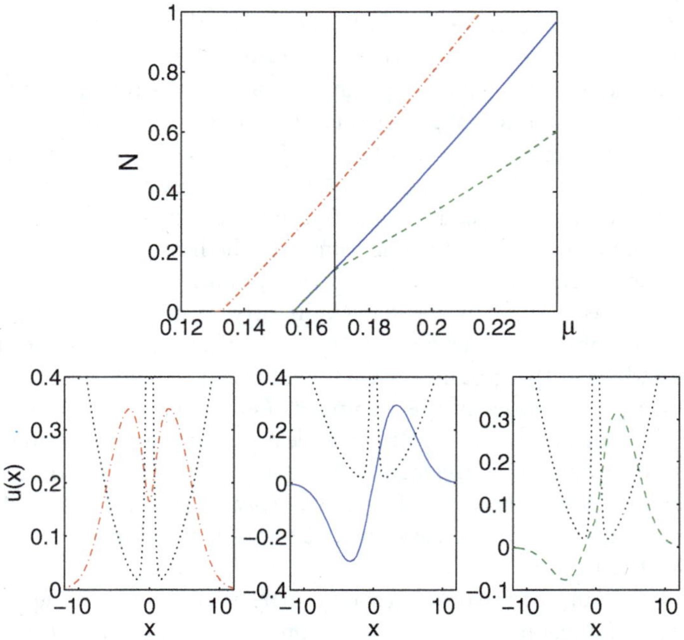

Figure 2.47. *Top panel: Same as in Fig. 2.45 but for the repulsive nonlinearity ($s = +1$). The red dashed-dotted line denotes the symmetric solution, the blue solid the antisymmetric one, while the green dashed line corresponds to the asymmetric solution generated from the pitchfork bifurcation occurring at $\mu_c \approx 0.168$. The analytical prediction (thin vertical black line) is $\mu_c = \omega_0 + 3BN = 0.16822$. Contrary to the case $s = -1$, the bifurcation originates from the antisymmetric branch. Bottom panels: The profiles of the wavefunctions corresponding to the symmetric (left), antisymmetric (middle), and asymmetric (right) branches for $\mu = 0.22$. Adapted with permission from Ref. [420].*

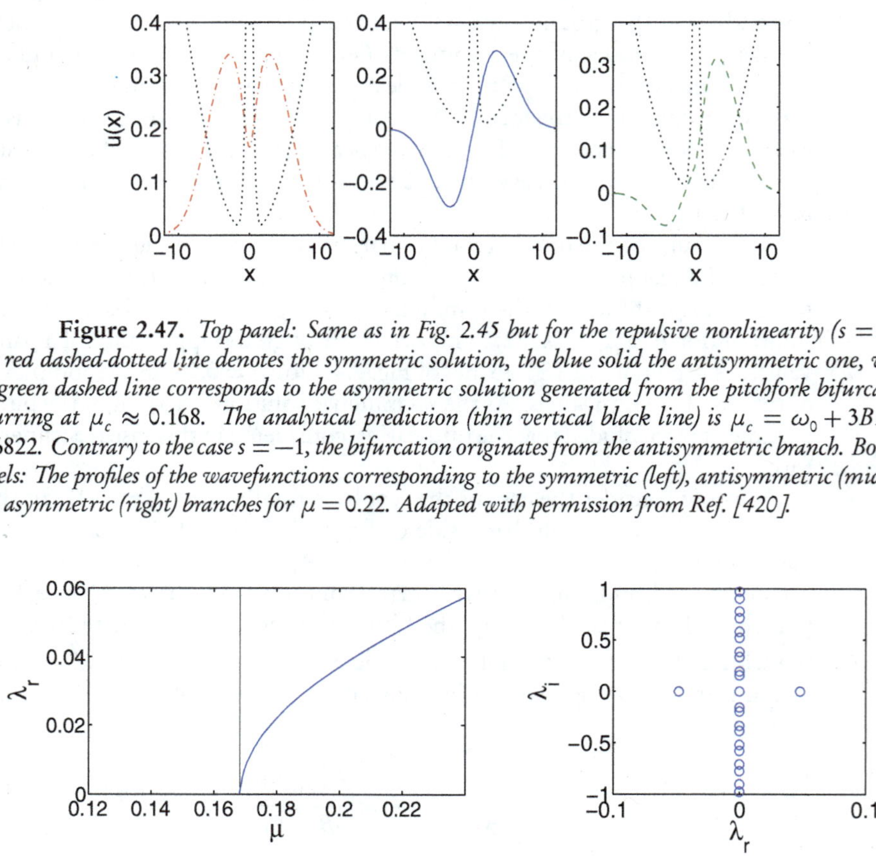

Figure 2.48. *Left panel: The maximal real eigenvalue associated with the linear stability analysis of the antisymmetric-solution branch. Right panel: The respective result of the linear stability analysis around the antisymmetric solution when $\mu = 0.22 > \mu_c$ in the complex plane. An eigenvalue with a positive real part implies instability of the solution. Adapted with permission from Ref. [420].*

2.5.3 ▪ Nonlinearity ("Feshbach Resonance") Management

Many of the fundamental properties of BECs (such as their shape, collective excitations, statistical fluctuations, etc.) depend on the sign and magnitude of the s-wave scattering length a, which controls the nonlinearity coefficient of the GPE. It is thus not surprising that much effort has been invested on the manipulations of matter waves by appropriate "adjustments" of the scattering length, leading to many interesting possibilities for "nonlinearity management" of BECs. On the other hand, in the context of optics, there exists a similar activity referring to the manipulation of nonlinear waves and solitons in nonuniform media, including artificial optical media, such as photonic crystals, and plasmonics (see the reviews [428, 429] and references therein). In this section, we will focus on the physics of BECs, and briefly describe results corresponding to the so-called *nonlinearity management* technique, which is also called "Feshbach resonance management" (FRM) one [271], for the reasons that will be better explained below.

A well-known technique to change and control the scattering length is to tune an external field in the vicinity of a *Feshbach resonance*; this external field may be a magnetic [430], an optical [431] or a dc-electric one [432]. In low-dimensional settings, one can also tune the effective scattering length by changing the BEC's transverse confinement [364] (see Sec. 1.2.4). The fields that control the scattering length may be time- or space-dependent and can act relatively far from the Feshbach resonance, so losses in the condensates (e.g., due to formation of molecules; see, e.g., [430]) are minimized. Generally, by employing such fields, one can control both the magnitude and sign of a, because the scattering length becomes zero relatively far from resonance, as in the case of ^{7}Li [95, 96] and ^{85}Rb [433] condensates.

To describe the basic idea of FRM, let us assume that the scattering length a is a function of a magnetic field B, i.e., $a = a(B)$, or an optical field of intensity I, i.e., $a = a(I)$. Then, considering that the external field, say A (corresponding to B or I) is time- or space-dependent, we may assume that $A = A_0 + A_1$, where A_0 is a constant value of the field (relatively far from resonance), and A_1 is a function describing the (temporal or spatial) change of A. Then, to a leading order approximation, one may express the scattering length—and, hence, the nonlinearity coefficient of the GPE model—as $a(t) = a_0 + a_1 A(t)$ (for *temporal FRM*) or $a(r) = a_0 + a_1 A(r)$ (for *spatial FRM*); here, a_0 is the value of a in the absence of the varying part of the external field A (i.e., $A_1 = 0$) and a_1 is a constant depending on the slope of a with respect to A, i.e., $a_1 = \partial_A a|_{A_1=0}$.

As far as the temporal FRM technique is concerned, temporal variations of the scattering were crucial to many important experimental discoveries, including the formation of bright matter-wave solitons [95, 96], the formation of molecular condensates [434], and probing the Bose–Einstein condensation in the Bardeen–Cooper–Schrieffer (BEC-BCS) crossover [435]. Additionally, theoretical studies have predicted that a time-dependent modulation of the scattering length can be used to stabilize attractive condensates against collapse in two [436, 437] and three [438] dimensions (in the latter case, the Feshbach resonance based technique is combined with a quasi-1D optical lattice), to create robust matter-wave solitons [271, 439, 440], periodic waves [441], or shock waves [442].

On the other hand, regarding the spatial FRM technique, the possibility of varying the scattering length *locally* (i.e., spatially) emerges as an interesting new direction for the manipulation of coherent matter waves. Such spatial dependence of a, which can be implemented upon utilizing a spatially inhomogeneous external magnetic field in the vicinity of a Feshbach resonance [250] renders the collisional dynamics inhomogeneous across the BEC. These so-called collisionally inhomogeneous BECs (according to the nomenclature of Ref. [250]) have also attracted much attention, as they are relevant to many interesting studies and applications: these include adiabatic compression of matter waves [250, 443], atomic soliton emission and atom lasers [444], enhancement of transmittivity of matter waves through barriers [445, 446], dynamical trapping of matter-wave solitons [445], dynamics of matter waves in the presence of periodic or random spatial variations of the scattering length [447], stable condensates exhibiting both attractive and repulsive interatomic interactions [448], and so on. Moreover, relevant studies in a quasi-2D setting were reported [449], while more rigorous results concerning the existence and the stability of solutions of the relevant GPEs also appeared [450].

Below, we will present some representative results for the temporal FRM, which correspond to the cases of periodic nonlinearity management and linear variation of the nonlinearity coefficient.

Periodic Nonlinearity Management

We start by considering a nearly 1D condensate, whose dynamics is described by the following normalized GPE:

$$iu_t = -\frac{1}{2}u_{xx} + a(t)|u|^2 u + V(x)u, \tag{2.271}$$

where $u(x,t)$ is the macroscopic wavefunction, $V(x) = (1/2)\Omega^2 x^2$ is the harmonic trap of strength Ω, and the Feshbach-managed nonlinearity coefficient (proportional to the scattering length) alternates between a_1 for $0 < t < \tau$ and a_2 for $\tau < t < T$ (with $a_2 < 0$ in most cases), which is repeated with a period T. The above "modulation map" defines the average ("dc") value of the nonlinear coefficient, as well as the ("ac") amplitude of the periodic variation, namely,

$$a_{\mathrm{dc}} \equiv \frac{a_1 + a_2}{2}, \qquad a_{\mathrm{ac}} \equiv \frac{a_1 - a_2}{2}.$$

Clearly, the modulation in time of the nonlinearity coefficient is reminiscent of the modulation of the group-velocity dispersion coefficient in optical models with dispersion management, where it is modulated as a function of the propagation distance [451]. Nevertheless, there is an essential difference from the dispersion management case, as in FRM the external parabolic trap is also incorporated; the latter plays an important role as will be seen below.

First, we note that FRM induces modulational instability (MI)—cf. the corresponding analysis in optical systems with dispersion management [452] or nonlinearity management [453], as well as relevant studies in BECs [454]—which leads to solitary waves. Interestingly, the corresponding equations feature a structure reminiscent of the Kronig-Penney model, well known in solid-state physics [455]. This can be seen upon observing at first that, in the absence of the trap ($\Omega = 0$), the uniform solution to Eq. (2.271) is $u_0(t) = A_0 \exp[iA_0^2 \int_0^t a(s)ds]$. Then, for the MI analysis (and rescaling, for simplicity, the amplitude to $A_0 \equiv 1$), a solution including an infinitesimal perturbation eigenmode is sought for as $u = u_0[1 + \epsilon w(t)\cos(kx)]$, where ϵ and k are the amplitude and wavenumber of the perturbation. This way, we obtain the following equation for $w_1 \equiv \mathrm{Re}(w)$:

$$\ddot{w}_1 = -k^2\left[k^2/4 + a(t)\right]w_1. \tag{2.272}$$

For a periodic function $a(t)$, Eq. (2.272) is Hill's equation [456] [if $a(t)$ is a harmonic function, Eq. (2.272) is the Mathieu equation]. For the proposed FRM scheme, the piecewise constant form of the function $a(t)$ allows an explicit solution, similar to that in the Kronig–Penney model [457]. In this case, Eq. (2.272) describes a harmonic oscillator with frequencies (which may be imaginary) $\omega_1 = \sqrt{k^2(k^2/4 + a_1)}$ for $0 < t < \tau$ and $\omega_2 = \sqrt{k^2(k^2/4 + a_2)}$ for $\tau \leq t \leq T$. The resulting solutions, taking into regard the continuity of the function w_1 and its derivative, lead to a 4×4 determinant condition (cf. Refs. [454, 453] for details) for the linear stability eigenfrequency ω; then, imaginary or complex values of ω give rise to MI. The final result, which was also confirmed in the experiment of Ref. [453],[16] is that the periodic nonlinearity leads to multiple instability bands, separated by stable regions, corresponding to the gaps and bands of the Kronig-Penney model, respectively [454, 453]. This result should be contrasted with the one corresponding to a uniform medium: indeed, thanks to FRM, there are multiple instability regions rather than just one as in a uniform medium [36].

As is expected, the MI in the nonlinearity managed system is responsible for the generation of strongly localized structures, to the consideration of which we now proceed. The most natural structure that is expected to be formed under the action of FRM, and in particular for small a_{dc}, is the one oscillating between the ground states that would exist at positive and negative constant values of a. In the former (repulsive) case, this state is well approximated by the TF wavefunction

$$u_{\mathrm{TF}} = \max\left\{\sqrt{\frac{2\mu-(\Omega x)^2}{2a}},0\right\}\exp(-i\mu t),$$

where μ is the chemical potential. In the latter (attractive) case, the Gaussian, i.e., the ground state of the quantum harmonic oscillator for small μ (while the solitonic waveform is relevant for large μ), is a natural shape (unless a is too large). Notice that the fundamental difference of this dynamical regime from its fiber-optic counterpart (the dispersion management soliton) is due to the presence of the confining potential, which supports the TF state.

The numerical simulations of Ref. [271] corroborate this assumption: starting with the initial TF state, prepared for $a(0) = 1$, simulations of Eq. (2.271) reveal persistent oscillations between the TF and Gaussian configurations, as shown in Fig. 2.49. For small values of a_1 and $|a_2|$ the oscillations are periodic; nevertheless, as a_1 and $-a_2$ increase and take values of order 1, more frequencies come into play and the oscillations become chaotic.

An overall description of the dynamics for the FRM-driven condensate is provided by a phase diagram which is displayed in the (a_{dc}, a_{ac}) plane in Fig. 2.50(a). The "breathing" state is stable beneath the solid curve, oscillating between the TF and Gaussian states quasiperiodically (area I) or chaotically (area II). In area III, the breather is naturally closer to the TF state, since, in this case $a_{dc} > a_{ac}$, i.e., the interaction is always repulsive. Finally, in area IV, the breathing soliton strongly resembles a well-known two-soliton solution of the underlying self-focusing NLS equation, which, however, is unstable in the NLS equation proper [36]; in accordance with that, in the absence of the trapping potential ($\Omega = 0$), this breather demonstrates an altogether different behavior: it emits radiation, losing $\approx 2\%$ of atoms and reorganizes itself into an ordinary fundamental NLS soliton shown in the inset of Fig. 2.50(b).

Another type of the FRM-controlled dynamical state can be obtained by embedding a dark soliton into the localized condensate, thus creating a dip in the center of the breather. This can be done upon using an initial condition of the form $u = u_{\mathrm{TF}} \tanh x$. Then, another stable state emerges, featuring drastically different dynamics: the central part of the condensate including the dark soliton remains virtually static even when $a(t)$ takes negative values, as can be seen in Fig. 2.51, and only "wings" oscillate quasiperiodically between states with smaller and larger curvature (which may be regarded, respectively, as remnants of the former TF and Gaussian states). This soliton-like state may be viewed

[16] In this work, optical beams propagated through alternating layers of glass and air, which form a multilayer "Kerr sandwich" [as each medium by itself had a different cubic (Kerr) nonlinearity].

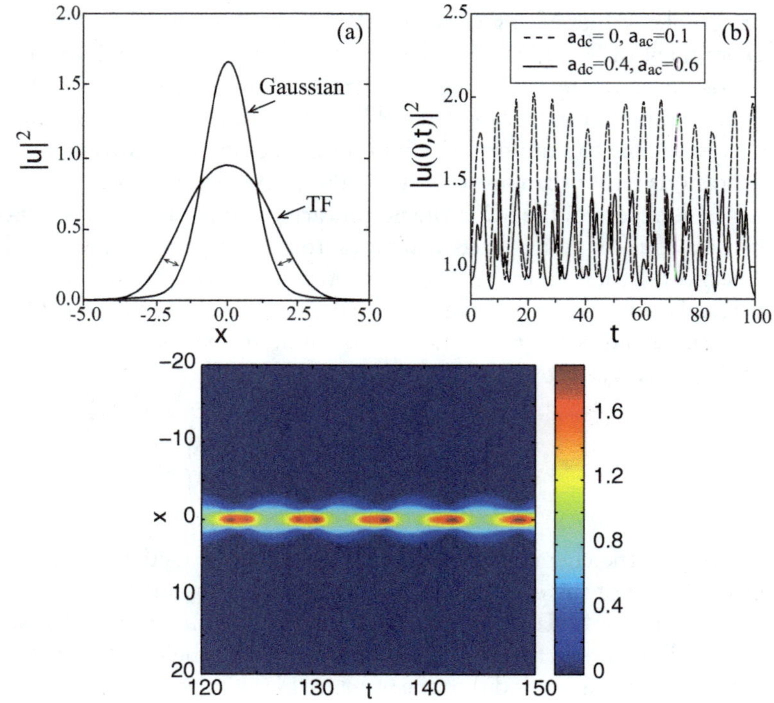

Figure 2.49. *Top panel:* (a) *Stable oscillations of a breather between the TF and Gaussian shapes, in the case* $T = 2$, $\tau = 1$, $a_{dc} = 0$, $a_{ac} = 0.1$. (b) *Time evolution of the field's amplitude for* $a_{dc} = 0$, $a_{ac} = 0.1$ *(dotted line), and* $a_{dc} = 0.4$, $a_{ac} = 0.6$ *(solid line). In all the cases,* $\Omega^2 = 0.25$. *In the former case, the oscillations are quasiperiodic, while in the latter case they are chaotic. Bottom panel: Spatio-temporal contour plot of the density,* $|u(x,t)|^2$, *for typical stable breather oscillations. Adapted with permission from Ref. [271].*

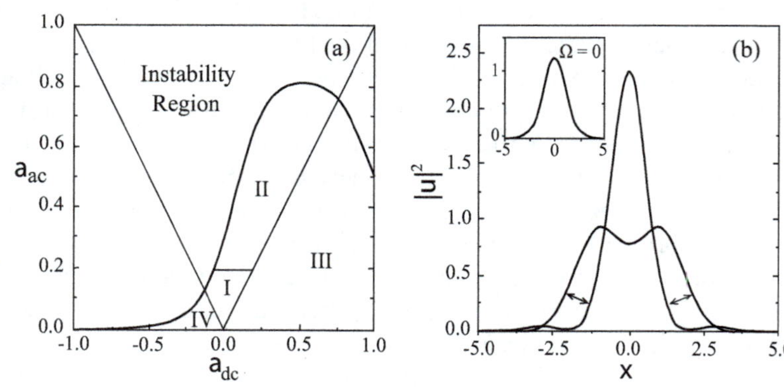

Figure 2.50. (a) *The phase diagram for dynamical states of the FRM-driven condensate for* $\tau = T/2 = 1$. (b) *Shape oscillations of the stable two-soliton state at* $a_{dc} = -0.4$ *and* $a_{ac} = 0.015$. *The inset shows the corresponding fundamental soliton stable state obtained with* $\Omega = 0$. *Adapted with permission from Ref. [271].*

as a nonlinear counterpart of the first excited state of the quantum harmonic oscillator [51] (see also next subsection below), and it also has its optical dispersion management counterpart in the form of an odd soliton [458]. However, there is a significant difference: as the bottom panel of Fig. 2.51 shows, the dark soliton-carrying condensate driven by

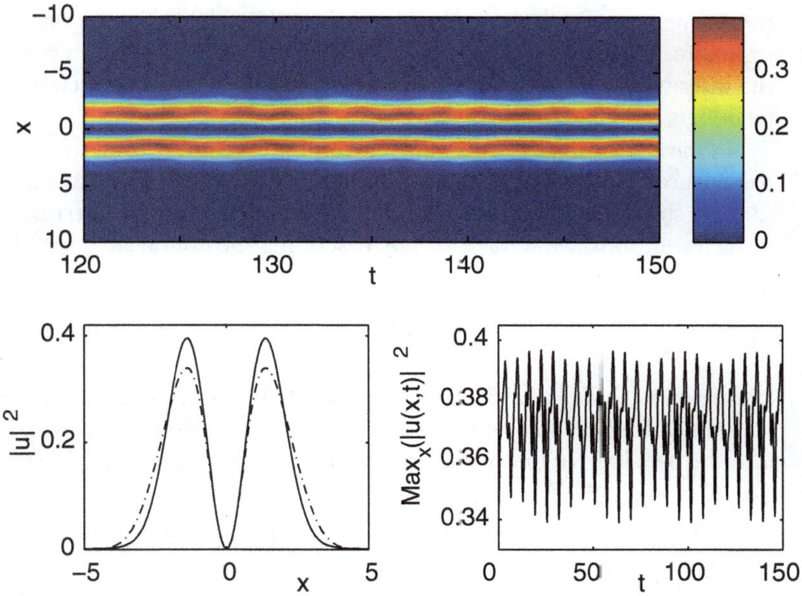

Figure 2.51. *Top panel: A dark soliton nested in the condensate in the FRM case with* $T = 2$, $\tau = 1$, $a_{\rm dc} = 0$, $a_{\rm ac} = 0.1$. *The top subplot shows the evolution of the density, the bottom-left one displays the density profiles at* $t = 139.6$ *and* $t = 142.8$ *(dashed dotted and solid lines, respectively), and the bottom right subplot shows the density maximum vs. time.*

FRM is *very* robust against perturbations, including even ones. On the contrary, the dispersion management odd soliton is indeed unstable against even perturbations [458].

We conclude this section by noting the following: in the case of the so-called strong nonlinearity management (similarly to the case of strong dispersion management [451]), i.e., in the case of $a(t) = \gamma_0 + (1/\varepsilon)\gamma(t/\varepsilon)$, where $\gamma_0 = \text{const.}$, ϵ is a small parameter, and $\gamma(t/\varepsilon)$ denotes the "dc" and rapidly oscillating "ac" parts of the time-dependent nonlinearity coefficient, it is possible to perform an averaging of the GPE model proper and derive an effective homogeneous equation. As was shown in Refs. [439, 440] for a harmonic γ, the averaged equation captures accurately the dynamics of the system and the shapes of the breathers presented above.

Linear Nonlinearity Management

We now consider another type of time-dependent nonlinearity, namely one described by a time-dependent nonlinearity $g(t)$. In particular, following the analysis of Ref. [52], our model is the following GPE:

$$i\psi_t = -\frac{1}{2}\psi_{xx} + g(t)|\psi|^2\psi + \frac{1}{2}\Omega^2 x^2 \psi, \tag{2.273}$$

with the wavefunction is normalized to unity, i.e., $||\psi||_2^2 = (\psi, \psi) \equiv \int \psi\psi^* dx = 1$ (for simplicity, below the integration limits are suppressed whenever the integration is over the whole real axis). Let τ be the characteristic temporal scale of $g(t)$, i.e., $|g_t|/|g| \sim 1/\tau$. Another characteristic time of the system is the one imposed by the trap potential through the linear oscillator period $2\pi/\Omega$. We will examine the competition between these time scales by showing characteristic cases with rapidly varying nonlinearity $\Omega\tau \ll \pi$ (in dimensionless units) and adiabatically varying nonlinearity $\nu\tau \gg \pi$. Notice that for the intermediate case the results are essentially interpolating between the ones that will be presented below for these two cases [52].

In the case of adiabatically varying nonlinearity it is natural to assume that the system passes through a sequence of states which at each moment of time are "close" to the respective stationary solutions. Both for that reason and per the intrinsic physical relevance of such solutions, we start by briefly describing the nonlinear eigenvalue problem

$$-u_{n,xx} + 2gu_n^3 + \Omega^2 x^2 u_n = \mu_n u_n, \tag{2.274}$$

associated with Eq. (2.273) for $\psi(x,t) = u(x)\exp(-i\mu t/2)$ (here g is considered as a constant). In particular, when the nonlinearity is zero ($g = 0$), Eq. (2.274) transforms into the well-known linear quantum harmonic oscillator eigenvalue problem

$$\mathscr{L} v_n \equiv -v_{n,xx} + \Omega^2 x^2 v_n = \mathscr{E}_n v_n, \tag{2.275}$$

with eigenvalues $\mathscr{E}_n = \Omega(2n+1)$ and it is assumed that $(u_n, u_n) = (v_n, v_n) = 1$.

One can now argue (based on continuity intuition) that for small enough $|g|$, the spectrum of the nonlinear eigenvalue problem (2.274) is located in some vicinity of the spectrum $\mathscr{E}_n$. Indeed, let us assume that $u_n(x)$ is an eigenfunction of (2.274) and that $\mathscr{E}_n$ is the closest eigenvalue of $\mathscr{L}$ to μ_n. Then, one can consider an auxiliary linear operator $\mathscr{L}_{g,n} = \mathscr{L} + 2gu_n^2(x)$ where $2gu_n^2(x)$ is considered as a perturbation. Clearly $\mathscr{L}_{g,n}$ is self-adjoint, $\mathscr{L}_{0,n} = \mathscr{L}$, and $\mathscr{L}_{g,n}u_n = \mu_n u_n$ (μ_n being a nondegenerate eigenvalue of the auxiliary eigenvalue problem). Then the spectrum of $\mathscr{L}_{g,n}$ belonging to some vicinity of $\mathscr{E}_n$ belongs asymptotically (when $g \to 0$) to the interval $(\mathscr{E}_n + g\mathscr{E}_n^{(1)} - o(g), \mathscr{E}_n + g\mathscr{E}_n^{(1)} + o(g))$ where $\mathscr{E}_n^{(1)} = (v_n, 2u_n^2 v_n)$ (see, e.g., [459, Theorem XII.23]).

Next, from Eqs. (2.274) and (2.275) we obtain $(\mathscr{E}_n - \mu_n)(v_n, u_n) = -2g(v_n, u_n^3)$. Thus, in the limit $g \to 0$,

$$\mu_n - \mathscr{E}_n = 2g(v_n, u_n^2 v_n) + o(g) = 2g(v_n^2, u_n^2) + o(g),$$

and we have $(v_n, u_n^3) = (v_n, u_n)(v_n^2, u_n^2) + o(g)$. Since (v_n^2, u_n^2) is positive definite, we conclude that in the general situation, when (v_n, u_n) does not go to zero when g goes to zero, $(v_n, u_n^3)\cdot(v_n, u_n) > 0$. The last results mean that the shift of the eigenvalues μ_n with respect to the eigenvalues $\mathscr{E}_n$ is defined by the type of the interatomic interactions: $\mathrm{sgn}(\mu_n - \mathscr{E}_n) = \mathrm{sgn}(g)$. Notice that for the ground states (having no nodes) one does not need to analyze the limit of small g but rather simply take into account that $(v_0, u_0^3)(v_0, u_0) > 0$. For the ground state and positive scattering length another proof of the same result can be found in [460].

The main conclusion of this analysis is the *existence* of a branch with "chemical potential" (nonlinear eigenvalue) larger than that of the linear problem in the repulsive interaction case. On the other hand, in the attractive case, the presence of the nonlinearity induces the opposite behavior (the bifurcation will occur with a smaller eigenvalue than that of the linear problem).

Subject to the condition of the conserved norm (number of particles) we can describe the transitions we are interested in, and we start with the background state $n = 0$. This state can be approximated in the different limits of g by the following known formulas:

$$u_0 \to \frac{1}{\sqrt{g}}\sqrt{\frac{(6g\Omega)^{2/3}}{4} - \Omega^2 x^2}, \qquad |x| < \mu_0, \quad \mu_0 \to (6g\Omega)^{2/3}/4, \tag{2.276}$$

$$u_0 = \frac{\Omega^{1/4}}{\pi^{1/4}} e^{-\Omega x^2/2}, \qquad \mu_0 = \mathcal{E}_0 = \Omega, \tag{2.277}$$

$$u_0 \to \frac{|g|^{1/2}}{2}\frac{1}{\cosh(|g|x/2)}, \qquad \mu_0 \to -\frac{g^2}{4}, \tag{2.278}$$

representing respectively a Thomas–Fermi approximation, the zero level of the linear oscillator, and approximate one-soliton solution when the soliton width is much less than the linear oscillator length (for positive, zero, and negative g, respectively). Thus, considering now the (temporal) variation of g from the linear limit of $g = 0$ to $g > 0$ and $g < 0$, one can interpret (2.276) and (2.278) as mapping the zero level of the linear oscillator (2.277) → Thomas–Fermi distribution (2.276) and the zero level of the linear oscillator (2.277) → bright soliton (2.278), respectively, providing one-to-one correspondence among these objects.

By analogy one can consider the first level which can be approximated in the different limits of g by the following formulas:

$$u_1 \to \sqrt{\frac{\mu_1 - \Omega^2 x^2}{2g}}\tanh\left(\sqrt{\frac{\mu_1}{2}}x\right), \qquad |x| < \frac{\sqrt{\mu_1}}{\Omega}, \quad \mu_1 \to (3g\Omega)^{2/3}, \tag{2.279}$$

$$u_1 = \frac{\sqrt{2}\Omega^{3/4}x}{\pi^{1/4}} e^{-\Omega x^2/2}, \qquad \mu_1 = \mathcal{E}_1 = 3\Omega, \tag{2.280}$$

$$u_1 \to \frac{|g|^{1/2}}{4}\left(\frac{1}{\cosh(|g|(x-L)/4)} - \frac{1}{\cosh(|g|(x+L)/4)}\right), \qquad \mu_1 \to -\frac{g^2}{16}. \tag{2.281}$$

Equation (2.279) represents a dark soliton on top of a background, providing a highly accurate description of the problem, as was justified in Ref. [173]. Equation (2.280) is the first excited state of the linear oscillator, while Eq. (2.281) is a two-soliton state, with the solitons having a π difference in their phases. As elaborated in the previous sections, such solitons repel each other and a stable state is possible due to the action of the parabolic trap acting in the opposite direction. It is clear that the equilibrium distance $2L$ between the solitons depends on the value of the nonlinearity. More specifically L grows with $|g|$ (see, e.g., Ref. [369] for details).

Once again here, the variation of the nonlinearity coefficient g from the linear limit provides a mapping of the first excited level of the quantum harmonic oscillator to the dark soliton (in a nonuniform background) in the repulsive case and in a two-soliton concatenation (the so-called twisted mode [369]) in the attractive case.

We have conjectured above that under adiabatic change of the nonlinearity, the system passes through the respective stationary states. This conjecture can be proved to be true upon examining the short-time behavior of the evolution problem: indeed, in Ref. [52], it was shown that if at $t = 0$ the nonlinearity coefficient is $g(t = 0) = g_1$ and the system is in the stationary state $\psi_1(x,t) = e^{i\mu_1 t}u_1(x)$, then the time divergence of the solution from this stationary state is small (for short times) in some appropriate sense. Hence, for adiabatic variation of the nonlinearity, the solution is guaranteed to stay close to its linear sibling for short times. As far as the longer term behavior is concerned, it was studied numerically in Ref. [52]: in this work, the numerical experiments illustrated the robustness of states with $n = 0$ and $n = 1$ nodes and the progressive destabilization of higher order excited states (which, however, may still be observable in physical experiments).

2.6 ▪ Dissipative Effects: Finite-Temperature Condensates

So far, we have considered the stability and dynamics of matter-wave dark solitons at zero temperature, $T = 0$. Nevertheless, as experiments are obviously performed at finite temperatures, it is relevant to consider the dissipative dynamics of ultracold quantum gases (see Ref. [461] for a recent review on this topic), and more specifically the instability of dark solitons induced by the thermal excitations that naturally occur. This problem was first addressed in Ref. [126], where a kinetic-equation approach, together with a study of the BdG equations, was used. In this work, it was found that the dark soliton center obeys an equation of motion which includes an *antidamping* term [similarly to Eq. (2.83)], which is nonzero (zero) for finite (zero) temperature. The behavior of the solutions of this equation of motion incorporating the antidamping term can be used to explain, at least qualitatively, the soliton dynamics observed in experiments: solitons either decay fast (for high temperatures) [46, 47, 50] or perform oscillations (for low temperatures) [68, 69, 70, 72] of growing amplitude and eventually decay, so that the system finally relaxes to its ground state (see also discussion below for the role of the antidamping term).

Dark soliton dynamics in BECs at finite temperatures was also studied in other works by means of different approaches. In particular, in Ref. [209], the problem was treated in the framework of a mean-field model, namely the so-called *dissipative GPE* (DGPE)—see below; this equation incorporates a damping term, first introduced phenomenologically [462] and later justified from a microscopic perspective (see, e.g., the review [463]). On the other hand, in Refs. [464, 465] the same problem was studied numerically, using coupled Gross–Pitaevskii and quantum Boltzmann equations, which include the mean-field coupling and particle exchange between the condensate and the thermal cloud; this treatment is based on the so-called Zaremba–Nikuni–Griffin (ZNG) approach [466, 467]. Furthermore, in Refs. [59, 468, 469], finite-temperature dynamics of dark solitons was studied by means of the so-called *stochastic GPE* (SGPE)—see, e.g., Refs. [470, 471, 472] and [461, 463] for reviews; on the other hand, in Ref. [473] quantum effects on dark solitons were additionally studied in the framework of the truncated Wigner approximation (see, e.g., Refs. [474, 475] for this approach); we also note that in the work [476], the dissipative dynamics of a dark soliton at temperatures T, lower than the chemical potential μ of the background Bose liquid, was studied.

In the works [468, 469], it was shown that for sufficiently low temperatures and certain parameter regimes, averaged dark soliton trajectories obtained by the stochastic GPE are in a very good agreement with results obtained by the dissipative GPE model. Thus, the results of Ref. [468, 469] indicate that the use of the dissipative GPE in studies of dark solitons in finite-temperature BECs (a) can reasonably be justified from a microscopic perspective, and (b) allows for an analytical description of the problem, by employing techniques exposed in Sec. 2.2.2, provided that the dynamics of the thermal cloud does not play the dominant role. For the above reasons, below we will focus on the SGPE and DGPE models and analyze the statics and dynamics of single- and multiple-dark-soliton states in these settings.

2.6.1 ▪ The SGPE and DGPE Models

Let us follow the analysis of Ref. [469] and consider a finite-temperature BEC confined in a highly elongated trapping potential. In this case, the low-energy modes of this effectively 1D system may be represented by the following stochastic differential equation [470, 471, 472], so-called stochastic GPE (SGPE):

$$i\hbar\frac{\partial\Psi(z,t)}{\partial t}=(1-i\gamma(z,t))\left[-\frac{\hbar^2}{2m}\frac{\partial^2}{\partial z^2}+V(z)+g|\Psi|^2-\mu\right]\Psi(z,t)+\eta(z,t), \tag{2.282}$$

where $\Psi(z,t)$ is a complex parameter describing the occupation of such low-lying modes, $V(z)=(1/2)m\omega_z^2z^2$ is the axial trapping potential, $g=2\hbar\omega_\perp a$ is the effective 1D interaction strength (with a being the s-wave scattering length), and $\eta(z,t)$ is a complex Gaussian noise term, with correlations given by the relation

$$\langle\eta^*(z,t)\eta(z',t')\rangle=2\hbar\gamma(z,t)k_BT\delta(z-z')\delta(t-t'), \tag{2.283}$$

where k_B is Boltzmann constant and T the BEC temperature. The strength of the noise, and damping, due to contact with higher energy thermal modes, is given by $\gamma(z,t)$.

While the soliton will reside in the low-lying modes, in arriving at Eq. (2.282) it is assumed that the high-energy thermal atoms in the system may be treated as though at equilibrium; this should be valid for small perturbations, e.g., when introducing a dark soliton into a large system (characterized by a Thomas–Fermi radius $R\gg\xi$, where the healing length $\xi=\hbar/\sqrt{mn_0g}$ sets the soliton width—with n_0 being the maximum condensate density) which would have little effect on the high-lying modes, which should instead remain close to equilibrium. In this picture, the high- and low-energy systems are assumed to be in diffusive and thermal equilibrium with a common temperature T and chemical potential μ. Moreover, due to the large mode occupations typical within degenerate Bose gases, we are well justified in the further assumption that the low-energy modes are sufficiently highly occupied for the classical field approximation to be valid (see, e.g., Ref. [471]).

Neglecting the noise term of Eq. (2.282) leads to the following dissipative GPE (DGPE) for the condensate wavefunction (see, e.g., the review of Ref. [463]):

$$i\hbar\frac{\partial\Psi(z,t)}{\partial t}=(1-i\gamma(z,t))\left[-\frac{\hbar^2}{2m}\frac{\partial^2}{\partial z^2}+V(z)+g|\Psi|^2-\mu\right]\Psi(z,t). \tag{2.284}$$

This is expected to be a good model when damping effects are dominant over diffusion, and representative of the mean-field soliton dynamics in this case. An advantage of deriving this model from the SGPE is that the damping may be calculated in an *ab initio* manner, via Eq. (2.285) as we will show below, rather than selected phenomenologically [477, 478].

Form of the Dissipation

According to the formulation given in Refs. [470, 471], the quantity parametrizing the strength of the noise and damping is the so-called Keldysh self-energy $\hbar\Sigma^K(z,t)$, which may be related to the dissipation term in Eq. (2.282) via $\gamma(z,t)=i\beta\hbar\Sigma^K(z,t)/4$ [471, 479]. We now follow the methods of Ref. [479] and assume that the system is close to equilibrium, so that the dissipation γ is for all practical purposes time-independent, i.e.,

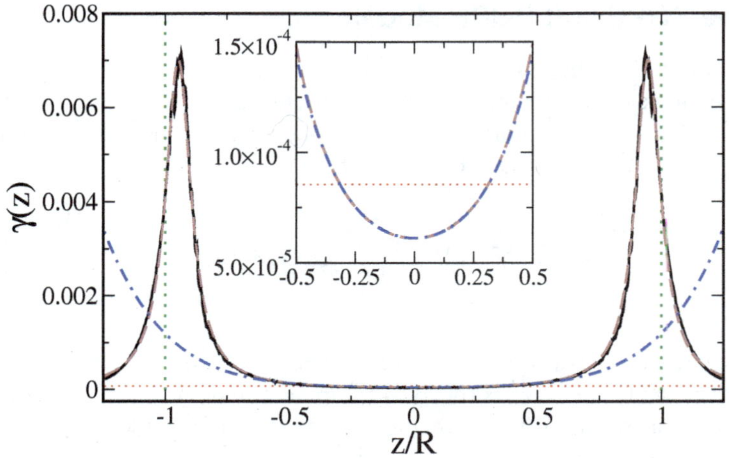

Figure 2.52. *Form of the equilibrium dissipation $\gamma(z)$ at $T = 150$ nK as found by Eq. (2.285) (solid black line), as fitted by Eq. (2.287) (dashed brown line), and as approximated by Eq. (2.288) (dot-dashed blue line), for $T = 150$ nK. The dotted red curve depicts the mean value of $\gamma(z)$, while the vertical dotted green lines indicate the points $z = \pm R$, which define the rims of the TF cloud. The inset shows a magnification of this figure closer to the trap center, i.e., for the spatial interval $[-R/2, R/2]$: there, the agreement between the fit to $\gamma(z)$ [Eq. (2.287); dashed brown line] and the approximate form of this [Eq. (2.288); dot-dashed blue line] is excellent, as the corresponding curves essentially coincide and become indistinguishable.*

$\gamma = \gamma(z)$; in such a case, the integral to be evaluated for the dissipation is

$$\gamma(z) = \frac{4\beta m a^2}{\pi \hbar^2} \int dE_2 \int dE_3 \; S(k_1, k_2, k_3)(1 + N_1) N_2 N_3, \tag{2.285}$$

where $0 < E_i < \infty$ and $N_i = [\exp(\beta(E_i + U(z) - \mu)) - 1]^{-1}$ is a Bose–Einstein distribution. Here, $U(z) = V(z) + 2g\langle|\Psi(z)|^2\rangle$, where $\langle\cdots\rangle$ denotes an ensemble average over many noise realizations, $E_1 = E_2 + E_3 + U(z) - \mu$, and

$$\begin{aligned} S(k_1, k_2, k_3) = \frac{1}{2}\Big[&\operatorname{sgn}(k_1 + k_2 - k_3) + \operatorname{sgn}(k_1 - k_2 + k_3) \\ &- \operatorname{sgn}(k_1 + k_2 + k_3) - \operatorname{sgn}(k_1 - k_2 - k_3)\Big], \end{aligned} \tag{2.286}$$

with $k_i = \sqrt{2mE_i/\hbar^2}$.

The damping $\gamma(z)$ may be evaluated numerically (e.g., using Gauss–Legendre quadrature), an example of which is shown in Fig. 2.52 for $T = 150$ nK (black solid line). A notable feature in the form of $\gamma(z)$ is the peak close to the Thomas–Fermi radius R, which indicates that the scattering rate is largest near the edge of the quasicondensate. It should be mentioned that the peaks in $\gamma(z)$ calculated using the SGPE density (black solid curve) are noticeably closer to the trap center than for the $T = 0$ Thomas–Fermi case at the same μ due to the effects of thermal depletion on the equilibrium density profile (see Fig. 1 in Ref. [469]). We also note that the shape of $\gamma(z)$ is qualitatively similar to the spatial form of the scattering rates found within numerical implementations of the ZNG approach [466, 467].

Importantly, the above form of dissipation $\gamma(z)$ can be characterized by means of a simple analytical fitting function. In fact, $\gamma(z)$ can be well approximated for temperatures ranging from $T = 125$ nK to $T = 300$ nK by the function

$$\gamma(z) = \frac{a}{(z+c)^2+d} + \frac{a}{(z-c)^2+d}, \tag{2.287}$$

which is the sum of two Lorentzians, centered near the edge of the quasicondensate. The parameters a, c, and d depend in general on temperature, T, the trapping frequencies, chemical potential, and atomic species (although all except T remain fixed in our analysis). Various values of these parameters, indicating their temperature dependence, can be found in Table I of Ref. [469]; for instance, we note that for $T = 150$ nK, we find that $a = 0.0219$, $c = 26.63$, and $d = 3.128$.

Based on the *ab initio* found form of $\gamma(z)$ [cf. Eq. (2.285)], and also on its approximate form [cf. Eq. (2.287)], it is now relevant to Taylor expand the function $\gamma(z)$ around the trap center ($z = 0$). This expansion reads

$$\gamma(z) \approx \gamma_0 + \gamma_2 z^2 + \gamma_4 z^4 + \gamma_6 z^6, \tag{2.288}$$

where the constant coefficients are given by

$$\begin{aligned}
\gamma_0 &= 2a/(c^2+d), \\
\gamma_2 &= 2a(3c^2-d)/(c^2+d)^3, \\
\gamma_4 &= 2a(5c^4-10c^2d+d^2)/(c^2+d)^5, \\
\gamma_6 &= 2a(7c^6-35c^4d+21c^2d^2-d^3)/(c^2+d)^7.
\end{aligned} \tag{2.289}$$

Using, as an example, the values of a, c, and d corresponding to $T = 150$ nK (see above), in Fig. 2.52 we compare the fit to $\gamma(z)$ given by Eq. (2.287) with the approximate expansion of Eq. (2.288). More specifically, we focus on the spatial interval of $[-R/2, R/2]$ (recall that R is the TF radius)—see inset of Fig. 2.52: this interval is particularly relevant for our analytical and numerical considerations (see below), as we are mainly concerned with solitons moving close to the trap center. As seen in the figure, in this regime, the approximate expression is almost identical to the more accurate one, thus justifying the degree of approximation used in Eq. (2.288). Here, it should also be noticed that, to the leading-order of approximation, and in the vicinity of the center of the trap, the dissipation function can also be assumed to be constant, i.e., $\gamma(z) \approx \gamma_C$, with $\gamma_0 \approx \int \gamma(z)dz/R$ over the interval $[-R/2, R/2]$; in such a case, γ_0 is found to scale with temperature as T^α, with $1/2 < \alpha < 4$ [468].

The above results suggest that, for sufficiently low temperatures, the *ab initio* calculation of the loss $\gamma(z)$ of the SGPE model can be used in the DGPE model. As we will show below, the latter can be treated analytically to provide the finite-temperature induced dissipation of dark solitons, with the results being in good agreement with full stochastic calculations in the framework of the SGPE [468, 469].

2.6.2 ▪ Dark Solitons in Single-Component Systems

Let us now focus on the dissipative GPE model (2.284). Introducing the transformation $t \to [1+\gamma^2(z)]t$, and taking into regard that in the considered parameter regime $d\gamma^2/dz \approx 0$, we can express Eq. (2.284) in the following dimensionless form [468, 469]:

$$[i-\gamma(z)]\partial_t \psi = \left[\frac{1}{2}\partial_z^2 + V(z) + |\psi|^2 - \mu\right]\psi, \tag{2.290}$$

where units are the same to the ones used for Eq. (2.1). Here, the external potential is assumed to be parabolic, i.e., $V(z)=(1/2)\Omega^2 z^2$ (where $\Omega=\omega_z/\omega_\perp \ll 1$ is the normalized trap strength), and the function $\gamma(z)$, which accounts for the dissipation, takes the form given in Eq. (2.288), as per the discussion in the previous section.

Below, we will present analytical results on the dynamics of dark matter-wave solitons that can be derived in the framework of Eq. (2.290). Our analysis will be based on the Hamiltonian approach of the perturbation theory for dark solitons (see Sec. 2.2.2). In particular, we will first discuss the relaxation dynamics to the ground state of the system and, then we will employ the formal perturbation theory for dark solitons.

Relaxation Dynamics to the Ground State of the System

As a first step of our approach, we will determine the evolution of the background carrying the soliton, namely the effective TF wavefunction. This is expected to be time-dependent due to the presence of the dissipative term $-\gamma(z)\partial_t\psi$ in Eq. (2.290). For simplicity, we will consider the case $\gamma=$ const.; in this case, we seek solutions of Eq. (2.290) in the form $\psi(z,t)=\psi_b(z,t)e^{-i\theta(t)}$, where the real functions $\psi_b(z,t)$ and $\theta(t)$ denote the profiles of the background wavefunction and phase, respectively. Then Eq. (2.290) is reduced to the following system:

$$\partial_t\psi_b+\gamma\psi_b\partial_t\theta=0, \tag{2.291}$$

$$\partial_t\psi_b+\frac{1}{2}\partial_z^2\psi_b-\psi_b^3+\mu\psi_b-\gamma\partial_t\psi_b-V(z)\psi_b=0. \tag{2.292}$$

In the case of a homogeneous BEC [i.e., $V(z)=0$], the above system admits a simple solution of the form $\psi_b(z,t)=\psi_0(t)$, which can be found as follows: using Eq. (2.291), and assuming that the parameter γ is sufficiently small (such that $\gamma^2\ll 1$), we put Eq. (2.292) in the form $d\psi_0/dt=\gamma\psi_0(\mu-\psi_0^2)$. This can readily be integrated, yielding the background density,

$$\psi_0^2(t)=\frac{\mu\psi_0^2(0)\exp(2\gamma\mu t)}{\mu-\psi_0^2(0)[1-\exp(2\gamma\mu t)]}, \tag{2.293}$$

while the phase is determined by means of Eq. (2.291) as

$$\partial_t\theta=-\mu+\psi_0^2(t). \tag{2.294}$$

Equations (2.293) and (2.294) describe analytically the evolution of a relaxation process: at $t=0$ the homogeneous condensate is described by the constant density $\psi_0^2(0)$ and a phase derivative $\partial_t\theta=-\mu+\psi_0^2(0)$, while in the limit $t\to\infty$ the constant density relaxes to a value equal to the chemical potential μ, while the time derivative of the phase becomes $\partial_t\theta=0$.

In the case of a confined BEC [i.e., $V(z)\neq 0$], it is clear that the BEC wavefunction will be inhomogeneous due to the presence of the trap. Taking into account that the maximum value of the wavefunction, occurring at $z=0$ [i.e., when $V(z=0)=0$], takes the value $\max\{\psi_b(z,t)\}=\psi_0(z=0,t)$, then it can readily be found that in the TF limit the background wavefunction has the form

$$\psi(z,t)=\max\left\{\sqrt{\psi_0(t)-V(z)},0\right\}\exp[-i\theta(t)], \tag{2.295}$$

where the functions $\psi_0(t)$ and $\theta(t)$ are respectively given by Eqs. (2.293) and (2.294). Similarly to the case of a homogeneous BEC, Eq. (2.295) provides the evolution of the

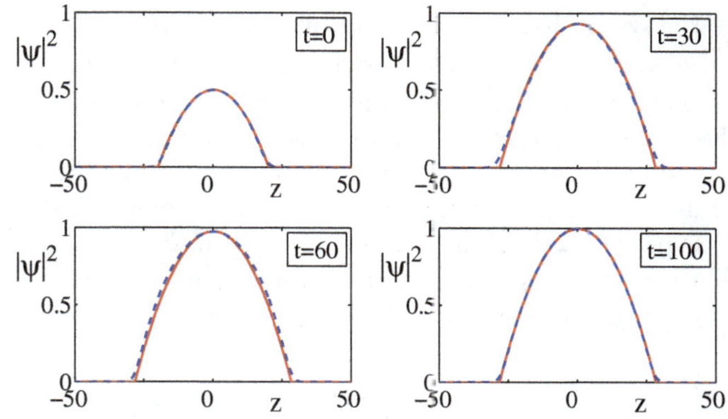

Figure 2.53. *Evolution of the density of a TF cloud characterized by an initial density at the trap center* $\psi_0^2(0) = 0.5$ *as obtained numerically (dashed blue line) and analytically (solid red line) in the framework of the GPE (2.290) with parameters* $\mu = 1$, $\gamma = 0.1$, *and* $\Omega = 0.05$. *Adapted with permission from Ref. [469].*

condensate wavefunction towards equilibrium. Notice that in the limiting case of $t \to \infty$, the condensate wavefunction (2.295) is reduced to the well-known form

$$\psi = \max\left\{\sqrt{\mu - V(z)}, 0\right\} \exp(-i\mu t).$$

Thus, Eq. (2.295) provides an approximate expression describing the relaxation to equilibrium of a TF cloud.

The above analytical results are compared to direct numerical simulations. In that respect, Fig. 2.53 shows the evolution of the density of a TF cloud characterized by an initial density at the trap center $\psi_0^2(0) = 0.5$ as obtained numerically (blue line) and analytically (red line) in the framework of the GPE (2.290) with parameters $\mu = 1$, $\gamma = 0.1$, and $\Omega = 0.05$. It is observed that both the evolution and the final spatial profile of the effective TF cloud in Eq. (2.295) (see the bottom right panel, at $t = 100$, where the system has reached equilibrium) are in excellent agreement with the respective numerical results.

It is clear that the above analysis leads to qualitatively similar results even in the case where γ is z-dependent, as long as $\gamma(z)$ is slowly varying with respect to z, which is indeed the case—at least close to the trap center, as indicated by the analysis of the previous section. Thus, below, we will use the analytical form of the TF cloud, and derive from Eq. (2.290) a perturbed NLS equation for the dark soliton wavefunction. The latter, will provide analytical results for the temperature-induced dissipative dynamics of dark solitons.

Perturbation Theory for Dark Solitons of the DGPE

We assume that the condensate dynamics involves a fast scale of relaxation of the background to the ground state, while the dark soliton subsequently evolves on top of the relaxed ground state. In particular, we seek for a solution of Eq. (2.290) in the form

$$\psi(z,t) = \psi_b(z,t)e^{-i\theta(t)}\upsilon(z,t), \tag{2.296}$$

where $\psi_b(z,t)$ and $\theta(t)$ obey Eqs. (2.291) and (2.292) and the unknown complex function $\upsilon(z,t)$ represents a dark soliton, which is governed by the following effective equation:

$$i\partial_t\upsilon + \frac{1}{2}\partial_z^2\upsilon - \psi_b^2(t)(|\upsilon|^2 - 1)\upsilon = -\frac{\partial_z\psi_b}{\psi_b}\partial_z\upsilon + \gamma\partial_t\upsilon. \tag{2.297}$$

When the system has reached equilibrium ($t \to \infty$), the nonlinearity coefficient $\psi_b^2(t)$ takes the asymptotic value $\psi_b^2 = \mu$ and Eq. (2.297) resembles a perturbed NLS model that has been used to describe the dynamics of dark matter-wave solitons in trapped BECs (see, e.g., Sec. 2.2.2). However, an important additional feature of this equation is the presence of the term $\gamma \partial_t \upsilon$ in the right-hand side which describes the temperature-induced dissipative dynamics of dark matter-wave solitons. Moreover, it should be noted that taking $\psi_b^2(t)$ as per Eq. (2.293), the effective NLS Eq. (2.297) can be also used to describe non-equilibrium dynamics of dark matter-wave solitons. In any case, at a first approximation, we use $\psi_b^2 = \mu$ and express Eq. (2.297) in the form

$$i\partial_t \upsilon + \frac{1}{2}\partial_z^2 \upsilon - (|\upsilon|^2 - 1)\upsilon = P(\upsilon), \tag{2.298}$$

where we have used the rescaling $t \to \mu t$ and $z \to \sqrt{\mu} z$, and the perturbation $P(\upsilon)$ has the form

$$P(\upsilon) = \frac{1}{2\mu}\left[2\left(1-|\upsilon|^2\right)V(z)\upsilon + \frac{dV}{dz}\partial_z \upsilon + 2\mu\gamma(z)\partial_t \upsilon\right]. \tag{2.299}$$

It is now possible to employ the Hamiltonian approach of the perturbation theory for dark solitons: following the methodology of Sec. 2.2.2, we consider a dark soliton of Eq. (2.298) of the form $\upsilon(z,t) = \cos\varphi \tanh Z + i \sin\varphi$ (where $Z = \cos\varphi\,[z - z_0(t)]$), whose center $z_0(t)$ and phase angle $\varphi(t)$ obeying the equations [see the respective Eqs. (2.65) and (2.66) in Sec. 2.2.2]

$$z_0(t) = \int_0^t \sin\varphi(t')dt', \tag{2.300}$$

$$\frac{d\varphi}{dt} = \frac{1}{2\cos^2\varphi \sin\varphi}\mathrm{Re}\left\{\int_{-\infty}^{+\infty} P(\upsilon)\partial_t \upsilon^* dz\right\}. \tag{2.301}$$

The integral in Eq. (2.301) involves three terms [cf. Eq. (2.299)]: the first two terms (accounting for the Hamiltonian part of the perturbation) can readily be evaluated upon expanding the potential $V(z)$ in Taylor series around the soliton's center, z_0; on the other hand, the third term (accounting for the dissipative part of the perturbation) can be evaluated by approximating the function $\gamma(z)$ by its Taylor expansion around the trap center; cf. Eq. (2.288). To this end, calculation of the integral in Eq. (2.301) leads to the following result:

$$\begin{aligned}\frac{d\varphi}{dt} = &-\frac{1}{2}\cos\varphi\frac{dV}{dz} + \gamma_0\frac{2}{3}\mu\cos\varphi\sin\varphi + \gamma_2\frac{\left(\pi^2-6\right)}{18}\mu\tan\varphi \\ &+ \gamma_4\frac{\pi^2\left(7\pi^2-60\right)}{360}\mu\tan\varphi\sec^2\varphi + \gamma_6\frac{\pi^4\left(31\pi^2-294\right)}{2016}\mu\tan\varphi\sec^4\varphi.\end{aligned}$$

Next, combining Eqs. (2.302) and (2.300), we obtain the following effective equation of motion for the dark soliton center,

$$\begin{aligned}\ddot{z}_0 = &\left[\gamma_0\frac{2}{3}\mu\dot{z}_0 - \left(\frac{\Omega}{\sqrt{2}}\right)^2 z_0\right]\left(1-\dot{z}_0^2\right) + \mu\dot{z}_0\left[\gamma_2\left(\frac{\pi^2-6}{18}\right)\right. \\ &\left. + \gamma_4\left(\frac{\pi^2(7\pi^2-60)}{360}\right)\left(1-\dot{z}_0^2\right)^{-1} + \gamma_6\left(\frac{\pi^4(31\pi^2-294)}{2016}\right)\left(1-\dot{z}_0^2\right)^{-2}\right].\end{aligned} \tag{2.302}$$

Notice that a variant of Eq. (2.302), corresponding to $\gamma_2 = \gamma_4 = \gamma_6 = 0$, was presented in Ref. [468], where the function $\gamma(z)$ was approximated by the constant value γ_0 (which is close to the mean value of $\gamma(z)$ in the interval $[-R/2, R/2]$; see Fig. 2.52). Nevertheless, here we are going to follow Ref. [469] and analyze the more general problem, taking into regard the spatially dependent profile of $\gamma(z)$.

Here, it should be mentioned that there appears to be a singularity in Eq. (2.302), corresponding to velocity values $\dot{z}_0 = 1$. This is also consistent with the analysis of Ref. [184], where it is shown that formal perturbation theory fails for extremely shallow dark solitons (with phase angles $\varphi \approx \pi/2$). In any case, in the physically relevant scenarios that we consider in our simulations below, solitons are observed to decay at the rims of the condensate at times smaller than the one needed for the soliton velocity to become $\dot{z}_0 = 1$.

Temperature-Dependent Damping Regimes

The equation of motion (2.302) is evidently nonlinear. Nevertheless, assuming that the dark soliton is close to a black one (i.e., φ is sufficiently small), Eq. (2.302) can be reduced to the following linearized form:

$$\ddot{z}_0 - \widetilde{\gamma}_0 \mu \, \dot{z}_0 + \left(\frac{\Omega}{\sqrt{2}}\right)^2 z_0 = 0, \tag{2.303}$$

where $\widetilde{\gamma}_0$ is given by

$$\widetilde{\gamma}_0 = \frac{2}{3}\gamma_0 + \frac{(\pi^2-6)}{18}\gamma_2 + \frac{\pi^2(7\pi^2-60)}{360}\gamma_4 + \frac{\pi^4(31\pi^2-294)}{2016}\gamma_6. \tag{2.304}$$

Equation (2.303) is similar to the equation of motion derived in Ref. [126] by means of a kinetic-equation approach. In the limiting case of zero temperature (i.e., for $\widetilde{\gamma}_0 = 0$), Eq. (2.303) recovers the result presented in Sec. 2.2.2, i.e., a dark soliton oscillates with constant amplitude and frequency $\Omega/\sqrt{2}$ in the harmonic trap $V(z) = (1/2)\Omega^2 z^2$. On the other hand, at finite temperatures (i.e., for $\widetilde{\gamma}_0 \neq 0$), the linearized equation of motion (2.303) additionally incorporates an antidamping term, $\propto -\dot{z}_0$ [with a coefficient that takes into account the spatial dependence of $\gamma(z)$]. Although it may sound counterintuitive, this antidamping term describes the dissipation of the dark soliton due to the interaction with the thermal cloud: in fact, this term results in the acceleration of the soliton towards the velocity of sound, i.e., the soliton becomes continuously grayer and, eventually, the soliton is expelled from the ground state of the condensate.

Clearly, the nature of the solutions of Eq. (2.303) depends on whether the roots of the auxiliary equation $s^2 - (2/3)\widetilde{\gamma}_0 \mu s + (\Omega/\sqrt{2})^2 = 0$ are real or complex. The roots are given by

$$s_{1,2} = \frac{1}{3}\widetilde{\gamma}_0 \mu \pm \left(\frac{\Omega}{\sqrt{2}}\right)\sqrt{\Delta}, \qquad \Delta = \left(\frac{\widetilde{\gamma}_0}{\gamma_{\rm cr}}\right)^2 - 1, \tag{2.305}$$

where $\gamma_{\rm cr} = (3/\mu)(\Omega/\sqrt{2})$, and the discriminant Δ determines the type of the motion. In the *supercritical* case of strong antidamping with $\Delta > 0$, i.e., for high temperatures such that $\widetilde{\gamma}_0 > \gamma_{\rm cr}$, the soliton trajectory is given by

$$z_0(t) = \frac{1}{s_2 - s_1}\left[(s_2 z_0(0) - \dot{z}_0(0))\exp(s_1 t) + (\dot{z}_0(0) - s_1 z_0(0))\exp(s_2 t)\right]. \tag{2.306}$$

In the *critical* case with $\Delta = 0$, i.e., $\widetilde{\gamma}_0 = \gamma_{\rm cr}$, the soliton trajectory is given as

$$z_0(t) = \left[z_0(0) + \left(\dot{z}_0(0) - \frac{1}{3}\widetilde{\gamma}_0 \mu z_0(0)\right)t\right]\exp\left(\frac{1}{3}\widetilde{\gamma}_0 \mu t\right). \tag{2.307}$$

Finally, in the *subcritical* case of weak antidamping with $\Delta < 0$, i.e., for sufficiently low temperatures such that $\tilde{\gamma}_0 < \gamma_{\rm cr}$, the soliton trajectory is given by

$$z_0(t) = \exp\left(\frac{1}{3}\tilde{\gamma}_0\mu t\right)\left[z_0(0)\cos(\omega_{\rm osc}t) + \omega_{\rm osc}^{-1}(\dot{z}_0(0) - \tilde{\gamma}_0\mu z_0(0))\sin(\omega_{\rm osc}t)\right], \quad (2.308)$$

where

$$\omega_{\rm osc} = \left(\frac{\Omega}{\sqrt{2}}\right)\left[1 - \left(\frac{\sqrt{2}}{3}\frac{\tilde{\gamma}_0\mu}{\Omega}\right)^2\right] \quad (2.309)$$

is the soliton oscillation frequency.

The above simple analysis shows that in the case of relatively high temperatures (with $\tilde{\gamma}_0 \geq \gamma_{\rm cr}$), the dark soliton will not "survive" long enough to oscillate in the trap, a result in qualitative agreement with experimental observations of dark matter-wave solitons in elongated Bose gases [46]. On the other hand, when the temperature is relatively low (i.e., $\tilde{\gamma}_0 < \gamma_{\rm cr}$), the dark soliton performs oscillations with an increasing amplitude and period—recall that the oscillation frequency is down-shifted as per Eq. (2.309) with respect to its value $\Omega/\sqrt{2}$, at zero temperature. These analytical results are in qualitative agreement with the numerical ones obtained in the framework of the ZNG model [464, 465].

Numerical Results and Comparison with the SGPE Model

We now compare analytical [Eqs. (2.302) and (2.304)] and numerical [Eqs. (2.290) with (2.287)] soliton trajectories, in each case making use of the approximate form for $\gamma(z)$ employed in the analytical approach. First, we consider Eq. (2.290), with $\gamma(z)$ given in Eq. (2.288), with an initial condition corresponding to a dark (black) soliton, initially placed off-center at $z_0(0) = 2$ [with zero initial velocity, $\dot{z}_0(0) = 0$], on top of a TF cloud, with $\mu = 1$, confined in a trap of strength $\Omega = 0.05$. The resulting soliton trajectories are shown in Fig. 2.54 for the supercritical ($\tilde{\gamma}/\gamma_{\rm cr} = 1.89$, top panel), critical ($\tilde{\gamma} = \gamma_{\rm cr} = 0.106$, middle panel), and subcritical ($\tilde{\gamma}/\gamma_{\rm cr} = 0.09$, bottom panel) cases. The black dashed lines in each plot depict the numerically obtained solutions of Eq. (2.302), while the white dashed lines show the corresponding analytical solutions of Eq. (2.303).

Generally, it is observed that the results based on the analytical approximations, i.e., the solutions of Eqs. (2.302) and (2.303)—which were derived by employing the approximate polynomial form of $\gamma(z)$ [cf. Eq. (2.288)]—are in good agreement with the numerical results obtained by the DGPE with the "exact" form of $\gamma(z)$ [cf. Eq. (2.287)]. Nevertheless, the solutions of the nonlinear equation of motion (2.302) are more accurate than the analytical solutions of Eq. (2.303) in capturing the soliton trajectories obtained by the DGPE. This behavior is more pronounced for longer times, where the soliton either decays (top two panels of Fig. 2.54) or performs large amplitude oscillations (bottom panel of Fig. 2.54): in fact, the solutions Eq. (2.302) are able to correctly predict the decay time of the solitons [which is underestimated by the solutions of Eq. (2.303)] in the supercritical and critical cases of strong dissipation, or follow quite accurately the DGPE trajectory in the subcritical case of weak dissipation. Notice that, in the latter case, the analytical solution of Eq. (2.303) underestimates (overestimates) the frequency (amplitude) of oscillation for longer times.

We now proceed with a systematic comparison between analytical approximations, focusing on the more accurate nonlinear equation of motion (2.302), and numerical (DGPE and SGPE) results. To ensure long soliton lifetimes for this comparison, we focus on one of the lower temperatures within our study, $T = 150$ nK, which nevertheless still

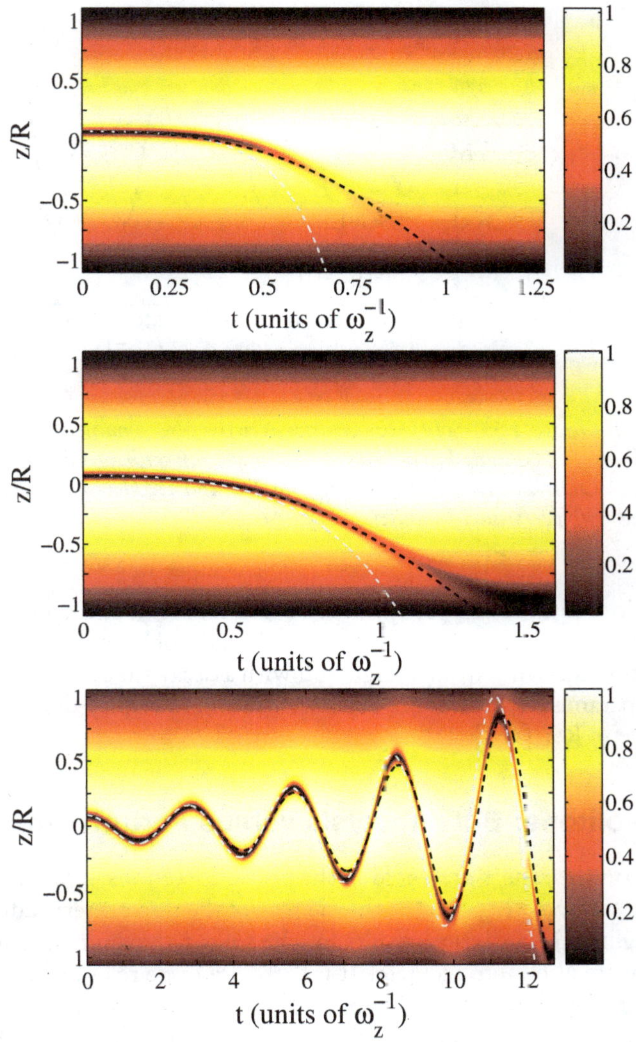

Figure 2.54. *Spatiotemporal evolution of the density of a BEC confined in an harmonic trap with $\Omega = 0.05$ and $\mu = 1$ [in the scaled units of Eq. (2.290)], with a dark (black) soliton initially placed at $z_0 = 2$ for the supercritical case with $\tilde{\gamma}_0/\gamma_{cr} = 1.89$ (top panel; $a = 84.44$, $c = 21.67$, and $d = 93.97$), critical case with $\tilde{\gamma}_0/\gamma_{cr} = 1$ (middle panel; $a = 43.67$, $c = 22.19$, and $d = 59.16$), and subcritical case with $\tilde{\gamma}_0/\gamma_{cr} = 0.09$ (bottom panel; $a = 4.53$, $c = 23.89$, and $d = 19.72$). Here $\gamma_{cr} = 0.106$. The white dashed lines across the soliton trajectories correspond to the analytical prediction of the linearized equation of motion Eq. (2.303), while the black dashed lines are obtained by numerically solving the nonlinear equation of motion Eq. (2.302). Adapted with permission from Ref. [469].*

corresponds to $k_B T = 0.8\mu$. We perform a direct numerical integration of the DGPE, Eq. (2.290) [with $\gamma(z)$ given by Eq. (2.287)] and compare this to respective results obtained via the analytic Eq. (2.302) and SGPE models.

In Fig. 2.55, we show the soliton trajectories found via the DGPE and the SGPE, as well the solutions of the nonlinear equation of motion (2.302) stemming from our analytical approximations. The initial condition takes the form of a dark soliton, initially placed at the trap center $z_0(0) = 0$ with initial velocity $\dot{z}_0(0) = 0.25$. It is clear that the numerical solution of the DGPE accurately captures that of the SGPE (similarly to the

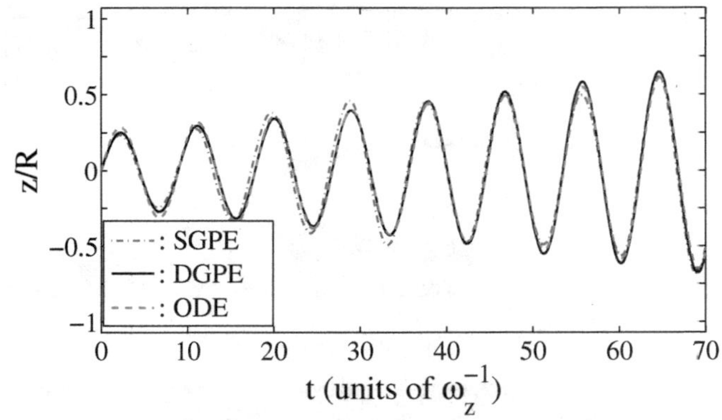

Figure 2.55. *Soliton trajectories as found by the numerical integration of the DGPE (solid black line), the SGPE (dashed-dotted red line; single run with a decay time from the mean bin), and the nonlinear ODE (2.302) (dashed green line), for $T = 150$ nK. Here $a = 0.0219$, $c = 26.63$, and $d = 3.128$. Adapted with permission from Ref. [469].*

behavior found in Ref. [468]), and also that the result of the analytical approximation is in very good agreement with the numerical results of both the DGPE and SGPE. The SGPE trajectory shown is from a single run with a decay time close to the ensemble average; for these parameters, such trajectories were found to display dynamics close to the ensemble average in Ref. [468].

2.6.3 ▪ Multisolitons: BdG Analysis versus Perturbation Theory Results

Based on the results presented in the previous sections, it is clear that the approximation of constant γ is very good in the vicinity of the trap center; additionally, in the same spatial region, the analytical results for the soliton dynamics obtained by the perturbation theory are also quite accurate. It is therefore quite relevant to compare the "eigenvalues associated with translation" provided by Eq. (2.305) with $\tilde{\gamma}_0 = \gamma =$ const., with the respective anomalous mode eigenvalues that can be found via the BdG analysis. Below, we will present such a study for both cases of single- and multisoliton states.

The One-Dark-Soliton State

First, we consider the one-dark-soliton state, for which we can determine the eigenvalue associated to the negative Krein signature eigenmode. This eigenmode can be derived as a solution of the eigenvalue problem [i.e., from the BdG equations arising from the GPE (2.290) for $\gamma(z) = \gamma =$ const.; cf. Sec. 2.2.3] stemming from a linearization around the stationary dark soliton, $u(x)$:

$$\psi(x,t) = u(x) + \epsilon[\exp(\lambda t)a(x) + \exp(\lambda^* t)b^*(x)], \tag{2.310}$$

where $\{\lambda \equiv \lambda_r + i\lambda_i, (a,b)^T\}$ is the eigenvalue-eigenvector pair, ϵ is a formal small parameter, and the asterisk denotes complex conjugate. Such a negative Krein sign eigenmode, namely the anomalous or negative-energy mode, is naturally expected in the case under consideration, because the dark soliton is actually an excited state of the condensate (albeit being dynamically stable for $\gamma = 0$). In fact, as discussed in Sec. 2.2.3, $\mathcal{N}$ such modes are

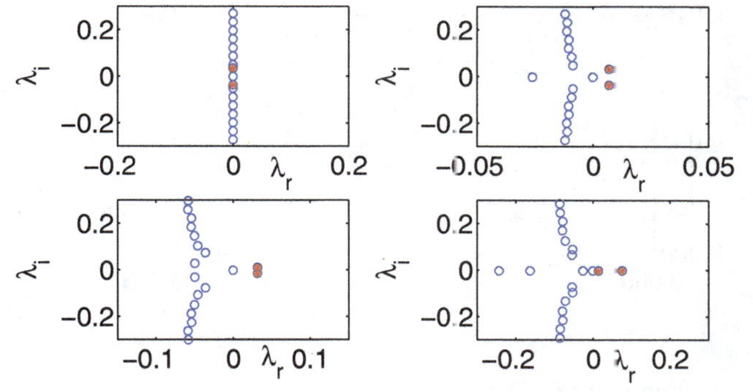

Figure 2.56. *Four spectral planes, corresponding to different values of γ, for the single-dark-soliton stationary state: in the top left panel $\gamma = 0$ (the zero-temperature Hamiltonian case), in the top right $\gamma = 0.02$, in the bottom left $\gamma = 0.1$, and in the bottom right $\gamma = 0.15$. The red stars highlight the negative Krein sign (in the Hamiltonian case) eigenvalues which lead to a Hopf bifurcation and an oscillatory instability of the dark soliton. Adapted with permission from Ref. [480].*

expected to exist for an $\mathcal{N}$-dark-soliton solution (i.e., a wavefunction with $\mathcal{N}$ nodes in its spatial profile). Here, it should be recalled that a remarkable feature of such eigenmodes (in comparison with the rest of the eigenmodes of positive Krein signature) is their behavior under the effect of the dissipative perturbation induced by the term proportional to γ. In particular, as was shown in Ref. [196], contrary to all other eigenmodes, negative Krein eigenmodes of the respective Hamiltonian case (i.e., $\gamma = 0$) bifurcate towards the right half plane of the excitation spectrum under the action of dissipation, while positive signature eigenmodes move to the left half-plane. This leads to *immediate* dynamical instability of the single- or multisoliton states. Furthermore, the more nodes (or dark solitons) the state has, the more unstable it will become, having $\mathcal{N}$ pairs of complex eigenvalues.

This situation is depicted in Fig 2.56, where the spectrum of linearization around a stationary single-dark-soliton state is shown for increasing values of γ (notice that such relatively large values of γ are used only for illustration purposes, since—as discussed above—only sufficiently small γ values are physically relevant). Clearly, as the parameter γ is increased, the departure from the Hamiltonian limit leads the eigenvalues with positive Krein signature to the left half-plane. On the other hand, the eigenvalue with negative Krein signature moves to the right half-plane with progressively decreasing imaginary part, until the relevant pair collides and splits along the real axis, giving rise to a purely exponential instability. Interestingly, this is mirrored in the frequency prediction of Eq. (2.305) (with $\tilde{\gamma}_0 = \gamma = \text{const.}$), since there exists a γ_{cr} above which the relevant pair of roots (eigenvalues) becomes purely real; recall that this was dubbed as a transition from an underdamped to an overdamped type setting.

It is now interesting to focus on a physically relevant value of γ, namely $\gamma = 0.00014$ (corresponding to the mean value $\int \gamma(z)dz/R$ over $[-R/2, R/2]$ for $T = 150$ nK; see Fig. 2.52) and compare the analytical result of Eq. (2.305) directly with the relevant (negative Krein sign) linearization eigenvalue of the full linear stability problem. This comparison is provided in Fig. 2.57, and the excellent agreement obtained clearly supports the explanation that this negative Krein signature eigenvalue stems from the breaking of translational invariance [196] (due to the presence of the trap) and is directly associated with the small-amplitude motion of the soliton, as also described from Eq. (2.304), for $\tilde{\gamma}_0 = (2/3)\gamma$. The above results (as well as the results that will be presented below) have been obtained for $\Omega = 0.05$ and $\mu = 1$.

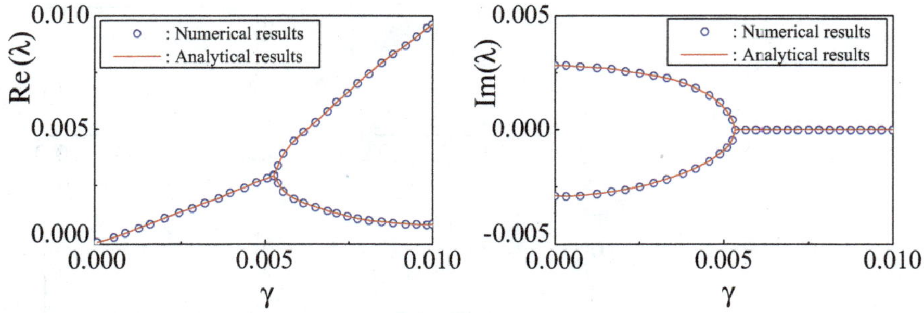

Figure 2.57. *Dependence of the real part (left; instability growth rate) and imaginary part (right; oscillation frequency) of the unstable eigenmode of the excitation spectrum on γ: analytical results (solid line) vs. DGPE numerics (circles), for $T = 150$ nK. For $\gamma < \gamma_{cr}$, the complex conjugate pair is responsible for the oscillatory growth dynamics, whereas beyond γ_{cr} the collision of the complex conjugate pair of eigenvalues creates a real pair (critical/supercritical exponential dynamics). Adapted with permission from Ref. [468].*

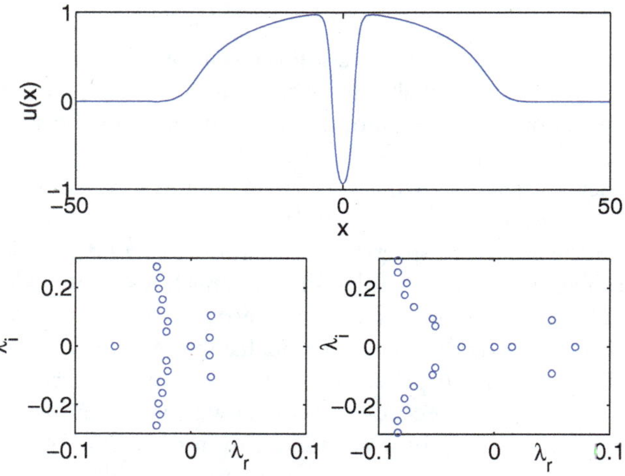

Figure 2.58. *The top panel shows the two-dark-soliton stationary state. Bottom panels show the pertinent linearization spectrum for $\gamma = 0.05$ (left) and $\gamma = 0.15$ (right). Adapted with permission from Ref. [480].*

The Two-Dark-Soliton State

We now proceed with the case of the two-dark-soliton solution, shown in the top panel of Fig. 2.58. In this case, as a natural generalization of the analysis for the single soliton, it is expected that for nonzero values of γ there will be *two* complex eigenvalue pairs. Both these pairs will eventually collide and lead to four real eigenvalues (and a purely exponential instability). The bottom panels of Fig. 2.58 illustrate the spectral plane for two values of γ, lending support to the above description.

It is also relevant to develop an analytical description for the relevant eigenvalue pairs, which are expected to be associated with the relative motion of the two dark solitons. To do so, we need to generalize the equation(s) of motion relevant to each of the solitons in order to include their pairwise interaction. This interaction was quantified in the Hamiltonian setting (corresponding to the zero-temperature limit of $\gamma = 0$) in Sec. 2.1.6, as stemming from a pair potential of the form $V(z_2 - z_1) = 1/\sinh^2(z_2 - z_1)$, where $z_{1,2}$ denote the two dark soliton centers. Given that we are interested in comparing eigenvalues of linearization (associated with small-amplitude motion around the equilibrium position and substantial separation between the two solitons), we will use the asymptotic (exponential) form of this interaction potential, that permits us to derive analytical expressions below. Thus, the two equations of motion for the dark solitons are, on the one hand, similar to Eqs. (2.104) and (2.105) as they incorporate the repulsive interaction potential and, on the other hand, they include the temperature-induced antidamping term—cf. Eq. (2.304):

$$\ddot{z}_1 = \frac{2}{3}\gamma \dot{z}_1 - \left(\frac{\Omega}{\sqrt{2}}\right)^2 z_1 - 8\exp(-(z_2 - z_1)), \tag{2.311}$$

$$\ddot{z}_2 = \frac{2}{3}\gamma \dot{z}_2 - \left(\frac{\Omega}{\sqrt{2}}\right)^2 z_2 + 8\exp(-(z_2 - z_1)). \tag{2.312}$$

The above system allows us to compute the equilibrium displacement between the two solitons as $\tilde{z} = z_1 = -z_2 = (1/4)W(64/\Omega^2)$, where $W(\eta)$ is Lambert's function defined as the inverse of $\eta(W) = We^W$. In the case of the two-soliton state, the distance $2\tilde{z}$ between the solitons is theoretically predicted to be 2×2.01, while numerically it is found to be 2×2.026, which once again attests to the accuracy of the relevant approximations. Considering now the linearization of the equations of motion (2.311) and (2.312) around this equilibrium distance yields for the relative displacements

$$\ddot{\eta}_1 = \frac{2}{3}\gamma \dot{\eta}_1 - \left(\frac{\Omega}{\sqrt{2}}\right)^2 \eta_1 + 16e^{-4\tilde{z}}(\eta_2 - \eta_1), \tag{2.313}$$

$$\ddot{\eta}_2 = \frac{2}{3}\gamma \dot{\eta}_2 - \left(\frac{\Omega}{\sqrt{2}}\right)^2 \eta_2 - 16e^{-4\tilde{z}}(\eta_2 - \eta_1). \tag{2.314}$$

To this end, adding or subtracting the two equations (2.313) and (2.314) from each other leads to the corresponding normal mode equations of the two-soliton system, yielding the corresponding eigenfrequencies

$$s_{1,2} = \frac{1}{3}\gamma\mu \pm \omega_i\sqrt{\Delta}, \qquad \Delta = \left(\frac{\gamma}{\gamma_{\rm cr}}\right)^2 - 1, \tag{2.315}$$

and, in this case, $\gamma_{\rm cr} = 3\,\omega_i$ (where $i = 1,2$), $\omega_1 = \Omega/\sqrt{2}$ (as before), while ω_2 is given by

$$\omega_2 = \sqrt{\frac{\Omega^2}{2} + 32e^{-4\tilde{z}}}. \tag{2.316}$$

As per the discussion of Sec. 2.2.3, the frequency ω_1 (ω_2) corresponds to the in-phase (out-of-phase) motion of the solitons. We can now once again compare the analytical expressions for the normal modes of the two-dark-soliton state with the corresponding dependence (on γ) of the negative Krein signature eigenvalues of the full linearization problem. This comparison is performed in Fig. 2.59, illustrating once again a very good agreement of the theoretical prediction with the full numerical result—especially for the physically relevant case of smaller values of the temperature-dependent parameter γ. It can be seen how two complex pairs arise for $\gamma \neq 0$, and how their subsequent collisions on the real axis lead to eventually four real eigenvalues. As a measure of quantitative comparison between the theoretical predictions and numerical findings, in the case of the

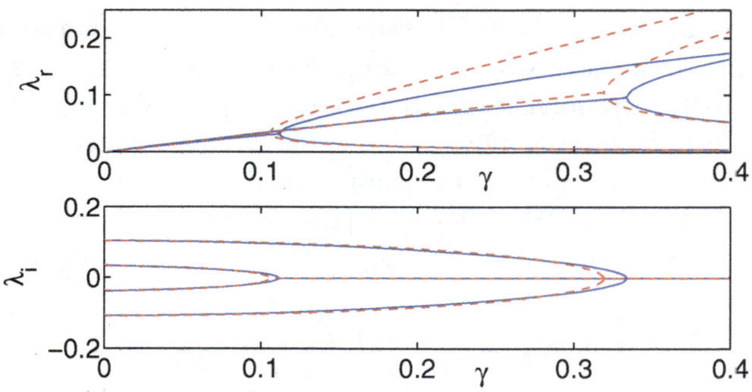

Figure 2.59. *The real (top panel) and imaginary (bottom panel) parts of the two negative Krein signature eigenvalue pairs for a stationary two-dark-soliton solution. Both lead to Hopf bifurcations, for $\gamma \neq 0$, and both pairs eventually collide and give rise to exponential instabilities through eigenvalues on the real axis. The solid lines yield the full numerical results, while the dashed lines provide the corresponding theoretical predictions. Adapted with permission from Ref. [480].*

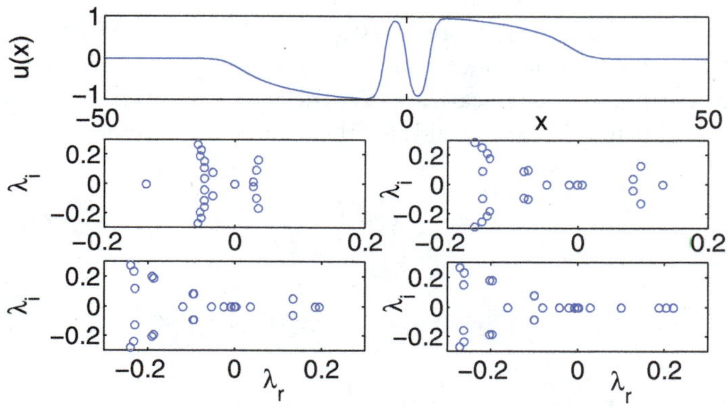

Figure 2.60. *The top panel shows the profile of the three-dark-soliton stationary state. The middle and bottom panels present the corresponding spectrum for the cases of $\gamma = 0.1$ (middle left), $\gamma = 0.3$ (middle right), $\gamma = 0.5$ (bottom left), and $\gamma = 0.6$ (bottom right). The different values of γ illustrate how the initially complex pairs of negative Krein signature eigenvalues gradually lead to real pairs through collisions on the real axis. Adapted with permission from Ref. [480].*

two solitons the first transition to a purely exponential instability arises for $\gamma > 0.106$ analytically, while it is computed to arise for $\gamma > 0.112$ numerically; on the other hand, the second transition is theoretically predicted to arise for $\gamma > 0.32$ analytically, while it emerges for $\gamma > 0.334$ numerically.

The Three-Dark-Soliton State

We now turn to the study of the three-dark-soliton case. The relevant state is illustrated in the top panel of Fig. 2.60, alongside its linearization spectrum for different values of γ, progressively increasing from $\gamma = 0.1$ to $\gamma = 0.6$. We observe the natural generalization of the scenario presented above. In particular, in this case, there exist three negative Krein sign eigenvalue pairs which, in turn, give rise to three complex pairs as soon as γ becomes nonzero. As γ is increased, the eigenvalues of each one of these pairs collide on the real axis, and progressively yield increasing numbers of real eigenvalues. The theoretical

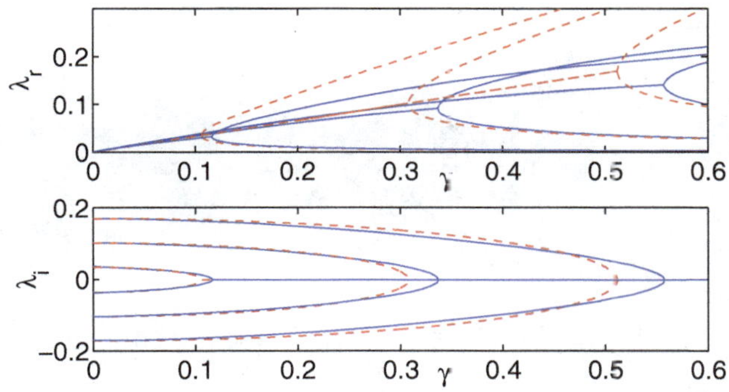

Figure 2.61. *Similar to Fig. 2.59, but now for the three-dark-soliton state. We observe that in this setting there are three negative Krein signature eigenvalue pairs, all of which acquire nonvanishing real parts, as soon as γ becomes nonzero (top panel). Once again, the theoretical predictions for the relevant eigenvalues are given by solid lines, both for the real (top) and imaginary (bottom) parts, while the corresponding theoretical predictions are demonstrated by dashed lines. Adapted with permission from Ref. [480].*

analysis can proceed along the same vein as before, taking into account now the pairwise interactions between the three different solitons. Writing down the equations of motion, it is straightforward to appreciate that symmetry will dictate that the equilibrium position of the middle soliton is at $z_2 = 0$, while $z_1 = -z_3$. Following a similar path as in the two-soliton case, we obtain that, at equilibrium, $z_1 = -z_3 = (1/2)W(32/\Omega^2)$. Using this expression for a quantitative comparison, we find that the theoretical prediction for the location of the solitons is at $z_1 = 3.725$, while they are numerically observed to be located at $z_1 = 3.75$.

We can now turn to the normal modes of the three-soliton state (which are again to be compared to the negative Krein sign eigenvalues of the linearization around that state in terms of the full numerical results). Writing down the equations for the three solitons, it is straightforward to show that the linear combinations that give rise to the normal modes will result in expressions for the corresponding eigenvalues, similar to the ones in Eq. (2.315). However, here the three (in this case) characteristic frequencies ω_i in the absence of dissipation will be given by

$$\omega_1 = \frac{\Omega}{\sqrt{2}}, \qquad \omega_2 = \sqrt{\frac{\Omega^2}{2} + 16e^{-2z_1}}, \qquad \omega_3 = \sqrt{\frac{\Omega^2}{2} + 48e^{-2z_1}}. \tag{2.317}$$

The comparison of the real and imaginary parts of the corresponding eigenvalues with the corresponding full numerical results is presented in Fig. 2.61. Once again, we can observe excellent qualitative and even good quantitative agreement between the analytical predictions for the complex (and subsequently real, upon collisions on the real axis) eigenvalue pairs (dashed lines) and the full numerical results (solid lines). As a measure of quantitative comparison of the relevant agreement, the first transition from a complex pair to a real one (upon collision of the relevant pair on the real axis) occurs for $\gamma = 0.106$ theoretically, and $\gamma = 0.117$ numerically. The second such transition occurs for $\gamma = 0.308$ theoretically, while it arises for $\gamma = 0.337$ numerically, while the last one emerges for $\gamma = 0.512$ theoretically and for $\gamma = 0.558$ numerically.

In Ref. [480], a direct numerical integration of the DGPE (2.290) with $\gamma(z) = \gamma =$ const., has been performed to study the dissipative dynamics of the two- and three-dark-soliton states. Results are shown in Fig. 2.62: in all cases, the full numerical results

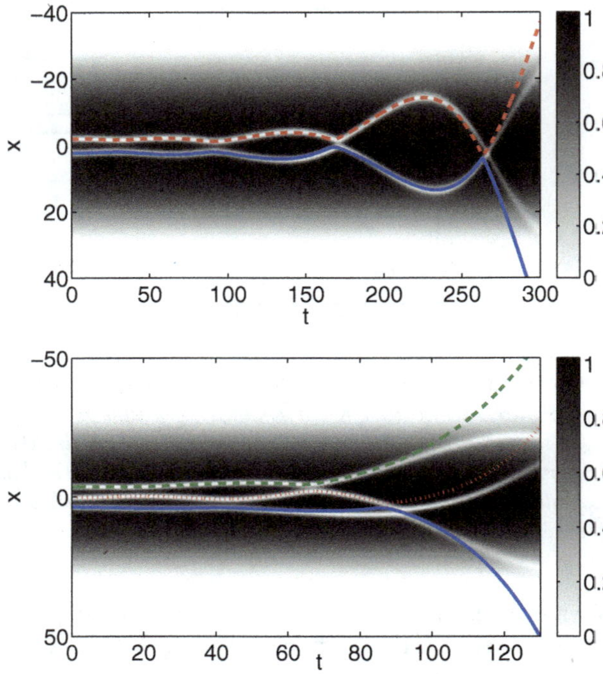

Figure 2.62. *Contour plot showing the space-time evolution of the BEC density (indicated by the colorbar) for the two- and three-dark-soliton cases. The top panel shows the case of two solitons—oscillating and interacting—in the case of* $\gamma = 0.05$ *(the eigenvalue of which is shown in the bottom left panel of Fig. 2.58). In this case, the solitons are initially placed at* $x_1 = 2.1$ *and* $x_2 = -2.2$*. The soliton evolution according to the solution of Eqs. (2.311) and (2.312) is shown by solid blue and dashed red lines. The bottom panel shows the case of three solitons, initially placed at* $x_1 = 3.65$, $x_2 = -3.7$, *and* $x_3 = 0.2$*. The results obtained by the corresponding equations of motion are shown by solid blue, dashed green, and dashed-dotted red lines. Adapted with permission from Ref. [480].*

involving the evolution of a slight displacement of the solitons from the corresponding equilibrium positions are directly compared with the ODE results for the evolution of the center of mass of the solitons. One can clearly observe the oscillatory evolution of the instabilities, associated with the complex eigenvalues identified previously. Despite slight quantitative discrepancies (associated, e.g., with multiple collision events for which the interaction potentials may be somewhat less accurate), we should also highlight the remarkable quantitative agreement between the reduced description of the simple ODEs describing the center of mass of the solitonic "particles" and the results of the full PDE model of Eq. (2.290).

It is thus clear that the effective particle description adopted above is able to capture—both qualitatively and quantitatively—the fundamental physical mechanisms arising in the setting under consideration: the antidamping effect due to the finite temperature, the oscillatory dynamics induced by the harmonic confinement, and the intersoliton interaction dynamics.

2.6.4 ▪ Dark-Bright Solitons at Finite Temperatures

Based on the considerations of Sec. 2.6, and also on the analytical tools developed in the present section, it is possible to study the dissipative dynamics of DB solitons due to finite-temperature effects. Our analysis will be based on the study of a system of two coupled DGPEs, which will be analyzed by means of the perturbation theory of Sec. 2.4.3. This way, we will derive an equation of motion for the soliton centers, and identify different temperature-dependent damping regimes, similar to those presented in Sec. 2.6.2. Our analysis will cover cases of both single- and multiple-DB-soliton states.

The System of Dissipative Gross–Pitaevskii Equations

We consider a quasi-1D (elongated along the x direction) two-component Bose gas at finite temperature, and assume that the thermal modes of energies $> \hbar\omega_\perp$ are at equilibrium (accounting for a heat bath in contact with the axial part of the gas), while the modes in the x direction are highly occupied. Then, extending the considerations of Sec. 2.6.1 to the two-component case under consideration, we may use the following set of two coupled 1D SGPEs to describe the axial modes of the system:

$$i\hbar\partial_t\psi_j = [1-\gamma_j(x,t)]\left(-\frac{\hbar^2}{2m}\partial_x^2\psi_j + V(x) - \mu_j + g\sum_{k=1}^{2}|\psi_k|^2\right)\psi_j + \eta_j(x,t).$$

Here, $\psi_j(x,t)$ ($j=1,2$) are complex order parameters characterizing each component of the binary Bose gas, m is the atomic mass, μ_j are the chemical potentials, while $V(x) = (1/2)m\omega_x^2x^2$ is the external trapping potential. Furthermore, $\eta_j(z,t)$ are complex Gaussian noise terms with correlations of the form $\langle\eta_j^*(x,t)\eta_j(x',t')\rangle = 2\hbar\gamma_j(x,t)k_BT\delta(x-x')\delta(t-t')$, where brackets denote averaging over different realizations of the noise. The strength of the latter can be calculated *ab initio* by the Keldysh self-energy [471]; for thermal clouds close to equilibrium, the relevant integrals determining the dissipation $\gamma_j(x,t)$ can be expressed as follows:

$$\gamma_j(x) = \pi^2\beta g^2\int\frac{dk_1}{2\pi}\int\frac{dk_2}{2\pi}\int\frac{dk_3}{2\pi}2\pi\delta(k_1-k_2-k_3)\delta(\epsilon_c^{(j)}+\epsilon_1^{(j)}-\epsilon_2^{(j)}-\epsilon_3^{(j)})$$
$$\times[N_1(1+N_2)(1+N_3)+(1+N_1)N_2N_3], \tag{2.318}$$

where $\beta = 1/k_BT$, $\epsilon_c^{(j)}$ are the condensate energies, $\epsilon_n^{(j)}$ are the energies of the nth excited states, $N_n^{(j)} = [\exp(\beta(E_n^{(j)}+V(x)+2g\sum_{k=1}^{2}\langle|\psi_k|^2\rangle-\mu_j))-1]^{-1}$ are Bose–Einstein distributions, while $E_n^{(j)}$ and $k_n = \sqrt{2mE_n/\hbar^2}$ denote, respectively, the kinetic energies and momenta of single particles in the nth excited state. Physically speaking, Eq. (2.318) describes the exchange of atoms between the thermal clouds and the condensates due to elastic collisions; notice that in the above description we have taken into regard exchanges up to the third excited state while, to leading-order approximation, we have omitted exchanges between the different hyperfine states (in other words, we have considered the simplest situation where each condensate component interacts with its own thermal cloud).

Under the above assumptions, the dissipation terms $\gamma_j(x)$ may in principle be calculated numerically, for several temperatures, as was done in the case of a single-component Bose gas in Sec. 2.6.1. Here, it should be recalled that, sufficiently close to the trap center (i.e., in the interval $[-R/2, R/2]$, where R is the Thomas–Fermi radius), the dissipation takes approximately constant values for a relatively wide range of temperatures (see, e.g., Fig. 2.52). Thus, below we will consider the situation where both dissipative terms γ_j are constant: such an assumption is consistent with our scope, i.e., to analyze the dynamics of the DB soliton near the center of the trap. Furthermore, based on the fact that simulations investigating soliton dynamics in the framework of the SGPE model were found to be in fairly good agreement with analytical and numerical results relying on the respective DGPE model, below we will omit the noise terms $\eta_j(x,t)$; this way, we will use the following system of two coupled DGPEs to describe the DB soliton dynamics in the two-component Bose gas at finite temperatures:

$$(i-\gamma_j)\hbar\partial_t\psi_j = \left(-\frac{\hbar^2}{2m}\partial_x^2\psi_j + V(x) - \mu_j + g\sum_{k=1}^{2}|\psi_k|^2\right)\psi_j. \tag{2.319}$$

Note that the above model was also used in Ref. [481], where the quantum Kelvin-Helmholtz instability of a two-component BEC was studied.

The system of Eqs. (2.319) can be expressed in dimensionless form as follows. Measuring the densities $|\psi_j|^2$, length, time, and energy in units of $2a$, $a_\perp = \sqrt{\hbar/\omega_\perp}$, $\omega_\perp^{-1}$, and $\hbar\omega_\perp$, respectively, Eqs. (2.319) become

$$(i-\gamma_d)\partial_t u_d = -\frac{1}{2}\partial_x^2 u_d + V(x)u_d + (|u_d|^2 + |u_b|^2 - \mu)u_d, \tag{2.320}$$

$$(i-\gamma_b)\partial_t u_b = -\frac{1}{2}\partial_x^2 u_b + V(x)u_b + (|u_b|^2 + |u_d|^2 - \mu - \Delta)u_b, \tag{2.321}$$

where we have used the notation $\psi_1 = u_d$ and $\psi_2 = u_b$, indicating—as per our considerations above—that the component 1 (2) is supposed to support a dark (bright) soliton. Note that the respective chemical potentials are now $\mu_1 = \mu_d = \mu$ and $\mu_2 = \mu_b = \mu + \Delta$, and we will assume that $\mu_d > \mu_b$, i.e., $\Delta = -|\Delta| < 0$). Finally, the external potential in Eqs. (2.320) and (2.321) takes the form $V(x) = (1/2)\Omega^2 x^2$, where $\Omega = \omega_x/\omega_\perp \ll 1$ is the normalized trap strength; the latter, along with the thermally induced damping parameters $\gamma_{d,b}$, are considered to be small parameters of the system (these will be treated as formal perturbation parameters in our analytical approximation—see below).

Relaxation to the Ground State of the System

As in the case of dark solitons in a single-component thermal BEC, we will first consider the dynamics of the background wavefunctions, namely a Thomas–Fermi (TF) wavefunction for the u_d component and a zero wavefunction for the u_b component. We will show that the coupled DGPEs Eqs. (2.320) and (2.321), similarly to their one-component counterpart (cf. discussion in Sec. 2.6.2), describe a *relaxation process*: the two components, starting (at $t=0$) from suitable initial conditions, will evolve so that, at sufficiently large times, u_d will converge towards a TF cloud with the prescribed value of the chemical potential μ, while u_b will vanish.

To show that this is the case indeed, we examine the peak amplitudes $U_{d,b}(t)$ of the wavefunctions $u_{d,b}(x=0,t)$, corresponding to their (absolute) values at the center of the trap [i.e., at $x=0$, where $V(x)=0$ as well], and assume respective phases $\theta_{d,b}(t)$. The evolution equations for $U_{d,b}(t)$ and $\theta_{d,b}(t)$, which can directly be obtained by introducing the ansatz $u_{d,b} = U_{d,b}(t)\exp[-i\theta_{d,b}(t)]$ into Eqs. (2.110) and (2.111), are of the form

$$\dot{U}_{d,b} + \gamma_{d,b}U_{d,b}\dot{\theta}_{d,b} = 0, \tag{2.322}$$

$$\gamma_d\dot{U}_d - \dot{\theta}_d U_d + (U_d^2 + U_b^2 - \mu)U_d = 0, \tag{2.323}$$

$$\gamma_b\dot{U}_b - \dot{\theta}_b U_b + (U_b^2 + U_d^2 - \mu - \Delta)U_b = 0, \tag{2.324}$$

where overdots denote time derivatives. Next, utilizing Eqs. (2.322), we obtain from Eqs. (2.323) and (2.324) the following system:

$$\dot{U}_d = -\tilde{\gamma}_d \left(U_d^2 + U_b^2 - \mu\right) U_d, \tag{2.325}$$

$$\dot{U}_b = -\tilde{\gamma}_b \left(U_d^2 + U_b^2 - \mu - \Delta\right) U_b, \tag{2.326}$$

where $\tilde{\gamma}_{d,b} \equiv \gamma_{d,b}/(1+\gamma_{d,b}^2)$. It is clear that the system of Eqs. (2.325) and (2.326) has a fixed point $(U_{d0}, U_{b0}) = (\sqrt{\mu}, 0)$ [a similar analysis can be done for the fixed point $(U_{d0}, U_{b0}) = (0, \sqrt{\mu+\Delta})$]. The evolution of small perturbations $U_{d1,b1}$ around this fixed point can then readily be found introducing the ansatz $U_{d0}(t) = \sqrt{\mu} + U_{d1}(t)$ and $U_{b0} = U_{b1}(t)$ into Eqs. (2.325) and (2.326) and linearizing with respect to $U_{d1,b1}$; this way, we can easily solve the equations for $U_{d1,b1}$ and finally obtain the following approximate expressions for the peak amplitudes of the wavefunctions:

$$U_d(t) \approx \sqrt{\mu} + (U_d(0) - \sqrt{\mu})e^{-2\tilde{\gamma}_d \mu t}, \tag{2.327}$$

$$U_b(t) \approx U_b(0)e^{-\tilde{\gamma}_b|\Delta|t}, \tag{2.328}$$

where $U_{d,b}(0)$ are initial conditions. Thus, at sufficiently large times, the peak amplitude of u_d will decay to the value $\sqrt{\mu}$, while the one of u_b will become zero. Accordingly, during the relaxation to equilibrium process, one may expect the following type of evolution towards relaxation. If the u_d component is initially a TF cloud of amplitude $U_d(0)$, its density will evolve as

$$|u_d(x,t)|^2 \approx \max\left\{U_d^2(t) - V(x), 0\right\}. \tag{2.329}$$

On the other hand, if the u_b component has initially the form of an arbitrary localized function, e.g., a Gaussian, of amplitude $U_b(0)$, it will asymptotically approach the trivial stationary state.

The above predictions can be directly compared to numerical simulations. In particular, in Fig. 2.63 we show the evolution of a state characterized by the initial densities $|u_d(x,0)|^2$ given by Eq. (2.329) and $|u_b(x,0)|^2 = U_b^2(0)\exp\left[-2(x/w)^2\right]$, with parameter values $U_d(0) = 0.86$, $U_b(0) = 0.6$, $\Omega = 0.05$, and $w = 10$; as found by direct numerical integration of Eqs. (2.320) and (2.321), with $\mu = 1.3$, $|\Delta| = 0.1$, and $\gamma_d = \gamma_b = 0.05$. The figure clearly shows the validity of our analytical approximations: the u_d component develops into a TF cloud with chemical potential $\mu = 1.3$, with the numerically found density profile (solid red line) being in fairly good agreement with the analytical prediction of Eq. (2.329) (dashed blue line); on the other hand, the u_b component (solid green line) vanishes at $t \approx 200$, a time consistent with the slow time scale $t_* \equiv (\tilde{\gamma}_b|\Delta|)^{-1} \approx 200$ suggested by Eq. (2.328).

2.6.5 ▪ Dissipative Dynamics of a Single Dark-Bright Soliton

Having studied the relaxation process described by Eqs. (2.320) and (2.321), we now proceed to investigate, in the same framework, the dissipative dynamics of DB solitons. We will assume that the dark soliton is on top of an already formed TF cloud with the equilibrium density $|u_{d,\mathrm{TF}}|^2 = \max\{\mu - V(x), 0\}$; this way, the density $|u_d|^2$ in Eqs. (2.320) and (2.321) is substituted by $|u_d|^2 \to |u_{d,\mathrm{TF}}|^2|u_d|^2$. Furthermore, we introduce the transformations $t \to \mu t$, $x \to \sqrt{\mu}x$, $|u_b|^2 \to \mu^{-1}|u_b|^2$, and cast Eqs. (2.320) and (2.321) into the form of Eqs. (2.130) and (2.131), but with the functional perturbations R_d and R_b now

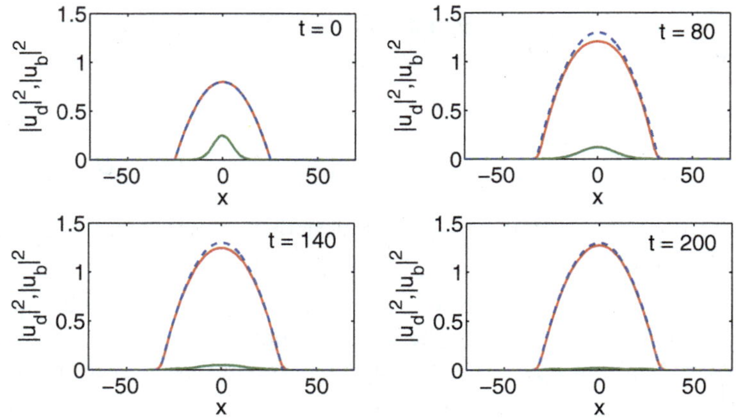

Figure 2.63. *Time evolution of a state characterized by the densities* $|u_d(x,0)|^2$ *given by Eq. (2.329) and* $|u_b(x,0)|^2 = U_b^2(0)\exp\left[-2(x/w)^2\right]$, *with parameter values* $U_d(0)=0.86$, $U_b(0)=0.6$, $\Omega = 0.05$, *and* $w = 10$. *The solid lines show the density of the dark (red) and bright (green) component, while the blue dashed line shows the analytical result of Eq. (2.327). The other parameter values used in Eqs. (2.320) and (2.321) are* $\mu = 1.3$, $|\Delta| = 0.1$, *and* $\gamma_d = \gamma_b = 0.05$. *Adapted with permission from Ref. [325].*

being given by

$$R_d \equiv (2\mu^2)^{-1}[2(1-|u_d|^2)V(x)u_d + V'(x)\partial_x u_d] + \gamma_d \mu^{-1}\partial_t u_d, \tag{2.330}$$

$$R_b \equiv \mu^{-2}[(1-|u_d|^2)V(x)u_b + \mu\gamma_b\partial_t u_b]. \tag{2.331}$$

Then, employing the Hamiltonian approach of the perturbation theory for DB solitons of Sec. 2.4.3, we consider the adiabatic evolution of a DB soliton, with parameters $D(t)$, $x_0(t)$, and $\phi(t)$. The evolution of $D(t)$ and $x_0(t)$ is respectively described by Eqs. (2.136) and (2.137), while the evolution of $\phi(t)$ (as found by the evolution of the DB soliton energy) is described by the equation

$$\begin{aligned} 4\dot{D}D^2 + \chi D\sec^2\phi(\dot{D} + D\dot{\phi}\tan\phi) = {} & \frac{1}{\mu^2}\left(2\cos^3\phi\sin\phi - \chi D\sin\phi\cos\phi\right)V'(x_0) \\ & - \frac{8}{3}\frac{\gamma_d}{\mu}D^3\sin^2\phi - \frac{2}{3}\frac{\gamma_b}{\mu}\chi D^4\tan^2\phi. \end{aligned} \tag{2.332}$$

We now proceed as in Sec. 2.4.3 and assume that the DB soliton moves in the vicinity of the trap center. We thus find the fixed points of Eqs. (2.136), (2.137), and (2.332) (which are identical to the ones found in the Hamiltonian case of Sec. 2.4.3), and assume small perturbations around these fixed points: $x_0 \to 0 + x_0$, $\phi \to 0 + \phi$, and $D \to D_{eq} + D_1$. Then, linearizing Eqs. (2.136) and (2.137) and Eq. (2.332) with respect to x_0, ϕ, and D_1, we derive again Eqs. (2.142), (2.144) and the following equation for the evolution of ϕ:

$$\dot{\phi} = \frac{\left(-2 + \chi D_{eq}\right)V'(x_0) + \frac{2}{3\mu}D_{eq}^3\left(4\gamma_d + \chi\gamma_b D_{eq}\right)\phi}{D_{eq}[-8D_{eq}\tilde{D} - \chi(2\tilde{D} - D_{eq})]}, \tag{2.333}$$

Now differentiating Eq. (2.144) with respect to time once, and using Eqs. (2.333) and (2.144), we obtain after some straightforward algebraic manipulations the following equation of motion for the DB soliton center x_0:

$$\ddot{x}_0 - a\,\dot{x}_0 + \omega_{\rm osc}^2 x_0 = 0, \tag{2.334}$$

where the oscillation frequency $\omega_{\rm osc}$ is given by Eq. (2.147), while the antidamping parameter a is given by

$$a = \frac{2}{3}\mu\left(\gamma_d - \frac{1}{8}\chi^2\gamma_b\right) + \frac{4}{3}\frac{\chi}{\chi_0}\mu\left(\gamma_b - \gamma_d + \frac{1}{8}\chi^2\gamma_b\right). \tag{2.335}$$

Notice that in the absence of dissipation ($a = 0$), Eq. (2.334) recovers the results of Sec. 2.4.3.

As in the case of dark solitons in single-component BECs, it is clear that the nature of the soliton trajectories $x_0(t)$ as predicted by Eq. (2.334) depends on whether the roots of the auxiliary equation $s^2 - as + \omega_{\rm osc}^2 = 0$ are real or complex. The roots are given by

$$s_{1,2} = \frac{1}{2}\left(a \pm \sqrt{a^2 - a_{\rm cr}^2}\right), \quad a_{\rm cr} \equiv 2\omega_{\rm osc}, \tag{2.336}$$

with the discriminant $\mathscr{D} \equiv a^2 - a_{\rm cr}^2$ determining the type of the motion. In particular, we identify different temperature-dependent damping regimes: the subcritical weak antidamping regime ($\mathscr{D} < 0, a < a_{\rm cr}$), the critical regime ($\mathscr{D} = 0, a = a_{\rm cr}$), and the supercritical strong antidamping regime ($\mathscr{D} > 0, a > a_{\rm cr}$). In the first regime the soliton performs oscillations of growing amplitude, with $x_0(t) \propto \exp(at)\cos(\omega_{\rm osc} t)$, while in the latter two regimes the soliton follows an exponentially growing trajectory, i.e., $x_0(t) \propto \exp(s_{1,2}t)$ (with $s_{1,2} \in \mathbb{R}$), and decays at the rims of the condensate cloud (see also below).

The above analytical predictions are supported by a linear stability analysis around the stationary DB soliton (with $\phi = 0$ and $x_0 = 0$). For such a state, the right-hand sides of Eqs. (2.320) and (2.321) still vanish and, thus, stationary DB solitons are exact solutions of the problem with $\gamma_{d,b} \neq 0$. We can obtain this solution by means of a fixed point algorithm and then find the linearization spectrum around the stationary DB soliton, as was done in Sec. 2.4.3.

In Fig. 2.64, we show an example of such a stationary DB soliton in a trap of strength $\Omega = 0.1$ (for simplicity, we consider the case with $\gamma_d = \gamma_b = \gamma$). Notice that upon the variations of γ (and hence of temperature) considered in the figure, the solution profile does not change, as mentioned above; however, the linearization problem and its eigenvalues significantly depend on the value of γ, as is shown in the four bottom panels of Fig. 2.64. In the zero-temperature (Hamiltonian) case of $\gamma = 0$, all eigenvalues are imaginary (see also Fig. 2.24), and the oscillatory motion of a single DB soliton is spectrally associated with the existence of a single anomalous mode (of negative Krein sign) in the spectrum. This anomalous mode (AM) possesses a frequency identical to the frequency of the DB soliton oscillation, i.e., $\omega_{\rm AM} \equiv {\rm Im}(\lambda_{\rm AM}) = \omega_{\rm osc}$.

Our analytical approximation for $\omega_{\rm osc}$ is tested against the numerical results for ${\rm Im}(\lambda_{\rm AM})$, both in the case examples of Fig. 2.64, as well as in the parametric dependence results of Fig. 2.65. It is clear from the spectral plots (middle and bottom rows of panels of Fig. 2.64) that, as soon as $\gamma \neq 0$, the relevant anomalous eigenmode (indicated by red stars in Fig. 2.65) becomes *complex*, leading to soliton oscillations of growing amplitude; this behavior, which corresponds to the "subcritical" regime mentioned above, is similar to the case of dark solitons in single-component systems, as discussed in Sec. 2.6.3. If γ

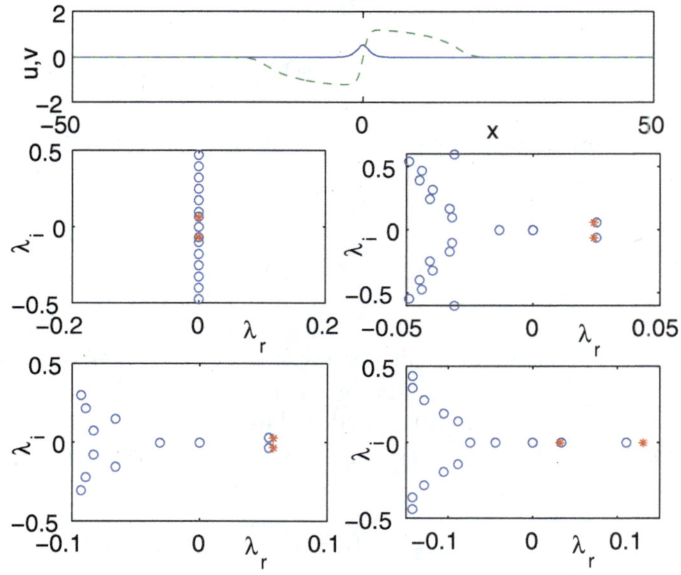

Figure 2.64. *The top panel depicts the stationary solution for a single DB soliton for $\mu = 1.5$, $|\Delta| = 0.6$, and $\Omega = 0.1$. The dark (bright) components are shown by the dashed green (solid blue) lines. The middle and bottom rows of panels depict four spectral planes, corresponding to different values of $\gamma = \gamma_d = \gamma_b$, for the single DB soliton stationary states: in the middle left panel $\gamma = 0$ (the zero-temperature Hamiltonian case), in the middle right $\gamma = 0.05$, in the bottom left $\gamma = 0.12$, and in the bottom right $\gamma = 0.17$. The red stars highlight the anomalous mode (of the Hamiltonian case) eigenvalues. Adapted with permission from Ref. [325].*

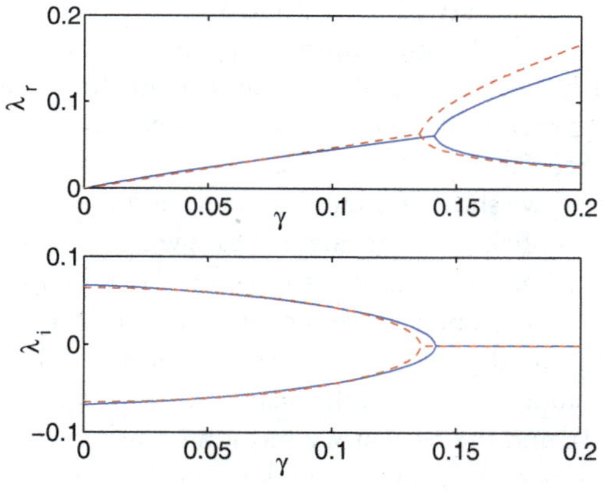

Figure 2.65. *The real part (top)—instability growth rate—and imaginary part (bottom)—oscillation frequency—of the unstable eigenmode of the linearization around a stationary DB soliton as a function of the parameter γ (parameter values are as in Fig. 2.64). Solid lines indicate the full numerical result, while dashed ones the analytical results of Eq. (2.336). For $\gamma < \gamma_{\rm cr} \approx 0.141$ (subcritical regime), the complex conjugate pair is responsible for soliton oscillations of growing amplitude. For $\gamma \geq \gamma_{\rm cr}$ (critical/supercritical regimes) the collision of the complex conjugate pair of eigenvalues creates a real pair, and the dynamics involves purely exponential growth. The parameter values are $\mu = 1.5$, $|\Delta| = 0.6$, and $\Omega = 0.1$. Adapted with permission from Ref. [325].*

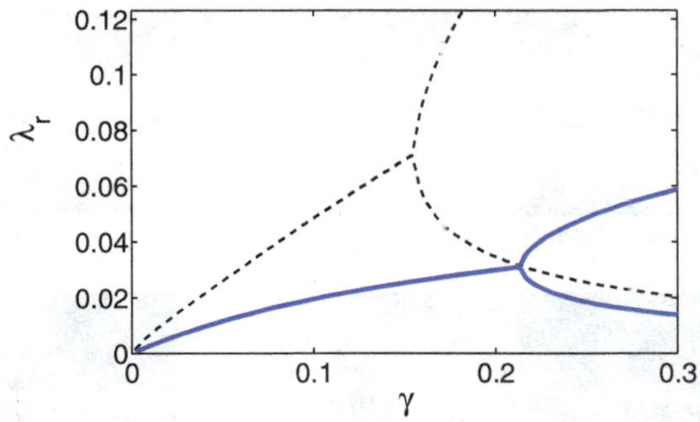

Figure 2.66. *Comparison of the bifurcation diagrams $\lambda_r(\gamma)$ for a stationary DB soliton (solid line) and a "bare" dark soliton (dashed line) as obtained by the BdG analysis. It is clear that the presence of the bright ("filling") component pushes $\gamma_{\rm cr}$ towards larger values, acquiring also smaller values of the instability growth rate λ_r as compared to the ones found for the dark soliton. The parameter values are $\Omega = 0.1$, $\mu = 1.5$, and $|\Delta| = 0.1$. Adapted with permission from Ref. [325].*

is increased beyond a critical point, $\gamma_{\rm cr} \approx 0.141$, the relevant eigenvalue pair collides with the real axis, leading to the emergence of a pair of real eigenvalues (cf. bottom right panels of Fig. 2.64 and Fig. 2.65). This corresponds to the "supercritical" regime mentioned above, where the divergence of the soliton from its center equilibrium is purely exponential. Notice that the analytical predictions for the relevant unstable eigenvalue (and the oscillatory or purely exponential divergence from the equilibrium position) in Figs. 2.64 and 2.65 are generally fairly accurate, especially for the physically relevant regime of small γ; mathematically, this can be understood by the fact that our analytical approximation relies on the smallness of γ, which was treated as a small parameter of the problem within our perturbation theory approach.

At this point, the role of the bright-soliton component in the dynamics should be highlighted in connection to the case of a dark soliton in a single-component condensate (where the bright soliton is absent). It can be directly seen from Eq. (2.335) that the anti-damping effect is always *weaker* for the DB soliton in comparison to the dark one (at least in the case $\gamma_d = \gamma_b = \gamma$ we consider herein). Hence, the lifetime of the DB soliton is always longer than that of the dark soliton and, in fact, it becomes larger, as the bright-soliton component "filling" of the dark one becomes stronger. This *partial stabilization* of the dark soliton evolution by means of its symbiotic second component is clearly illustrated in Fig. 2.66, where the bifurcation diagrams for the DB soliton are directly compared to the ones corresponding to the "bare" dark soliton. It is clear that the whole bifurcation diagram for the DB soliton is "drifted" towards larger values of γ (e.g., $\gamma_{\rm cr} = 0.212$ for the DB soliton and $\gamma_{\rm cr} = 0.155$ for the dark soliton), acquiring also smaller values of the instability growth rate λ_r as compared to the ones found for the dark soliton for the same values of the temperature parameter γ. This clearly indicates that the "filled" dark soliton in a two-component BEC is *more long-lived* in the presence of finite temperature than a "bare" dark soliton in a single-component BEC.

Our analytical predictions can also be tested against direct numerical simulations illustrating the evolution of the single DB soliton, both for the subcritical case of oscillatory growth (see left panels of Fig. 2.67), and for the supercritical case of purely exponential growth (see right panels of Fig. 2.67). In both cases it can be seen that the dashed line corresponding to the analytical solution of the ODE (2.334) accurately tracks the evolution of the center of the DB soliton, which progressively loses its contrast and eventually

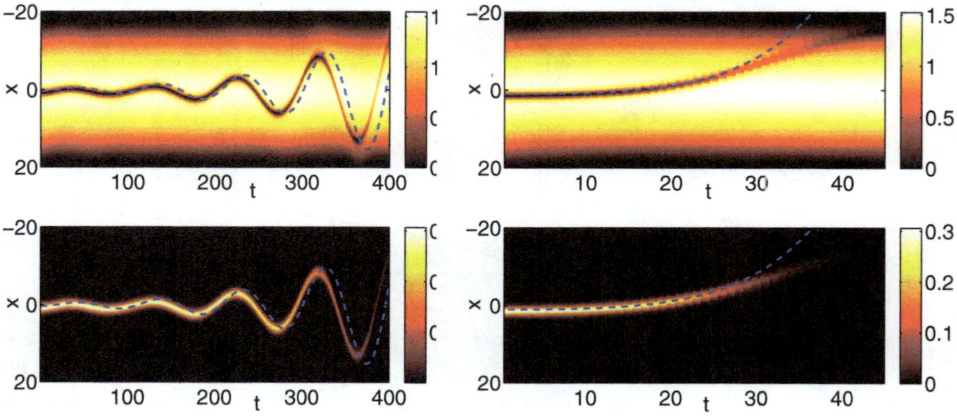

Figure 2.67. *Contour plots showing the space-time evolution of the density of a single DB soliton, initially placed at* $x_0(0) = 0.4$*, in a trap of strength* $\Omega = 0.1$*. The left panels correspond to the subcritical regime with* $\gamma = 0.02$*, while the right panels to the supercritical case of* $\gamma = 0.15$*. Top panels represent the dark solitons and the bottom ones the bright solitons. The blue dashed lines represent the analytical results of Eq. (2.334), namely* $x_0(t) = x_0(0)\exp[(a/2)t]\cos(\omega_{\rm osc} t)$ *(left panels) and* $x_0(t) = \frac{x_0(0)}{s_2 - s_1}[s_2\exp(s_1 t) - s_1\exp(s_2 t)]$ *(right panels). The parameter values are* $\mu = 1.5$ *and* $|\Delta| = 0.6$*. Adapted with permission from Ref. [325].*

disappears in the condensate background, with the system converging to its ground state (see Sec. 2.6.4).

2.6.6 ▪ Two-Dark-Bright-Soliton States at Finite Temperatures

We now turn our attention to the study of DB soliton "molecules" composed of two-DB-soliton states. Our analysis relies on the generalization of the considerations of Sec. 2.4.5 for the finite-temperature induced dissipative case under consideration. In particular, combining the thermal effect on each DB soliton in the trap, represented by Eq. (2.334), and interaction effects included in Eq. (2.160), we may use the following approximation to describe the motion of the centers of the two DB solitons:

$$\ddot{x}_0 - a\,\dot{x}_0 - (F_{\rm tr} + F_{\rm int}) = 0. \tag{2.337}$$

The equilibrium points $x_{\rm eq}$, can easily be found as solutions of the transcendental equation resulting from Eq. (2.337) letting $\dot{x}_0 = \ddot{x}_0 = 0$, for cases of both in- and out-of-phase DB solitons. To study the stability of these equilibrium points in the framework of Eq. (2.337), we may again use the ansatz $x_0(t) = x_{\rm eq} + \delta(t)$, and obtain a linear equation for the small-amplitude perturbation $\delta(t)$, namely, $\ddot{\delta} - a\dot{\delta} + \omega_1^2\delta = 0$, where the frequency ω_1 is given by

$$\omega_1^2 = \omega_{\rm osc}^2 - \left.\frac{\partial F_{\rm int}}{\partial x_0}\right|_{x_0 = x_{\rm eq}}, \tag{2.338}$$

where $\omega_{\rm osc}^2$ and a are respectively given by Eq. (2.147) and Eq. (2.335).

We now test the relevant predictions against BdG simulations, first for in-phase DB solitons in Fig. 2.68, and then for out-of-phase DB solitons in Fig. 2.69. As expected, in the case of the in-phase configuration the BdG analysis reveals the existence of two

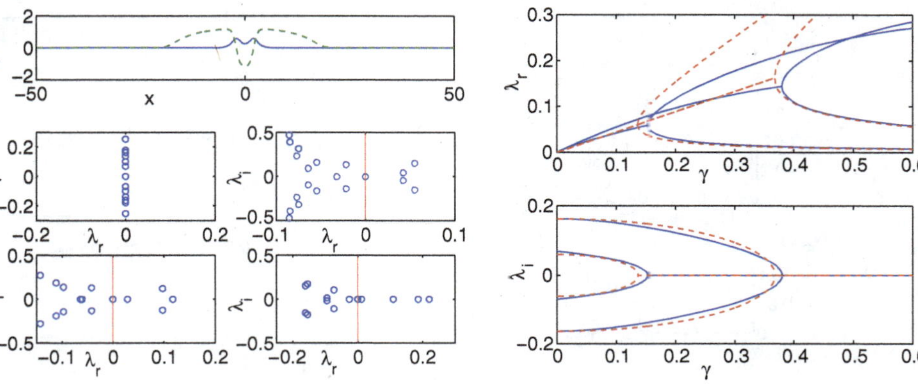

Figure 2.68. *Left set of panels: Top panel depicts a stationary in-phase two-DB-soliton state for* $\mu = 1.5$, $|\Delta| = 0.6$ *and* $\Omega = 0.1$ *[dark (bright) components are shown by the dashed green (solid blue) lines]; the middle and bottom panels show spectral planes for* $\gamma = 0$ *[middle left (the zero-temperature Hamiltonian case)],* $\gamma = 0.1$ *(middle right),* $\gamma = 0.2$ *(bottom left), and* $\gamma = 0.4$ *(bottom right). Right set of panels: The real (top) and imaginary (bottom) parts of the two anomalous mode eigenvalues for an in-phase two-DB-soliton state; solid blue lines show the numerical results, while the dashed red lines show the corresponding theoretical predictions of Eqs. (2.337) and (2.338). Adapted with permission from Ref. [325].*

anomalous modes: the one with the smaller (larger) eigenvalue—in the zero-temperature case—corresponds to an in- (out-of-) phase motion of the two DB solitons, similarly to the case of a two-dark-soliton state in a single-component BEC (see Sec. 2.6.3). These two anomalous mode pairs lead to complex eigenfrequencies for $\gamma \neq 0$, and the two-DB-soliton state performs oscillations of growing amplitude. Similarly to the case of the single DB soliton, as γ is increased these pairs collide pairwise on the real axis at two critical points, namely $\gamma_1 \approx 0.153$ and $\gamma_2 \approx 0.38$. Beyond the second critical point γ_2, the growth of the trajectory of the DB soliton center becomes purely exponential. The theoretical approximation of the relevant complex (and subsequently real) eigenvalues depicted by the red dashed lines in the right panels of Fig. 2.68 is again fairly accurate (but becomes progressively worse as γ increases).

A similar phenomenology arises in the case of out-of-phase two-DB-soliton states, as shown in Fig. 2.69. However, there exists a rather nontrivial twist in comparison to the previous case. In particular, a third pair of complex eigenvalues emerges due to the fact that a third anomalous mode exists for $\gamma = 0$. This mode is no longer a *translational* one associated with the in- or out-of-phase motion of the two soliton centers (as before and as shown in the bottom left eigenmodes of Fig. 2.69). It is instead a mode associated with the π relative phase of the peaks: if we add the eigenvector of this unstable (for $\gamma \neq 0$) mode to the two-DB-soliton out-of-phase solution, we observe that while the center location of the state remains intact, the relative heights of the two solitons are affected, leading to a symmetry breaking of the configuration. We will not consider this unstable mode further, since its induced instability is weaker than those of the (in-phase and out-of-phase) translations. Nevertheless, we note that all three pairs of modes eventually collide on the real axis, eventually leading to pairs of purely real eigenvalues.

In Fig. 2.70, we show results of direct numerical simulations, for both the in-phase and the out-of-phase two-DB-soliton states for the low-γ, oscillatory growth (sub-critical) regime. Despite the complexity of the resulting system and of the DB soliton interactions, it can still be clearly observed that the ODE (2.337) can be used to capture fairly

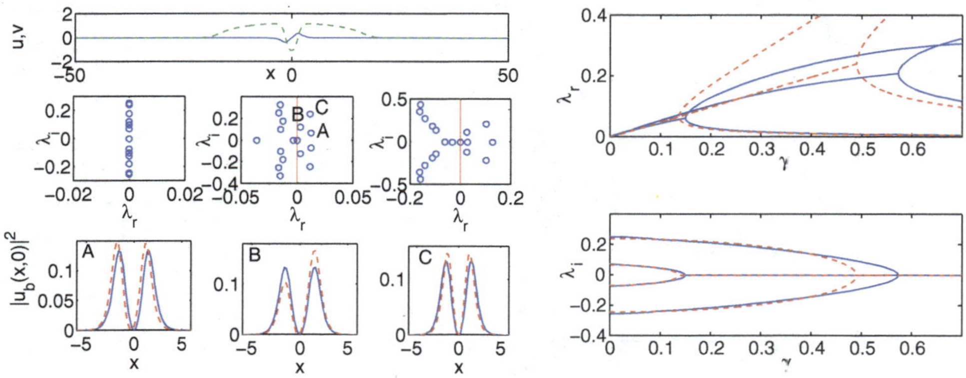

Figure 2.69. *Left set of panels: Top panel depicts a stationary out-of-phase two-DB-soliton, with $\mu = 1.5$, $|\Delta| = 0.6$, and $\Omega = 0.1$ [dark (bright) components are shown by the dashed green (solid blue) lines]; middle panels show three spectral planes for $\gamma = 0$, $\gamma = 0.02$, and $\gamma = 0.2$ (from left to right); bottom panels show the bright soliton components (solid blue line) against the perturbed states (dashed red line) obtained by adding to them the respective BdG eigenfunctions; "A", "B", and "C" correspond to the eigenfunctions of the three anomalous modes' eigenvalues in ascending order (see text). Right set of panels: Real (top) and imaginary (bottom) parts of the two anomalous mode eigenvalues (corresponding to "A" and "C" in the left middle panel); solid blue lines depict numerical results, while the red dashed lines provide the corresponding theoretical predictions of Eqs. (2.337) and (2.338). Adapted with permission from Ref. [325].*

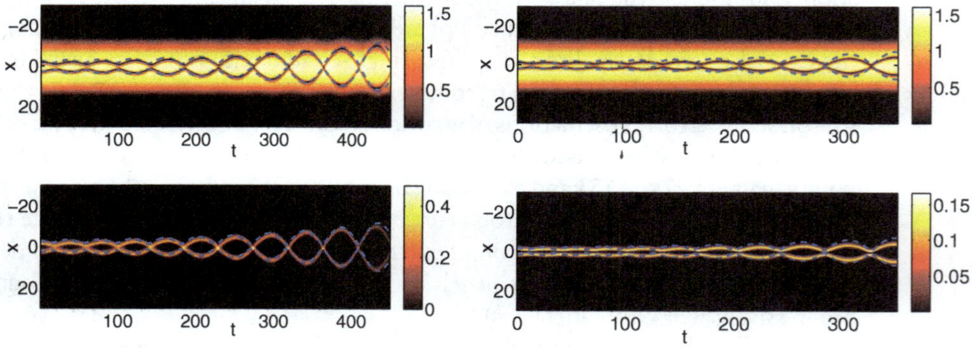

Figure 2.70. *Contour plot showing the space-time evolution of the density of in-phase (left panels) and out-of-phase (right panels) two-DB-soliton states for $\gamma = 0.01$; top (bottom) panels shows the dark (bright) soliton. The in-phase (out-of-phase) solitons are initially placed at $x_1 = 2.75$ and $x_2 = -2.75$ ($x_1 = 1.76$ and $x_2 = -1.76$). The blue dashed lines show the results obtained by numerical solution of Eq. (2.337). Adapted with permission from Ref. [325].*

accurately the relevant dynamics even for the long-time evolutions considered in the figure. Here, it should be mentioned that the temperature-induced dissipation results in an interesting effect. Particularly, as observed in Fig. 2.70, for short times, the individual DB solitons clearly behave like repelling particles, which can always be characterized by two individual density minima—even at the collision point. Nevertheless, for longer times, the nature of their interaction changes: due to dissipation, they gain kinetic energy and completely overlap at the collision point (see also bottom panel of Fig. 2.2).

Chapter 3
The Two-Dimensional Case

3.1 ▪ Ground State, Fundamental Vortex, and Other Radially Symmetric Solutions: Existence, Stability, and Dynamics

3.1.1 ▪ Model Setup

Our starting point for two-dimensional (2D) considerations is the Gross–Pitaevskii equation (GPE) in its two-dimensional reduction form. Such an "effective" 2D GPE applies to situations where the condensate has a nearly planar, i.e., "pancake" shape; see, e.g., Ref. [482] and references therein. We express the equation in harmonic oscillator units [483] in the form

$$i u_t = -\frac{1}{2}\Delta u + \frac{\Omega^2}{2}(x^2+y^2)u + g\,|u|^2 u, \tag{3.1}$$

where $u = u(x,y,t)$ is the 2D wavefunction (the condensate order parameter) and $\rho = |u|^2$ is the local condensate density. The external potential (also predominantly in what follows) consists of the harmonic trapping potential $V(r) = \Omega^2 r^2/2$, where Ω is the effective frequency of the parabolic trap which can be expressed as $\Omega \equiv \omega_r/\omega_z$ (with $r^2 \equiv x^2+y^2$), and $\omega_{r,z}$ are the confinement frequencies in the radial and axial directions respectively. It is assumed that $\Omega \ll 1$ for the pancake geometry considered herein. Finally, $g = \pm 1$ expresses a normalized measure of the interatomic interactions (nonlinear coefficient in the GPE in harmonic oscillator units) and is fixed to $g = -1$ for attractive interactions and to $g = +1$ for repulsive interactions. Accordingly, the squared L^2 norm is

$$\iint_{-\infty}^{+\infty} |u|^2\, dx\, dy = N, \tag{3.2}$$

where N is the number of atoms.

In order to numerically identify stationary nonlinear solutions of Eq. (3.1), we use the standing wave ansatz $u(x,y,t) = v(r)\exp[i(\mu t + S\theta)]$, where we assume the density of the solution is radially symmetric with chemical potential μ and topological charge (vorticity) S. In this section, we restrict our considerations to radially symmetric density solutions, although in later sections of this text, we will lift this constraint. The above ansatz results in the steady-state equation

$$\mu v = -\frac{1}{2}\left(v_{rr} + \frac{1}{r} v_r - \frac{S^2}{r^2} v \right) + \frac{\Omega^2}{2} r^2 v + g\, |v|^2\, v. \tag{3.3}$$

Equation (3.3) exhibits infinite branches of nonlinear bound states, each branch stemming from the corresponding modes of the underlying linear problem. Although the linear problem possesses both radially symmetric and non-radially-symmetric modes, we will focus on the former in the present section.

These branches are constructed and followed using the method of pseudo-arclength continuation [484, 485], in order to obtain subsequent points along a branch, once a point on it has been identified. The first point on each branch is found using a bound state of the underlying linear equation

$$E v = -\frac{1}{2}\left(v_{rr} + \frac{1}{r} v_r - \frac{S^2}{r^2} v \right) + \frac{\Omega^2}{2} r^2 v. \tag{3.4}$$

The linear state corresponds to a solution of Eq. (3.3) with parameters $\mu = E$, vorticity S, radial zero crossings n_r and normalization $\int r|v|^2 dr$ tending to zero; in that limit, the nonlinearity becomes irrelevant and the linear solutions are the well-known Gauss–Laguerre modes with $E = (2n_r + S + 1)\Omega$. Using such an initial seed, a nontrivial solution is found through a fixed-point iterative scheme for a slightly perturbed value of the chemical potential.

The pseudo-arclength method, used to trace subsequent solutions, works via the introduction of a pseudo-arclength parameter s and an additional equation $F(v, \mu, s) = 0$ such that $F(\bar{v}, \bar{\mu}, 0) = 0$ where $(\bar{v}, \bar{\mu})$ is a solution of Eq. (3.3). We used for F

$$F(v, \mu, s) = |v - \bar{v}|^2 + |\mu - \bar{\mu}|^2 - s^2. \tag{3.5}$$

Lastly, the new expanded system of equations can be solved, e.g., using a predictor-corrector method.

The linear stability of the solutions is analyzed by using the following ansatz for the perturbation in Eq. (3.1):

$$u(r,t) = e^{i(\mu t + S\theta)}\left[v(r) + a(r) e^{iq\theta + \lambda t} + b^*(r) e^{-iq\theta + \lambda^* t} \right], \tag{3.6}$$

where the asterisk stands for the complex conjugation and θ is the polar angle. One solves the resulting linearized equations for the perturbation eigenmodes $(a(r), b(r))^T = (a_q(r), b_q(r))^T$ and eigenvalues $\lambda = \lambda_q$ associated with them. The key observation that allows one to carry this task through is that the subspaces of the different azimuthal perturbation eigenmodes, i.e., subspaces of different q, decouple [486, 266], leading each eigenmode $e^{iq\theta}$ to be coupled only with its complex conjugate. This, in turn, allows the examination of the stability of the 2D problem in the form of a denumerably infinite set of 1D radial eigenvalue problems that are solved on the same numerical (Chebyshev) grid as the original existence problem of Eq. (3.3).

3.1.2 ▪ Principal Results

In this section, following Ref. [269], we present the principal results for the radial nonlinear states that bifurcate from the *linear* spectrum of the 2D problem. This linear limit

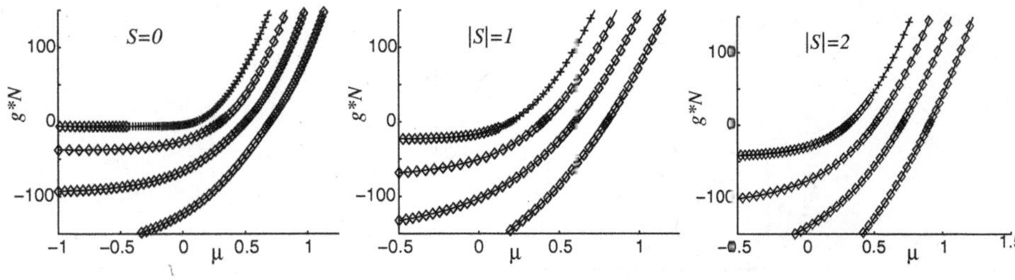

Figure 3.1. *Linear stability analysis along the first three branches: ground state (no radial nodes), first, and second excited states ($n_r = 1$ and $n_r = 2$ radial nodes respectively) depicted, respectively, in the curves from left to right in each panel. The left, middle, and right panels display g times the norm of the solution as a function of the chemical potential for $S = 0$ (no topological charge), $|S| = 1$ (singly charged) and $|S| = 2$ (doubly charged) solutions. Each point on the branch is depicted by crosses for stable solutions and diamonds for unstable solutions. The areas of stability for $|S| = 0, 1$ appear only for $n_r = 0$ and only for lower powers for $g = -1$ and all powers of the ground state for $g = 1$. In the case of $|S| = 2$, the ground state with $n_r = 0$ is only linearly stable for larger powers in the case of $g = 1$. Adapted with permission from Ref. [269].*

is one of the analytically (or semi-analytically) tractable limits of the problem that we use to obtain a handle on the emergent nonlinear states thereof.

Our steady state and stability results are summarized in Fig. 3.1. A partial set of branch profiles, linearization eigenvalues, and dynamical evolutions is given in Figs. 3.4–3.8 and is briefly touched upon in what follows. Figure 3.1 shows the continuation from the linear limit of the relevant states. The nonlinearity leads to a decrease of the chemical potential with increasing power for the focusing case and to a corresponding increase for the defocusing case. Notice that some of the branches presented here have also appeared in the earlier work of Ref. [487]. The spectral stability of the obtained states is encapsulated straightforwardly in the symbols associated with each branch. The plus symbols denote branches which are stable, while the diamond symbols correspond to unstable cases. We observe that the only truly stable solution corresponds to the ground state with $n_r = 0$ of the defocusing problem, which for large number of atoms can be well approximated by the 2D Thomas–Fermi state given by $|u|^2 = \max(0, \mu - V(r))$, in the presence of the potential $V(r)$ [43, 42]. Furthermore, this is true only for topological charges $S = 0$ and $|S| = 1$; for higher topological charges $|S| \geq 2$, there are regions of stability for $n_r = 0$, as was originally shown in Ref. [488]; however, outside these regions these excited states tend to break up into lower charge vortices. All higher excited states with $n_r > 0$, i.e., with one, two, or more nodes are also directly found to be unstable, even close to the corresponding linear limit of these states.

Briefly, for completeness—and also for comparison purposes—here we also address the corresponding stability properties of the focusing case of attractive condensates. In the latter, the situation is even more unstable due to the catastrophic effects of self-steepening and wave collapse [489]. In particular, even the ground state with $n_r = 0$ may be unstable due to collapse (represented by the mode with $q = 0$ in this case), although the instability growth rate may be very small, as is the case for $S = 0$; see, e.g., Fig. 3.2. Excited states are always unstable for the focusing nonlinearity as well; in fact, they are more strongly so than in the defocusing case, again due to the presence of the collapse-inducing $q = 0$ mode. More generally, this self-attractive case in higher dimensions is always faced with the potential for a collapse instability and hence rather restricted in the number of atoms that can be used and the phenomenology that can be observed without being subject to self-focusing effects.

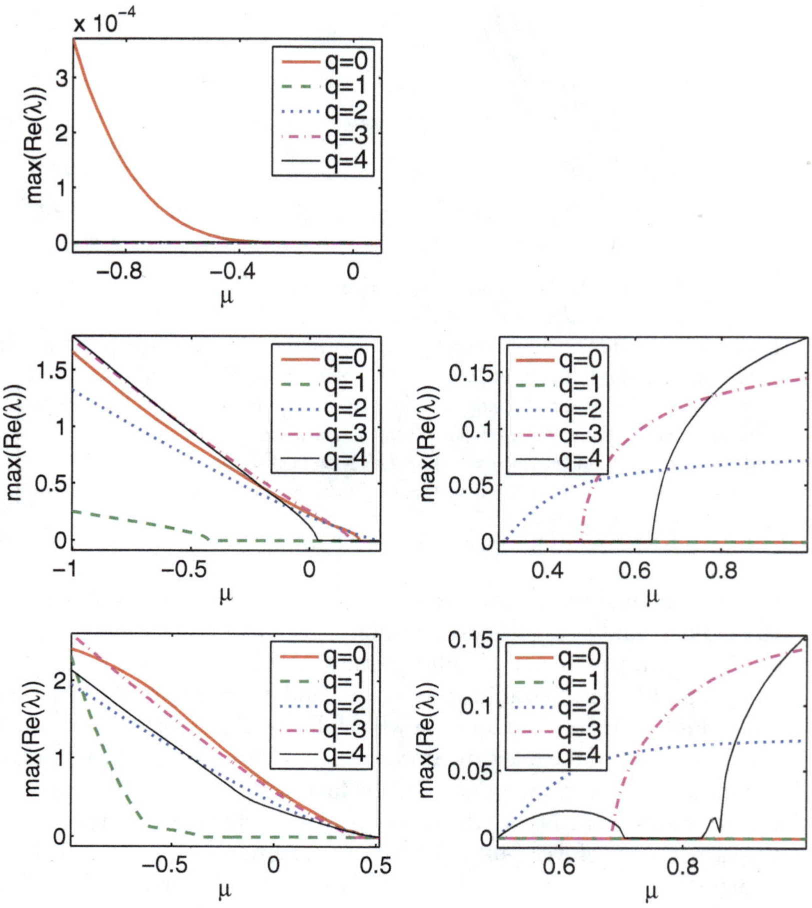

Figure 3.2. *Stability eigenvalues for chargeless ($S = 0$) solutions. The left column of plots corresponds to the real part of the primary eigenvalue for $q = 0, 1, 2, 3, 4$ for the ground state with $n_r = 0$ (top panel) and first and second excited states with $n_r = 1$ and 2 (middle and bottom panels, respectively) for $g = -1$; while the right column of plots corresponds to the case of $g = +1$. The ground state for $g = +1$ is omitted since the entire branch is stable. Adapted with permission from Ref. [269].*

Having offered an outline of the stability in Fig. 3.1, we now examine its details in what follows. In Figs. 3.2 and 3.3 we depict the real part of the primary ($q = 0, 1, 2, 3, 4$) eigenvalues for $S = 0$ and $S = 1$ (similar results in Ref. [269] are given also for $S = 2$) and for the states with $n_r = 0, 1$, and 2 in each case (i.e., the ground state and first two excited states in the top, middle, and bottom panel of each figure). The stability results showcase the presence of at least one unstable eigenvalue. In fact, there is only one instability eigenvalue for the ground state of the attractive case, and a larger number of such eigenvalues for excited states. The presence of such eigenvalues, whose larger magnitude is tantamount to more rapid dynamical development of the instability, indicates that the dynamical evolution of such stationary states—when small perturbations are added to them—will manifest the presence of such unstable eigenmodes through the deformation and likely destruction of the initial wave profile.

This is shown in detail for $S = 0$ in Figs. 3.4 and 3.5 (for $n_r = 0$ and $n_r = 1$). For the profile considered in Fig. 3.4 for $S = 0$, it is clear that the solution for the repulsive

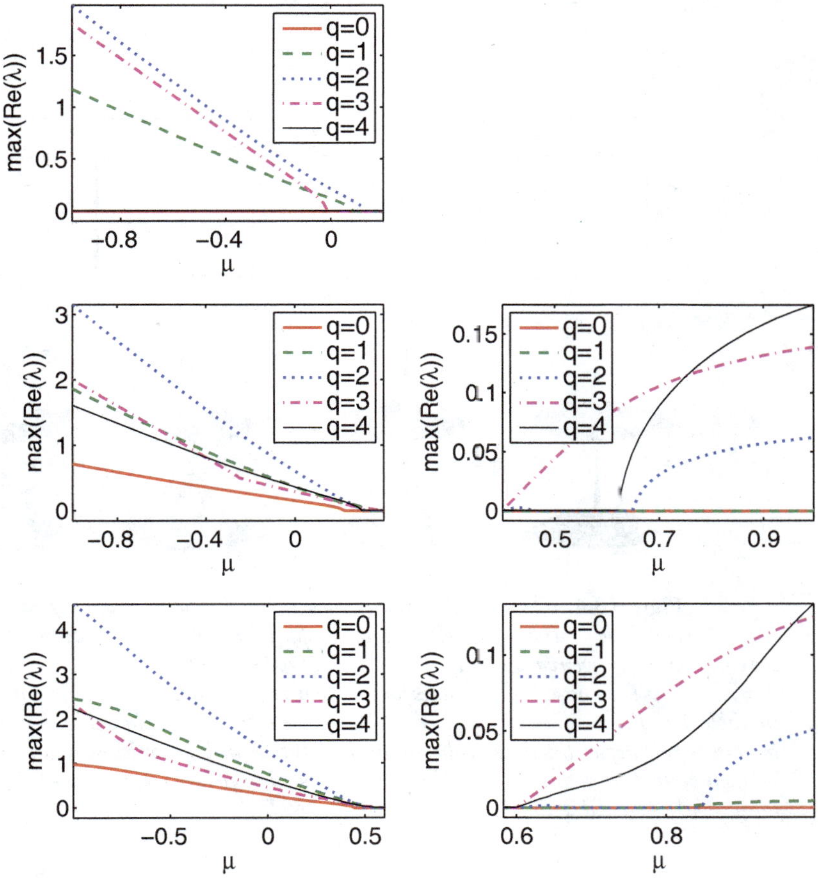

Figure 3.3. *Same as Fig. 3.2 for singly charged ($|S| = 1$) solutions. Adapted with permission from Ref. [269].*

case ($g = +1$) is stable since it corresponds to the ground state of the repulsive system. This can be contrasted to the solution for the attractive case ($g = -1$) which is weakly unstable, due to the $q = 0$ mode and the associated collapse, as discussed above (see also the bottom left panels of Fig. 3.4).

In Fig. 3.5 we present the equivalent results for the *first* excited states ($n_r = 1$ and $S = 0$); higher excited states were also systematically considered in Ref. [269]. In these cases the instability manifests itself with the presence of a richer scenario of unstable eigenvalues. Let us explain in detail the first excited state in Fig. 3.5 for the repulsive case, as the attractive case is subject not only to azimuthal modulations but also once again to catastrophic collapse. For $g = 1$, $S = 0$, and $n_r = 1$, the $q = 0$ eigenvalue is not unstable as there is no collapse in the repulsive case. The dynamic evolution of the instability in this case leads to the competition between the unstable eigenvalues with $q = 2, 3$, and 4 in this particular example. This is seen to be dominated by the most unstable mode $q = 3$, as can be evidenced by the three humps (surrounding the central peak) displayed at $t = 150$. Interestingly, this state is exactly the ring dark soliton (RDS) mentioned in Sec. 2.3.3 of the previous chapter. A slightly different perspective on the unstable evolution of this RDS is given in Fig. 3.6. There it can be seen that the RDS is indeed subject to snaking instability (due to the $q = 2, 3, 4$ modes described above), leading to the formation of

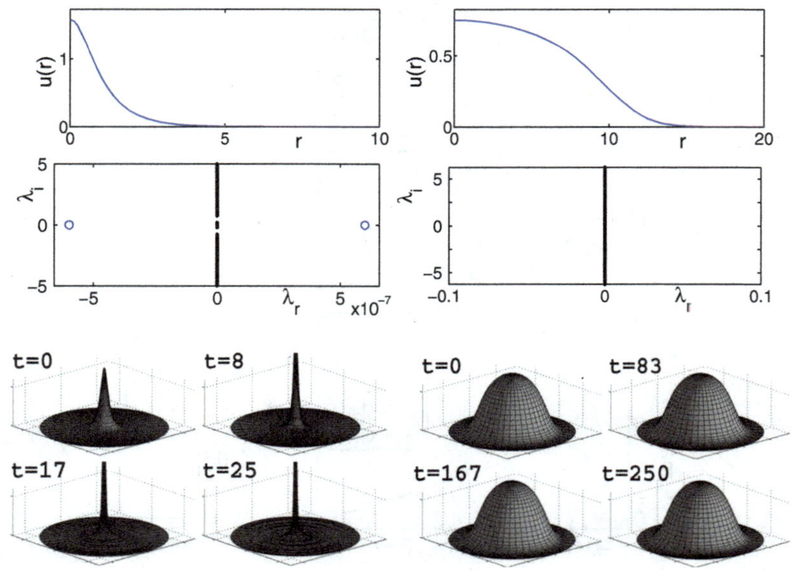

Figure 3.4. *Data for the ground state ($n_r = 0$) for $S = 0$ for $g = -1$ (left panels) and $g = +1$ (right panels). The top panels show the profile of the solution and the middle panel panels show the corresponding eigenvalues for $q = 0, \ldots, 50$ in Eq. (3.6) on the complex plane (λ_r, λ_i) of eigenvalues $\lambda = \lambda_r + i\lambda_i$. The unstable eigenvalues for different values of q are given different symbols as described in Table 3.1. Eigenvalues with* $\mathrm{Re}(\lambda) < 10^{-7}$ *or for $q > 11$ are plotted with a small dark dot. The bottom panels depict the time evolution of the solution in harmonic oscillator units after an initial random perturbation of* 10^{-2}. *All solutions correspond to $\mu = -0.5$ for $g = -1$ and norm $N|g| = 100$ for $g = +1$. Adapted with permission from Ref. [269].*

Table 3.1. *Table of symbols used for the unstable eigenvalues in Figs. 3.4–3.8. For eigenvalues smaller than* 10^{-7} *or $q > 11$ we use a small black dot. Adapted with permission from Ref. [269].*

q value	0	1	2	3	4	5	6	7	8	9	10	11	12–50
symbol	○	×	+	∗	□	◊	▽	△	◁	▷	⬠	⬡	•

vortex-antivortex pairs in multiples of four, which are initially set along a ring, forming a so-called vortex necklace. Eventually, this pattern relaxes to a set of four pairs located on a ring, which oscillates in the radial direction between the same limits which confined the oscillations of the original RDS; simultaneously, the pairs perform oscillatory motion along the ring [265].

In general, for the repulsive cases where collapse is absent, we observe that in all cases the solution is subject to symmetry-breaking azimuthal modulations that produce coherent structures somewhat reminiscent of the "azimuthons" of Ref. [490]. In the attractive cases, on the other hand, collapse appears to ubiquitously occur due to the mode with $q = 0$ and additionally to that azimuthal modulations appear to emerge.

The above discussion describes in detail the stability and dynamics for the non-topologically-charged ($S = 0$) solutions. Let us now describe the case when the solutions have an intrinsic topological charge, namely $|S| > 0$. In Fig. 3.7 we display the results for $n_r = 0$ in the presence of unit topological charge (or vorticity), namely $S = 1$; states with $n_r = 1$ and $n_r = 2$ (for $S = 1$) are presented in Ref. [269]. It can be directly observed that this

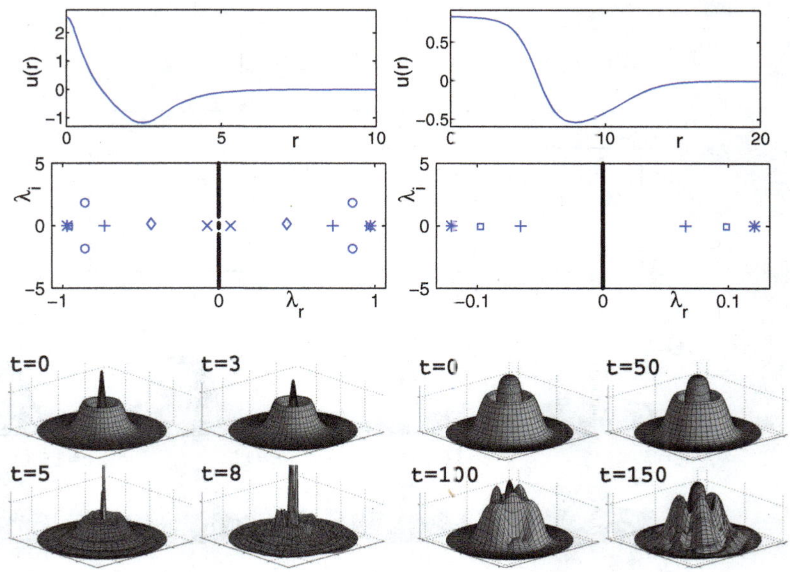

Figure 3.5. *Same as Fig. 3.4 ($S = 0$) for the first excited state ($n_r = 1$). Adapted with permission from Ref. [269].*

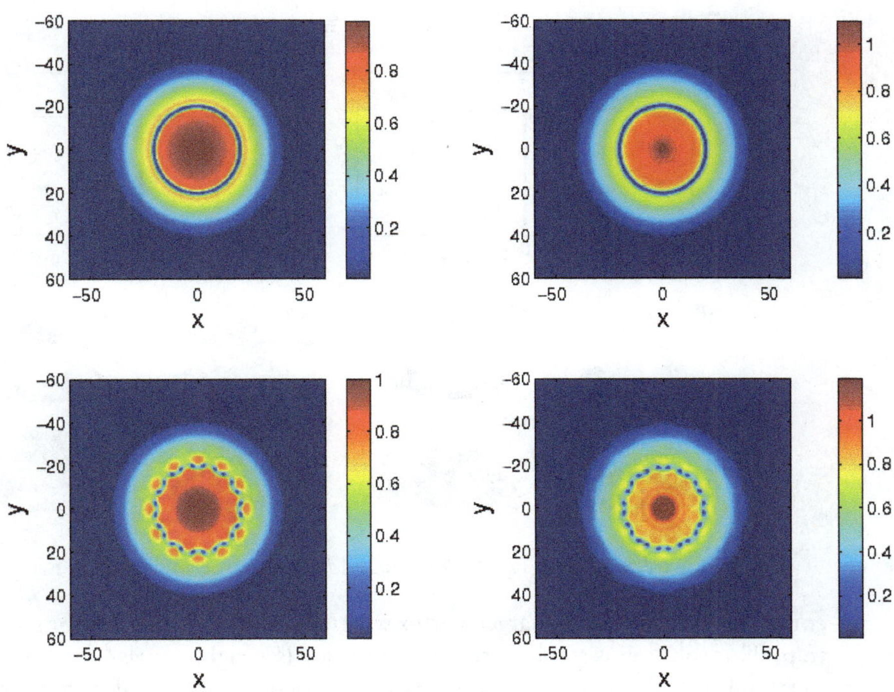

Figure 3.6. *Contour plots of the density of a disk-shaped condensate carrying a stationary RDS for $\mu = 1$ and $\Omega = 0.0035$, which develops the snaking instability. The top left and top right panels show, respectively, the initial condition ($t = 0$) and a snapshot at $t = 40$, while the bottom left and right panels show, respectively, the onset of the snaking instability ($t = 80$) and the formation of the vortex necklace ($t = 100$). Adapted with permission from Ref. [75].*

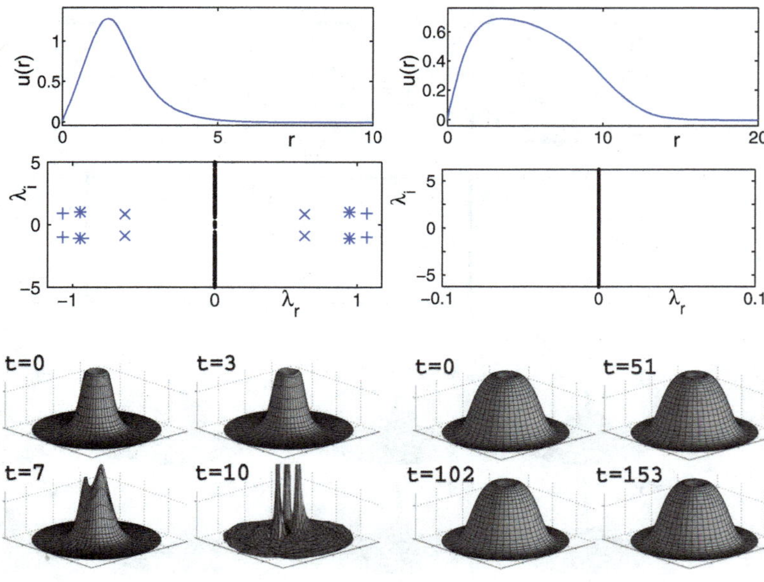

Figure 3.7. *Same as Fig. 3.4 for* $S = 1$. *Adapted with permission from Ref. [269].*

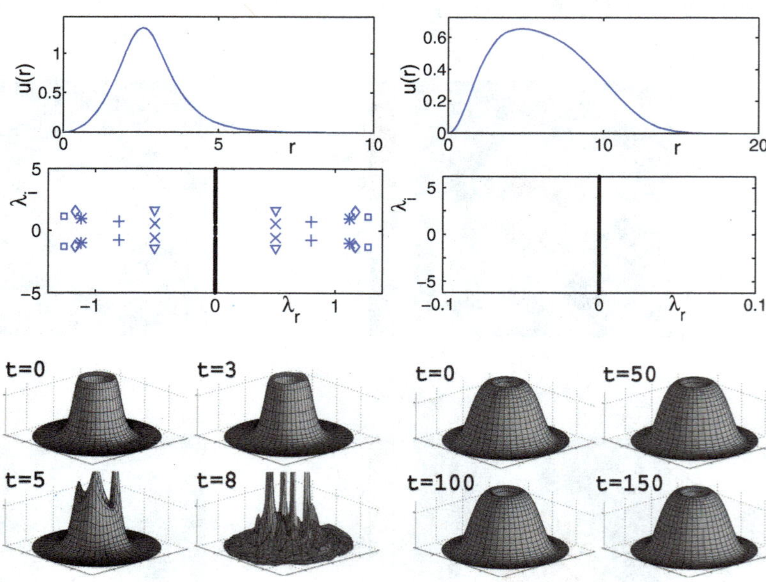

Figure 3.8. *Same as Fig. 3.4 for* $S = 2$. *Adapted with permission from Ref. [269].*

corresponds to the singly charged vortex solution supported at the bottom of the parabolic trap. This solution is stable in the repulsive case (see right panels—further discussion on the stability of this state will follow in the next section) and unstable in the attractive case (see left panels). It is interesting to note that the instability of the vortex solution in the attractive case is not driven by collapse (at least not through the $q = 0$ eigenvalue) since the vorticity tends to push the mass away from the center ($r = 0$) of the trap. This is a general feature of the cases with $|S| > 0$, where collapse appears to arise in the rings of the cloud, rather than at its center. Finally, in Fig. 3.8 we display similar results for the doubly charged case $S = 2$ (but again only for $n_r = 0$, while $n_r \neq 0$ are omitted and can be found

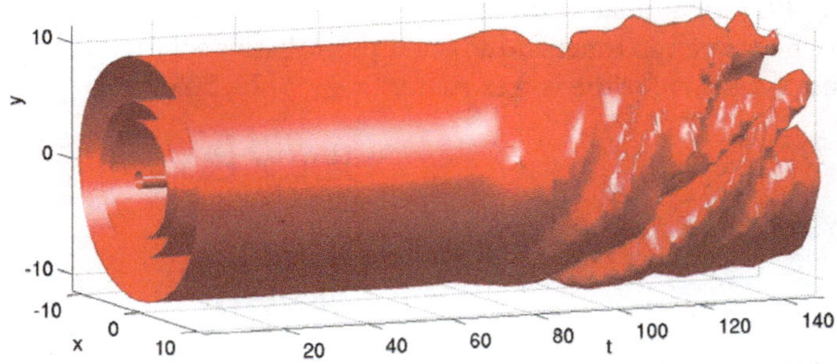

Figure 3.9. *Spatio-temporal isodensity contour for the first excited state $n_r = 1$ for $g = +1$, and $S = 2$. Clearly observable is the rotation of the unstable $q = 4$ mode due to the intrinsic vorticity ($S = 2$) of the solution. Adapted with permission from Ref. [269].*

for completeness in Ref. [269]). The latter is the first example of a vortex that is unstable in suitable intervals of atom numbers, even in the case of repulsive condensates, an instability which—for this, as well as higher charges—has also been considered experimentally, as we will see in more detail below.

It is worth mentioning that the unstable modes for vorticity-carrying solutions ($|S| > 0$) tend to rotate as they are growing. This effect can be clearly seen in the 3D isodensity contour plot depicted in Fig. 3.9 corresponding to the first excited state with $S = 2$ and $g = +1$. Interestingly, a single solution with multiple radial nodes (excited states) is able to pick out more than one growing mode since each ring might be affected by a different q mode. From here on, however, we focus our attention (for the most part in what follows) on the most fundamental solutions with $n_r = 0$.

3.2 ▪ Stability and Dynamical Properties of a Single-Charge Vortex

We now turn to a more detailed examination of the basic stability and dynamical properties of the quintessential nonlinear (topological) state of the 2D condensates, namely the vortex. Vortices are one of the most thoroughly studied type of excitations in the realm of BECs. In two- (but also often in higher-) dimensional settings these structures are also of particular interest not only in nonlinear optics [491, 492] (see also the reviews [35, 493]), but more broadly in nonlinear field theories in various branches of science [54]. Nevertheless, BECs represent a pristine setting where numerous features of the exciting nonlinear dynamics of single- and multicharge vortices, as well as of vortex crystals and vortex lattices, can be not only theoretically studied, but also experimentally observed. More specifically, from the viewpoint of experiments, the first experimental observation of vortices [146] by means of a phase-imprinting method between two hyperfine spin states of a ^{87}Rb BEC [494] paved the way for a systematic investigation of their dynamical properties. Stirring the BECs [495] above a certain critical angular speed [496, 497, 498] led to the production of few vortices [498] and even of robust vortex lattices [499]. Other vortex-generation techniques were also used in experiments, including the breakup of the BEC superfluidity by dragging obstacles through the condensate [248], as well as nonlinear interference between condensate fragments [158]. In addition, apart from unit-charged vortices, higher-charged vortex structures were produced [500, 501] and their dynamical (in)stability was examined.

In the presence of a parabolic trap as in our considerations herein, in order to study the effect of the potential on the vortex, one can follow the lines of Ref. [502] (see also

Ref. [491] for similar work in the context of optics) and use a matched asymptotics approach between an inner and an outer perturbative solution leading to the following equations of motion for the vortex center (x_v, y_v) [76, 503]:

$$\dot{x}_v = \frac{\Omega^2}{2\mu}\log\left(A\frac{\mu}{\Omega}\right)y_v, \tag{3.7}$$

$$\dot{y}_v = -\frac{\Omega^2}{2\mu}\log\left(A\frac{\mu}{\Omega}\right)x_v, \tag{3.8}$$

where A is an appropriate numerical factor (detailed comparison with numerics yields very good agreement in the Thomas–Fermi regime, e.g., for $A \approx 8.88 \approx 2\sqrt{2}\pi$). These results suggest a precession of the vortex in the harmonic trap with a frequency

$$\omega_{\rm pr} = \frac{\Omega^2}{2\mu}\log\left(A\frac{\mu}{\Omega}\right), \tag{3.9}$$

where the precession frequency $\omega_{\rm pr}$ corresponds to the "anomalous" (mode) as we will discuss in what follows. In the context of trapped BECs this precession frequency calculation was first developed in the work of Fetter; see, e.g., the early review [76] and for a recent exposition, including extensions to multicomponent systems, see Ref. [504]. A more mathematical analysis of the relevant estimates can be found in Ref. [505].

Naturally, and as may be expected based also on the results of the previous section, the position $(0,0)$, i.e., the center of the trap constitutes an equilibrium [a stable one according to the dynamics of Eqs. (3.7) and (3.8)]. It is this stationary equilibrium state $u_0(x,y)$ that we identified in the previous section. The linearization around such a state performed through the Bogoliubov–de Gennes analysis [43, 42, 44] involves the ansatz

$$u = u_0(x,y) + \left[a(x,y)e^{i\omega t} + b^*(x,y)e^{-i\omega^* t}\right], \tag{3.10}$$

where we now use eigenfrequencies (ω) instead of eigenvalues (λ) as in Eq. (3.6). Subsequent derivation of the linearized equations and the ensuing Bogoliubov–de Gennes (BdG) eigenvalue problem yields the eigenfunctions $(a(x,y), b(x,y))^T$ and eigenfrequencies ω. As concerns the latter, we note that due to the Hamiltonian nature of the system, if ω is an eigenfrequency of the Bogoliubov spectrum, so are $-\omega$, ω^* and $-\omega^*$; see, e.g., Ref. [506]. Notice that a linearly stable configuration is tantamount to $\mathrm{Im}(\omega) = 0$, i.e., all eigenfrequencies being real.

An important quantity resulting from the BdG analysis is the amount of energy carried by the normal mode with eigenfrequency ω, namely,

$$E = \iint (|a|^2 - |b|^2)\omega \, dx \, dy. \tag{3.11}$$

The sign of this quantity, known as *Krein sign* [195], is a topological property of each eigenmode. In the case of the vortex, for one of the eigenvalues with $\omega > 0$, this sign is negative. The corresponding mode is called *anomalous mode* [43], or *negative energy mode* [507], or mode with *negative Krein signature* [195] (see also a relevant discussion in Sec. 2.2.3 for the case of dark solitons). If such a mode becomes resonant with a mode with positive Krein signature then, typically, complex frequencies appear in the excitation spectrum, i.e., a dynamical instability arises [195]. For a general discussion of the properties of this signature from a topological side see, e.g., Ref. [195], while for a discussion from a perspective of applications to GPE systems, see, e.g., Ref. [507].

We note in passing that similar setups were investigated in Ref. [508] for finite displacements of the vortex from the center of the trap. Also, in Ref. [509] a variational approximation was employed and the frequency of the anomalous mode was derived with a similar functional form. However, in Ref. [509], the case of a BEC unbounded in the axial direction ($\omega_z = 0$) was considered and, as a result, the constant in Eqs. (3.7) and (3.8) [see also Eq. (3.9)] was found to take a different value, $A = 2$. It is also worth noting that the connection between quantum fluctuations and anomalous modes of matter-wave vortices under Magnus forces was considered in Ref. [510].

It is important, at this stage, to make a few comments regarding the nature of the Bogoliubov spectrum resulting from the linearization around the vortex. The system at hand, namely the disk-shaped condensate carrying the vortex, is *not* in the ground state (a similar situation occurs in the 1D analogue of the system, namely a quasi-1D BEC carrying a dark soliton—cf. Sec. 2.2.3). The existence of the anomalous mode, characterized by negative energy, indicates that the vortex (and the dark soliton in the 1D case) is thermodynamically unstable and, in the presence of dissipation, the system is driven towards a lower energy configuration, namely the ground state. Also, as the chemical potential is increased, the eigenfrequency of the anomalous mode of the vortex (similarly to the case of the dark soliton of the 1D GPE) deviates from its value $\omega = \Omega$ in the linear limit. In the latter limit, the vortex is represented by the linear superposition $|1,0\rangle + i|0,1\rangle$, where $|m,n\rangle$ denotes the mth linear eigenstate of the quantum harmonic oscillator along the x direction and nth one along the y direction (see also Fig. 3.10). Generally, the anomalous mode (in both 1D and 2D cases) is the *lowest* excitation frequency of the system and the only one below the trap frequency, which is associated with the doubly degenerate (in the two-dimensional case) Kohn mode corresponding to the dipolar motion of the condensate [42, 43]. Since this anomalous mode is related to the thermodynamic instability of the vortex that, if activated, will drive it away from the center, it is expected that this mode is inherently associated with the motion of the vortex. Then, given that this is an oscillatory mode associated with the motion of the vortex and since we saw that when displaced from the center, the vortex executes precessional motions with a particular frequency ω_{pr}, it is natural to expect that the frequency of the anomalous mode is indeed the frequency of near-equilibrium precessional motions. However, in the case of the vortex (and contrary to what is the case for the dark soliton), there is one more frequency, which may be smaller than ω_{pr}, at least for small chemical potentials. However the corresponding mode grows monotonically away from the origin and thus becomes larger for increasing chemical potential than the anomalous mode and the Kohn mode. This frequency was described through a small parameter expansion in Sec. V.B of Ref. [511]. The relevant eigenfrequency is given (in our units) by the expression

$$\omega \approx \mu - 2\Omega, \tag{3.12}$$

which becomes increasingly more accurate as $\mu \to 2\Omega$. This is in contrast to the case for the expression of the precession frequency Eq. (3.9), which should be increasingly more accurate in the Thomas–Fermi (TF) limit, corresponding to large μ.

In Fig. 3.10 we show the numerically obtained eigenfrequency ω of the Bogoliubov spectrum as (blue) solid lines as a function of the chemical potential μ and the analytical predictions of Eq. (3.9) (green) dashed-dotted line and Eq. (3.12) (red) dashed line. The numerically obtained frequencies are real, denoting that the system is dynamically stable. We observe that in accordance with the analytical predictions, the lowest modes are (i) the one monotonically increasing away from zero and (ii) the anomalous mode, connected to the vortex precession (see previous section), bifurcating from the constant dipolar mode

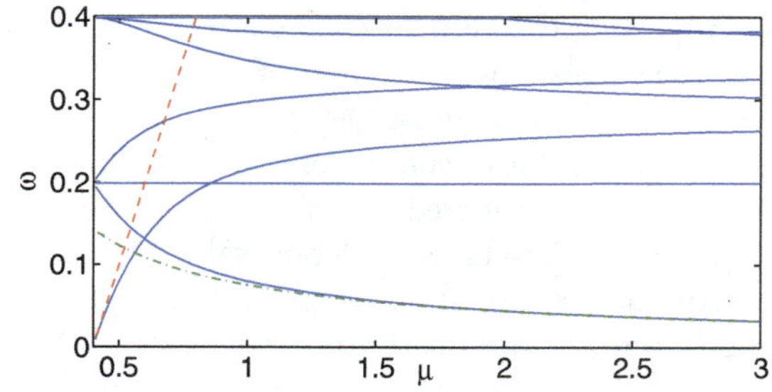

Figure 3.10. *The eigenfrequencies ω of the Bogoliubov spectrum for a harmonically confined singly charged vortex as a function of the chemical potential μ (for a trap strength $\Omega = 0.2$). Theoretical predictions are given by $\omega = (\mu - 2\Omega)$ (red dashed line) for the mode monotonically increasing from zero and $\omega = (\Omega^2/(2\mu))\log(A\mu/\Omega)$ (green dashed-dotted line); the constant A was chosen to be $2\sqrt{2}\pi \approx$ 8.886. Adapted with permission from Ref. [503].*

in the linear limit. For the monotonically increasing mode, we notice that the nonradial nature of the solutions at hand (due to their phase profile) leads to the absence of additional symmetries of the eigenvalue problem away from the linear limit. The only symmetry generally present is that of the phase or gauge invariance, associated with the conservation of the number of particles. This sustains a pair of linearization eigenfrequencies at $\omega = 0$, but as discussed in Ref. [511], at the linear limit the dimension of the corresponding kernel is 4, hence an eigenfrequency pair should depart from the origin (at least for small μ) according to Eq. (3.12). As observed in Fig. 3.10, the predictions for the linear mode (3.12) especially for small μ and the anomalous mode (3.9) especially for large μ are in good agreement with the numerical results. Naturally, deviations for the linear mode are observed for larger chemical potentials. On the other hand, as concerns the precession frequency (3.9), we notice its monotonically decreasing dependence on μ for given Ω (the latter was set to $\Omega = 0.2$ in Fig. 3.10), its bifurcation from the Kohn mode eigenfrequency limit and its excellent agreement with the theoretical prediction in the TF limit (for all $\mu > 1$). We note in passing that in this 2D case, a pair of Kohn modes can be seen to be preserved at $\omega = \Omega = 0.2$, being associated with the dipole oscillations of the condensate along the two spatial directions, while the fourth mode at 0.2 in the linear limit results in a monotonically growing eigenfrequency as μ is increased.

We have also tested the validity of the analytical predictions concerning the two lowest eigenfrequencies for different values of Ω, and as a function of μ (see Fig. 3.11). Once again, a very good agreement of the two asymptotic theoretical descriptions in their respective limits is found (for more details see Ref. [503]).

These results for the linear stability and the dynamics of the vortex can be corroborated by means of direct numerical simulations of the 2D GPE for initial states containing a single vortex (displaced from its center of the trap equilibrium). In order to determine the position of the vortex as a function of time we first compute the fluid velocity [512]

$$\mathbf{v}_s = -\frac{i}{2}\frac{u^*\nabla u - u\nabla u^*}{|u|^2}. \tag{3.13}$$

The fluid vorticity is then defined as $\omega_{\text{vor}} = \nabla \times \mathbf{v}_s$. Due to our setup, the direction of the fluid vorticity is always the z direction and, therefore, we can treat this quantity as a scalar. Furthermore, we investigate single vortex states leading to a single maximum of the fluid vorticity at the position of the vortex. This allows us to determine the position of the vortex by determining the center of mass of the vorticity ω_{vor}.

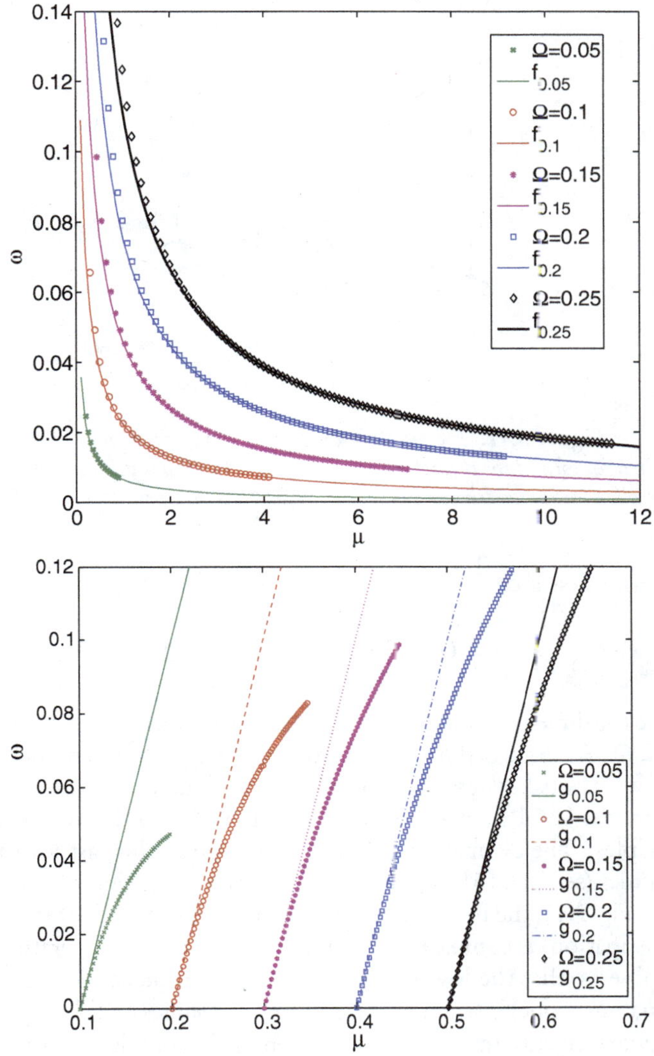

Figure 3.11. *Top: Dependence of the anomalous mode eigenfrequency on the chemical potential for different trap frequencies* Ω. *The data points are interpolated by the functions* $f_\Omega(\mu) = \omega_{\rm pr}$ *of Eq. (3.9). Bottom: Similar to top panel, but for the mode bifurcating from zero, as compared to the theoretical prediction* $g_\Omega(\mu) = \mu - 2\Omega$ *of Eq. (3.12). Adapted with permission from Ref. [503].*

Figure 3.12 shows the evolution of a single vortex for $\mu = 3$. We displaced the vortex initially from the center of the trap to $(x_0, y_0) = (-1.5, 0)$ and propagated the state numerically using the 2D GPE. These results are compared to the solutions of Eqs. (3.7) and (3.8) $x = x_0 \cos(\omega_{\rm pr} t)$ and $y = y_0 \sin(\omega_{\rm pr} t)$ and the initial position (x_0, y_0). The theoretical predictions agree very well with our numerical findings: the vortex oscillates around the center of the trap with constant frequency and radius.

A final topic that we would like to touch upon in the context of the stability and dynamics of a single vortex is the effect of dissipative perturbations that may stem, e.g., from the coupling of the BEC to the thermal cloud. In this case, a pertinent model to consider is the dissipative GPE (cf. Sec. 2.6.1), which can be expressed in the following

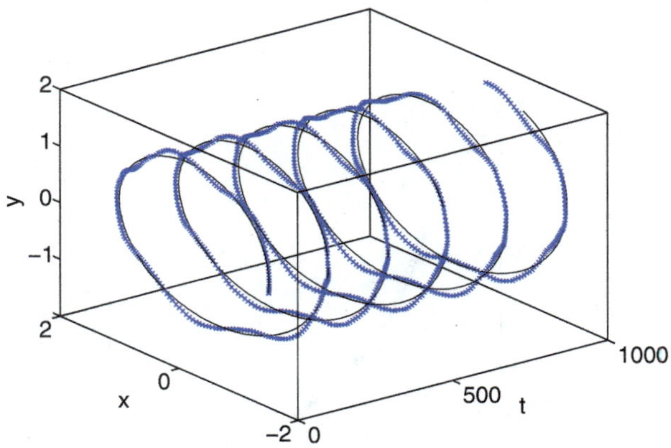

Figure 3.12. *The panel shows the trajectory of the vortex for $\mu = 3$ obtained by direct integration (crosses) and the theoretical prediction (solid line) obtained by solving Eqs. (3.7) and (3.8). Notice the excellent agreement between the two. Adapted with permission from Ref. [503].*

dimensionless form:

$$(i-\gamma)u_t = -\frac{1}{2}\Delta u + V(r)u + |u|^2 u - \mu u, \tag{3.14}$$

where the dimensionless parameter γ can be associated with the system's temperature (see Sec. 2.6). As discussed in the 1D setting, this type of phenomenological model has been used in reasonable agreement with the full dynamics of more complex models such as the stochastic GPE, incorporating the effect of coupling to the thermal cloud from first principles. The dissipative GPE has been used in the past for the analysis of the vortex dynamics (see Ref. [513]); here, we follow the exposition of Ref. [503].[17]

In Fig. 3.13, the BdG spectrum of a vortex for the case of $\gamma = 0$ (zero temperature, i.e., no dissipation) is compared to that of $\gamma = 0.2$ (finite temperature, i.e., dissipation). One can observe that the lowest frequency mode of the condensate (which for $\mu = 1$ is the anomalous mode) is the one that, for nonzero values of γ, immediately acquires a positive imaginary eigenfrequency becoming genuinely complex, contrary to what is the case for all other eigenmodes of the system, which are, in fact, decaying. This property was rigorously proved in Ref. [196] for negative Krein signature (i.e., negative energy) eigenmodes, namely that their bifurcation upon such dissipative perturbations happens *oppositely* to that of all other modes (of positive energy) of the system. This remarkable feature is directly consonant with the property of this excited state of the system resulting, via the effect of dissipation (and through the complex nature of the relevant eigenmode), eventually into the ground state of the system. Moreover, notice that this complex eigenmode also implies the combination of growing amplitude with the previously analyzed precessional motion, leading to the spiraling of the vortex core toward the edges of the TF cloud (see Fig. 3.14) and its eventual disappearance, in favor of the ground state of the system. This is also related to the notion of the dynamic stability but *thermodynamic instability* of the vortex at $T = 0$. In the latter case, the solution bears no dynamically unstable eigenmode, but the existence of the anomalous mode signals its excited nature. Hence, should

[17] Note that in the above dissipative model the interaction between the thermal cloud and the condensate is only modeled by particle exchange resulting in the dissipative factor γ; we should note that physically, the relevant case is that of $\gamma \ll 1$, although for illustration purposes we explore the phenomenology also far from that regime (as in the 1D case studied in Sec. 2.6).

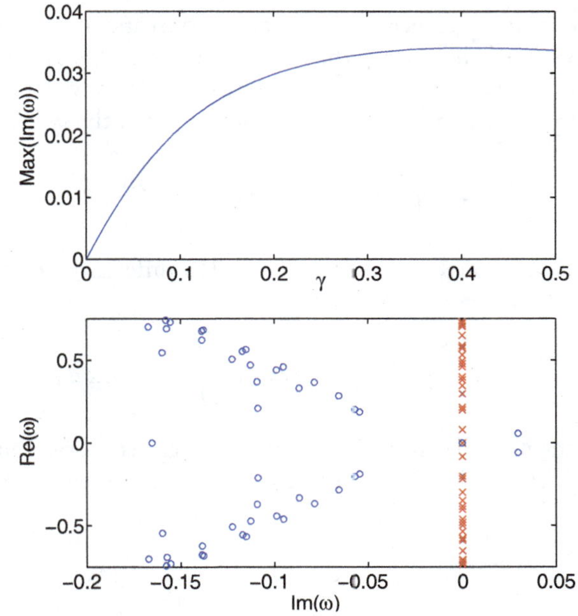

Figure 3.13. *The top panel shows the immediate acquisition of a nonvanishing imaginary part of the eigenfrequency associated with the anomalous mode, as soon as the temperature-dependent dissipative prefactor γ becomes nonzero. The bottom panel highlights the special behavior of the anomalous mode by illustrating the BdG spectrum for the cases of $\gamma = 0$ (red crosses) and that of $\gamma = 0.2$ (blue circles). Parameters $\mu = 1$ and $\Omega = 0.2$ were used. Adapted with permission from Ref. [503].*

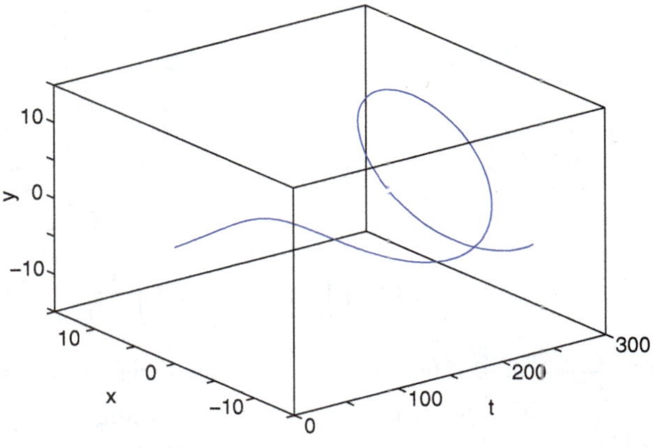

Figure 3.14. *Trajectory of the vortex for the case taking into account dissipation, namely integrating Eq. (3.14) with $\gamma = 0.2$. Other parameters are $\mu = 1$ and $\Omega = 0.2$. Adapted with permission from Ref. [503].*

some dissipative channel become available in the system, it would enable the possibility of a dynamic instability, which is prohibited at $T = 0$, due to the conservative nature of the system. This instability, in turn, renders possible the return of the system from this excited state to its ground state through a suitable shedding of the vortex excitation (i.e., its decay in the periphery of the cloud).

In a recent work, following up on the important earlier studies of Ref. [514], an analytical calculation was presented [515] that captures the main features of the vortex dynamics in the combined presence of the parabolic trap and also of effective, thermally induced dissipation. We briefly recap the main elements of this calculation in order to connect it to the above numerical observations (both at the steady state and at the dynamical level). Starting from the conserved, $\gamma = 0$ limit, energy of the system,

$$E = \frac{1}{2}\iint \left\{\left(|u_x|^2 + |u_y|^2\right) + (|u|^2 - \mu)^2 + 2V|u|^2\right\} dx\, dy, \tag{3.15}$$

we can derive through a direct calculation [by differentiating and substituting the dynamical Eq. (3.14)] the moment identity

$$\frac{dE}{dt} = -2\gamma \iint |u_t|^2 dx\, dy, \tag{3.16}$$

which is the natural generalization of the energy conservation of the $\gamma = 0$ case. Using a variational calculation for the energy of a vortex presented earlier in Ref. [516], we have that

$$\frac{dE}{dt} \approx -\omega_{\mathrm{pr}}(2x_0\dot{x}_0 + 2y_0\dot{y}_0) \tag{3.17}$$

for a vortex centered at $(x_0(t), y_0(t))$. On the other hand, evaluating the right-hand side of the moment identity of Eq. (3.16), we find that

$$\begin{aligned} -2\gamma \iint |u_t|^2\, dx\, dy &= -2\gamma \iint |u_t(x - x_0(t), y - y_0(t))|^2 dx\, dy \\ &= -2\gamma \iint \left(\frac{\partial v}{\partial \xi}(-\dot{x}_0) + \frac{\partial v}{\partial \eta}(-\dot{y}_0)\right) \\ &\qquad \times \left(\frac{\partial v^*}{\partial \xi}(-\dot{x}_0) + \frac{\partial v^*}{\partial \eta}(-\dot{y}_0)\right) d\xi\, d\eta \\ &= -2\gamma\left(\dot{x}_0^2 s + \dot{y}_0^2 s\right), \end{aligned} \tag{3.18}$$

where

$$s = \iint \frac{\partial v}{\partial \xi}\frac{\partial v^*}{\partial \xi} = \iint \frac{\partial v}{\partial \eta}\frac{\partial v^*}{\partial \eta} \tag{3.19}$$

and $\mathrm{Re}(\iint \frac{\partial v}{\partial \eta}\frac{\partial v^*}{\partial \xi}) = 0$. Here, we have used the variables $\xi = x - x_0(t)$ and $\eta = y - y_0(t)$, so that $v(\xi, \eta) = u(x - x_0(t), y - y_0(t))$. The resulting integral constant s can be directly evaluated [515], e.g., using Padé approximations to the vortex profile.

Using the above results, it is possible to combine the left- and right-hand sides of Eq. (3.16) and obtain

$$\omega_{\mathrm{pr}}(x_0\dot{x}_0 + y_0\dot{y}_0) = s\gamma\left(\dot{x}_0^2 + \dot{y}_0^2\right). \tag{3.20}$$

From this, we can infer the equations of motion of the single vortex structure. Looking for equations of motion that correspond to rotation with antidamping, we add a term proportional to $\dot{x}_0\dot{y}_0$ to both sides of Eq. (3.20) and split it into the following two equations:

$$\omega_{\mathrm{pr}}x_0 + \dot{y}_0 = s\gamma\dot{x}_0, \tag{3.21}$$

$$\omega_{\mathrm{pr}}y_0 - \dot{x}_0 = s\gamma\dot{y}_0. \tag{3.22}$$

Then, the analytical expression for the complex eigenfrequency $\omega = \omega_r + i\,\omega_i$ is

$$\omega_i = \mathrm{Im}(\omega) = \frac{\omega_{\mathrm{pr}} s\gamma}{1+(s\gamma)^2}, \tag{3.23}$$

$$\omega_r = \mathrm{Re}(\omega) = \frac{\omega_{\mathrm{pr}}}{1+(s\gamma)^2}. \tag{3.24}$$

In turn, at this level of approximation, this leads to an explicit analytical form of the above presented outward spiraling vortex trajectories as

$$x = e^{\omega_i t}\left[y_0 \sin(\omega_r t) + x_0 \cos(\omega_r t)\right], \tag{3.25}$$

$$y = e^{\omega_i t}\left[y_0 \cos(\omega_r t) - x_0 \sin(\omega_r t)\right]. \tag{3.26}$$

Hence, through the above analysis, we can fully corroborate both the steady state properties (i.e., the vortex spectrum and its complex eigenmode in the presence of dissipation), as well as their corresponding dynamical implications (the outward vortex center spiraling, as seen in the example of Fig. 3.14). It should be noted in passing that, in Ref. [515], extensions of these considerations to multivortex dynamics were given; additionally, an effort was made to render the relevant estimates rigorous. Furthermore, very good agreement was found between the above prediction and the corresponding linearization, as well as the full dynamical GPE results for a single vortex.

3.3 ▪ Dark Soliton Stripes

3.3.1 ▪ Transverse (Modulational) Instability of Dark Solitons in Higher Dimensions and How to Avoid It

In the absence of the external potential, $V(x,y)=0$, the 2D GPE admits an exact analytical solution, which is in fact a 1D dark soliton. Assuming that its nodal plane is oriented along the x direction, this solution is (see previous chapter for more details)

$$u(x,y,t) = \sqrt{\mu}\{\mathscr{B}\tanh[\sqrt{\mu}\mathscr{B}(x-vt)] + i\mathscr{A}\}e^{-i\mu t}. \tag{3.27}$$

Here, $\sqrt{\mu}\exp(-i\mu t)$ is the stationary background with density μ (which is the chemical potential) supporting the dark soliton. Further, $\sqrt{\mu}\mathscr{B}$ and $v=\sqrt{\mu}\mathscr{A}$ represent, respectively, the soliton's depth (or inverse width) and velocity, with $\mathscr{A}$ and $\mathscr{B}$ subject to the constraint $\mathscr{A}^2+\mathscr{B}^2=1$. Recall that the dark soliton with $\mathscr{A}=0$ ($\mathscr{A}\neq 0$) is stationary (moving), and is usually called a "black" ("gray") soliton.

Unfortunately, in the homogeneous medium these dark soliton stripe solutions are unstable against the effect of transverse modulations, which induce a snake deformation [35], causing the nodal plane to decay into vortex pairs. This instability, first revealed in the homogeneous limit of the 2D GPE in the work of Ref. [253], has been examined in the context of BECs in Refs. [170, 126], and the Bogoliubov spectrum of the dark soliton has been obtained; in this context, the relevant imaginary modes were identified to transfer the energy of the condensate to collective excitations parallel to the nodal plane destroying the configuration. The work of Ref. [517] examined the criterion for the transverse instability in the presence of the trap, testing it against direct numerical simulations, and exposing the possible dynamical scenarios and their dependence on the trap parameters. It is that latter approach that we follow in our exposition below.

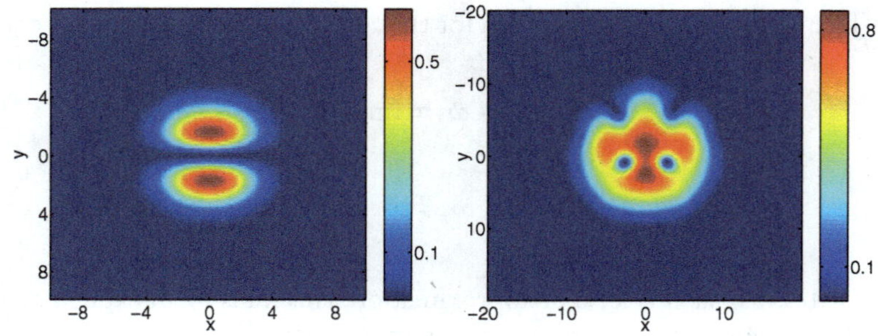

Figure 3.15. *The panels show the contour plots of the density* $|u|^2$ *for* $\Omega = 0.35$ *at* $t = 1000$ *(left) and* $\Omega = 0.15$ *at* $t = 190$ *(right). In the first case the transverse instability is clearly suppressed, while in the second it sets in, giving rise to a formation of a vortex pair. Adapted with permission from Ref. [517].*

In the absence of the potential, the transverse instability occurs for perturbation wavenumbers

$$k < k_{\mathrm{cr}} \equiv \left[2\sqrt{\sin^2\phi + u_0^{-2}\sin\phi + u_0^{-4}} - \left(2\sin\phi + u_0^{-2}\right)\right]^{1/2}, \tag{3.28}$$

where $\sin\phi$ is the dark-soliton velocity [35] and u_0 is the amplitude of the homogeneous background, connected with the chemical potential through $u_0^2 = \mu$. In the case of stationary (black) solitons of interest here, $\sin\phi = 0$ and hence $k_{\mathrm{cr}} = u_0^{-1}$. On the other hand, for $V(r) = \Omega^2 r^2/2$, the characteristic length scale of the BEC (i.e., the diameter of the TF cloud) is $\lambda_{\mathrm{BEC}} \approx 2\sqrt{2\mu}/\Omega$, i.e., the condensate extends over a region of (maximal) length of twice the Thomas–Fermi radius. Then, in order to suppress this infrared catastrophe, i.e., this short-wavenumber/long-wavelength type of transverse instability, we need to demand that the scale of the BEC be shorter than the minimal one for the instability. The corresponding condition reads

$$\Omega > \frac{\sqrt{2\mu}}{\pi u_0}. \tag{3.29}$$

To obtain the minimum value of Ω we need to know how μ is connected with u_0. As a first guess, in the absence of the dark soliton, one can assume $u_0^2 \approx \mu$ (close to the center of the BEC), which yields $\Omega > \sqrt{2}/\pi = 0.45$. Hence, stronger (the above threshold) trapping should "drown" the transverse instability and preserve dark soliton stripes on top of the TF cloud (i.e., stable "dipole" solutions). Note that in terms of real physical units, the above mentioned critical value of Ω may correspond, e.g., to a weakly interacting ^{87}Rb pancake condensate, containing $\approx 10^3$ atoms, confined in a trap with $\omega_r = 2\pi \times 5$Hz and $\omega_z = 2\pi \times 50$Hz.

The numerical tests of Ref. [517] indicated that the above threshold is an *overestimate* of the critical trapping frequency for the transverse instability, which is $\Omega_{\mathrm{cr}} \approx 0.31$. Figure 3.15 shows a dynamical evolution example of the dipolar solution, for $\Omega = 0.35$ (left) and $\Omega = 0.15$ (right). Clearly, in the former case the transverse instability is suppressed, while in the latter a vortex pair has been formed demonstrating the dynamical instability of the configuration. This type of numerical simulation naturally motivates the study of such vortex pairs and their connection to the planar dark soliton, a theme that we will explore below.

We have monitored the asymptotic, long-time evolution of the dipole and have observed the following phenomenology: for $0.18 \lesssim \Omega \lesssim 0.31$, while the stripe is dynami-

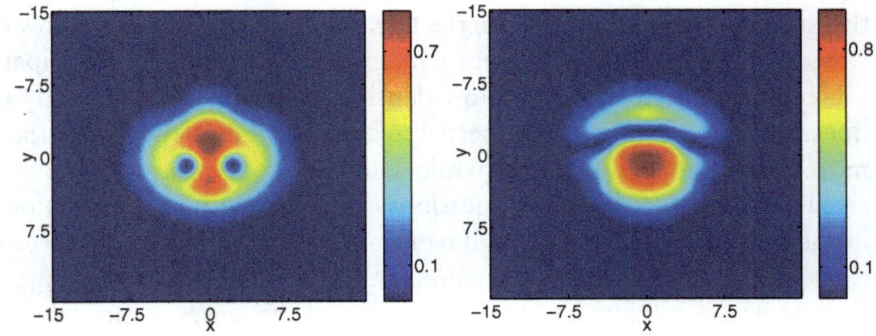

Figure 3.16. *Snapshots of a vortex-pair evolution in a case where snaking instability has set in ($\Omega = 0.2$). In the left panel ($t = 200$) the formed vortex pair is shown, while the right panel ($t = 220$) shows the recombination of the two vortices, resulting in the regeneration of a dark stripe structure. The latter is unstable and decays at longer times ($t \approx 400$). Adapted with permission from Ref. [517].*

cally unstable, there is not sufficient "space" for the instability-induced vortices to fully develop; as a result, after their formation, they subsequently recombine and disappear. This behavior is shown in Fig. 3.16 (for $\Omega = 0.2$): the two vortices formed are shown at $t = 200$ (left panel), while at $t = 220$ they recombine to form a transient dark stripe (right panel). This configuration is unstable and it subsequently breaks up to a new vortex pair (not shown here), which eventually recombines at longer times ($t \approx 400$). It is interesting to note that as, in the present case, the available size of the condensate is of the order of a few healing lengths ξ (in our units, $\xi = 1/\sqrt{2}$), the two vortices formed hardly fit the condensate size (recall that the vortex core is of order of ξ [76]). This is a possible qualitative explanation of this recombination, whose origin is the competition of the length scales available in the BEC. Finally, it should be noted that for $\Omega \lesssim 0.18$, the vortices will survive in the asymptotic evolution of the system, and naturally the weaker the trapping the larger the number of "engulfed" vortices generated due to the stripe breakup.

A question that naturally arises in the results above concerns the disparity between the critical point theoretical estimate for the transverse instability and the corresponding numerical finding. To address this issue, it was argued in Ref. [517] that the theoretical stability analysis of Ref. [35] is performed for the infinite homogeneous medium (the dark-soliton pedestal), in the absence of a magnetic trap. In the presence of the trap on the one hand, the background is inhomogeneous, while, on the other hand, for tight traps resulting in small condensate sizes, the presence of the dark soliton at the BEC center, significantly modifies the maximum density. Thus, one should expect that the relation $u_0^2 = \mu$ should be modified as $\delta u_0^2 = \mu$, where the "rescaling" factor $\delta < 1$. Based on the results of the numerical simulations, it was argued that a reasonable estimate is $\delta = 1/2$, leading to a new Ω minimum, i.e., $\Omega = 1/\pi \approx 0.318$. This modified criterion for the suppression of the transverse instability, namely $\Omega > 0.318$, was found to be in very good agreement with the above mentioned numerically found condition $\Omega > 0.31$.

Finally, it is worth mentioning that it is also possible to stabilize dark soliton stripes with localized stripe-type potentials as it shown in Sec. 3.9.1 below.

3.3.2 ▪ A Complementary Perspective: Bifurcation of the Dark Soliton Stripe and Its Instability in the Vicinity of the Linear Limit

Our exposition so far has viewed the instability of the planar dark soliton and the concomitant formation of vortices in the large-density Thomas–Fermi limit. Here we will take a different (yet complementary in a number of ways) perspective. We will explore this state as one bifurcating from the linear limit of the problem at low density. These two ways of approaching the problem will, in fact, constitute our principal analytical tools in much of what follows. In the low density limit, we will generically use the underlying linear states to construct nonlinear states and to study bifurcations therefrom that lead to more complex daughter states (while destabilizing the parent states from which these derived). Few mode Galerkin truncations of the same type as used in double-well potentials in the previous (1D) chapter will be implemented in an attempt to capture such bifurcations. On the other hand, the complementary perspective for large densities and high nonlinearities will involve the examination of the waves as individual particles (solitons, or most typically in 2D as vortices) which are moving (e.g., precessing) and interacting within the BEC.

The discussion, more specifically, of the cascade of bifurcations arising in the context of the dark soliton stripe will follow the analysis of Ref. [518], in which the trap strength was fixed (to $\Omega = 0.2$) and the variable parameter over which the relevant bifurcation structure was unfolded was the chemical potential μ. In the noninteracting (linear) case, the Hamiltonian describing the stationary states can be written as $\hat{H} = \hat{H}_x + \hat{H}_y$, with $\hat{H}_x$ and $\hat{H}_y$ being the Hamiltonian of a single particle in a harmonic trap oriented along the x and y directions, respectively. Therefore, the linear states factorize as well, and can be written as $\Psi_{nm}(x,y) = \psi_n(x)\psi_m(y)$ with $\psi_n(x)$ and $\psi_m(y)$ being the states associated with the nth and mth eigenvalues of $\hat{H}_x$ and $\hat{H}_y$, respectively. The resulting states $\Psi_{nm}(x,y)$ with energy $\mu = (n+m+1)\Omega$ denote the eigenstates of the linear limit. Due to the rotational symmetry of the problem, one can rotate any solution around the center of the trap obtaining another solution. Hence, when we consider in the following a solution with a distinct symmetry direction, it should be kept in mind that this direction may be chosen arbitrarily.

The top left panel of Fig. 3.17 shows the number of atoms N as a function of the chemical potential for different branches of possible solutions. The bottom left panel in this figure shows, for better clarity of exposition, the atom difference $\Delta N = N - N_{\text{ds}}$ between the atom number N of a particular branch and the atom number N_{ds} for the planar dark soliton branch. We have not shown the branch of the ground state starting at $\mu = \Omega = 0.2$ from the linear limit, as the stable ground state of the system will not be relevant in our discussion hereafter.

The first branch, with the largest atom number (among the ones considered therein) for fixed chemical potential, is the single vortex (sv) state located at the center of the trap. As we discussed above and as is well known [43, 42, 44, 76], this solution is dynamically stable despite of the fact that it possesses an anomalous mode in its excitation spectrum. In the linear limit, this state is given by $\Psi_{\text{sv}}(x,y) = \psi_1(x)\psi_0(y) + i\psi_0(x)\psi_1(y)$. Therefore, it is "initialized" at an energy of $\mu = 2\Omega$. For increasing chemical potential the atom number increases monotonically. Furthermore, no other states bifurcate from this one (in consonance also with its absence of stability changes).

The second (and most critical, for our discussion) branch starts from $\mu = 2\Omega$ in the linear limit as well. The wavefunction of this branch is given in the linear limit by $\Psi_{\text{ds}}(x,y) = \Psi_{10}(x,y) = \psi_1(x)\psi_0(y)$ (without loss of generality we assumed the relevant stripe to be oriented along the y axis). This state, which has a density minimum along the y axis and a phase jump across this minimum, is precisely the planar dark soliton (ds) stripe with the line symmetry that characterizes such a 2D generalization of the 1D dark soliton.

The essence of our near-linear limit discussion lies in the (symmetry-breaking) bifurcations arising from this planar dark soliton stripe solution. In particular, in the case of $\Omega = 0.2$ considered in Fig. 3.17, and for $\mu \approx 0.68$, a bifurcation of a novel state involving

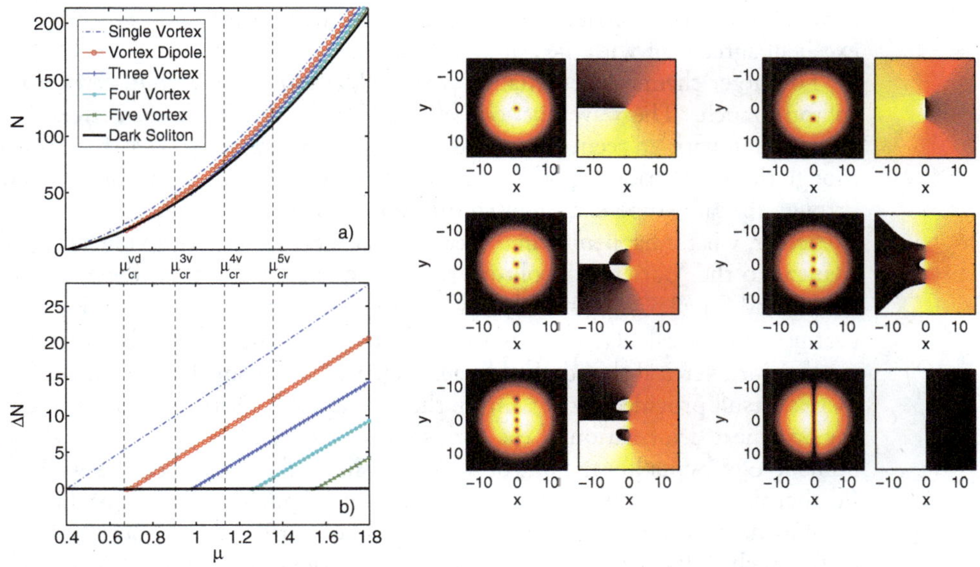

Figure 3.17. *Bifurcations arising from a dark soliton stripe state. Top left panel* (a): *Number of atoms as a function of the chemical potential for the different states for* $\Omega = 0.2$ *bifurcating from the dark soliton stripe. Bottom left panel* (b): *Corresponding atom number difference with respect to the dark soliton stripe branch. Bifurcation of vortex multipole states arises when their atom number difference from the dark soliton stripe vanishes. The corresponding theoretical predictions of such bifurcations are shown by the vertical lines. Right panels (from left to right and top to bottom): Density and phase profiles (left and right subpanels respectively) corresponding to bifurcating states from the dark soliton stripe (bottom right panel): the single vortex state, the aligned two- (vd), three- (3v), four- (4v), and five-vortex (5v) states. Adapted with permission from Ref. [518].*

the vortex dipole state arises. This vortex dipole is exactly the same type of pair of two oppositely topologically charged vortices that we observed manifested in the dynamical simulations of the previous subsection. The bifurcation of a stationary state involving this vortex pair constitutes a supercritical pitchfork bifurcation, as there are two potential installments of such a dipole with opposite, between them, relative positions of their $S = 1$ and $S = -1$ vortex constituents. The location of this bifurcation is sufficiently close to the linear limit so that it can be predicted by using a Galerkin-type approach [420] (see also Refs. [419, 519] for a general discussion). The dark soliton state is given, close to the linear limit, by $\Psi_{ds}(x,y) = \psi_1(x)\psi_0(y)$. The vortex dipole (vd) state occurs due to an admixture of the $\Psi_{02}(x,y) = \psi_0(x)\psi_2(y)$ state to Ψ_{ds}, but importantly with a phase difference of $\pi/2$ with respect to the Ψ_{ds}. Hence, the bifurcation theory developed in Ref. [420] can be directly applied but with the crucial difference of the relative phase of the symmetry-breaking admixing state (with respect to the original state) of $\phi = \pi/2$. In that case, one can repeat the calculations of Ref. [420] [cf. Eq. (12) therein] to obtain the relevant bifurcation point as

$$N_{cr} = \frac{\omega_{10} - \omega_{02}}{I - I_0}, \tag{3.30}$$

where ω_{10} and ω_{02} are, respectively, the eigenvalues corresponding to the states $\Psi_{10}(x,y)$ and $\Psi_{02}(x,y)$, $I_0 = \int \Psi_{10}^4 dx\,dy$, and $I = \int \Psi_{10}^2 \Psi_{02}^2 dx\,dy$. One can also determine the critical chemical potential at which the symmetry breaking is expected to occur, namely $\mu_{cr} = \omega_{10} + I_0 N_{cr}$ [420]; for these parameter values, this leads to $\mu_{cr}^{vd} = \frac{10}{3}\Omega$ (= 2/3 for

$\Omega = 0.2$; see second vertical thin dashed line in the case shown in Fig. 3.17), which is in excellent agreement with the numerical findings.

For larger chemical potential, at $\mu \approx 0.98$, another branch bifurcates from the dark soliton branch. The wavefunctions of this branch contain three vortices (3v) oriented along a line with adjacent vortices having opposite vorticity. It is remarkable that although this bifurcation happens quite far from the linear limit, one can still understand it through the same supercritical pitchfork (symmetry-breaking) theoretical framework. In particular, what is happening in this case is that the Ψ_{03} state is providing the "admixture" element to the "pure" dark soliton stripe state. In light of that, the analysis presented above [and Eq. (3.30)] still holds but now $\omega_{02} \to \omega_{03} = 3\Omega$ and $I = \int \Psi_{10}^2 \Psi_{03}^2 dx\, dy$. Repeating the same calculation as above, one finds $\mu_{cr}^{3v} = 86\Omega/19$ (which in this case is ≈ 0.91, see the third vertical thin dashed line in Fig. 3.17, in reasonable agreement with the numerical result provided above). Given the substantial departure from the linear limit and the nonlinear deformation of the dark soliton stripe (still represented in the above phenomenology by its linear limit of Ψ_{10}), this slight discrepancy can be well understood. Furthermore, the symmetry-breaking, supercritical pitchfork nature of the bifurcation is again evident since two installments of the same "vortex tripole" (as it will be called below) branch may emerge, one with two $S = 1$ vortices and an $S = -1$ centrally separating them and one with two $S = -1$ vortices with an $S = 1$ between them.

At $\mu \approx 1.26$ a four-vortex (4v) state (an aligned quadrupole), with the vortices oriented along a line, bifurcates from the dark soliton state. At this point, the sequence of symmetry breaking events is clear: now the Ψ_{04} state is offering the $\pi/2$ out-of-phase admixture leading to a critical point through the same procedure and the corresponding reformulation of the above constants of $\mu_{cr}^{4v} = 890\Omega/157$; in connection to the $\Omega = 0.2$ case of Fig. 3.17, we find $\mu_{cr}^{4v} \approx 1.13$ (see fourth vertical thin dashed line), with the slight progressive degradation of the critical point attributable again to the further departure from the linear limit. In this case also, the vortex parity variation of $S \to -S$ for such a four-vortex line configuration whose adjacent vortices possess opposite charges, yields the two bifurcating branches associated with this critical point.

Interestingly (a point we will revisit later), a state with four vortices exists even in the linear limit as is shown by the branch starting at $\mu = 3\Omega$ in the linear limit: this represents a four vortex state, with vortices located at the vertices of a square, i.e., a vortex quadrupole (vq). This is among the states previously considered in Refs. [520, 521], as well as independently proposed as a member of a larger class of so-called vortex necklace states in Ref. [522]. At the linear limit, this state can be represented as $\Psi_{vq}(x,y) = \psi_2(x)\psi_0(y) + i\psi_0(x)\psi_2(y)$ and it emerges at $\mu_{cr}^{vq} = 3\Omega$ ($= 0.6$ in the case of Fig. 3.17, cf. first vertical thin dashed line in the figure). With increasing chemical potential the number of atoms of the state monotonically increases. At $\mu \approx 1.4$ this branch crosses the branch representing the dark soliton stripe. However, it is important to note that due to the structural disparity of those states, the crossing of the number of atoms (contrary to the previous cases considered above) does *not* constitute a bifurcation and, therefore, does not lead to any stability change.

Nevertheless, at higher values of μ (e.g., for $\mu \approx 1.54$ in the case of $\Omega = 0.2$ of Fig. 3.17), a five-vortex (5v) state (a vortex quintupole) bifurcates from the dark soliton stripe. Once again, the vortices are oriented along a line and adjacent vortices have opposite vorticities. Furthermore, the bifurcation shares the standard supercritical pitchfork character of the earlier ones, and its characterization through the above analytical tools reveals a critical point at $\mu_{cr}^{5v} = 726\Omega/107$ (which, for $\Omega = 0.2$, leads to $\mu_{cr}^{5v} \approx 1.36$; see last vertical thin dashed line in Fig. 3.17). Typical case examples of all these different vortex states associated with the planar dark soliton are presented in the right panels of Fig. 3.17.

At this point, let us add a short note about the physical intuition of the general picture. What is happening in the present setting is that there exists a sequence of supercritical pitchfork bifurcations in which the Ψ_{10} gets mixed progressively with each of the Ψ_{0m} modes. In so doing, the m-fold symmetry of the latter breaks the line symmetry of the former. In addition, the $\pi/2$ relative phase between them introduces a picture which locally at each perpendicular intersection of the m nodal lines of the latter with the nodal line of the former resembles a $\Psi_{10} \pm i\Psi_{01}$ (with the pluses alternating with the minuses due to the field structure of the Ψ_{0m}). As a result, at each of these nodal line intersections, entities with alternating (between adjacent ones) vortices emerge, progressively formulating the dipole, tripole, aligned quadrupole, quintupole, and so on. Importantly, also, general bifurcation theory can be used to identify the stability characteristics of the resulting states. In particular, since the stripe is dynamically stable as it emerges from the linear limit, the vortex dipole state that arises from it upon the first symmetry-breaking "event" ($m = 2$) should inherit this stability. However, now, once the stripe has become unstable, all higher bifurcations with $m \geq 3$ will *necessarily* result into dynamically unstable states. Moreover, the tripole will inherit one such instability, the aligned vortex quadrupole will inherit two instabilities, the quintupole three, and so on.

Having explored the bifurcations from the linear limit for the setting of the dark soliton stripe, let us present in more detail the resulting nonlinear states (i.e., few-vortex clusters) in the next section. We will also generalize this idea to bifurcations from other linear states (or combinations thereof) to illustrate its power and generality.

3.4 ▪ Few-Vortex Clusters/Crystals: The Low Density Limit of Bifurcation from Linear States

Numerical results on the symmetry-breaking bifurcations resulting in the emergence of vortex cluster states from the first excited state (single dark soliton stripe) are summarized in Fig. 3.17 for $\Omega = 0.2$. The emergence of $1 \times m$ ($m = 2,3,4,5,\ldots$) vortex states can be observed to occur respectively at $(0.68, 0.98, 1.26, 1.54)$ while the corresponding theoretical predictions are $(\mu_{\text{cr}}^{\text{vd}}, \mu_{\text{cr}}^{3\text{v}}, \mu_{\text{cr}}^{4\text{v}}, \mu_{\text{cr}}^{5\text{v}}) = (2/3, 0.91, 1.13, 1.36)$, as shown in the previous section. Clearly, the two-mode approach captures the fundamental phenomenology, despite a slight progressive degradation of the agreement on the critical point arises due to the increasing departure from the linear limit. Additionally, the linear stability properties of the resulting states directly reflect the theoretical stability expectations discussed above; see also Ref. [518]. We should note here that the notation $1 \times m$ reflects the existence of m vortices located along a single line. For m vortices along two lines, we will use the notation $2 \times m$ etc. In exploring the different vortex clusters below, we will follow the presentation of Ref. [523].

The advantage of our bifurcation method and of the wealth of states that can be derived from it is unveiled, e.g., when considering the next set of excited states, namely the combinations with $n + m = 2$, i.e., the degenerate states u_{20}, u_{02}, and u_{11}. In this setting, already a vortex *quadrupole* can be formed at the linear limit as $u_{20} + i u_{02}$ [522]. Interestingly, so can a doubly charged vortex through $u_{20} - u_{02} + 2i u_{11}$. However, these states do not present symmetry breaking bifurcations and, therefore, are not further detailed in what follows. Focusing on the states that do present relevant bifurcations (see Fig. 3.18), some prototypical examples are

(i) the solitonic state consisting of two stripes (i.e., a two-dark-soliton state), i.e., the u_{20} state,

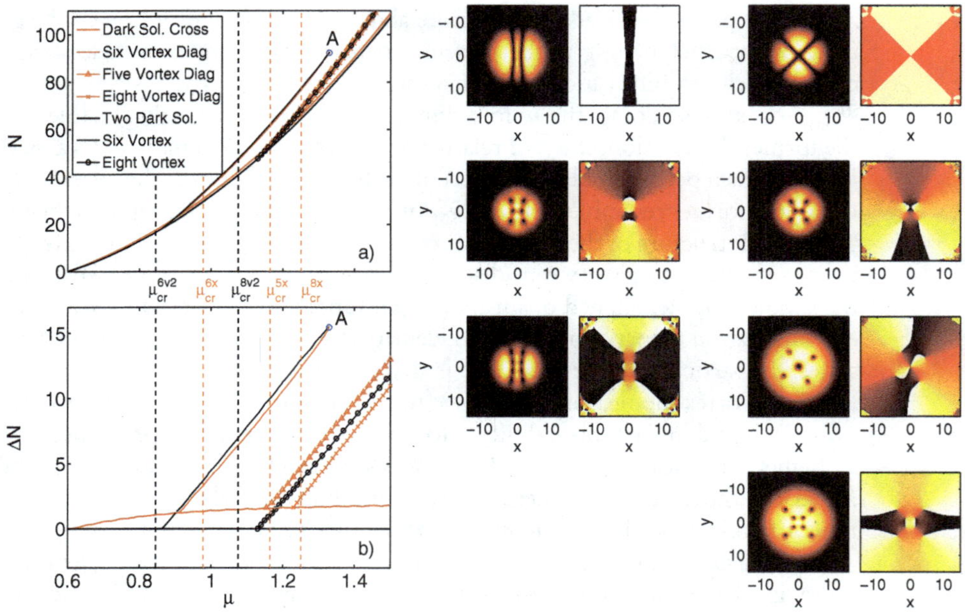

Figure 3.18. *Bifurcations arising from two-dark-soliton stripes and X-shaped dark soliton states. Top left panel* (a): *Number of atoms as a function of the chemical potential for the different states bifurcating from the two-dark-soliton stripes (black lines) and from the dark soliton cross (orange lines) for* $\Omega = 0.2$. *Bottom left panel* (b): *Corresponding atom number difference with respect to the two-dark-soliton stripes branch. The occurrence of bifurcations at zero crossings of* ΔN *and the corresponding theoretical vertical line predictions thereof are similar to the previous (and following) figures. Middle column of panels (from top to bottom): Density and phase profiles (left and right subpanels respectively) corresponding to bifurcating states from the two-dark-soliton stripes (top): the aligned six- (6v2) and eight-vortex states (8v2). Right column of panels (from top to bottom): Density and phase profiles (left and right subpanels respectively) corresponding to bifurcating states from the X-shaped dark soliton cross (top): the diagonal six-vortex state (6x), the four and doubly charged vortex state (5x), and the diagonal eight-vortex state (8x). Adapted with permission from Ref. [523].*

(ii) an X-shaped dark soliton cross emerging from $u_{20} - u_{02}$, as well as

(iii) a ring dark soliton state (see also Refs. [265, 266, 269, 522]) arising from $u_{20} + u_{02}$.

The u_{20} state is one for which the generalization of the phenomenology of Fig. 3.17 is most straightforward as with $\phi_0 = u_{20}$ and $\phi_1 = u_{0m}$ states with $2 \times m$ are formed with $m > 2$. For example, for $\phi_1 = u_{03}$ and $\phi_1 = u_{04}$, such bifurcations are predicted at $\mu_{\rm cr}^{6v2} = 283\Omega/67$ and $\mu_{\rm cr}^{8v2} = 2965\Omega/551$ (see thick black vertical dashed lines in Fig. 3.18), respectively, leading to $6 = 2\times 3$ and $8 = 2\times 4$ two-line vortex clusters. On the other hand, the X-shaped dark soliton pair is subject to similar symmetry-breaking bifurcations/destabilizations due to $\phi_1 = u_{03}, u_{13}, u_{04}$, etc. These bifurcations, occurring in turn at $\mu_{\rm cr}^{6x} = 33\Omega/7$, $\mu_{\rm cr}^{5x} = 99\Omega/17$, and $\mu_{\rm cr}^{8x} = 369\Omega/59$ (see thin orange vertical dashed lines in Fig. 3.18), lead to a diagonal state with six vortices, a doubly charged vortex in the center together with a four-vortex quadrupole around it, and an eight-vortex cluster of near-diagonal vortices.

Finally, the ring dark soliton gets mixed with states of the form $\phi_1 = q_l(r)\sin(l\theta)$, where $q_l(r) = \sqrt{\frac{2}{2!\pi}}(r^l)\exp(-\Omega r^2/2)$, for $l > 3$. Remarkably, these symmetry-breaking events, which can be predicted to occur, e.g., at $\mu_{\rm cr}^{6r} = 5\Omega$ and $\mu_{\rm cr}^{8r} = 59\Omega/9$ (see,

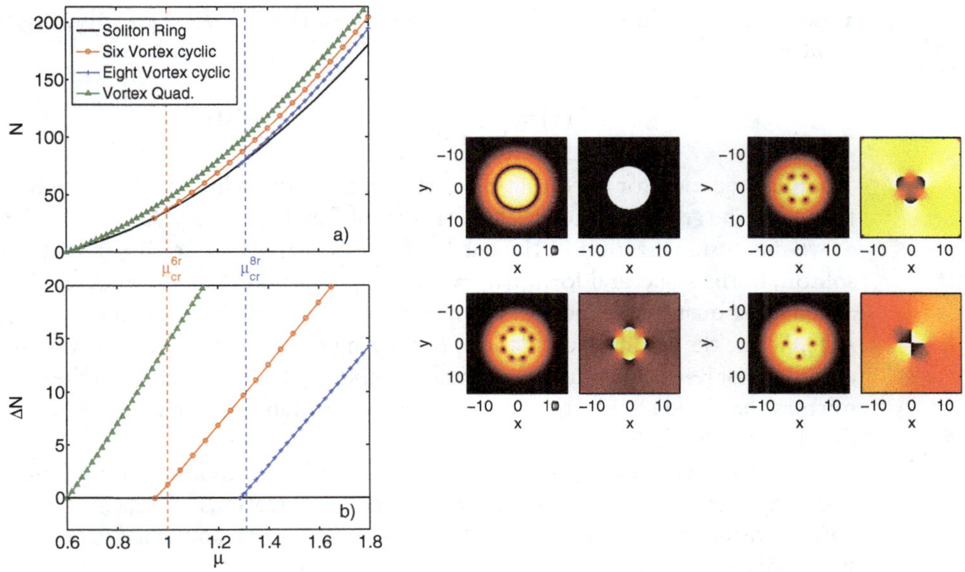

Figure 3.19. *Bifurcations arising from the ring dark soliton. Top left panel* (a)*: Number of atoms as a function of the chemical potential for the different states bifurcating from the ring dark soliton for* $\Omega = 0.2$*. Bottom left panel* (b)*: Corresponding atom number difference with respect to the ring dark soliton. Right panels (from left to right and top to bottom): Density and phase profiles (left and right subpanels, respectively) corresponding to bifurcating states from the ring dark soliton state (top left): the hexagonal (6r) and octagonal (8r) vortex states. Bottom right panel: vortex quadrupole state (note that this state does not bifurcate from the ring dark soliton). Adapted with permission from Ref. [523].*

respectively, orange and blue vertical dashed lines in Fig. 3.19), for $l = 3$ and 4, give rise to polygonal vortex configurations with the vortices now placed on the periphery of the circle. This way, vortex hexagons, octagons, decagons, etc. can be systematically constructed at will.

It is interesting to mention here that this ring state has been of significant attention in its own right. An earlier discussion of it within the context of radially symmetric states appeared in the beginning of this chapter and more specifically in Fig. 3.5 (cf. also with the perturbative analysis of the stability of rings in the previous chapter). There, it could already indeed be observed that it is the modes of $q = 2$, $q = 3$, and $q = 4$ that are the predominant causes of the destabilization of the ring and thus it is expected that it is precisely these modes that will arise in the respective dynamics. Indeed, the $q = 2$ mode pertains to the stable vortex quadrupole state $u_{20} + i u_{02}$, which is always detrimental (immediately off of the linear limit) to the stability of the ring state. However, for larger values of the chemical potential, we observe in Fig. 3.5 additional instability bifurcations of the $q = 3$ mode (leading to the vortex hexagon above) and of $q = 4$ (leading to the vortex octagon above). It is these types of instabilities that gave rise to the vortex necklaces first observed in Ref. [265]; see also the recent work of Ref. [524]. Complex multivortex patterns were seen to emerge through the relevant azimuthal instability, and relaxational processes that would expunge some of the vortices were also found to be possible in the relevant dynamics. Nevertheless, a feature of these ring dark solitons that was also of particular interest in Refs. [265, 524] was the (only metastable, given the above mechanism) dynamics of the radius of the ring. This was found to be captured by a radial form of the dark soliton perturbation theory (cf. Sec. 2.3.3 and Ref. [265]), which led to a dynamical equation of motion for the radius of the ring $R(t)$ [see also Eqs. (2.97) and (2.98)]:

$$\ddot{R}(t) = \left(-\frac{1}{2}\frac{dV}{dR} + \frac{1}{3R(t)}\right)\left(1 - \dot{R}(t)^2\right). \tag{3.31}$$

A similar equation for the evolution of the dark soliton radius was found in Ref. [524], by following the adiabatic perturbation theory of Ref. [174, 175]. According to Ref. [524], a two-dimensionalization of the energy $E_{1D} = (4/3)(\mu - V(x) - \dot{x}^2)^{3/2}$ of the 1D dark soliton, in the structural form $E_{2D} = 2\pi R E_{1D}$, and a differentiation of E_{2D} should yield a vanishing quantity, which, in turn, permits a derivation of a dynamical equation for $\ddot{R}(t)$. While the resulting oscillations of the ring's radius could be observed (see, e.g., the discussion of Ref. [265]), and an anomalous mode in the BdG analysis of the ring spectrum mirrors that mode, given the strong dynamical instability of the ring state, we will not dwell on this effect further.

We also mention in passing that interesting additional bifurcation events (collisions and disappearances into "blue sky," saddle-node, bifurcations) arise, e.g., between the six-vortex cluster derived from the X-shaped dark soliton cross and that derived from the two-stripe soliton (see blue circle denoted by A in Fig. 3.18). On the other hand, in Fig. 3.19 we depict the states that bifurcate from the vortex ring: vortex hexagons (6r) and octagons (8r). In this figure we also depict the vortex quadrupole state (bottom right panel) which does not bifurcate from the soliton ring. In all the cases, good quantitative agreement is found on the critical points predicted by the theory and those observed numerically.

Figures 3.20 and 3.21 complement the above existence picture with a systematic linear stability analysis of each of the corresponding states for Figs. 3.18 and 3.19 respectively. The two-stripe soliton is dynamically stable near the linear limit, a stability inherited by its first symmetry-breaking offspring, the six-vortex (2×3) state; while the aligned eight-vortex case (2×4) is unstable from its existence onset (see left column of panels in Fig. 3.20). Similarly, the X-shaped dark soliton cross state also bears imaginary eigenfrequencies for all values of μ and hence its derivative states inherit its dynamical instability (see right column of panels in Fig. 3.20). On the other hand, the ring dark soliton is, as was also found earlier [265, 266, 269, 522], always unstable, hence the polygonal vortices that derive from it inherit this instability (see Fig. 3.21). However, it should be noted that the instability of such states weakens as the chemical potential μ increases. Note that the vortex quadrupole is stable in the linear limit. Since no state bifurcates from the vortex quadrupole, this stability is generally preserved (apart from a small window of oscillatory instabilities).

A general comment on the observed BdG spectra is due here. We can see that the spectra bear a large number of positive Krein sign modes (see blue points in all BdG spectra) which appear to be asymptoting to appropriate corresponding values in the large chemical potential limit. In addition, there is a number of negative Krein sign eigenfrequencies (see orange points is all BdG spectra) which may lead to oscillatory (upon collision with positive sign ones) or exponential instabilities. Generally, we can comment that the positive Krein sign eigenmodes correspond to the ground state "background" on which a particular solitonic or vortex (or mixed) state exists. Furthermore, these eigenfrequencies have a well-defined limit when μ is large, as discussed, e.g., in Ref. [525] (see also references therein). On the other hand, the negative Krein sign modes reflect the excited state nature of the considered soliton or vortex (or mixed) states and are the ones which bear the potential for dynamical instabilities.

Importantly, the present approach can be generalized to higher excited states (see Fig. 3.22). As merely a small sample of further exotic configurations that can emerge

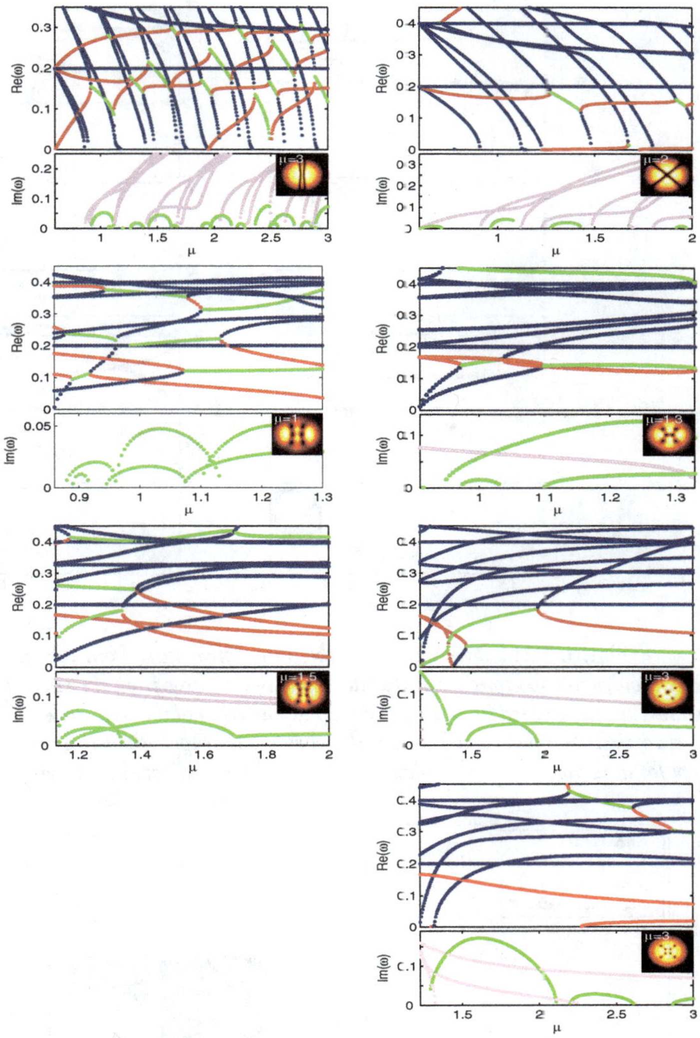

Figure 3.20. *BdG spectra for the two-dark-soliton stripe and its bifurcating states (left column) and for the dark soliton cross and its bifurcating states (right column) corresponding to the states described in Fig. 3.18. The inset in each of the spectra depicts a typical configuration at the chemical potential indicated. In all BdG spectra we depict the real (top subpanel) and imaginary (bottom subpanel) parts of the eigenfrequencies as a function of the chemical potential* μ *using the following color coding: blue* $\rightarrow$ *positive energy (Krein sign) modes, orange* $\rightarrow$ *negative energy modes, green* $\rightarrow$ *oscillatory unstable modes, and pink* $\rightarrow$ *nonoscillatory unstable modes. The presence of nonvanishing imaginary parts is an indication of instability. Note that the only stable solutions are the two-dark-soliton stripe (top left panel) and its first symmetry breaking offspring, the six-vortex state (left panel of second row), which are stable only for low enough chemical potentials. Adapted with permission from Ref. [523].*

(some of which can even be structurally robust), we mention a nine-vortex cluster (a square grid of 3×3 vortex "particles"), which emerges from the linear limit as $u_{30} + i u_{03}$ and can, therefore, be regarded as a higher excited analogue to the single vortex and the vortex quadrupole states, and is expected to be stable in the vicinity of that limit [at least with respect to purely imaginary (i.e., nonoscillatory) instabilities, see below]. On the other hand, there are bifurcations from that state including the bifurcation of a ring with

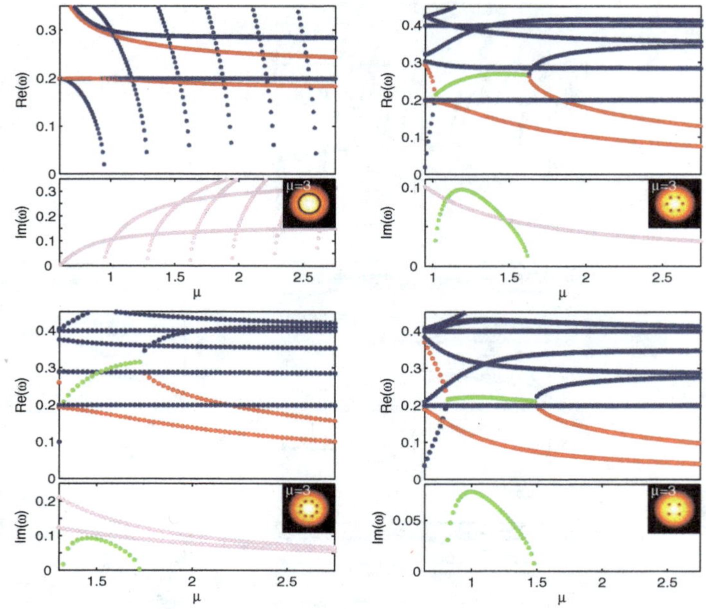

Figure 3.21. *BdG spectra for the states bifurcating from the ring dark soliton and for the vortex quadrupole (bottom right) that does not itself emerge from the ring. These spectra correspond to the bifurcations depicted in Fig. 3.19. For an explanation of the color codes see Fig. 3.20. From the above it is clear that the most robust state is the vortex quadrupole (which is dynamically stable for small* μ *except for a narrow interval of oscillatory instabilities). Adapted with permission from Ref. [523].*

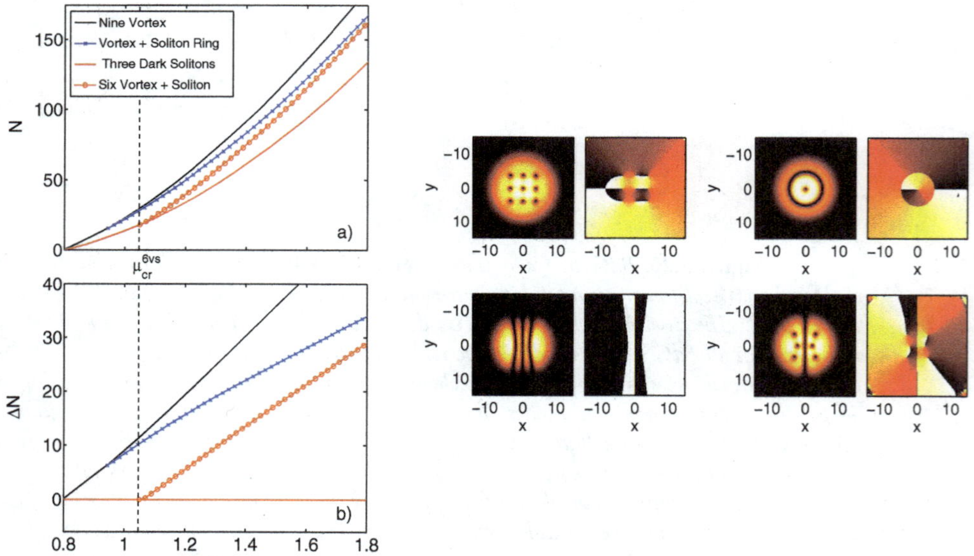

Figure 3.22. *Bifurcations for higher order states. Top left panel* (a): *Number of atoms as a function of the chemical potential for the different states bifurcating from* $\mu = 4\Omega$ *for* $\Omega = 0.2$. *Bottom right panel* (b): *Corresponding atom difference with respect to the three-soliton state. Right panels (from left to right and top to bottom): Density and phase profiles of the nine-vortex state, the vortex-soliton ring, the three-soliton state and the mixed state consisting of a dark stripe and six vortices. Adapted with permission from Ref. [523].*

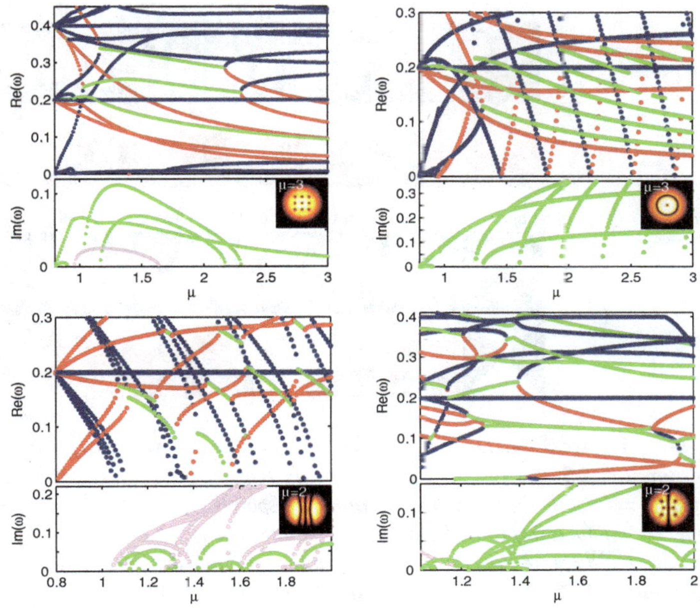

Figure 3.23. *BdG spectra for the bifurcating states depicted in Fig. 3.22. From left to right and top to bottom: The nine-vortex state, the ring dark soliton plus vortex, the three-dark-soliton stripe (which is stable for low values of the chemical potential), and the mixed soliton-vortex state. For an explanation of the color codes see Fig. 3.20. Adapted with permission from Ref. [523].*

a vortex in its center (which was again discussed, e.g., in Ref. [269]). Another state that exists and should be robust near the linear limit is a three-soliton stripe. In fact, from such a state bifurcations again emerge due to the mixture with states such as $\phi_1 = u_{0m}$ with $m \geq 4$, but also with $\phi_1 = u_{1m}$ with $m \geq 3$. We focus briefly on the latter, which is theoretically predicted to occur at $\mu_{\rm cr}^{\rm 6vs} = 1250\Omega/239$ (see vertical dashed line in Fig. 3.22), because it gives rise to yet another novel type of state, namely a mixed state between vortices and a dark soliton: this soliton-vortex mixed state has a dark soliton stripe nodal line plus two additional lines over each of which three vortices reside. Similar states with eight, ten, etc. vortices beside the soliton also exist, arising through subsequent bifurcations.

To corroborate the above theoretical predictions in Figs. 3.22 and 3.23 some prototypical examples are shown for states that emerge from the third excited linear branches and their corresponding BdG stability spectra. The nine-vortex "crystal" is stable close to the linear limit in the sense that its BdG spectrum possesses no imaginary mode. However, there is a complex mode inducing an oscillatory instability. At $\mu \approx 0.95$ the vortex plus ring soliton state bifurcates from the nine-vortex "crystal" inducing a complex mode. However, the latter is small and vanishes again at $\mu \approx 1.6$. The BdG spectrum of the vortex plus ring soliton state contains no purely imaginary mode but many complex modes. Note that the eigenvalue spectrum looks fairly similar to the ring dark soliton case, but the imaginary part of the modes is not due to imaginary modes which are created by bifurcations. In this case, modes with positive energy cross zero and thus obtain a negative energy. These negative energy modes then collide with modes with positive energy and create the complex modes. Moreover, we show the three-soliton stripe, as well as the mixed soliton–six-vortex cluster state emerging from its first bifurcation event in Fig. 3.22. The three-soliton state is stable near the linear limit (of $\mu = 0.8$) but becomes destabilized at $\mu = 1.05$ (and then further so at 1.06, 1.1, and 1.3). The first bifurcation gives rise

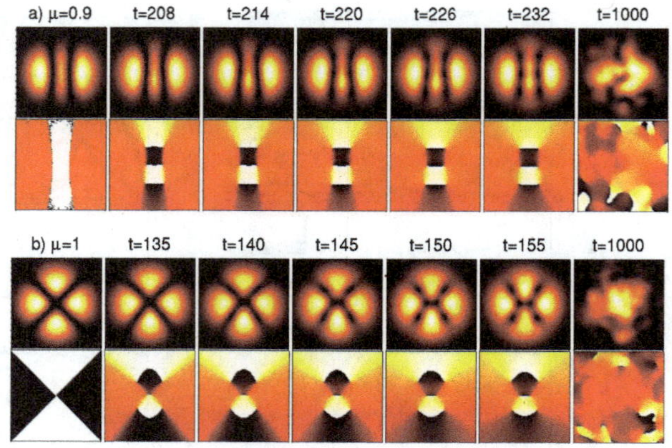

Figure 3.24. *Dynamics for* (a) *the unstable two dark-soliton-stripe and* (b) *the dark soliton cross for $\mu = 0.9$ and $\mu = 1$ respectively. The left-most column corresponds to the stationary steady state at $t = 0$ while the right-most column corresponds to the configuration after a relatively long time span. The top (bottom) rows depict the condensate density (phase) at the times indicated. To initiate the instability of the state, we added to the initial steady state configuration a spatial random perturbation of magnitude* 10^{-8}. *Adapted with permission from Ref. [523].*

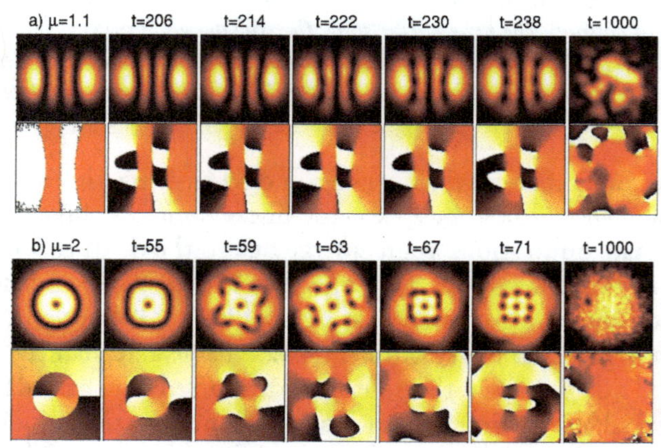

Figure 3.25. *Same as in Fig. 3.24 for* (a) *the three-dark-soliton stripe with $\mu = 1.1$ and* (b) *the vortex-soliton ring for $\mu = 2$. Adapted with permission from Ref. [523].*

to the very weakly unstable soliton-vortex state predicted above; the critical point for this bifurcation is found to be at $\mu = 1.034$, once again in very good agreement with the full numerical result. The present approach can naturally be extended to a multitude of additional states which, however, are expected to be dynamically unstable; therefore, having presented the most fundamental ones, and having indicated how the relevant phenomenology proceeds towards higher excited state clusters, we will limit ourselves to the few-vortex crystals presented above.

To complement the stability (i.e., stationary) picture of the bifurcating solutions, we depict in Figs. 3.24 and 3.25 the typical destabilization dynamics for some of the modes studied above. In particular, we focus on the configurations that after an initial random destabilization evolve temporarily into other configurations on nearby bifurcation branches. Specifically, in Fig. 3.24(a) we depict the destabilization dynamics for the two dark soliton stripe where we choose a chemical potential $\mu = 0.9$ that is inside the instability window for this configuration (cf. Fig. 3.20). The figure shows how the two-dark-soliton stripe is reconfigured, through the snaking instability of its constituent two-dark-soliton stripes (a generalization of the mechanism presented in the previous section), into the six vortex state (6v2) configuration whose branch bifurcates precisely from the former stripe state at $\mu = \mu_{\rm cr}^{6v2}$ that is slightly smaller than the value $\mu = 0.9$ that we chose for this example. This reconfiguration is possible since the two branches are close to each other both in terms of the number of atoms and in terms of the actual density configuration. It is important to mention that this transient six-vortex state is unstable (cf. Fig. 3.20) and therefore it eventually destabilizes also giving rise to a disordered state [see the last snapshot in Fig. 3.24(a)]. A similar scenario for the dark soliton cross is depicted in Fig. 3.24(b) where the configuration destabilizes and transitions to the nearby configuration branch of the diagonal six-vortex state (6v). Again this transient configuration is unstable and eventually destabilizes towards a disordered sate. It is worth noting that the amount of vortices created in the two examples depicted in Fig. 3.24 depends on the length of the dark soliton stripes and the density of the background field where they are supported. This, in turn, depends on the total mass contained in the original configuration: the larger the mass (i.e., the larger the chemical potential μ), the more vortices will be nucleated by the snaking instability; see, e.g., Ref. [526, 523]. As is the case for the soliton stripes and crosses, the ring dark soliton transiently passes through a neighboring branch configuration (a number of vortex pairs arranged in a necklace configuration) which in turn is unstable and eventually gives rise to a disordered state. The number of remaining vortices for long times depends on the annihilation of vortex pairs (vortex-antivortex pairs) as also observed in the dynamics of merging BEC fragments in Ref. [159].

Finally, in Fig. 3.25, we depict the destabilization of a three-soliton stripe and the vortex soliton ring. The former transiently passes through the mixed soliton-vortex state configuration and the latter passes through the nine-vortex state that eventually tends, in turn, to a disordered dynamical state. It is worth mentioning that all the configurations described above have zero total angular momentum except the vortex-soliton ring [see Fig. 3.25(b)]. That is the reason why a single vortex remains for long times in the latter case (see the small black hole in the density and its 2π phase winding for $t = 1000$) while vortex-antivortex annihilations eliminate all vortices in all the other cases (except, as explained above, for the soliton ring for relatively large chemical potentials).

3.5 ▪ Few-Vortex Clusters/Crystals: The High Density Limit of Interacting Particles

The above discussion and bifurcation picture is especially relevant close to the point of emergence of these multivortex solution branches. A complementary aspect of our description concerns the case (for larger chemical potentials) where the vortices are well separated, individual interacting entities which can be regarded as interacting particles subject to the precessional effect due to the parabolic confinement. In that picture, it is relevant to develop (kinematic) equations of motion characterizing the vortex dynamical evolution, as well as their potential equilibrium positions and their near-equilibrium behavior, similarly to what was done, e.g., for characterizing the oscillations and interactions of dark solitons in quasi-1D settings in Ref. [70] (see also Ref. [347]). Our presentation of these features of the particle-equation based vortex clusters will follow the recent exposition of Ref. [527].

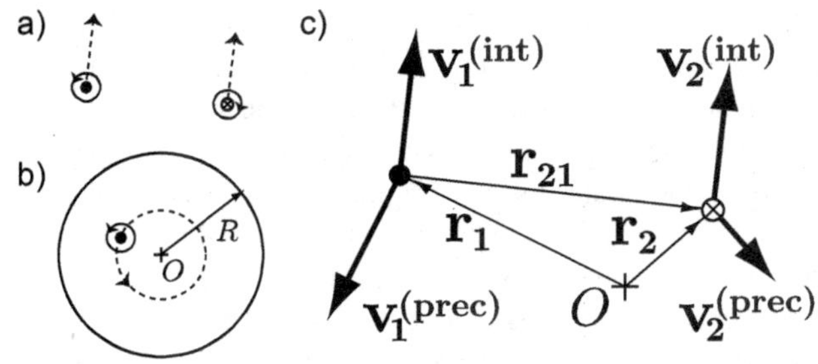

Figure 3.26. *Elements of the vortex particle model.* (a) *In an unbounded homogeneous fluid, two counter-circulating vortex lines move together (dashed lines) in the direction of the flow between them.* (b) *A single vortex line in a bounded condensate precesses (dashed line) about the center of the condensate O in the same sense as its circulation. The radius R may be either the Thomas–Fermi radius of the condensate or the wall of the confining cylinder.* (c) *Two counter-circulating vortex lines in a bounded condensate experience instantaneous velocities* $\mathbf{v}_{1,2}^{(\mathrm{prec})}$ *and* $\mathbf{v}_{1,2}^{(\mathrm{int})}$ *associated with precession and interactions, respectively (thick arrows). The thin arrows define the displacements of the two vortices with respect to one another* ($\mathbf{r}_{21}$) *and the center of the condensate* ($\mathbf{r}_{1,2}$). *Adapted with permission from Ref. [527].*

3.5.1 ▪ Model Setup

Our considerations assume straight line vortices in a BEC with cylindrical symmetry. The most prominent physical example arises in the context of harmonically confined, oblate BECs [528], but the same general equations and conclusions may be drawn in the case of a hard-wall container as well [529]. We begin by considering an ensemble of singly quantized vortices, and in the following section specialize to the case of the vortex dipole.

The motion of a vortex line involves its interactions with other vortex lines, as well as the effect of "boundary conditions" associated with the fluid confinement [518]. We begin with the interactions. The angular velocity of the fluid flow a distance r from a singly quantized vortex core in a homogeneous fluid is

$$\Phi(r) = \frac{\hbar}{m r^2}, \tag{3.32}$$

where m is the atomic mass. This flow pattern determines the motion of a second vortex line, which moves with the flow at its center (core) [530]. The motion of the first vortex line is similarly determined by the flow pattern established by the second. If the separation between the jth and kth vortex lines is r_{jk}, then the two same-charge vortex lines orbit one another at angular frequency $\Phi(r_{jk})$. On the other hand, if the vortices are counter-circulating (vortices of opposite charge), then the vortex lines move together with linear speed $v_{jk} = r_{jk}\Phi(r_{jk})$ in the direction of the flow between them [Fig. 3.26(a)].

The motion of the kth vortex line in either a hard-wall or a disk-shaped harmonic potential also involves gyroscopic precession about the condensate center at fixed distance r_j [Fig. 3.26(b)]. The precession frequency may be expressed in both cases by [76, 529, 531]

$$\Omega(r_k) = \frac{\omega_{\mathrm{pr}}}{1 - r_k^2/R^2}, \tag{3.33}$$

with

$$\omega_{\text{pr}} = \begin{cases} \dfrac{\hbar}{mR^2} & \text{hard-wall,} \\ \dfrac{2\hbar\omega_r^2}{8\mu}\left(3+\dfrac{\omega_r^2}{5\omega_z^2}\right)\ln\left(\dfrac{2\mu}{\hbar\omega_r}\right) & \text{harmonic,} \end{cases} \tag{3.34}$$

where μ is the chemical potential, ω_r and ω_z are the confining radial and axial frequencies of the harmonic trap, respectively, and R is the Thomas–Fermi radius of the condensate or the radius of the hard-wall cylinder [529]. The common dependence on r_k is what draws together these two different physical situations.

Summing these two velocities for the kth vortex in a "gas" of n vortices gives, in complex coordinates,

$$i\dot{z}_k = -S_k\Omega(r_k)z_k + \frac{b}{2}\sum_{j\neq k}^{n} S_j\Phi(r_{jk})(z_k - z_j), \tag{3.35}$$

where (x_k, y_k) is the position of the kth vortex and $z_k = x_k + iy_k = r_k e^{i\theta_k}$, $r_k = |z_k|$, and $r_{jk} = |z_k - z_j|$. The topological charge of the kth vortex is $S_k = \pm 1$, with the positive (negative) sign referring to counterclockwise (clockwise) circulation as viewed from the positive z axis. The constant parameter b modifies the interaction strength slightly from the homogeneous case in a harmonic trap; for experimentally relevant parameter values, it has been argued that to a good approximation it is given by $b = 1.35$ [518], while for the hard-wall potential it is $b = 2$. This value of b essentially takes into consideration the effect of screening due to the fact that interaction of the vortices arises in an inhomogeneous background. A more general approach to account for the inhomogeneity is presented in Ref. [532]. However, in what follows, we consider an averaged "effective" value of b as accounting for the density decrease due to the trap.

Figure 3.26(c) shows the relevant vortex velocities and coordinates for a vortex dipole. Since the model is only valid for straight line vortices, we adopt the convention of referring to the location of a vortex line as the location of the vortex (particle).

3.5.2 ▪ Conserved Quantities and Integrability

The system of interest has a Hamiltonian structure. To verify this, let us write the differential equations in Cartesian coordinates as

$$\begin{aligned} \dot{x}_k &= -S_k\Omega(r_k)y_k - \frac{b}{2}\sum_{j\neq k}^{n} S_j\Phi(r_{jk})(y_k - y_j), \\ \dot{y}_k &= S_k\Omega(r_k)x_k + \frac{b}{2}\sum_{j\neq k}^{n} S_j\Phi(r_{jk})(x_k - x_j). \end{aligned} \tag{3.36}$$

For simplicity, we work henceforth in dimensionless distance units setting $R = 1$. Let us define the function

$$H(z_1,\ldots,z_n) = -\frac{\omega_{\text{pr}}}{2}\sum_{k=1}^{n}\ln(1-r_k^2) - \frac{\Phi_R}{4}\sum_{k=1}^{n}\sum_{j\neq k}^{n} S_j S_k \ln(r_{jk}^2), \tag{3.37}$$

where $r_k = |z_k| = \sqrt{x_k^2 + y_k^2}$ and $\Phi_R \equiv b\Phi(R) = \hbar b/m$ with $R = 1$. Then, it can readily be found that

$$S_k \dot{x}_k = -\frac{\partial H}{\partial y_k},$$
$$S_k \dot{y}_k = \frac{\partial H}{\partial x_k} \tag{3.38}$$

for every $k = 1, \ldots, n$, which means that H is a first conserved quantity, i.e., the first integral or Hamiltonian along the orbits of the system. Given initial positions $z_k(0)$, then

$$H(z_1(t), \ldots, z_n(t)) = E_0 \qquad \forall t, \tag{3.39}$$

for some suitable constant E_0.

It is directly verifiable that a second conserved quantity is

$$L = \sum_{k=1}^{n} S_k r_k^2, \tag{3.40}$$

which represents the angular momentum of the system; see, e.g., Ref. [533] for a discussion about conservation laws (in the absence of the precessional terms).

The existence of two conserved quantities guarantees integrability in the classical Liouville sense [506] for the case $n = 2$, whether the vortices are co- or counter-rotating. This implies that the energy level sets are compact and the phase space is foliated by invariant tori. On each of these, the motion is quasiperiodic with two frequencies. In the following, we consider other dynamical aspects of the $n = 2$ case for vortex dipoles with $S_1 = 1$ and $S_2 = -1$. It is worthwhile to mention that, in the absence of the trap, even the three-vortex system is integrable due to the existence of a third conserved quantity, the linear momentum:

$$P = \sum_{k=1}^{n} S_k z_k. \tag{3.41}$$

However, due to the presence of the trap, this conserved quantity ceases to exist, rendering the two-vortex system the maximal one (in terms of number of constituent vortex particles) that preserves the integrable structure.

3.5.3 ▪ The Case of Two Counter-Rotating Vortices

No Collisions

The two vortices in the framework of our above mathematical model never collide. To see this, we exploit the fact that the Hamiltonian is constant along orbits. Taking exponentials on the Hamiltonian, we have

$$(1 - r_1^2)^{\omega_{\rm pr}} (1 - r_2^2)^{\omega_{\rm pr}} r_{12}^{\Phi_R/2} = C^2 > 0 \qquad \forall t, \tag{3.42}$$

where $C^2 = (1 - r_1(0)^2)^{\omega_{\rm pr}} (1 - r_2(0)^2)^{\omega_{\rm pr}} r_{12}(0)^{\Phi_R/2}$. Note that $0 < C^2 < 2^{\Phi_R/2}$ because $0 \leq r_i(0)^2 < 1$ $(i = 1, 2)$ and $0 < r_{12}(0) < 2$. A first consequence of Eq. (3.42) is

$$r_{12}(t) > C^{4/\Phi_R} \qquad \forall t, \tag{3.43}$$

that is, the vortices are separated by a computable minimal distance that depends on the initial position of the vortices.

It is worth reiterating that the model considers the dynamics of straight vortex lines only, such as those occurring in oblate BECs, and does not therefore preclude the possibility of collisions and possible reconnection phenomena when the vortex lines are tilted or bent.

No Ejections

The vortices never reach the edge of the fluid, remaining confined instead within a computable inner circle. Assuming $r_{12}(t) < 2$ for all t (i.e., the distance between the two vortices cannot exceed twice the radius, $R = 1$, of the condensate), we have

$$(1-r_1^2)^{\omega_{\rm pr}}(1-r_2^2)^{\omega_{\rm pr}} = \frac{C^2}{r_{12}^{\Phi_R/2}} > \frac{C^2}{2^{\Phi_R/2}} \qquad \forall t. \tag{3.44}$$

In consequence,

$$r_i^2(t) < 1-h, \tag{3.45}$$

where $h = (C^2 2^{-\Phi_R/2})^{1/\omega_{\rm pr}} < 1$.

Again, it is worth noting that at finite temperature the dissipative presence of thermal atoms is expected to cause the vortices to leave the fluid. These effects are not considered in the present model, although as shown in the single vortex discussion they can, in principle, be accounted for; see, e.g., Ref. [514, 515].

3.5.4 ▪ The Stationary Two-Vortex Equilibrium and its Stability

We now prove the existence of a stationary equilibrium of the vortex pair dynamics and illustrate its stability.

Existence

Lemma 3.1. *There is an equilibrium, unique up to rotations, given by*

$$(x_1^0, y_1^0) = \left(\sqrt{\frac{\Phi_R}{4\omega_{\rm pr}+\Phi_R}}, 0\right),$$

$$(x_2^0, y_2^0) = \left(-\sqrt{\frac{\Phi_R}{4\omega_{\rm pr}+\Phi_R}}, 0\right).$$

Proof. Recall that $n = 2$ and $S_1 = -S_2 = 1$. The system (3.35) in complex notation is

$$\begin{aligned} i\dot{z}_1 &= -\Omega(r_1)z_1 + \frac{\Phi_R}{2r_{12}^2}(z_1 - z_2), \\ i\dot{z}_2 &= +\Omega(r_2)z_2 - \frac{\Phi_R}{2r_{12}^2}(z_2 - z_1). \end{aligned} \tag{3.46}$$

Here $z_1 = r_1\exp(i\theta_1)$ and $z_2 = r_2\exp(i\theta_2)$ are, in general, time dependent, but we look for equilibria, i.e., constant solutions. By direct substitution of the given solution above, one sees that it is indeed a constant solution. Let us prove that this is unique. Naturally, r_i, θ_i denote polar coordinates (or equivalently an amplitude-phase decomposition). We can find that the evolution of such polar coordinates reads

$$\begin{aligned}
\dot{\theta}_1 &= \Omega(r_1) - \frac{\Phi_R}{2r_{12}^2}\left[1 - \frac{r_2}{r_1}\cos(\theta_1 - \theta_2)\right],\\
\dot{\theta}_2 &= -\Omega(r_2) + \frac{\Phi_R}{2r_{12}^2}\left[1 - \frac{r_1}{r_2}\cos(\theta_1 - \theta_2)\right],\\
\dot{r}_1 &= \frac{\Phi_R r_2}{2r_{12}^2}\sin(\theta_1 - \theta_2),\\
\dot{r}_2 &= \frac{\Phi_R r_1}{2r_{12}^2}\sin(\theta_1 - \theta_2).
\end{aligned} \tag{3.47}$$

Let us call $\theta = \theta_1 - \theta_2$. From the last equation, $\sin\theta = 0$. Then, $\theta = 0$ or $\theta = \pi$. It is straightforward to show that the first option must be discarded. In particular, if $\theta = 0$, from the first equation

$$0 = \Omega(r_1) - \frac{\Phi_R}{2r_{12}^2}\left[1 - \frac{r_2}{r_1}\right].$$

Since $\Omega(r_1) > 0$,

$$1 - \frac{r_2}{r_1} = \frac{2r_{12}^2}{\Phi_R}\Omega(r_1) > 0,$$

and then $r_1 > r_2$. Arguing analogously with the second equation, one gets $r_2 > r_1$ and thus we reach a contradiction. In conclusion, it must be the case that $\theta = \pi$. Now, we prove that $r_1 = r_2$. On the contrary, let us assume that $r_1 > r_2$ (the remaining case is analogous). Adding the two first equations

$$\Omega(r_1) - \Omega(r_2) = \frac{\Phi_R}{2r_{12}^2}\left[\frac{r_2}{r_1} - \frac{r_1}{r_2}\right] < 0,$$

which leads to a contradiction because $\Omega(r)$ is strictly increasing.

In conclusion, $\theta = \pi$, $r_1 = r_2$ and we can find the exact value $r_1 = r_2 = \sqrt{\frac{\Phi_R}{4\omega_{pr}+\Phi_R}}$ by solving the first equation of the system (3.47).

To emphasize the rotational invariance of our system, it is convenient to work with the variables (r_1, r_2, θ). The (unique) equilibrium then reads

$$(r_1^0, r_2^0, \theta_0) = \left(\sqrt{\frac{\Phi_R}{4\omega_{pr} + \Phi_R}}, \sqrt{\frac{\Phi_R}{4\omega_{pr} + \Phi_R}}, \pi\right).$$

Stability

Theorem 3.2. *The equilibrium (r_1^0, r_2^0, θ_0) is stable (in the sense of Lyapunov).*

Proof. In the coordinates (r_1, r_2, θ), the system of equations (3.46) reads

$$\begin{aligned}
\dot{\theta} &= \Omega(r_1) + \Omega(r_2) - \frac{\Phi_R}{2r_{12}^2}\left[2 - \frac{r_1}{r_2} - \frac{r_2}{r_1}\right]\cos\theta, \\
\dot{r}_1 &= \frac{\Phi_R r_2}{2r_{12}^2}\sin\theta, \\
\dot{r}_2 &= \frac{\Phi_R r_1}{2r_{12}^2}\sin\theta.
\end{aligned} \tag{3.48}$$

Of course, the Hamiltonian H is still a conserved quantity; in the new variables, it reads

$$H(r_1, r_2, \theta) = -\frac{1}{2}\left[\omega_{\mathrm{pr}}\ln(1 - r_1^2) + \omega_{\mathrm{pr}}\ln(1 - r_2^2) + \frac{\Phi_R}{2}\ln(r_1^2 + r_2^2 - 2r_1 r_2\cos\theta)\right].$$

Note that basic trigonometric considerations give $r_{12}^2 = r_1^2 + r_2^2 - 2r_1 r_2\cos\theta$.

The equilibrium (r_1^0, r_2^0, θ_0) is a critical point of H. The Hessian matrix evaluated at (r_1^0, r_2^0, θ_0) is

$$\frac{1}{8}\begin{pmatrix} 7\Phi_R + 12\omega_{\mathrm{pr}} + \frac{\Phi_R{}^2}{\omega_{\mathrm{pr}}} & \Phi_R + 4\omega_{\mathrm{pr}} & 0 \\ \Phi_R + 4\omega_{\mathrm{pr}} & 7\Phi_R + 12\omega_{\mathrm{pr}} + \frac{\Phi_R{}^2}{\omega_{\mathrm{pr}}} & 0 \\ 0 & 0 & \Phi_R \end{pmatrix}.$$

One can easily prove that this matrix is positive-definite by Sylvester's criterion [534]. Hence, H is a Lyapunov function that attains a minimum at (r_1^0, r_2^0, θ_0), and the equilibrium is stable.

The stationary equilibrium is relevant to vortices that are symmetric with respect to a line passing through the condensate center. We illustrate several typical orbits in Fig. 3.27. This class of orbits has been predicted theoretically [535, 518] and observed experimentally [536, 537] (see also the detailed discussion below). For relatively small perturbations from the symmetric equilibrium, the (linear) rotational frequency (i.e., the imaginary part of the stability eigenvalue associated with the equilibrium configuration) of these orbits is in reasonable agreement with those found experimentally [537].

3.5.5 ▪ Guiding Center Two-Vortex Equilibria and Their Stability

Finally, we prove the existence of a class of guiding center equilibria and illustrate their stability.

Existence

Consider two vortex positions (x_1, y_1) and (x_2, y_2), or in complex notation, $z_1 = r_1\exp(i\theta_1)$ and $z_2 = r_2\exp(i\theta_2)$, where all of the r's and θ's are time dependent. The system under

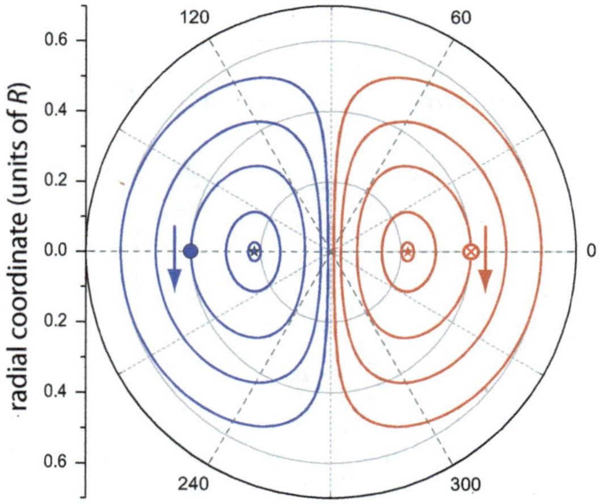

Figure 3.27. *Typical computed orbits associated with symmetric vortex configurations, determined by integrating the equations of motion [Eqs. (3.46)]. The vortex on the left has a counterclockwise sense of circulation; the vortex on the right has a clockwise sense of circulation. The stationary equilibrium positions of the vortices are indicated by stars. These orbits arise through competition between the effects of interaction and precession [535]. Adapted with permission from Ref. [527].*

study is

$$i\dot{z}_1 = -\frac{\omega_{pr}}{1-r_1^2}z_1 + \frac{\Phi_R}{2r_{12}^2}(z_1 - z_2), \tag{3.49}$$

$$i\dot{z}_2 = \frac{\omega_{pr}}{1-r_2^2}z_2 - \frac{\Phi_R}{2r_{12}^2}(z_2 - z_1). \tag{3.50}$$

We seek precessing guiding centers about which the vortices oscillate. For the special case of guiding centers equidistant from the center of the fluid, we expect the precession frequency ω to vanish, reproducing the stationary equilibrium case above. Given a vortex positioned at one guiding center, and a (counter-circulating) antivortex at the other, we expect that each vortex will stay at its own guiding center; that is, if the guiding centers are themselves precessing, each vortex orbits the center of the condensate at the precession frequency of the guiding centers. From symmetry, we also expect the angular positions of the guiding centers to be π radians apart on a line that passes through the center of the condensate, i.e., on opposite sides of the center.

Therefore, we adopt a trial solution of the form

$$z_1(t) = r_1 \exp(i\omega t),$$

$$z_2(t) = r_2 \exp(i\omega t + i\pi) = -r_2 \exp(i\omega t),$$

where r_1 and r_2 are now constant, as is ω, the precession frequency of the guiding center. Note that in this case, $r_{12} \equiv r_1 + r_2$ is the constant separation distance between the two guiding centers.

It is now convenient to introduce the notation

$$\alpha = \frac{1}{1-r_1^2}, \quad \beta = \frac{1}{1-r_2^2}, \quad \text{and} \quad \gamma = \frac{1}{2r_{12}^2},$$

with α, β, and γ all time-independent constants, by assumption.

Since the time dependence in $z_1(t)$ and $z_2(t)$ is now explicit, we take the derivatives directly:

$$\dot{z}_1 = i\omega r_1 \exp(i\omega t) \quad \text{and} \quad \dot{z}_2 = -i\omega r_2 \exp(i\omega t),$$

whereupon the differential equations become

$$\omega r_1 = \alpha\omega_{\text{pr}} r_1 - \gamma\Phi_R(r_1 + r_2), \tag{3.51}$$

$$\omega r_2 = -\beta\omega_{\text{pr}} r_2 + \gamma\Phi_R(r_1 + r_2). \tag{3.52}$$

Recalling $r_{12} = r_1 + r_2 = (2\gamma)^{-1/2}$ as a final (as yet unused) constraint, we now have three equations and three unknowns (r_1, r_2, and ω); that is, upon choosing r_{12}, the guiding center locations r_1 and r_2 can be identified in terms of r_{12}, as well as the precession frequency of the guiding centers ω—provided that there are solutions to this set of equations. Given the nature of the ensuing algebraic equations (see below), such solutions will generically exist. We derive the relevant equations below.

We start with the precession frequency ω. From the system of Eqs. (3.51) and (3.52), we find

$$\omega = \frac{1}{2}\left[\omega_{\text{pr}}(\alpha - \beta) + \gamma\Phi_R\left(\frac{r_1}{r_2} - \frac{r_2}{r_1}\right)\right]. \tag{3.53}$$

Note that if $r_2 = r_1$, then $\alpha = \beta$ and $\omega = 0$, recovering the expected stationary equilibrium.

Furthermore, from the same Eqs. (3.51) and (3.52) we immediately obtain

$$\omega_{\text{pr}}(\alpha + \beta) - \gamma\Phi_R\left(2 + \frac{r_1}{r_2} + \frac{r_2}{r_1}\right) = 0. \tag{3.54}$$

Substituting in for α, β, and γ, in terms of r_1 and r_2, and simplifying, one finds

$$\omega_{\text{pr}}\left(\frac{1}{1-r_1^2} + \frac{1}{1-r_2^2}\right) - \frac{\Phi_R}{2r_1 r_2} = 0. \tag{3.55}$$

Once the position of r_2 is fixed, the location of the first vortex $r \equiv r_1$ can be found from the third order polynomial

$$\beta r^3 - \frac{\Phi_R}{2r_2} r^2 - \left(\omega_{\text{pr}} + \beta\right) r + \frac{\Phi_R}{r_2} = 0. \tag{3.56}$$

An alternative possibility is to fix r_{12}, then, the location of the first vortex $r \equiv r_1$ can be found from the quartic algebraic equation

$$c_4 r^4 + c_3 r^3 + c_2 r^2 + c_1 r + c_0 = 0, \tag{3.57}$$

where the coefficients c_j ($j = 0, 1, 2, 3$) are given by

$$\begin{aligned} c_0 &= B(r_{12}^2 - 1), \\ c_1 &= 2r_{12} - r_{12}^3 - 2r_{12}B, \\ c_2 &= 3r_{12}^2 - 2 - B(r_{12}^2 - 2), \\ c_3 &= -4r_{12} + 2r_{12}B, \\ c_4 &= 2 - B, \end{aligned} \tag{3.58}$$

and $B \equiv \Phi_R/(2\omega_{\text{pr}})$.

Stability

In the following we prove the stability of the guiding center of the form

$$z_1(t) = r_1 \exp(i\omega t), \qquad z_2(t) = -r_2 \exp(i\omega t), \tag{3.59}$$

obtained in the previous subsection. To this purpose, we pass to a co-rotating frame by making the change of variables $\tilde{z}_i = z_i \exp(-i\omega t)$ on the original system (3.46). Keeping the more convenient z_i notation for the state variables in a slight abuse of notation, the resulting system is

$$\begin{aligned} i\dot{z}_1 &= \omega z_1 - \Omega(r_1) z_1 + \frac{\Phi_R}{2r_{12}^2}(z_1 - z_2), \\ i\dot{z}_2 &= \omega z_2 + \Omega(r_2) z_2 - \frac{\Phi_R}{2r_{12}^2}(z_2 - z_1). \end{aligned} \tag{3.60}$$

Note that a rotating solution like (3.59) of the original system (3.46) is equivalent to the statement that $(r_1, -r_2)$ is an equilibrium of the new system (3.60).

In Cartesian coordinates, the system (3.60) reads

$$\begin{aligned} \dot{x}_1 &= \omega y_1 - \Omega(r_1) y_1 + \Phi_R \frac{y_1 - y_2}{2r_{12}^2}, \\ \dot{y}_1 &= -\omega x_1 + \Omega(r_1) x_1 - \Phi_R \frac{x_1 - x_2}{2r_{12}^2}, \\ \dot{x}_2 &= \omega y_2 + \Omega(r_2) y_2 - \Phi_R \frac{y_2 - y_1}{2r_{12}^2}, \\ \dot{y}_2 &= -\omega x_2 - \Omega(r_2) x_2 + \Phi_R \frac{x_2 - x_1}{2r_{12}^2}, \end{aligned} \tag{3.61}$$

and passing to polar coordinates, we obtain

$$\begin{aligned} \dot{\theta}_1 &= -\omega + \Omega(r_1) - \frac{\Phi_R}{2r_{12}^2}\left[1 - \frac{r_2}{r_1}\cos(\theta_1 - \theta_2)\right], \\ \dot{\theta}_2 &= -\omega - \Omega(r_2) + \frac{\Phi_R}{2r_{12}^2}\left[1 - \frac{r_1}{r_2}\cos(\theta_1 - \theta_2)\right], \\ \dot{r}_1 &= \frac{\Phi_R r_2}{2r_{12}^2}\sin(\theta_1 - \theta_2), \\ \dot{r}_2 &= \frac{\Phi_R r_1}{2r_{12}^2}\sin(\theta_1 - \theta_2). \end{aligned} \tag{3.62}$$

Of course, when $\omega = 0$ we recover Eqs. (3.47).

Now the Hamiltonian is

$$\begin{aligned} H(r_1, r_2, \theta) = -\frac{\omega}{2}(r_1^2 - r_2^2) - \frac{1}{2}\Big[& \omega_{\mathrm{pr}} \ln(1 - r_1^2) + \omega_{\mathrm{pr}} \ln(1 - r_2^2) \\ & + \frac{\Phi_R}{2}\ln(r_1^2 + r_2^2 - 2r_1 r_2 \cos\theta)\Big], \end{aligned} \tag{3.63}$$

where again $\theta = \theta_1 - \theta_2$. However, here, the additional difficulty is the absence of a simple explicit expression for the solution.

Lemma 3.3. *Any equilibrium of system (3.62) satisfies* $\theta = \pi$.

Proof. From the last equation, $\theta = \pi$ or $\theta = 0$. By contradiction, let us assume that $\theta = 0$. Then from the first two equations of system (3.62)

$$-\omega + \Omega(r_1) - \frac{\Phi_R}{2r_{12}^2}\left[1 - \frac{r_2}{r_1}\right] = 0, \tag{3.64}$$

$$-\omega - \Omega(r_2) + \frac{\Phi_R}{2r_{12}^2}\left[1 - \frac{r_1}{r_2}\right] = 0,$$

subtracting one from the other leads to

$$\Omega(r_1) + \Omega(r_2) = \frac{\Phi_R}{2r_{12}^2}\left[2 - \frac{r_2}{r_1} - \frac{r_1}{r_2}\right] > 0,$$

then

$$2 - \frac{r_2}{r_1} - \frac{r_1}{r_2} > 0,$$

and multiplying by $r_1 r_2$, one gets $-(r_1 + r_2)^2 > 0$, which is a contradiction.

Now it is ensured that any equilibrium should be of the form (r_1, r_2, π). We can then compute the Hessian matrix of the Hamiltonian above, evaluated at the point (r_1, r_2, π) and obtain

$$\begin{pmatrix} \Omega(r_1) + \delta_1 + \delta_{12} - \omega & \delta_{12} & 0 \\ \delta_{12} & \Omega(r_2) + \delta_2 + \delta_{12} + \omega & 0 \\ 0 & 0 & \delta_{12} r_1 r_2 \end{pmatrix},$$

where $\delta_i \equiv \frac{2\omega_{\mathrm{pr}} r_i^2}{(1-r_i^2)^2} > 0$ and $\delta_{12} \equiv \frac{\Phi_R}{2(r_1+r_2)^2} > 0$. Here it is difficult to see that the matrix is positive-definite, but, from Eqs. (3.64), the frequency ω satisfies

$$\omega = \Omega(r_1) - \delta_{12}\left(1 + \frac{r_2}{r_1}\right) = -\Omega(r_2) + \delta_{12}\left(1 + \frac{r_2}{r_1}\right).$$

Inserting this information into the Hessian yields

$$\begin{pmatrix} \delta_1 + \delta_{12}\left(2 + \frac{r_2}{r_1}\right) & \delta_{12} & 0 \\ \delta_{12} & \delta_2 + \delta_{12}\left(2 + \frac{r_1}{r_2}\right) & 0 \\ 0 & 0 & \delta_{12} r_1 r_2 \end{pmatrix},$$

and now one easily realizes that this matrix is positive-definite again by Sylvester's criterion [534]. Hence, once again the relevant guiding-center equilibrium is a stable one, just as it is in the special case in which $\omega = 0$ and $r_1 = r_2$.

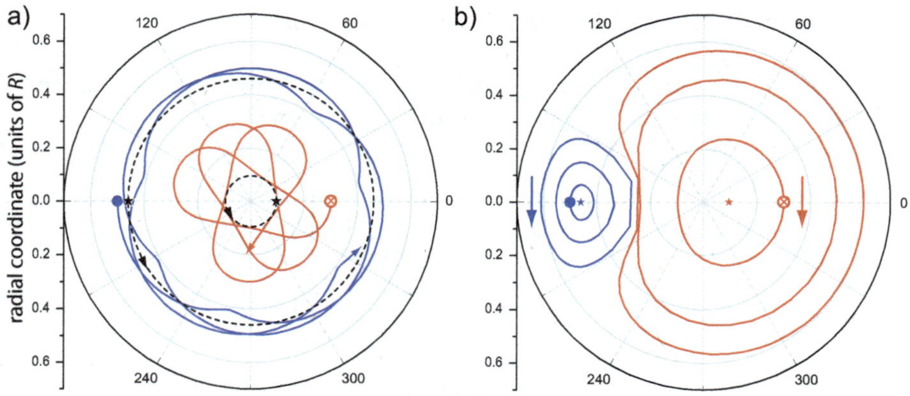

Figure 3.28. *Typical computed orbits associated with asymmetric vortex configurations, determined by integrating the equations of motion [Eqs. (3.46)].* (a) *Motion of the vortices in the lab frame, where the outer vortex (blue) has a counterclockwise sense of circulation and the inner vortex (red) has a clockwise sense of circulation. The guiding center equilibria are indicated by stars, and their trajectories follow circles in the same sense as the rotation of the outermost vortex (black dashed lines).* (b) *Motion of vortices in a frame co-rotating with the guiding centers. Several representative orbits with the same guiding centers are shown, with the circle symbols attached to the curves corresponding to the motion in* (a). *Adapted with permission from Ref. [527].*

Rotating guiding center equilibria are relevant when the vortices are asymmetrically located with respect to a line through the condensate center. A representative example is given in Fig. 3.28; such behavior has been observed in experiments [537], to the discussion of which we now turn.

3.5.6 ▪ Connection with Experimental Results

A comparison of the above theoretical predictions with experiments was offered in Ref. [537]. The apparatus and experimental procedure used therein were also used earlier and described in detail in Refs. [279, 531]. The experiments involved an oblate ^{87}Rb condensate of $\sim 6 \times 10^5$ atoms in the $|F = 1, m_F = -1\rangle$ state, confined in a TOP magnetic trap [538] with effective frequencies $\{\omega_r, \omega_z\} = 2\pi \times \{35.8(2), 101.2(5)\}$ Hz. The Thomas–Fermi radius of the condensate is $R_\perp \sim 17$ μm. Vortex dipoles arise spontaneously [57] but infrequently ($\sim$ 3% of the time) during the final stage of evaporative cooling. To observe the vortex dynamics a real-time imaging method was used [531] that repeatedly extracts, expands, and absorptively images small (1% or 5%) fractions of the atomic sample. A final time-of-flight image of the remaining atoms is taken at the end of the experiment.

Image sequences of the motion of the vortex lines are shown in Figs. 3.29 and 3.30. A surface fit to each image determines the coordinates of each vortex and the condensate center. The number of atoms was estimated based on the fraction extracted for each image, including an extra reduction (typically 1%) between images to reflect atomic losses due to three-body recombinations [539]. The sudden reduction in the number of atoms following expansion is expected to excite a small-amplitude (a few percent) breathing mode, with only slight alteration of vortex trajectories [540]. Experimentally, neither the anticipated excitation nor any effect on the motion of a solitary vortex line [531] was observed, further suggesting that any effect on a dipole is small.

The few-vortex dynamics described by the ODEs presented above provide a number of experimentally accessible predictions for the observed symmetric and asymmetric vor-

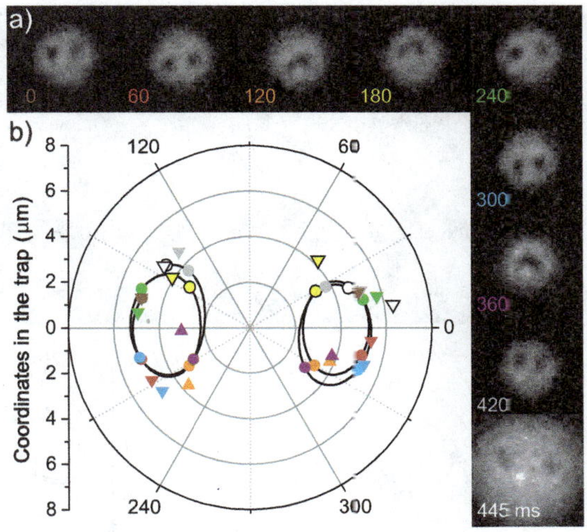

Figure 3.29. *Symmetric vortex configuration.* (a) *Experimental images of the atomic density (optical depth), 1% extraction.* (b) *Polar plot of the experimental data (triangles) and the theoretical prediction of the ODEs (solid line). The colored circles correspond to the calculated vortex coordinates at the same times as the experimental data. Only the inner region of the condensate is shown. Adapted with permission from Ref. [537].*

tex dipole configurations. First, an explicit prediction can be made about the stationary equilibrium positions of the constituent vortices. The velocities of the two vortices are zero for all times at a separation distance

$$d_{\rm sep} = 2R_\perp \sqrt{\frac{b\Phi_R}{4\omega_{\rm pr} + b\Phi_R}}, \tag{3.65}$$

with $\omega_{\rm pr} \equiv \Omega(0)$ and $\Phi_R \equiv \Phi(R_\perp)$. Equation (3.65) yields an average equilibrium distance of $0.436R_\perp$, in remarkable agreement with the experimental value $0.432(17)R_\perp$.

The frequency of the epicyclic motion observed in Fig. 3.29 is calculated by linearizing the two-vortex equations around $d_{\rm sep}$. We obtain two pairs of frequencies, $\omega_{1,2} = 0$ and

$$\omega_{3,4} = \pm\sqrt{2}\omega_{\rm pr}\left(1 + \frac{b\Phi_R}{4\omega_{\rm pr}}\right)^{3/2}. \tag{3.66}$$

The first pair illustrates the rotational invariance (around the center of the condensate) of the dipole [521], whereas the second pair corresponds to the anomalous modes of the vortex system [76], similar to those of dark solitons [70, 347]. These anomalous modes are responsible here for the counter-rotating, oscillatory vortex motion about the equilibrium points [518], which can be understood intuitively as the result of competition between the precession and interaction velocities [535]. The measured oscillation frequency is $1.66(5)\overline{\Omega}_0$, where the overbar denotes the average value during the experiment. This compares reasonably with the prediction of $1.52\overline{\Omega}_0$ from Eq. (3.66), given the relatively large amplitude of the observed vortex motion. The initially cyclic vortex motion observed in Ref. [536] is also of this persuasion, albeit with larger amplitude. We note here in passing that the experiment in Ref. [536] involved the dragging of an (optical) defect through the condensate, as a superfluid analogue of the classical flow past a cylinder experiment (see Sec. 2.5.1). Naturally, past a critical speed (for a given density) or a crit-

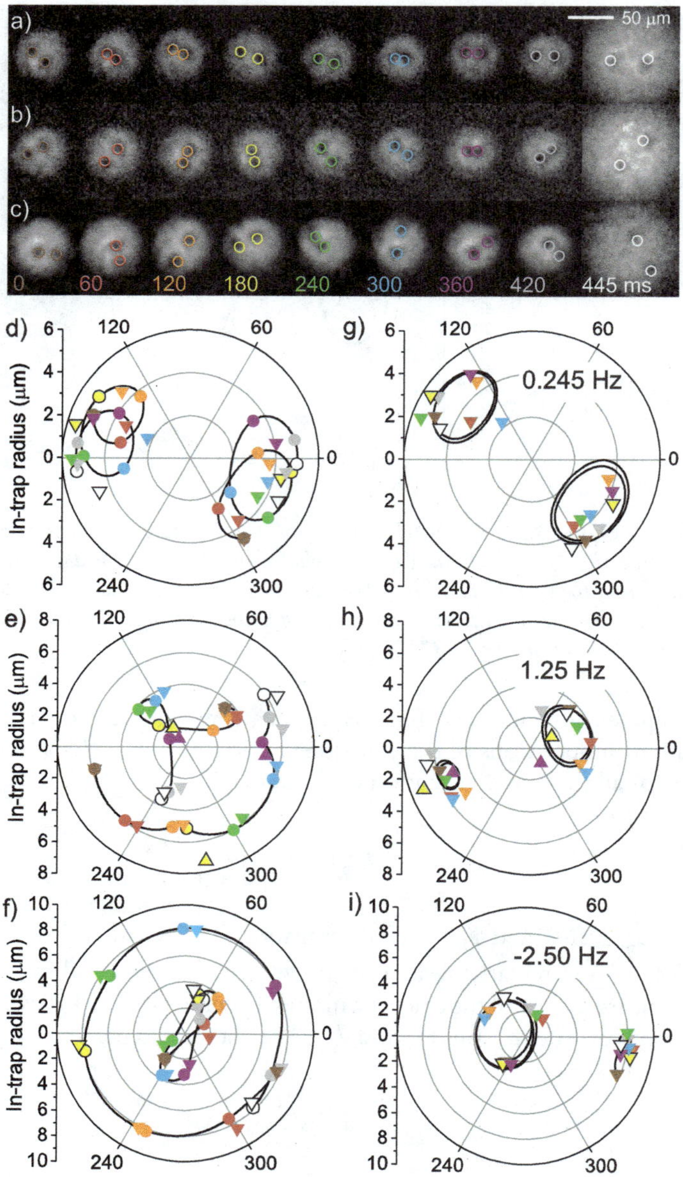

Figure 3.30. *Asymmetric vortex configurations.* (a)–(c) *Experimental images of the atomic density overlaid with calculated vortex positions from the theoretical model (colored rings).* (d)–(f) *Trajectories for the data in panels* (a)–(c), *using the notation of Fig. 3.29.* (g)–(i) *Trajectories for the data in panels* (d)–(f) *in the indicated rotating frames. In* (a)–(b) *the extracted fraction is 1%, whereas in* (c) *it is 5% and the image intensity is scaled by a factor of 2. Adapted with permission from Ref. [537].*

ical density (for a given speed), the superfluid becomes unstable towards the emission of excitations in the form of vortices in such oblate BECs. These, both in experiments and in the numerical simulations of Ref. [536] seem to follow the predicted epicyclic form of motion (although with much larger amplitude).

A direct comparison between experimental trajectories and those predicted by the model is made possible by integrating the equations of motion [as shown in 3.29(b)]. It is convenient to make this comparison by rescaling the post-expansion data by the ratio $R_\perp/R_\perp^*$, where $R_\perp^*$ is the measured condensate radius after expansion, and $R_\perp$ is the theoretical in-trap Thomas–Fermi radius calculated for a condensate of $N(t)$ atoms. The theoretical points are calculated from the ODE "particle" model, using values for the chemical potential that are also determined from $N(t)$, and are updated after each measurement in the subsequent time-evolution. In Ref. [537], the initial positions of the modeled vortices were chosen to coincide with those of the experiment at $t = 0$. As a sole free parameter, a small ($\lesssim 1\,\mu$m) offset (i.e., uncertainty) of the condensate center $(\delta x, \delta y)$ with respect to all of the measured data points was also included. The offset is attributed to a technical effect arising from atomic micromotion in the time-averaged potential of the magnetic trap; similar shifts have been observed in condensates with single vortex lines [531]. With these parameters and initial conditions, good agreement was obtained between the experiment and the results of the numerical integration [see Fig. 3.29(b)] for symmetric displacements of the vortices from the equilibrium points.

Vortex trajectories arising from asymmetric displacements of the vortex lines were also considered in Ref. [537]. In a sense, neither the equilibrium positions (i.e., zero-frequency motions), nor the epicyclic precessions around them (i.e., one-frequency motions) constitute the generic motion of vortices in this system; it is this generic asymmetric motion that we would like to now explore based on both the "top level" of experimental description and the "bottom level" of ODE-particle reduction (which also agrees with the "middle level" of the partial differential equation). Three typical image sequences are shown in Figs. 3.30(a)–(f). In Fig. 3.30(d), cyclic motion akin to that of Fig. 3.29 is superposed upon a slow precession. With increasing asymmetric displacements, Figs. 3.30(e) and (f), one vortex moves in loops near the center of the condensate, while the other moves closer to the periphery. The agreement between the simulation and the experiment ranges between fair (d) and quite good (e) and (f), with more pronounced discrepancies when the vortices are close together and therefore are difficult to resolve.

Remarkably, all three data sets of Fig. 3.30 exhibit essentially identical epicyclic behavior, as is easily seen by transforming in each case to a suitable co-rotating reference frame [Figs. 3.30(g)–(i)]. We therefore arrive at a unified and intuitive general understanding of the vortex dipole behavior: the vortex lines oscillate in phase about stable equilibrium points (or "guiding centers") that precess rigidly about the center of the condensate. Both the precession frequency and the cyclic frequency (in the rotating frame) increase with displacement from the condensate center, and the amplitude of the outer vortex orbit becomes smaller than that of the inner vortex; this last fact is due to the faster precession of the outer vortex chiefly influencing the co-rotating frame frequency. Ultimately, the stationary vortex dipole, as well as the oscillating dipole in Fig. 3.29, can be perceived as special cases of this general motion in which guiding centers are stationary in the laboratory frame. These are precisely the *stable* guiding center solutions that were analyzed in the previous subsection. Notice that this conclusion is perfectly in line with the structure of a two degrees-of-freedom system, which possesses two integrals of the motion—namely, the Hamiltonian H and the angular momentum. This, in turn, guarantees the integrability in the classical Liouville sense [506], and the phase space is foliated by invariant tori, on each of which the motion is quasiperiodic with two frequencies: the precession frequency of the equilibrium points, and the oscillation frequency about the precessing equilibria.

3.5.7 ▪ The Case of $n > 2$ Vortices

We now further explore the particle model in the presence of a higher number of vortices of nonuniform charge, starting with the case of $n = 3$. In this exposition, we follow the discussion of Ref. [541]. Notice in that light that for simplicity in this context, we will assume that $\Omega(r)=\Omega=$ constant in the discussion of this section, in order to avoid complications due to the radial dependence of the single vortex precession frequency. While quantitative details of our conclusions (such as the exact location of the fixed points) will be modified by a radial distance dependence of the precession frequency, the qualitative properties thereof (including the stability characteristics) will not.

In the case of $n=3$, the only fixed point of the particle system is a collinear configuration. To prove this fact, we start by assuming without loss of generality, fix $y_1=0$. If $(x_1,0,x_2,y_2,x_3,y_3)$ with charges S_1, S_2, and S_3 is a fixed point of the system with $S_1=1$, the equations of motion read

$$\frac{S_2 y_2}{r_{12}^2} = \frac{-S_3 y_3}{r_{13}^2}, \tag{3.67}$$

$$\frac{S_3(y_2-y_3)}{r_{23}^2} = -S_2\Omega y_2 - \frac{y_2}{r_{12}^2}, \tag{3.68}$$

$$\frac{S_2(y_3-y_2)}{r_{23}^2} = -S_3\Omega y_3 - \frac{y_3}{r_{13}^2}. \tag{3.69}$$

It is then straightforward to see that for any combination of S_2 and S_3 within the values ± 1, these lead to $y_2=-y_3$. Finally, substituting in the equations for $\dot{y}_i$, we obtain that for $\Omega=1$, $y_2=-y_3=\sqrt{2}/2$. A direct calculation of the stability of this configuration yields $\lambda_{1,2}=\pm i\sqrt{5}$, $\lambda_{3,4}=\pm\sqrt{7}$ and $\lambda_{5,6}=0$. The last pair indicates the rotational invariance of the three-vortex configuration (due to the isotropic nature of the trap). However, and while the first pair is associated with a vibrational mode, the second one leads to an exponential instability. This naturally corroborates the near linear picture discussed above, whereby the bifurcation of the tripole arose from the (already unstable) planar dark soliton, clearly inheriting its symmetry breaking instability (through the corresponding supercritical pitchfork bifurcation). Previous theoretical discussions of this tripole configuration can be found in Refs. [520, 521, 523], while a recent experimental observation has been reported in Ref. [542].

The case $n=4$ is more subtle, as the system exhibits more than one fixed point, as we show analytically below. We will start by conjecturing that all the fixed points of the vortex equations of motion in our (nonuniform charge) setting are symmetric about the origin. Assuming the conjecture is true, one can prove that for $n=4$, the only fixed points of the vortex particle equations are square or collinear (i.e., one in which all vortices are located on a line going through the origin) configurations. However, it is presently unclear whether this conjecture may be true (in that it is solely based on numerical investigations).

The proof proceeds as follows. Choosing $\mathbf{x}_1=(x_1,0)$ implies that $\mathbf{x}_2=(-x_1,0)$. Then, it must be that $x_3=-x_4$ and $y_3=-y_4$. If $y_3=0$, the configuration is collinear, so assume $y_3\neq 0$. Then, the equation for $\dot{x}_1$ yields

$$\frac{S_3}{r_{13}^2} = \frac{S_4}{r_{14}^2}.$$

Thus $r_{13}=r_{14}$, but since the configuration is symmetric about the origin this implies that it is a square.

Fixing $\mathbf{x}_1=(1,0)$, $\mathbf{x}_2=(-1,0)$, $\mathbf{x}_3=(0,1)$, $\mathbf{x}_4=(0,-1)$, $S_1=S_2=1$, and $S_3=S_4=-1$ leads to a square fixed point configuration with $\Omega=1/2$. Linearizing about this equilibrium yields the following eigenvalues:

$$\lambda_{1,2} = \pm\frac{i}{\sqrt{2}}, \qquad \lambda_{3,4} = \lambda_{5,6} = \pm\frac{i}{4}, \qquad \lambda_{7,8} = 0. \tag{3.70}$$

Using the same value of $\Omega = 1/2$, we find that the collinear fixed point satisfies $\mathbf{x}_1 = (c,0) = -\mathbf{x}_3$, $\mathbf{x}_2 = (d,0) = -\mathbf{x}_4$, where

$$\begin{aligned} c &= \sqrt{1+\sqrt{2(-1+\sqrt{2})}}, \\ d &= \frac{1}{2}\left[10\sqrt{1+\sqrt{2(-1+\sqrt{2})}} - 9\left(1+\sqrt{2(-1+\sqrt{2})}\right)^{3/2}\right. \\ &\quad \left. +4\left(1+\sqrt{2(-1+\sqrt{2})}\right)^{5/2} - \left(1+\sqrt{2(-1+\sqrt{2})}\right)^{7/2}\right]. \end{aligned} \tag{3.71}$$

This configuration has eigenvalues $\lambda_{1,2} \approx \pm 2.15$, $\lambda_{3,4} \approx \pm 1.43i$, $\lambda_{5,6} \approx \pm .914$, $\lambda_{7,8} = 0$, and thus has $n-2 = 2$ directions of instability.

It is important to note that these results are consonant with the observations of the GPE, as presented above. In particular, the (rotationally symmetric, hence the vanishing eigenvalue pair) vortex quadrupole configuration of Fig. 3.19 is never exponentially unstable; it only features an interval of oscillatory instability which cannot be observed at the particle level above (as this instability emerges from the resonance of the localized vortex modes with the background extended ones ignored in our particle model). Furthermore, the aligned vortex quadrupole emerges from the next sequential instability of the dark soliton stripe, as discussed earlier in the context of Fig. 3.17. As a result, it possesses an additional instability in comparison to the tripole of $n = 3$ and hence has $n-2 = 2$ exponentially unstable modes, as expected.

Turning now to the case of $n = 5$, we again confine our considerations to cases symmetric around the origin. Thus, fixing $(x_1, y_1) = (0,0)$ and $(x_2, y_2) = (r_1, 0)$, we obtain by symmetry $(x_3, y_3) = (-r_1, 0)$. Then if we set $(x_4, y_4) = (r_2\cos\theta_2, r_2\sin\theta_2)$, $0 \le \theta_2 < \pi$, we have that $(x_5, y_5) = (-r_2\cos\theta_2, -r_2\sin\theta_2)$. We then compute

$$\dot{x}_1 = \frac{(S_3 - S_5)\sin\theta_2}{r_2}$$

and so either $\sin\theta_2 = 0$ or $S_3 = S_5$. Assuming the latter holds we find from the equation for y_1 that $S_2 = S_4$. Then from the equation for x_2 it follows that $\sin(2\theta_2) = 0$.

If we choose the root $\theta_2 = \frac{\pi}{2}$, then the configuration must be a square or rhombus centered on the origin. Here, two cases are physically relevant. In the first, the vortices alternate in sign in the counterclockwise direction with the central vortex being of either sign. In the second, the outer vortices have the same sign which, in turn, is opposite to the sign of the vortex at the origin. In the first case, we find that the system has no fixed points. In the second, we find that if the central vortex has charge $M > 0$ and the outer vortices have charge -1, then the system has a fixed point if and only if $r_1 = r_2$. As an example, when the radius is one and $M = 2$ we compute the configuration to be unstable with eigenvalues $\lambda_{1,2} \approx 3.14i$, $\lambda_{3,4} = \sqrt{6}$, $\lambda_{5,6} \approx -1.15 \pm 1.07i$, $\lambda_{7,8} \approx 1.15 \pm 1.07i$, and $\lambda_{9,10} = 0$. More generally, considering the algebraic constraints, one finds that the configuration with a square surrounding the central vortex can be realized provided suitable constraints connecting the central vortex charge, the surrounding charges, the precession frequency and the square's radius. The appropriate condition in this case reads

$$r_1^2 = \frac{2M - 3\widetilde{M}}{2\Omega\widetilde{M}}, \tag{3.72}$$

where $-\widetilde{M}$ are the surrounding charges to the central one. It is thus clear that such configurations will only exist if $\widetilde{M} \times (2M - 3\widetilde{M}) > 0$ and assuming $\widetilde{M} > 0$ without loss of generality, this leads to $M > 3\widetilde{M}/2$. In the case of $\widetilde{M} = 1$, the lowest charge that will work is $M = 2$. This configuration can be found to be definitely unstable due to a real pair $\lambda = \pm 2\widetilde{M}^{3/2}\Omega\sqrt{4M - 2\widetilde{M}}/(2M - 3\widetilde{M})$. Such a configuration was indeed identified in Ref. [523], and was found for large μ to contain a complex eigenfrequency quartet, as well as a real eigenvalue pair as manifestations of its instability. These sources of instability are consonant with the ones identified by the particle model. Additionally, we should note that this is an example of a $4+1$ vortex configuration. Such configurations will be revisited later in this chapter.

In the case of $\theta_2 = 0$, we have a collinear fixed point, which only exists when the vortices have alternating charges. This fixed point can be described by the configuration $(0,0)$, $(\pm r_1, 0)$, $(\pm r_2, 0)$ where

$$\begin{aligned} r_1 &= \sqrt{\frac{2-\sqrt{3}}{2\Omega}}, \\ r_2 &= \sqrt{\frac{\sqrt{3}}{2\Omega}}. \end{aligned} \tag{3.73}$$

This configuration was computed to be unstable with eigenvalues $\lambda_{1,2} \approx \pm 6.95\Omega$, $\lambda_{3,4} \approx \pm 6.69\Omega$, $\lambda_{5,6} \approx \pm 3.64i\Omega$, $\lambda_{7,8} \approx \pm 2.8\Omega$, $\lambda_{9,10} = 0$. In this case too, there are $n-2=3$ directions of real instability, as well as a neutral direction. These conclusions agree with the ones of Refs. [523, 518] in the setting of large chemical potential.

Lastly from within the setting of small n's, we explore the case of $n = 6$, in which case two fixed points are well known, namely the hexagonal and the collinear configurations. In the former case, the eigenvalues for $\Omega = 1/2$ read

$$\lambda_{1,2} = \pm\frac{3i}{\sqrt{2}}, \quad \lambda_{3,4} = \lambda_{5,6} = \pm i, \quad \lambda_{7,8} = \lambda_{9,10} = \pm\frac{1}{\sqrt{2}}, \quad \lambda_{11,12} = 0,$$

confirming its instability according to the discussion around Fig. 3.19 (and its emergence through a supercritical pitchfork bifurcation from the already unstable ring dark soliton [523]).

On the other hand, the collinear configuration can be represented by $(\pm a, 0)$, $(\pm b, 0)$, $(\pm c, 0)$ where $a \approx 0.23$, $b \approx 0.67$, and $c \approx 1.58$ were computed numerically (see, however, also below for a semi-analytical calculation thereof). The eigenvalues of the linearization about this fixed point were computed to be $\lambda_{1,2} \approx \pm 4.88$, $\lambda_{3,4} \approx \pm 4.86$, $\lambda_{5,6} \approx \pm 2.65$, $\lambda_{7,8} \approx \pm 2.2i$, $\lambda_{9,10} \approx \pm 0.96$, and $\lambda_{11,12} = 0$ and so this configuration also has $n-2=4$ directions of instability.

As regards the collinear configurations, we can clearly discern the emergence of a pattern regarding their progressively increasing unstable eigendirections as n is increased. In particular, such configurations emerge from a cascade of bifurcations of the planar dark soliton, each adding one additional unstable eigendirection to the emergent structure. As a result, we can conjecture that the collinear fixed point is unstable for $n \geq 3$ with $n-2$ real directions of instability. While this is true for all the cases that we are aware of, no general proof of this fact has been given yet, to the best of our knowledge. The existence of such a collinear configuration of vortices, for any n, suggests that there are always real numbers $x_1, x_2, \ldots, x_n$ such that for every $j = 1, \ldots, n$, $\sum_{k=1,k\neq j}^{n} (-1)^{k-j}/(x_j(x_k - x_j)) = 1$. In the case of even n, our computations indicate a symmetric configuration of the form $-x_{n/2}, -x_{(n-2)/2}, \ldots, -x_1, x_1, \ldots, x_{(n-2)/2}, x_{n/2}$, while for n odd, one of the vortices is always placed at 0. In the latter case, one can rephrase the statement as $\sum_{k=1,k\neq j}^{n} (-1)^{k-j}/(x_k - x_j) = x_j$. The remaining vortices are placed symmetrically around 0 as above. It is clear that this is essentially an algebraic problem. Effectively, this can be rewritten as a set of n polynomial equations in n unknowns. As such, it defines a variety, and the question is whether this variety always has a real point $(x_1, \ldots, x_n)$. Again, our analysis for different cases of finite n and computations indicate that this conjecture should be true, but a proof of this statement is still missing.

Another important class of stationary vortex configurations in the case of alternating charges is that of a polygonal ring with n vortices (sometimes referred to as n-gon). In the work of Ref. [541], it was shown that such n-gons only exist in the case of n even and are only linearly stable if $n = 2$ (abusing notation to consider the dipole as a special case of an n-gon) and $n = 4$.

Without loss of generality, we can assume the vortices of the n-gon as living on the unit circle as $z_1 = e^{2\pi i/n}, z_2 = e^{4\pi i/n}, \ldots, z_n = 1$, accordingly fixing the rotation frequency Ω. The proof of (non)existence for the different n-gons is most straightforward to present in polar coordinates. When n is even, by symmetry $\dot{r}_j = 0$ for each j. For $r_j = 1$, one finds

$$\dot{\theta}_j = (-1)^{j+1}\Omega + \frac{1}{2}\sum_{k\neq j}(-1)^{k+1} = \begin{cases} \Omega - \frac{1}{2} & \text{if } j \text{ is odd}, \\ -\Omega + \frac{1}{2} & \text{if } j \text{ is even}, \end{cases} \tag{3.74}$$

and so the n-gon is a fixed point when $\Omega = 1/2$. On the other hand, when n is odd we have

$$\dot{\theta}_j = (-1)^{j+1}\Omega + \frac{1}{2}\sum_{k\neq j}(-1)^{k+1} = \begin{cases} \Omega & \text{if } j \text{ is odd}, \\ \Omega + 1 & \text{if } j \text{ is even}, \end{cases} \tag{3.75}$$

and hence the n-gon cannot be a fixed point for any value of Ω.

As regards the stability conclusions, once again it is preferential to calculate the relevant stability matrix in the polar coordinates:

$$M := \begin{pmatrix} A & B \\ C & D \end{pmatrix}.$$

This is a $2n \times 2n$ block matrix made up of $n \times n$ blocks with elements $a_{j,k}$, $b_{j,k}$, $c_{j,k}$, and $d_{j,k}$ which can be computed explicitly. If $\theta_{jk} = \theta_j - \theta_k$ and $r_{jk} = |z_j - z_k|$, then

$$a_{j,k} = \frac{S_j(r_k^2 - r_j^2)\sin(\theta_{jk})}{r_{jk}^4}, \qquad j \neq k,$$

$$a_{j,j} = \sum_{k\neq j} \frac{2r_k S_k(r_j - r_k\cos(\theta_{jk}))\sin(\theta_{jk})}{r_{jk}^4}.$$

In the case of the n-gon where $S_k = (-1)^{k+1}$ and $r_j = 1$ for all j we see that the above entries are identically zero, hence $A = 0$. Matrix B has elements

$$b_{j,k} = \frac{r_j S_j((r_j^2 + r_k^2)\cos(\theta_{jk}) - 2r_j r_k)}{r_{jk}^4}, \qquad j \neq k,$$

$$b_{j,j} = \sum_{k\neq j} \frac{r_k S_k(2r_k r_j - (r_j^2 + r_k^2)\cos(\theta_{jk}))}{r_{jk}^4}.$$

These expressions simplify for $r_j = 1$ to

$$b_{j,k} = -\frac{S_j}{r_{jk}^2},$$

$$b_{j,j} = \sum_{k\neq j} \frac{S_k}{r_{jk}^2}.$$

The matrix C is

$$c_{j,k} = \frac{S_k((r_j^2 + r_k^2)\cos(\theta_{jk}) - 2r_j r_k)}{r_k r_{jk}^4}, \qquad j \neq k,$$

$$c_{j,j} = -\sum_{k\neq j} \frac{S_k(r_j^2 - 2r_j r_k \cos(\theta_{jk}) + r_k^2 \cos(2\theta_{jk}))}{r_j r_{jk}^4} - \sum_{k\neq j} \frac{S_k(r_j - r_k\cos(\theta_{jk}))}{r_j^2 r_{jk}^2}.$$

Again setting $r_j = 1$, one obtains after some simplification

$$c_{j,k} = -\frac{S_j}{r_{jk}^2},$$

$$c_{j,j} = \sum_{k\neq j} \frac{S_k}{r_{jk}^2} - \sum_{k\neq j} S_k,$$

where we have used that $r_{jk}^2 = 2 - 2\cos(\theta_{jk})$. Finally, for matrix D,

$$d_{j,k} = \frac{S_j r_j (r_j^2 - r_k^2)\sin(\theta_{jk})}{r_k r_{jk}^4}, \qquad j \neq k,$$

$$d_{j,j} = \sum_{k\neq j} \frac{S_k r_k (r_k^2 - r_j^2)\sin(\theta_{jk})}{r_j r_{jk}^4},$$

which again vanishes for $r_j = 1$. We are interested in the eigenvalues of M, hence the vanishing of the determinant

$$\begin{pmatrix} -\lambda I & B \\ C & -\lambda I \end{pmatrix}.$$

This, in turn, leads to

$$\det(M - \lambda I_{2n\times 2n}) = \det(\lambda^2 I_{n\times n} - BC).$$

For $\Omega = 1/2$, the resulting matrix BC whose eigenvalues need to be computed is circulant and hence $\lambda^2 \in R$, which implies that λ is either real or imaginary (but not complex). It is worthwhile to note here that complex instabilities may arise in the full GPE model in the case of the quartet [518], but they only stem from resonances of the vortex "localized" (negative Krein signature) modes with those of the background condensate.

Given the circulant nature of the matrix $\widetilde{B} = BC$, it is sufficient to know the first row of the matrix, in order to determine its eigenvalues. In particular, for our setting of $S_k = (-1)^{k+1}$, we have

$$\tilde{b}_{1,1} = \left(\sum_{k\neq 1} \frac{(-1)^{k+1}}{r_{1k}^2} \right)^2 + \sum_{k\neq 1} \frac{(-1)^{k+1}}{r_{1k}^2} + \sum_{k\neq 1} \frac{(-1)^{k+1}}{r_{1k}^4},$$

$$\tilde{b}_{1,m} = (-1)^{m+1} \frac{1}{r_{1m}^2} \left[\sum_{k\neq 1} \frac{(-1)^{k+1}}{r_{1k}^2} + \sum_{k\neq m} \frac{(-1)^{k+1}}{r_{mk}^2} + (-1)^{m+1} \right] + \sum_{k\neq 1,m} \frac{(-1)^{k+1}}{r_{1k}^2 r_{mk}^2}.$$

The eigenvalues of $\widetilde{B}$ are given by

$$\gamma_j = \sum_{k=1}^{n} \tilde{b}_{1,k} \omega_j^{k-1}, \qquad j = 1, \ldots, n$$

and the eigenvectors are

$$v_j = \{1, \omega_j, \omega_j^2, \ldots, \omega_j^{n-1}\},$$

where $\omega_j = e^{2\pi i j/n}$ is the jth root of unity.

It was found in Ref. [541] that for $j = n-2$, the corresponding $\gamma_{n-2} = (\sqrt{2}n/4 - \sqrt{2})^2$, which gave rise to a marginally stable configuration for $n = 4$ and an unstable one for any even $n > 4$. This, in turn, explains (in the large density, highly nonlinear limit) the instability of configurations such as the vortex hexagon, octagon, etc. Indeed, these states were identified as bifurcating from states of the linear limit (such as the ring dark soliton) in Fig. 3.19 and the above conclusions corroborate their BdG analysis in Fig. 3.21. In fact, the numerical conclusions lead also a conjecture (again worth investigating in future work) regarding the stability of such n-gons as follows. For $n \geq 6$, the n-gon has $\frac{n}{2} - 2$ distinct pairs of real eigenvalues.

After presenting the case of co-rotating vortices, and especially the method of generating functions, below we will return to the study of the existence problem for the counter-rotating vortex setting.

3.5.8 ▪ The Case of Co-Rotating Vortices

While the consideration of constant vortex precession frequency already yields much of the interesting phenomenology (and stationary states) discussed above for the case of alternating sign vortices, there are some features which are particular byproducts of the radial dependence of the precession frequency. In particular, as was originally analyzed by Ref. [76] and was subsequently corroborated by Ref. [531], the radial dependence of the precession frequency as per $f(r) = \omega_{\text{pr}}^0/(1 - r^2)$ is more accurate for vortices precessing at radii closer to the Thomas–Fermi radius R_{TF}.[18] It is then natural to inquire whether in addition to rendering more quantitatively accurate the vortex dipole phenomenology, as

[18] In fact, our very recent effort to identify a formula accurate—from a fitting perspective—for all $r < R_{\text{TF}}$ numerically has suggested that a dependence $\propto 1/(1-\alpha r^2)$ with $\alpha \approx 0.78$ provides the best simple fit throughout the Thomas–Fermi region.

was shown, e.g., in Ref. [537], this functional form of the precession frequency also introduces novel features. This question is, perhaps, most accurately addressed in the positive by the case of co-rotating vortices that we now explore.

When examining the large chemical potential limit of "vortex particles," it is straightforward to observe from the pertinent equations of motion that while in the case of opposite charges the precession and intervortex interaction counter each other (potentially yielding equilibria such as the vortex dipole), this is no longer true for co-rotating vortices. In the latter case, the cooperation of the two effects naturally gives rise to states that are solidly rotating with a constant angular momentum. It is these states that we will examine in what follows assuming the general framework of n vortices.

Using again complex notation, whereby $z_j = x_j + iy_j$, the ODEs for the vortex positions can be written as

$$\dot{z}_j = if\left(\left|z_j\right|\right) z_j + ic \sum_{k \neq j} \frac{z_j - z_k}{\left|z_j - z_k\right|^2}, \quad j = 1, \dots, n. \tag{3.76}$$

According to the above discussion, we seek relative (i.e., in a rotating frame) equilibria of the vortex configuration, upon using $z_j(t) = e^{i\omega t} r_j$; this leads to the steady state equation

$$\omega r_j = f\left(\left|r_j\right|\right) r_j + c \sum_{k \neq j} \frac{r_j - r_k}{\left|r_j - r_k\right|^2}, \tag{3.77}$$

which will be the main focus of our considerations in what follows.

As a (partially numerical) aside, we will also consider the following "aggregation" equation:

$$\dot{r}_j = \left(f(\left|r_j\right|) - \omega\right) r_j + c \sum_{k \neq j} \frac{r_j - r_k}{\left|r_j - r_k\right|^2}. \tag{3.78}$$

By construction, the relative equilibrium $z_j(t) = e^{i\omega t} r_j$ of Eq. (3.76) corresponds to the relaxational equilibrium $r_j(t) = r_j$ of Eq. (3.78) and vice versa. Moreover there is an intimate connection between the stability of the two models that we will use intensely below. More specifically, in our discussion of this case, we will follow closely the arguments of Ref. [543], starting from the following theorem (discussed in the appendix of Ref. [543]):

Theorem 3.4. *Suppose that an equilibrium $r_j(t) = \xi_j$ of Eq. (3.78) is stable. Then the relative equilibrium $z_j(t) = e^{i\omega t}\xi_j$ of Eq. (3.76) is (neutrally) stable. The converse is also true in the following two cases:* (i) $f(r) = \text{const.}$ *or* (ii) $|\xi_j| = \text{const.}$ *for all* j.

3.5.9 ▪ Stability of the Vortex Ring Configuration

Given the above comment about the cooperative effect of precession and interaction, the prototypical configuration that one expects to observe as a relative (rotating) equilibrium in the case of small vortex number n is the "ring" configuration, with the vortices sitting at the vertices of a canonical polygon. This is also, in part, motivated by the corresponding result in the absence of precession [544, 545]. Assuming such a relative equilibrium to Eq. (3.76) with radius R, we can express the vortex positions as

$$z_j(t) = R \exp(i\omega t) \exp\left(\frac{2\pi i}{n} j\right). \tag{3.79}$$

Then, the interaction term becomes

$$\sum_{k\neq j}\frac{z_j-z_k}{\left|z_j-z_k\right|^2}=\frac{1}{R}\exp\left(i\omega t+\frac{2\pi i}{n}j\right)\frac{n-1}{2}, \tag{3.80}$$

by using the identity

$$\sum_{k=0}^{n-1}\frac{1}{1-\exp\left(-\frac{2\pi i}{n}k\right)}=\frac{n-1}{2}.$$

Therefore the radius R satisfies

$$\omega=f(R)+\frac{c(n-1)}{2R^2}. \tag{3.81}$$

Using the above Theorem 3.4 and the analysis of the aggregation equation (cf. for a similar analysis Refs. [546, 547]), the following stability theorem, generalizing the classical result of Ref. [544] was proved in Ref. [543].

Theorem 3.5. *Consider the ring solution for Eq. (3.76), of radius R as given by Eq. (3.79), where the frequency ω is given by Eq. (3.81). Suppose that n is odd. Then the ring is stable provided that*

$$f'(R)R+\frac{c}{8R^2}(n-1)(n-7)<0,$$

and is unstable if the inequality is reversed. Suppose that n is even. Then the ring is stable provided that

$$f'(R)R+\frac{c}{8R^2}\left(n^2-8n+8\right)<0,$$

and is unstable if the inequality is reversed.

The ring is generically unstable if $n\geq 7$ and $f(r)$ is an increasing function.

Notice that, remarkably, the generalization of the classic result of Ref. [544] is due to the feature that, here, the precession frequency, denoted by f in the above statement, is not constant but rather an increasing function of the radius.

Proof. It is sufficient to consider the stability of the steady state $r_j(t)=R\exp(2\pi i k/n)$ of the aggregation equation (3.78). Consider small perturbations of this state, in the form

$$r_k(t)=R\exp(2\pi i k/n)(1+h_k(t)),\quad |h_k|\ll 1,$$

where h_k is a small, complex-valued perturbation whose evolution is given by

$$\frac{dh_k}{dt}=\left(f'(R)\frac{R}{2}+f(R)-\omega\right)h_j+f'(R)\frac{R}{2}\bar{h}_j+c\sum_{k\neq j}\frac{\exp(2\pi(k-j)/n)\bar{h}_j-\bar{h}_k}{4R^2\sin^2\left(\frac{\pi(k-j)}{n}\right)}, \tag{3.82}$$

where the overbar denotes complex conjugation. Using the Fourier mode decomposition for the perturbation,

$$h_j(t)=\xi_+(t)e^{im2\pi j/n}+\xi_-(t)e^{-im2\pi j/n},\quad m\in\mathbb{N}, \tag{3.83}$$

and collecting like terms in $e^{im2\pi j/n}$ and $e^{-im2\pi j/n}$, the system (3.82) decouples into a sequence of 2×2 subproblems

$$\xi_+' = \left(f'(R)\frac{R}{2} + f(R) - \omega\right)\xi_+ + f'(R)\frac{R}{2}\bar{\xi}_- + \sigma_+\bar{\xi}_-, \tag{3.84}$$

$$\xi_-' = \left(f'(R)\frac{R}{2} + f(R) - \omega\right)\xi_- + f'(R)\frac{R}{2}\bar{\xi}_+ + \sigma_-\bar{\xi}_+, \tag{3.85}$$

where

$$\sigma_+ \equiv c\sum_{k,k\neq j}\frac{e^{i2\pi(k-j)/n} - e^{im2\pi(k-j)/n}}{4R^2\sin^2(\pi(k-j)/n)},$$
$$\sigma_- \equiv c\sum_{k,k\neq j}\frac{e^{i2\pi(k-j)/n} - e^{-im2\pi(k-j)/n}}{4R^2\sin^2(\pi(k-j)/n)}.$$

Using the identity

$$\sum_{k=1}^{n-1}\frac{e^{\pm im2\pi k/n}}{\sin^2(\pi k/n)} = 2\left(m-\frac{n}{2}\right)^2 - \frac{n^2}{6} - \frac{1}{3}, \qquad m=0,\dots,n, \tag{3.86}$$

it is possible to write

$$\sigma \equiv \sigma_+ = \sigma_- = \frac{c}{2R^2}(m-1)(n-m-1). \tag{3.87}$$

Taking the conjugate of Eq. (3.85), the system can be written as

$$\begin{pmatrix}\xi_+' \\ \bar{\xi}_-'\end{pmatrix} = \begin{pmatrix} f'(R)\frac{R}{2} + f(R) - \omega & f'(R)\frac{R}{2} + \sigma \\ f'(R)\frac{R}{2} + \sigma & f'(R)\frac{R}{2} + f(R) - \omega\end{pmatrix}\begin{pmatrix}\xi_+ \\ \bar{\xi}_-\end{pmatrix}, \tag{3.88}$$

whose eigenvalues are given by

$$\lambda_\pm(m) = f'(R)\frac{R}{2} + f(R) - \omega \pm \left(f'(R)\frac{R}{2} + \frac{c}{2R^2}(m-1)(n-m-1)\right), \quad m=0,\dots,n-1.$$

Using Eq. (3.81) this simplifies to

$$\begin{aligned}\lambda_-(m) &= \frac{c}{2R^2}[-(n-1) - (m-1)(n-m-1)],\\ \lambda_+(m) &= f'(R)R + \frac{c}{2R^2}[(m-1)(n-m-1) - (n-1)].\end{aligned} \tag{3.89}$$

Note that $\lambda_-(0) = 0$ due to rotation invariance. Moreover, $(m-1)(n-m-1)$ is positive for all integers $1 \le m \le n-1$ and attains its max at $m = n/2$. Therefore, $\lambda_-(m)$ correspond to stable eigendirections and the instability threshold is obtained by setting $\lambda_+(\lfloor n/2 \rfloor) = 0$, where $\lfloor\cdot\rfloor$ denotes the floor function.

On the other hand, if n is odd we substitute $m=(n-1)/2$ into Eq. (3.89) to obtain

$$\lambda_+((n-1)/2)=f'(R)R+\frac{c}{8R^2}(n-1)(n-7), \quad n \text{ odd.}$$

If n is even we substitute $m=n/2$ which yields

$$\lambda_+(n/2)=f'(R)R+\frac{c}{8R^2}\left(n^2-8n+8\right), \quad n \text{ even.}$$

From the above, we can infer stability of the ring in the aggregation equation (and hence in the vortex model) if the right-hand side of the above two equations, respectively, for n odd and even, is negative and unstable otherwise. If $f'=0$, this recovers the $n=7$ threshold, well known from the classical work of Ref. [544]. This concludes the proof.

3.5.10 ▪ Special Cases for Small Vortex Numbers n

We now explore a number of special cases of small vortex numbers which have been examined recently both experimentally [548] and by means of Monte Carlo numerical methods [549].

We start with the $n=7$ case, which, as shown in the previous section, is the *critical* one in the absence of precession, with the corresponding eigenvalue being neutral. Here, however, the presence of precession adds a crucial positive part [for $f(R)$ increasing as is the case for vortices in BECs] to the pertinent eigenvalue rendering the corresponding configuration *generically* unstable.

For the cases with $n<7$, the corresponding eigenvalue can be written as $\lambda=f'(R)R+\frac{c}{8R^2}\left(n^2-8n+s\right)$, where $s=7$ for n odd and $s=8$ for n even. A key observation now (which directly connects the present work with the experimental observations of Ref. [548] and the computational/theoretical analysis of Ref. [549]) is that although for these $n<7$ cases the configuration is *not* generically unstable, nevertheless it *may* become unstable for sufficiently large R, or equivalently for sufficiently large angular momentum L of the vortex system. The latter in our case is simply $L=nR^2$. Then, for the specific dependence suggested for $f(r)\sim 1/(1-r^2)$ by the analysis of Ref. [76] and the experiments of Ref. [531], one obtains that

$$\lambda=\frac{2L}{n(1-\frac{L}{n})^2}+\frac{cn}{8L}\left(n^2-8n+s\right).$$

This results in the critical angular momentum given by setting $\lambda=0$ satisfying

$$16L^2+c(n-L)^2(n^2-8n+s)=0,$$

resulting in

$$L_{\text{cr}}=\frac{cn(n^2-8n+s)+4n\sqrt{c(8n-n^2-s)}}{16+c(n^2-8n+s)}.$$

This general expression yields the relevant critical angular momenta for all the earlier experimentally and numerically considered cases of $n=2$ to $n=5$ of Refs. [548, 549] Using the notation

$$\rho_1^2\equiv\frac{\sqrt{c}}{\sqrt{c}+2},\quad \rho_2^2\equiv\frac{\sqrt{c}}{\sqrt{c}+\sqrt{2}}, \tag{3.90}$$

we have that

$$\begin{aligned}
L_{\mathrm{cr},n=2} &= 2\rho_1^2,\\
L_{\mathrm{cr},n=3} &= 3\rho_2^2,\\
L_{\mathrm{cr},n=4} &= 4\rho_2^2,\\
L_{\mathrm{cr},n=5} &= 5\rho_2^2,\\
L_{\mathrm{cr},n=6} &= 6\rho_1^2.
\end{aligned} \tag{3.91}$$

It is rather intriguing that the critical angular momentum for the different cases exhibits an apparent pattern although not a clearly definite one. In the special case of $n=7$, while the dominant eigenvalue is always positive (as per the above discussion distinguishing it from its critical behavior in the absence of a radially monotonically increasing precession frequency [544]), even the next one $\lambda_{(n-3)/2}$ can be seen to cross 0 at $L_{\mathrm{cr},n=7}=7\rho_2^2$, extending this pattern.

An interesting question that now spontaneously arises is that of the fate of the ring configuration past this critical instability threshold. This question has been addressed in Refs. [548] and [549], through a combination of experimental observations and numerical computations that we now discuss. It was generally found that the above analyzed instability of the ring configuration is associated with a symmetry breaking pitchfork bifurcation that gives rise to the birth (or dynamical manifestation) of an asymmetric rotating vortex state.

We start the relevant analysis by writing the system of vortex particle differential equations in a radial-polar form:

$$\begin{aligned}
\dot{r}_i &= -c\sum_{i\neq j} S_j \frac{r_j}{\rho_{ij}^2}\sin(\theta_i-\theta_j),\\
\dot{\theta}_i &= \frac{S_i}{1-r_i^2} + c\sum_{i\neq j} S_j\left[\frac{1}{\rho_{ij}^2} - \frac{r_j}{r_i\rho_{ij}^2}\cos(\theta_i-\theta_j)\right],
\end{aligned} \tag{3.92}$$

for the corresponding coordinates (r_i,θ_i). The constant c here is an adimensional parameter created by dividing the equations of motion by the precession frequency of the vortex at (or, in fact, very near) the center referred to above as ω_{pr}. This constant was found (up to logarithmic corrections in atom numbers) to assume values of $c\approx 0.11$ in the experiments of Ref. [548]. The conserved quantities of the system are once again the Hamiltonian and the angular momentum of the vortex particles.

Case of $n=2$

The case of $n=2$ is the simpler to analyze theoretically and given its integrable character (again two independent degrees of freedom with two conservation laws, as in the counter-rotating vortex case considered earlier), it is amenable to a fully analytical treatment. In particular, it was found in Ref. [548] that the pitchfork bifurcation is a supercritical one leading past the bifurcation point to the existence of an *asymmetric* co-rotating pair of vortices with relative angle $\theta_{mn}=\pi$ and equilibrium distances of the vortices from the trap center $r_1^*\neq r_2^*$ satisfying

$$-r_1^* r_2^*(r_1^*+r_2^*)^2 + c\left(1-r_1^{*2}\right)\left(1-r_2^{*2}\right)=0.$$

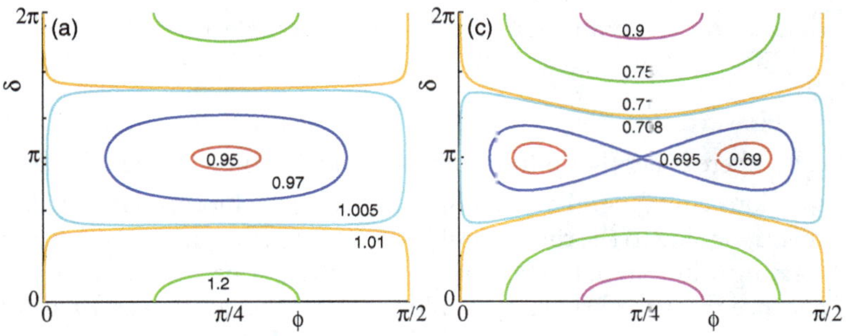

Figure 3.31. *Contours from the reduced Hamiltonian of the vortex system (3.37) in a subcritical case of $L < L_{cr}$ (left) and a supercritical case of $L > L_{cr}$ (right). Adapted with permission from Ref. [548].*

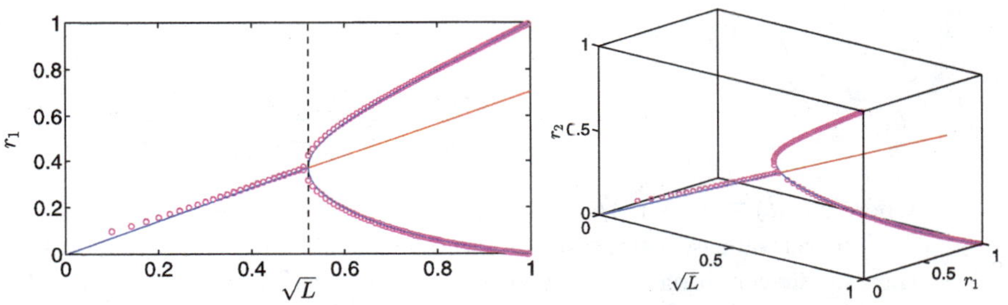

Figure 3.32. *Bifurcation scenario for the case of $n = 2$. The first panel shows the relevant bifurcation diagram in the plane of r_1 (the distance of the first vortex from the trap center) versus $\sqrt{L}$ (i.e., the square root of the angular momentum), while the second one shows the three-dimensional plane $(r_1, r_2, \sqrt{L})$. Adapted with permission from Ref. [549].*

The two-degree-of-freedom nature of the configuration (with the corresponding conservation laws) enabled additionally a simple and informative visualization of the system's dynamics in an effective phase plane with the variables $\phi = \tan^{-1} r_2/r_1$ and $\delta = \theta_2 - \theta_1$. Equi-energetic trajectories (according to the Hamiltonian energy quantity of 3.37) are shown in the panels of Fig. 3.31.

The Monte Carlo (MC) analysis of Ref. [549] offered a complementary view of the same bifurcation event. By using a combination of hyperspherical coordinates, a Metropolis step at a very low $kT = 10^{-6}$ and a final deterministic local search at $T = 0$, this work ensured the convergence to energy minima for different vortex configurations. The output of this search was compared to the comprehensive (still feasible for small n) analysis of the fixed points of the vortex system using the bifurcation package AUTO [550]. The result in the case of $n = 2$ is shown in Fig. 3.32. There, it can be directly seen that the MC results follow the bifurcation, by converging to the symmetric branch for $L < L_{cr}$ and converging to the asymmetric co-rotating configuration for $L > L_{cr}$.

Case of $n = 3$

In this case, the ring assumes the form of an equilateral triangle with relative angles between adjacent vortices of $\theta_{ij} = 2\pi/3$. In this case too, as per the general considerations above, the symmetric state becomes unstable at $r = r_{\text{cr}}^{(3)} \equiv \sqrt{\sqrt{c}/(\sqrt{c}+\sqrt{2})}$ while $\sqrt{L_{\text{cr},1}^{(3)}} = \sqrt{3}r_{\text{cr}}^{(3)}$.

However, more generally, the $n = 3$ case is richer than the $n = 2$ one. More specifically, there is another bifurcation. which takes place at the critical point $L_{\text{cr},2}^{(3)}$ with new solutions emerging. These constitute a pair of solutions with a trilateral symmetry (C_3) corresponding to the three possible *isosceles triangles* of vortices that can emerge as rigidly rotating solutions in the system. For these solutions, one of the vortices is at, say a longer distance from the origin, r_1, while the other two are at a shorter distance r_2. Then the angular momentum constraint reads $L = r_1^2 + 2r_2^2$, while the following conditions specify the relevant solution in analytical form:

$$\begin{aligned}
\theta_{12} &= \theta_{31} = \frac{d \pm \sqrt{d^2+8}}{4},\\
\theta_{23} &= 2\pi - 2\delta_{12}, \qquad\qquad (3.93)\\
c &= \frac{3r_1^4 r_2^2 + 3r_1^2 r_2^4 - a\, r_1^3 r_2^3}{(1-r_1^2)(1-r_2^2)(r_1^2 - r_2^2 - a\, r_1 r_2)},
\end{aligned}$$

where $\cos(d) \equiv (r_1^2 + r_2^2)/(2r_1 r_2)$, and $a^2 \equiv r_1^4 + 34 r_1^2 r_2^2 + r_2^4$.

The relevant isosceles solutions were identified also numerically in Ref. [549], where their saddle-center nature was revealed. Namely, each of the three (rotated by $2\pi/3$) triangles comes with an "unstable partner." This saddle-center bifurcation arises numerically at $L_{\text{cr},2}^{(3)}$, while very close to this value the MC computation jumps at the newly arising (stable) branch of solutions which suggests that it quickly becomes the global minimum of the energy surface. Interestingly, the unstable partner of these isosceles saddle-center pairs results in a collision with the symmetric equilateral solutions at $L_{\text{cr},1}^{(3)}$ and, hence, becomes responsible for their destabilization. This phenomenology is illustrated in Fig. 3.33.

Case of $n = 4$

Once again in the case of $n = 4$, we encounter the instability of the ring configuration in the form, now, of a square with $\theta_{ij} = \pi/2$. This now arises at $r = r_{\text{cr},1}^{(4)} \equiv \sqrt{\sqrt{c}/(\sqrt{c}+\sqrt{2})}$, which corresponds to the higher angular momentum $L_{\text{cr},1}^{(4)} = 4\sqrt{c}/(\sqrt{c}+\sqrt{2})$.

It is interesting to note that in this case one can seek reduced symmetry configurations such as the rhombic one with $r_1 = r_3$ and $r_2 = r_4$ (yet with the same adjacent vortex relative angles as for the square). In Ref. [549] such configurations were identified with $L = 2(r_1^2 + r_2^2)$ and $r_2 = \sqrt{c(1-r_1^2)/(2r_1^2 + c(r_1^2 - 1))}$. It is exactly this rhombic branch that emerges through a supercritical pitchfork bifurcation from the square branch at $r = r_{\text{cr},1}^{(4)}$.

However, in Ref. [549], both the MC computation and the direct solution of the corotating steady state equations revealed additional complexity and further bifurcations. In particular, it was illustrated that, after the destabilization of the squares towards the rhombi, a saddle center bifurcation arises whereby novel states consisting of a pair of asymmetric (so-called irregular) quadrilaterals of vortices may emerge with unequal sides. Shortly after their emergence, the stable solutions of such a branch become the global

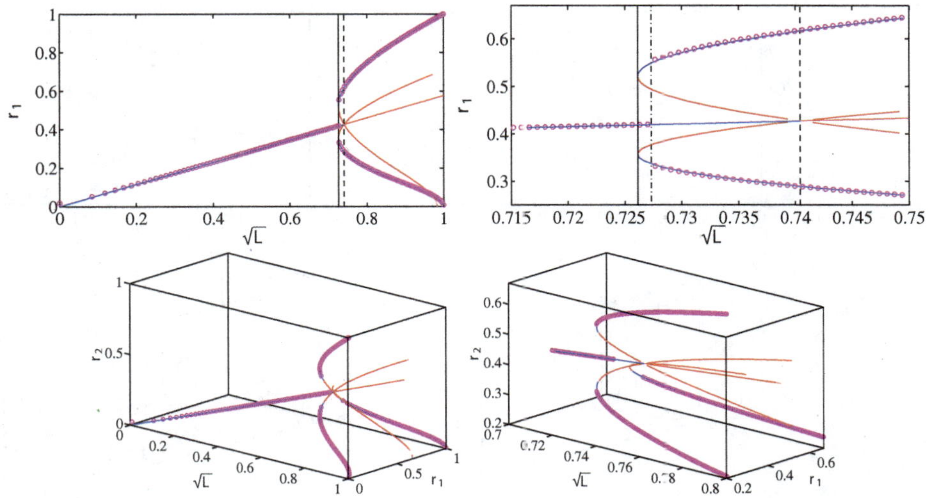

Figure 3.33. *Bifurcation scenario for the case of $n = 3$. The left panels correspond to the entire domain $0 \leq L \leq 1$ while the right panels depict a zoomed in version near the bifurcation region. The top panels show the bifurcation diagrams in a planar view, namely r_i vs. $\sqrt{L}$. The bottom panels depict the bifurcation diagram in the three dimensional space ($\sqrt{L}, r_1, r_2$) where the C_3 symmetry of the solution is clearly visible. The thin dashed vertical line corresponds to the symmetry breaking bifurcation where the symmetric (equilateral triangle) loses its stability. The thin solid vertical line depicts the location of the saddle-center bifurcation of collisions between asymmetric (isosceles triangle) solutions. The thin dash-dotted vertical line corresponds to the location where the MC simulation switches branches (i.e., location where the energy minimum configuration switches from an equilateral to an isosceles one). Adapted with permission from Ref. [549].*

energy minimum and hence the MC computation robustly converges to such states (instead of the rhombi). These quadrilaterals approximately satisfy $r_1 = r_3$, have r_2 close to r_1, r_3 (but not equal to them), while r_4 is much larger (of course, rotated versions of such states also exist from symmetry). The relevant dynamical features, as well as the full set of states and their stability are illustrated in Fig. 3.34.

Experimental Results for Small Vortex Numbers n

The experiments of Ref. [548] corroborated the above theoretical predictions for settings involving $n = 2$, 3, and 4 vortices. Following these experiments, we show two figures below from 8 out of 52 overall examined experimental time series. Four of these in Fig. 3.35 involve the case of 2 co-rotating vortices, while two are for 3 and two for 4 vortices in Fig. 3.36. Each of the runs is shown via 8 snapshots spanning a total time of 210 to 420 ms [see panels (a)–(d) in both figures]. The vortex positions were extracted via a least-squares algorithm, normalized to the BEC radius (i.e., TF units) and the corresponding angular momentum L_0 and Hamiltonian H were computed for each frame [panels (e)–(h) in the figures]. For each series, the average angular momentum $\overline{L}_0$ and Hamiltonian $\overline{H}$ are determined (horizontal dashed lines in the middle panels in Fig. 3.35). Using $\overline{L}_0$, the experimental points representing each orbit were positioned on the (ϕ, δ) plane to examine their correspondence to the isocontours of H corresponding to $\overline{H}$, as shown in the right column of panels in Fig. 3.35. A fairly reasonable agreement between the two can be observed, confirming that the vortices appear to closely follow the trajectories

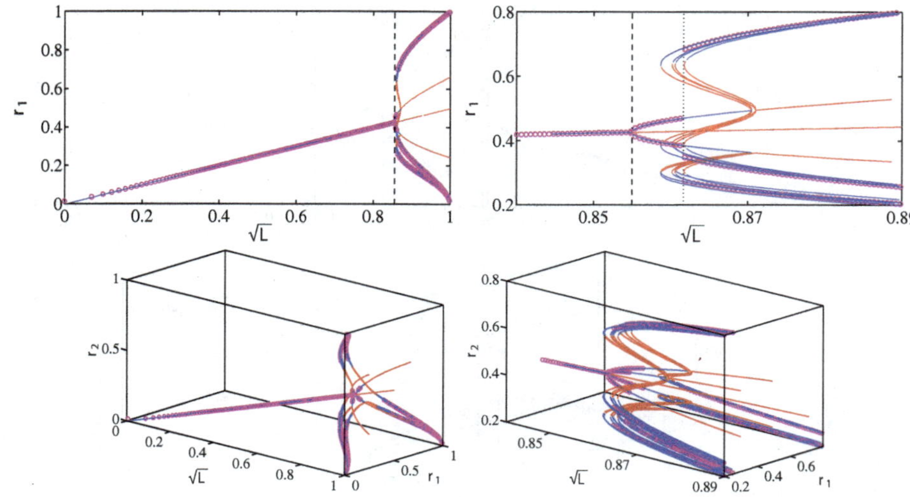

Figure 3.34. *Bifurcation and MC results for the case of $n = 4$. The top panels illustrate planar representations of the solutions, while the bottom ones three-dimensional examples of $(r_1, r_2, \sqrt{L})$. The blue lines are stable branches, while the red lines represent the unstable branches and the MC data are overlaid using small magenta circles. This conveys not only how new branches (such as the rhombus and general quadrilateral) emerge through suitable bifurcations (supercritical pitchfork or saddle center, respectively), but also when they become the global energy minimizers and hence are followed by the MC simulation. The thin dashed vertical line corresponds to the symmetry breaking bifurcation where the symmetric (square) configuration loses its stability towards the newly created rhombus configuration, while the thin dashed-dotted vertical line corresponds to the location where the MC simulation switches branches (i.e., transitions from the rhombic configuration to a more general quadrilateral without any apparent symmetry). Adapted with permission from Ref. [549].*

predicted by the effective particle model (despite the substantial simplifications involved in the relevant reduction).

The different qualitative cases that we observe as displayed in Fig. 3.35 may be grouped as follows:

- For $\overline{L}_0 < L_{\rm cr}$ and relatively small $\overline{H}$, the experiment displays symmetric orbits (rows a and e in Fig. 3.35).
- For $\overline{L}_0 > L_{\rm cr}$ and moderate $\overline{H}$, the experiment displays (i) orbits where both vortices (in the rigidly rotating frame) are approximately on the same side of the cloud chasing each other on the same path [rows (b) and (f) in Fig. 3.35] or (ii) asymmetric orbits [rows (c) and (g) in Fig. 3.35]. The selection between (i) and (ii) is determined by the initial conditions. Initial conditions inside the area delimited by the separatrix [red double-loop curve in the right panels of Figs. 3.35(f) and 3.35(g)] emanating from the saddle point $(\phi, \delta) = (\pi/4, \pi)$ give rise to asymmetric orbits.
- For $\overline{L}_0 > L_{\rm cr}$ and large $\overline{H}$, the experiment displays orbits in which one vortex remains close to the center while the other orbits around it close to the periphery of the cloud [rows (d) and (h) in Fig. 3.35].

Similar conclusions can be drawn from trios and quartets, as shown in Fig. 3.36. The figure shows an example in each case, where the angular momentum is subcritical [(a) and (e) for trios, and (c) and (g) for the quartets] and one which is supercritical [(b) and (f) for

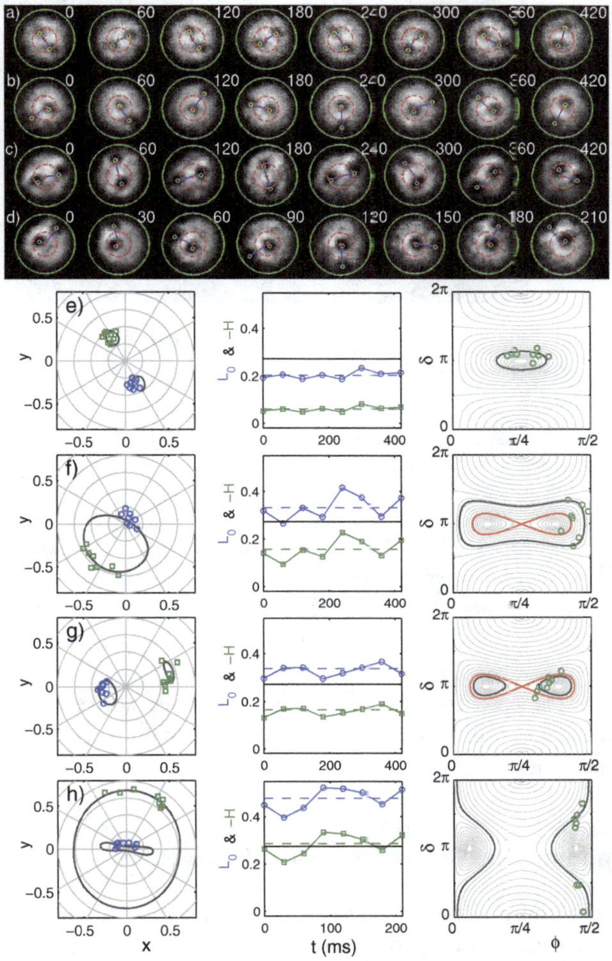

Figure 3.35. (a)–(d) *Typical experimental series for the dynamics of two co-rotating vortices (time shown in ms). The large (green) circles and the (red) crosses represent, respectively, the fitted TF radius and center of the cloud while the small (yellow) dots depict the fitted vortex centers. The (red) dashed circles represent the critical radius,* r_{cr}, *above which symmetric orbits become unstable.* (e)–(h) *Manifestation of the pitchfork bifurcation for the experimental series depicted in panels* (a)–(d) *which correspond to* $c = 0.1$. *Left column: Experimental vortex positions and their corresponding orbit from the reduced ODE model (solid line), in TF units in the rigidly rotating frame. Middle column: Corresponding* L_0 *(blue circles) and* $-H$ *(green squares) and their averages (horizontal dashed lines) as well as the critical* L_0 *(solid horizontal line), so as to deduce the subcritical* (e) *or supercritical* (f)–(h) *nature of the configurations. Right column: Corresponding orbits in the* (ϕ, δ) *plane along with isocontours (thin gray lines) for constant* H *(highlighted is the isocontour corresponding to the average* H *and in red is the separatrix delimiting the area containing asymmetric orbits). Adapted with permission from Ref. [548].*

trios, and (d) and (h) for quartets]. In the former case, it is clear that the configuration preserves its roughly equilateral triangle shape (in the case of trios) and nearly square shape (in the case of quartets). In the latter case, the configuration definitely features an asymmetric profile corroborating the symmetry breaking expectation. A small comment should be added here in that the configurations cannot be *a priori* seeded in particular locations in the actual physical experiment and, as such, cannot be initialized at the (relative) steady

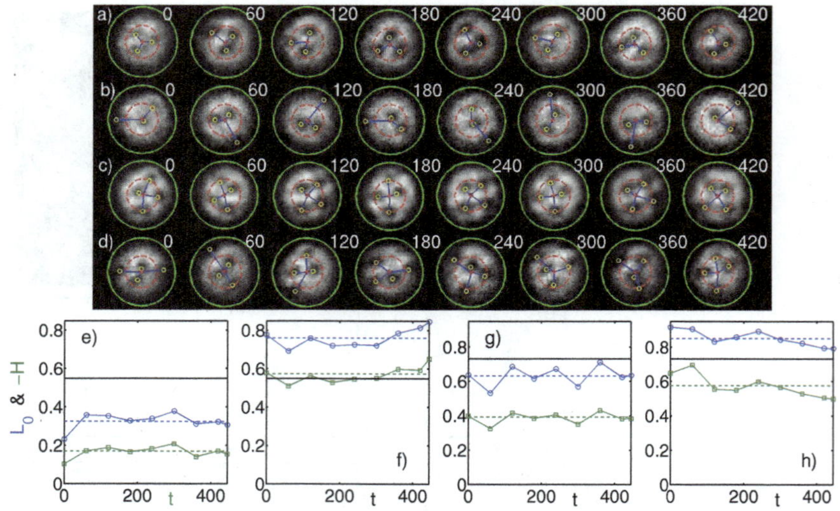

Figure 3.36. *Experimental series for trios [*(a) *and* (b)*] and quartets [*(c) *and* (d)*] below [*(a) *and* (c)*] and above [*(b) *and* (d)*] the critical threshold.* (e)–(h) *Corresponding time series for L_0 and $-H$. Same notation and units as in Fig. 3.35. Adapted with permission from Ref. [548].*

state, i.e., the co-rotating equilibrium prescribed above. For this reason, we do not expect an exact equilibrium in these dynamical evolution experiments, but only something proximal to it.

Other Configurations: The Case of the $n + 1$ Vortex Ring

The above considerations suggest that our setting is fundamentally distinct from the classical one, e.g., of Ref. [545], where the precession term is absent. Yet, this raises the important question (discussed above in the context of $n = 2$, 3, and 4) of the nature of the potential ground state configuration as n is increasing. This is an especially interesting problem both for intermediate values of n and even for large n where vortex lattice configurations arise [551]; see also Ref. [529] for a recent review. The case of large n is briefly touched upon below. Here, we would like to add some comments on the case of intermediate values of n.

Unfortunately, these asymmetric configurations which are stabilized by the presence of the precession term (in its absence such asymmetric configurations are unstable, as discussed, e.g., in Ref. [552]), do not have a general closed form that would enable their stability analysis. Nevertheless, there exists another symmetric configuration that emerges as relevant for the ground state of the system, at least in the case of $n = 5$ examined in Ref. [549] (cf. Fig. 7 therein) and potentially also for cases of larger n. This is the so-called $n + 1$ vortex configuration, consisting of n vortices on the polygonal ring and one more at the center. For the classical vortex problem in the absence of precession, the stability of this configuration was analyzed in Ref. [553]; see also Ref. [554]. In Ref. [543], this analysis was generalized to the case where precession is present yielding the following result.

Theorem 3.6. *Consider the $n+1$ configuration of Eq. (3.76) consisting of n vortices uniformly distributed on a ring of radius R and angular velocity ω given by Eq. (3.79) plus a vortex at* *the origin. Then R satisfies*

$$\omega = f(R) + \frac{c}{2R^2}(n+1). \tag{3.94}$$

Let

$$\lambda_+^*(m) = \lambda_+(m) - 2c/R^2, \tag{3.95}$$

where $\lambda_+(m)$ as given by Eq. (3.89) and let

$$M_0 = \begin{pmatrix} f'(R)\frac{R}{2} - \frac{c(n+1)}{2R^2} & f'(R)\frac{R}{2} - \frac{c}{R^2} & \frac{c}{R^2} \\ f'(R)\frac{R}{2} - \frac{c}{R^2} & f'(R)\frac{R}{2} - \frac{c(n+1)}{2R^2} & 0 \\ \frac{cn}{R^2} & 0 & f(0) - f(R) - \frac{c(n+1)}{2R^2} \end{pmatrix}. \tag{3.96}$$

The $n+1$ configuration is stable if and only if $\lambda_+^*(\lfloor n/2 \rfloor) < 0$ and all eigenvalues of M_0 are negative.

In particular, this theorem shows that the $n+1$ configuration is stable if the n-ring is stable and if in addition the matrix M_0 is negative definite.

When $f = 0$, both $9+1$ and $3+1$ configurations are marginally stable [the former due to $\lambda_+^*(4) = 0$ when $n = 9$, the latter due to the eigenvalue of M_0 crossing zero when $n = 3$]; the $n+1$ configuration is stable for $3 < n < 7$ and is unstable for $n > 9$ or $n < 3$. This is in agreement with the results in Ref. [553]. When the precession term f is increasing as in our BEC setting, both $9+1$ and $3+1$ configurations lose their marginal stability and become unstable, so that $n+1$ configuration becomes unstable for any $n \geq 9$ or $n \leq 3$. Hence, it is natural to further explore in the future the potential relevance of such a configuration for the cases $n = 5$ (as in Ref. [549]), 6, 7, and 8 and even for $n = 4$.

An Alternative Approach: Generating Functions and Classical Polynomials

The examination of vortex "particles" is a remarkable theme of research in its own right, both for smaller and even for larger values of the number of vortices n, that has stirred considerable interest in the fluid dynamics engineering, as well as mathematical community. It is in that light that it has been characterized as a "classical mathematics playground" [555]. One of the particularly enticing aspects of this research activity has been the connection with the theory of classical (such as the Hermite) and even more modern (so-called Adler–Moser) polynomials. An excellent summary of this link that interweaves applied mathematics with algebra and the theory of polynomials is given in Ref. [556].

In the present section, we briefly touch upon this theme, connecting it to the considerably more complex setting considered herein whereby the vortices feature precession in addition to their pairwise interactions. Perhaps the most elementary yet particularly elegant calculation associated with both settings can be given in the case of n vortices of the same charge, when considering solidly rotating solutions of aligned vortices in the absence of precession. In that case, the equation of motion reads

$$\Omega x_i = \sum_{j=1, j\neq i}^{n} \frac{1}{x_i - x_j}. \tag{3.97}$$

In this setting, the precession frequency Ω effectively renormalizes the vortex positions (which scale as $\propto 1/\sqrt{\Omega}$), hence we will set it to unity in what follows. Importantly, we should note that, even in the presence of precession, Eq. (3.97) still holds provided that the precession frequency is constant. Then, that constant merely renormalizes the precession frequency Ω.

The remarkable idea that apparently dates back to the work of Stieltjes [557] (subsequently developed by Marden [558], and Szegö [559], among others) was to consider a so-called generating function in the form of

$$P(x)=\prod_{i=1}^{n}(x-x_i)=(x-x_1)\ldots(x-x_n). \tag{3.98}$$

Then, straightforward differentiation leads to the identities

$$P'(x)=P(x)\sum_{i=1}^{n}\frac{1}{x-x_i}, \quad P''(x)=P(x)\sum_{i=1}^{n}\sum_{j=1,j\neq i}^{n}\frac{1}{(x-x_i)(x-x_j)}. \tag{3.99}$$

A key observation then is that the product involved in the second derivative can be simply written as

$$\frac{1}{(x-x_i)(x-x_j)}=\left(\frac{1}{x-x_i}-\frac{1}{x-x_j}\right)\frac{1}{x_i-x_j}.$$

But then, the sum over j in the term $1/[(x-x_i)(x_i-x_j)]$ can be performed yielding for the first term $x_i/(x-x_i)$ according to Eq. (3.97). The second term is handled in the same manner (the summation over i is performed), and exchanging indices we get $P''(x) = 2P(x)\sum\frac{x_i}{x-x_i}$. Of course this is true *only* provided that the points x_i are the equilibrium vortex configuration locations. Then, rewriting $x_i = -(x-x_i)+x$, we obtain a summation of (−) unity [yielding a factor of $-n$ as multiplying $2P(x)$], as well as a factor of $2xP'(x)$, from the identity of Eq. (3.99) for $P'(x)$. Combining all the above, the differential equation

$$P''(x)=-2nP(x)+2xP'(x) \tag{3.100}$$

arises for the generating function which is *exactly* the differential equation satisfied by the nth order Hermite polynomial. In light of the above analysis, this implies that the roots of a rotating equilibrium configuration involving n vortices will lie exactly at the root locations of the nth order Hermite polynomial.

We now turn to a variant of this analysis that is relevant to our setting involving vortices also featuring precession. Perhaps the most appropriate context for illustrating relevant ideas is the case that we discussed above in Sec. 3.5.7 (also in the context of Ref. [541]): this concerns equilibria of precessing vortices involving opposite charges. In this case, we will start by assuming that there are n_1 vortices of one charge and n_2 vortices of the other charge. Furthermore, in what follows, we will assume a constant precession frequency. This is a reasonable assumption within, roughly, half of the BEC cloud [548]. Extending considerations to the case of a spatially (i.e., radially) dependent precession frequency is a challenging generalization worth considering in its own right. Our discussion of these configurations will follow the recent exposition of Ref. [560].

In this setting, the steady state equations for the n_- "odd" charges (we assume $s_j = -1 = s_{2k+1}$ without loss of generality) reads

$$x_{2k+1}=\sum_{l=1}^{n_+}\frac{1}{x_{2k+1}-x_{2l}}-\sum_{l=1,l\neq k}^{n_-}\frac{1}{x_{2k+1}-x_{2l+1}}. \tag{3.101}$$

Similarly, for the n_+ "even" positive charges, we have

$$x_{2k} = -\sum_{l=1, l\neq k}^{n_+} \frac{1}{x_{2k} - x_{2l}} + \sum_{l=1}^{n_-} \frac{1}{x_{2k} - x_{2l-1}}. \tag{3.102}$$

We can now define $P(x) = \prod_{l=1}^{n_+}(x - x_{2l})$ and $Q(x) = \prod_{l=1}^{n_-}(x - x_{2l+1})$, which satisfy identities similar to Eq. (3.99) (since these do not involve in any way the equations of motion). Then, using the same partial fraction decomposition as above, as well as the equations of motion (3.101) and (3.102), we obtain

$$P''(x) = 2P(x)\left[\sum_{l=1}^{n_+} \frac{-x_{2l}}{x - x_{2l}} + \sum_{l=1}^{n_+}\sum_{k=1}^{n_-} \frac{1}{(x - x_{2l})(x_{2l} - x_{2k+1})}\right], \tag{3.103}$$

$$Q''(x) = 2Q(x)\left[\sum_{k=1}^{n_-} \frac{-x_{2k+1}}{x - x_{2k+1}} + \sum_{l=1}^{n_+}\sum_{k=1}^{n_-} \frac{1}{(x - x_{2k+1})(x_{2k+1} - x_{2l})}\right]. \tag{3.104}$$

Multiplying $Q\times$ (3.103) $+P\times$ (3.104), and once again reshuffling according to our well-known by now partial fraction identity, we retrieve a differential equation combining the generating functions P and Q in the form

$$PQ'' + QP'' = 2P'Q' + 2(n_+ + n_-)PQ - 2x(PQ)'. \tag{3.105}$$

Remarkably, such equations in the simplest setting of vortices purely interacting (but not precessing) had been developed previously, stemming already from the (unpublished thesis) work of Tkachenko in 1964; see the relevant discussion in Ref. [556]. However, the more standard case of purely interacting vortices provides a much simpler form of Eq. (3.105) as $P''Q + PQ'' = 2P'Q'$.

Let us now seek special case example solutions of Eq. (3.105) to explore the power of the generating function formalism in comparison to the brute force calculations, based on the equations of motion in Sec. 3.5.7. The simplest case to examine is that of monomials with $P(x) = (x - a)$ and $Q(x) = (x + a)$, as per the vortex dipole solution examined previously; see also the recent experiments of Ref. [537]. In this case, we have $Q(x) = -P(-x)$; in fact, more generally for even vortex number with $n_+ = n_- = n$, we have that $Q(x) = (-1)^n P(-x)$, based on the symmetries discussed in Sec. 3.5.7. A direct substitution of P and Q into Eq. (3.105) yields immediately $a = \pm 1/\sqrt{2}$. This, in turn, implies that $P(x)Q(x) = (x^2 - a^2) = (x^2 - 1/2) \equiv H_2(x)$; i.e., it appears that the Hermite polynomials resurface, albeit now in the form of $P(x)Q(x)$. Unfortunately, this is only a fortuitous coincidence, as in this case of monomials $P'' = Q'' = 0$ and hence one can proceed to show that

$$(PQ)'' = 2P'Q' = 2x(PQ)' + 4PQ. \tag{3.106}$$

This, in turn, implies that $PQ = H_2(x)$ according to the equation for the relevant polynomials [cf. Eq. (3.100)].

In the case of odd $n_+ + n_-$, the above symmetry of the generating functions is broken and we have, e.g., for $n_+ = 1$ and $n_- = 2$ that $P(x) = x$ and $Q(x) = x^2 - a^2$. Once again, a direct substitution yields $a = \pm 1/\sqrt{2}$.

On the other hand, for $n_+ = n_- = 2$, we have a choice of $P(x) = (x - a)(x - b)$ and $Q(x) = (x + a)(x + b)$. In this case, direct substitution of P and Q in Eq. (3.105) yields the algebraic conditions

$$a^2+b^2=1, \quad ab=\frac{1\pm\sqrt{2}}{2}, \tag{3.107}$$

which is in line with Eq. (3.71) in Sec. 3.5.7 above.

In the case of $n_+=3$ and $n_-=2$, the generating functions assume the following form $P(x)=x(x^2-a^2)$ and $Q(x)=x^2-b^2$. Here, the direct substitution yields anew a system for a and b according to

$$a^2+b^2=1, \quad 2a^2-6b^2+8a^2b^2=0. \tag{3.108}$$

This, in turn, results in $a^2=\sqrt{3}/3$ and $b^2=(2-\sqrt{3})/2$, which we again find to be in agreement with Eq. (3.73).

The last two cases that we wish to consider following this method are $n_+=3$ and $n_-=3$, as well as $n_+=3$ and $n_-=4$. In the former case, $P(x)=(x-a)(x-b)(x-c)$ and $Q(x)=-P(-x)$. This results in three algebraic equations of the following form:

$$a^2+b^2+c^2=\frac{3}{2}, \tag{3.109}$$

$$12a^2b^2c^2-2(a^2b^2+a^2c^2+b^2c^2)-8abc(a+b+c)=0, \tag{3.110}$$

$$4(a+b+c)^2-8(a^2b^2+b^2c^2+a^2c^2)=0. \tag{3.111}$$

The solution of the algebraic system of Eqs. (3.109)–(3.111) is $a=0.1623$, $b=-0.4765$, and $c=1.1165$, which indeed, when scaled by $\Omega=1/2$ (as used in Ref. [541]), yields values identical to the ones reported therein.

Finally, the case of $n_+=3$ and $n_-=4$ yields $P(x)=x(x^2-a^2)$ and $Q(x)=(x^2-b^2)(x^2-c^2)$. Here, the algebraic equations lead to

$$a^2+b^2+c^2=\frac{3}{2}, \tag{3.112}$$

$$4(b^2+c^2-a^2)-8(a^2b^2+a^2c^2+b^2c^2)=0, \tag{3.113}$$

$$12a^2b^2c^2-2(a^2b^2+a^2c^2-3b^2c^2)=0. \tag{3.114}$$

These yield $a=0.5546$, $b=0.2594$, and $c=1.0607$ (again potentially scalable by $\sqrt{1/\Omega}$).

It is interesting to add here that observation of the above patterns of algebraic equations suggests that it may be possible to derive general algebraic conditions that the roots of the polynomials may satisfy irrespectively of n_+ and n_-. For instance, we can see that both for $n_+=2=n_-$ and for $n_+=3=n_-+1$, it is true that $\sum_{i=1}^{n_++n_-}x_i^2=1$, while the same sum becomes $3/2$ for $n_+=3=n_-$ and for $n_+=4=n_-+1$. Motivated by this finding, we have more generally examined the resulting powers of the polynomials arising in both the left- and the right-hand sides of Eq. (3.105). As a result, we have found that at power $x^{n_++n_--2}$, the prefactors of the two sides are respectively $n_+(n_+-1)+n_-(n_--1)$ and $-4\sum_{i=1}^{n_++n_-}x_i^2+2n_+n_-$, leading eventually to the identity

$$\sum_{i=1}^{n_++n_-}x_i^2=\frac{1}{4}\big[2n_+n_--n_+(n_+-1)-n_-(n_--1)\big]. \tag{3.115}$$

Such identities (verified by all of the above considered cases) have also been considered in the absence of precession in Ref. [556] [see, e.g., Eq. (10) therein], although they are obtained at the level of the dynamical equations for the vortices rather than those of the generating function ODEs.

We now turn our attention to generalizing the above considerations of the Tkachenko equation to the 2D complex plane. In this case, if we take n_+ vortices $z_1,\ldots,z_{n_+}$ with charge $+1$ and n_- vortices $\zeta_1,\ldots,\zeta_{n_-}$ with charge -1, the relevant stationary equations will be of the form

$$z_j = -\sum_{k=1,k\neq j}^{n_+} \frac{z_j - z_k}{|z_j - z_k|^2} + \sum_{k=1}^{n_-} \frac{z_j - \zeta_k}{|z_j - \zeta_k|^2}, \tag{3.116}$$

$$\zeta_j = -\sum_{k=1,k\neq j}^{n_-} \frac{\zeta_j - \zeta_k}{|\zeta_j - \zeta_k|^2} + \sum_{k=1}^{n_+} \frac{\zeta_j - z_k}{|\zeta_j - z_k|^2}. \tag{3.117}$$

We again define the generating functions P and Q in the same way as before, namely $P(z) = \prod_{i=1}^{n_+}(z - z_i)$ and $Q(z) = \prod_{j=1}^{n_-}(z - \zeta_j)$. While Eqs. (3.99) remain the same in this case, the main change arises at the level of Eqs. (3.103) and (3.104). Here, the identities become

$$P''(z) = P(z) \sum_{i=1}^{n_+} \frac{2}{z - z_i} \left[-\bar{z}_i + \sum_{k=1}^{n_-} \frac{1}{z_i - \zeta_k} \right], \tag{3.118}$$

$$Q''(z) = Q(z) \sum_{j=1}^{n_-} \frac{2}{z - \zeta_j} \left[-\bar{\zeta}_j + \sum_{k=1}^{n_+} \frac{1}{\zeta_j - z_k} \right]. \tag{3.119}$$

Paralleling the derivation of Eq. (3.105), we form $P''(z)Q(z) + Q''(z)P(z)$. Then the second term in the brackets within Eqs. (3.118) and (3.119) will again yield $2P'(z)Q'(z)$, but the first term is more complicated. In particular, for $z_j = x_j + iy_j$ and $\zeta_j = u_j + iv_j$, we have $\bar{z}_j = z_j - 2iy_j$ and $\bar{\zeta}_j = \zeta_j - 2iv_j$. The first term (involving z_j and ζ_j) is now entirely analogous to the real case and will provide the terms $2(n_+ + n_-)P(z)Q(z) - 2z(P(z)Q(z))'$. However, the last term provides an additional term $4iP(z)Q(z)(\sum_{j=1}^{n_+}\frac{y_j}{z-z_j} + \sum_{j=1}^{n_-}\frac{v_j}{z-\zeta_j})$ (cf. the term of the right-hand side of Eq. (12) of Ref. [556]). The resulting final form of the Tkachenko equation is

$$\begin{aligned} P''Q + PQ'' &= 2P'Q' + 2(n_+ - n_-)PQ - 2z(PQ)' \\ &\quad + 4iPQ\left(\sum_{j=1}^{n_+} \frac{y_j}{z - z_j} + \sum_{j=1}^{n_-} \frac{v_j}{z - \zeta_j} \right). \end{aligned} \tag{3.120}$$

Generally, Eq. (3.120) is far more difficult to solve than Eq. (3.105). However, some special case examples can still be extracted from it. Perhaps a prototypical one is that of two interlaced polygonal vortex rings (with the same radius) consisting of oppositely charged vortices. Nonstationary variants of such configurations were observed to spontaneously emerge through the instability of ring dark solitons in Ref. [265]. Stationary variants of such states were identified in Ref. [522] and a systematic bifurcation analysis thereof connecting them to the instabilities of the ring dark soliton was given in Ref. [523]. From our above stability analysis in Sec. 3.5.7 it was found there that the only (non-collinear) such configuration which is stable at the "particle level" is the so-called vortex quadrupole. On the other hand, configurations such as the vortex hexagon, octagon, decagon, etc. are always unstable, arising through a cascade of progressive bifurcations from the ring dark soliton. Now, we turn to the identification of such states through the Tkachenko equation approach of Eq. (3.120).

We assume that $n_+ = n_- \equiv n$ and then the two polygonal rings, possessing the same radius, can be represented in the complex plane by $P(z) = z^n - R^n$ and $Q(z) = z^n - R^n e^{in\phi}$. In the second expression the roots (i.e., vortex locations) are assumed to be (uniformly) shifted by an angle ϕ. This achieves the interlaced vortex ring configuration of interest. Direct substitution of the P and Q in Eq. (3.120) yields numerous cancellations and only two conditions, due to the vanishing of the polynomial prefactors of z^{2n-2} and of z^{n-2}.[19] These two conditions are

$$R^2 = \frac{1}{2}, \tag{3.121}$$

$$e^{in\phi} = -1. \tag{3.122}$$

The former condition represents an algebraic constraint either on the radius of the vortex ring or, equivalently (for arbitrary vortex ring radius) on the precession frequency, while the latter condition suggests that the two rings are displaced with respect to each other by an angle of $\phi = \pi/n$. This is consonant with the numerical observations of the PDE model, as, e.g., in the case of the vortex quadrupole of $n = 2$, the rotation is by $\pi/2$, in that of the hexagon [522, 523] of $n = 3$, it is by $\pi/3$, and so on. Moreover, since the vortices of a single charge are rotated with respect to their neighbors of the same sign by $2\pi/n$ (due to the structure of the complex roots of unity), this suggests that in the interlaced polygonal configuration (as expected by symmetry) the vortices are positioned in an alternating charge fashion at equal angles of π/n.

Beyond the single vortex polygon one can consider two or more nested polygons. Two rings, each possessing 4 vortices, were observed numerically in the PDE model as bifurcating from the X-shaped dark soliton cross [523]. Like in the case of a single polygon, we find that there are no fixed equilibria consisting of nested polygons with odd numbers of vortices. With this as motivation we investigate two polygons (each interlaced as above) having an equal, even number of vortices. We consider Eq. (3.120) with

$$P(z) = (z^n - R_1^n)\left(z^n - R_2^n e^{in\phi}\right), \tag{3.123}$$

$$Q(z) = (z^n - R_2^n)\left(z^n - R_1^n e^{in\phi}\right), \tag{3.124}$$

and obtain a polynomial in z which, after factoring out z^{n-2}, has only four (potentially) nonzero terms:

$$4n(-1 + R_1^2 + R_2^2) z^{4n}, \tag{3.125}$$

$$-n(1 + e^{in\phi})(R_2^n(-3 + n + 4R_1^2 + 2R_2^2) + R_1^n(-3 + n + 2R_1^2 + 4R_2^2)) z^{3n}, \tag{3.126}$$

$$\begin{aligned} &2ne^{in\phi}((-1 + 2R_1^2)R_2^{2n} + R_1^{2n}(-1 + 2R_2^2) \\ &+ 2R_1^n R_2^n(-1 + 2n + R_1^2 + R_2^2 + (-1 + R_1^2 + R_2^2)\cos(n\phi)) z^{2n}, \end{aligned} \tag{3.127}$$

$$-e^{in\phi}(1 + e^{in\phi}) n R_1^n R_2^n((-1 + n + 2R_1^2)R_2^n + R_1^n(-1 + n + 2R_2^2)) z^n. \tag{3.128}$$

Setting the first expression equal to zero we find the relationship between the two radii $R_1^2 + R_2^2 = 1$. Using this relationship and noting that $n \geq 2$ forces us to choose $\phi = \pi/n$

[19] It should be added as an aside that properties of the complex roots of unity and geometric series are used in order to evaluate the last term in Eq. (3.120); cf. Eq. (13) of Ref. [556].

from the second equation. This choice also satisfies the fourth equation, and we now have two equations which can be used to determine admissible radii for the rings:

$$1 = R_1^2 + R_2^2, \tag{3.129}$$
$$0 = 4nR_1^n R_2^n + (-1 + 2R_1^2)R_2^{2n} + R_1^{2n}(-1 + 2R_2^2). \tag{3.130}$$

Interestingly, we are able to modify the forms of P and Q slightly to deduce the existence of another family of equilibria made up of two interlaced n-gons with different radii and a vortex at the center of the configuration. A configuration from this family can be found in [523, Fig. 6(b)]. Without loss of generality, we choose the vortex at the center to have charge $+1$ and set

$$P(z) = z(z^n - R_1^n), \tag{3.131}$$
$$Q(z) = \left(z^n - R_2^n e^{in\phi}\right). \tag{3.132}$$

Equation (3.120) is now a polynomial with two nonzero coefficients corresponding to z^n and z^{n-1}. The coefficients are zero provided that

$$1 = R_1^2 + R_2^2, \tag{3.133}$$
$$0 = -e^{in\phi}(1 + n + 2R_1^2)R_2^n - R_1^n(-3 + n + 2R_2^2). \tag{3.134}$$

Separating real and imaginary parts in the second equation, we again find $\phi = \pi/n$. For a given n the remaining equations can be used to determine R_1 and R_2. Finally, these types of considerations can also enable the capturing of other symmetric configurations such as the 5x (4 vortices of the same charge $S = 1$, surrounding a higher opposite charge $S_0 = -2$ at the center) of Fig. 3.18; details can again be found in Ref. [560], but we do not expand on this further, as this configuration is typically unstable due to the (split) instability of the charge-2 vortex at the center.

The above perspective provides an arguably more elegant and gratifying complementary viewpoint to the direct analysis of Ref. [541]. Yet, there are perhaps numerous aspects that are worthy of additional investigation in the realm of oppositely charged vortices in the presence of precession. Such an interesting avenue of potential further analysis involves a recursion formula can be set up in the classical Tkachenko case of stationary opposite charge vortices without precession [556] that provides a direct connection between the polynomials P_{n-1}, P_n, and P_{n+1} arising for different vortex numbers. This recursion relation is associated (in the absence of precession) to the so-called Adler–Moser polynomials [561], that surprisingly also arise in the study of rational solutions of the Korteweg–de Vries (KdV) equation. Identifying the relevant recursion relation and connecting it (potentially) to particular polynomial properties would be a separate topic worth investigating in its own right.

The Large n Case: Coarse Graining and the Continuum Limit

Having considered the case of small n, it is worthwhile to now turn to the case of large n. Here, we expect that a substantially different vortex distribution will arise. This is confirmed by the well-known vortex lattice observations of, e.g., Refs. [562, 563], which reveal a triangular pattern in the form of the well-known Abrikosov lattice [564]. This is also illustrated by the direct numerical evolution of the aggregation equation (3.78); see the left panel of Fig. 3.37. Recall that the aggregation equation has the benefit of relaxing to equilibrium attractors corresponding to the marginally stable equilibria of Eq. (3.76).

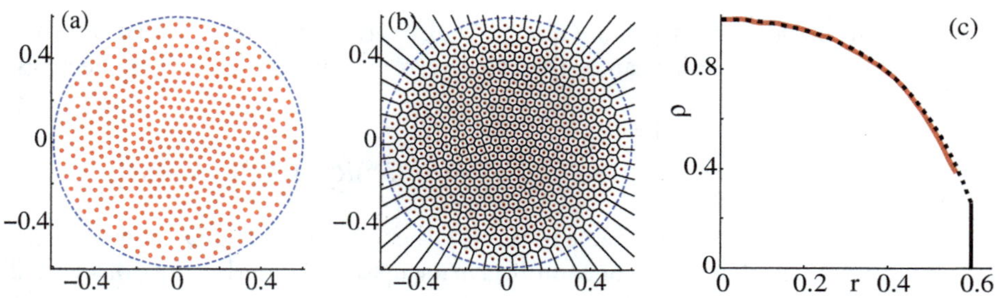

Figure 3.37. (a) *Typical stable equilibrium of Eq. (3.78) for* 500 *vortices The dashed circle is the asymptotic boundary whose radius* $R = 0.6$ *is the physically relevant solution to Eq. (3.144).* (b) *Voronoi tessellation used to compute the 2D density distribution* ρ. (c) *Average of* $\rho(|x|)/\rho(0)$ *as a function of* $r = |x|$. *The red solid curve corresponds to the numerical computation, while the black dashed one is the formula (3.145), with the boundary* $r = R$ *denoted by a vertical line. Adapted with permission from Ref. [543].*

A natural quantitative diagnostic in this case is to attempt to infer the distribution of such a large number of vortices. Here, we will follow the exposition of Ref. [543] and techniques similar to Ref. [565] (see also Ref. [566]) in that direction.

To coarse-grain the vortex system, we define the particle density as

$$\rho(x) = \sum_{k=1}^{n} \delta(x - x_k). \tag{3.135}$$

Then, the aggregation Eq. (3.78) can be rewritten as $\dot{x}_j = v(x_j)$, where we define the continuum limit of the velocity as

$$v(x) \equiv (f(r) - \omega)x + c \int_{\mathbb{R}^2} \frac{x-y}{|x-y|^2} \rho(y)\,dy.$$

Here, $r = |x|$ represents the radial variable, and the density normalization condition to the number of vortices n reads

$$\int_{\mathbb{R}^2} \rho(x)dx = n.$$

For large n, the conservation of mass yields the continuity equation

$$\rho_t + \nabla \cdot (v\rho) = 0, \tag{3.136}$$

which can be deemed as the continuum limit of the aggregation equation.

Using, for any smooth radial function $g(r)$, the identity

$$\int_{\mathbb{R}^2} \frac{x-y}{|x-y|^2} g(|y|)\,dy = x\frac{2\pi}{r^2}\int_0^r g(s)s\,ds,$$

we obtain for $\rho(y)$

$$v = \left(f(r) - \omega + \frac{2\pi c}{r^2}\int_0^r \rho(s)s\,ds \right)x. \tag{3.137}$$

Let $V(r)$ be the bracketed expression in Eq. (3.137) so that $v = V(r)x$ and note that

$$\nabla \cdot (v\rho) = \nabla \cdot (V(r)\rho x) = \frac{1}{r}\left(V\rho r^2\right)_r$$

so that $\rho_t + \nabla \cdot (v\rho) = 0$ becomes

$$(r\rho)_t + (Vr\rho r)_r = 0. \tag{3.138}$$

Now define the integrated density in the form

$$u(r) = \int_0^r \rho(s)s\,ds.$$

Integrating Eq. (3.138) we obtain

$$u_t + Vru_r = 0. \tag{3.139}$$

Recall that we have

$$Vr = r(f(r) - \omega) + \frac{2\pi c}{r}u. \tag{3.140}$$

Thus, we obtain the following characteristics for Eq. (3.139) where u is a constant:

$$\frac{dr}{dt} = r(f(r) - \omega) + \frac{2\pi c}{r}u, \tag{3.141}$$

Then, assuming that the initial density is radially symmetric and has finite support of radius R, we have

$$u(R) = \int_0^R r\rho\,dr = \frac{n}{2\pi}. \tag{3.142}$$

Thus, the corresponding characteristic $r = R$ evolves according to

$$\frac{dR}{dt} = R(f(R) - \omega) + \frac{cn}{R}. \tag{3.143}$$

In particular at the steady state $t \to \infty$ and with $f(R) = a/(1-R^2)$, we obtain the equation for the support radius,

$$\omega = \frac{a}{1-R^2} + \frac{cn}{R^2}. \tag{3.144}$$

From Eq. (3.141), at the steady state we have $u = -r^2(f(r) - \omega)/(2\pi c)$. From Eq. (3.142) we then obtain $\rho = -\frac{1}{2\pi c}\frac{1}{r}\frac{\partial}{\partial r}\left(r^2[f(r) - \omega]\right)$ or

$$\rho = \frac{1}{\pi c}\left(\omega - \frac{a}{(1-r^2)^2}\right). \tag{3.145}$$

Note that for fixed ω, Eq. (3.144) has either zero or (one at the critical point or) two solutions for R. If it has two solutions $R_- < R_+$, then R_- is stable and R_+ is unstable, as is easily deduced from Eq. (3.143). The threshold occurs by setting $\partial\omega/\partial R = 0$ to obtain

$$\omega_c = \left(\sqrt{a} + \sqrt{cn}\right)^2, \qquad R_c^2 = \frac{\sqrt{cn}}{\sqrt{a} + \sqrt{cn}}.$$

Thus, two solutions exist when $\omega > \omega_c$. The one with smaller support is stable, and the one with bigger support is unstable. Interestingly, the density $\rho(R)$ vanishes precisely at $R = R_c$. Hence, among the two solutions only the stable one with $R_- < R_c$ is physically observable, while the unstable one with $R_+ > R_c$ cannot be computationally obtained (and presumably physically observed, since it would involve negative vortex densities for $R > R_c$).

To complement these analytical findings, in Ref. [543] a computation of the steady-state distribution of the vortices was obtained by evolving Eq. (3.78) until it settled to an equilibrium state [Fig. 3.37(a)]. Computing the Voronoi tessellation of the plane [Fig. 3.37(b)] each vortex x_j was assigned a region (Voronoi cell) which consists of all points in the plane that are closer to x_j than to any other vortex. The density distribution at x_j can then be approximated by $\rho(x_j)=1/\text{area}_j$ where area_j is the area of the Voronoi cell associated with x_j. The resulting radial density can be plotted as in Fig. 3.37(c), computed by taking the average of $\rho(x)$ along $|x|=r$. This compares very favorably with the distribution obtained from the analytical expression of Eq. (3.145) as can be seen in this panel.

A natural followup question at the level of the particle model concerns the identification of the vibrational or so-called Tkachenko modes of the vortex lattices; see, e.g., Refs. [562, 567, 568, 569]. Attempting to obtain such modes based on the particle equations using analytical coarse-graining approaches similar to the ones above would yield considerable insight on these modes and would be of particular interest for future studies. Some of the above works have tackled this question analytically or semi-analytically for small n [568], while others have addressed the case of large n numerically [567, 569], attempting to connect with the experimental data [562].

3.6 ▪ Using the Handle of Anisotropy: Modifications of Existence, Stability, and Dynamics of Vortices in Anisotropic BECs

3.6.1 ▪ A Vortex Particle-Picture in Anisotropic Traps

Let us now turn to the case of a harmonic trapping potential, expressed in dimensionless form as

$$V(x,y)=\frac{1}{2}(\omega_x^2x^2+\omega_y^2y^2),$$

where the two trapping frequencies ω_x and ω_y can be different, i.e., in the potential presence of *anisotropy*. The parameter $\alpha=\omega_y/\omega_x$ will be used to quantify the anisotropy of the trap. The presentation of this anisotropic setting will follow directly the analysis of Ref. [570]. For simplicity, we restrict the discussion to singly charged vortices with charge $S=\pm1$, although generalizations to multiply quantized vortices are available [76].

We start by mentioning that in Ref. [76] a set of vortex precession ODEs was derived within the full 3D Gross–Pitaevskii framework, which also holds in the anisotropic regime of $\omega_x\neq\omega_y$. Matching this result to the expressions obtained in Ref. [518] for the 2D GPE in the isotropic limit, we find the following set of ordinary differential equations (ODEs) governing the precessional dynamics of a single vortex of charge S:

$$\begin{aligned}\dot{x}&=-S\omega_y^2Q\,y,\\ \dot{y}&=\ \ S\omega_x^2Q\,x.\end{aligned}\tag{3.146}$$

where $Q(\mu,\omega_x,\omega_y)=\ln(A\mu/\omega_{\text{eff}})/(2\mu)$, $\omega_{\text{eff}}=\sqrt{(\omega_x^2+\omega_y^2)/2}$, and the numerical constant $A\approx 8.88$. These equations describe elliptical vortex orbits in the anisotropic trap, where the anisotropic precession frequency is given by $\omega_{\text{pr}}=\omega_x\omega_yQ$. Our discussion above for the case of isotropic traps has in a number of cases (e.g., in the instability of symmetric configurations of co-rotating vortices) highlighted the importance of the radial dependence of the precession frequency. Nevertheless, given the additional layer of complication imposed by anisotropy as regards this (now, no longer radial) dependence, we will assume the precession frequency roughly constant (i.e., independent from the distance from the trap center). This, in turn, suggests that we restrict our considerations to vortices close(r) to the trap center (than to the Thomas–Fermi radius).

In the presence of more than one vortex, there is, as before, an additional interaction contribution to the equations of motion: Each vortex moves with the local velocity field created by all the other vortices [77, 533]. Neglecting modifications of the velocity field profiles due to the inhomogeneous condensate background in the presence of the trap, we can employ the interaction term that has also been used in Refs. [518, 537], yielding

$$\begin{aligned} \dot{x}_k &= -S_k \omega_y^2 Q y_k + B \sum_{j \neq k} S_j \frac{y_j - y_k}{2\rho_{jk}^2}, \\ \dot{y}_k &= S_k \omega_x^2 Q x_k - B \sum_{j \neq k} S_j \frac{x_j - x_k}{2\rho_{jk}^2}, \end{aligned} \tag{3.147}$$

where $\rho_{jk}^2 = (x_j - x_k)^2 + (y_j - y_k)^2$. Here, B is a numerical constant, which in earlier works for isotropically trapped quasi-2D condensates of aspect ratio $\omega_r/\omega_z = 0.2$ (which coincides with the isotropic limit in our simulations) has been found to be $B \approx 1.35$ from fits of the dipole's equilibrium position [518, 537]. We will use this value of B whenever numerically evaluating results from the above ODE system.

The "particle picture" of Eqs. (3.147) will be used throughout this section to study equilibrium positions, linearization frequencies and dynamics of few-vortex arrangements in anisotropic traps.

3.6.2 ▪ Aligned Vortex States in the Particle Picture

One particular class of stationary vortex clusters in Bose–Einstein condensates, that has received considerable attention in the past years, consists of configurations where a number of singly charged vortices are located along one of the symmetry axes of the trap, and the sign of the vortex charges is alternating between adjacent vortices—see, e.g., Refs. [520, 521, 540]. We will refer to these solutions of the Gross–Pitaevskii equation as "aligned vortex states." In the following, we will apply the particle picture ODEs to determine equilibrium positions and linearization frequencies of the aligned vortex cluster states. In contrast to numerous previous works on the theme of counter-rotating vortices (see Ref. [518] for a recent discussion of the relevant literature), we will not restrict the analysis to isotropic traps, but allow for different trapping frequencies in the x and y directions. This will turn out to have important consequences for the stability of the vortex clusters. It should be noted, on the other hand, that the subject of co-rotating vortices in the presence of anisotropy has been considered in some detail in a series of works [532, 571, 572].

Single Vortex

To start out, we numerically determine the stationary single vortex solution to the full Gross–Pitaevskii equation for different values of $\alpha = \omega_y/\omega_x$ by representing the Laplacian in terms of finite differences on a spatial grid and employing a Newton–Krylov method as described in Ref. [573]. Technically, to scan α, we fix $\omega_x = 0.2$ throughout this work and vary ω_y. The chemical potential is held fixed at $\mu = 2.5$, high enough for the Thomas–Fermi large-density approximation to be valid. As expected, and predicted

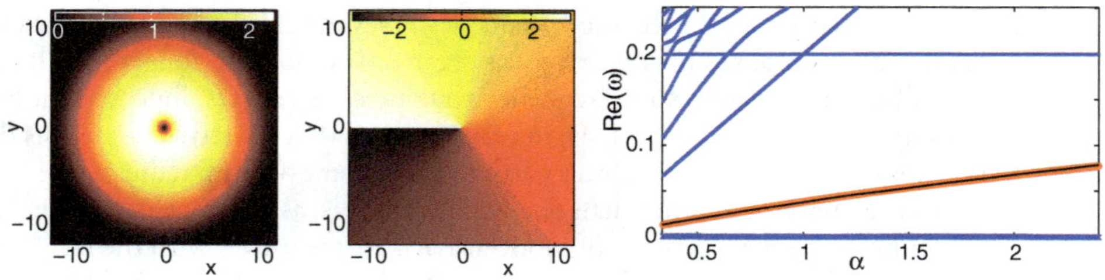

Figure 3.38. *Profiles (density in the left panel and phase in the middle panel) and BdG spectrum (right panel) of the single vortex state. In the BdG spectrum, the precession frequency from the particle picture is depicted in black, coinciding with the anomalous BdG mode (thick orange line). Chemical potential fixed at $\mu = 2.5$. Adapted with permission from Ref. [570].*

by the particle picture, the stationary single vortex is located at the trap center $x = y = 0$ for any value of α. A typical profile of the state's density and phase structure are shown in the left panels of Fig. 3.38.

Having numerically identified the vortex solution in various anisotropic settings, we calculate its BdG spectrum as a function of α by diagonalizing the ensuing BdG matrix. The resulting spectrum is shown in the right panel of Fig. 3.38, together with the linearization frequency calculated from the particle picture (black line). The spectrum contains one anomalous (negative Krein) mode indicated by the thick orange line. This anomalous mode in the BdG spectrum is connected to the precessional motion of the vortex, as before. Exciting it slightly shifts the vortex from its equilibrium position and makes it precess around the trap center. Obviously, this mode's numerically found dependence on α is very well described by the precession frequency of the particle picture, both in the isotropic limit and in anisotropic settings. Finally, we stress that the BdG spectrum contains no eigenfrequencies with nonzero imaginary part, indicating that the single vortex state is dynamically stable in arbitrary anisotropic traps.

Apart from the anomalous mode responsible for vortex precession, the full BdG spectrum exhibits a large number of "background modes," which are not captured by the vortex particle picture. The $\omega = 0$ mode, present for any value of α, can be identified as the Goldstone mode related to the $U(1)$ invariance of the Gross–Pitaevskii equation. In addition, there is always a so-called Kohn or dipolar mode with frequency ω_x (which assumes the value 0.2 independent of α in Fig. 3.38). Similarly, there is a dipole mode (linear in α in Fig. 3.38) with frequency ω_y. These modes involve a collective oscillation of the entire cloud around the center of the trap in each of these two directions.

Vortex Dipole

Let us perform the same analysis for the vortex dipole. The particle picture predicts two equilibrium positions of the vortices along the y axis, namely,

$$x_1 = 0,\ y_1 = \pm\sqrt{B/(4\omega_y^2 Q)} \quad \text{and} \quad x_2 = 0,\ y_2 = \mp\sqrt{B/(4\omega_y^2 Q)}.$$

Naturally, the solutions associated with the two different signs can be transformed into each other by simply interchanging the positive- and the negative-charge vortices.

The middle panel of Fig. 3.39 compares the numerically found equilibrium positions of the vortices forming the stationary dipole to this prediction. Technically, from the nu-

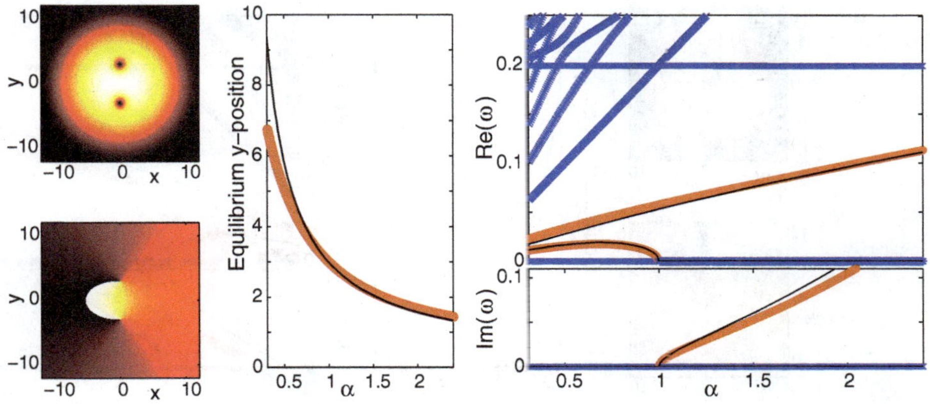

Figure 3.39. *The left panels show a typical example of the density and phase of a vortex dipole for an equilibrium position of the vortices within the PDE. The middle panel shows the equilibrium position of the dipole configuration as a function of the 2D aspect ratio: numerical data (thick orange line) and ODE prediction (black line). The right panel shows the results of the linearization around such a dipole as a function of α. For the anomalous (internal) modes of the two vortices the prediction from the particle equation theory is given again by the solid black lines, while the numerical findings are given by thick orange lines. Purely imaginary modes are also denoted by thick orange lines. Notice the instability emerging for $\alpha > 1$. The chemical potential is fixed at $\mu = 2.5$ throughout. Adapted with permission from Ref. [570].*

merically calculated wavefunctions the vortex locations are extracted by evaluating the z component of the superfluid vorticity. The vortices then show up as sharp, well localized extrema. It can be observed that the agreement between the numerical data and the ODE prediction is better in the $\alpha > 1$ regime, and for small α the errors become larger. Partially, this inaccuracy of the particle picture in the small α regime can presumably be attributed to the dependence on the distance from the center of the vortex precession frequency which we do not take into account. Another factor that should be taken into consideration here, and accounts for the observed discrepancy, is the modification of the vortex-vortex interaction due to the nonhomogeneous, modulated TF density, especially near the condensate boundaries. The latter effect has been implicitly included in the equations through a shift of the B factor away from its background value. However, the deviations in the middle panel of Fig. 3.39 suggest that including effects due to the presence of the trap (and the ensuing nonhomogeneous condensate background) on the vortex-vortex interaction by using an effective B is not sufficiently accurate for full quantitative agreement. A refined description of vortex interaction in the trapped condensate, taking into account not only the vortices' positions, but also the nontrivial shape of the background density distribution and the ensuing deformation of the velocity fields around the vortices, would be an interesting direction for further studies; when included into our particle picture equations, this could (presumably) significantly improve their quantitative predictions—see for a relevant discussion in the case of anisotropic vortices the work of Refs. [532, 574].

The dipole's linearization frequencies are found to be

$$\omega_{1,2} = \pm\sqrt{2}\omega_{\text{pr}}, \quad \omega_{3,4} = \pm\omega_{\text{pr}}\sqrt{1-\alpha^2}.$$

A key consequence of this prediction is that for $\alpha > 1$, the spectrum of the vortex dipole exhibits a purely imaginary mode, indicating instability in this regime. On the other hand, for $\alpha \leq 1$ the dipole is stable. In this stable interval, the spectrum contains two

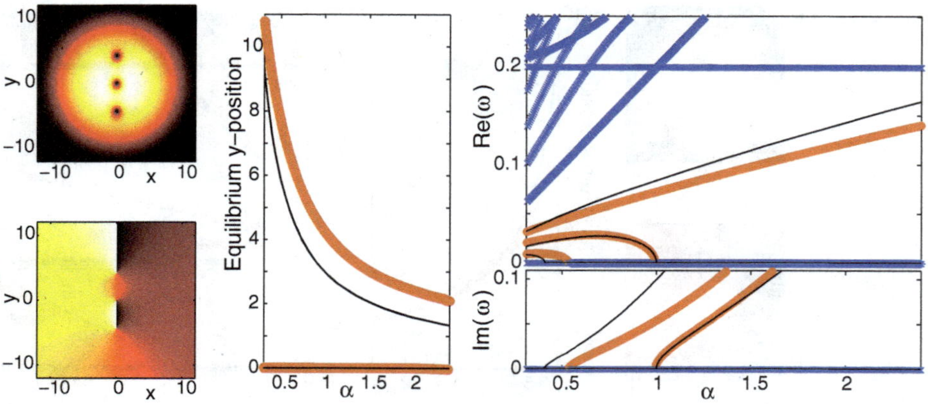

Figure 3.40. *The left panels show a typical example of the density and phase of an equilibrium vortex tripole. The position of the vortices in the tripole configuration as a function of the 2D aspect ratio is shown in the middle panel with numerical data shown as thick orange lines and the ODE prediction shown in black solid line. The right panel shows the eigenmodes of linearization around the tripole: ODE theoretical predictions represented by solid black lines and anomalous modes and purely imaginary modes by thick orange lines. Chemical potential $\mu = 2.5$. Adapted with permission from Ref. [570].*

anomalous modes (indicated by the thick orange lines in the right panel of Fig. 3.39) whose functional dependence is well described by the linearization frequencies predicted in the particle picture. At $\alpha = 1$, one of these two anomalous modes vanishes. The existence of such a zero mode in the isotropic limit is a general feature that will be found for all subsequent aligned vortex states: One can think of the presence of the aligned vortex configuration breaking the rotational symmetry of the $\alpha = 1$ system, which was responsible in the isotropic case for the emergence of a corresponding Goldstone mode. For $\alpha > 1$, this former Goldstone mode in the dipole's spectrum becomes a purely imaginary eigenfrequency, which again is predicted correctly by the particle picture. Intuitively, in the isotropic setting of $\alpha = 1$ the dipole as a whole can be arbitrarily rotated. This neutrality is represented by the zero mode in the BdG spectrum. When $\alpha > 1$, in our current setup, the vortex pair is compressed along the axis of the vortex dipole, which favors "buckling" of the axial structure and leads to instability, while $\alpha < 1$ has the opposite effect.

Vortex Tripole

Let us now turn to the next aligned vortex state, consisting of three vortices of alternating charge, i.e., $S_1 = \pm 1$, $S_2 = \mp 1$, $S_3 = \pm 1$. As for the vortex dipole, we take the stationary vortices to be aligned along the y axis, i.e., $x_{1,2,3} = 0$, and for their equilibrium y coordinates we consider the symmetric ansatz $y_2 = 0$, $y_1 = -y_3 = y$. Inserting into the particle picture ODEs yields a fixed point for $y = \sqrt{B/(4\omega_y^2 Q)}$. Thus, the particle picture predicts a stationary "vortex tripole" state, where two vortices of the same charge are placed along one of the trap's main axes, while the oppositely charged third vortex rests between them, at the trap center.

Again, we can compare the fixed point coordinates calculated from the vortex particle picture to the vortex positions obtained from numerically identifying the vortex tripole solution of the stationary GPE. The results are shown in Fig. 3.40.

Interestingly, we observe that while choosing $B = 1.35$ led to good agreement for the dipole configuration, this is no longer the case for the tripole. Here, better quantitative agreement is achieved if the interaction constant B is taken at its background value $B = 1.95$, valid for vortex interaction in a homogeneous condensate [518] (data not shown). This discrepancy may again be regarded as a warning sign that accounting for condensate inhomogeneities due to the trap by rescaling B is insufficient in the general case, and it is certainly desirable to take into account the effects of the nonhomogeneous background, as discussed above.

Next, we calculate the linearization frequencies around the tripole equilibrium. From the particle picture ODEs we obtain

$$\omega_{1,2} = \pm\sqrt{2}\omega_{\rm pr}\sqrt{1-\alpha^2}, \quad \omega_{3,4,5,6} = \pm\omega_{\rm pr}\sqrt{4-5\alpha^2 \pm \sqrt{9+2\alpha^2+25\alpha^4}}.$$

In the isotropic limit of $\alpha = 1$, these equations reproduce the result of Ref. [518]. Concerning stability, the most important conclusion to be drawn is that below the critical anisotropy $\alpha_{\rm cr} = 1/\sqrt{6} \approx 0.408$, all linearization frequencies become real, i.e., the tripole can be completely stabilized by means of strong enough transverse confinement. Comparing with the numerically found BdG spectrum of the tripole essentially confirms this prediction. The particle picture still captures the overall behavior of the relevant vortex modes. In detail, however, the predictions are less accurate than for the vortex dipole. In particular, while stabilization in general is correctly predicted by the particle picture, the critical value of α for which it occurs is found to be $\alpha_{\rm cr} \approx 0.53$. It should also be mentioned that similarly to the dipole case, the neutral mode present for $\alpha = 1$ (due to isotropic rotation of the tripole) can be tipped towards stability (for $\alpha < 1$) or instability (for $a > 1$), depending on the direction of anisotropic compression of the condensate with respect to the axis of the multivortex state (perpendicular, or parallel, respectively).

Having discussed both cases of the vortex dipole and the vortex tripole should render evident the fact that anisotropy presents a remarkable handle for controlling the stability and dynamics of multivortex clusters at will away from the isotropic limit. In particular, it is evident that configurations such as the vortex dipole, which are structurally robust in the isotropic limit, can be immediately rendered unstable when departing from that limit for values of $\alpha > 1$. On the flip side, any configuration which is more highly excited and unstable in the isotropic limit can instead be stabilized when operating in a sufficiently anisotropic regime for $\alpha < 1$. Examples of all four of these scenarios—perturbed but stable isotropic dipole, perturbed unstable anisotropic dipole for $\alpha > 1$, perturbed unstable isotropic tripole and finally, perturbed but stable sufficiently anisotropic tripole for $\alpha \ll 1$—are shown systematically in Fig. 3.41.

Aligned Vortex Quadrupole

As a final example, let us consider the case of four vortices with $S_1 = S_3 = \pm 1$, $S_2 = S_4 = \mp 1$. An ansatz where these four vortices are aligned along the y axis, symmetrically with respect to the origin, allows us to determine the following equilibrium positions of the aligned vortex quadrupole along the y axis:

$$\begin{aligned} y_1 = -y_4 &= \sqrt{\frac{B}{4\omega_y^2 Q}} \cdot \sqrt{1+\sqrt{2\sqrt{2}-2}}, \\ y_2 = -y_3 &= \sqrt{\frac{B}{4\omega_y^2 Q}} \cdot \sqrt{1-\sqrt{2\sqrt{2}-2}}. \end{aligned} \tag{3.148}$$

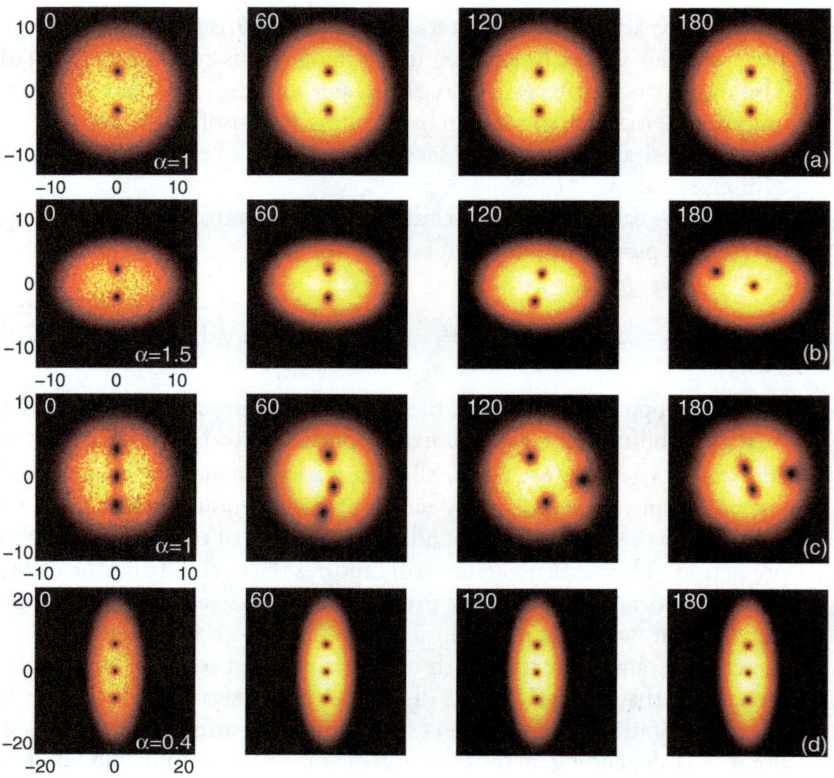

Figure 3.41. *Dynamical evolution of aligned vortex states perturbed by white noise for different values of* α*:* (a) $\alpha = 1$ *dipole,* (b) $\alpha = 1.5$ *dipole,* (c) $\alpha = 1$ *tripole,* (d) $\alpha = 0.4$ *tripole. All plots show* $|\psi|^2$*, and the elapsed dimensionless time is given in the upper left corners. Adapted with permission from Ref. [570].*

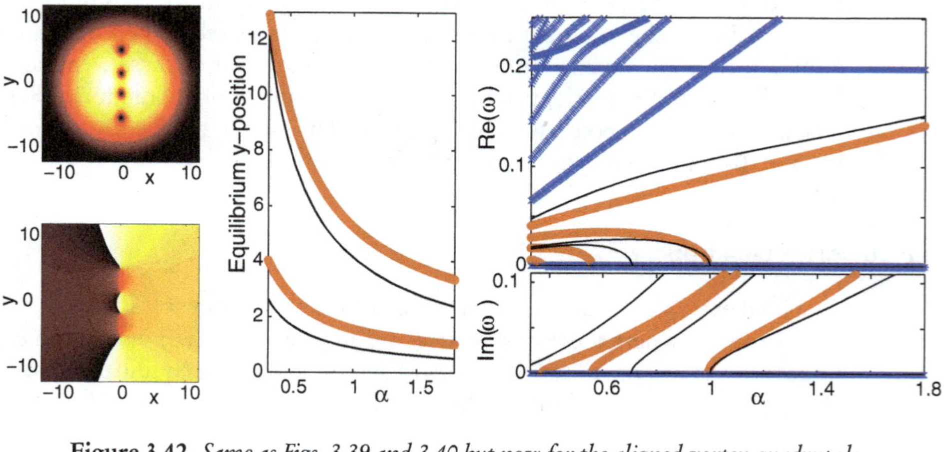

Figure 3.42. *Same as Figs. 3.39 and 3.40 but now for the aligned vortex quadrupole.*

The middle panel of Fig. 3.42 compares these predictions to the numerically calculated equilibrium positions. Again, qualitative (and to some extent also quantitative) agreement is very good.

The linearization frequencies from the particle equations of motion could not be obtained analytically in this case. The results calculated by numerical diagonalization of the ODE system's Jacobian at the aligned quadrupole fixed point, together with the full BdG spectrum, are shown in Fig. 3.42, following the analysis of Ref. [570].

Once again, we observe the onset of stabilization for small enough values of α. This feature is contained in the particle picture predictions as well, but the critical value $\alpha_{\rm cr}$ below which stabilization occurs is predicted too low. From the BdG spectrum we obtain $\alpha_{\rm cr} \approx 0.37$. In summary, quantitative agreement between our particle picture predictions and the BdG modes calculated from the full Gross–Pitaevskii theory somewhat deteriorates as more vortices are considered. We mainly assume two effects responsible for this: On the one hand, the more vortices are present, the further the vortex cluster stretches out into the condensate and the off-center correction term due to the precession frequency which we neglect becomes increasingly important. On the other hand, our simple, semiclassically justified modeling of vortex interaction is not fully accurate in that it does not adequately account for density inhomogeneities due to the trap i.e., the background-induced effect mentioned above. The higher the number of vortices, the further outward the vortex cluster extends and hence the more the density variation at the rims of the cloud affects the result. Nevertheless, for all the cases considered the qualitative agreement between the conclusions of the particle picture and those of the full PDE is very good.

3.6.3 ▪ Aligned Vortex States: Bifurcations

We will now study branches of vortex cluster (and solitonic) solutions to the stationary anisotropic GPE, varying the chemical potential μ and monitoring the number of particles $N = \int |\psi|^2\, dxdy$, analogously to what we did in the isotropic case in Sec. 3.4. Relevant bifurcations are identified and related to the stability properties of the different branches. This will prove to be a useful complementary tool to better understand the changes in stability induced by anisotropy that we observed for the vortex states of Sec. 3.6.2. This approach will complement the particle theory perspective given so far in Sec. 3.6.1.

Extending the weakly nonlinear analysis to the anisotropic regime is essentially straightforward. Let us in the following explain how insight into qualitative changes in the bifurcation diagram due to $\alpha \neq 1$ (and the ensuing implications for stability) can be gained using the Galerkin approach. Linear eigenfunctions in the anisotropic trap still factorize according to $\gamma_{mn}(x,y) = \gamma_m(x)\gamma_n(y)$, where the one-dimensional modes now read

$$\gamma_m(x) \propto H_m(\sqrt{\omega_x}x)\exp(-\omega_x x^2/2), \tag{3.149}$$

$$\gamma_n(y) \propto H_n(\sqrt{\omega_y}y)\exp(-\omega_y y^2/2). \tag{3.150}$$

Normalization constants have been omitted and H_n denotes the nth Hermite polynomial. The energy eigenvalue of state γ_{mn} is given by

$$E_{mn} = (m+1/2+\alpha(n+1/2))\,\omega_x.$$

Thus, for the two linear modes used to predict the bifurcation of aligned vortex clusters, γ_{10} and γ_{0n}, the difference between the eigenenergies is $E_{0n} - E_{10} = (n\alpha - 1)\omega_x$. This energy difference crucially enters the expression for the critical particle number of the bifurcation predicted by the Galerkin approach, and in turn the position of the critical point is controlled by the anisotropy parameter α.

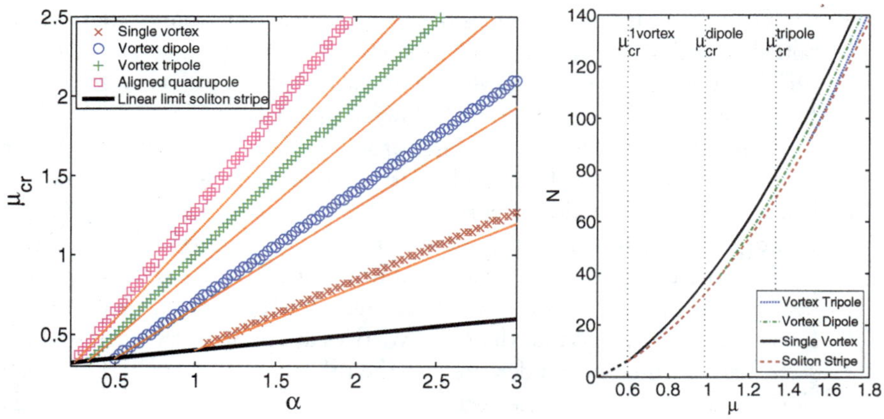

Figure 3.43. *(Left) Critical values of μ at which bifurcations leading to aligned vortex states occur, as a function of the anisotropy parameter α: predictions of the Galerkin approach (thin orange lines) and numerically found values (symbols as shown in legend). (Right) Example of an $N(\mu)$ bifurcation diagram, $\alpha = 1.5$, predicted critical values indicated by dotted vertical lines. Adapted with permission from Ref. [570].*

The theoretical predictions for $\mu_{\rm cr}$ are shown in Fig. 3.43, together with the bifurcation points obtained from our numerical simulations. We find that for any of the bifurcations considered, the Galerkin approach gives correct results as long as the bifurcation happens sufficiently close to the linear limit. As α is increased, the bifurcation points are shifted to higher values of the particle number N, and the Galerkin approximation is less accurate.

Intuitively, for $\alpha > 1$, i.e., for $\omega_y > \omega_x$, the energy of the γ_{0n} $(n \geq 1)$ states is higher than in the isotropic limit, while the energy of the γ_{10} state is only weakly affected. In particular, the degeneracy of γ_{10} and γ_{01} (that constitute the one-vortex state) is lifted: in contrast to the isotropic case, the admixture of a γ_{01} component to the γ_{10} soliton state is suppressed, and thus the bifurcation of the single vortex is shifted away from the linear limit to a nonzero value of N. By the same reasoning, all the other aligned vortex states bifurcate further away from the linear limit, too. We note that this has implications also for the stability of the higher aligned multivortex states, since the only stable bifurcating state will be the single charge vortex inheriting the (dark) soliton's initial stability and the remaining aligned states will, by necessity, be more unstable (by one eigenmode) than before, a feature corroborated in the previous section by the particle picture for $\alpha > 1$.

On the other hand, for $\alpha < 1$ the energy of the γ_{0n} states is lower than in the isotropic case, that is their admixture to the γ_{10} soliton state is favored. For values of α just below 1, the consequences are most drastic for the single vortex state. As the energy of the γ_{01} state is now lower than that of the γ_{10} soliton state, it is no longer the case that the vortex emerges by an admixture of $\pm i\gamma_{01}$ to γ_{10}. Rather, this picture is reversed, with the vortex emerging by an admixture of $\pm i\gamma_{10}$ to the (energetically favorable) γ_{01} state. In other words, the vortex branch now bifurcates from the soliton stripe oriented along the x axis. Similar arguments apply to the other aligned vortex states if the value of α is further decreased. For $\alpha = 1/2$, the energies of the γ_{10} soliton state and the γ_{02} state (whose admixture leads to the vortex dipole) are the same, and thus the vortex dipole emerges from the linear limit. For even smaller values of α, the vortex dipole no longer bifurcates from the single soliton, but rather from the two-soliton state γ_{02} parallel to the x axis (through an admixture of $\pm i\gamma_{10}$). Obviously, this goes for any aligned vortex state:

For $\alpha = 1/n$, the states γ_{10} and γ_{0n} are degenerate and the n-vortex state emerges from the linear limit. For smaller values than this, the bifurcation picture is reversed, and the vortex state no longer bifurcates from the single dark soliton stripe along the y axis, but rather γ_{0n} becomes the lower energy state to which $\pm i\gamma_{10}$ gets admixed.

Having understood the structural dependence of the bifurcation diagram on α, let us now summarize the conclusions on the stability properties of aligned vortex states in the presence of anisotropy.

We first consider the regime of $\alpha > 1$. As in the isotropic case, the soliton stripe is stable when it emerges from the linear limit. The first bifurcation (which now leads to the single vortex state) renders it unstable, with the vortex inheriting the soliton's stability. The vortex dipole then bifurcates from this already unstable soliton stripe and is thus unstable as well (in contrast to the isotropic case). The higher aligned vortex states (tripole, quadrupole, ...) are all more unstable by the same reasoning. Recall that, as stated above, (further) destabilization of the dipole, tripole, and aligned quadrupole for $\alpha > 1$ is also predicted by the particle picture and confirmed by our numerical computations.

We can also draw some conclusions on the stability of the soliton stripe itself. For $\alpha > 1$, we find that the length of the interval between the emergence of the soliton from the linear limit and the first bifurcation point increases as a function of α. The Galerkin approach predicts a linear increase. This corresponds to a growing range of values of the chemical potential for which the soliton stripe is stable. This, in turn, reflects the fact that for $\alpha \gg 1$ we progressively approach the 1D regime where the soliton stripe—ultimately, the 1D dark soliton—is stable for all values of μ for which it exists (which is consonant with the prediction of the 1D Gross–Pitaevskii theory exposed in the previous chapter).

On the other hand, for $\alpha < 1$ the aligned vortex clusters with the vortices located along the y axis tend to get stabilized. In the interval $1/2 \leq \alpha \leq 1$ the vortex dipole is the first state bifurcating from the soliton and is thus stable, while the soliton stripe gets destabilized. For $1/3 \leq \alpha \leq 1/2$, the vortex dipole no longer bifurcates from the soliton, and the tripole takes its place (and its stability properties) as the first emerging state. Thus, for $\alpha \leq 1/2$ the vortex tripole (which is unstable in isotropic traps) is expected to be stabilized. By the same reasoning, the aligned quadrupole gets stabilized for $\alpha \leq 1/3$, the vortex quintupole for $\alpha \leq 1/4$, and so on. It can further be argued that the aligned vortex states are still stable when α is so small that they no longer bifurcate from the soliton stripe along the y axis. In this case (as discussed above), the state with n vortices aligned bifurcates from the γ_{0n} branch through an admixture of $\pm i\gamma_{10}$. Similarly to the γ_{01} soliton stripe branch, the general γ_{0n} solitonic branch can be expected to be stable when it emerges from the linear limit, and the admixture of $\pm i\gamma_{10}$ will induce the first bifurcation from it, thus leading to a stable vortex cluster state (while the n-soliton state gets destabilized). Therefore, the states with n vortices aligned along the y axis are expected to be stable for any $\alpha \leq 1/(n-1)$, which is in very good agreement with our numerical results for the single vortex, vortex dipole, tripole and quadrupole shown above. It has been demonstrated numerically in Ref. [570] that even large aligned vortex clusters of up to $n = 17$ vortices can be stabilized by strong enough transverse confinement, and that the anomalous BdG eigenmodes of such a linear cluster are reminiscent of standing waves on a classical string; see Fig. 3.44. Interestingly, in such a "vortex string," there will be n (i.e., in the above example 17) internal (anomalous) modes of vibration in its spectrum, which will, in turn, correspond to the n normal modes of such a vortex lattice.

Let us also remark that our results are consistent with previous results on the stability of the one-soliton state in anisotropic settings [192]. In this work it was found (employing box boundary conditions and keeping the particle density fixed) that in the regime corresponding to our $\alpha \gg 1$ relaxing the confinement in the y direction (i.e., approaching

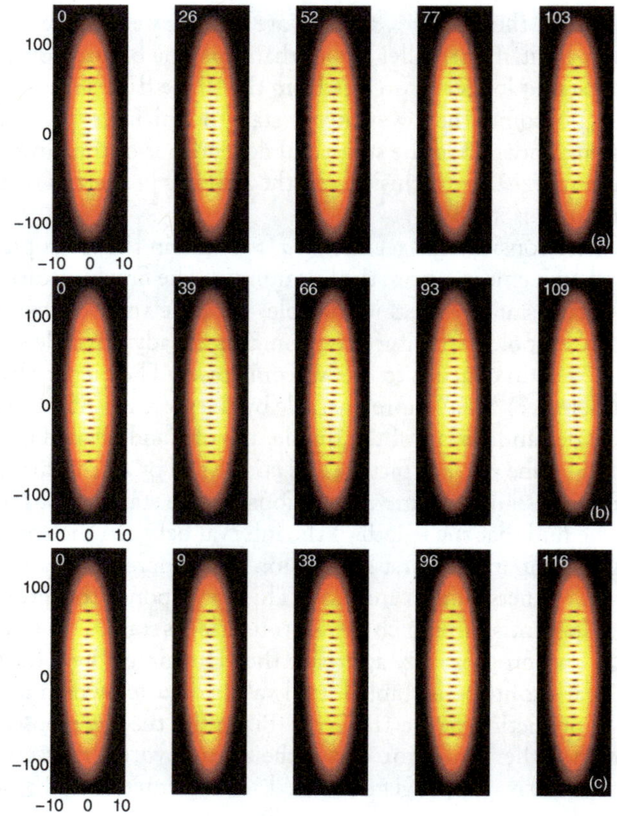

Figure 3.44. *Anomalous modes of a stabilized 17-vortex cluster, resembling the fundamental and first harmonic modes of a classical string. Note the different scaling in the x and y axes. Adapted with permission from Ref. [570].*

$\alpha = 1$) opens up an increasing number of decay channels for the soliton, with the first one leading to a single vortex, the second one leading to a vortex dipole, and so on. In our analysis, these "decay channels" correspond to the imaginary modes in the soliton's BdG spectrum, induced by supercritical pitchfork bifurcations from the soliton branch having happened at lower values of N (or μ) than the one under consideration.

Furthermore, the authors of Ref. [192] report the numerical observation of a "solitonic vortex" solution to the GPE in anisotropic settings, i.e., a stationary state with its density and phase properties in between those of a soliton stripe and a single vortex. This is reported to bifurcate from the dark soliton stripe as the anisotropy of the confinement is varied. Even though different boundary conditions are employed, this agrees well with our findings: If $\alpha \neq 1$, the single vortex state bifurcates from the soliton stripe at a finite particle number (due to the nondegeneracy of γ_{01} and γ_{10}). This, in turn, necessarily means that close to the bifurcation point the vortex state will still show some similarity to the soliton solution from which it just bifurcated. Figure 3.45 shows an example of the single vortex state close to its bifurcation from the soliton stripe. Clearly, the density shows remnants of the soliton stripe. The phase runs from $-\pi$ to $+\pi$ continuously, a characteristic of the singly charged vortex, however the azimuthal phase gradient is not constant (as it is for the known vortex solution in isotropic settings). Rather, the phase changes very sharply in the region around the y axis, again owing to the state's former solitonic properties.

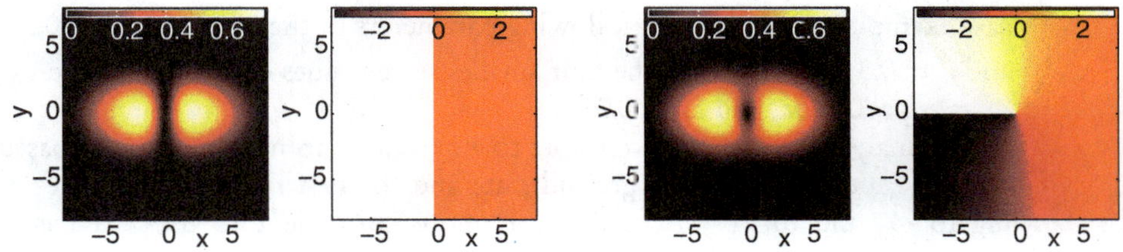

Figure 3.45. *Soliton stripe (left panels) and single vortex state (right panels) profiles at* $\mu = 1$, $\alpha = 2.1$ *("solitonic vortex").*

The interested reader is referred to Ref. [570] for a systematic analysis of the impact of anisotropy in nonaligned vortex clusters, such as regular (i.e., nonaligned) vortex quadrupoles, among others. Having given a flavor of the impact of anisotropy on the most prototypical (dark soliton, vortex stripe, and aligned vortex states), we will now turn our attention to a different theme, namely 2D realizations of double-well potentials.

3.7 ▪ Two-Dimensional Few-Well Potentials

The aim of this section is to generalize the double-well setting that we encountered in the 1D chapter into a corresponding "few-well" configuration within a 2D trap. In our presentation, we will chiefly follow the approach of Refs. [575, 576]. In these works, a configuration bearing four wells in the 2D setting was considered, through the combination of a parabolic magnetic trap or an optical dipole trap and a cosinusoidal optical lattice in BECs. The same model can be realized in optics too, using a bulk nonlinear medium with a symmetric set of four embedded waveguiding rods.

As suggested by numerous works that analyzed the spontaneous symmetry breaking (SSB) and related effects in 1D settings in the previous chapter, a natural approach to the analysis of the four-well configuration may be based on a Galerkin-type few-mode truncation that reduces the 2D GPE to a discrete system, cf. Ref. [218]. In the framework of this approach, we generate a bifurcation diagram, with the aim to predict a global set of possible stationary states of the underlying four-well system. Then, applying numerical methods to the GPE, we verify that all the states predicted by the four-site discrete-nonlinear-Schrödinger (DNLS) reduction indeed exist in the underlying continuum model. Furthermore, their stability—in the limit of strong nonlinearity—coincides with what is expected from the DNLS model. Notice that, in this context too, we take a weakly nonlinear approach, which uses as a "basis" for the reduction at hand the one of the lowest linear eigenmodes of the system.

3.7.1 ▪ Analytical Approach

Our starting point is the GPE in its usual rescaled form,

$$i\partial_t u = \hat{L}u + |u|^2 u - \mu u, \tag{3.151}$$

with the operator $\hat{L} = -(1/2)\Delta + V(x,y)$ (Δ is the 2D Laplacian) and the potential

$$V(x,y) = \frac{1}{2}\Omega^2 r^2 + V_0[\cos(2kx) + \cos(2ky)],$$

where $r^2 = x^2 + y^2$. The parameter μ represents, as usual, the chemical potential (or the propagation constant in optics). The sign in front of the cubic term corresponds to the self-repulsion of the interatomic interactions within our model. In the present case example, we assume the following parameters of the potential: $\Omega = 0.21$, $V_0 = 0.5$, and $k = 0.3$, in which case the four smallest eigenvalues of operator $\hat{L}$ are $\omega_0 = 0.3585$, $\omega_1 = \omega_2 = 0.3658$, and $\omega_3 = 0.3731$.

To analyze the system, we resort to a natural four-mode reduction, based on eigenstates u_0 and $u_{1,2,3}$, i.e., the ground state and the first three excited states corresponding to ω_0 and $\omega_{1,2,3}$. Actually, it is more convenient to use a transformed basis, $(\Phi_0 \ \ \Phi_1 \ \ \Phi_2 \ \ \Phi_3) = (u_0 \ \ u_1 \ \ u_2 \ \ u_3)\mathbf{T}$, with an appropriate matrix $\mathbf{T}$. This basis is formed by an orthonormal set of functions Φ_j localized in each of the wells which constitute the four-well potential. Thus, we adopt the following decomposition of the full solution: $u(x,y,t) = \sum_{j=0}^{3} c_j(t)\Phi_j(x,y)$.

Substituting this expression into Eq. (3.151) and projecting onto the orthonormal basis $\{\Phi_0, \Phi_1, \Phi_2, \Phi_3\}$, we derive a system of four ODEs for the complex amplitudes $c_j(t) \equiv \rho_j(t)e^{i\varphi_j(t)}$, $j = 0,1,2,3$:

$$\begin{aligned} i\dot{c}_j &= \tilde{\omega}_j + A_j|c_j|^2 c_j + \sum_{k\neq j} B_{jk}(2|c_k|^2 c_j + c_k^2 c_j^*) \\ &+ \sum_{k\neq j}[D_{kj}|c_k|^2 c_k + D_{jk}(2|c_j|^2 c_k + c_j^2 c_k^*)] + \sum_{k\neq l\neq j\neq k} E_{kjl}(2|c_k|^2 c_l + c_k^2 c_l^*) \\ &+ \sum_{k\neq l\neq j\neq k} E_{jkl}(c_j^* c_k c_l + c_j c_k^* c_l + c_j c_k c_l^*) + G \sum_{\substack{k\neq l\neq m\neq k \\ k,l,m\neq j}} c_k^* c_l c_m, \end{aligned} \tag{3.152}$$

with $k,l,m = 0,1,2,3$, where

$$\begin{aligned} \tilde{\omega}_0 &\equiv (1/4)[c_0(\gamma_0 - 4\mu) + c_1\gamma_1 + c_2\gamma_3 + c_3\gamma_2], \\ \tilde{\omega}_1 &\equiv (1/4)[c_0\gamma_1 + c_1(\gamma_0 - 4\mu) + c_2\gamma_2 + c_3\gamma_3], \\ \tilde{\omega}_2 &\equiv (1/4)[c_0\gamma_3 + c_1\gamma_2 + c_2(\gamma_0 - 4\mu) + c_3\gamma_1], \\ \tilde{\omega}_3 &\equiv (1/4)[c_0\gamma_2 + c_1\gamma_3 + c_2\gamma_1 + c_3(\gamma_0 - 4\mu)], \end{aligned}$$

and

$$\begin{aligned} \gamma_0 &\equiv \omega_0 + \omega_1 + \omega_2 + \omega_3, \\ \gamma_1 &\equiv \omega_0 + \omega_1 - \omega_2 - \omega_3, \\ \gamma_2 &\equiv \omega_0 - \omega_1 + \omega_2 - \omega_3, \\ \gamma_3 &\equiv \omega_0 - \omega_1 - \omega_2 + \omega_3. \end{aligned}$$

Notice that $\gamma_1 = \gamma_2 = \omega_0 - \omega_3$ and $\gamma_3 \approx 0$, for the above-mentioned particular values of the eigenfrequencies. The above ODEs also involve a number of overlap integrals, *viz.*,

$$\begin{aligned} G &\equiv \iint \Phi_0\Phi_1\Phi_2\Phi_3\, dx\, dy, \\ A_n &\equiv \iint \Phi_n^4\, dx\, dy, \quad n = 0,1,2,3, \\ B_{mn} &\equiv \iint \Phi_m^2\Phi_n^2\, dx\, dy, \quad m,n = 0,1,2,3, \quad m \neq n, \\ D_{mn} &\equiv \iint \Phi_m^3\Phi_n\, dx\, dy, \quad m,n = 0,1,2,3, \quad m \neq n, \\ E_{lmn} &\equiv \iint \Phi_l^2\Phi_m\Phi_n\, dx\, dy, \end{aligned}$$

with mutually different l,m,n. For our parameters (and, in fact, also for other cases corresponding to well-separated potential wells), other overlap integrals are much smaller than the A_n's. Then, neglecting these small overlap terms leads to an approximation of the tight-binding, alias discrete NLS, type, which may actually be quite accurate, in comparison with numerical solutions of the underlying continuous NLS equation.

The resulting set of ODEs gives rise to both real and complex-valued solutions. Our analysis of the discrete model indicates that principal complex solutions in this setting correspond to discrete vortices with phase sets $\varphi_j = j(\pi/2)$ [577]. This solution has been studied in detail elsewhere [578, 511], and we will touch upon it in our discussion of vortices within optical lattices below. Instead, here we focus on real solutions and a bifurcation diagram for them, based on the four algebraic equations derived from Eq. (3.152) by setting $\varphi_j = j\pi$. We will also compare this approximation with numerical solutions of the full GPE, and report results for the numerical linear-stability analysis for such solutions.

3.7.2 ▪ Numerical Results

The principal bifurcation diagram, shown in Fig. 3.46, displays the squared L^2 norm (the number of atoms in BEC, or the power in optics), denoted by N, as a function of chemical potential μ in Eq. (3.151). The version of the diagram shown in the right panel Fig. 3.46 is obtained from the algebraic equations stemming from Eq. (3.152), and demonstrates very good agreement between the full numerical and semi-analytical results. Different branches labeled in the bifurcation diagram are displayed in detail in Fig. 3.47. We have also developed a symbolic representation in the form of 2×2 matrices, labeling different waveforms that arise in the diagram, as follows:

$$\mathrm{A1} \equiv \begin{pmatrix} -1 & 1 \\ 1 & -1 \end{pmatrix}, \quad \mathrm{A2} \equiv \begin{pmatrix} 1 & 1 \\ -1 & -1 \end{pmatrix}, \quad \mathrm{A3} \equiv \begin{pmatrix} 1 & 1 \\ 1 & 1 \end{pmatrix}, \quad \mathrm{A4} \equiv \begin{pmatrix} 0 & 1 \\ -1 & 0 \end{pmatrix},$$

$$\mathrm{B1} \equiv \begin{pmatrix} -1 & 1 \\ \varepsilon & -\varepsilon \end{pmatrix}, \quad \mathrm{B2} \equiv \begin{pmatrix} -\varepsilon & 1 \\ 1 & -\varepsilon \end{pmatrix}, \quad \mathrm{B3} \equiv \begin{pmatrix} 1 & 1 \\ -\varepsilon & -\varepsilon \end{pmatrix},$$

$$\mathrm{C1} \equiv \begin{pmatrix} 1 & -\varepsilon \\ -\varepsilon & \varepsilon \end{pmatrix}, \quad \mathrm{C2} \equiv \begin{pmatrix} -\varepsilon & 1 \\ 1 & -1+\varepsilon \end{pmatrix}, \quad \mathrm{C3} \equiv \begin{pmatrix} -\varepsilon & 1 \\ 1 & 1+\varepsilon \end{pmatrix},$$

$$\mathrm{C4} \equiv \begin{pmatrix} -1+\varepsilon & 1 \\ 1 & 1+\varepsilon \end{pmatrix}, \quad \text{and} \quad \mathrm{D1} \equiv \begin{pmatrix} 1+\varepsilon & 1 \\ -\varepsilon & -1+\varepsilon \end{pmatrix}.$$

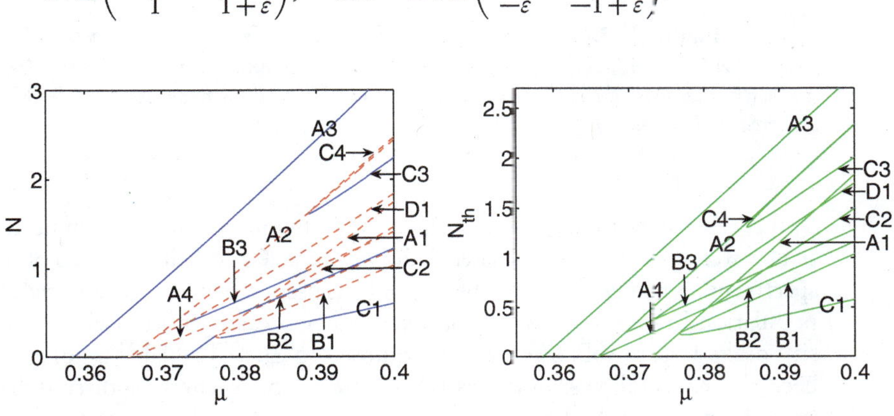

Figure 3.46. *Norm N of the numerically found solutions of Eq. (3.151) (left), and their counterparts predicted by the four-mode approximation (right), as a function of μ. The branches are explained in the text and their profiles and stability are detailed in subsequent figures. Solid lines and dashed lines denote stable and unstable solutions, respectively. Adapted with permission from Ref. [575].*

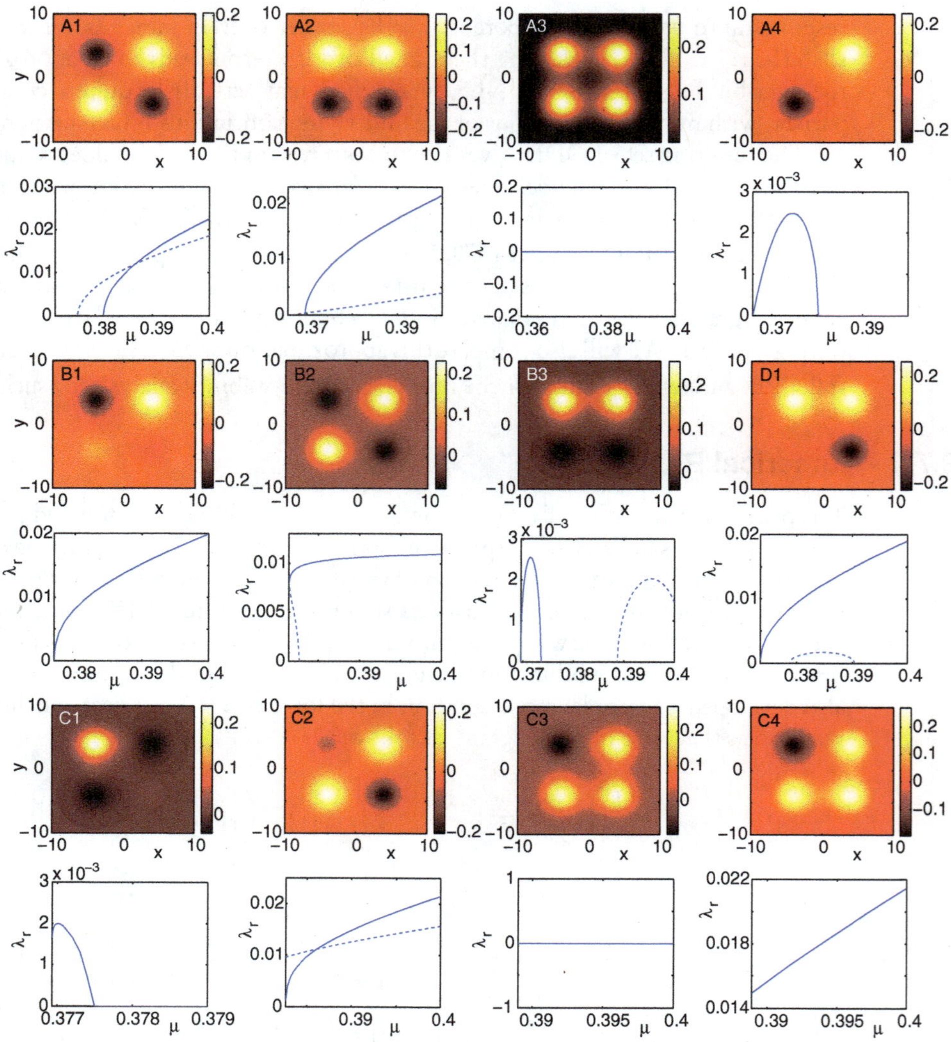

Figure 3.47. *The first, third, and fifth rows show the wavefunctions of the bifurcation branches A1, A2, A3, A4, B1, B2, B3, D1, C1, C2, C3, and C4 at* $\mu = 0.395$. *The second, fourth, and sixth rows show the real parts of unstable eigenvalues of the respective branches as a function of* μ. *Adapted with permission from Ref. [575].*

In this representation, 1, −1, and 0 have the obvious meaning, by indicating that a particular well is or is not "populated," and its phase (0 and π in the cases of +1 and −1, respectively), when populated. The symbol ε is used to denote either a small (but nonzero) population in one of the wells, or a symmetry-breaking effect (when some of the density peaks feature values $\pm 1 \pm \varepsilon$, as opposed to ± 1). The labeling of the branches has been chosen as follows: branches A1–A4 are the ones emerging directly from the linear limit; branches B1–B3 bifurcate from those of the A1–A4 types and feature two pairs of peaks with different amplitudes; branches C1–C4 either bifurcate from B1–B3, or arise in a saddle-node bifurcation, but their four peaks have three different amplitudes, while branch D1 has all four of its peaks different. Clearly, branches B, C, and D are results of an elaborate sequence of symmetry-breaking bifurcations.

Examining the emergence of new branches and their stability in more detail, we conclude the following. Branch A3 is the stable ground state of the system. Branch A2 is immediately unstable, upon the departure from the linear limit, having a real eigenvalue pair. A supercritical pitchfork, leading to the emergence of branch B3, adds a second pair of real eigenvalues to the linearization around this branch. On the other hand, states A4 bifurcating from the same linear mode as A2, are unstable due to a quartet of eigenvalues near the linear limit, but become stabilized at higher values of chemical potential μ. On the contrary, A1 is stable near the linear limit, but becomes unstable due to a supercritical *and* a subcritical pitchfork, which involve, respectively, branches B1 and C1, apparently endowing branch A1 with *two* identical pairs of real eigenvalues. A subsequent bifurcation of branch B2 from A1 (at larger values of N) leads to yet another pair of real eigenvalues for the linearization around A1 (i.e., three pairs in total for N large enough). Although B2 remains unstable with two real pairs of eigenvalues, a subsequent bifurcation of C2 from it leads to a reduction of the number of such pairs to one, while the resulting branch C2 inherits the other two real pairs. Furthermore, C3 and C4 arise through a saddle-node bifurcation near $\mu = 0.389$, and one of them carries a real eigenvalue pair, while the other one does not (due to the nature of the bifurcation). Lastly, branch D1 bifurcates from B3 and always possesses a real eigenvalue pair, as well as potentially an eigenvalue quartet.

Note that many of the general stability characteristics of the above-mentioned branches, which are valid at *large* N, can be understood on the basis of a few simple principles developed in the context of discrete systems (from the so-called anticontinuum (AC) limit [577]) with the self-defocusing nonlinearity [579]. In particular, branches of states which feature a single-site shape in the AC limit should be stable, those based on a set of two nearest neighbors with opposite signs produce real eigenvalue pairs, while in-phase sets may lead to complex eigenvalue quartets. Finally, sets of in-phase next-nearest neighbors lead to real pairs, while sets of the same type but with the out-of-phase arrangement, may produce quartets.

Naturally, these considerations do not apply to some of the asymmetric branches, such as C3 or D1, which cannot be examined in the AC setting. Nevertheless, this analysis provides a useful set of guidelines in understanding most of the stability features of the fundamental branches.

A similar picture but for the attractive case was painted in Ref. [576]. There, the DNLS based interactions and stability are, arguably, even simpler to understand as the excitation of any two sites with in-phase eigenstates produces spectral instability due to a real eigenvalue pair, while the case of out-of-phase excitations may yield oscillatory instability (for suitable parametric intervals). For more details on this latter case, we refer the interested reader to Ref. [576].

3.8 ▪ A Generalization: Two-Dimensional Nonlinear States in a Combination of a Magnetic Trap and an Optical Lattice

In this section, we will use both semi-analytical and numerical techniques to offer some insights on how the properties of the coherent structures are modified in the additional presence of an optical lattice (i.e., a 2D egg carton) potential. In particular, our first line of attack will involve the familiar weakly nonlinear technique of bifurcation of nonlinear states from the underlying linear limit. The basic idea of such an approach is that in the absence of interactions the GPE is reduced to a linear Schrödinger equation for a confined single-particle state; in this limit, and in the case of, e.g., a harmonic external potential, the linear problem becomes the equation for the quantum harmonic oscillator characterized by discrete energies and corresponding eigenmodes. Exploiting this simple physical picture, one may then use analytical and/or numerical techniques for the *continuation* of these linear eigenmodes supported by the particular type of the external trapping potential into nonlinear states as the interactions become stronger.

Following the approach of Ref. [580], we will examine the structure and the stability of a pancake-shaped condensate confined by the combination of a harmonic trap and a periodic potential, with periodicity in two orthogonal directions. We will adopt the above-mentioned approach of continuation of linear states to nonlinear ones, thus providing a host of interesting solutions that should be tractable within presently available experimental settings. In particular, our analysis starts by first considering the noninteracting limit. In this regime, we employ a multiscale perturbation method (which uses the harmonic trap strength as a formal small parameter) to find the discrete energies and the corresponding eigenmodes of the pertinent single-particle Schrödinger equation with the combined harmonic and periodic potential. We then use this linear limit as a starting point for initializing a 2D "nonlinear solver" that identifies the relevant stationary nonlinear eigenstates as a function of the chemical potential (i.e., the nonlinearity strength) and of the optical lattice depth.

Once the basic structure of the condensate is found, we subsequently perform a linear stability analysis of the nonlinear modes that can be initiated by the noninteracting ground and first few excited states. The linear stability analysis will indicate the *dynamical* instability (which is associated to exponential growth of small perturbations) of the solution When nonlinear states are found to be unstable, we use direct numerical simulations to study their dynamics and monitor the evolution of the relevant instability. Essential results that will be presented below are the following: for a fixed harmonic trap strength, there exist certain regions in the parameter plane defined by the chemical potential and the optical lattice depth, where not only the ground state, but also excited states are stable or only weakly unstable. Particularly, an excited state with a shape resembling an out-of-phase matter-wave soliton pair (for attractive interactions) is found to persist for long times, being stable (weakly unstable) for attractive (repulsive) interatomic interactions. Thus, the ground state and the aforementioned excited state have a good chance to be observed in a real experiment with either attractive or repulsive pancake BECs. Similar conclusions can be drawn for more complex states, such as a quadrupolar one which may also be stable in the attractive case; however, higher excited states are typically more prone to instabilities, as is shown in the numerical computations. We will present below only the repulsive (defocusing) nonlinear case in detail (based on the discussion of Ref. [580]), while results for the focusing case can be found in Ref. [580].

The dimensionless model of interest in this setting reads

$$i\partial_t u = -\frac{1}{2}\Delta u + g|u|^2 u + V(x,y)u. \tag{3.153}$$

Here, the normalized trapping potential $V(x,y)$ is now given by

$$V(x,y) = V_{\mathrm{MT}}(r) + V_{\mathrm{OL}}(x,y) \equiv \frac{1}{2}\Omega^2 r^2 + V_0(\cos(2x) + \cos(2y)). \tag{3.154}$$

In the above equation, length is measured in units of $a_L = d/\pi$, time in units of $\omega_L^{-1} = \hbar/E_L$, and energy in units of $E_L = 2E_{\mathrm{rec}} = \hbar^2/ma_L^2$ [where E_{rec} is the lattice recoil energy and the periodicity of the optical lattice is $d = (\lambda/2)/\sin(\theta/2)$, where θ is the angle between the two beams and λ is their wavelength]. Furthermore, the normalized lattice depth V_0 is measured in units of $4E_{\text{rec}}$, while the normalized harmonic trap strength is given by

$$\Omega = \frac{a_L^2}{a_\perp^2} = \frac{\omega_\perp}{\omega_L}. \tag{3.155}$$

3.8.1 ▪ Linear Analysis

We start by considering the noninteracting limit $g \to 0$, in which the GPE becomes a linear Schrödinger equation. Then, seeking stationary localized solutions of the form $u(x,y,t) = \exp(-iE_{m,n}t)u_{m,n}(x,y)$ [where $E_{m,n}$ are discrete energies and $u_{m,n}(x,y)$ are the corresponding linear eigenmodes], and rescaling spatial variables by $\sqrt{\Omega}$, we obtain the following equation:

$$\begin{aligned}\frac{E_{m,n}}{\Omega}u_{m,n} = &-\frac{1}{2}\Delta u_{m,n} + \frac{1}{2}(x^2+y^2)u_{m,n} \\ &+\frac{V_0}{\Omega}\left[\cos\left(\frac{2x}{\sqrt{\Omega}}\right)+\cos\left(\frac{2y}{\sqrt{\Omega}}\right)\right]u_{m,n}.\end{aligned} \tag{3.156}$$

The next step is to separate variables through $u(x,y) = u_m(x)u_n(y)$ and split the energy into $E_{m,n} = E_m + E_n$, to obtain two 1D eigenvalue problems of the same type as above, namely,

$$-\frac{1}{2}\frac{d^2u_m}{dx^2} + \frac{1}{2}x^2 u_m + \frac{V_0}{\Omega}\cos\left(\frac{2x}{\sqrt{\Omega}}\right)u_m = \frac{E_m}{\Omega}u_m, \tag{3.157}$$

and a similar one for y (with x replaced by y and the subscript m replaced by n).

We now restrict ourselves to the physically relevant regime of $0 < \Omega \ll 1$ as discussed above. In this case, we may use $\nu \equiv \sqrt{\Omega}$ as a formal small parameter and develop methods of multiple scales and homogenization techniques [511] in order to obtain analytical predictions for the linear spectrum. In particular, introducing the fast (i.e., rapidly varying) variable $X = x/\nu$ and rescaling the energy as $\varepsilon_m = E_m/\Omega$, the eigenvalue problem of Eq. (3.157) is expressed as follows:

$$\left(\nu^2\mathcal{L}_{\text{HT}} - \nu\frac{\partial^2}{\partial x \partial X} + \mathcal{L}_{\text{OL}}\right)u_m = \nu^2\varepsilon_m u_m, \tag{3.158}$$

where

$$\mathcal{L}_{\text{HT}} = -\frac{1}{2}\frac{\partial^2}{\partial x^2} + \frac{1}{2}x^2, \tag{3.159}$$

$$\mathcal{L}_{\text{OL}} = -\frac{1}{2}\frac{\partial^2}{\partial X^2} + V_0\cos(2X), \tag{3.160}$$

[note that V_0/Ω was treated as an $\mathcal{O}(1)$ parameter]. Additionally, we consider a formal series expansion (in ν) for u_m and ε_m [511], namely,

$$u_m = u_0 + \nu u_1 + \nu^2 u_2 + \cdots, \tag{3.161}$$

$$\varepsilon_m = \varepsilon_{-2}\nu^{-2} + \varepsilon_{-1}\nu^{-1} + \varepsilon_0 + \varepsilon_1\nu + \cdots. \tag{3.162}$$

To this end, substitution of this expansion in the eigenvalue problem of Eq. (3.158) and use of the solvability conditions for the first three orders of the expansion [i.e., $\mathcal{O}(1)$, $\mathcal{O}(\nu)$ and $\mathcal{O}(\nu^2)$] yields the following results for the eigenvalue problem of the original operator. The energy of the mth mode can be approximated by

$$E_m = -\frac{1}{4}V_0^2 + \left(1 - \frac{1}{4}V_0^2\right)\Omega\left(m + \frac{1}{2}\right), \tag{3.163}$$

while the corresponding eigenfunction is given by

$$u_m(x) = c_m H_m\left(\frac{x}{\sqrt{1-\frac{V_0^2}{4}}}\right)\exp\left(-\frac{x^2}{2-\frac{V_0^2}{2}}\right) \times \frac{1}{\sqrt{\pi}}\left[1 - \frac{V_0}{2}\cos\left(\frac{2x}{\sqrt{\Omega}}\right)\right], \tag{3.164}$$

where $c_m = (2^m m! \sqrt{\pi})^{-(1/2)}$ is the normalization factor and $H_m(x)$ are the Hermite polynomials.

Combining the results in the two orthogonal directions x and y yields a total energy eigenvalue

$$E_{m,n} = -\frac{1}{2}V_0^2 + \left(1 - \frac{1}{4}V_0^2\right)\Omega(n + m + 1), \tag{3.165}$$

and a corresponding eigenfunction which, up to normalization factors, can be written as

$$u_{m,n}(x,y) \propto H_m\left(\frac{x}{\sqrt{1-\frac{V_0^2}{4}}}\right)\left[1 - \frac{V_0}{2}\cos\left(\frac{2x}{\sqrt{\Omega}}\right)\right] \times H_n\left(\frac{y}{\sqrt{1-\frac{V_0^2}{4}}}\right)\left[1 - \frac{V_0}{2}\cos\left(\frac{2y}{\sqrt{\Omega}}\right)\right] \times \exp\left(-\frac{r^2}{2-\frac{V_0^2}{2}}\right). \tag{3.166}$$

The particularly appealing feature of this expression is that it allows us to combine various (ground or excited) states in the x direction with different ones along the y direction. Here, we will focus on the simplest possible combinations of $m, n \in \{0,1,2\}$ and examine the various states generated by the combinations of these, which we hereafter denote $|m,n\rangle$. In particular, below we will focus on the ground state $|0,0\rangle$ and the excited states $|1,0\rangle$, $|1,0\rangle + |0,1\rangle$, $|1,1\rangle$, and $|2,0\rangle$ (although we note that topological states such as $|1,0\rangle + i|0,1\rangle$ also exist [581] and will be considered separately below). Our aim is to investigate which of these states persist in the nonlinear regime in the case of repulsive interatomic interactions (similar considerations for the attractive case can be found in Ref. [580]), and study the dynamical stability of these states in detail. The states can be experimentally examined in expansion or an interference-type of experiment, and to observe stability one would just examine the state for longer times.

We start by examining the validity of the above analytical predictions concerning the linear limit of the problem, namely, Eqs. (3.165) and (3.166). These results are summarized in Fig. 3.48. Panel (a) shows the 1D harmonic oscillator energy spectrum (for $\Omega = 0.1$)

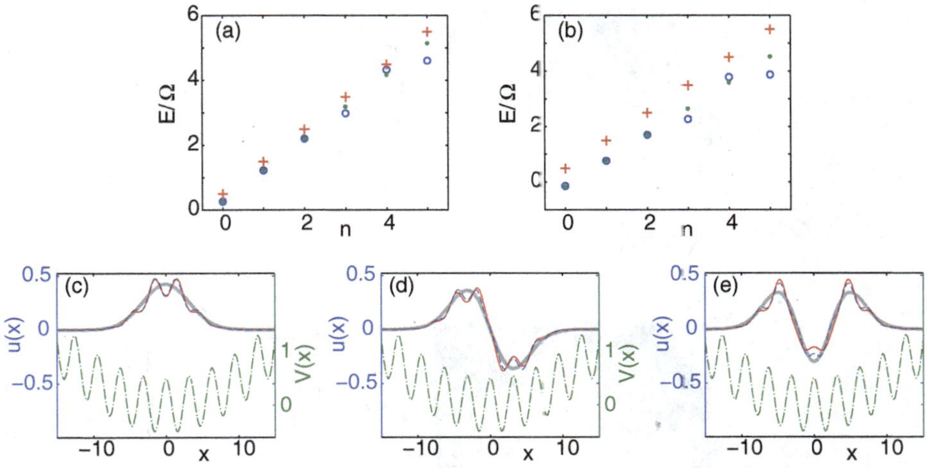

Figure 3.48. *Panel* (a) *shows the energy spectra corresponding to a purely parabolic potential [i.e., $E/\Omega = (n+1/2)$, depicted by pluses], and also how these eigenvalues move when an optical lattice is introduced [numerically (blue circles) and as predicted analytically (green dots)]. Shown are only the first six eigenvalues for $\Omega = 0.1$ and $V_0 = 0.3$. A similar result is demonstrated in panel* (b) *but for a larger lattice depth, $V_0 = 0.5$ (and the same value of Ω). For the latter case ($V_0 = 0.5$), panels* (c), (d), *and* (e) *show the first few eigenmodes for the parabolic potential (thick solid gray lines), parabolic and lattice potential computed numerically (thin solid red lines), and the same ones given by Eq. (3.164) (dashed blue lines). The potential (rescaled for visibility) is shown by the dashed-dotted green line. Adapted with permission from Ref. [580].*

and compares it with the energy spectrum obtained from numerical and approximate theoretical [see Eq. (3.163)] solutions of the combined harmonic and optical lattice potential for $V_0 = 0.3$. Panel (b) offers a similar comparison but for a larger lattice depth, namely $V_0 = 0.5$. One can clearly see that the theoretical calculation approximates very accurately the numerical results for the first few states (i.e., $n = 0, 1, 2$), while deviations become more significant for higher-order excited states. Panels (c), (d), and (e) show the zeroth, first, and second eigenfunctions of the purely harmonic potential (thick solid gray line), as well as of the harmonic trap and optical lattice potential as found numerically (thin solid red line) and analytically [given by Eq. (3.164)]. The green dashed-dotted line represents the form of the combined potential. Notice the good agreement of the analytical results in comparison with the full numerical computation.

3.8.2 ▪ Repulsive interatomic interactions

Existence and Stability

Now we will investigate the results pertaining to repulsive interatomic interactions ($g > 0$) for the same linear states examined above. We start with the ground state $|0,0\rangle$ branch shown in Fig. 3.49. The top left panel of Fig. 3.49 shows the number of atoms $N(V_0, \mu)$. However, the linear stability $\sigma(V_0, \mu)$ (i.e., the maximum real part of the stability spectrum) for this case is omitted because, as may be expected, this branch is stable throughout the parameter space, in contrast to its attractive counterpart (which is potentially subject to collapse).

We now turn to excited states with $m+n = 1$. Figure 3.50 shows features similar to the previous one, but now for the state $|1,0\rangle$. This branch is found to always be unstable due

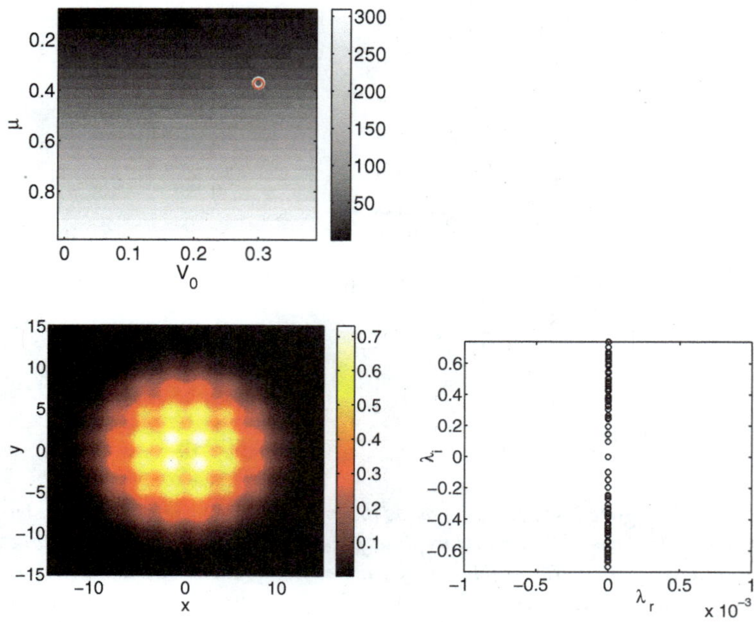

Figure 3.49. *The ground state* $|0,0\rangle$ *solution in the case of repulsive interatomic interactions. The top left panel depicts the number of atoms* $N(V_0,\mu)$. *The bottom panels depict the density of the configuration (left) and its corresponding spectrum (right) for the parameter values* $(V_0,\mu)=(0.3,0.37)$ *(see red point in the top left panel). The stability of the ground state persists over the entire parameter space, and hence the stability surface is omitted from this set. Adapted with permission from Ref. [580].*

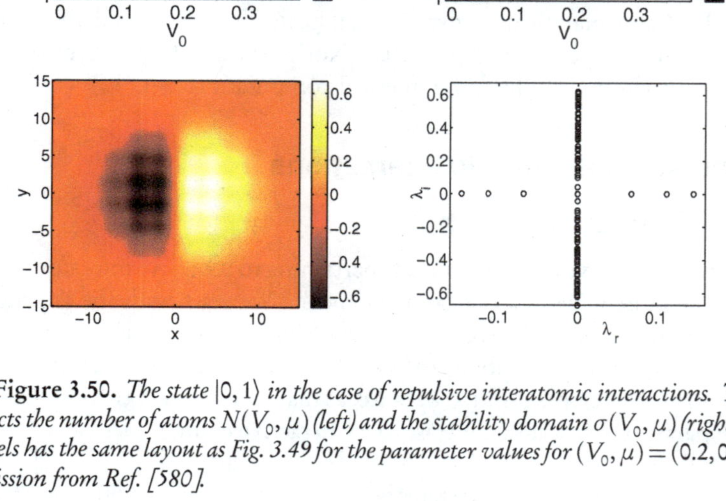

Figure 3.50. *The state* $|0,1\rangle$ *in the case of repulsive interatomic interactions. The top row of panels depicts the number of atoms* $N(V_0,\mu)$ *(left) and the stability domain* $\sigma(V_0,\mu)$ *(right). The bottom row of panels has the same layout as Fig. 3.49 for the parameter values for* $(V_0,\mu)=(0.2,0.47)$. *Adapted with permission from Ref. [580].*

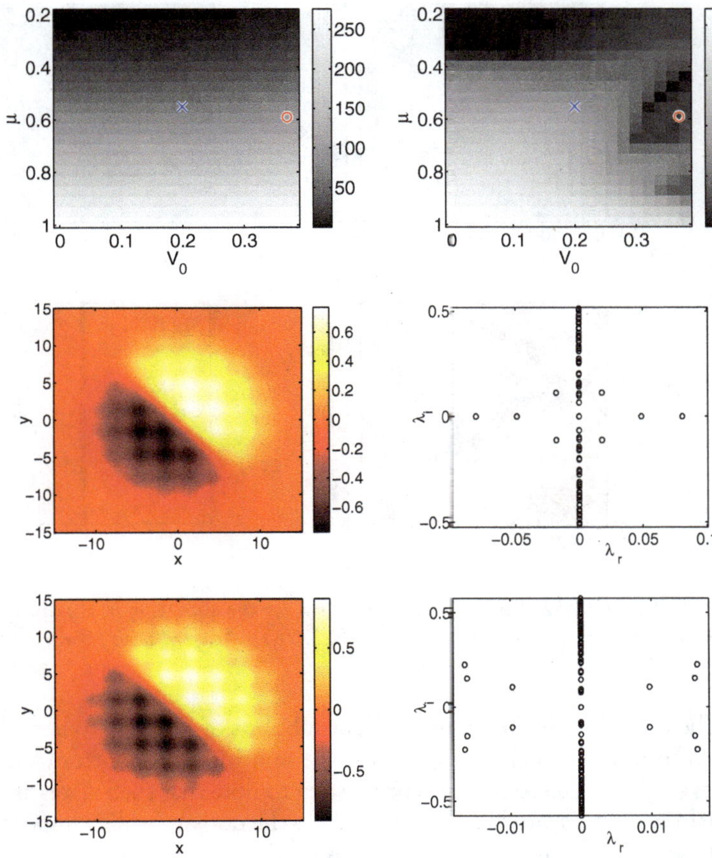

Figure 3.51. *Same as the previous figures but for the* $|1,0\rangle+|0,1\rangle$ *state in the case of repulsive interatomic interactions. This state is always unstable due to at least an eigenvalue quartet and up to two other real pairs. Note that there exists a region of weak instability, where only small magnitude quartets are present. The middle and bottom panels show the contour plots of this state and its linearization spectrum for* $(V_0,\mu)=(0.2,0.55)$ *(see blue cross in top panels) and* $(0.37,0.59)$ *(see red circle in top panels), respectively. Adapted with permission from Ref. [580].*

to the appearance of up to three real eigenvalue pairs (for the parameter range considered herein). The top panels depict the dependence of the number of atoms $N(V_0,\mu)$ (left), and instability growth rate $\sigma(V_0,\mu)$ (right) on the lattice depth V_0 and the chemical potential μ. A sample profile and its eigenvalue spectrum are given in the bottom panels, indicating the presence, in this case, of multiple real eigenvalue pairs.

Next, we consider the state $|1,0\rangle+|0,1\rangle$, which is presented in Fig. 3.51. This state always possesses a quartet of complex eigenvalues, and up to two additional pairs of real eigenvalues, and is thus unstable for all μ. It is worth noticing, however, that the instability weakens to relatively benign small magnitude complex quartets for intermediate values of the chemical potential, roughly $\mu\in(0.4,0.9)$, and large lattice depths, $V_0>0.3$. This suggests that such a configuration should be long-lived enough that it could be observable in experiments with repulsive condensates.

We now consider the states with $n+m=2$ (again for repulsive interatomic interactions), starting with the $|1,1\rangle$ branch in Fig. 3.52. The branch is also always unstable, possessing a complex quartet and then up to four additional real pairs for larger values of

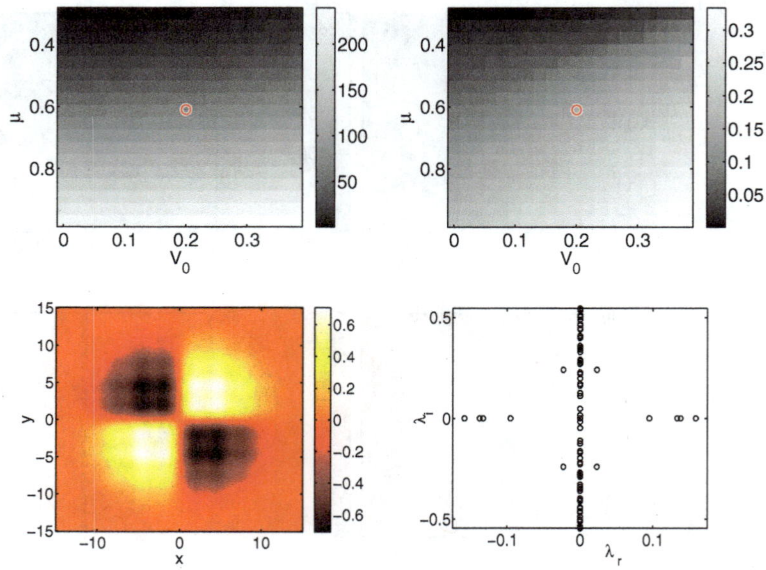

Figure 3.52. *Same as in Fig. 3.50, but for the state* $|1,1\rangle$ *in the case of repulsive interatomic interactions for parameter values* $(V_0, \mu) = (0.2, 0.61)$. *Adapted with permission from Ref. [580].*

μ. The instability is shown in the right subplots of Fig. 3.52, where the spectral plane of the bottom right panel corresponds to the solution of the bottom left one, for parameter values $(V_0, \mu) = (0.2, 0.61)$. It is interesting to note that such states are reminiscent of the domain walls presented in Ref. [251] (here the domain wall is imposed by the difference in phase), which however were found as potentially stable structures in multicomponent condensates (and will be discussed in the corresponding section below).

Next, the case of the $|2,0\rangle$ state is shown in Fig. 3.53. Here, there are up to three complex quartets along with three real pairs of eigenvalues (for the considered parameter range). It is notable that for higher values of the lattice depth, these states are deformed as the lattice "squeezes" the central maximum separating the two minima (see middle left panel of the figure). The contour plot shown in the bottom right panel suggests that further increase of lattice depth may lead to a new configuration altogether when the two local maxima eventually pinch off of each other. This deformation is a direct consequence of the presence of the (repulsive) nonlinearity, which results in drastically different configurations as compared to the linear limit of the structure.

Dynamics

We performed numerical simulations to investigate the evolution of typical unstable states in the case of repulsive interatomic interactions. Apart from the ground state, all excited states presented in the previous section were predicted to be unstable, and this is confirmed in this section. In the particular case of the state $|1,0\rangle + |0,1\rangle$ which was found to be weakly unstable (see bottom row of Fig. 3.51), the instability takes a considerable time to manifest itself. The evolution of this state is depicted in panel (b) of Fig. 3.54; it is clearly seen that the state persists up to $t \approx 300$. On the other hand, panel (a) shows the evolution of the state $|1,0\rangle$ which persists up to $t \approx 40$, while the next two panels show the dynamics of the (c) $|1,1\rangle$ state and of the (d) $|2,0\rangle$ state, both persisting also up to $t \approx 40$. All of these excited states degenerate into ground-state-like configurations. Notice that transient vortex-like

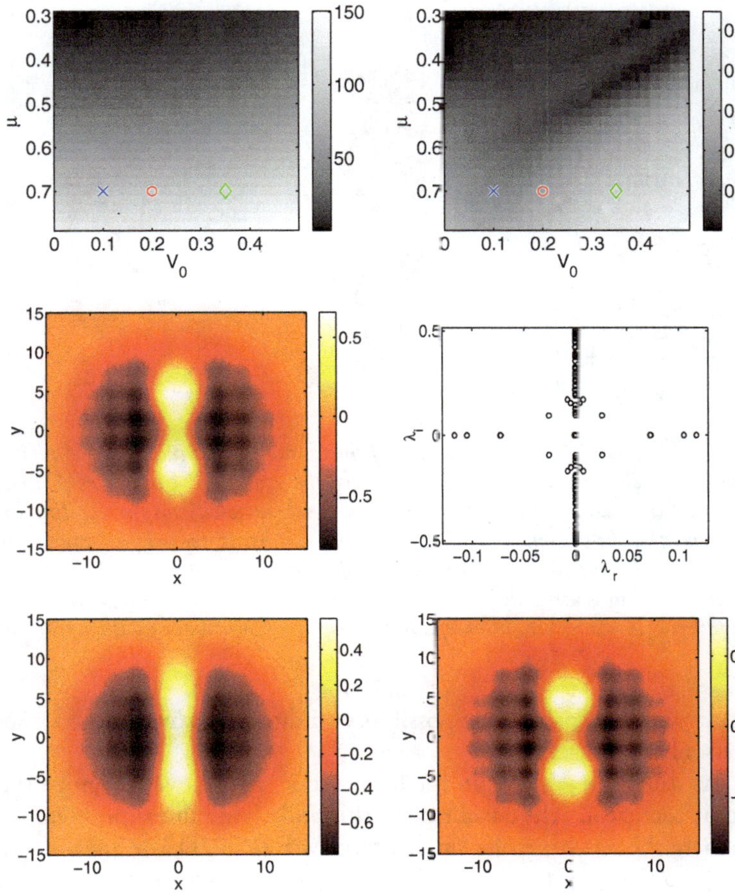

Figure 3.53. *Same as in Fig. 3.50, but for the state* $|2,0\rangle$ *(in the case of repulsive interatomic interactions) with parameter values* $(V_0, \mu) = (0.2, 0.7)$ *(see red circle in top panels). The bottom row shows profiles for smaller,* $V_0 = 0.1$ *(left; see blue cross in top panels), and larger,* $V_0 = 0.35$ *(right; see green diamond in top panels), values of the optical lattice depth. Adapted with permission from Ref. [580].*

structures seem to appear during this process, but they do not persist in the eventual dynamics and are hence not further discussed here. Interestingly, vortices emerge for all cases in which the structures from the higher excited states transform themselves into the ground state.

3.8.3 ▪ A Case Example of a Highly Nonlinear State in the Combined Potential: Vortex Existence, Stability, and Dynamics

Having examined the different real solutions of the underlying linear problem (in the presence of both a parabolic trap and a periodic potential) and their nonlinear continuations, we now wish to turn to a set of topologically charged states, namely the vortices and their existence, stability, and dynamics, when subjected to a general optical lattice (OL) potential of the form

$$V_{\mathrm{OL}} = V_0[\cos^2(k_x x + \phi) + \cos^2(k_y y + \phi)]. \tag{3.167}$$

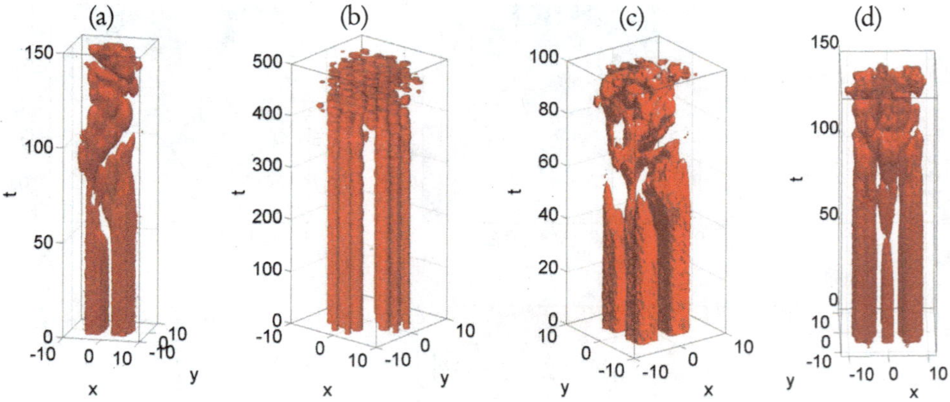

Figure 3.54. *The dynamics of the unstable states in the case of repulsive interatomic interactions. Panels* (a) *and* (b) *show, respectively, the evolution of the states* $|1,0\rangle$ *and* $|1,0\rangle + |0,1\rangle$*. It is clear that the state* $|1,0\rangle + |0,1\rangle$ *is subject to a weaker oscillatory instability for the parameter values mentioned in the bottom row of Fig. 3.51 and, as a result, the original configuration persists for a long time. Panels* (c) *and* (d) *show the dynamics of the states with* $m+n=2$*, namely* (c) $|1,1\rangle$*, and* (d) $|2,0\rangle$*. All these solutions ultimately degenerate into ground-state-like configurations. The density isosurfaces are taken at* $a=0.5$ *with the exception of* (d) *at* $a=0.4$*. Adapted with permission from Ref. [580].*

where we have enabled the possibility of different wavenumbers and phases for the lattice's different directions.

As usual, for $V_0 = 0$, in the linear limit, $s \to 0$, we recognize the 2D quantum harmonic oscillator. The advantage of this underlying linear problem is that it allows us to generate the structure of a single charge vortex in 2D as a linear combination of $|1,0\rangle = H_{1,0}e^{-\Omega r^2/2}$ and $|0,1\rangle = H_{0,1}e^{-\Omega r^2/2}$, where $H_{1,0}=x$ and $H_{0,1}=y$, according to $w(x,y) = |1,0\rangle + i|0,1\rangle$. Similarly, a linear, doubly charged vortex-type state can be constructed as $w(x,y) = |2,0\rangle - |0,2\rangle + 2i|1,1\rangle$ and such constructions can also be produced for higher charges, as well. These can be used as good initial guesses for the fixed point iteration, towards producing a solution of the full nonlinear problem.

In the next subsections, following the work of Ref. [578] (see also the earlier work of Ref. [581]) we will present a two-parameter continuation in μ, which corresponds to the chemical potential of the BEC (and is related to its number of atoms), and in V_0 which parametrizes the strength of the OL potential. A uniform space mesh with $\Delta x = \Delta y = 0.15$ is used for these results to prevent contamination of spurious modes, which may occur when the spatial scale of the unstable bound state in the vortex core competes with the grid size. Once the solution of the nonlinear problem has been identified to a prescribed accuracy (typically 10^{-7}–10^{-8}), we solve the linearized eigenvalue problem of the Bogoliubov–de Gennes equations to determine the stability of each solution.

$S = 1$ Vortices

We start by investigating the stability of the single quantized vortex state in the cases of $\phi = 0$ and $\phi = \pi/2$ over a range of chemical potentials μ between 0.2 (the linear quantum harmonic oscillator limit for such a state with $V_0 = 0$) and 1.2. For $V_0 = 0$, our results confirm previous ones [488] in that the single quantized vortex is stable in the presence of a pure parabolic potential over all values of μ. Perhaps more interestingly, we found the $\phi = 0$ case to be stable over all V_0 as well, in agreement with the dynamical observations of

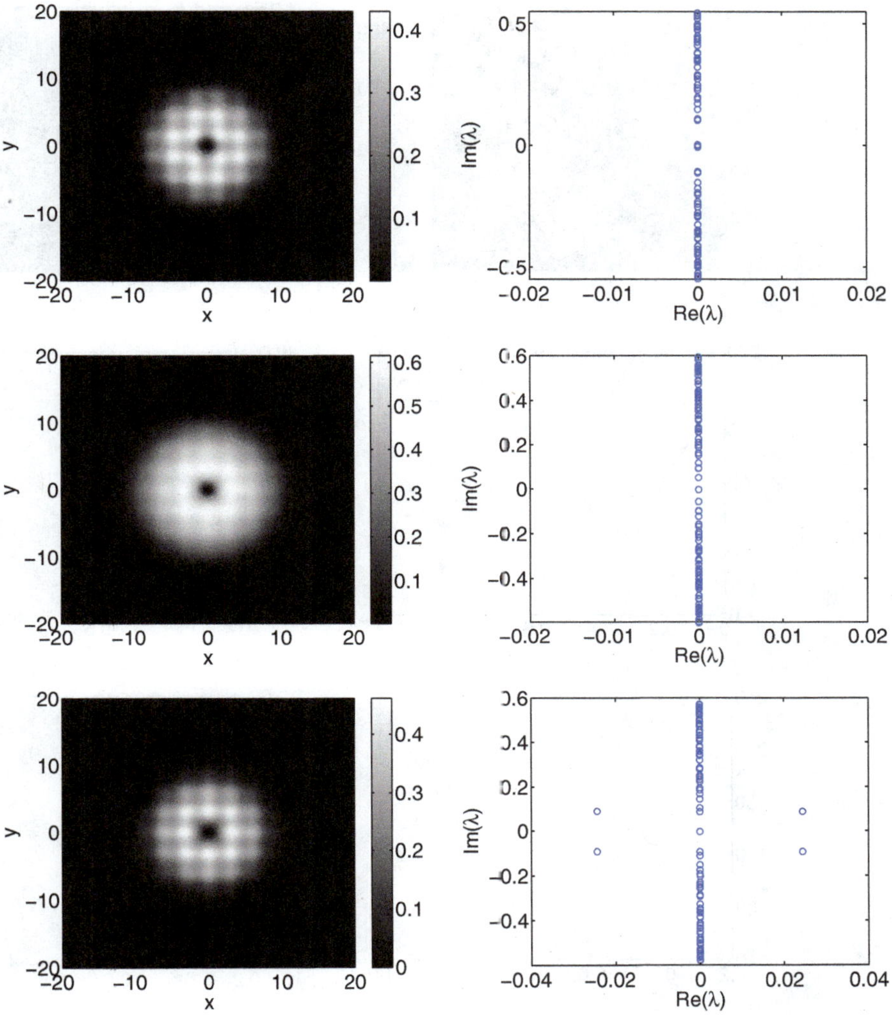

Figure 3.55. *Top: Profile (left) and spectrum (right) for a stable vortex with* $\phi = 0$ *at* $(V_0, \mu) = (0.3, 0.8)$*. Middle: Stable vortex with* $\phi = \pi/2$ *at the point* $(V_0, \mu) = (0.1, 0.8)$ *in our parameter plane. Bottom: Same for the unstable case at the intersection of the slices from the previous panels:* $(V_0, \mu) = (0.3, 0.8)$*. Adapted with permission from Ref. [578].*

Ref. [581], where such a solution was observed to remain at its location, when initialized in the center of such a potential. An example density profile and spectrum are presented in the top panels of Fig. 3.55.

On the other hand, when the vortex of single charge is located at a local minimum ($\phi = \pi/2$), instability sets in for a sufficiently modulated trap (i.e., for $V_0 \approx 0.2$) and the solutions remain unstable essentially throughout their interval of existence. Notice that for each value of μ, the solution can only sustain a certain level of modulation from the optical lattice, before it finally degenerates into the zero solution (beyond a certain critical, μ-dependent OL strength). Typical examples of the subcritical and supercritical profiles and their corresponding spectra for the relevant $S = 1$ mode are shown in Fig. 3.55. It is evident that in this case, beyond the critical V_0, a mode becomes unstable through an oscillatory instability associated with an eigenvalue quartet. Both the existence regime

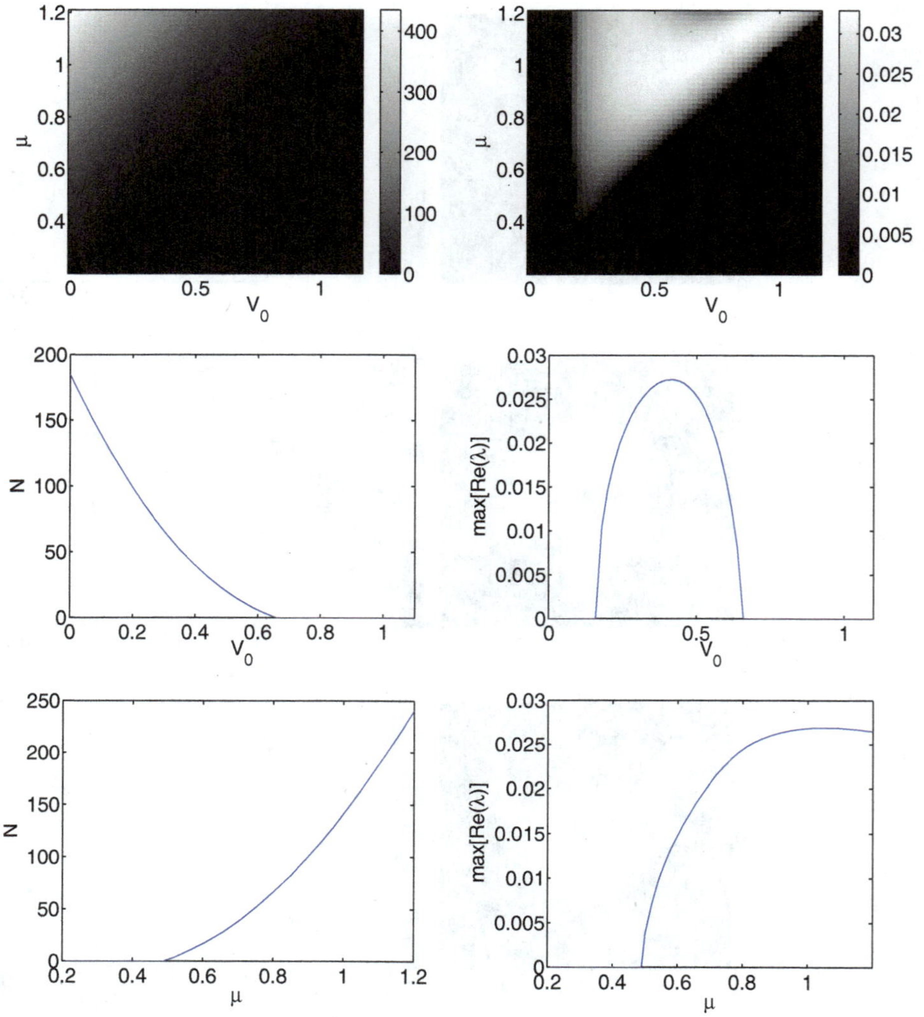

Figure 3.56. *Top: The (dimensionless) number of particles, $N = \int |u|^2 dx dy$, is shown as a contour plot in the two-parameter plane of (V_0, μ) in the left panel, while the maximum real part of λ is shown in a similar way in the right panel for a single charged vortex with $\phi = \pi/2$. Middle: Slices of $N(V_0)$ (left) and* $\max[\mathrm{Re}(\lambda(V_0))]$ *(right) for $\mu = 0.8$. Bottom: Slices of $N(\mu)$ (left) and* $\max[\mathrm{Re}(\lambda(\mu))]$ *(right) for $V_0 = 0.3$. Adapted with permission from Ref. [578].*

(top left panel) and the stability regimes (top right panel) of such waveforms are illustrated in Fig. 3.56 (the figure also contains relevant one-parameter cross sections of the number of atoms and of the eigenvalue of the most unstable mode of the linearization).

Our stability computations above are corroborated by direct numerical experiments involving the dynamical evolution of the vortices. We perturb the stationary state at unstable parameter values (for $\phi = \pi/2$) in the direction $\tilde{u}$ of its unstable eigenmode. The instability is clearly present and the vortex spirals out from the center of the condensate within the first 100 time units, as is shown in Fig. 3.57. We note in passing that this result is consonant with the findings of Ref. [581].

It is interesting to also note that for $\phi = \pi/2$ the instability growth rate appears to be zero for small V_0, while for larger V_0 it becomes nonzero—in fact increasing initially

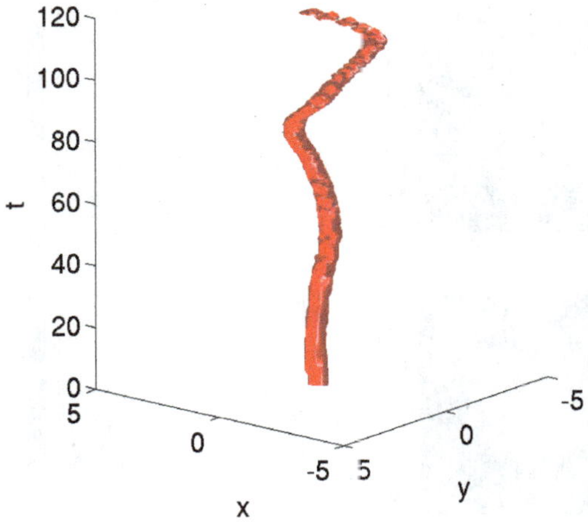

Figure 3.57. *The dynamics of the unstable mode when $\phi = \pi/2$ for parameter values* (0.3, 0.8). *We show an isosurface of the vorticity. This clearly illustrates the winding out of the vortex towards the periphery of the BEC cloud. Adapted with permission from Ref. [578].*

as a function of V_0 as seen in Fig. 3.56 (before it eventually decreases back to zero along with the solution, for very strong lattices). This indicates that the vortices should spiral outward faster for larger V_0 (with the exception of the case of very strong lattices), which is perhaps counterintuitive. This fact is confirmed by observing the trajectory $\mathbf{p}(t)$ of the initially stationary vortex slightly perturbed by a random noise of amplitude $\pm 0.05\%$, for different values of V_0 in the runs of Fig. 3.58. In these runs, the location of the vortex center is obtained by

$$\mathbf{p}(t) = \frac{1}{\iint ||\omega_{\mathrm{vor}}||^2 \, dx \, dy} \left(\iint x ||\omega_{\mathrm{vor}}||^2 \, dx \, dy, \iint y ||\omega_{\mathrm{vor}}||^2 \, dx \, dy \right), \tag{3.168}$$

where $\omega_{\mathrm{vor}} = \nabla \times \mathbf{v}_s$ is the fluid vorticity obtained from curl of the fluid velocity [see Eq. (3.13)]. A related quantitative diagnostic is defined as

$$t^* = \min\{t; ||\mathbf{p}(t)|_2 > 0.5\}, \tag{3.169}$$

illustrating the time at which the vortex arrives at a distance of 0.5 from its initial central location. In the absence of external potential, the perturbed vortex remains at the center of the harmonic trap for the entire simulation time of 450 units (see top row of Fig. 3.58). For a slightly larger value of $V_0 = 0.2$ it becomes slightly unstable due to the small magnitude of the eigenvalue quartet [with max(Re(λ)) = 0.0165] which emerges. This is seen in the middle row of Fig. 3.58; notice the (red) dot in the center image, presenting the vortex coordinates at $t = t^* = 300$, which represents the first time the vortex leaves the radius of 0.5 from its original position. Note that subsequently the perturbed vortex remains within the local minimum of the lattice for the duration of 450 time units. As V_0 is increased further to the value of $V_0 = 0.4$, depicted at the bottom row of the Fig. 3.58, it moves away from the initial location faster, having a critical time of $t^* = 134$. This is in line with the fact that this case corresponds to a larger growth rate with max(Re(λ)) = 0.0307.

An intuitive explanation that one can give about this effect (larger growth rate as V_0 increases) stems from a comparison with the work of Ref. [582], where the effect of sound

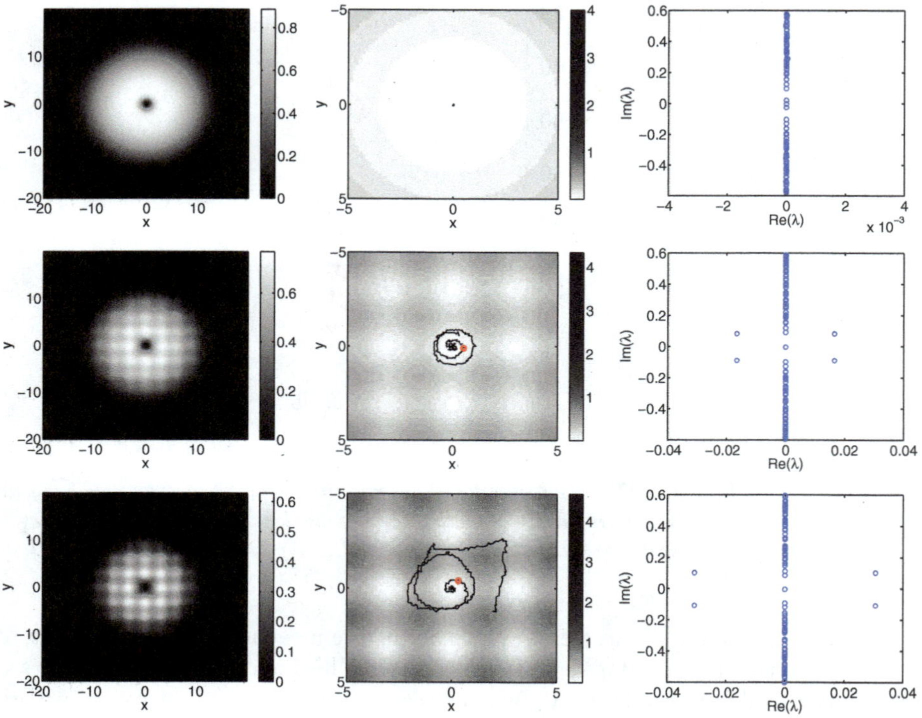

Figure 3.58. *The vortex trajectory for 450 time units for charge $S = +1$ solutions (central panels) with the respective stationary solutions in the left panels, and spectra in the right panels. All solutions are with $\mu = 1$ and are overlaid on top of the inverse image of the external potential in order to clarify the setting. The top row is for $V_0 = 0$, middle is $V_0 = 0.2$, and the bottom row is for $V_0 = 0.4$. The red dot represents $p(t^*)$ [see Eqs. (3.168) and (3.169)]. The value of t^* is 300 for the middle case and 134 for the bottom. Adapted with permission from Ref. [578].*

orresponding spiraling of the vortex away from the center of a magnetic trap with an additional dimple potential is considered. In particular:

- For small V_0, the vortex is close to the stable case of a magnetic trap and encounters a potential that is still close to a parabolic one, leading to a recombination effect of the emitted sound waves [582] with the vortex and a near closed trajectory (see the middle panel of Fig. 3.58 and notice its similarity to the deep dimple case in Fig. 3 of Ref. [582]).

- For large V_0, the potential no longer resembles that of a magnetic trap due to the strong modulation by the optical lattice. As a result, in this case the sound waves escape without recombining with the emitting vortex, allowing it to rapidly spiral out from the central location. This situation has a much more rapidly spiraling trajectory (shown in the bottom panel of Fig. 3.58, which is similar to the case of the shallow dimple trap in Fig. 3 of Ref. [582]).

$S = 2$ Vortices

We now turn to the case of the doubly quantized vortex. Again, following the presentation of Ref. [578], when $V_0 = 0$, we observe the solutions disappear at the linear quantum harmonic oscillator limit of $\mu = 0.3$. Our results in the $V_0 = 0$ continuation over μ are

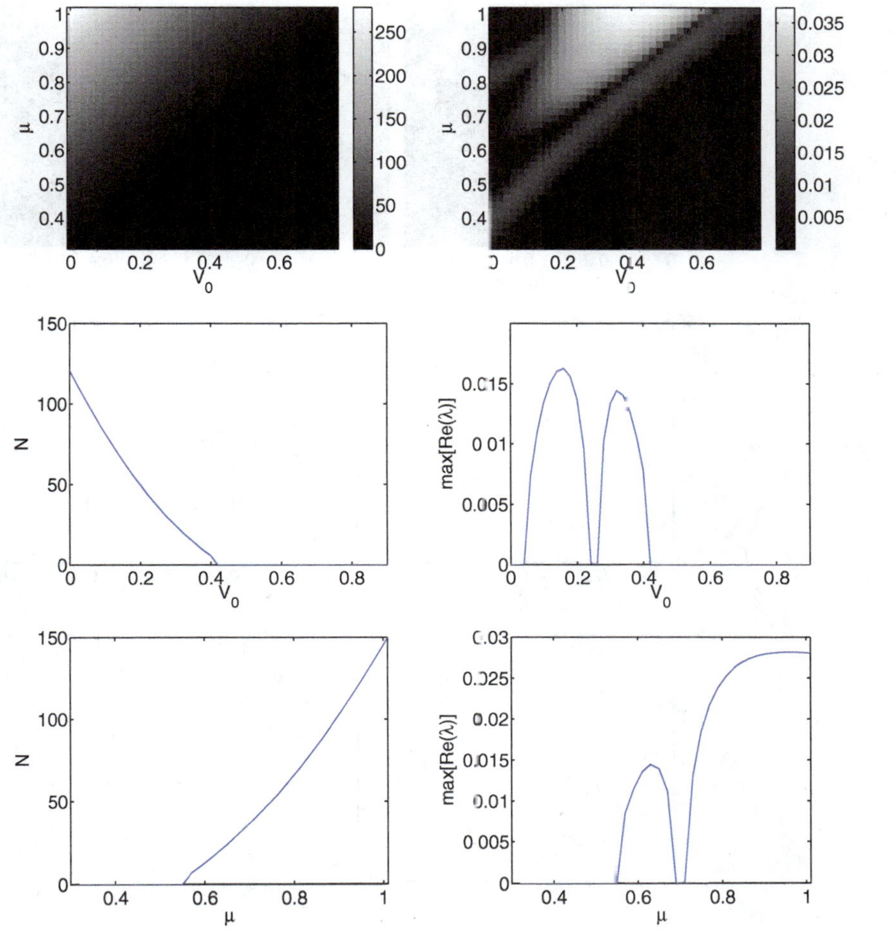

Figure 3.59. *The corresponding images to Fig. 3.56 for a double charge vortex with $\phi = \pi/2$ in the same layout as for the single charge case in Fig. 3.56. V_0 slices are for $\mu = 0.71$ and μ slices are for $V_0 = 0.24$. Adapted with permission from Ref. [578].*

consistent with those of Ref. [488]. In particular, there are windows of stability along this branch for μ in the intervals (0.45, 0.75) and (0.85, 1). Then, as we fill out the surface of solutions in V_0, we observe in Fig. 3.59 a similar behavior of N (as for the $S = 1$ vortices), while the stability intervals appear to narrow as V_0 increases. Furthermore, the stability landscapes appear to be quite similar for $\phi = 0$ and $\phi = \pi/2$, as can be observed in the comparison of the upper right corners of Figs. 3.59 and 3.60. While the stability intervals quickly narrow as V_0 increases, it is worthwhile to note that stable solutions with a considerable amount of modulation by the optical lattice can still be identified.

The dependence of the structure of the solutions on the phase ϕ of the OL becomes apparent in the case of these higher-charged vortices. In the case $\phi = \pi/2$, the vortex is centered at a local minimum and therefore, when it merges with its neighboring local minima, forms either an "X" or a square, depending on the relative size of the condensate (μ) and the lattice (V_0), as can be seen in Fig. 3.61. In the case of $\phi = 0$, where the vortex merges with its neighboring local minima, it assumes the form of a cross, as is shown in Fig. 3.62.

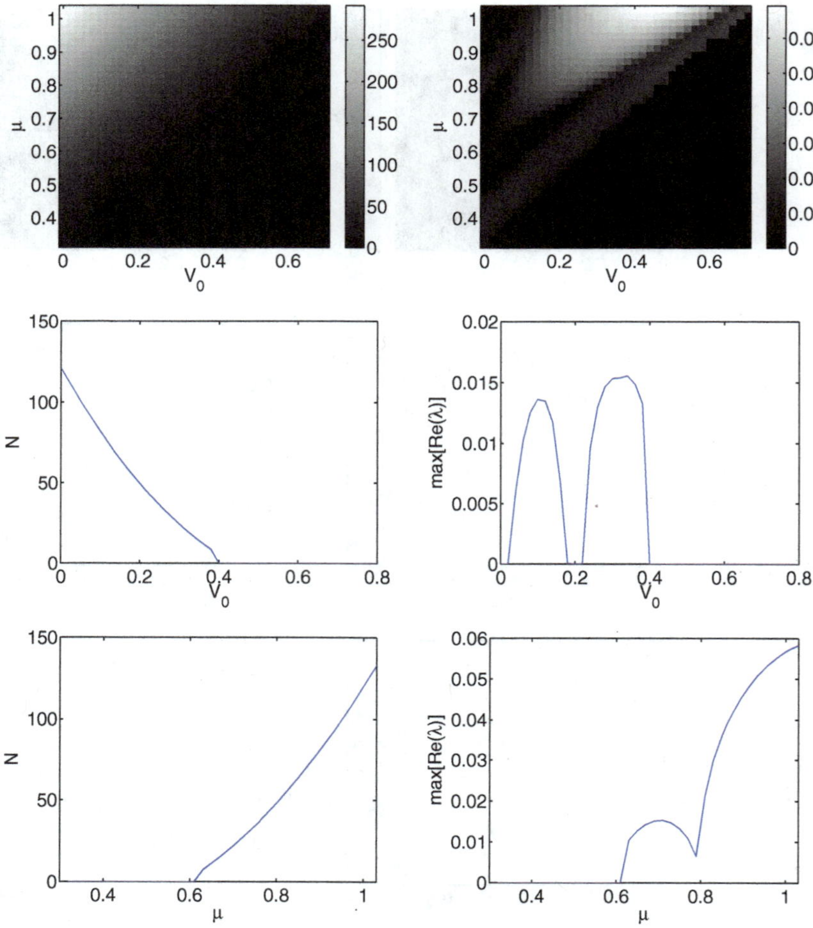

Figure 3.60. *Same as Fig. 3.59 except with $\phi = 0$. V_0 slices are for $\mu = 0.71$ again and μ slices are for $V_0 = 0.3$. Adapted with permission from Ref. [578].*

Finally, we examine the dynamical evolution of vortices in the case of $\phi = \pi/2$ and $\phi = 0$. Over time, both double charge vortices split into a pair of single charge vortices which follow one another around in an oscillatory spiral motion. Also, it is clear that the vortex located at the maximum (Fig. 3.62) has typically somewhat larger instability growth rates and in fact may possess two unstable eigenvalue quartets in its spectrum. This can also be observed to have a bearing on the evolution images, particularly as is highlighted in the contrast between the left panel of Fig. 3.63, for parameter values for which the solution has a small instability growth rate, and the right panel of Fig. 3.63, for parameter values for which the solution has a strong growth rate. The spiraling of the former case is slower and less violent than that of the latter.

Dynamics were also examined in Ref. [578] for equal parameter values of $(0.25, 0.8)$ and the vortex trajectory for 450 time units along with the profiles and spectra are presented in Fig. 3.64. The magnitude of the real part of the single quartet in each case is comparable, as are the corresponding dynamics. For $\phi = 0$, $\max(\text{Re}(\lambda)) = 0.036$, while for $\phi = \pi/2$, $\max(\text{Re}(\lambda)) = 0.027$. While this slight discrepancy is hard to detect from the trajectories shown in Fig. 3.64, we can define a similar diagnostic as in the single charge case. Here, we define $\mathbf{p}_1$ and $\mathbf{p}_2$ as in Eq. (3.168) and $t^{**} = \min\{t; \max(|\mathbf{p}_1 - \mathbf{p}_2|) > 1.5\}$,

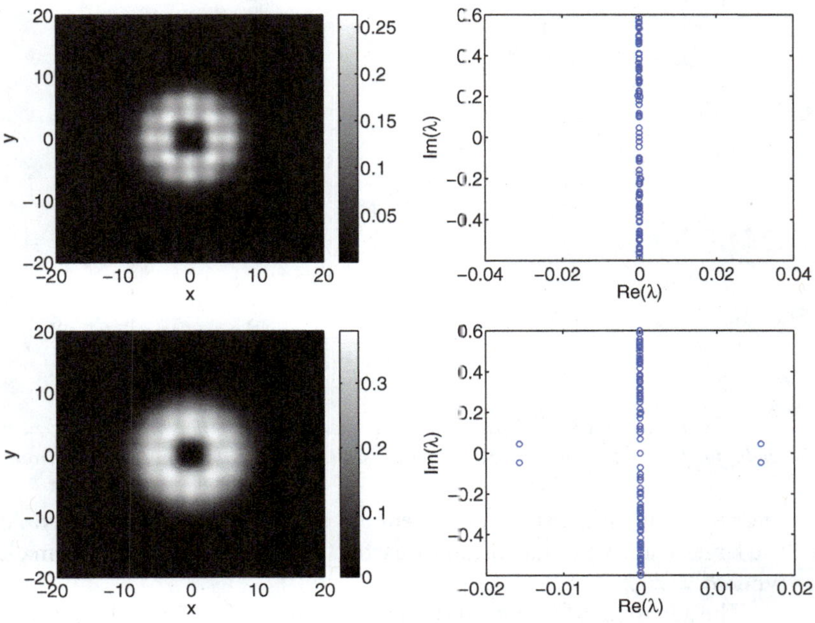

Figure 3.61. *Stable (top,* (0.25, 0.71)*) and unstable (bottom,* (0.13, 0.71)*) double charge vortices (both their profiles and their corresponding spectral planes of eigenvalues) with $\phi = \pi/2$. Adapted with permission from Ref. [578].*

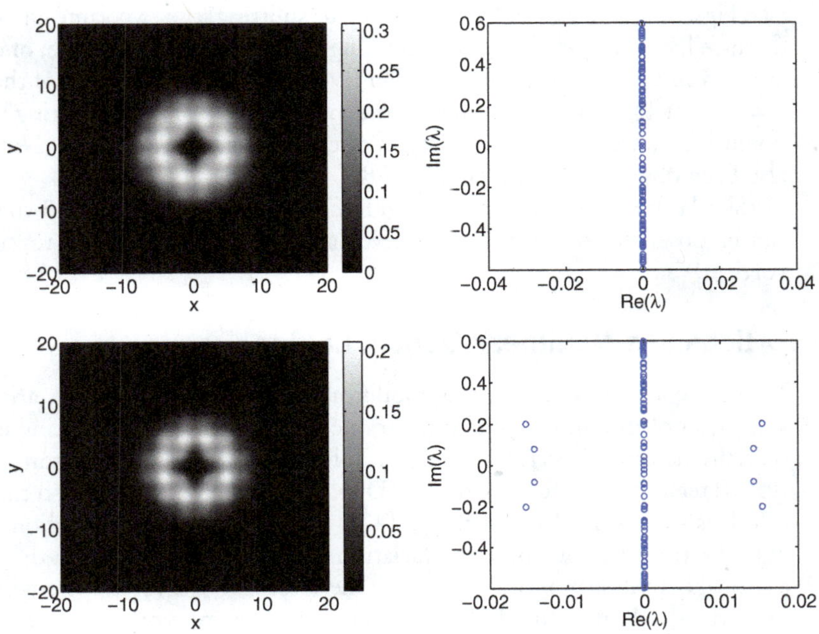

Figure 3.62. *Similarly to Fig. 3.61, stable (top,* (0.21, 0.71)*) and unstable (bottom,* (0.3, 0.71)*) double charge vortices but now for $\phi = 0$. Adapted with permission from Ref. [578].*

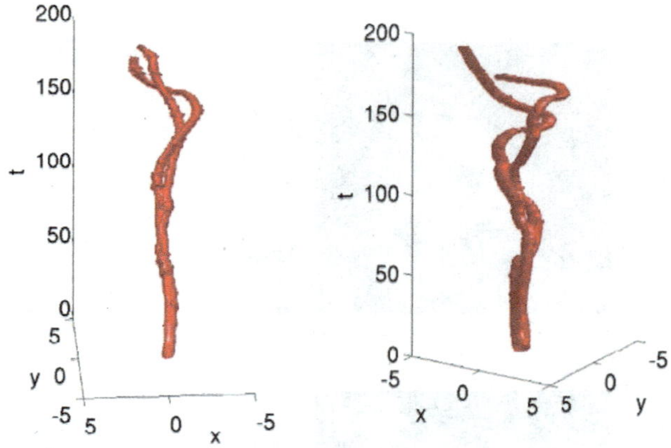

Figure 3.63. *The space-time dynamics of the unstable mode when $\phi = \pi/2$ (left) and $\phi = 0$ (right) for parameter values* (0.3, 0.71) *is illustrated. Adapted with permission from Ref. [578].*

in order to investigate this fundamentally different type of instability. As we expect from the larger real part of its unstable quartet, $t^{**} = 152$ when $\phi = 0$ is smaller than $t^{**} = 216$ when $\phi = \pi/2$.

The apparent difference of the above result for $S = 2$ and the case of the $S = 1$ vortex, where the potential with $\phi = 0$ induced stability, while that of $\phi = \pi/2$ could potentially lead to instability. However, an important clarification should be made here. The instability of the vortices with $S = 1$ and that of the vortices with $S = 2$ are *fundamentally* different. In particular, the former concerns the spiraling of the vortices away from the center of the trap (see Figs. 3.57 and 3.58). On the other hand, the latter predominantly involves the effect of *splitting* of the vortex of charge 2 into two vortices of charge $S = 1$ (see Figs. 3.63 and 3.64). As a result of this splitting, the two same-charge vortices (as is discussed, e.g., in Ref. [77]), start rotating around each other (while, of course, spiraling away from the center of the trap). This type of effect (the breakup of the $S = 2$ into two $S = 1$ vortices) may even happen in the magnetic trap alone, rendering vortices of $S = 2$ potentially unstable in that setting as discussed, e.g., in Ref. [488]. Notice that this has also been observed experimentally in Ref. [501].

Similar findings were identified in Ref. [578] for the case of $S = 3$ vortices. The latter can be observed to also naturally develop the splitting instability towards three $S = 1$ vortices.

3.8.4 ▪ Vortices in a Nonlinear Optical Lattice

We now consider an interesting modification to the above picture, arising from a consideration of inhomogeneous interatomic interactions. Such a setting is described by a spatially dependent scattering length, in line with the Feshbach-resonance management (FRM) technique discussed—in the 1D setting—in Sec. 2.5.3 (see also the special volume [583], as well as the detailed review [429] for more extensive discussions). Here, we will consider the effect of a periodic variation of the nonlinearity strength, $g \equiv g(x,y)$ (i.e., a sort of a nonlinear optical lattice) on the spectrum of a vortex. We will also attempt to draw the similarities and differences to the above picture of a vortex in a linear periodic potential. Our discussion below will follow the exposition of Ref. [503].

In Fig. 3.65, we study the case of a cosinusoidal variation of the nonlinearity strength, namely,

$$g(x,y) = 1 + s\left(\cos^2(\pi x/4) + \cos^2(\pi y/4)\right).$$

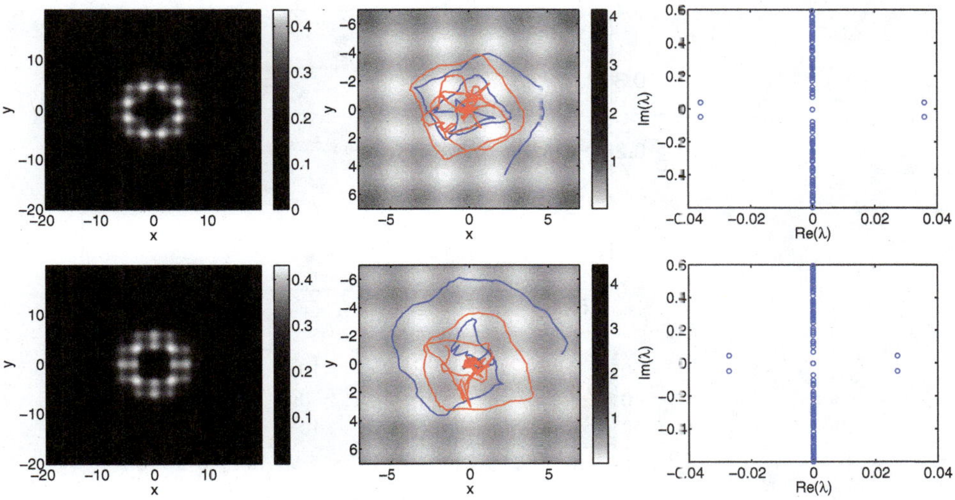

Figure 3.64. *The top row contains the profile (left) spectrum (right) and temporal evolution for 450 time units of the charge 2 vortex with* $\phi = 0$*, where one can visualize the splitting into two charge 1 vortices in blue and red. The bottom row is the same for the case* $\phi = \pi/2$*. Parameter values are equal in both cases,* $(0.25, 0.8)$*. Adapted with permission from Ref. [578].*

We monitor the vortex spectrum as a function of the chemical potential, where s is the strength of the oscillation. In the same figure, the typical form of the density and phase of the wavefunction (the former showcasing the spatial variation dictated by the corresponding variation of the scattering length, and the latter demonstrating the vortex structure of the configuration), as well as the Bogoliubov excitation spectrum, are also illustrated. Most of the relevant eigenfrequencies are only very weakly affected by the spatially dependent nonlinearity. Yet, the one which is most dramatically affected is, in fact, the anomalous mode. The comparison of the blue solid line with $s = 0$ case with the red dashed line of $s = 0.1$ and the green dashed-dotted line of $s = 0.3$ illustrates that the latter two not only approach zero (as $s = 0$ does asymptotically), but rather cross it at a finite value of μ. For $s = 0.1$, the anomalous mode hits the origin of the spectral plane at $\mu = 2.61$, while for $s = 0.3$ at $\mu = 1.56$. However, it is perhaps even more remarkable that this collision does not produce an instability through an imaginary eigenfrequency (real eigenvalue) pair, but rather maintains the stability of the configuration (the eigenfrequencies appear to go through each other). Generally, it can be seen that the trend of increasing the oscillation strength in the cosinusoidal case leads to a more rapid decrease of the anomalous mode eigenfrequency with μ and an "earlier" collision (i.e., occurring for smaller μ) with the spectral plane origin. In these considerations, we focus on a periodicity (wavelength) of the nonlinearity that is larger than the size (core) of the vortex. Throughout this long-wavelength regime the spectral results are qualitatively the same. The case pertaining to wavelengths of the nonlinearity comparable or smaller than the size of the vortex is certainly a worthwhile theme of study that should be considered separately. Nonetheless, it can be anticipated that for small enough wavelengths compared to the core of the vortex, the spatial modulation of the nonlinearity will effectively, through spatial homogenization, act as a constant nonlinearity (possibly shifted from its original $g = 1$ value) in a manner akin to the effects of (linear) periodic potentials generated by optical lattices acting on harmonically trapped dark solitons [208].

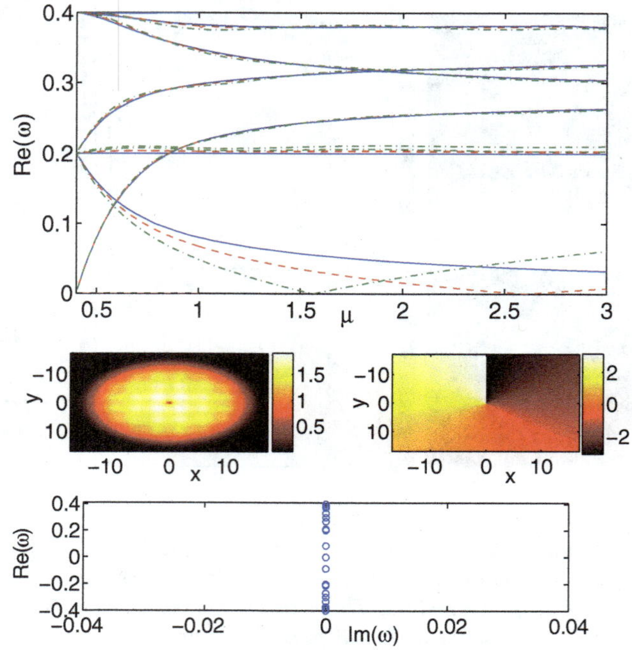

Figure 3.65. *The case of a cosinusoidal variation of the nonlinearity strength for a trap strength $\Omega = 0.2$. The top panel shows the eigenfrequency ω of the Bogoliubov spectrum as a function of the chemical potential μ. Comparison of $g(x,y) = 1 + s\left(\cos^2(\pi x/4) + \cos^2(\pi y/4)\right)$, with $s = 0.1$ (red dashed line) and $s = 0.3$ (green dashed-dotted line), with the case of $s = 0$ (blue solid line). The middle panels show contour plots of the density (left) and phase (right) of the wavefunction, while the bottom panel shows the respective Bogoliubov excitation spectrum (real vs. imaginary parts of the eigenfrequency ω, where instability would correspond to the existence of eigenfrequencies with $Im(\omega) \neq 0$). The chemical potential is $\mu = 4$ (approaching the Thomas–Fermi limit). Adapted with permission from Ref. [503].*

Extending the above considerations, we also examine a sinusoidal modulation of the nonlinearity strength, namely,

$$g(x,y) = 1 + s\left(\sin^2(\pi x/4) + \sin^2(\pi y/4)\right).$$

In this case, as observed in Fig. 3.66, the fundamental difference is that the anomalous mode is larger than that of the homogeneous interactions case ($g = 1$). More importantly perhaps, its dependence can also be nonmonotonic, resulting in the increase of the corresponding eigenfrequency for a chemical potential $\mu \gtrsim 1$. This raises the possibility of collision of the relevant eigenmode with other modes bifurcating from $\omega = \Omega$ for sufficiently large μ (see [green] cross for $\mu \approx 2.11$ in the case of $s = 0.3$ considered in the figure). This, in turn, produces an instability due to the opposite signature of the colliding modes, yielding a quartet of complex eigenfrequencies. The (positive) imaginary part of the quartet, is shown in the bottom panel of Fig. 3.66; see also the second and third rows of panels for a typical profile and spectral plane of the relevant configuration.

The above features of the anomalous mode seem structurally similar to the linear periodic potential case, where again the cosinusoidal case was found to be dynamically stable, while the sinusoidal one was unstable beyond a critical lattice strength [578]. More specifically, independently of the strength of the amplitude parameter of the cosinusoidal lattice (s in the above nonlinear case and V_0 in the linear case of Ref. [578]) it can be seen that no eigenvalue collisions or quartet bifurcations arise. On the other hand, in the sinusoidal

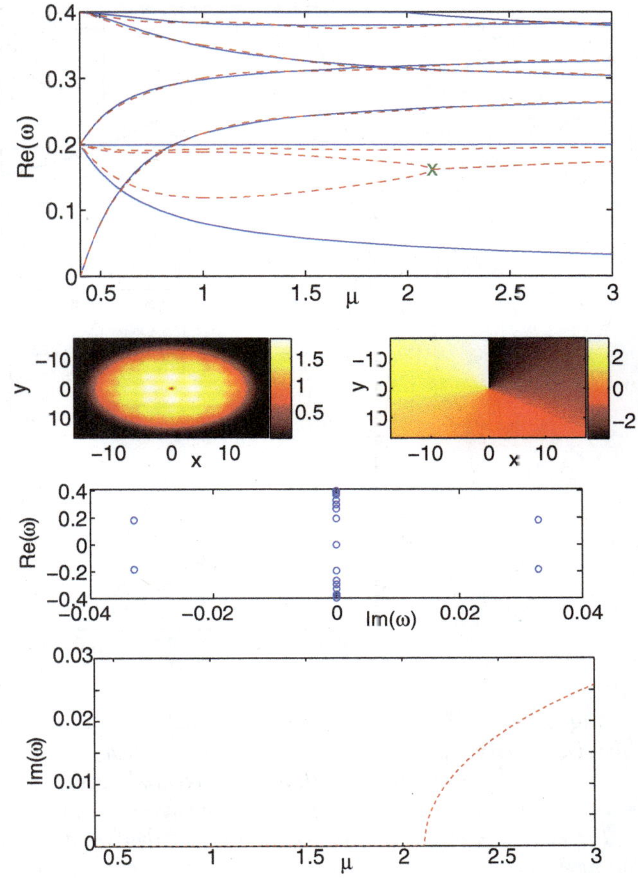

Figure 3.66. *Similar to Fig. 3.65, but for the case of a sinusoidal variation of the nonlinearity strength. The first three rows of panels are analogous to the panels of Fig. 3.65. The case of* $g(x,y) = 1+s\left(\sin^2(\pi x/4)+\sin^2(\pi y/4)\right)$ *for* $s = 0.3$ *(red dashed line) is compared to that of* $s = 0$ *(blue solid line). The bottom panel shows the imaginary part of the complex eigenfrequency; the oscillatory instability arises for* $\mu > 2.11$. *Adapted with permission from Ref. [503].*

case within both the linear and the nonlinear setting such collisions (of anomalous modes with other modes of the eigenfrequency spectrum) arise which, in turn, lead to windows of oscillatory instability.

It is also worthwhile to investigate the effect of a periodic modulation of the nonlinearity on the stability of the system for the finite-temperature case. The relevant model of Eq. (3.14), including phenomenological dissipation, in the presence of a spatially varying scattering length then becomes

$$(i-\gamma)u_t = -\frac{1}{2}\Delta u + V(r)u + g(x,y)|u|^2 u - \mu u. \tag{3.170}$$

For a periodic cosinusoidal modulation of the nonlinearity, the imaginary part of all eigenfrequencies vanishes and the system is stable for $\gamma = 0$ as discussed above. The top panel of Fig. 3.67 shows the maximal imaginary part of the eigenfrequency as a function of γ for different chemical potentials μ for $g(x,y) = 1 + s\left(\cos^2(\pi x/4)+\cos^2(\pi y/4)\right)$, with $s = 0.3$. For $\gamma = 0$ the imaginary part of the eigenfrequency vanishes for all cases.

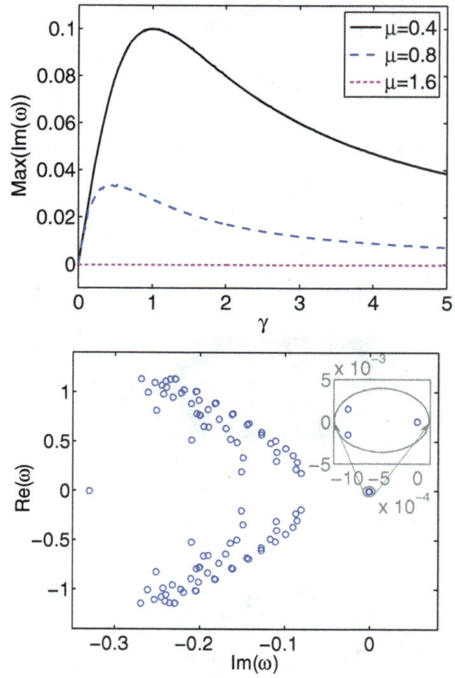

Figure 3.67. *The top panel shows the immediate acquisition of a nonvanishing imaginary part of the eigenfrequency associated with the anomalous mode, as soon as the temperature-dependent dissipative prefactor* γ *is nonzero for small chemical potentials, but no acquisition of an imaginary part for large chemical potentials. The bottom panel gives an overview of the eigenfrequencies for* $\gamma = 0.2$ *and* $\mu = 1.6$*. The maximum imaginary part is zero. The inset shows that the eigenfrequencies corresponding to the (formerly) anomalous mode were shifted into the negative imaginary half-plane. Adapted with permission from Ref. [503].*

However, for small μ the imaginary part of the eigenfrequency becomes nonzero immediately, similar to the case of a constant nonlinearity strength. On the other hand, for large chemical potential the maximal imaginary part of the eigenfrequencies remains zero *independent* of γ. Thus, the system remains stable even in the presence of dissipation.

This behavior can be understood by a closer inspection of Fig. 3.65. The occurrence of a positive imaginary part of the eigenfrequencies is due to the fact that the anomalous mode is of negative Krein signature. However, for the case of a cosinusoidal modulation of the nonlinearity, one observes that the value of the frequency of the anomalous mode decreases with increasing μ and, finally, even crosses the origin. At that critical point, the frequency curve shows a crossover with its opposite-value companion (of the same pair). The latter mode, however, has a positive Krein sign and, therefore (since all negative signatures arise for negative frequencies, and positive signatures for positive frequencies), the imaginary parts of the eigenfrequencies become negative in the case of nonzero γ. The bottom panel in Fig. 3.67 provides an overview of the eigenfrequencies for $\gamma = 0.2$ and $\mu = 1.6$. All eigenfrequencies were shifted further into the negative imaginary half-plane. Importantly, the eigenfrequencies corresponding to the (formerly) anomalous mode were shifted into the negative imaginary half-plane as is shown in the inset, given that the effect of the nonlinear optical lattice has, indeed, changed the nature of the mode.

We now turn to some prototypical results of the dynamical evolution of the vortices in the presence of a spatially dependent nonlinearity (without and with the effect of dissipa-

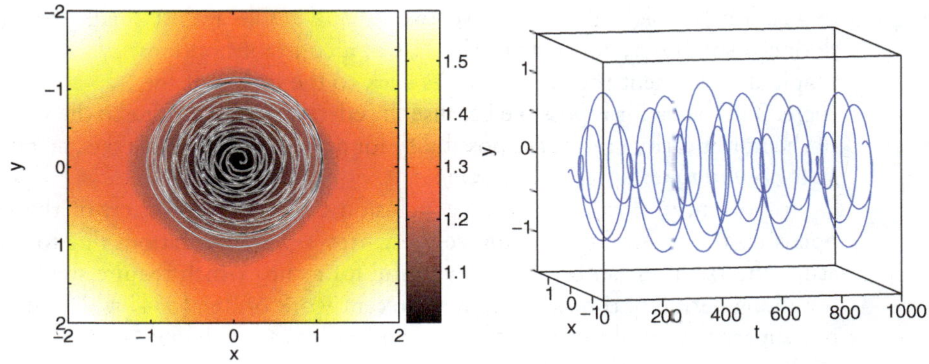

Figure 3.68. *The trajectory of a vortex for* $\mu = 3$ *for the case* $g(x,y) = 1 + s\left(\sin^2(\pi x/4) + \sin^2(\pi y/4)\right)$ *with* $s = 0.3$. *In the left panel the trajectory is plotted on top of the profile of the coupling* $g(x,y)$, *whereas the right panel shows the trajectory as a function of time. Adapted with permission from Ref. [503].*

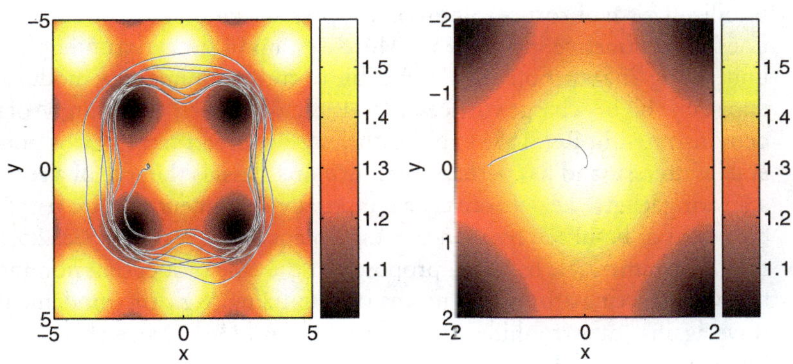

Figure 3.69. *The trajectory of the vortex for* $\mu = 3$ *for the case* $g(x,y) = 1 + s\left(\cos^2(\pi x/4) + \cos^2(\pi y/4)\right)$ *with* $s = 0.3$ *without dissipation (left) and with dissipation (right) with* $\gamma = 0.2$. *The trajectories are plotted on top of the profile of the coupling* $g(x,y)$. *Notice the fast inward vortex motion in the presence of dissipation. Adapted with permission from Ref. [503].*

tion). Figure 3.68 shows the trajectory of a vortex for the case of a periodically modulated sinusoidal nonlinearity, $g(x,y) = 1 + s\left(\sin^2(\pi x/4) + \sin^2(\pi y/4)\right)$. The initial configuration is a slightly perturbed eigenstate leading to a small shift of the position of the vortex. Due to the instability of the sinusoidal $g(x,y)$ landscape, the vortex spirals outwards initially, but then spirals inwards after reaching a region with approximately constant nonlinearity. Subsequently the vortex follows a series of such alternating (spiraling first outwards, and then inwards) cycles in an apparently quasiperiodic orbit.

Figure 3.69 shows the trajectory of a vortex for the case of a periodically modulated cosinusoidal nonlinearity, $g(x,y) = 1 + s\left(\cos^2(\pi x/4) + \cos^2(\pi y/4)\right)$, without dissipation (left panel) and with dissipation (right panel). In this case, small perturbations do not get amplified since the system is stable. However, a macroscopic displacement of the vortex to $(x_0, y_0) = (-1.5, 0)$ leads to the trajectories shown in the figure (see left panel). In the case without dissipation the vortex moves outwards (reaching a region outside the "square" of the first minima of the nonlinearity) and oscillates around the center on a trajectory with roughly constant nonlinearity. For the case with cosinusoidal nonlinearity *and* dissipation (see right panel), the vortex remains stable against small perturbations and does not spiral out, contrary to the case of a constant nonlinearity. Even for a macroscopic displacement the vortex moves back to the center of the trap and remains stable there. This behavior is possible because the effective potential due to the spatial variation of the nonlinearity creates the possibility for a dynamically stable (i.e., attracting) vortex state *even in the presence of dissipation.*

Importantly, this leads to the conclusion that the modulation of the nonlinearity opens up the possibility to stabilize the vortex against excitations due to finite temperature effects. This can be extremely useful for setups which require stable vortex states for a long period of time as, e.g., in the recent work of Ref. [584] which suggests the use of a superposition of two counter-rotating BECs as a gyroscope.

Before we turn to a different theme (namely the role of impurities in 2D BECs), we would like to add some further remarks regarding this realm of modulated scattering lengths. In particular, in the above we have considered only spatial modulations of the scattering length. This theme has been receiving increasing amounts of attention not only in the atomic but also in the optical physics community. A recent example of this mounting interest is the review of Ref. [429], which covers themes such as the shaping and stabilized form of vortex solitons in the presence of optical lattices of spatially modulated nonlinear refractive index [585]. Moreover, upon suitably tuning the potential (and the underlying linear problem), it is possible to create nonlinearity modulations that may sustain stable higher charge vortices, as well as radial states. An example of this form has been demonstrated in Ref. [586]. Nevertheless, a related theme that is potentially of considerable interest (and has perhaps been explored less) is the role of a temporally dependent scattering length, i.e., effectively a temporally dependent nonlinearity, as discussed under the realm of Feshbach resonance management in the realm of Sec. 2.5.3. It is well known that this framework has been proposed in the case of focusing nonlinearities and attractively interacting atoms as a means of stabilizing, e.g., the fundamental bright solitary wave against catastrophic collapse instabilities [436]. Nevertheless, considerably less is known about its role in the dynamics of vortices in repulsively interacting atomic BECs. This may be an interesting theme for further study along the lines of the kinematics and dynamics of vortices and 2D BECs that we have described above.

3.9 ▪ The Role of Impurities in Two-Dimensional BECs: Some Case Examples

We will now turn to a partial examination of the role of other forms of the external potential, including a stripe-type quasi-1D "impurity" potential (and more complex multistripe generalizations thereof in some cases—see below), as well as localized in both directions blob-type impurity potentials and their interaction with coherent structures such as vortices.

3.9.1 ▪ Stripe-Type Potential Barrier and Its Role in Controlling Transverse Instabilities

We start by considering the standard 2D GPE with an external potential of the form

$$V(x,y) = V_{\rm MT}(x,y) + V_{\rm LS}(x), \tag{3.171}$$

that, in addition to the standard magnetic (harmonic) trap

$$V_{\rm MT}(x,y) = \frac{1}{2}\Omega^2\left(x^2+y^2\right), \tag{3.172}$$

includes the potential barrier (corresponding to a far-detuned laser sheet illuminating the BEC) with the quasi-1D Gaussian profile

$$V_{LS}(x) = A \exp\left[-x^2/\left(2\sigma^2\right)\right]. \tag{3.173}$$

Here A and σ measure the height and width of the barrier, with $A > 0$ ($A < 0$) corresponding to the blue- (red-) detuned laser beam, which repels (attracts) atoms in the condensate (note that $A < 0$ corresponds to a potential trough of depth $|A|$, rather than a barrier). In exploring the role of this barrier within the condensate, we will follow the analysis of Ref. [526].

It is worth noticing that the model under consideration can also be deemed as having an interpretation in the context of nonlinear optics: there, it governs the evolution of the local amplitude of the electromagnetic wave in the self-defocusing bulk waveguide, with t replaced by the propagation distance z, while x, y are the transverse coordinates, and $V(x,y)$ describes a local modulation of the refractive index. The barrier potential then corresponds to a slab with a lower value of the refractive index embedded into the waveguide, while the potential (3.172) approximates the global waveguiding structure.

When $V(x,y) = 0$, the 2D GPE admits an exact analytical solution for the dark soliton stripe (DSS). Assuming that its nodal plane is oriented along the x direction, to be aligned with the potential barrier, when it is switched on, the DSS solution is

$$u(x,y,t) = \sqrt{\mu}\{\mathcal{B}\tanh[\sqrt{\mu}\mathcal{B}(x - vt)] + i\mathcal{A}\}e^{-i\mu t}. \tag{3.174}$$

Here, as in Chapter 2, $\sqrt{\mu}\mathcal{B}$ and $v = \sqrt{\mu}\mathcal{A}$ represent, respectively, the soliton's depth (or inverse width) and velocity, with $\mathcal{A}$ and $\mathcal{B}$ subject to the constraint $\mathcal{A}^2 + \mathcal{B}^2 = 1$. The DSS with $\mathcal{A} = 0$ ($\mathcal{A} \neq 0$) is stationary (moving), and is called a "black" ("gray") dark soliton. Below, we first employ the variational approximation to find an approximate stationary DSS solution to the 2D GPE in the presence of the potential barrier (3.173), and then perform the linear-stability analysis to study the transverse modulational instability (MI).

Dark-Soliton Stripe Steady State in the Presence of the Potential Barrier

In the absence of the harmonic trapping potential (3.172), but in the presence of barrier (3.173), the profile of a stationary DSS solution is sought for as

$$u(x,y,t) = U(x)e^{-i\mu t}, \tag{3.175}$$

where μ is the chemical potential (or $-\mu$ is the propagation constant, in terms of the optical model), and $U(x)$ is a real function. The resulting ordinary differential equation for u reads

$$\frac{1}{2}U'' + \left[\mu - Ae^{-x^2/(2\sigma^2)}\right]U - U^3 = 0. \tag{3.176}$$

As Eq. (3.176) describes both the dark soliton and the uniform background, one should subtract the contribution of the latter into the respective Lagrangian, which will be used as the basis of our variational approximation. To this end, we follow Ref. [26] (see also Ref. [184]), and write the Lagrangian as follows:

$$L = (U')^2 - 2\left[\mu - Ae^{-x^2/(2\sigma^2)}\right](U^2 - \mu) + (U^4 - \mu^2). \tag{3.177}$$

Then, to approximate the stationary dark-soliton solution of Eq. (3.176), we use the following ansatz [cf. Eq. (3.27)]:

$$U(x;A,\sigma) = \sqrt{\mu}\tanh(\beta(A,\sigma)x), \tag{3.178}$$

where the inverse width of the soliton, β, is a variational parameter, while the density μ is the fixed background chemical potential. Inserting (3.178) into the Lagrangian (3.177) and performing the integration over $-\infty < x < +\infty$, we arrive at the effective (averaged) Lagrangian

$$L_{\text{eff}} = \frac{4}{3}\mu\left(\beta + \frac{\mu}{\beta}\right) - 2A\mu\int_{-\infty}^{+\infty} e^{-x^2/(2\sigma^2)}\,\text{sech}^2(\beta x)dx. \tag{3.179}$$

Further, we use the Euler–Lagrange equation, $\partial L_{\text{eff}}/\partial\beta = 0$, to derive the following implicit equation for the inverse soliton's width β:

$$\beta^2 = \mu - 3A\beta^2 J(\beta,\sigma), \tag{3.180}$$

where

$$J(\beta,\sigma) \equiv \int_{-\infty}^{+\infty} x e^{-x^2/(2\sigma^2)}\,\text{sech}^2(\beta x)\tanh(\beta x)dx. \tag{3.181}$$

Using the rescaling $x \to \beta x$ we conclude that $J(\beta,\sigma) = J(1,\overline{\sigma})/\beta^2$, where $\overline{\sigma} \equiv \beta\sigma$, hence one can rewrite Eq. (3.180) as

$$\beta^2 = \mu - 3AJ(1,\beta\sigma). \tag{3.182}$$

Apparently, Eq. (3.182), with J being a function of the single argument, is simpler than Eq. (3.180).

In Fig. 3.70 we corroborate the results of the variational approximation by a direct comparison of the width extracted from the numerically found steady state, and from the variational equation (3.182) [thick (blue) and thin (red) lines, respectively], for different values of the barrier's parameters. As seen in the figure, for $A > 0$ (i.e., for the repulsive potential barrier), the variational approximation is very accurate, with the steady-state solution closely resembling the tanh profile of ansatz (3.178) [see panel (b) in the figure]. On the other hand, as seen in panel (c), for $A < 0$ (the attractive potential trough, instead of the barrier) the approximation deteriorates, especially for large widths of the trough ($\sigma > 0.6$, which exceeds the width of the DSS). This observation is explained by the fact that the steady-state profile develops a localized hump induced by the attractive trough, which is not captured by the proposed ansatz.

Transverse Modulational Instability of the Dark-Soliton Stripe in the Presence of the Barrier

In the case where the potential barrier is present, the steady-state (black) DSS may be approximated by

$$u(x,y,t) = u_0(x,y)e^{-i\mu t} = \sqrt{\mu}\tanh(\beta x)e^{-i\mu t}, \tag{3.183}$$

where β is determined by Eq. (3.182). To analyze the MI of this DSS, we follow the approach of Ref. [253] and consider small transverse perturbations. The result of the analysis [526] is that the critical wavenumber which defines the width of the instability band for solution (3.175) is

$$k_{\text{cr}} = \beta, \tag{3.184}$$

where β is determined by Eq. (3.182). That is, the DSS (3.183) is unstable against transverse perturbations with wavenumbers $k < k_{\text{cr}} = \beta$. When this instability sets in, the DSS undergoes a snake-like transverse deformation, which eventually breaks it into vortex-antivortex pairs (see discussion in Sec. 2.3.2). This outcome of the evolution can be observed in Fig. 3.71.

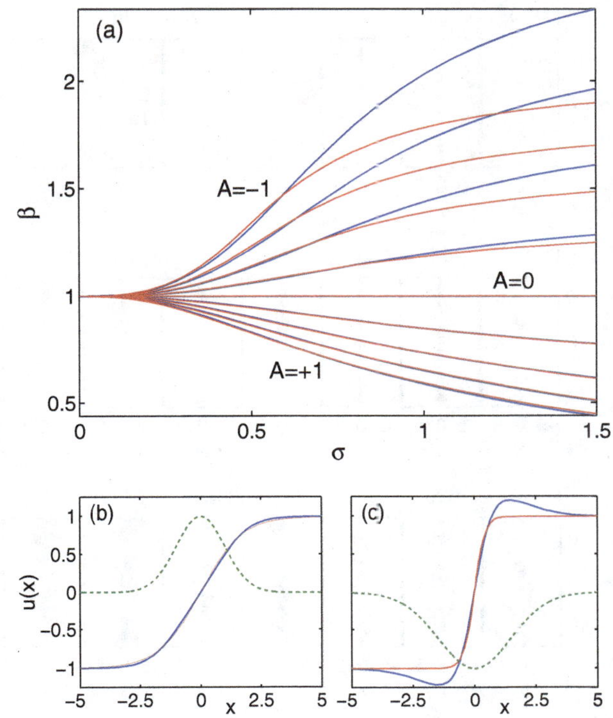

Figure 3.70. (a) *Inverse width β of the dark soliton as a function of width σ of the potential barrier, with height A ranging from $A=-1$ (top curves) to $A=+1$ (bottom curves) in steps of* 0.25. *The thick (blue) curves depict the width of the numerical solution, obtained by fitting its profile to* $\tanh(\beta x)$. *The thin (red) lines represent the results predicted by the variational approximation; see Eq. (3.182). Panels* (b) *and* (c) *depict, respectively, two solutions for $(A,\sigma)=(1,1)$ and $(A,\sigma)=(-1,1.5)$, the latter case corresponding to the potential trough, rather than a barrier. The thick (blue) line depicts the numerically found steady state, while the thin (red) line represents the fitted* $\tanh(\beta x)$ *profile. The transverse profile of the potential barrier is depicted by the dashed (green) line. The chemical potential is fixed to be $\mu=1$. Adapted with permission from Ref. [526].*

From Eq. (3.184) we can derive the critical number of vortices in the chain produced by the splitting of the DSS (3.183):

$$N_{\rm cr} = \beta L/\pi, \tag{3.185}$$

where L is the length of the DSS. Since β is implicitly determined by height A and width σ of the potential barrier (3.173), as per Eq. (3.182), it is possible to control the critical wavenumber by appropriately adjusting the barrier's parameters. This, in turn, offers a control of the upper bound of the vortex pairs that are nucleated by the transverse instability of the DSS. Below, it will be shown that it is indeed possible to control the number of the nucleated vortex pairs.

Numerical Results

We first examine the results in the absence of the harmonic trap. Typical examples of the snaking instability are depicted in Fig. 3.71. The figure depicts the evolution of the density $|u(x,y,t)|^2$ at different times, starting from a stationary DSS with a small initial random perturbation added to it. The middle panel in the figure depicts the case of a

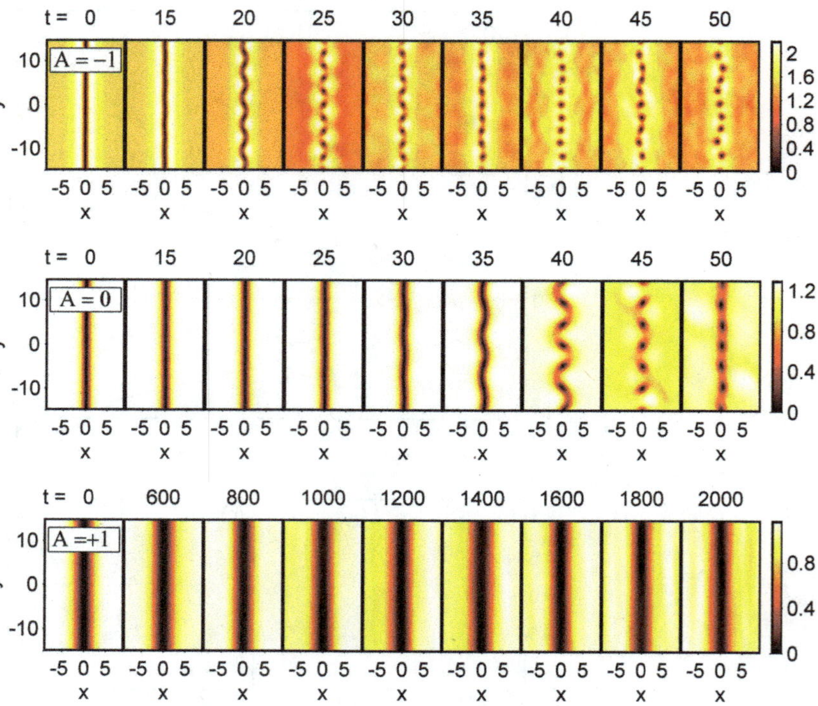

Figure 3.71. *The dynamics induced by the transverse modulational instability of a dark-soliton stripe in the absence of the external harmonic trap. Shown are different density snapshots at indicated moments of time. The initial condition corresponds to a stationary stripe with an added random perturbation of relative size* 10^{-5}. *The top panel: The case of the attractive potential trough, with* $A = -1$ *and* $\sigma = 1.5$. *The middle panel: No potential barrier or trough (*$A = 0$*). The bottom panel: The case of the repulsive potential barrier, with* $A = 1$ *and* $\sigma = 1.5$. *Note the faster destabilization and nucleation of a larger number of vortices (see dark circular spots for later times) in the case of the attractive potential trough (the top panel), in comparison to the case when the trough is absent (the middle panel), and the complete stabilization of the dark-soliton stripe by the repulsive potential barrier in the bottom panel. The simulations in this figure were carried out in the domain of* $(x,y) \in [-15,15] \times [-15,15]$. *Adapted with permission from Ref. [526].*

quiescent (black) DSS ($\beta = 1$), with no potential barrier $A = 0$, where the instability manifests itself with wavenumber $k \approx 2\pi/10 = 0.628$ (wavelength $\simeq 10$), which is smaller than the predicted critical wavenumber, $k_{\rm cr} = \beta = 1$. In fact, this unstable wavenumber is the most unstable one, $k_{\rm max}$, for this configuration.

In order to quantify the instability and to compare it to the analytical results of the previous section, we compute the stability spectra of the stationary solutions $u_0(x,y)$ through the standard BdG analysis for the eigenfrequencies ω, connected to the eigenvalues through $\omega = i\lambda$.

The MI spectrum for the DSS in the case of no potential barrier ($A = 0$) is depicted by the (green) squares in all panels in Fig. 3.72. The most unstable wavenumber, $k_{\rm max}$, corresponds to the location of the maxima of $\mathrm{Re}(\lambda) > 0$. Note that, as predicted by the linear stability analysis, all modes with $k > k_{\rm cr} = \beta = 1$ are stable. It is also interesting to note that for strong enough attractive ($A < 0$) potential barriers the whole spectrum is shifted to the right, so that small wavenumbers become stable [see (blue) circles in Fig. 3.72]. Furthermore, it is important to note that the results presented in Fig. 3.72 were obtained in a

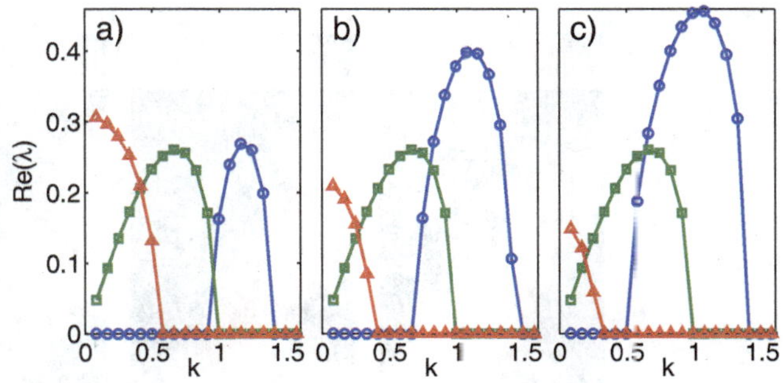

Figure 3.72. *The growth rate of the transverse modulational instability of dark-soliton stripes at different values of the height and width of the potential barrier, A and σ. The panels correspond to the increasing width:* (a) $\sigma = 0.5$, (b) $\sigma = 1$, *and* (c) $\sigma = 1.5$, *as indicated by the dashed lines in Fig. 3.73. Each panel displays the real part of the eigenvalue, for the potential trough/barrier with heights* $A = -1$ *(blue circles),* $A = 0$ *(green squares), and* $A = 1$ *(red triangles). The computations corresponding to this figure were carried out in the domain of* $(x, y) \in [-50, 50] \times [-50, 50]$. *Adapted with permission from Ref. [526].*

large box, $(x, y) \in [-50, 50] \times [-50, 50]$, in order to safely capture small wavenumbers. In contrast, the simulations depicted in Fig. 3.71 were performed in a smaller domain box, $(x, y) \in [-15, 15] \times [-15, 15]$ so that the case for $A = 1$ (the bottom row in the figure) is rendered stable since the corresponding unstable wavenumbers do not fit into the integration box. This effect, discussed further below, becomes important when considering trapped condensates that inherently possess a finite size.

The next step involves examining the effect of the potential barrier on the stability (and dynamics) of the DSS. The top panel in Fig. 3.71 depicts the snaking instability corresponding to the attractive potential trough with strength (depth) $A = -1$. As seen in the figure, the attractive trough naturally enhances the instability, in comparison to the free-standing DSS, in two ways: (i) the MI sets in earlier, and (ii) the most unstable wavenumber is right-shifted ($k_{\max} \approx 2\pi/6 = 1.0472$). The corresponding instability spectrum for this case is depicted by the (blue) circles in Fig. 3.72(c). In fact, comparing the (green) squares and the (blue) dots in this panel, it is evident that the net effect of the attractive potential trough is the right-shift of the instability spectrum. Note that the spectrum is located to the left of the threshold, $k_{\rm cr} = \beta \approx 1.78$, which is predicted by the linear-stability analysis.

In contrast to the destabilizing effects of the attractive potential, the repulsive barrier stabilizes the DSS, as might be expected and is shown in the bottom panel of Fig. 3.71. The spectrum for this case is depicted by the (red) triangles in Fig. 3.72(c). Note that, although some perturbation modes are still unstable in this case, their wavelengths are too large to fit the integration domain used in the simulations shown in the bottom panel of Fig. 3.71 and, thus, the respective sufficiently strongly repulsive barrier renders the DSS effectively stable (in a still larger domain, we do observe the development of the weak transverse MI in this case, which is not shown here). The same stabilizing effect is also observed for other values of the widths of the barrier, as it can be clearly seen in all spectra depicted in Fig. 3.72. This figure depicts, for three values of width σ, the shift of the MI spectrum for the attractive (blue circles), zero (green squares), and repulsive (red triangles) channel potentials. The figure suggests that, as mentioned above, for a sufficiently strong repulsive potential barrier, only very small wavenumbers may be unstable. Therefore, if the longitudinal size of the condensate (set by the external trap) or optical waveguide, into which the DSS is embedded, is not too large, the instability might be fully suppressed, as

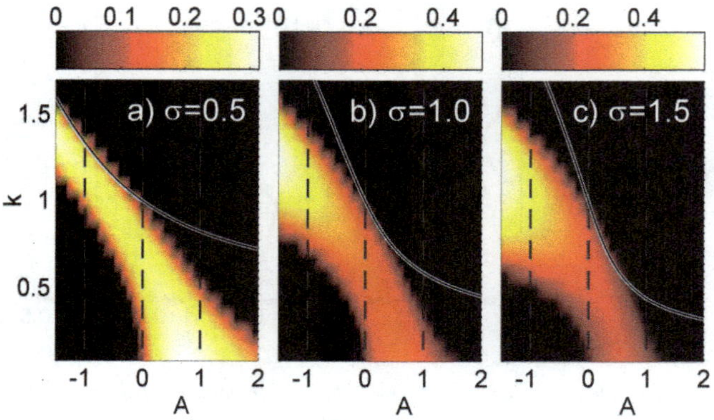

Figure 3.73. *Regions of the transverse modulational instability of the dark-soliton stripe for different parameters of the potential barrier. Depicted is the real part of the eigenvalue (the lighter color represents the stronger instability, while black corresponds to the stability) as a function of wavenumber k and the barrier's height A. Different panels correspond to increasing width of the barrier:* (a) $\sigma = 0.5$, (b) $\sigma = 1$, *and* (c) $\sigma = 1.5$. *The black and white solid lines depict the prediction of the critical wavenumber as per Eq. (3.184). Vertical dashed lines represent the cuts depicted in Fig. 3.72. Adapted with permission from Ref. [526].*

the unstable wavelengths would be too large to comply with the longitudinal size (see also Ref. [517] and the earlier discussion of Sec. 3.3.1).

On the other hand, a sufficiently strong repulsive barrier may completely suppress all instabilities, including those with very small wavenumbers. In Fig. 3.73, we depict the instability spectra for three different widths of the barrier, as a function of the sign and strength of the channel potential. The light-colored areas in the panels correspond to unstable modes, with the shading scale corresponding to the associated instability growth rate for each mode. As shown in the figure, for sufficiently wide potential channels ($\sigma \geq 1.0$, see the two right panels), there is a positive value of the barrier's height above which all wavenumbers are stable (the zero real part of the eigenvalues corresponds to the black shade in the panels). Therefore, it is possible to completely suppress the snaking instability and render the DSS stable with the appropriate choice of the potential barrier.

For example, for the domain used in Figs. 3.72 and 3.73 and a potential barrier width of $\sigma = 1$ ($\sigma = 1.5$) and height larger or equal to $A = 1.65$ ($A = 1.35$) the DSS is completely stable. We also plot in Fig. 3.73 (see solid black and white lines) the critical wavenumber (3.184). This approximate result (valid for $A = 0$) captures the qualitative behavior of the critical wavenumber, but fails to give its precise location.

Another important feature of the control over the location of the most unstable mode $k_{\max}$ is that it allows one to precisely manipulate the number of nucleated vortices per unit length which emerge from the snaking instability. For example, the destabilizing (attractive) channel potential, with $(\sigma, A) = (1.5, -1)$, enhances the outcome by producing $N_v = 11$ vortices in the domain of length $L = 30$, namely the vortex density $\rho_v = 10/30 = 0.33\ldots$. On the other hand, the black DSS ($A = 0$) is responsible for the nucleation of $\rho_v = 6/30 = 0.2$ vortices per unit length. Thus, one can manipulate the density of the nucleated vortices, selecting it from zero to a maximum value which is inversely proportional to the healing length of the condensate (recall that two single-charged vortices cannot coexist at distances smaller than this length).

We now consider the combined effects of the external harmonic trap and the channel potential. As mentioned above, this combination can produce an effective suppression of the snaking instability if the trap's strength is such that the corresponding TF radius of the

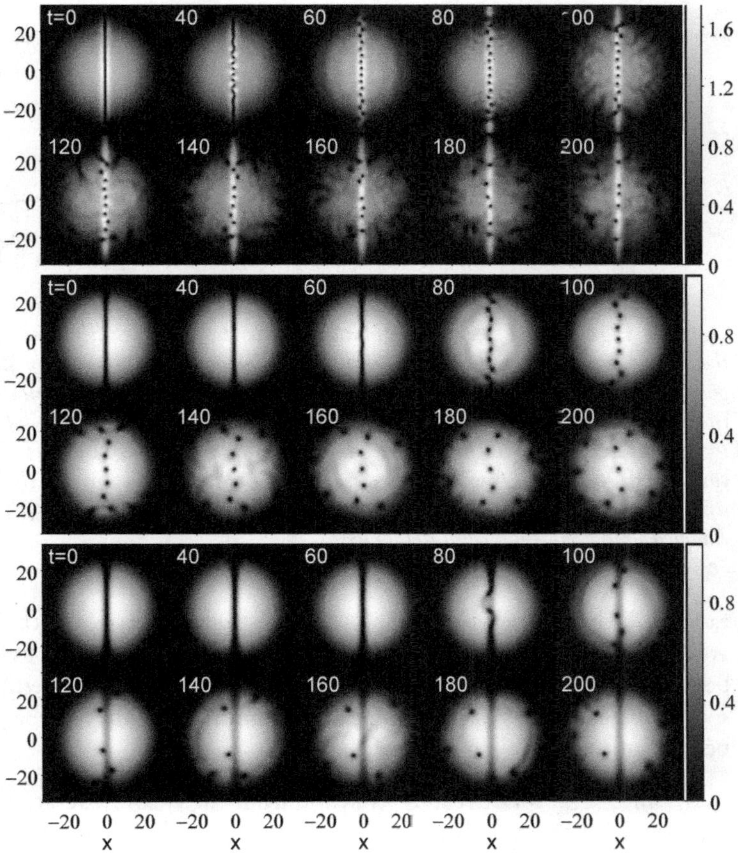

Figure 3.74. *Dynamics of the transverse modulational instability of a dark-soliton stripe in the BEC loaded into the harmonic trap of strength* $\Omega = 0.05$. *Shown are different snapshots of the density at the times indicated in the figure. The initial condition corresponds to a stationary stripe to which a random perturbation of relative size* 10^{-5} *was added. The top panel shows the case with the attractive potential trough of depth* $-A = 0.8$ *and width* $\sigma = 1.5$. *The middle panel: the situation in the absence of the channel potential (*$A = 0$*). The bottom panel: the case of the repulsive potential barrier of height* $A = 0.4$ *and width* $\sigma = 1.5$. *Note that the number of nucleated vortices can be controllably varied from about 18 for* $A = -0.8$ *(in the top panel) to none for* $A \gtrsim 1$ *(not shown here). Adapted with permission from Ref. [526].*

BEC, $R_{\mathrm{TF}} = \sqrt{2\mu}/\Omega$, is smaller than the smallest wavelength of unstable perturbations. In Fig. 3.74 the development of the DSS' snaking instability and the concomitant vortex nucleation are displayed for the BEC confined in the harmonic trap of strength $\Omega = 0.05$. The three panels correspond, from top to bottom, to channel potentials with width $\sigma = 1.5$ and strength $A = -0.8$, 0, and 0.4. It is difficult to precisely measure the number of nucleated vortices because some of them are nucleated at the rims of the configuration and are almost invisible. Therefore, we herein focus on measuring the number of nucleated vortices, N_v, inside the circle of the TF radius. It is clear that, for $A = -0.8$, at least $N_v = 16$ vortices are nucleated, while $N_v = 10$ for $A = 0$ (without the channel potential), and only $N_v = 5$ vortices emerge for $A = 0.8$.

Figure 3.75 shows the number of nucleated vortices for two different values of the trapping frequency. The curve corresponding to $\Omega = 0.05$ has been scaled by factor 1/2 for a better comparison with the one corresponding to $\Omega = 0.1$, since the former has a TF

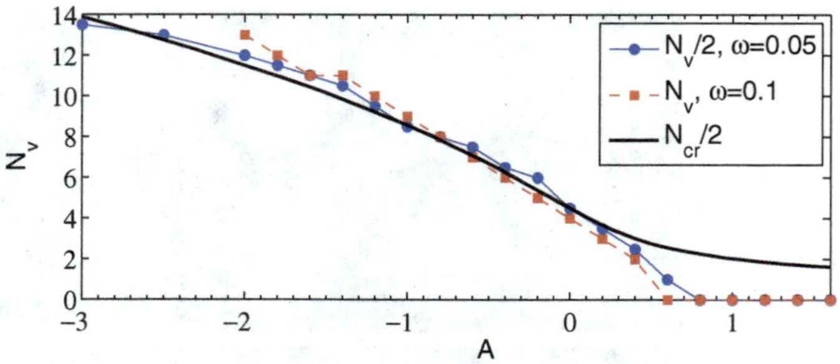

Figure 3.75. *The number of nucleated vortices (N_v) resulting from the snaking instability of the dark-soliton stripe in the BEC loaded into the harmonic trap, versus the strength (A) of the channel potential. The number of vortices is measured by direct count of the number of empty cores present in the density profile after the snaking instability is fully developed. The solid line depicts half of the theoretical prediction of Eq. (3.186) [i.e., the estimation given in Eq. (3.187)]. Circles: The trap's strength corresponds to $\Omega = 0.05$ and depicted is the half number of the vortices ($N_v/2$). Squares: The trap's strength corresponds to $\Omega = 0.1$. The width of the potential channel is $\sigma = 1.5$ for both data sets. Adapted with permission from Ref. [526].*

radius which is twice as large. In order to estimate analytically the number of nucleated vortices inside the BEC cloud, we first establish an upper bound by considering the number of nucleated vortices in a DSS of length $L = 2R_{\text{TF}}$ with the threshold wavenumber $k = k_{\text{cr}}$. Clearly, this estimation is an upper bound for the number of nucleated vortices because of the following reasons:

(i) since $k_{\max}$ is not readily available from the theory, we are using k_{cr} ($k_{\max} < k_{\text{cr}}$), which corresponds to the instability threshold and not to the wavenumber with the maximum growth rate;

(ii) at the periphery of the TF cloud, it is difficult to pick the number of vortices through the direct count of the number of empty cores in the density profiles;

(iii) as it can be noticed from Fig. 3.73, the analytical prediction for k_{cr} is always an overestimation of the true wavenumber threshold.

Thus, an upper bound for the number of vortices can be safely given by

$$N_{\text{cr}} = \frac{\beta L}{\pi} = \frac{\beta 2R_{\text{TF}}}{\pi} = \frac{2\beta\sqrt{2\mu}}{\pi\Omega}. \tag{3.186}$$

In connection with item (iii) above, it is worth mentioning that it is in principle possible to measure the number of vortices by analyzing the vorticity (curl of the superfluid velocity) of the condensate. This method reveals (results not shown here) that, indeed, some vortices created at the periphery of the cloud are missed by the direct empty-core counting method, based on the inspection of the density profiles. Nonetheless, we have observed that these peripheral vortices are quickly pushed out and always stay at the periphery. Therefore, such fringe vortices do not induce a considerable change in the long-term BEC dynamics.

It is also worth pointing out that counting vortices by the number of empty cores in the density profile is tantamount to what is used in current BEC experiments. Of course, in order to image the vortices in real experiments, the BEC cloud has to be left to expand and this could potentially modify the density distribution and "push" vortices inwards or outwards the BEC cloud. However, since the expansion time is usually short, it is likely that any dynamics during expansion will be rather slow and, thus, no change in the measurable number of vortices should be expected. As concluded above, the observable number of vortices nucleated by the DSS, N_v, will be smaller that $N_{\rm cr}$ defined in Eq. (3.186), hence a relation of the type $N_v = \delta N_{\rm cr}$, with $\delta < 1$, should be more appropriate.

Motivated by the observation that the most unstable growth rate $k_{\max}$ is approximately half of the critical threshold $k_{\rm cr}$ when no potential barrier is present [see (green) squares in Fig. 3.72], we choose to use $\delta = 1/2$ as a first approximation. In fact, as it can be noticed from Fig. 3.75—see the solid black line—we have found that the choice of $\delta = 1/2$ gives a good match to the actual number of observed vortices in the numerical experiments for a wide range of values of the potential barrier intensity A.[20] Therefore, with this *a posteriori* estimation for the value of δ, we give the following estimate for the number of nucleated vortices generated by the snaking instability of the DSS confined in a combined harmonic trap and potential barrier:

$$N_v \approx \frac{1}{2} N_{\rm cr} = \frac{\beta\sqrt{2\mu}}{\pi\Omega}, \tag{3.187}$$

where Ω is the harmonic trap strength, and β is given by Eq. (3.182) for the potential barrier (3.171) with width σ and strength A.

3.9.2 ▪ The Case of a Fully Localized 2D "Impurity" Interacting with a Vortex: Capture, Dragging, and Manipulation

We now turn to the examination of the effect of an impurity of the form

$$V_{\rm Imp}(x,y,t) = V_{\rm Imp}^{(0)} \exp\left(-\frac{[x-x_a(t)]^2+[y-y_a(t)]^2}{\varepsilon^2}\right), \tag{3.188}$$

following the work of Ref. [587]. In this equation, and similarly to the previous section, $V_{\rm Imp}^{(0)}$ is proportional to the peak laser intensity divided by the detuning of the laser from the atomic resonance, and $\varepsilon = w_0/\sqrt{2}$ where $2w_0$ is the adimensional Gaussian beam width. A positive (negative) $V_{\rm Imp}^{(0)}$ corresponds to the intensity of a blue- (red-) detuned, repulsive (attractive) potential. The fundamental difference in this case is that the impurity is indeed localized in both spatial directions (rather than only in one as before).

Steady-state solutions of the GPE are obtained through the standard separation of the spatial and temporal dependencies as $u(x,y,t) = \Psi(x,y)\, e^{-i\mu t}$.

The initial condition used in Ref. [587] was one that closely approximates a vortex solution of unit charge $S = \pm 1$ centered at (x_0, y_0):

$$\begin{aligned}\Psi(x,y) = \Psi_{\rm TF}(x,y) \tanh[(x-x_0)^2+(y-y_0)^2] \\ \times \exp\left[iS \tan^{-1}\{(y-y_0)/(x-x_0)\}\right],\end{aligned} \tag{3.189}$$

where $\Psi_{\rm TF}(x,y) = \sqrt{\max(\mu - V(x,y), 0)}$ represents the shape of the TF cloud formed in the presence of the considered external potential. Subsequently, this approximate initial condition was allowed to converge to the numerically "exact" solutions by means of a fixed point iteration.

[20] Incidentally, this choice of δ that takes into account the inhomogeneity of the condensate's background is precisely the same choice of δ to again account for the inhomogeneity when computing the critical threshold for instability of dark soliton stripes presented at the end of in Sec. 3.3.1.

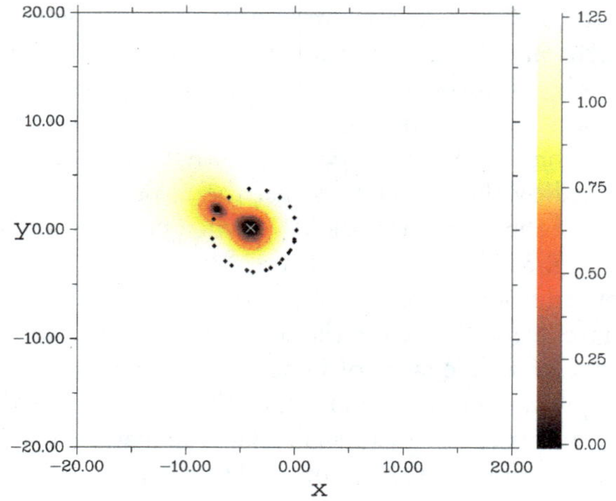

Figure 3.76. *Density plot showing a snapshot of the interaction of the vortex with a localized impurity in the absence of the harmonic trap (i.e., $\Omega = 0$). The presence of the impurity (wider field depression) induces a clockwise rotation of the vortex (narrower field depression) along a path depicted by the dark dots. The parameters are as follows: $(\mu, \Omega, s, V_{\text{Imp}}^{(0)}, \varepsilon) = (1, 0, 1, 5, 1)$. The colorbar shows the condensate density in adimensional units (i.e., in units of $\hbar\omega_z / g_{2D}$). Distances are given in units of the transverse harmonic oscillator length a_z. Adapted with permission from Ref. [587].*

The Static Picture: Vortex Pinning and the Bifurcations Beneath

As discussed earlier in this chapter, a vortex interacting with a harmonic trap undergoes a precession, whose characteristics depend on the healing length and the trap parameters. Since we are introducing a localized impurity into the trap it is worthwhile to first observe the behavior of a vortex interacting with only the localized impurity, in the absence of the harmonic potential. We note that in this case, the parameters x_a and y_a may be neglected, and the parameters x_0 and y_0 can be interpreted as the coordinates of the vortex relative to the impurity.

By symmetry, a vortex placed at the center of an impurity (i.e., $x_0 = y_0 = 0$) will result in a steady state without precessing. However, a vortex placed off center with respect to the impurity will precess at constant speed around the impurity due to the gradient in the background field induced by the impurity [491]. In order to study this behavior in a simple physically meaningful setting, we examine a positive-charge ($S = 1$) vortex without the impurity and then the impurity is adiabatically switched on at a prescribed distance away from the center of the vortex. We find that for $V_{\text{Imp}}^{(0)} > 0$ the vortex then begins to precess around the impurity in a clockwise direction. Reversing the sign of $V_{\text{Imp}}^{(0)}$ in order to create an attractive impurity induces a counterclockwise precession with respect to the impurity. An example of the vortex precession induced by the impurity is shown in Fig. 3.76. It is crucial to note that if the impurity is turned on "close enough" to the steady-state vortex such that the impurity is within the vortex funnel then the vortex would begin its usual rotation but would be drawn into the center of the impurity, effectively pinning the vortex. This effective attraction is related to the emission of sound by the vortex when it is inside the funnel of the impurity, as described in Ref. [582].

Throughout this vortex-impurity interaction characterization, we will follow the center of the vortices by detecting the extrema of the superfluid vorticity $\omega_{\text{vor}} = \nabla \times \mathbf{v}_s$, where the fluid velocity is given by Eq. (3.13) [44, 512].

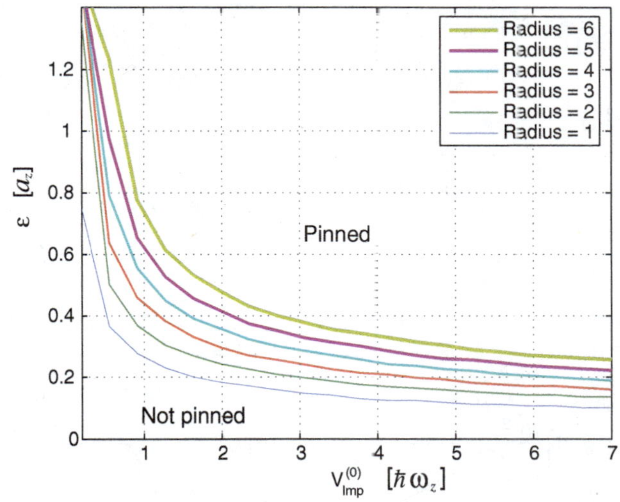

Figure 3.77. *Phase diagram depicting the vortex pinning by the localized impurity of strength $V_{\rm imp}^{(0)}$ (in units of $\hbar\omega_z$) and width ε (in units of transverse harmonic oscillator length a_z). Each curve represents a different pinning location at the indicated radii (i.e., distance from the center of the harmonic trap). Parameters are as follows:* $(\mu, \Omega, s) = (1, 0.065, 1)$. *Adapted with permission from Ref. [587].*

Next, the net effect of the pinning induced by the impurity and the precession induced by the harmonic trap are considered. Since the aim of Ref. [587] was to identify conditions needed for the manipulation of vortices using the repulsive impurity, a minimum requirement identified therein was that the impurity's pinning strength be sufficient to overcome the precession inside the trap and thus pin the vortex very close to the location of the impurity [i.e., $(x_0, y_0) \approx (x_a, y_a)$]. Therefore, we seek to find the minimum conditions that an off-center vortex, at a particular radius measured from the center of the harmonic trap, could be pinned by a localized impurity at that same location. For certain combinations of beam parameters (strong beam intensity, or large beam widths), the vortex will remain localized near this point. For other parameters (weak intensity, small beam widths), the beam cannot overcome the vortex precession induced by the harmonic trap and the vortex would be driven away from the beam position. This would give us a lower bound for the possible beam intensities and widths for which a vortex might be dragged to the corresponding position within the BEC. The existence of such pinned states was identified by searching in the impurity parameter space (strength $V_{\rm imp}^{(0)}$ and width ε) for several off-center radii (i.e., distances measured from the center of the trap). The results are shown in the phase diagrams of Fig. 3.77, where each curve corresponding to a different radius (decreasing from top to bottom) depicts the boundary in $(V_{\rm imp}^{(0)}, \varepsilon)$ parameter space for which pinning is possible. In other words, for the points in parameter space below a given curve, one gets primarily vortex precession dynamics induced by the harmonic trap, whereas above these curves (i.e., for strong or wide enough impurities), the vortex is trapped by the impurity and stays very close to it.

We now further elaborate our investigation of the pinning statics and the associated dynamical stability picture. In particular, we thoroughly analyze the bifurcation structure of the steady states with single-charge vorticity in the setting investigated above including their stability. The latter will be examined by the eigenvalues of the linearization around the steady state. Upon obtaining a steady state solution Ψ and considering, as

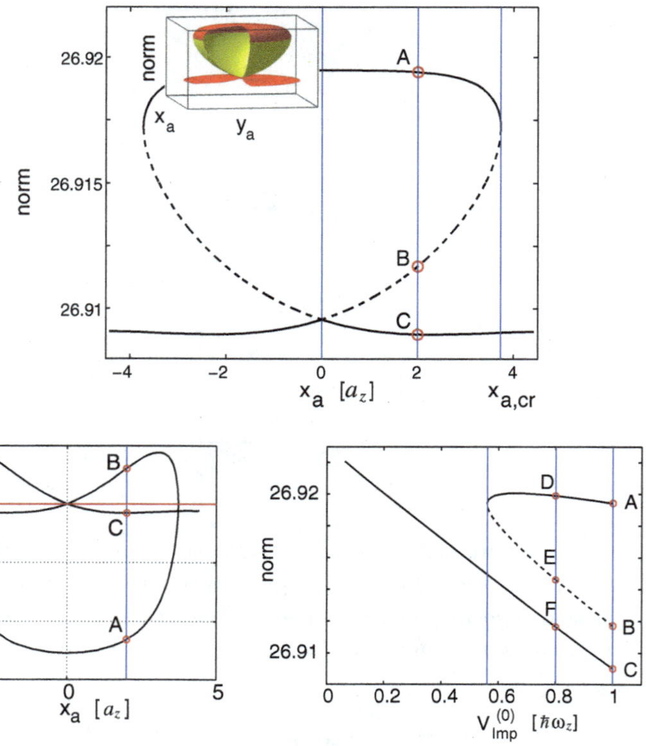

Figure 3.78. *The top panel is a 1D slice (for $y_a = 0$) of the solution surfaces (represented in the inset panel) as a function of the position of the impurity (x_a, y_a) for fixed $V_{\rm Imp}^{(0)}$. The vertical axis corresponds to the L^2 norm squared of the solution (i.e., the normalized number of atoms). The dashed lines [yellow (lighter) surface] correspond to unstable solutions, while the solid lines [red (darker) surfaces] correspond to stable solutions. The critical value of the radius, $x_{a,\rm cr} = 3.72$ as well as the characteristic value $x_a = 2$ and $x_a = 0$ are represented by vertical lines. The bottom left panel shows the squared bifurcating eigenvalues, λ^2, along these branches ($\lambda^2 > 0$ corresponds to an instability), while the bottom right shows the saddle-node bifurcation as a function of $V_{\rm Imp}^{(0)}$ for $x_a = 2$ fixed (there are lines at $V_{\rm Imp}^{(\rm cr)} = 0.57$, $V_{\rm Imp}^{(0)} = 0.8$, and $V_{\rm Imp}^{(0)} = 1$). The solutions and spectra for each of the three branches, represented by circles (and letters) for each of the characteristic values of x_a and $V_{\rm Imp}^{(0)}$ are presented in Fig. 3.79. For these branches $(\mu, s, \varepsilon) = (1, 1, 0.5)$. Adapted with permission from Ref. [587].*

usual, a separable complex valued perturbation $\tilde{u} = a(x,y)e^{\lambda t} + b^*(x,y)e^{\lambda^* t}$ of the steady state, we obtain the standard BdG eigenvalue problem for the growth rate, λ, of the perturbation.

As evidenced by the bifurcation diagrams presented in Fig. 3.78, there exist three solutions (i.e., steady-state vortex positions) for any impurity displacement radius in the interval $(0, x_{a,\rm cr})$, letting $y_a = 0$ without any loss of generality. At the low end of the interval [i.e., for an impurity at the center of the trap, $(x_a, y_a) = (0, 0)$] there is a transcritical bifurcation where the left (i.e., ones for negative x_a) and right (i.e., for positive x_a) solutions collide (see below for further explanation) and exchange stability. At the other end of the interval [i.e., for an impurity at $(x_a, y_a) = (x_{a,\rm cr}, 0)$] there exists a saddle-node bifurcation where two new steady-state vortex solutions can be thought of as emerging as x_a decreases to values $x_a < x_{a,\rm cr}$, or conversely can be thought of as disappearing as x_a increases to values $x_a > x_{a,\rm cr}$. Among them, one stable vortex position is found very close to the impurity (see cases "A" and "D" in Figs. 3.78 and 3.79), and another unstable vortex

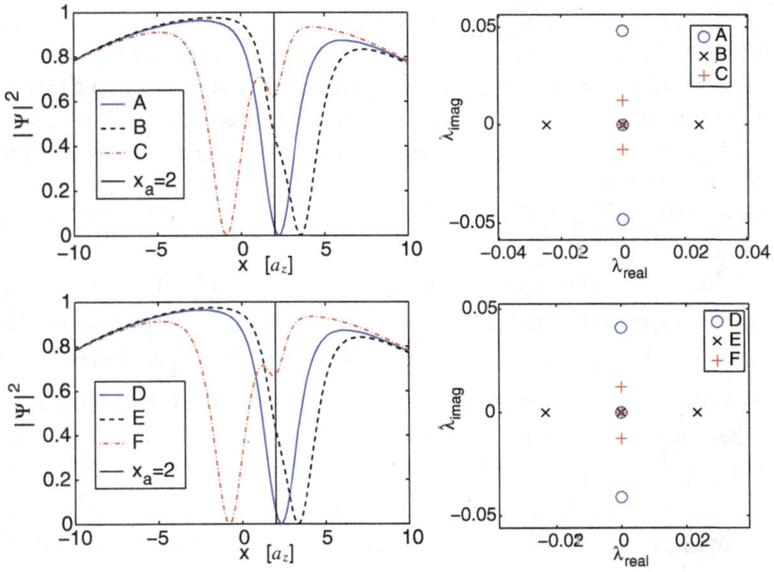

Figure 3.79. *The solutions (left, zoomed in to the region where the vortices "live") and corresponding linearization spectra in a neighborhood of the origin (right), for the various parameter values indicated by the circles in Fig. 3.78. The depicted solutions correspond to cuts along $y = 0$ of the 2D density $|\Psi(x,y)|^2$. The top (bottom) row corresponds to $V_{\rm Imp}^{(0)} = 1$ ($V_{\rm Imp}^{(0)} = 0.8$). Adapted with permission from Ref. [587].*

position further away (see cases "B" and "E" in Figs. 3.78 and 3.79). Considering only x_a, with $y_a = 0$, two solution branches are stable: one with the vortex sitting very close to the impurity (see cases "A" and "D" in Figs. 3.78 and 3.79), and one with the vortex close to the center of the harmonic trap (see cases "C" and "F" in Figs. 3.78 and 3.79). The other solution is unstable (see cases "B" and "E" in Figs. 3.78 and 3.79), with the vortex sitting in the small effective potential minimum on the side of the impurity opposite the center of the harmonic trap. This branch of solutions collides with the one that has a vortex pinned at the impurity and the two disappear in a saddle-node bifurcation as the attraction of the impurity becomes too weak to "hold" the vortex (this bifurcation corresponds to the curves of critical parameter values in Fig. 3.77). At $x_a = 0$, the potential becomes radially symmetric and the solution with the vortex on the outside of the impurity becomes identical (up to rotation) to the one close to the origin. Indeed, at this point there is a single one-parameter family of invariant solutions with the vortex equidistant from the origin (i.e., for any angle in polar coordinates), in addition to the single solution with the vortex centered at $(0,0)$. The solutions in this invariant family, not being radially symmetric and being in a radially symmetric trap, necessarily have an additional pair of zero eigenvalues in the linearization spectrum to account for the additional invariance, i.e., they have 4 instead of 2 zero eigenvalues (note that the solution with the vortex in the center for $x_a = 0$ only has a single pair of zero eigenvalues due to its radial symmetry). For $x_a < 0$ the branches exchange roles (transcritical bifurcation) as the previously imaginary pair of eigenvalues for the stable $x_a > 0$ branch (see cases "C" and "F" in Figs. 3.78 and 3.79) emerges on the real axis and the pair of previously real eigenvalues from the unstable $x_a > 0$ branch (see cases "B" and "E" in Figs. 3.78 and 3.79) emerges on the imaginary axis, i.e., the branches exchange their stability properties, becoming the reflected versions of one another; see the bottom left panel of Fig. 3.78 and the right column of Fig. 3.79.

In summary, for an impurity that is strong enough and close enough to the trap center (specifically, $0 \le x_a < x_{a,\mathrm{cr}}$, where $x_{a,\mathrm{cr}}$ depends of the strength of the impurity) it is possible to induce the stable pinning of a vortex very close to the impurity. However, if the impurity is too far away from the trap's center (specifically, $x_a > x_{a,\mathrm{cr}}$) the vortex can no longer be pinned by the impurity. The top image of Fig. 3.78 depicts the bifurcations via the L^2 norm squared (i.e., the normalized number of atoms) of the solution as a function of the radius, x_a (for $y_a = 0$), and as a function of arbitrary impurity location (x_a, y_a) (inset).

In fact, the above picture holds for any fixed, sufficiently large $V_{\mathrm{Imp}}^{(0)}$. Conversely, the same bifurcation structure can be represented by a continuation in the amplitude of the impurity, $V_{\mathrm{Imp}}^{(0)}$ (see bottom right panel of Fig. 3.78). In particular, for a fixed radius of $x_a = 2$, a continuation was performed and a saddle-node bifurcation appears for $V_{\mathrm{Imp}}^{(0,\mathrm{cr})} \approx 0.57$ where an unstable and a stable branch emerge (corresponding exactly to those presented in the continuation in x_a). One can then infer that the critical $V_{\mathrm{Imp}}^{(0)}$ decreases as x_a decreases.

Lastly, it should be noted here that recent work in Ref. [588] examined the possibility of an impurity to trap not only vortices of $S = 1$ but also vortices of higher charge in the form of persistent current. Both numerical computations and experimental results corroborated the possibility of trapping such higher charges, which would split into multiple vortex filaments upon release from the trap and expansion.

The Dynamical Picture: Dragging and Capturing

The next step is to take a pinned vortex and adiabatically drag it with the impurity in a manner akin to what is has been proposed for bright [589, 590, 231] and dark solitons [208, 224, 225] in the quasi-1D setting. Manipulation of the vortex begins with the focused laser beam at the center of the vortex. The laser is then adiabatically moved to a desired location while continually tracking the position of the vortex. Adiabaticity for the motion of the impurity is controlled by the adiabaticity parameter τ controlling the acceleration of the center $(x_a(t), y_a(t))$ of the impurity as

$$\begin{aligned} x_a(t) &= x_i - \frac{1}{2}(x_i - x_f)\left[1 + \tanh\left(\frac{t-t^*}{\tau}\right)\right], \\ y_a(t) &= y_i - \frac{1}{2}(y_i - y_f)\left[1 + \tanh\left(\frac{t-t^*}{\tau}\right)\right], \end{aligned} \tag{3.190}$$

where the initial and final positions of the impurity are, respectively, (x_i, y_i) and (x_f, y_f). We will assume $y_a(t) = 0$ (i.e., $y_i = y_f = 0$) for the discussion below. The instant of maximum acceleration is

$$t^* = \tanh^{-1}\left(\sqrt{1-\delta\tau}\right)\tau, \tag{3.191}$$

where δ is a small parameter, $\delta = 0.001$, such that the initial velocity of the impurity is negligible and that $x_a(0) \approx x_i$ and $x_a(2t^*) \approx x_f$ (and the same for y). This condition on t^* allows for the reduction of parameters and allows us to ensure that we begin with a localized impurity very close to the center of the trap [i.e., $(x_a(0), y_a(0)) \approx (0,0)$] and that we will drag it adiabatically to (x_f, y_f) during the time interval $[0, 2t^*]$. The next objective is to determine the relation between adiabaticity and the various parameters such as strength ($V_{\mathrm{Imp}}^{(0)}$) or the width (ε) of the impurity in order to successfully drag a vortex outward to a specific distance from the center of the harmonic trap. Following

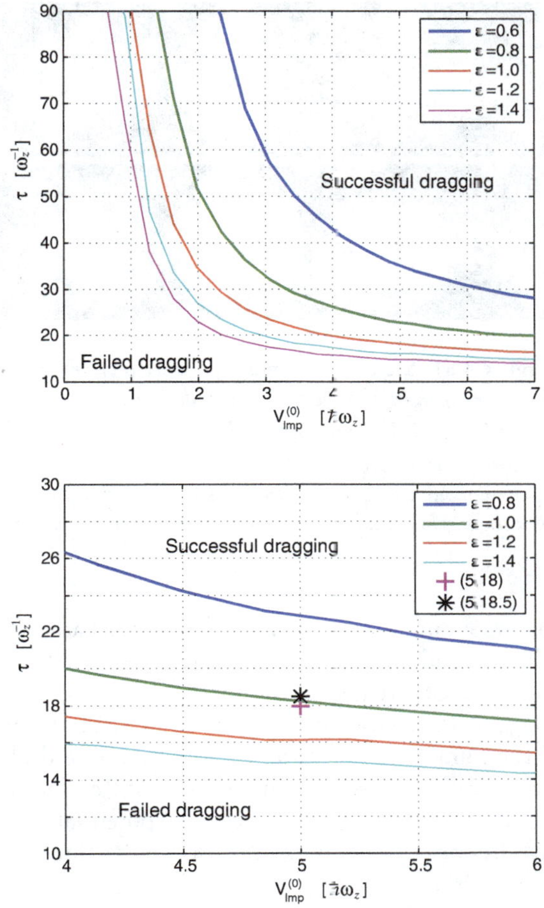

Figure 3.80. *Parameter regions for successful manipulation of a single vortex inside a harmonic trap. The area above each curve corresponds to the successful dragging region. The bottom panel depicts a zoomed region of the top panel where the asterisk and cross correspond, respectively, to the manipulation success and failure depicted in Fig. 3.81. These panels indicate that higher intensity beams, and broader beams, can successfully drag vortices over shorter timescales than weaker beams. Adapted with permission from Ref. [587].*

Ref. [587], we set this distance to be half of the radius of the cloud (half of the Thomas–Fermi radius). We use the value t^* to also define when to stop dynamically evolving our system. In particular, we opt to continue monitoring the system's evolution until $t_f = 3t^*$. This choice ensures that a vortex that might have been lingering close to the impurity at earlier times would have either been "swallowed up" by the impurity and remain pinned for later times, or will have drifted further away due to the precession induced by the trap.

Applying this technique, along with a bisection method (successively dividing the parameter step in half and changing the sign of the parameter stepping once the threshold pinning value is reached) within the span of relevant parameters yields the phase diagram depicted in Fig. 3.80. The various curves in the figure represent the parameter boundaries for successful dragging of the vortices for different impurity widths (increasing widths from top to bottom). All the curves for different widths are qualitatively

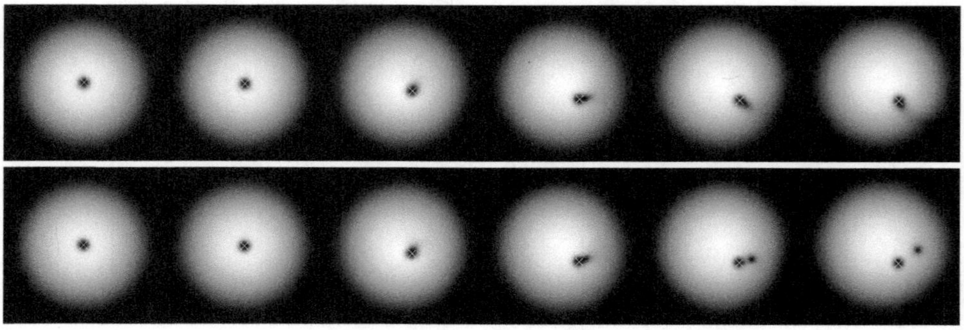

Figure 3.81. *Successful (top row) and failed (bottom row) vortex dragging cases by a moving laser impurity corresponding to the parameters depicted, respectively, by an asterisk and a cross in the bottom panel of Fig. 3.80. In both cases, the laser impurity, marked by a cross, with* $(V_{\rm Imp}^{(0)}, \varepsilon) = (5, 1)$ *is moved adiabatically from* $(0, 0)$ *to* $(5.43, -5.43)$ *and the snapshots of the density are shown every* $0.5t^*$. *The top row corresponds to a successful manipulation for* $\tau = 18.5$ *while the bottom row depicts a failed dragging for a slightly lower adiabaticity of* $\tau = 18$. *Adapted with permission from Ref. [587].*

similar corresponding to higher values of the adiabaticity parameter as the width is decreased. This trend continues as ε approaches the existence threshold established in Fig. 3.77. In Fig. 3.81 we depict snapshots for the two cases depicted by an asterisk (successful dragging) and a cross (failed dragging) in the lower panel of Fig. 3.80.

All of the numerical simulations discussed above deal with dragging the vortex by means of the localized impurity. In the work of Ref. [587], it was also attempted to produce similar results via an optical lattice (OL) potential generated by counter-propagating laser beams. In one dimension, the case of bright solitons manipulated by OLs was studied earlier in Refs. [590, 231] while the dark soliton case has been treated in Refs. [208, 224, 225]. For a 1D OL, the potential minima (or maxima) are isolated from each other providing good effective potential minima for pinning and dragging. On the other hand, each 2D minimum (or maximum) is no longer isolated, and that between two minima (or maxima) there are areas for which the vortex can escape (near the saddle points of the potential). This is exactly what was observed in Ref. [587] when attempting to drag a vortex using the 2D OL without sufficient adiabaticity. The vortex would meander around the various facets of the lattice outside of our control. To overcome this one needs to displace the potential with a high degree of adiabaticity. In doing so, dragging the vortices was successful under some constraints (relatively small displacements from the trap center). It was thus concluded that, for sufficient adiabaticity, with a strong enough OL and for small displacements from the trap center, it is possible to successfully drag the vortex.

A natural extension of the above results is to investigate whether it is possible to capture a vortex that is already precessing by an appropriately located and crafted impurity. This idea of capturing, paired with the dragging ability, suggests that a vortex created off-center, which is typically the case in an experimental setting, can be captured, pinned, and dragged to a desired location either at the center of the trap or at some other distance off-center. We now give a few examples demonstrating that it is indeed possible for a localized impurity to capture a moving vortex. The simulation begins with a steady-state solution of a vortex pinned by an impurity at a prescribed radius and a second impurity on the opposite side of the trap at a different radius.

As is shown in Fig. 3.82, the capturing impurity must be located sufficiently lower (i.e., closer to the trap center) than the trapping impurity in order for the vortex to be

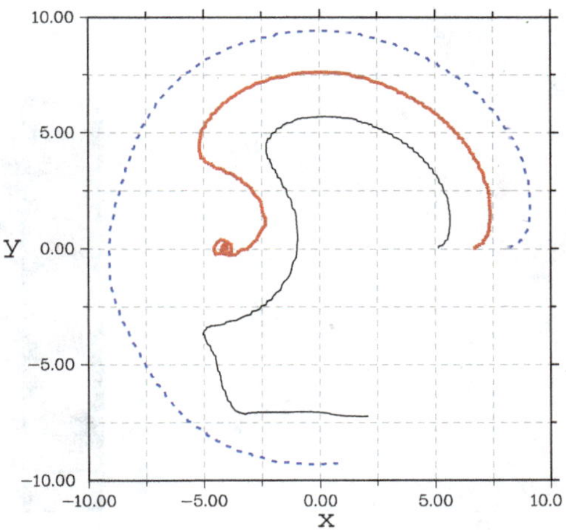

Figure 3.82. *Capturing a precessing vortex by a stationary impurity for* $(V_{\rm Imp}^{(0)}, \varepsilon, \Omega) = (5, 1, 0.065)$. *The different paths correspond to isolated vortices that are released by adiabatically turning off a pinning impurity at the following off-center locations:* $(5,0)$ *(thin black line),* $(6.5,0)$ *(thick red line), and* $(8,0)$ *(blue dashed line). The capturing impurity is located at* $(-4,0)$. *The first and third cases fail to produce capturing while the second case manages to capture the vortex. One interesting feature that we observed in the case of successful capture is that, before its gets captured, the vortex gets drawn into the impurity potential, but then almost gets knocked back out by the phonon radiation waves created from the capture which bounce around within the condensate. Adapted with permission from Ref. [587].*

pulled from its precession and be captured by the impurity. Intuitively one might come to the conclusion that if the vortex and impurity were located the same distance away from the center of the trap, then the vortex should be captured. But due to the interaction between the vortex and the impurity that was discussed earlier, as the vortex approaches the impurity, it begins to interact with it by precessing clockwise around the impurity. Thus, the orientation of the vortex and impurity with respect to the trap center greatly determines the dynamics. This combination of the interactions of the vortex with the trap and the vortex with the impurity then dictates the following: for a vortex to be captured by the impurity while precessing in the trap and rotating around the impurity, the impurity must be positioned at least closer to the trap center than the initial distance between the vortex and the trap center.

3.9.3 ▪ Formation of Vortices Through Interference of Fragments Separated by Quasi-1D Impurities

We will now use the quasi-1D impurities previously used to examine the modification of the properties of the dark soliton stripes for a different purpose, namely to perform numerical interference-type experiments. This is the same vein as the earlier proposal of Ref. [151], according to which such nonlinear interference in quasi-1D BECs leads to the formation of dark solitons (cf. Sec. 2.1.5); this proposal had been utilized in the experiments of Ref. [70], as discussed in Sec. 2.3.5.

We start with two different experimentally feasible situations (which, however, have not been realized experimentally—see below). In the more standard one, i.e., the direct analogue of the 1D case of Ref. [151], the potential in the 2D GPE reads

$$V(x,y) = V_{\rm MT}(x,y) + V_{\rm LS}(x,y), \tag{3.192}$$

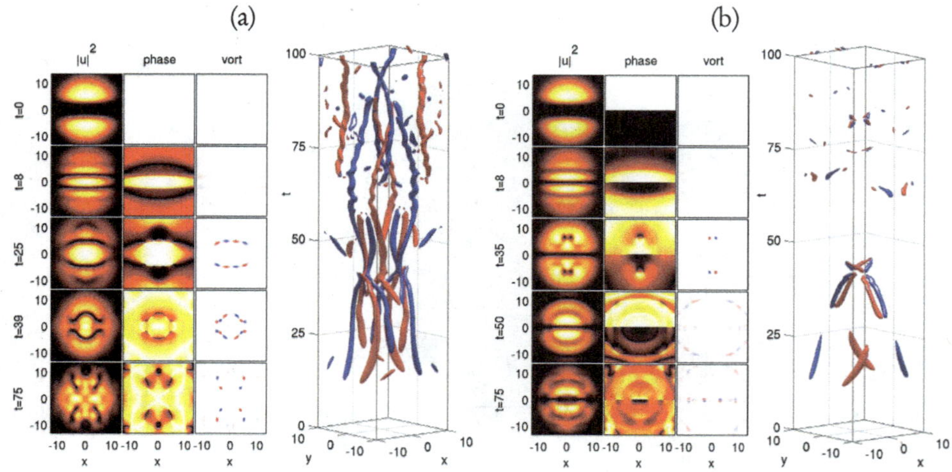

Figure 3.83. *Collision of two condensate fragments originally separated by a quasi-1D light sheet lifted at $t=0$. Panels* (a) *correspond to fragments with same initial phase ($\Delta\phi = 0$) while panels* (b) *correspond to $\Delta\phi = \pi$. For each case, the first column of panels depicts the density of the condensate $|u(x,y,t)|^2$ at the times indicated (brighter regions indicate higher densities). The result of the release of the light sheet is the interference between the two fragments, see $|u(t=8)|^2$, that eventually leads to the nucleation of vortex pairs. The second column of panels depicts the corresponding phases (brighter regions indicate phases close to $-\pi$ and $+\pi$ while dark regions correspond to zero phase). The third column of panels depicts the corresponding fluid vorticity (blue/red indicates positive/negative vorticity); see text. The fourth panel depicts the spatio-temporal evolution of the vorticity by showing a contour slice at a tenth of the maximum vorticity. The evolution is responsible for the nucleation and annihilation of vortex pairs resulting in the spatio-temporal vortex filaments shown in the panel. In all the 2D simulations, we use, for the spatial variable, a discretization of 301×301 sites centered around $(x,y)=(0,0)$. Also, the trap parameters for all 2D simulations are $\Omega = 0.1$, $V_0 = 5$, and $b = 1$. Adapted with permission from Ref. [156].*

where

$$V_{\rm MT}(x,y) = \frac{1}{2}\Omega^2(x^2+y^2), \tag{3.193}$$

$$V_{\rm LS}(x,y) = V_0\,\text{sech}(by). \tag{3.194}$$

Here, $V_{\rm LS}$ is a localized repulsive potential describing the "light sheet" (with V_0 and b representing the normalized intensity and inverse width of the laser beam).

For this trap, the ground state of the system has the form shown in the top left panel in Fig. 3.83(a): namely, two identical condensate fragments separated by the light sheet. This ground state is obtained by relaxation (imaginary time integration) starting from a Thomas–Fermi approximate profile with chemical potential μ that will be set to unity in what follows. The proposed numerical experiment, as it was carried out in Ref. [156], assumes that the condensate, confined in the potential of Eq. (3.192), is in the ground state at $t=0$, and that both fragments share the same phase (see below for a description when fragments have different phases). Then, at that time, $t=0$, we "lift" (i.e., switch off) the light sheet by setting $V_0 \to 0$, and subsequently let the system evolve according to the 2D GPE. During the evolution, the two fragments originally constituting the condensate expand, and eventually interfere [see later times in the left column of Fig. 3.83(a)]. In order to quantify the amount of vorticity generated by the collision of the different fragments we monitor the vorticity $\omega_{\rm vor}(\mathbf{r},t)$. The typical numerical experiment showing the collision between two fragments with same initial phase ($\Delta\phi = 0$, see below) shown in Fig. 3.83(a) is performed for $\Omega = 0.1$, $V_0 = 5$, and $b = 1$.

Through the interference, quasi-1D "nonlinear" fringes are formed, i.e., dark stripes [see atom density at $t = 8$ in the left panel of Fig. 3.83(a)] resembling dark solitons (in direct analogy to the 1D case), but now in the 2D setting. However, as discussed previously in this chapter, such 1D stripes are unstable in 2D (and 3D) towards transverse modulations [253, 526]. As a result of the ensuing snaking instability, such stripes break up into vortex-antivortex pairs—because the original zero vorticity of the solution needs to be preserved. This is precisely what happens in our case as well. The bending of the stripes [see the atom density at $t = 25$ in Fig. 3.83(a)] is responsible for the break up into four vortex pairs, as is shown in vorticity plot at $t = 25$. As the dark stripes continue bending and mixing, a series of vortex pair nucleations and annihilations occur.

To follow this evolution, we depict in the right panel of Fig. 3.83(a) a space-time plot of the vorticity. The main results stemming from the evolution of the vorticity can be summarized as follows. Between $t \approx 10$ and $t \approx 15$ vortex patches start forming, from which 4 vortices nucleate at $t \approx 15$. Two more quartets of vortices are nucleated at $t \approx 22$ and $t \approx 35$. At $t \approx 45$ the first quartet merges and is annihilated. The second quartet merges at $t \approx 58$. This process of merging and spontaneous nucleation of vortex pairs and quartets remains active for long times (data not shown here). It is interesting to note that, after $t \approx 60$, the apparent superfluid turbulence present in the condensate makes the vortices "wiggle" about their original location.

Another possibility to perform this type of experiment is to grow a condensate in the harmonic trap alone and then split the condensate into fragments by adiabatically ramping up the light sheet. Depending on the degree of adiabaticity and, more importantly, on the time the fragments are kept separate before releasing the light sheet, the phases of the different fragments will evolve independently and assume, effectively, random phases. As a result, it is worthwhile to explore the effects of such phase differences between the condensate fragments. We therefore consider the case where just before the light sheet release the phase of the top (bottom) fragment is ϕ_1 (ϕ_2). Without loss of generality, we fix $\phi_1 = 0$ and we focus on the phase difference between the condensates $\Delta\phi \equiv \phi_2 - \phi_1 = \phi_2$. In Fig. 3.83(b), we depict the collision of two fragments as in Fig. 3.83(a), but with a different initial relative phase between the condensates of $\Delta\phi = \pi$. In the simulations, initial configurations with arbitrary phase differences $\Delta\phi$ were achieved by using the steady state solution found for $\Delta\phi = 0$ as before, then applying a phase shift of $\Delta\phi$ to the bottom fragment (i.e., for $y < 0$), and running again the relaxation scheme (imaginary time integration) until convergence.

The results presented in Figs. 3.83(a) and 3.83(b) indicate that the detailed evolution of the vortex formation and merging depends on the relative phases of the condensates. In fact, further numerical results with phase differences between $\Delta\phi = 0$ and $\Delta\phi = \pi$ show similar vortex formation and annihilation with vortex "activity" decreasing (both in terms of the number of vortices produced and the persistence time of the structures) as $\Delta\phi$ was increased from 0 to π (results not shown here). It is interesting to note that the dynamics for phase differences different from zero loses its fourfold symmetry when compared to the $\Delta\phi = 0$ case. In all cases that were examined, vortex pairs are nucleated at some point in the simulation irrespective of $\Delta\phi$. However, it is evident from the figures that the smaller the phase difference between the fragments, the more vortex structures are nucleated and the longer they live. In fact, the extreme case of a $\Delta\phi = \pi$ between fragments barely produces any vortex pairs. This is due to the fact that the initial condition is close to a steady-state domain wall (dark soliton stripe separating out-of-phase domains) and, thus, there is little extra energy for the collision of the fragments. Indeed, as it can

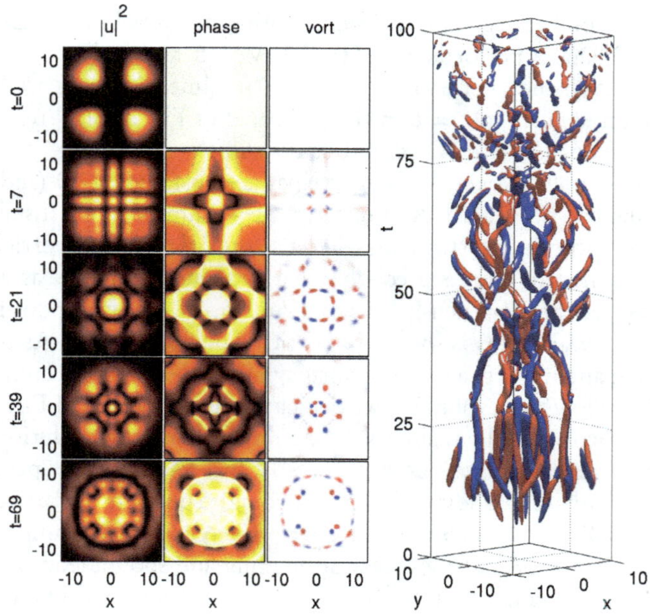

Figure 3.84. *Similar to Fig. 3.83*(a), *but for an initial condition that contains four fragments (see* $|u(t=0)|^2$*), being the ground state of the potential of Eq. (3.195). Notice the fourfold symmetry of the interference in this case and the eventual formation of a complex pattern of vortex pairs. Same parameters as in Fig. 3.83 with* $a = b = 1$*. Adapted with permission from Ref. [156].*

be evidenced from Fig. 3.83(b) (where $\Delta\phi = \pi$), the condensate always maintains a configuration similar to a domain wall with some perturbations: the density always shows a nudge at $y = 0$ and there is a predominant phase difference of π between the upper and lower half-planes.

It should be noted that in the work of Ref. [156], additionally, the effects of a phenomenological damping (representing the coupling to the thermal cloud) and of the ramp-down time of the laser sheet separating the BEC fragments were considered. Both the increase of damping (due to its detrimental effect on vortices) and the effect of increase of the ramp-down time (due to its increased adiabaticity and enabling of "healing" of the phase differences between adjacent BEC fragments) were found to have a reduced number of resulting vortices within the numerical experiments.

A more genuinely 2D installment of the same "experiment" can also be envisioned and is proposed in Fig. 3.84 (again per the work of Ref. [156]). In this case, the light-sheet potential is of the form

$$V_{\rm LS} = V_0\,[\mathrm{sech}(ax) + \mathrm{sech}(by)], \tag{3.195}$$

resulting in a ground state which contains four in-phase condensate fragments (as is shown by the initial density $|u(t=0)|^2$ in Fig. 3.84). Subsequently, as the "light cross" is lifted at $t = 0$, the fragments attempt to fill in the empty space, resulting in an interference pattern with fourfold symmetry. The oscillation and ensuing bending of the resulting dark stripes results in a rich evolution of vortex pairs as it can be evidenced by the space-time plot of the vorticity in the right panel of Fig. 3.84. Notice that for this initial condition, more exotic patterns, including *vortex necklaces* (cf. also Ref. [265]) and structures with a higher number of vortices are formed. However, these patterns are considerably less long-lived

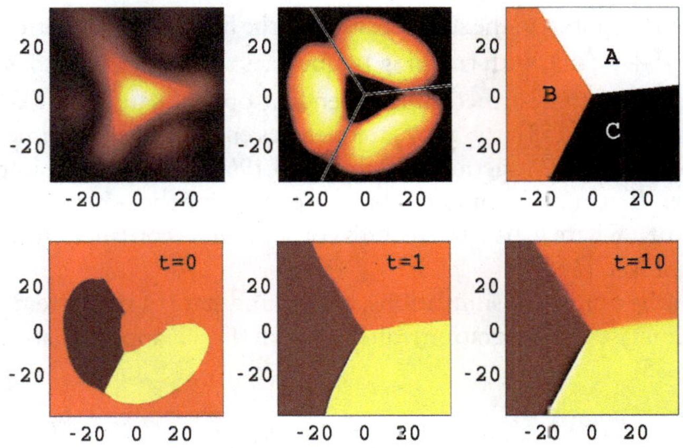

Figure 3.85. *Top-left: Intensity profile of the optical potential responsible for segmenting the potential well into three local minima. Top-center: Thomas–Fermi approximation used as an initial condition for the relaxation method to obtain the ground state of the system. Top-right: Regions A, B, and C with respective phases ϕ_1, ϕ_2, and ϕ_3. The bottom row of panels depicts the evolution of the phase during the relaxation (imaginary time relaxation) towards the initial steady state with different phases for the fragments. This example shows the phase for the case $\phi_k = 2\pi k/3$ at the times indicated. In all panels, the field of view is approximately 70 μm per side. The axis numbers indicate x and y coordinates relative to the center of the unsegmented harmonic trap. Adapted with permission from Ref. [159].*

as compared to the ones observed in the two in-phase fragments numerical experiment described above.

Results pertaining to the generation of vortex rings in 3D settings by collision of multiple BEC fragments is covered in Sec. 4.3.1.

3.9.4 ▪ Formation of Vortices Through Interference of Three Fragments and Connections with Experiment

A proposal similar in spirit to the ones above was performed in the work of Ref. [158], where a three-armed potential was experimentally implemented. This induced the separation of the condensate into three fragments (see top-left and top-center panels in Fig. 3.85). Subsequently, this optical potential barrier was lifted and the interference and controlled merging of the three isolated fragments (which are essentially independent BECs) was observed. Such a process has close ties to the one of topological defect formation in-phase transitions, as proposed by Kibble [55] and Zurek [56]. We should note that the Kibble-Zurek mechanism was responsible for the vortex formation process by means of a rapid quenching through the quantum phase transition of Bose–Einstein condensation as was originally proposed in the pioneering work of Ref. [57]. The latter mechanism was, in fact, the one utilized to produce the vortices and vortex dipoles of Ref. [531, 537]. We should also add that alternative methods proposed for nucleating vortices include vortex formation by Bragg reflection from an optical lattice [65], and vortex formation in turbulent-like regimes in stirred condensates [591]. In considering the interference effects of the three merging condensates, we follow the presentation of Ref. [159].

We will once again employ as our fundamental model the 2D GPE, where the potential $V(\mathbf{r};t)$ is taken to be of the form

$$V(\mathbf{r};t) = V_{\rm MT}(\mathbf{r}) + \alpha(t)\, V_{\rm L}(\mathbf{r}). \tag{3.196}$$

This corresponds to the superposition of the harmonic magnetic trap, $V_{\mathrm{MT}}(\mathbf{r}) = \frac{1}{2}m(\omega_x^2 x^2 + \omega_y^2 y^2 + \omega_z^2 z^2)$, with trapping frequencies $\omega_x = \omega_y = 2\pi \times 7.4$ Hz and $\omega_z = 2\pi \times 14.1$ Hz, and the three-armed time-dependent optical barrier, $\alpha(t)V_{\mathrm{L}}(\mathbf{r})$, used in the experiments of Ref. [158] (the above frequencies emulate the experimentally used ones as well).

Note that the function $\alpha(t)$ in Eq. (3.196) describes the ramping down of the optical barrier. The maximum initial barrier energy for the potential is taken to be $\alpha_0 \equiv \alpha(0) = 26\,k_B$ nK, where k_B is Boltzmann's constant, in accordance with the experimental parameters of Ref. [158].

In the numerical simulations, the ground state of the system is obtained by relaxation (imaginary time integration) initiated with the TF approximation (see top-center panel in Fig. 3.85)

$$\psi(\mathbf{r},0) = \sqrt{\max\{0, \mu - V(\mathbf{r})\}} \times \phi(\mathbf{r}), \tag{3.197}$$

where $\mu = 8\,k_B$ nK (again using the same value as in the experiments of Ref. [158]) is the chemical potential, and $\phi(\mathbf{r})$ contains the chosen phase of the different wells separated into the three regions A, B, and C depicted in the top-right panel of Fig. 3.85. The notation ϕ_1, ϕ_2, and ϕ_3 is used to denote the initial phases in regions A, B, and C, respectively. Notice that while in the experiments of Ref. [158] the different fragments have uncorrelated (random) phases, in the numerical simulations of Ref. [159], the initial phases of the fragments (and more importantly their relative phases) can be controlled.

The output of the relaxation was used in Ref. [159] as a suitable initial condition for the 2D GPE. The bottom row of Fig. 3.85 depicts the evolution of the phase for the case $\phi_k = 2\pi k/3$ during relaxation. As can be observed from the figure, the initial "seed" ($t = 0$) starts with sharp phase boundaries around the localized fragments. As the relaxation procedure evolves, the phase boundaries become smoother and adjust to the boundaries for the top-right panel of the figure. The phase profile seems to settle after 10 ms of imaginary time relaxation. In the simulations, 50 ms of imaginary time relaxation was used to ensure proper convergence to the steady-state solution.

We now explore the effect of the ramp-down time of the potential barrier on the formation of vortices through the merging of the different fragments of the condensate. For this purpose we use a linear ramp:

$$\alpha(t) = \max\left\{\frac{\alpha_0}{t_b}(t_b - t), 0\right\}, \tag{3.198}$$

where $\alpha_0 \equiv \alpha(0)$ is the maximum barrier energy as defined above and t_b is the ramping time (in ms) of the barrier. This is a ramp directly related to the one used experimentally in Ref. [158].

The formation of vortices, as well as the overall vorticity of the system was examined in Ref. [159] using various vorticity-related diagnostics. The results for the merger of the three BEC fragments with relative phases $\phi_k = 2\pi k/3$, $k \in \{1,2,3\}$, for different merging times are depicted in Fig. 3.86. The number of vortex pairs nucleated by the merger is extremely sensitive to the ramping time t_b. Shorter ramping times give rise to an extremely rich vorticity pattern as the fragments merge [see, for example, the top two rows in Fig. 3.86 corresponding to the instantaneous ($t_b = 0$) removal of the barrier], including the appearance of structures resembling vortex streets.

However, for longer ramp times, i.e., slower ramping, just a handful of vortices are nucleated. In fact, for $t_b > 100$ ms (results not shown here), the only vortex that is nucleated is the central one. It is evident that, independently of the ramping time, the central vortex is always formed for this pair of relative phases between the fragments.

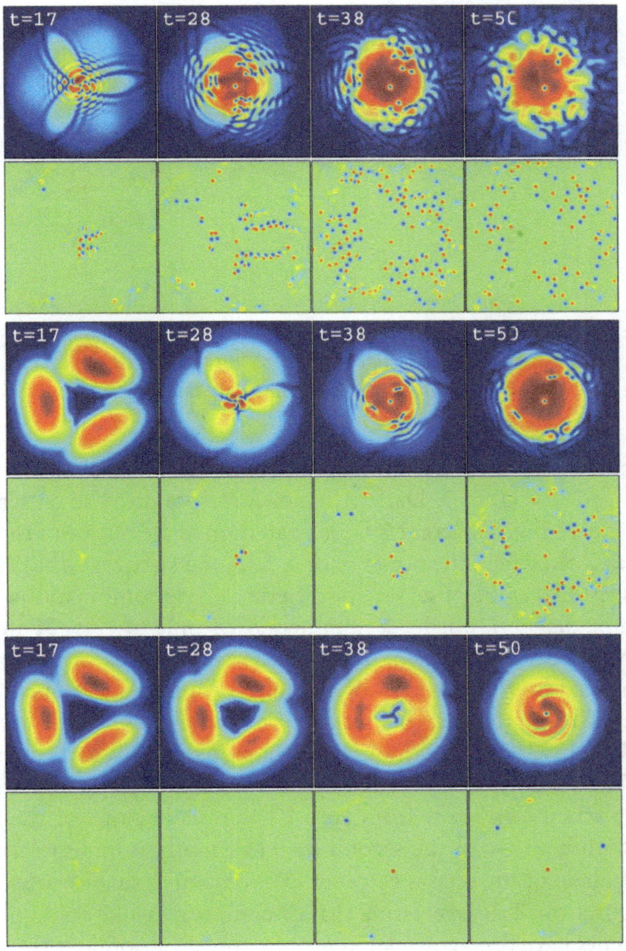

Figure 3.86. *Evolution of the 2D condensate density (respective top series of panels) and vorticity (respective bottom series of panels) for three different ramp-down times of the optical potential barriers. From top to bottom, the three sequences correspond to $t_b = 0$ ms (first and second rows), $t_b = 25$ ms (third and fourth rows), and $t_b = 50$ ms (fifth and sixth rows). The times are indicated in the panels in ms and the field of view is approximately 70 μm per side. For the initial conditions in all cases, the different condensed fragments have relative phases of $2\pi/3$, namely, $\phi_k = 2\pi k/3$ ($k \in \{1,2,3\}$). Adapted with permission from Ref. [159].*

Figure 3.87 shows the spatio-temporal rendering of the vortex formation for the middle row example of Fig. 3.86 (i.e., phases given by $\phi_k = 2\pi k/3$ and a ramp-down time of $t_b = 25$ ms). In this figure, a space-time contour plot of the vorticity is presented where blue/red contours correspond to negatively/positively charged vortices, respectively. The figure clearly shows the formation of pairs of vortices with opposite charge, some of which self-annihilate at later times, while others oscillate together with the cloud: upon formation, they expand to the rims of the cloud and are subsequently reflected from its outskirts and contract anew, following the density profile oscillations resulting from the merging process.

In order to follow in more detail the formation of vortices in the central portion of the cloud as a function of the relative initial phases of the different fragments, systematic simulations were performed in Ref. [159] for a large set of relative initial phases. The

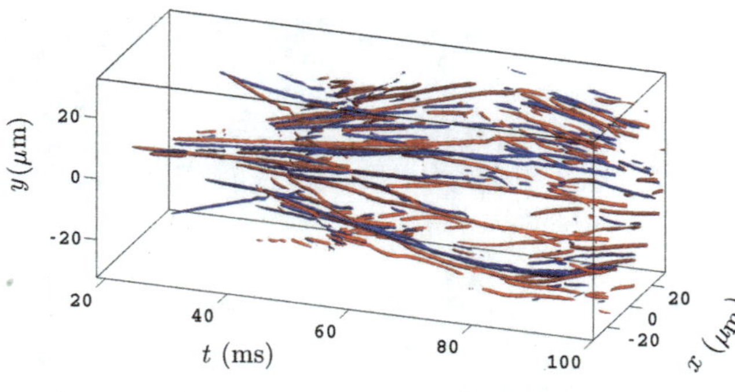

Figure 3.87. *Evolution of the vortex structures in the 2D condensate density for $t_b = 25$ ms and $\phi_k = 2\pi k/3$ (cf. middle rows in Fig. 3.86). Adapted with permission from Ref. [159].*

results are presented in Fig. 3.88, where the existence of vortices is depicted as a function of (ϕ_2, ϕ_3) for $\phi_1 = 0$. Darker shades correspond to the presence of more vortices. The top panel of Fig. 3.88 corresponds to an immediate release of the barrier (ramp-down time of $t_b = 0$ ms) where the presence of 0, 1, 2, or 3 vortices (white, light gray, gray, and black, respectively) can be observed for different phase combinations. The middle panel depicts the same diagram, but for a ramping-down time of $t_b = 50$ ms. It is clear that, for this relatively slow ramp-down, the formation of a vortex in the central region is exclusively determined by the vorticity of the initial configuration. Namely, if any of the relative phases is larger than π, the initial configuration resembles more that of a *discrete vortex* (see Refs. [592] and [493] for reviews). In particular, each fragment can be thought of a "unit" and the whole configuration corresponds to a discrete vortex with three units with a net vorticity different from zero. The middle panel of the figure clearly reveals that *only* within an arc of the second (and essentially symmetrically of the fourth) quadrant of the plane of the phases $(\phi_2 - \pi, \phi_3 - \pi)$ will a single vortex form at the center of the configuration. It is interesting that discrete vortex-like configurations consisting of three fragments have been considered in the context of nonlinear optics in Ref. [593] and even in genuinely discrete systems as, e.g., in Refs. [594, 595]. In the bottom panel of Fig. 3.88 we depict the difference between the top and middle panels to show the amount of vortices that are created exclusively by the fragment collision and not by the intrinsic vorticity of the initial configuration. It is important to mention that some of the vortices counted in the top panel come from vortices that are created outside the central region but that migrate towards the center as time progresses.

In comparing the above results to the experiments of Ref. [158], we indicate that in the latter an optical potential was used to segment a harmonic trap before condensation was achieved; with additional evaporative cooling, three initially isolated and mutually independent (i.e., uncorrelated phases between the different fragments) condensates were created. The initial condition of incoherent condensate fragments served as the conceptual basis behind the motivation to impose and examine various relative phases between the fragments in the numerical simulations of Refs. [159]. The optical barrier was experimentally ramped off approximately linearly over time scales between 50 ms and 3 s, significantly longer than the time scales considered in the simulations. At the end of the merging process, immediately after the barrier was completely removed, the fully merged BEC was released from the harmonic trap and allowed to ballistically expand for 56 ms to enable imaging of vortex cores. Nevertheless, it should be added that the simulations do not involve an expansion stage (although the latter is also not expected to significantly affect the vortex dynamics). In both experimental results and in the simulations, as dis-

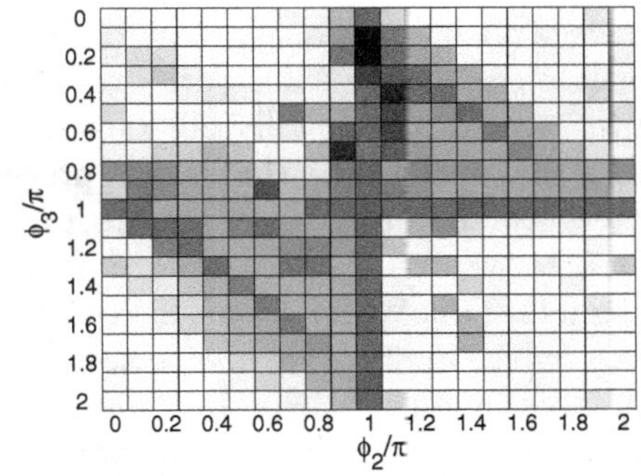

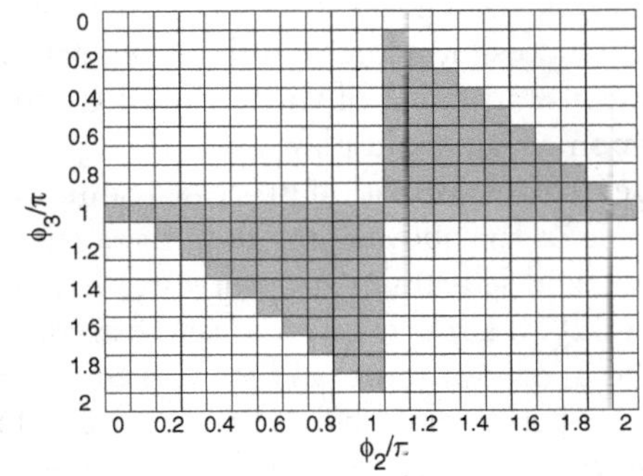

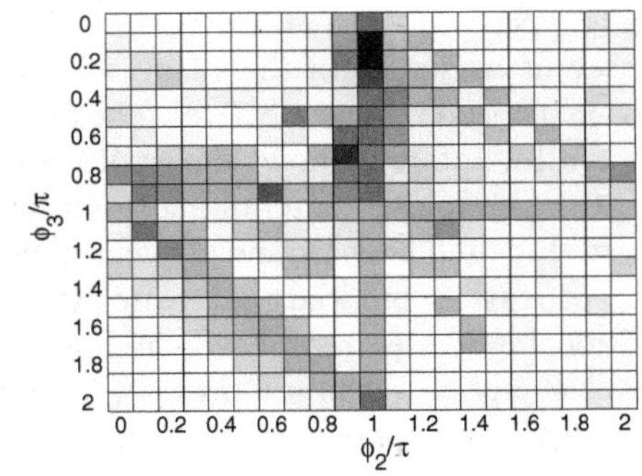

Figure 3.88. *Phase diagram for vortex formation at the trap center region as a function of the relative phases of the merging fragments (where ϕ_1 is assigned the value of 0). Top: Ramp-down time of $t_b = 0$ ms (i.e., instantaneous removal) of the barrier. The different shades of gray indicate the amount of vorticity contained in the central region after 100 ms. White, light gray, gray, black correspond, respectively, to 0, 1, 2, and 3 vortices. Middle: Same as above for a ramp-down time of $t_b = 50$ ms. This case only produces no vortices (white) or one vortex (gray). Bottom: Difference between top and middle diagrams. This corresponds to the vortices formed by the collision of the fragments and not by the intrinsic vorticity of the initial configuration (which depends on the relative phases of the different fragments). Adapted with permission from Ref. [159].*

cussed in Ref. [159], a series of important similarities could be identified. First, it is clear that vortex cores may be formed during the merging process. Second, as noted in the experimental work, the vortex formation process should depend upon relative phases between the condensates. The simple analysis of a slow merging process leading to a 25% probability of vortex formation, as presented in the experimental work, matches the results of the simulations summarized in Fig. 3.88. Finally, experimental and simulated results show that faster merging leads to more vortices initially created, but these vortices may self-annihilate over time by holding the fully merged BEC in the trap.

However, there is one notable quantitative difference between the experimental and numerical results. In the experimental work, for barrier ramp-down times longer than 1 s, single vortices were experimentally observed in approximately 25% of the images obtained directly at the conclusion of the merging process. Multiple vortex cores were not observed under these conditions. For faster ramps, experimental images containing either single or multiple cores were more often obtained, with significantly more than 25% of the images containing at least one vortex core. The results of the numerical data show that the slow merging limit (where multiple vortices cease to be created during merging) is reached for merging times that are much shorter than in the experiment. In other words, images with multiple vortices are seen in the experiment under conditions where the numerical results would suggest that a given BEC should have at most one vortex.

There are a few possible sources of this discrepancy. At first glance, it might appear that the spontaneous formation of vortices in BECs during evaporative cooling in an axisymmetric harmonic trap could play a role in higher percentage of vortices seen in the experiment. Such vortex formation processes cannot be described by the GPE and thus are not observable in the above computational analysis. However, due to angular momentum damping and self-annihilation of vortices in the asymmetric local potentials in which the three BEC fragments grow, it can be argued that vortices that might be spontaneously created in one (or more) of the three BECs are unlikely to survive at rates that would affect the experimental observations of Ref. [158]. This possible source for the quantitative discrepancy could be tested, for example, with simulation methods based on the stochastic GPE [596].

Perhaps a more likely source of the quantitative discrepancy may lie in the optical potential energy or shape; differences between the experiment and simulations regarding barrier heights, widths, and ramp-down trajectory might induce more vortices to form during merging. For example, if center-of-mass oscillations of the cloud were induced in the experiment, atomic fluid flow around the central portion of the optical barrier could induce formation of vortices and lead to increased vortex observation rates in the experiment. Such additional processes may be worthwhile to be studied in GPE simulations in order to further characterize dynamical processes that may be involved in vortex formation. Experimentally, the use of oscillatory excitations through external time-dependent magnetic fields has been a particularly fruitful technique over the last few years, as it has been implemented to produce vortex [597] and multivortex configurations [542], and perhaps even more importantly to generate tangles of vortices and superfluid quantum turbulence in atomic BECs [598]. It might also be possible that imperfections in the true optical barrier used in the experiment could pin vortices for a portion of the barrier ramp process, and significantly alter the vortex formation and annihilation process. Finally, much of the present work focused on the central portion of the trap, a region that encompasses the center of the merged BEC. In the experimental work; however, vortices were most often seen further from the BEC center.

3.9.5 ▪ Formation of Vortices through the Breakdown of Superfluidity: Defect Dragging

As we saw previously in 1D BECs (cf. Sec. 2.5.1), the dragging of an obstacle through a condensate beyond a critical speed, leads to the breakdown of superfluidity and the spontaneous emission of coherent structures[21] as a nonlinear analogue of Čerenkov radiation. In 2D BECs, the corresponding phenomenology has also attracted considerable atten-

[21] This emission assumes in 1D BECs the form of dark solitons; cf. details in Sec. 2.5.1.

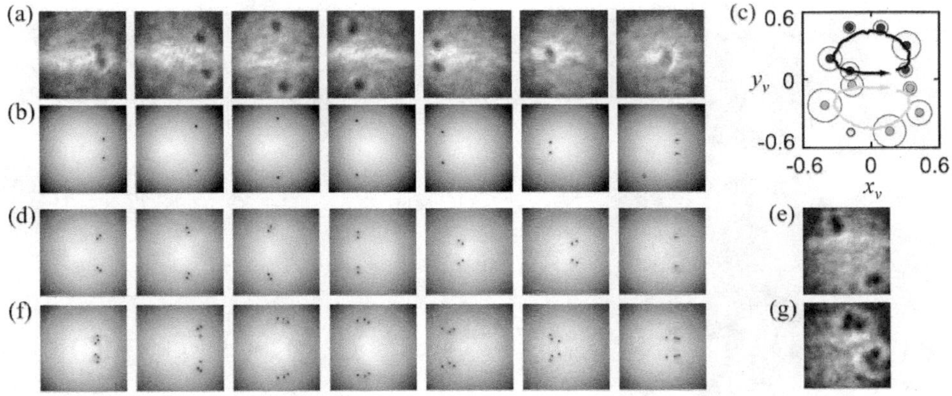

Figure 3.89. *Panel* (a) *shows an experimental sequence of snapshots from dragging a blue-detuned laser beam through an oblate condensate. Panel* (b) *shows a corresponding numerical sequence, while* (c) *summarizes the (epicyclic; cf. with the internal modes of the vortex dipole) motion of the vortices in the dipole. Panels* (d) *and* (e) *refer to a theoretical sequence and a single experimental snapshot for a larger speed inducing a pair of vortex dipoles, while* (f) *and* (g) *do the same but now for a triplet of vortex dipoles. Adapted with permission from Ref. [536].*

tion [391]. In particular, the GPE model has been used to study the flow of a BEC around an obstacle, leading to vortex emission and formation of bow waves [599, 512, 600]. Importantly, these theoretical studies spurred a series of early experiments [390, 248], demonstrating the onset of dissipation induced by the motion of an impurity (in the form of a blue detuned laser beam) through the condensate for velocities only above a critical velocity. In these pioneering experiments, indirect methods were used for the identification of the breakdown of superfluidity (and the consequent formation of vortices). In particular, in Ref. [390], a strong heating of the condensate and a calorimetric method were utilized for the identification of the critical point, while in the later work of Ref. [248], the onset of the formation of a pressure gradient was used for the relevant observations. Importantly, a decade later, the improvement of experimental techniques enabled a *direct observation* of the formation of one or multiple vortex dipoles through the dragging of a blue-detuned laser beam in the experiments of Ref. [536].

A principal figure representing the experimental findings is shown in Fig. 3.89. The figure shows snapshots from the experiment in the top row and corresponding different sets of numerical simulations in (b), (d), and (f) with progressively increasing speeds. The snapshots for the larger speeds illustrate the possibility both numerically [(d) and (f)] and experimentally [(e) and (g)] for a doubly charged [(d) and (e)] and even a triply charged [(f) and (g)] vortex dipole. It is intriguing to point out the connection of the epicyclic motion of the vortices in this dynamics with the epicyclic mode of the linearization around the stationary vortex dipole analyzed previously in this Chapter.

Although we will explore some case examples of multicomponent 2D BECs in a separate section below, it is worthwhile to note that the dragged defect problem and the associated breakdown of superfluidity is also of particular interest in the context of multicomponent miscible condensates. This has been examined in the case of pseudospinor (two-component) condensates in the work of Ref. [326] and in the case of spinor (three-component) BECs in Ref. [410]. The remarkable feature in this case, as explained in Sec. 2.5.1, is the presence of multiple distinct speeds of sound, which in the context of an infinitesimally weak defect can be analytically calculated and expressed as a function of the interaction strengths and the individual components' chemical potentials/densities [326].

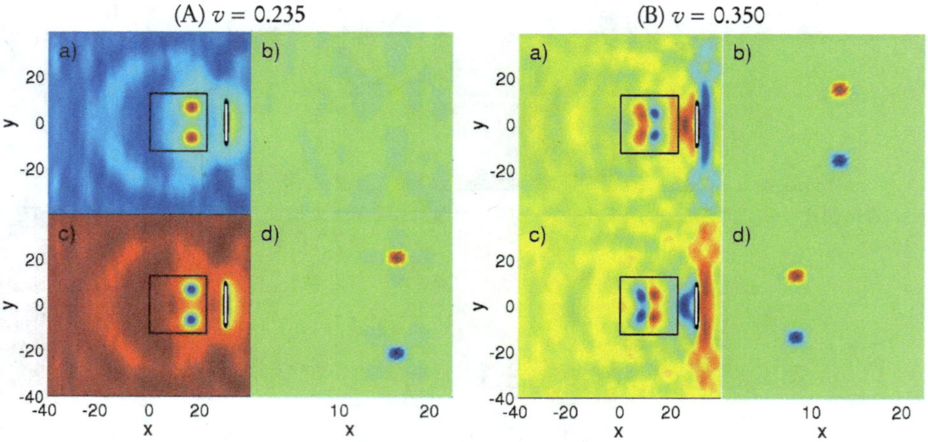

Figure 3.90. *Snapshots after vortex nucleation for different velocities of a running defect of size* $w = 10$ *[elongated vertical bar in panels* (a) *and* (c)*].* (A) $v = 0.235$ *is subsonic for the first component and supersonic for the second component.* (B) $v = 0.350$ *is supersonic for both components. The top (bottom) panel corresponds to the first (second) component. Left panels* (a) *and* (c) *show the square modulus of the solution together with the moving impurity. The right panels* (b) *and* (d) *show the vorticity* ω_{vor} *in the rectangular area depicted in the left panel counterparts. Red/blue (top/bottom spots) corresponds to regions with positive/negative vorticity. Adapted with permission from Ref. [326].*

Assuming the case of two components for simplicity, if the dragging speed is subcritical with respect to both sound speeds, there are no excitations arising. If the speed is supercritical with respect to both speeds, then vortices are emitted in both components. However, the most intriguing case is the intermediate one where the defect is supercritical with respect to one component, but subcritical with respect to the other. In that case, a vortex dipole will emerge in the supercritical component, but no vortices can arise in the subcritical second one. Instead, what we observe in Fig. 3.90 is the emission of a pair of "lumps" in the second component that "accompany" the vortex dipole.

More specifically, in the case shown in the figure, the impurity potential has a 2D form [326]:

$$V_1(x,y) = \frac{A}{4}\exp\left[-(x-vt)^2/2\epsilon^2\right][\tanh(y+w/2)+1][\tanh(-y+w/2)+1],$$

modeling a light sheet of strength A, width ϵ, and size w [see elongated vertical bar in panels (a) and (c) of Fig. 3.90]. In the 2D simulations shown, parameter values are $A = 0.9$, $\epsilon = 0.5$, and $w = 10, 15, 20$. In Fig. 3.90 we illustrate the two regimes leading to vortex nucleation (the trivial regime for subcritical velocity in both components is not shown here):

(A) $v = 0.235$ is subcritical in the first component, but supercritical in the second, resulting for the latter in a vortex state which is coupled to a lump (a 2D structure on a finite background) in the first component.

(B) $v = 0.350$ is supercritical for *both* components. This results in the formation of a dipole state which contains a vortex-lump pair, coupled to a lump-vortex pair, in a form reminiscent of the molecules reported in Ref. [601]. These molecular states also appear to represent a two-dimensional generalization of the dark-dark soliton state discussed in Sec. 2.4.6. Further elucidation of this connection would certainly be worthwhile; see also the discussion in Sec. 3.11.1 below.

It is worth mentioning that similar findings were reported in the case of spinor condensates in Ref. [410].

3.10 ▪ Quantum Turbulence in Two-Dimensional BECs

We now turn to a brief examination of the potentially turbulent dynamics of one-component 2D BECs. Our exposition will follow the path of Ref. [602]; see also the relevant discussion of Ref. [603].

A distinctive feature of 2D flows subject to small-scale forcing is that energy flux is blocked through the small length scales and, instead, it is transferred towards larger scales. This constitutes the so-called inverse energy cascade of 2D turbulence [604, 605]. The nature of 2D turbulence in quantum fluids is less clear than its classical fluid counterpart. Progress in 2D quantum turbulence (2DQT) may offer innovative routes to understanding quantum fluid dynamics [606, 607] and aspects of its potential universality. The aim of the present section (and of the work of Ref. [602]) is to describe an experimental and numerical study of forced and decaying 2DQT in a dilute-gas BEC. The relevant study revealed three key characteristics of 2D turbulence that may simultaneously appear in systems exhibiting 2DQT:

(i) emergence of large-scale flow from small-scale forcing, seen experimentally and numerically,

(ii) numerical observation of the formation of coherent vortex structures accompanying approximate enstrophy conservation, and

(iii) numerical observation of an incompressible kinetic energy spectrum with $k^{-5/3}$ dependence for low wavenumbers k and k^{-3} dependence for high k.

These observations are consistent with the notion that an inverse energy cascade can exist in this system, although admittedly the relevant issue has not been decisively settled.

The theme of 2D turbulence and its quantum ramifications has a rich history. Analyzing point vortex motion in a bounded domain, Onsager proposed in 1949 that long-lived vortices may develop via mergers of smaller vortices in turbulent flows of a 2D fluid, enabling the remaining vortices to move more freely and thereby maximize system entropy [608, 609]. He also proposed that vortices in superfluids have quantized circulation, and implied that turbulent 2D vortex dynamics might be ideally studied in superfluids. However, experimental demonstration of 2DQT has not been reported and has only recently been addressed numerically [610, 611, 612, 613, 614, 615].

To experimentally reach the 2DQT regime, optical and magnetic confinement was utilized in Ref. [602] to create highly oblate BECs [536]. A harmonic potential with radial (r) and axial (z) trapping frequencies $(\omega_r/2\pi, \omega_z/2\pi) = (8, 90)$ Hz confines BECs of up to $\sim 2\times 10^6$ ^{87}Rb atoms having radial and axial radii of 52 μm and 5 μm respectively. For these conditions, vortex bending and tilting are suppressed [616], leading to 2D vortex dynamics. An annular trap was created with a 23 μm radius, blue-detuned, Gaussian laser beam directed axially through the trap center; the beam creates a barrier of height $U_0 \sim 1.5\mu_0$, where $\mu_0 \sim 8\hbar\omega_z$ is the BEC chemical potential in the purely harmonic trap. Relative to the phase transition temperature $T_c \sim 116$ nK, the initial temperature was $T \sim 0.9T_c$.

At time $t = 0$ of each experimental run, a magnetic bias field moves the center of the harmonic trap, but not the central barrier, in one complete 5.7 μm diameter circle over 333 ms. This motion induces small-scale forcing and nucleation of numerous vortices in a highly disordered distribution, which we identify with 2DQT much as the notion of a "vortex tangle" is identified with 3D quantum turbulence [606, 499, 598]. Afterwards,

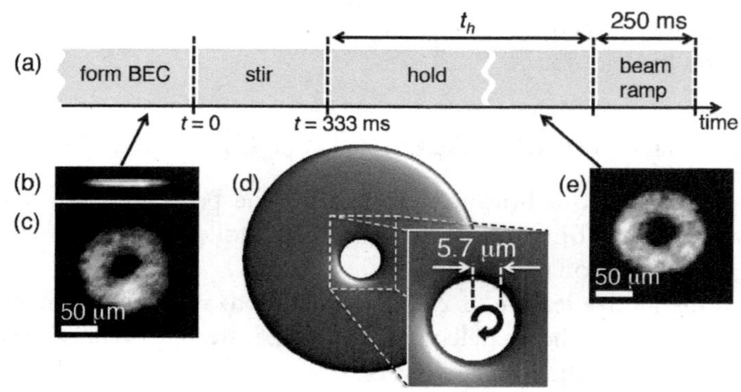

Figure 3.91. (a) *Timing sequence used to study 2DQT in Ref. [602].* (b) *and* (c) *Experimental in situ column-density images of the BEC immediately prior to the stir, as viewed* (b) *in the plane of 2D trapping and* (c) *along the z axis. Lighter grayscale shades indicate larger column densities, as in subsequent experimental and numerical density data.* (d) *Illustration of stirring; the black arrow shows the trajectory of the harmonic trap center relative to the larger fluid-free region created by the laser barrier.* (e) *In situ image of the BEC 10 s after stirring; vortices are not observable, necessitating an expansion stage to resolve them. Adapted with permission from Ref. [602].*

the BEC remains in the annular trap for a variable hold time t_h up to 50 s while the 2DQT decays. At t_h =1.17 s, the system temperature is reduced to $\sim 0.6T_c$ in order to decrease rates of thermal damping and vortex-antivortex recombination. At the end of the hold period, the central barrier is ramped off over 250 ms, the BEC is released from the trap, and ballistic expansion of the BEC enlarges the vortex cores such that they can be resolved by absorption imaging. Figure 3.91 illustrates this sequence and shows images of the trapped BEC.

Two experimental time sequences of post-stir dynamics are shown in rows (a) and (b) of Fig. 3.92. One can clearly observe the microscopic variability of vortex distributions. Just after the stir ($t_h = 0$ ms) a disordered vortex distribution appears. Large-scale superflow is evident after $t_h \approx 0.33$ s and with increasing t_h, as indicated by the large fluid-free hole in the middle of the expanded BEC; this flow evolves into a persistent current by $t_h \approx 8.17$ s. An optional 3 s hold between barrier ramp-down and BEC release gives the vortices pinned by the central barrier time to separate enough to determine the circulation winding number about the barrier. The experiment, therefore, demonstrates that under suitable conditions of forcing and dissipation, a highly disordered vortex distribution can evolve into a large-scale flow in an annular trap. However, measuring kinetic energy spectra and in situ vortex dynamics remain highly nontrivial experimental problems, due to the necessity to acquire phase information which is not directly available in a distributed field form, at present (as is, e.g., the density). For that reason, the work of Ref. [602] employed numerical modeling and analysis for further characterizing 2DQT in a stirred, trapped BEC.

As a first-principles theoretical approach that is numerically tractable, the modeling of Ref. [617] was used. The physical system of interest consists of a large noncondensate component close to thermal equilibrium and a BEC responding both to external forcing and to damping by the noncondensate component. Numerically, we focus on the dynamics of just the BEC. The experimental stirring procedure was simulated using the dissipative Gross–Pitaevskii theory [462]. The parameters most readily measured are the total atom number N and temperature T. An efficient Hartree–Fock scheme was used for determining the chemical potential $\mu(N,T)$ and reservoir cutoff energy $\epsilon_{\text{cut}}(N,T)$ in

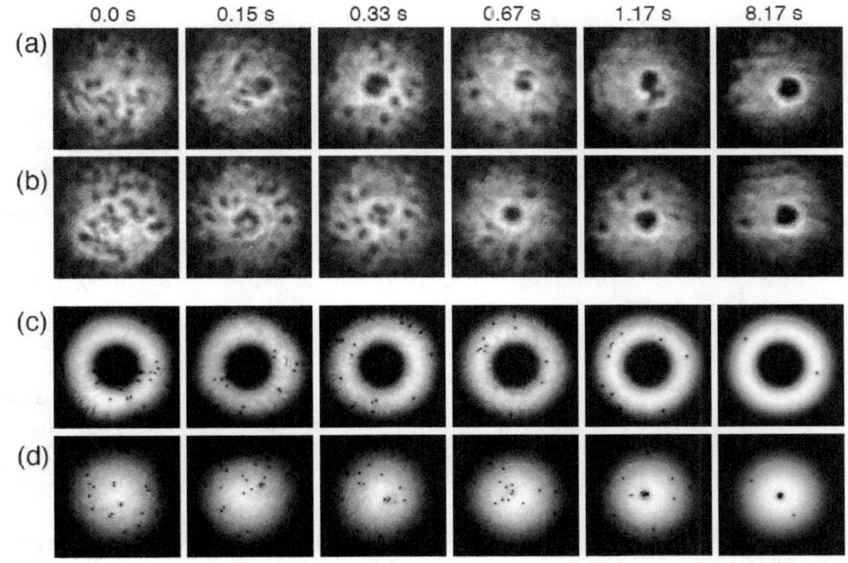

Figure 3.92. (a) *and* (b) *200-μm-square experimental column-density images acquired at the hold times t_h indicated above the images. Each BEC undergoes ballistic expansion immediately after the central barrier ramp-down in order to resolve the vortex cores. Each image is acquired from a separate experimental run.* (c) *In situ numerical data (96-μm-square images) for the hold times indicated. For each state represented in* (c), *ramping off the laser barrier in 250 ms gives the data shown in* (d). *Adapted with permission from Ref. [602].*

Ref. [618]. This was adapted to the present experiment, accounting for the shift in the trap minimum caused by the central barrier.

Row (c) in Fig. 3.92 shows simulations that correspond to experimental observations. Here, too, vortices become pinned to the central barrier to form a persistent current; at $t_h = 8.17$ s, three vortices are pinned to the barrier, as indicated by the corresponding quantum phase profile (not shown here). Ramping off the obstacle beam in the simulation (over 250 ms as in the experiment) gives the column densities shown in row (d) of Figure 3.92, with density distributions more readily compared to rows (a) and (b). Development of superflow at $t_h = 8.17$ s in row (c) leads to a large region of low density in the trap center after the barrier is ramped down, as seen in rows (a), (b), and (d).

In order to further analyze the numerical simulations and characterize 2DQT, the wavenumber spectrum of the system and its temporal evolution need to be studied. This leads to the observation of two distinct dynamical features of the system evolution, namely the development of a logarithmically bilinear kinetic energy spectrum, and the formation of tightly bound, long-lived clusters of vortices with the same sign of circulation. To examine numerically the dependence of the kinetic energy on the wavenumber k at any instant in time, we use the techniques of previous studies [611, 612, 613, 614, 615, 619] for extracting $E^i(k)$, the portion of the kinetic energy spectrum that corresponds to an incompressible superfluid component, derived by extracting the divergence-free density-weighted velocity field that embeds vorticity; the curl-free part of this field corresponds to sound waves and acoustic energy.

The spectra of Fig. 3.93 are obtained from various times of the simulation and calculated using spatial grids of 1811×1811 points separated by $\xi/4 = 0.1$ μm, where the ξ is the healing length. Each curve shows the spectrum of a 2D slice through $z = 0$, although the spectra are only weakly affected by averaging slices through the BEC. The ultravio-

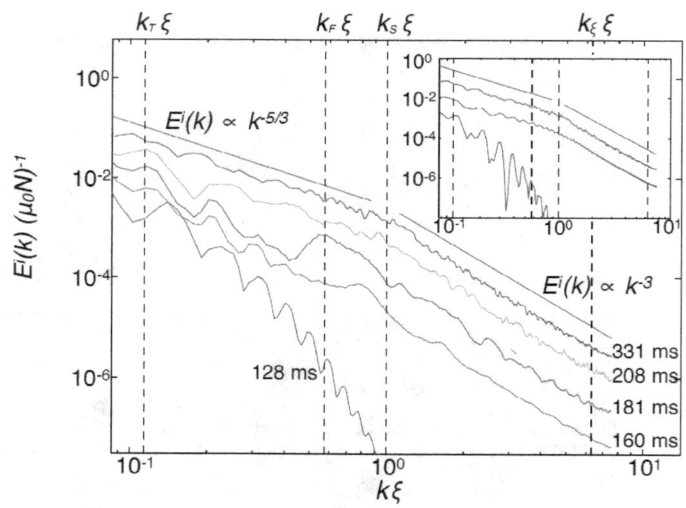

Figure 3.93. *Log-log plots of* $E^i(k)$ *(per atom) for times* (128, 160, 181, 208, 331) *ms over which forcing occurs, plotted against* $k\xi$ *with healing length* $\xi = 0.42\ \mu m$*. Vertical dashed lines indicate* k_T, k_F, k_s, *and* k_ξ, *defined in the text. Red and blue lines indicate* $E^i(k) \propto k^{-5/3}$ *and* k^{-3}, *respectively. A spectral peak at* 181 *ms appears at* $k_F \approx 2\pi/(11\xi)$*. Inset: log-log plot of* $E^i(k)$ *vs.* $k\xi$ *(labels omitted) over the same domain as the main plot. From top to bottom, curves show* $E^i(k) \propto k^{-5/3}$ *and* k^{-3} *(solid lines), and* $E^i(k)$ *at* 331 *ms, after* 14 *s of free decay, and for a charge-3 persistent current. Adapted with permission from Ref. [602].*

let (large k) $E^i(k) \propto k^{-3}$ region of the spectrum is a conspicuous feature once vortices are present. This power law is a universal property of a quantized vortex core in a compressible 2D quantum fluid, as analyzed in Ref. [603], occurring for $k > k_s \equiv \xi^{-1}$. The associated length scale $\sim 2\pi\xi$ thus serves to distinguish between scales where the system's physical characteristics are dominated by motion of point-like vortices ($k < k_s$), and those where characteristics derive from the structure of individual vortex cores ($k > k_s$). The ultraviolet power law plays a role in the energy spectrum through its amplitude, which is proportional to the total vortex number [603]. The only mechanisms that can appreciably change the incompressible energy for $k > k_s$ are creation and loss of free vortices.

As stirring injects kinetic energy into the system a Kolmogorov $k^{-5/3}$ power-law spectrum develops in the $k < k_s$ region, and is determined by the vortex configuration [603]. This spectrum spans a decade in k-space and is established by the end of the stir ($\sim$ 331 ms) which is also when total incompressible kinetic energy is maximal. Logarithmically bilinear spectra are also obtained after ramping off the central barrier (not shown). After the end of the stir, Fig. 3.93 (inset) indicates a slow loss of energy with approximate preservation of the Kolmogorov power law. Eventually the system spectrally condenses, leading to the generation of a metastable persistent current with three units of circulation.

Three additional wavenumbers are indicated in Fig. 3.93. The cross-sectional radial thickness of the BEC approximately corresponds to the length scale $25\ \mu m = 2\pi/k_T$. At high wavenumbers, $k_\xi = 2\pi/\xi$ corresponds to the scale of the healing length $\xi = 0.42\ \mu m$, the approximate size of the smallest features (e.g., vortex cores) supported by a BEC. Finally, a wavenumber k_F is associated with the peak in the incompressible spectrum at $\sim$ 181 ms, as was described in Ref. [602].

In the classical theory of forced 2D turbulence [605], spectrally localized forcing is related to the injection rates of enstrophy (η) and energy (ϵ) density as $k_F = \sqrt{\eta/\epsilon}$. In our numerical results, the forcing that precedes the full development of the logarithmically bi-

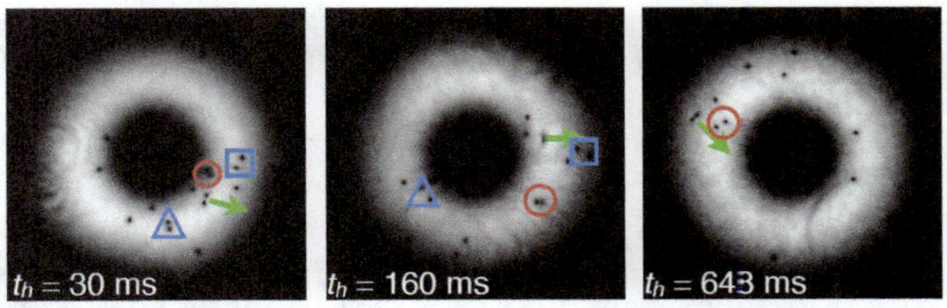

Figure 3.94. *Numerically obtained BEC column density shown for three hold times. In each 96-μm-square image, symbols indicate clusters of same-sign vortices, either co-rotating (red), or counter-rotating (blue) with the stir. Each shape represents the same cluster at different times. Vortex dipoles and their propagation directions are indicated by green arrows. At $t_h = 648$ ms, one cluster remains after having traveled clockwise halfway around the BEC. Adapted with permission from Ref. [602].*

linear spectrum is associated with the spectral peak at 181 ms in Fig. 3.93. At 208 ms the peak has dispersed and the spectrum is already approximately logarithmically bilinear. We estimate the location of k_F from the numerically computed changes in total incompressible kinetic energy, $\Delta E^i \approx 2.9 \times 10^{-3}\, \mu_0 \cdot N$, and enstrophy, $\Delta\Omega \approx 0.963 \times 10^{-3}\, \mu_0 \cdot N/\xi^2$, occurring during this 27 ms time interval. We find $k_F \equiv \sqrt{\Delta\Omega/\Delta E^i} = 0.57\xi^{-1} \simeq 2\pi/(11\xi)$, shown in Fig. 3.93, and coinciding with the spectral peak.

The physical mechanism for injection of energy and vorticity into the BEC involves coupling between pairs of opposite-circulation vortices and acoustic energy. In the stirring process, density waves are first generated in the BEC, most prominently where the fluid density approaches zero. These density waves decay to vortices in two characteristic ways. First, a density wave can develop into a localized dark soliton-like entity and subsequently decay to a vortex dipole. A second injection mechanism involves the decay of a density wave near the central barrier into a single vortex within the fluid and a partner antivortex pinned by the barrier.

Empirically we find that the length scale $\sim 2\pi/k_F$ coincides with the separation of phase singularities created from the decay of a localized sound pulse into a vortex dipole. Examining the instances of vortex dipole creation from sound during the stir period, we find dipole lengths d in the range 6.7ξ to 11ξ, suggesting an injection of incompressible energy near a wavenumber $k \sim 2\pi/d$. During the stir, there is one case of a vortex dipole annihilating irreversibly to sound at $t = 0.23$ s, where the dipole length is $d \sim 6.7\xi$. Two transient events during the nominal constant enstrophy period discussed below correspond to dipole annihilation at $\sim 6.7\xi$ and $\sim 8.9\xi$. Furthermore, the superfluid density modulations preceding vortex nucleation at the 181 ms spectral peak have a length scale of approximately $11\xi \simeq 2\pi/k_F$. Taken together, these observations indicate that forcing involves efficient energy and enstrophy transfer from the compressible to the incompressible fluid components for wavenumbers $k_F \lesssim k \lesssim k_s$.

The conservation of enstrophy in 2D turbulence corresponds to conservation of vortex number in 2DQT [612, 603]. Between ~300 ms and ~600 ms after the stir begins, the total vortex number is nominally constant in our simulation. When vortex-antivortex annihilation occurs, the resulting sound pulses are quickly (within ~20 ms) refocused by the inhomogeneous density, regenerating the vortices. Thus, this 300 ms period of nominally constant vortex number is identified with enstrophy conservation. During this period, four instances of the formation of two-vortex clusters (same-sign vortices) were observed in Ref. [602]. Figure 3.94 shows three two-vortex clusters and a vortex dipole (opposite-sign vortices) present 30 ms after the end of stir. This dipole exists for 20 ms, while the longest-lived of the two-vortex clusters exists for 630 ms. The vortices of this cluster orbit each other 15 times, travel together halfway around the BEC, and eventually

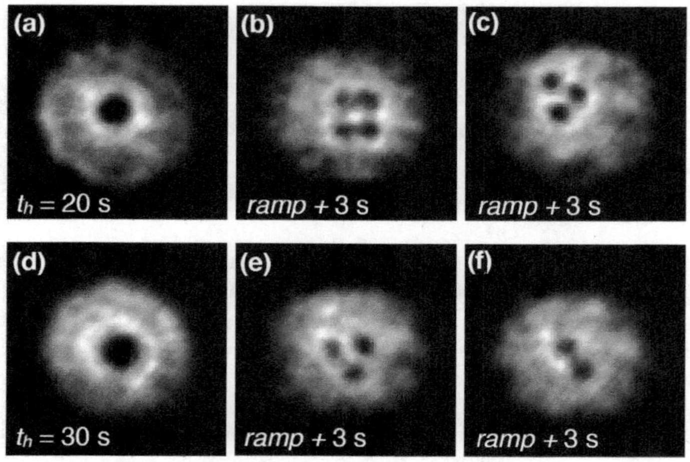

Figure 3.95. *Vortices observed in experimental data. Here, BECs are held in the annular trap for 20 s prior to the central barrier ramp down, trap release, expansion, and imaging. In* (a) *the expansion and imaging procedure takes place immediately after the barrier ramp. The dark region in the center of the BEC corresponds to multiple units of circulation, and is not due to the presence of the barrier itself; i.e., with no circulation about the barrier, a hole would not appear in the BEC after ballistic expansion. In* (b) *and* (c), *BECs are held for an additional 3 s in the trap before the expansion imaging procedure. This additional hold time allows the vortices that comprise the region of vorticity shown in* (a) *to dissociate and become experimentally observable.* (d)–(f) *show similar images for* $t_h = 30$ *s. Adapted with permission from Ref. [602].*

dissociate upon colliding with a vortex dipole; see Fig. 3.94. As seen in Fig. 3.92(a) and (b), there exist large regions free of vortices, and regions where many vortices are densely packed, which is indicative of clustering, but no further direct measures of this feature were developed. Clustering was, however, experimentally observed in Ref. [536] in the form of dipolar clusters reproducibly generated by a moving obstacle. In the present case of circular forcing, numerical observations indicated the intermittent emergence of long-lived clusters within otherwise irregular flows.

Over the tens of seconds after stirring stops, free vortices decay either by being pinned to the obstacle beam, or by damping at the outer BEC boundary, or by annihilating with free or pinned vortices of opposite sign. This results in the formation of a persistent current, reflecting the net angular momentum imparted by the stirring. The development of this superflow represents the growth of large-scale flow originating from small-scale forcing in an annular geometry, and serves as a check of our numerical procedures and interpretations. It should be reminded here that multiquantum vortices are energetically unstable. For this reason, the loose clustering of multiple quanta is favorable to perfect colocation of multiple vortices. We make use of this energetic instability in order to determine the size of the persistent current formed and its subsequent time evolution in the experiment. This is shown in more detail in Fig. 3.95. The extra hold time between beam ramp down and expansion allows for a count of the total number of free and pinned vortices for any hold time. For long-enough hold times where the number of free vortices has dropped to much less than one per image (on average), the observed vortices can be attributed to current pinned at the barrier, particularly if they appear clustered about the position of the barrier. This feature is directly demonstrated in Fig. 3.95, where one can clearly discern 4 and 3 vortices, respectively, in panels (b) and (c) after a hold time of 20 s, while 3 and 2 vortices, arise, respectively, in panels (e) and (f) after a hold time of 30 s.

Earlier investigations of 2DQT centered on numerically obtaining kinetic energy spectra, but the possibility of an inverse energy cascade in 2DQT, conservation of enstrophy, and correspondence between spectra and vortex dynamics, was not conclusively established. In the study of Ref. [602], simulation of the experimentally realized forcing showed the development of an inertial range. Additionally, small-scale forcing within the trap enabled the clustering of vortices; as clustering suppresses vortex-antivortex annihilation, it provides a mechanism to enable enstrophy conservation in a compressible superfluid. Regarding the possibility of a compressible superfluid supporting an inverse energy cascade, energy flux calculations provided the most direct route to analyzing cascades, although such an approach for a trapped BEC with an inhomogeneous density distribution is still an open problem. Nevertheless, the following observations have been found to be consistent with the existence of an inverse energy cascade near the end of the stir:

(i) vortex dipole recombination was suppressed and thus there was little dissipation over a forcing range k_F to k_s;

(ii) $E^i(k) \propto k^{-3}$ for $k > k_s$, a range that cannot support incompressible energy flux [603];

(iii) enstrophy was nominally conserved, and kinetic energy spectral developments occurred primarily for $k < k_s$;

(iv) $E^i(k) \propto k^{-5/3}$ for $k < k_s$, a signature of conserved energy transfer across the associated scales.

These observations indicate that characteristics of forced and decaying 2DQT in compressible quantum fluids may be analogous to those of 2D turbulence in incompressible classical fluids. In particular, growth of large-scale flow, aggregation of vorticity, nominal enstrophy conservation, and an energy spectrum with $k^{-5/3}$ and k^{-3} spectral features occur with suitable forcing. It would certainly be worthwhile that these observations motivate further investigations of 2DQT, with future work focusing on energy fluxes, the roles of dissipation and boundary conditions, and direct experimental observations of turbulent vortex dynamics of a 2D quantum fluid (which at the moment appear to be somewhat beyond the present state of the art).

3.11 ▪ Multicomponent Two-Dimensional BECs

3.11.1 ▪ Vortex-Bright and Vortex-Vortex Solitons

In this section, following the analysis of Ref. [620], we will study a prototypical excitation in 2D, two-component settings, namely the so-called vortex-bright solitary wave (note that such solutions were also proposed in Ref. [507]). This study was initially motivated as a 2D (more generally, multidimensional) generalization of the recent experimental studies of Refs. [68, 333, 334, 335, 336] which considered DB solitons in two-component, quasi-1D BECs (see details in Sec. 2.4.2). Note that, as in the case of a dark soliton in the quasi-1D setting, a vortex in the quasi-2D setting creates an "effective potential well" in which the bound bright soliton resulting in the second component (which is normally not supported in the presence of repulsive interactions) can be perceived as a nonlinear bound state. Thus, similarly to DB solitons, vortex-bright solitary waves can be considered as "symbiotic" (perhaps a more appropriate term is "parasitic") structures.

For our present considerations, we will use a nondimensional Hamiltonian for the two-component condensate in the form

$$H=\int d\mathbf{r}(\nabla\Psi)^{\dagger}(\nabla\Psi)+\Psi^{\dagger}V(\mathbf{r})\Psi+\frac{1}{2}|\Psi|^{2\dagger}U|\Psi|^{2}-\Psi^{\dagger}M\Psi, \tag{3.199}$$

where $\Psi(\mathbf{r}) \in \mathbb{C}^2$ is the pseudospinor order parameter, $|\Psi|^2 = (|\Psi_1|^2, |\Psi_2|^2)^{\dagger}$, $M = \text{diag}\{\mu_1, \mu_2\}$ is the diagonal matrix of chemical potentials associated with the conservation of the number of atoms $N_1 = \int d\mathbf{r}|\Psi_1|^2$ and $N_2 = \int d\mathbf{r}|\Psi_2|^2$; notice that a diagnostic that was also used in Ref. [620] is $R = N_1/(N_1+N_2) = N_1/N$. Finally, in Eq. (3.199), U is a 2×2 matrix accounting for the nonlinear interatomic interactions. For the $|1,-1\rangle$ and $|2,1\rangle$ components of ^{87}Rb, one has [279] $U_{11} = 1.03$, $U_{12} = U_{21} = 1$, and $U_{22} = 0.97$. These determine, through the negative sign of $\det(U)$, the immiscible nature of the interactions leading to phase separation [277, 279]. The confining potential is

$$V(\mathbf{r},z)=\underbrace{\frac{\omega_r^2}{4}|\mathbf{r}|^2+\frac{\omega_z^2}{4}z^2}_{V_{\text{MT}}}+\underbrace{A[\sin^2(2\omega_r x)+\sin^2(2\omega_r y)]}_{V_{\text{OL}}}, \tag{3.200}$$

where, as usual, V_{MT} denotes the parabolic (magnetic trap) component and V_{OL} the optical lattice component. Further details about the spatial and temporal rescalings can be found in Ref. [620].

The equations of motion, $(\Psi_t, \text{c.c.})^T = J\sigma(\delta H/\delta\Psi, \text{c.c.})^T = J\sigma DH$, where $J = \text{diag}(-iI, iI)$ and σ interchanges rows $(3,4)$ with $(1,2)$, can be expressed as

$$i\Psi_t=-\nabla^2\Psi+V(\mathbf{r})\Psi+U|\Psi|^2\Psi-M\Psi. \tag{3.201}$$

Then, the stability of stationary solutions is determined by the eigenvalues of the Hessian of the Hamiltonian, σD^2H, and of $J\sigma D^2H$. Here it should be recalled that negative eigenvalues of σD^2H indicate energetic instability, since "dissipative" perturbations (e.g., from exchanges of atoms with the thermal cloud if the temperature deviates from zero) in the system can render them dynamically unstable, as can collisions with other eigendirections even in the pure Hamiltonian (zero-temperature) system. The linear stability of the latter system is examined (as before) through the eigenvalues $\lambda = \lambda_r + i\lambda_i$ of $J\sigma D^2H$; instability arises when $\lambda_r \neq 0$ since, due to the Hamiltonian structure, the eigenvalues are symmetric over both the real and imaginary axes.

The variation can be posed in the $\{\Psi, \Psi^*\}$ or the $\{\Psi_{\text{real}}, \Psi_{\text{imag}}\}$ basis. The former is useful when the potential is axisymmetric, since then small excitations to a stationary solution, bearing charges S_1 and S_2, $\Psi = (\Psi_1(r)e^{iS_1\theta}, \Psi_2(r)e^{iS_2\theta})$ of the form $\psi = (a_1, a_2)^T(\mathbf{r})e^{\lambda t} + (b_1, b_2)^T(\mathbf{r})e^{\lambda^* t}$ will have definite angular momentum $\alpha_j(r,\theta) = \tilde{\alpha}_j(r)e^{ix_{\alpha_j}\theta}$. If we set $x_{a_1} = x$ then $x_{b_1} = x-2S_1$, $x_{a_2} = x-S_1+S_2$, and $x_{b_2} = x-S_1-S_2$, so a single index x will indicate the angular momentum of the excitation with given eigenvalue λ. Hence, the spectrum of eigenvalues $\{\lambda\}$ can be decomposed as the union of the spectra $\{\lambda_x\}$ pertaining to angular momentum x. We will also assume $S_2 = 0$, so that $S_1 = S$ and $x_{a_2} = x_{b_2}$. It has been shown numerically [488] and analytically [621] that instability windows arise in a single component with topological charge S only for wavenumbers with $|x| < S$ (see also discussion in Sec. 3.1). The null eigenvalues corresponding to gauge invariance appear in the spectrum of $x = S$. For a single component in a parabolic trap, an anomalous mode for $x = S-1$ converges to zero as $\omega_r \to 0$, accounting for translational invariance and leading to the energetic stability of the $S = 1$ vortex without external potential. For each $0 \le x < S-1$ $(S > 1)$ an anomalous mode leads to windows of instability

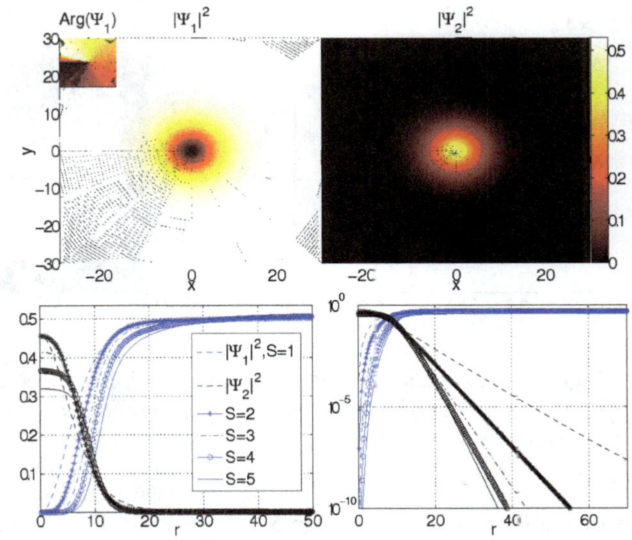

Figure 3.96. *The energetically stable $S = 1$ vortex-bright state without external potential for $R = 0.99$, $r_{max} = 60$, and $N = 5900$ (top row). The bottom row shows radial profiles of unit and higher charge vortex-bright solitons in a homogeneous medium on a regular (left) and logarithmic (right) scale. All profiles are for $N = 10000$, $R = 0.99$, and $r_{max} = 80$. Adapted with permission from Ref. [620].*

[488, 621]. As demonstrated in Ref. [620], these can be significantly suppressed, although the $S-1$ spectrum occasionally leads to small instability windows.

The numerical methods developed in Ref. [620] extended those used earlier in, e.g., Refs. [622, 623]. The spatial discretization in (r, θ, z) employed Chebyshev polynomials to represent the r dependence [624]. The Fourier modes representing θ and z make the Laplacian operators diagonal in these directions. To identify stationary states of Eq. (3.201), an initial estimate was obtained via imaginary time (i.e., replacing $t \to -it$) integration using a first-order implicit/explicit Euler scheme with $\Delta t = 10^{-2}$. Subsequently, the solution was refined using Newton's method. The linear system arising at each Newton step was solved using the matrix-free IDR(s) algorithm [625, 626], which requires only the action of the Hessian. To accelerate inversion, the system was preconditioned with the inverse Laplacian, using its block diagonal structure. Hence, the system solved involved $\nabla^{-2}D^2H(\Psi_n)\Delta_n = \nabla^{-2}DH(\Psi_n)$ and the update $\Psi_{n+1} = \Psi_n - \Delta_n$ for $n = 0, 1, \ldots$. Fewer than 5 Newton iterations were typically needed to achieve an accuracy of $||\nabla^{-2}DH(\Psi)||_{l^2}/||\Psi||_{l^2} < 10^{-12}$.

For each stationary solution Ψ, the matrix-free implicitly restarted Arnoldi algorithm was used to iteratively compute the eigenpairs of the linearization $J\sigma D^2H(\Psi)$ to a specified tolerance [627]. In order to find the desired eigenvalues inverse iteration was utilized, with the IDR(s) method and inverse Laplacian preconditioning to solve the linear systems, as above. Here, the preconditioner was taken to be $[J\sigma(\nabla^2)]^{-1}$, so that each iteration solves $\nabla^{-2}D^2H(\Psi)v_{n+1} = -\nabla^{-2}\sigma J v_n$. A resolution in (r, θ, z) of $40 \times 64 \times 80$ was used to represent nonaxisymmetric solutions and eigenvectors. For axisymmetric solutions, quantitative accuracy required only 30 radial modes for $N < 1000$, but up to 200 modes for larger N. For eigenvectors, only $S+1$ modes in θ were used and identified quantitatively all expected invariant and negative directions and windows of instability, using as a benchmark the pioneering work of Ref. [488].

In 2D ($\omega_z \to \infty$) for $\omega_r = A = 0$, we first demonstrate in Fig. 3.96 the existence of an energetically stable (and hence also dynamically stable) vortex-bright soliton state.

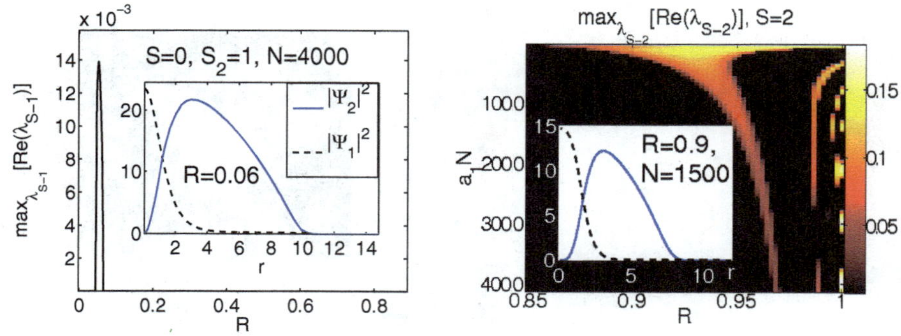

Figure 3.97. *Growth rate of the $S-1$ mode as a function of R with $N=4000$ for $S=0$ and $S_2=1$ (left), and growth rate of the $S-2$ mode as a function of $R\times a_1N$ for $S=2$ (right). Adapted with permission from Ref. [620].*

This is so for all of the (R,N) values that we have sampled. A similar state exists for vortices of higher topological charge S, as shown in Fig. 3.96 for $S\leq 5$. In this case, the negative energy modes [488] in the spectra associated to $0\leq \varkappa < S-1$ may lead to dynamical instability from complex quartets of eigenvalues as a result of Hamiltonian–Hopf bifurcations which can occur upon collision with positive modes.

In a parabolic trap, the $S=1$ vortex-bright structure remains dynamically stable. However, the breaking of the translational invariance produces a negative energy mode in the $\varkappa = S-1$ spectrum. Thus, an additional control parameter (through the atom number of the bright component) may lead to collisions of this mode with positive energy modes and, hence, rare isolated windows of instability arising for large $R<1$ and large N. A similar conclusion was also reached in Ref. [507]. This happens when $S_1=S=0$ and $S_2=1$ for sufficiently small values of R. A similar (oscillatory instability) feature has recently been shown in the DB 1D analogue of the vortex-bright states [335].

For higher-charge vortex-bright structures, the same situation holds for the $\varkappa = S-1$ spectrum, while for $R=1$ windows of instability arise from the negative modes in the spectra of $0\leq \varkappa < S-1$. As R decreases, however, these windows of instability are generically suppressed by the increasing presence of the second bright component. Figure 3.97 (right) depicts the growth rate of the $S-2$ spectrum for $S=2$ over (R,N) parameter space. An example of an unstable solution perturbed in the growing excitation direction is depicted in Fig. 3.98. The vortices first split from the center, forming a pair of vortex-bright solitons (to which we will return below) and begin to part and precess, but once they are far enough and the bright component is bimodal, they approach again and the bright component resumes uni-modality. The sequence repeats, similarly to single-component $S=2$ vortices [628].

When we impose an additional sinusoidal lattice potential, $A>0$, the one-component ($R=1$) $S=1$ vortex may become unstable (due to resonant eigenvalue collisions and ensuing oscillatory instabilities), at least for A sufficiently large [578]. The same holds for a large mass ratio $R<1$. However, below a critical R, once again the bright component has a stabilizing influence.

An additional nonlinear excitation that can exist in this two-component system is the so-called vortex-vortex solution. What this explores is that the above two-component system of equations in the special case where $g_{11}=g_{22}=g_{12}$ is invariant under SU(2) rotations. In particular, let us first recall that a general matrix element of SU(2) takes the form

$$\mathscr{U}=\begin{pmatrix} \alpha & -\beta^* \\ \beta & \alpha^* \end{pmatrix},$$

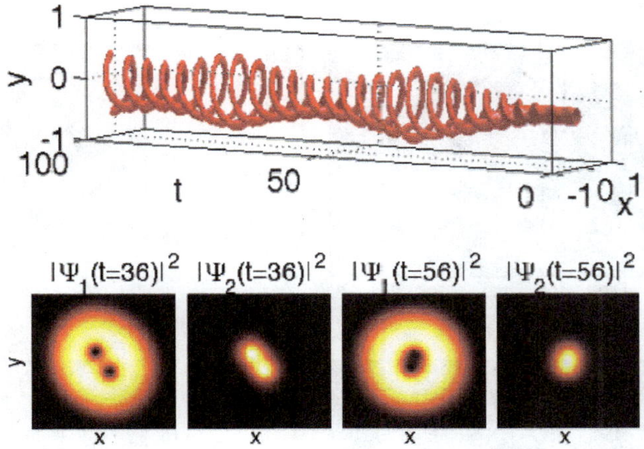

Figure 3.98. *The evolution in time of the unstable $S = 2$ solution for $(R,N) = (0.93, 200)$ perturbed in the direction of the growing excitation. Top: Approximate vorticity density isocontours show trajectories of the two vortices. Bottom: First splitting and rejoining. Adapted with permission from Ref. [620].*

where α and β are complex constants such that $|\alpha|^2 + |\beta|^2 = 1$. Then it can easily be shown that if $(u_1, u_2)^T$ are solutions of the nonlinear two-component system, then

$$\begin{pmatrix} u_1' \\ u_2' \end{pmatrix} \equiv U \begin{pmatrix} u_1 \\ u_2 \end{pmatrix} = \begin{pmatrix} \alpha u_1 - \beta^* u_2 \\ \beta u_1 + \alpha^* u_2 \end{pmatrix},$$

are also solutions of the same equations. As we saw in the previous chapter for the 1D BECs (see Sec. 2.4.2), this suggests that we may start from the exact DB soliton solution which exists in the absence of an external potential and derive the beating dark-dark (DD) soliton solution. To reduce the number of independent variables we can consider for simplicity (as in the 1D setting), the case of SO(2) rotated solitary waves. In this case, the corresponding orthogonal matrix is given by

$$U = \begin{pmatrix} \cos(\chi) & -\sin(\chi) \\ \sin(\chi) & \cos(\chi) \end{pmatrix}, \tag{3.202}$$

where χ is an arbitrary angle. In the present 2D, two-component realm, the canonical entity to rotate through the above mechanism is the vortex-bright solitary wave. In analogy to the case of the 1D BECs, where the rotated—beating in time—waves are termed dark-dark solitons (cf. Sec. 2.4.2), in the 2D case the relevant states are called vortex-vortex solitary waves. A prototypical example of such a vortex-vortex state for a $\chi = \pi/4$ rotation of a vortex-bright waveform is illustrated in Fig. 3.99.

3.11.2 ▪ A Radial Excitation: The Dark-Bright Ring

In addition to examining a two-component state involving vortices, in the present subsection we will also consider a two-component generalization of another principal state of the system, namely of the ring dark soliton. In particular, following the exposition of Ref. [629], we now discuss the existence of a radially symmetric 2D generalization of the DB soliton, the dark-bright ring soliton, a density profile of which is shown in Fig 3.100. For such a state, in the first, "dark" component the nodal line of zero density is accompanied by a phase jump of π in the order parameter. One-component ring dark solitons have

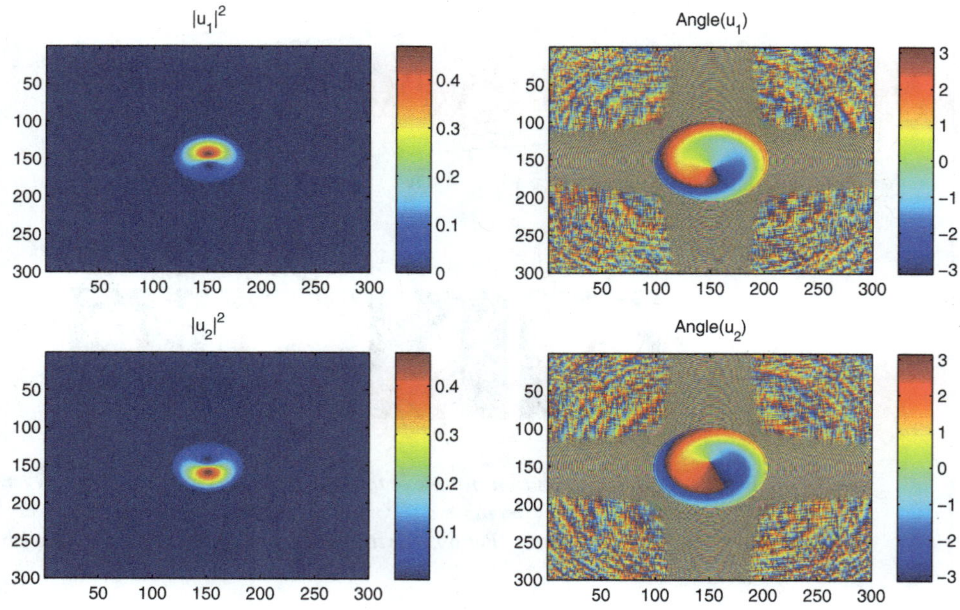

Figure 3.99. *A typical snapshot of a vortex-vortex solution created by the SO(2) rotated vortex-bright solitary wave with a $\chi = \pi/4$. The left panels illustrate the complementary densities of the two components, while the right panels show the corresponding phases. Notice that, by analogy to the 1D DD solitons, the vortex cores are displaced in the two components. Instead of oscillating around the common center as the 1D DD solitons do, the 2D vortices precess around the common core.*

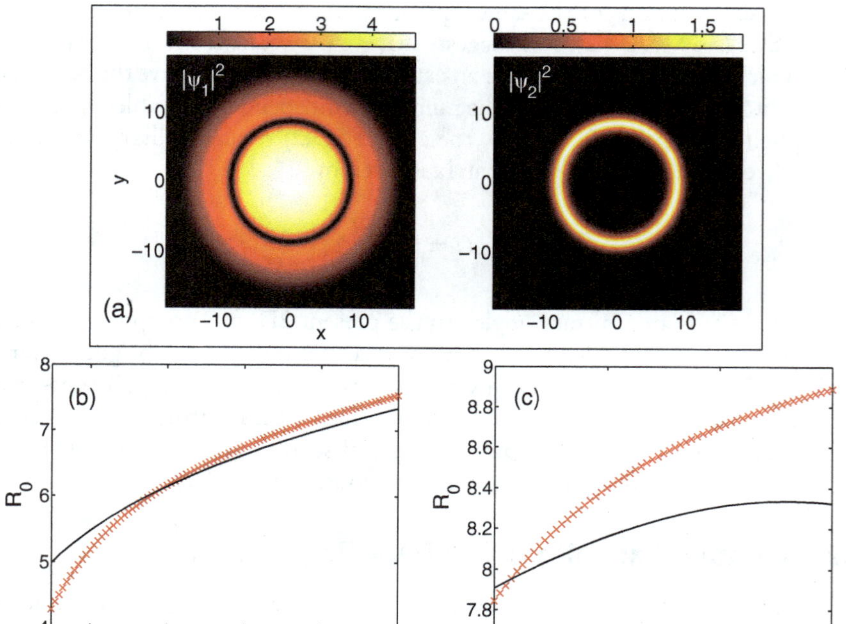

Figure 3.100. (a) *DB ring soliton's profile at $\Omega = 0.2$, $\mu_1 = 5$, $\mu_2 = 4.07$, $N_1 = 1616$, and $N_2 = 150$.* (b) *The ring's radius as a function of N_1 at $N_2 = 100$ fixed, as calculated from Eq. (3.206) (black) and numerical data (red crosses).* (c) *The ring's radius as a function of N_2 at $\mu_1 = 5$ fixed. Adapted with permission from Ref. [629].*

been studied, e.g., in Ref. [182, 266, 524] (see also Sec. 3.1 for details) and were found to be always unstable towards the formation of vortices. The DB ring soliton expands the idea of "symbiotic" states put forth in the discussion of DB and vortex-bright solitons, enabling the formation of a bright ring, due to the dark ring operating as an effective potential for the bright one. More importantly, as the work of Ref. [629] showed, the presence of such a second, "bright" component has a stabilizing effect on the symbiotic state.

Once again, the canonical mean-field model of interest is provided by the following 2D coupled normalized Gross–Pitaevskii equations:

$$i\partial_t \psi_1 = \left[-\frac{1}{2}\Delta + V + g_{11}|\psi_1|^2 + g_{12}|\psi_2|^2\right]\psi_1, \tag{3.203}$$

$$i\partial_t \psi_2 = \left[-\frac{1}{2}\Delta + V + g_{22}|\psi_2|^2 + g_{12}|\psi_1|^2\right]\psi_2. \tag{3.204}$$

Here, as in previous cases, we focus on the experimentally relevant case of ^{87}Rb hyperfine states $|F=1, m_F=-1\rangle$ and $|F=2, m_F=1\rangle$, where all scattering lengths are positive and nearly equal, leading to dimensionless coupling constants $g_{11}=1.03$, $g_{22}=0.97$, and $g_{12}=+1$ [279]. The (isotropic) harmonic potential of interest reads $V(x,y)=\Omega^2(x^2+y^2)/2$ and the numerical computations were performed for $\Omega \equiv \omega_r/\omega_z = 0.2$. As usual, stationary solutions are obtained by factorizing $\psi_j(x,y,t)=\exp(-i\mu_j t)\phi_j(x,y)$, where μ_1 and μ_2 denote the two components' chemical potentials.

The theoretical methodology used in Ref. [629] generalized the earlier approaches of Refs. [524, 174, 145] in order to construct an approximately conserved adiabatic invariant of the dark-bright ring, from which a description of its radial dynamics can be inferred. For a highly localized ring soliton, i.e., for one whose width is much smaller than its radius R, such an approximate constant of the motion in the regime of large μ_1 and small N_2 is given by

$$E_{\rm DB}(R) = 2\pi R\left(\frac{4}{3}\left[\mu_1 + \frac{N_2^2}{128\pi^3 R^2} - V(R)\right]^{3/2} + \frac{N_2}{\sqrt{2\pi}4\pi R}V(R)\right) - 4\pi R\dot{R}^2\sqrt{\mu_1 + \frac{N_2^2}{128\pi^3 R^2} - V(R)}. \tag{3.205}$$

To construct the ring energy $E_{\rm DB}$, we multiply the energy of a 1D DB soliton [145][22] placed R units away from the trap center by its circumference $2\pi R$ [524]. In this expression, we take into account a scaling correction and the rescaled atom numbers $N_{1,2} = \iint |\psi_{1,2}(x,y)|^2\,dx\,dy$. By using energy conservation arguments (i.e., $dE_{\rm DB}/dt=0$) one obtains a radial equation of motion which for a harmonic potential $V(R)=\omega^2R^2/2$ and for low velocities reads

$$\ddot{R}(R) = \frac{\mu_1}{3R} - \frac{N_2^2}{192\pi^3R^3} - \frac{2}{3}\omega^2 R + \frac{N_2\omega^2}{16\pi\sqrt{2\pi}\sqrt{\mu_1 + \frac{N_2^2}{128\pi^3R^2} - \frac{\omega^2}{2}R^2}}. \tag{3.206}$$

The validity of Eq. (3.206) is constrained within radii satisfying $\mu_1 - V(R) > 0$. This equation of motion predicts a stable (with respect to radial motions) DB ring equilibrium

[22] In Eq. (3.205), we have used the values $g_{11}=g_{22}=g_{12}=1$, for mathematical simplicity. We have checked that our numerical results are essentially unchanged, if these values are used instead of the experimentally relevant parameters.

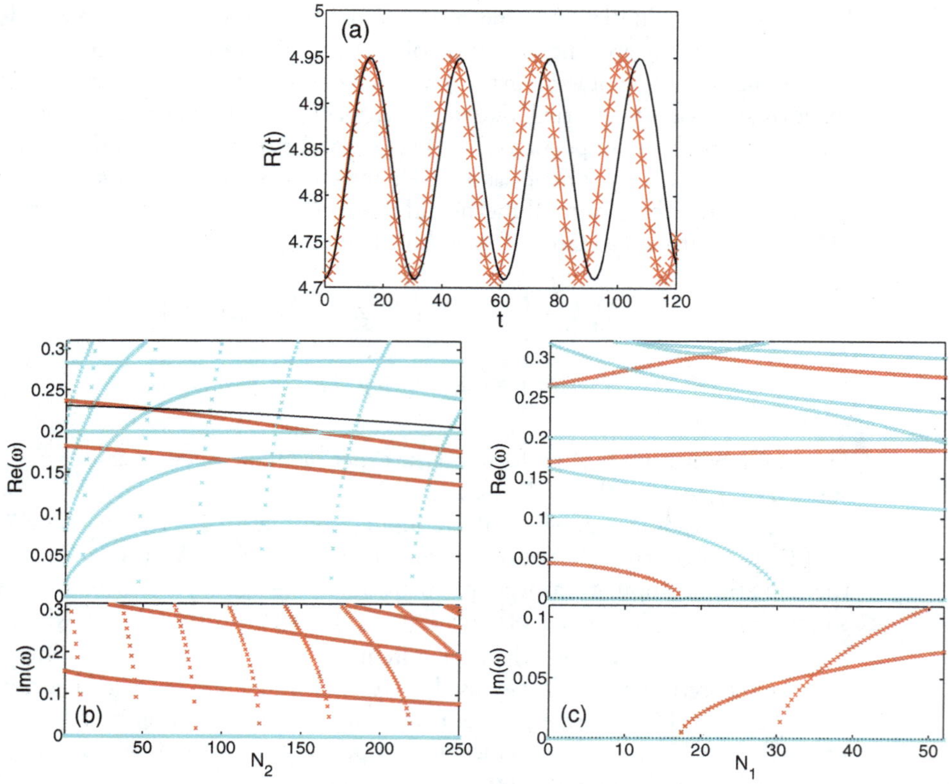

Figure 3.101. (a) *Radial oscillations of the DB ring when excited with the breathing mode for* $\mu_1 = 2$ *and* $N_2 = 100$. *Numerical data and harmonic oscillation at the predicted frequency are depicted, respectively, in red and black.* (b) *BdG spectrum as a function of* N_2 *for* $\mu_1 = 5$ *fixed. The prediction for the breathing frequency is depicted by the thin black line. Anomalous modes in the real part are shown in red, background modes in blue; purely imaginary modes in red.* (c) *BdG spectrum as a function of* N_1 *at* $N_2 = 20$ *fixed, exhibiting a stable parameter regime at small* N_1. *Adapted with permission from Ref. [629].*

at $R = R_0$ [cf. panels (b) and (c) of Fig. 3.100] and a breathing radial motion around this equilibrium [cf. Fig. 3.101(a)].

To complement this approach valid in the regime of large N_1, the DB ring was also studied in Ref. [629] near the linear limit of $N_1 \to 0$, $N_2 \to 0$, where the coupled GPEs reduce to two independent harmonic oscillator Schrödinger equations. Stationary solutions of the GPEs reduce to harmonic oscillator eigenstates in this limit. More specifically, to examine the ring, linear combinations of the form

$$\psi_1(x,y,t) = c_0(t)\varphi_0(x,y) + c_1(t)\varphi_1(x,y)$$

were considered for the dark component, where the linear modes are $\varphi_0 = |2,0\rangle$ and $\varphi_1 = |0,2\rangle$. Here, $|n_x, n_y\rangle$ denotes the 2D harmonic oscillator eigenstates and $c_{0,1}(t)$ are complex time-dependent prefactors. This two-mode approximation has been shown to be useful in describing one-component ring dark solitons in Ref. [523]. For the bright component of the ring, we numerically find that it reduces to the oscillator ground state $|0,0\rangle$ in the linear limit, and therefore for small enough atom numbers we tentatively approximate the order parameter as $\varphi_2 = \psi_2 = \sqrt{N_2}|0,0\rangle$. Note that in this near-linear regime the bright component does not yet "fill" the dark ring's zero density region; the ring's characteristic density profile only emerges in the nonlinear regime of larger atom numbers. Following an approach similar to Ref. [420], we substitute the above harmonic oscillator ansätze into Eq. (3.203), and use the amplitude-phase decomposition $c_k = \beta_k e^{i\alpha_k}$, to obtain the consistency condition

$$\begin{aligned} 0 = \dot{\alpha} = &- g_1\left[\beta_1^2 A_{1111} - \beta_0^2 A_{0000} + 3A_{0011}(\beta_0^2 - \beta_1^2)\right] \\ &- \frac{g_1}{\beta_0\beta_1}\left[A_{0111}\beta_1^2(3\beta_0^2 - \beta_1^2) + A_{0001}\beta_0^2(\beta_0^2 - 3\beta_1^2)\right] \\ &+ A_{0022} - A_{1122} + \frac{1}{\beta_0\beta_1}A_{0122}(\beta_1^2 - \beta_0^2), \end{aligned}$$

where $\alpha = \alpha_0 - \alpha_1$ and the nonlinear overlap integrals are denoted $A_{ijkl} = \int \varphi_i\varphi_j\varphi_k\varphi_l\, dx\, dy$. This algebraic equation for small $N_1 = \beta_0^2 + \beta_1^2$ has a single root at $\beta_0 = \beta_1 = \sqrt{N_1/2}$, corresponding to the DB ring approximately given by $\psi_1 \propto |0,2\rangle + |2,0\rangle$ and $\psi_2 \propto |0,0\rangle$ in this regime. Beyond a critical N_1^{cr}, two more roots emerge, which quickly tend to $(\beta_0, \beta_1) \approx (0, \sqrt{N_1})$ and $(\beta_0, \beta_1) \approx (\sqrt{N_1}, 0)$, respectively, as N_1 is further increased. These new solutions correspond to branches of states exhibiting two parallel DB soliton stripes that bifurcate from the ring branch (cf. Fig. 3.102). The effect of this symmetry-breaking bifurcation on the stability of the DB ring and its parametric dependence on the bright soliton's atom number N_2 is discussed below.

The above theoretical predictions were subsequently compared with numerical computations in Ref. [629] yielding the following results. In Figs. 3.100(b) and (c) the dependence of the DB ring's equilibrium radius R_0 is shown as a function of N_1 and N_2, respectively. It can be seen that the relevant prediction obtained from Eq. (3.206) is accurate in the proper regime and has the right functional dependence on $N_{1,2}$. Inaccuracies in the regime of small N_1 are attributable to a breakdown of the TF approximation. For large N_2, we expect repulsive interaction effects between the elementary bright solitonic components (within the ring) to become important which are not accounted for in our expression for E_{DB}. The radial breathing of the DB ring is shown in Fig. 3.101(a) in fair agreement with the theoretical prediction for the relevant frequency of linearization around $R = R_0$ in Eq. (3.206). Figures 3.101(b) and (c) illustrate the dependence on $N_{1,2}$ of the frequencies of the BdG analysis around a stationary ring. When one of these frequencies possesses a nonvanishing imaginary part, the corresponding rings are dynamically unstable. A key observation here is that for fixed N_2, the critical N_1 for which imaginary eigenfrequencies arise is larger, the larger N_2 is. In fact, for $N_2 = 0$, the dark ring is always unstable [266], while the presence of the second component yields a stabilization window at small N_1; see Fig. 3.101(c). This effect can also be observed in Fig. 3.101(b), showing the BdG frequency spectrum's N_2 dependence, where a weakening of the dynamical instabilities [i.e., decrease of the corresponding growth rates $\mathrm{Im}(\omega)$] and stabilization of a number of unstable modes is seen as N_2 is increasing. In these BdG plots, the radial breathing is represented by the "anomalous mode" of the DB ring, whose frequency is in fairly good agreement with the corresponding analytical prediction.

For small atom numbers, we can relate the imaginary modes arising in the ring's BdG spectrum, as N_1 is increased, to symmetry-breaking bifurcations leading to new states. A typical $N_1(\mu_1)$ bifurcation diagram for a relatively small bright atom number of $N_2 = 20$ is shown in Fig. 3.102(a). As the dark component's atom number increases, branches of two DB soliton stripes bifurcate from the ring, destabilizing it. This class of bifurcations can be understood within the near-linear framework of the two-mode approximation. Both

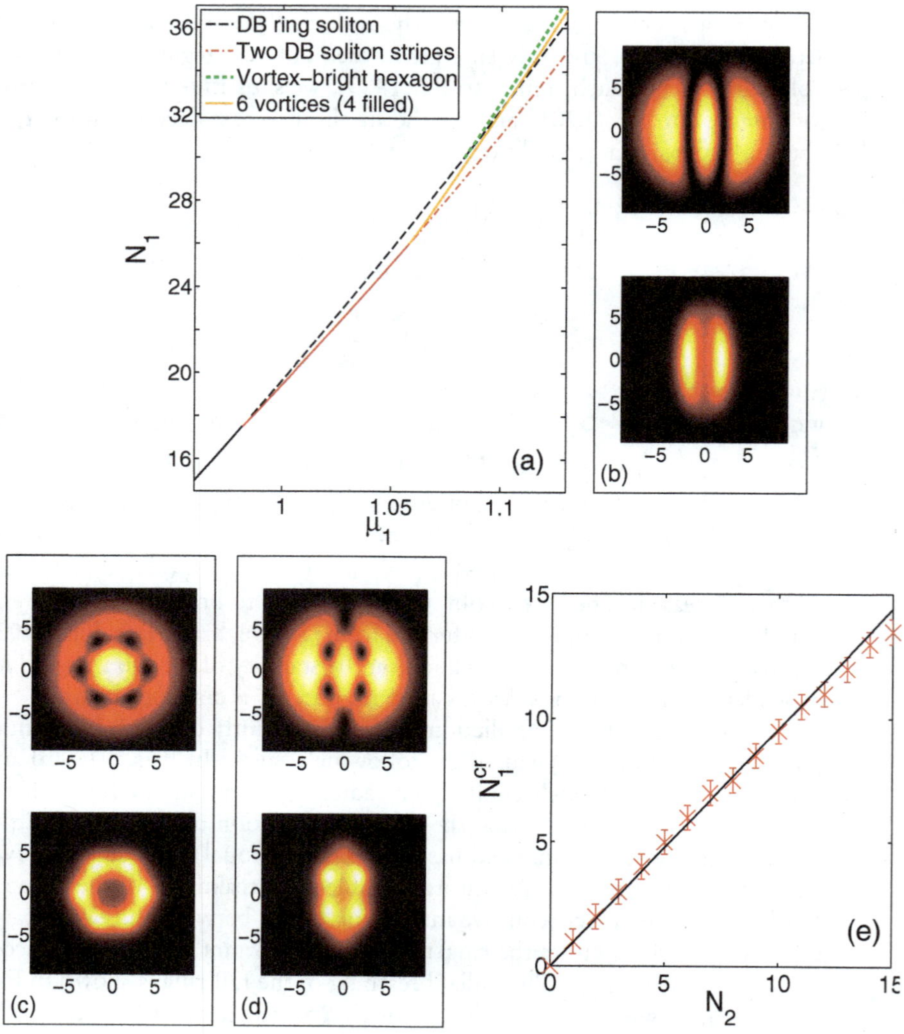

Figure 3.102. (a) $N_1(\mu_1)$ *bifurcation diagram at* $N_2 = 20$. *Solid (dashed) lines denote stable (unstable) branches.* (b)–(d) *Density profiles of the relevant branches [two DB soliton stripes* (b), *vortex-bright hexagon* (c), *filled-empty six vortex state* (d)*] at* $N_1 = 100$, $N_2 = 20$. (e) *Critical particle numbers at which the two DB soliton branches bifurcate from the DB ring branch as a function of* N_2*: two-mode approach predictions (black) and numerical data (red crosses). The error bars indicate the step size of the branches' numerical continuation in* N_1. *Adapted with permission from Ref. [629].*

in this simplified model and in full numerical simulations of the GPE we find that the critical value N_1^{cr} where the soliton stripes' bifurcation occurs increases as a function of N_2; see the very good agreement of these predictions in Fig. 3.102(e). This critical point marks (for small N_2) the termination of the window of stability of the DB ring due to the second component referred to above and is associated with the emergence of imaginary modes in its BdG spectrum.

Subsequent bifurcations from the ring lead to a sequence of polygonal vortex-bright clusters, starting with a hexagonal necklace state shown in Fig. 3.102(c). These configurations are natural two-component generalizations of the polygonal states discussed in Ref. [523]. For small N_2, they are found to be unstable, as they bifurcate from the destabilized ring branch. The DB soliton stripes also lose their stability to a bifurcating vortex cluster, namely the six-vortex state shown in Fig. 3.102(d). Interestingly, in this cluster the bright component only fills four of the six vortices, while two of the cores remain empty. The one-component analogs of these states were also discussed above and identified in Ref. [523]. For $N_2 \gtrsim 42$ we observe a change in the bifurcation order, with the vortex-bright hexagon branch now bifurcating before the DB soliton stripes and thus inheriting the ring's initial stability.

The principal conclusion that can be drawn from the above findings is that the presence of the second component parametrically "delays" the instability inducing bifurcations in the case of the ring dark soliton. This type of phenomenology is not only present in the case of the ring-like state, but can also be seen to persist for other states such as the crossed (diagonal) DB solitons. This can be thought of as a two-component generalization of the always unstable diagonal dark solitons of Ref. [523]. For this configuration too, a BdG analysis reveals that for fixed $N_2 > 0$ there exists a range of N_1 for which it is dynamically stable, while beyond the first critical point N_1^{cr}, it becomes unstable towards the formation of a vortex-bright quadrupole cluster.

3.11.3 ▪ Multi-Vortex-Bright Solitons and Their Interactions

We will now examine the possibility of two-component condensates to sustain multiple vortex-bright solitary wave structures, in the form of dipoles. This will be in the same spirit as the earlier consideration of single-component vortex dipoles in this chapter and that of multiple DB solitons in the previous chapter (see, in particular, Sec. 3.5). The vortex-bright dipoles, will consist of a vortex pair of opposite circulation in the first component, and a corresponding trapped bright soliton pair in the second component. It is relevant to mention here that particular realizations (i.e., ones where the bright solitons are in-phase) of such states have been very recently proposed as realizable by means of dragging numerical experiments in the immiscible regime of the pseudospinor system in Ref. [630]. The excitations resulting by dragging a laser beam through the two-component system had earlier been explored in the miscible regime in Ref. [326].

Following the exposition of Ref. [631], we will analyze the relevant case of binary condensates composed of ^{87}Rb atoms in the two spin states $|F=1, m_F=-1\rangle$ and $|F=2, m_F=1\rangle$, leading to the (by now) customary dimensionless coupling constants of approximately $g_{11} = 1.03$, $g_{22} = 0.97$, and $g_{12} = +1$. Once again, we will consider the harmonic potential $V(x,y) = \Omega^2(x^2+y^2)/2 \equiv \Omega^2 r^2/2$, with $\Omega = 0.2$ as before, while stationary solutions will be obtained by factorizing $\psi_j(x,y,t) = \exp(-i\mu_j t)\phi_j(x,y)$, $j = 1,2$.

The prototypical stationary solution of the GPE that will be the building block for our considerations is the single vortex-bright (VB) soliton state, analyzed in Sec. 3.11.1 (see also Refs. [620, 507]). One important observation to make here is that there is a single negative energy mode [507] in the spectrum of the vortex for small (or vanishing) $N_2 = \int |\psi_2|^2 dx\, dy$. In the case of the VB soliton, it is noteworthy that this mode decreases in frequency due to the presence of the bright component and ultimately crosses the zero frequency point. Yet, this crossing does *not* produce an instability; the relevant eigenfrequency pair remains real, but now the energy of the mode is positive, signaling the transition of the vortex state from a saddle in the energy landscape into a local minimum thereof (this is a setting analogous to what is observed for a vortex in the presence of rotation; see, e.g., the relevant discussion of Ref. [632]). It is important to note that this observation is in agreement with the results described in Ref. [633], where it was observed that filled vortex cores exhibit slower precessions, and thus a decrease of the

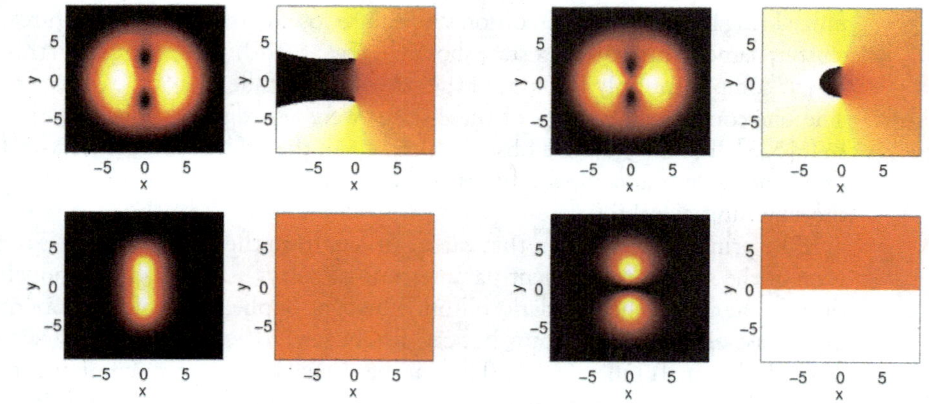

Figure 3.103. *Left panels: Density (first column) and phase (second column) of an in-phase vortex-bright dipole. Right panels: Density (third column) and phase (fourth column) of the out-of-phase vortex-bright dipole. Both examples are for $N_1 = 70$ and $N_2 = 20$. Adapted with permission from Ref. [631].*

precession frequency is expected. It is relevant to mention that a similar phenomenology was also found in the case of DB solitons (see details in Sec. 2.4.3), both in theory [145] and experiment [335].

Vortex-Bright Soliton Dipole States

We now look for bound states containing multiple such entities. In the same spirit as in the work of Ref. [518] (see also, e.g., Ref. [535] and references therein), the prototypical relevant bound state is the VB dipole. We have been able to identify two such states, which are both shown in Fig. 3.103 for $N_1 = 70$ and $N_2 = 20$. In both cases the first component contains two vortices located symmetrically with respect to the trap center, which are filled by the second component. The difference between the two dipoles is evident in the phase profile: the two bright solitons filling the vortex cores can have either the same phase, or there may be a phase difference of π between them.

The emergence of the dipole state branches can be thought of in terms of a bifurcation picture in an equivalent way to the one described for one-component BECs in Sec. 3.4 (see also Ref. [518]). The two possible dipole branches, distinguished from the presence or absence of the phase jump of π in the bright component, arise at critical values of N_1 from two states, which can be regarded as generalizations of the dark soliton stripe observed in one-component condensates. The density and phase profiles of these two solitonic states are shown in Figs. 3.104, where it is evident that the two states are again distinguished by the presence or absence of a π-phase jump in the bright component.

In the first case of the in-phase (IP) VB dipole, the parent state can be seen to consist of a dark soliton stripe in the first component accompanied by a bright stripe in the second component, which is trapped inside the dark stripe. In the notation of the corresponding linear limit of the two single-particle Hamiltonians, the relevant state of the first component is the $|1,0\rangle$ state (i.e., the first excited state in the x-direction) while that of the second component is the $|0,0\rangle$ ground state of the latter. More generally, the symbolism $|m,n\rangle$ will be again used to denote the states $H_m(\sqrt{\Omega}x)H_n(\sqrt{\Omega}y)e^{-\Omega r^2/2}$ where H_m stands for the mth Hermite polynomial and normalization constants have been omitted. These are stationary states of the underlying linear problem with eigenvalue $\mu = \Omega(m+n+1)$, from which the soliton stripe solutions arise in the presence of the effective nonlinear-

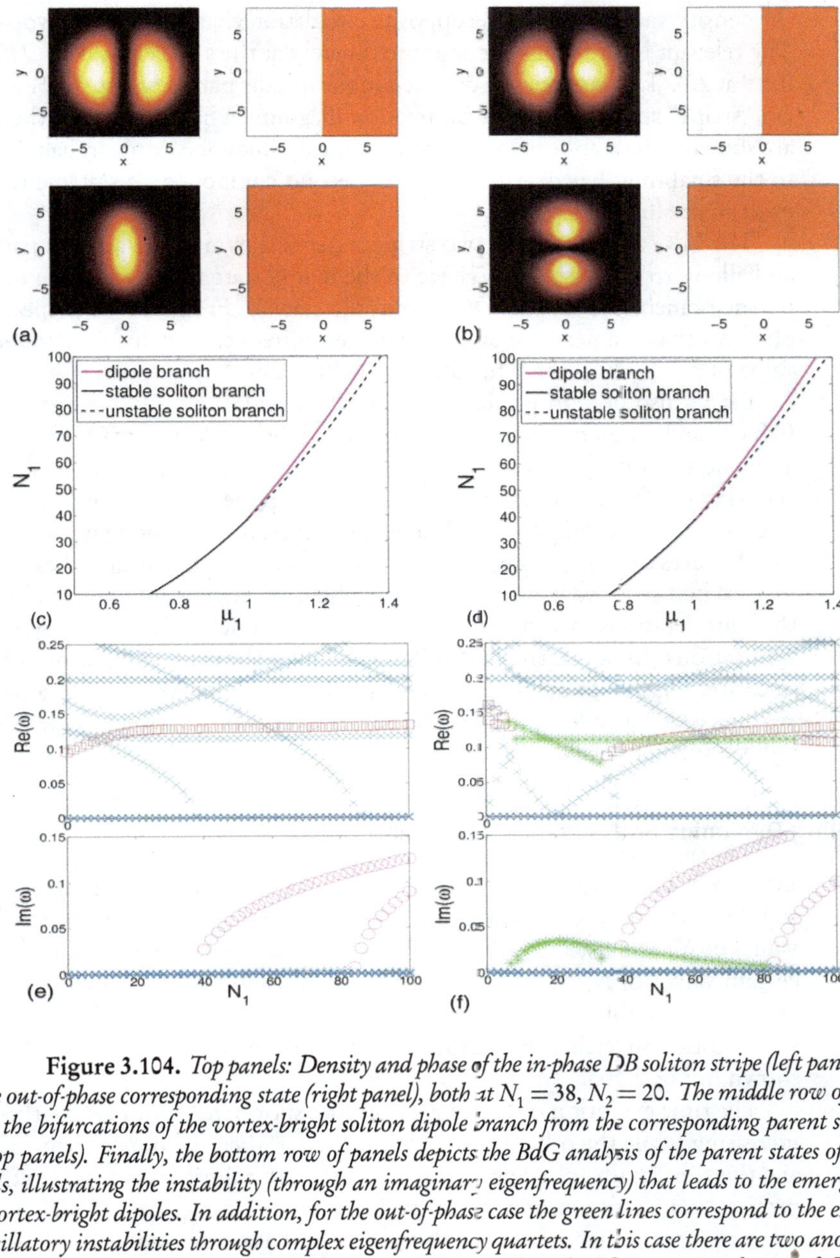

Figure 3.104. *Top panels: Density and phase of the in-phase DB soliton stripe (left panels) and of the out-of-phase corresponding state (right panel), both at $N_1 = 38$, $N_2 = 20$. The middle row of panels show the bifurcations of the vortex-bright soliton dipole branch from the corresponding parent states (of the top panels). Finally, the bottom row of panels depicts the BdG analysis of the parent states of the top panels, illustrating the instability (through an imaginary eigenfrequency) that leads to the emergence of the vortex-bright dipoles. In addition, for the out-of-phase case the green lines correspond to the existence of oscillatory instabilities through complex eigenfrequency quartets. In this case there are two anomalous modes, while there is only one such for the in-phase stripe. Adapted with permission from Ref. [631].*

ity induced by interatomic interactions. It is the nonlinearity and immiscibility of the two components that leads the $|0,0\rangle$ state of the second component to become elongated (as opposed to circularly symmetric) in the DB stripe of the top left panel of Fig. 3.104. These are rectilinear variants of the states considered previously in Sec. 3.11.2 (see also Ref. [629]). In the case of the top right panel of Fig. 3.104 and the out-of-phase (OOP) VB dipole, the parent state is the $|1,0\rangle$ of the first component coupled to the $|0,1\rangle$ of the second component.

While the bright component remains roughly invariant as the transition point is crossed, the dark component develops two singular points in the phase corresponding to the density vanishings and the opposite circulation characteristic of a vortex dipole [518]. The relevant bifurcations occur approximately at the same critical N_1 ($N_1^{\rm cr} \approx 40$, with N_2 fixed at 20). This is explicitly depicted in the middle panels of Fig. 3.104 for the two different "stripe" states, where the bifurcation diagrams which determine the relation $N_1(\mu_1)$ are shown. These approximately equal critical values of N_1 can, arguably, be expected due to the small number of particles of the second component, a statement which becomes exact in the limit of $N_2 \to 0$.

The BdG analysis of the two stripe states is shown in the bottom panel of Fig. 3.104 and allows to relate the emergence of the dipole state to the changing of stability of the parent branch (the IP and OOP solitonic states). From the BdG spectra, in fact, one observes the emergence of an imaginary eigenfrequency pair for $N_1^{\rm cr} \approx 40$, which is the value of N_1 at which the bifurcation actually occurs.

Let us also note in passing that the IP stripe bears only one negative energy mode (which can be seen to lead to oscillatory instabilities), while the OOP one bears two such anomalous modes. This is intuitively expected by the out-of-phase or excited state nature that the former bears only in the first component, while the latter has that type of structure in both components. Namely, in the second component, the former configuration features a ground state, symmetric waveform, while the latter has an antisymmetric first excited state waveform. This type of characteristics will be of critical relevance to the considerations that follow below, regarding tunneling effects and symmetry-breaking bifurcations. Importantly, the OOP mode, given the present repulsive nature of the nonlinearity, suggests an attractive interaction between the two atomic blobs trapped by the vortices and hence hints at the possibility of tunneling between these two "wells" as we will see further below.

Tunneling Dynamics and Symmetry-Breaking Bifurcations

Let us now come back to the vortex-bright soliton dipoles and especially their dynamical stability through the BdG analysis. The IP VB dipole appears to be stable for arbitrary values of N_1 and N_2, while the OOP VB dipole shows purely imaginary modes arising at certain values of N_1 and N_2 (when the state is scanned over N_1 or N_2, respectively). Exciting this unstable mode for the OOP dipole with $N_1 = 220$ and $N_2 = 1$, by adding some white noise and letting the perturbed states propagate in time, gives rise to the evolution shown in Fig. 3.105.

The time evolution of the density profiles shows the size of one of the bright solitons increasing while the other one is getting smaller, then the opposite situation is observed and this trend with enlarging and shrinking bright solitons is repeated in the course of the propagation. This observed time evolution has an immediate physical interpretation as concerns the bright component. A fraction of the corresponding particles within the second component moves from one vortex core to the other during the time propagation of the system. This can be explained by considering the dark component's density as acting in the form of an effective potential for particles of the bright one. Namely, the particles of the second component can be considered as tunneling within the vortex-induced (i.e., formed by the vortex cores) double-well potential. Within this very potential, this suggests the possibility of a spontaneous symmetry breaking bifurcation, as responsible for the observed phenomenology.

In order to validate this assumption, we consider the analytical model, developed in Ref. [420], based on a two-mode expansion, which is used to determine stationary states and to study the symmetry-breaking bifurcations occurring in 1D, single-component repulsive BECs confined in double-well potentials. The bifurcation emerges from (and

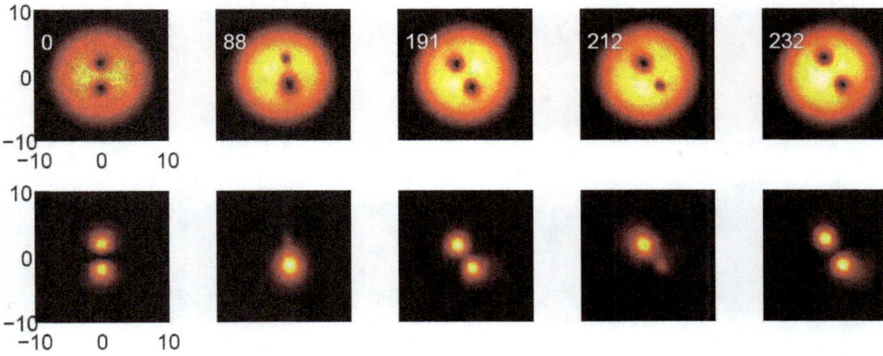

Figure 3.105. *Time evolution of the perturbed OOP VB soliton dipole state at* $N_1 = 220$ *and* $N_2 = 1$. *The top and bottom lines depict the density profiles of the dark and bright components, respectively, evaluated at five moments of the simulated evolution. The elapsed (dimensionless) time is indicated in the panels' upper left corners. The substantial addition of white noise (that seeds the instability) can be seen in the pixelization of the corresponding dynamical panels. Adapted with permission from Ref. [631].*

destabilizes) the antisymmetric, first-excited state in our repulsive interaction case (while it stems from and destabilizes the symmetric ground state in the attractive interaction case). This is consonant with the principal observation above of the generic stability of the IP VB, which has a symmetric second component and the symmetry breaking associated destabilization of the OOP VB with the antisymmetric second component.

The key quantitative observation now is that if we freeze the first component (assuming that it forms the vortex-core double well potential), then the theory of Ref. [420] can be applied directly but for the effective double-well potential in the form $V_{\text{eff}}(x,y) = V(x,y)+|\psi_1(x,y)|^2$, where $V(x,y)$ is the trapping harmonic potential. The equation for the second component can then be rewritten as

$$i\partial_t \psi_2(x,y) = \left[-\frac{1}{2}\Delta_{2D} + V_{\text{eff}}(x,y) + g_2|\psi_2(x,y)|^2\right]\psi_2(x,y), \tag{3.207}$$

which upon the rescaling $\tilde{\psi}_2(x,y) = \sqrt{g_2}\psi_2(x,y)$ leads to a standard single-component (2D) double-well setting. It should be noted that the potential of the double-well clearly depends on N_1 (through its dependence on $|\psi_1(x,y)|^2$). For each N_1 (in multiples of 5) in the interval $[0,300]$ the density profile of the dark component of the VB dipole state with $N_2 = 0$ has been utilized (to form the corresponding effective potential). For each considered value of N_1, one can diagonalize the obtained Hamiltonian $H = -\frac{1}{2}\Delta_{2D} + V_{\text{eff}}$, and keep the two lowest energy eigenstates, namely the symmetric ground state and the antisymmetric first excited state of the single particle operator with the effective potential. These, denoted hereafter as $u_0(x,y)$ and $u_1(x,y)$, will be used for the two-mode reduction.

The fundamental analytical prediction of the two-mode approximation is that as N_2 (and μ_2) is increased, a critical point will be reached, given by

$$N_{2,\text{cr}} = \frac{\Delta\omega}{(3B - A_1)g_2}, \tag{3.208}$$

whereafter the antisymmetric branch (interpreted here as the OOP VB dipole) will become unstable. Past this critical point, an asymmetric branch (i.e., an asymmetric VB dipole) should emerge as a stable configuration. In the expression of Eq. (3.208) $A_1 =$

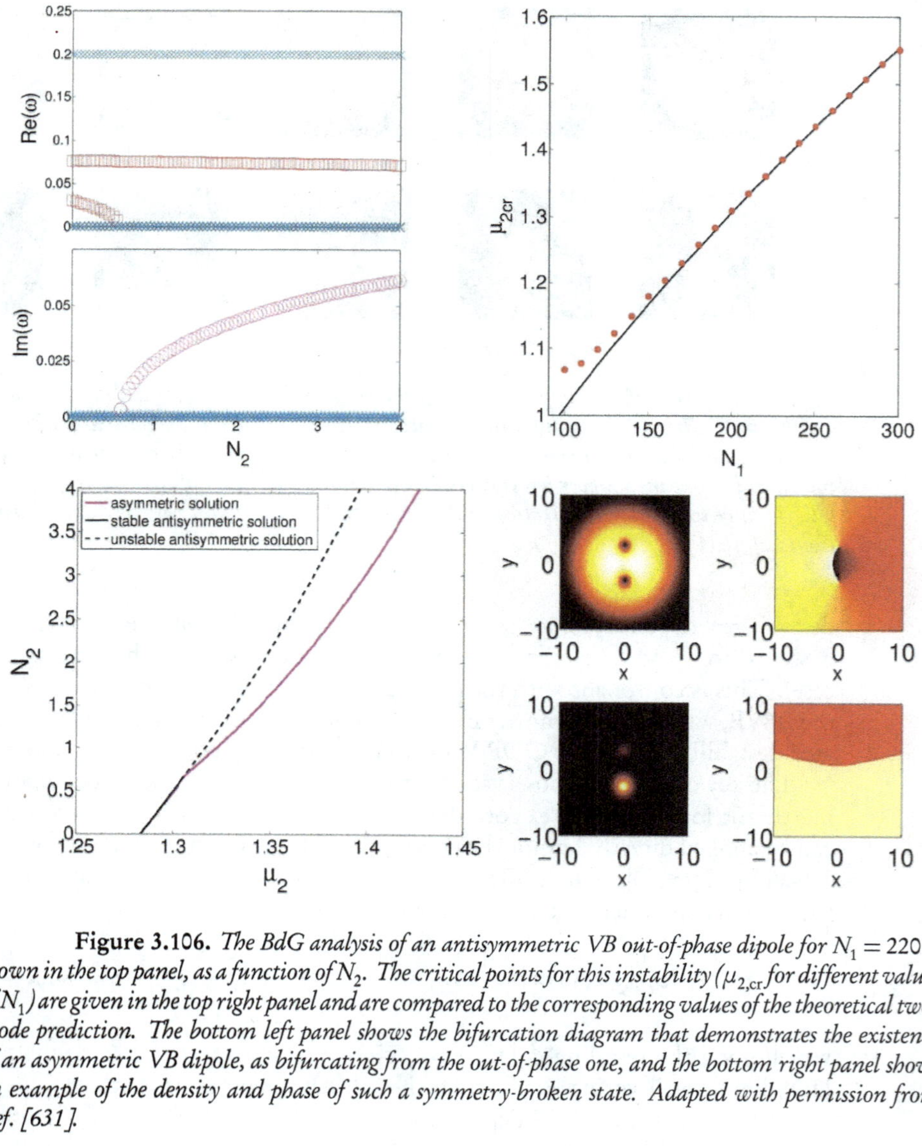

Figure 3.106. *The BdG analysis of an antisymmetric VB out-of-phase dipole for $N_1 = 220$ is shown in the top panel, as a function of N_2. The critical points for this instability ($\mu_{2,\mathrm{cr}}$ for different values of N_1) are given in the top right panel and are compared to the corresponding values of the theoretical two-mode prediction. The bottom left panel shows the bifurcation diagram that demonstrates the existence of an asymmetric VB dipole, as bifurcating from the out-of-phase one, and the bottom right panel shows an example of the density and phase of such a symmetry-broken state. Adapted with permission from Ref. [631].*

$\int u_1^4\,dx\,dy$ and $B = \int u_0^2 u_1^2\,dx\,dy$ are overlap integrals of the two lowest energy eigenstates of the linear Schrödinger problem and $\Delta\omega = \omega_1 - \omega_0$, where ω_0 and ω_1 are the energy eigenvalues corresponding to u_0 and u_1. The presence of the factor g_2 in the denominator ensures consistency with the scaling of ψ_2 discussed above. The critical chemical potential is calculated from the particle number by making use of the expression $\mu_{2,\mathrm{cr}} = \omega_0 + 3BN_{2,\mathrm{cr}}$ [420].

Repeating this procedure for each considered N_1, we get the critical values of μ_2, which are compared with the numerically obtained data; see the top right panel of Fig. 3.106. These latter were obtained in Ref. [631] by making a scan within a suitable interval of N_2 of the OOP VB dipole, and then performing the BdG analysis of the resulting branch of solutions. In the ensuing BdG spectrum, the bifurcation point can be identified as the onset of instability of the antisymmetric VB dipole. For a relevant example, see the top left panel of Fig. 3.106 for the scan obtained for $N_1 = 220$. There we can see the imaginary mode emerging at $N_2 \approx 0.6$ (generally, the relevant critical values $N_{2,\mathrm{cr}}$ are of order unity). At this critical value of N_2, for which the antisymmetric dipole becomes unstable, we expect the bifurcation of a stable asymmetric state, as was numerically verified in Ref. [631].

From the top right panel of Fig. 3.106, it is evident that the agreement between analytical predictions and numerical results is very good as concerns $\mu_{2,\mathrm{cr}}$ for $N_1 \gtrsim 120$. For lower values of N_1 numerics and analytical predictions start to be in disagreement with each other and this has the following explanation by decreasing N_1, the height of barrier between the two wells forming the effective potential $V_{\mathrm{eff}}(x,y) = V(x,y) + |\psi_1(x,y)|^2$ is also decreased. Thus, for low N_1, the density $|\psi_1(x,y)|^2$ provides a much lower contribution to the potential, and the effect of the harmonic trap is much more appreciable. This influences the form of the energy spectrum: the gap separating the almost equal lowest energy eigenvalues ω_0 and ω_1 from the larger eigenvalues $\omega_2, \omega_3, \ldots$, which is typically large for a high barrier, if the latter is lowered becomes smaller and smaller, until the equidistant eigenvalue spectral limit of the harmonic oscillator, is reached. But, if this is the case, the two-mode approximation that one makes by projecting the problem on the eigenmodes u_0 and u_1 fails as the contribution of the other eigenstates cannot be neglected anymore. Another intuitive argument to explain the disagreement between predicted and numerical results observed with lowering N_1 involves the fact that the effective potential $V_{\mathrm{eff}}(x,y) = V(x,y) + |\psi_1(x,y)|^2$ has been calculated deriving the term $|\psi_1(x,y)|^2$ from the density profile of the dark component of the VB dipole state with $N_2 = 0$ and is considered to be fixed. The above approach seeks a bifurcation occurring at a finite value of N_2, but neglects that for this value the density profile for the dark component is different from the one with $N_2 = 0$. Thus, in doing this approximation, we do not take into account the effect of the second component on the first due to the interaction between them. Considering V_{eff} independent of N_2 is expected to be valid if N_1 is large enough, as the dark component will then have a robust configuration with respect to variations due to the intercomponent interaction. But, with decreasing N_1, varying N_2 will start to influence both components in a sensitive way and, in particular, an evident variation of the density profile of the dark component is expected. Therefore, for low N_1, the assumption that $|\psi_1(x,y)|^2$ is independent of N_2 is no longer appropriate and the model described above cannot be expected to be valid anymore. Hence, in the top right panel of Fig. 3.106, values of N_1 smaller than 100 are not taken into account.

Let us now come back to the expected bifurcation from the antisymmetric branch, once its destabilization occurs, towards an asymmetric VB dipole solution. Indeed, an example of this bifurcation is illustrated in the bottom left panel of Fig. 3.106. The bottom right panel of the figure shows a prototypical example of the daughter state in the form of an asymmetric VB dipole. The robustness of the latter, combined with the Hamiltonian dynamics of the system, is what gives rise to the tunneling observations of Fig. 3.105 (since the anti-symmetric solution becomes a saddle, and the asymmetric one a center around which the dynamics revolves). Naturally, there are two such asymmetric states, depending on which vortex-induced well picks up part of the bright mass of the other, justifying the pitchfork character of the bifurcation. The BdG spectrum of the asymmetric state (not shown here) reveals the complete stability of the latter, in accordance with the expectations from the above bifurcation theoretic arguments. Although the above exposition was primarily concentrated on the tunneling dynamics arising in the limit of small N_2, in Ref. [631] it was illustrated that the relevant tunneling oscillation (albeit not necessarily in exactly the same periodic form) persists also in the case of large(r) N_2. In the latter case, it may also be accompanied by a rotational motion of the vortices during propagation.

3.11.4 ▪ Domain Walls, Propellers, Spirals, and Other Exotic Two-Component Entities

Theoretical Setup

We now turn to the consideration of various entities that may stem from the immiscibility of the two-components within multi-species BECs. In particular, we observe that for the case of ^{87}Rb, the scattering lengths a_{11}, a_{12}, and a_{22} are such that

$$\Delta = a_{11}a_{22} - a_{12}^2 = -0.0009 \leq 0. \qquad (3.209)$$

Hence ^{87}Rb is very weakly on the immiscible side. Furthermore, for a mixture of the different spin states of ^{23}Na, the immiscibility parameter is $\Delta_{\mathrm{Na}} = -0.036$ which also indicates that ^{23}Na is prone to phase separation; see the corresponding discussion in Refs. [43, 42, 44]. In light of these observations, it is relevant to examine domain-wall type entities where the two components phase separate due to the intercomponent interaction being stronger than their self-action.

We study such coherent structures in the context of the two-component 2D GPE model of the form

$$i\frac{\partial u_j}{\partial t} = \left[\hat{H}_{2\mathrm{D}} + \sum_{k=1,2} \alpha_{jk}|u_k|^2\right] u_j, \quad j = 1,2, \qquad (3.210)$$

$$\hat{H}_{2\mathrm{D}} \equiv (-1/2)\nabla_\perp^2 + V(r) - \Omega_L L_z, \qquad (3.211)$$

where $\nabla_\perp^2$ is the 2D Laplacian, and

$$V(r) \equiv (1/2)\Omega^2 r^2, \quad L_z \equiv i(x\partial_y - y\partial_x), \quad \Omega_L \equiv \omega_L/\omega_z. \qquad (3.212)$$

Notice that further to the standard kinetic and potential terms in the linear part of the Hamiltonian, we have added a rotational term parametrized by the frequency ω_L (normalized as all frequencies of the system to the z component of the trapping strength ω_z), in order to examine the role of rotation in the dynamics of the relevant nonlinear waveforms. The nonlinearity coefficients in the 2D system (3.210) are denoted here as $\alpha_{11} = 1$, $\alpha_{12} = a_{12}/a_{11}$, and $\alpha_{22} = a_{22}/a_{11}$. Then, the numbers of atoms in the species are $N_j = (4\sqrt{2}\pi^{3/2})^{-1}\sqrt{\hbar/m\omega_z}Q_j$, where $Q_j \equiv \int |u_j|^2 d^2\mathbf{r}_\perp$ are the squared norms of the 2D wavefunctions. In our exposition of the relevant structures below, we follow the presentation of Ref. [251].

At this point, it is worth mentioning in passing that the spatial evolution of bimodal optical beams in twisted photonic crystal fibers is also described by Eqs. (3.210) for the beam amplitudes u_j of the two components; in this case, t is replaced by the propagation distance z and $V(r)$ in Eq. (3.211) is a potential function describing the fiber cladding. In this context, N_j are total powers of the two components of the beam, and the nonlinear coefficients take values close to $\alpha_{12} = 2\alpha_{11} = 2\alpha_{22}$; the latter values definitely guarantee the strong "immiscibility" and, consequently, the mutual repulsion of the two optical modes.

Solutions to Eqs. (3.210) in the form of either a simple rectilinear domain wall (DW), first without the rotation ($\Omega_L = 0$), are constructed as follows. We start with the TF configuration for the two components with chemical potentials μ_j,

$$\left(u_j\right)_{\mathrm{TF}} = e^{-i\mu_j t}\sqrt{(\max\left[\mu_j - V(r), 0\right]/\alpha_{jj}}, \qquad (3.213)$$

if $\mu_j - V(r) > 0$, and $(u_j)_{\mathrm{TF}} = 0$ otherwise (in the case of photonic crystal fibers, $-\mu_j$ are the propagation constants of the optical modes). Then, to construct a rectilinear DW, we effectively remove each component from a half-plane, upon multiplying $(u_j)_{\mathrm{TF}}$ by

$$f_j(x) \equiv \frac{1}{2}\left[1-(-1)^j \tanh x\right]. \tag{3.214}$$

The resulting function u_1 (u_2) is positive in the left (right) half of the (x, y) plane.

To construct a DW cross in a similar fashion, one can also start from the TF ansatz (3.213), and then remove two quarter-planes in each component, multiplying the expression (3.213) by $(1/2)[f_j(x)f_j(y) + f_j(-x)f_j(-y)]$, where the functions $f_{1,2}$ are defined as per Eq. (3.214). As a result, the wavefunctions u_1 and u_2 have support, respectively, practically only in the quadrants (sectors) 1 and 3 (i.e., $xy > 0$), and 2 and 4 (i.e., $xy < 0$). It is also possible to multiply the entire configuration by $(-1)^j \tanh x$, hence the functions u_1 and u_2 will be, respectively, positive (negative) in the sectors 1 (3) and 2 (4). This introduces a phase shift of π between the two quadrants filled by each species (in other words, we aim to construct a DW cross which, simultaneously, is a "latent dark soliton" in each component). Without these phase shifts, the cross will be obviously unstable against splitting into a set of two approximately parallel quasirectilinear DWs, as a pair of the sectors filled by the same species without the phase shift between them will tend to merge into a single stripe-like pattern.

After the prototype rectilinear or cross DW configuration was constructed as described above, Eqs. (3.210) were numerically integrated in Ref. [251] in imaginary time, to let the system relax into a stable stationary state. For the DW-cross case, the relaxation in imaginary time did not essentially alter the imprinted pattern, always converging to a numerically exact stable cross pattern, provided that the immiscibility condition was met. This outcome of the imaginary time integration took place for arbitrarily small values of the immiscibility coefficient $|\Delta|$, including the case when Δ was set precisely equal to 0. With $\Delta > 0$, the integration always showed that two species would completely mix into a uniform state.

In the case of the rectilinear DW, the imaginary time integration converged to a stable quasi-1D DW pattern only if the condition $\Delta \leq \Delta_{\mathrm{rect}}^{(\mathrm{cr})} = -0.061$ was satisfied, which is more restrictive than the general immiscibility criterion. For smaller values of $|\Delta|$, stable rectilinear DWs could not be found; instead, the two species would completely mix up, even if Δ was negative. Thus, the DW-cross pattern, although being more complex than its simplest rectilinear counterpart, is a much more robust one, and has a much better chance to be created in experiments with the currently available weakly immiscible spin-state mixtures. The enhanced robustness of the DW-cross structure may be explained by the additional π phase shifts lent to its both components, as described above.

As an example, Fig. 3.107 shows the result of the relaxation for each type of the prototype pattern, with the value of Δ taken as for the spin-state mixture in ^{87}Rb. This was realized by setting $\alpha_{11} = 1$, $\alpha_{22} = 0.94$, and $\alpha_{12} = 0.97$ in Eq. (3.210). As is seen, the rectilinear DW pattern cannot be formed indeed, preferring to mix itself up into a uniform state [panels (a) and (b)]. On the contrary, the initial (prototype) DW-cross pattern readily relaxes into a configuration of exactly the same type.

In the example shown in Fig. 3.107, the normalized magnetic-trap strength is $\Omega = 0.05$, and both chemical potentials are set equal to 1. In terms of the real-world parameters, this choice corresponds to a mixture of two spin states in the ^{87}Rb condensate in a disk-shaped trap with

$$\omega_r = 2\pi \times 6 \text{ Hz}, \qquad \omega_z = 2\pi \times 120 \text{ Hz}, \tag{3.215}$$

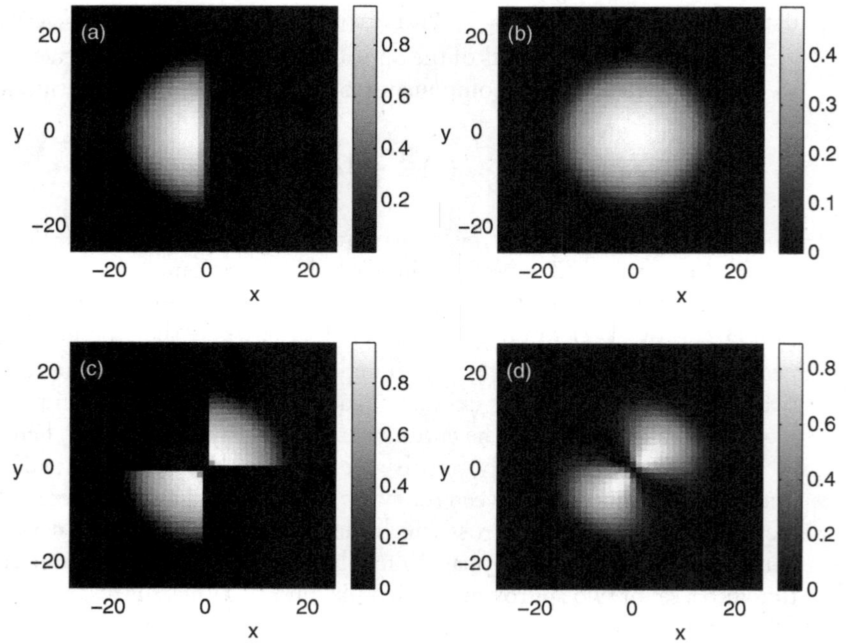

Figure 3.107. *Grayscale plots showing the distribution of the density $|u_1|^2$ of one species (the density $|u_2|^2$ is complementary to $|u_1|^2$) in the initial configuration* (a) *corresponding to the prototype rectilinear DW, and in the corresponding final counterpart* (b), *which is generated by the numerical integration of the GPE in imaginary time, in the case of weak immiscibility, with $\Delta = -9 \times 10^{-4}$ (corresponding to $\alpha_{11} = 1$, $\alpha_{22} = 0.94$, $\alpha_{12} = 0.97$) and $\Omega = 0.05$. Panels* (c) *and* (d) *display the same for the initial and final configurations in the case of the DW cross. Adapted with permission from Ref. [251].*

the initial TF radius and numbers of atoms in each species being

$$R = 34\ \mu\text{m}, \qquad N = 3.6 \times 10^3. \tag{3.216}$$

Furthermore, in the work of Ref. [251], larger values of $|\Delta|$ were considered that could not be directly found in one of the above atomic gases, but which may in principle be attainable through the use of the Feshbach resonance and its ability to tune scattering lengths. This enabled the generation of patterns that are much sharper, while qualitatively they are completely tantamount to those found at smaller (more directly realizable) values of the immiscibility parameter. In that light the parameters

$$\alpha_{11} = 1, \quad \alpha_{22} = 1.01, \quad \alpha_{12} = 1.52 \tag{3.217}$$

were used, which correspond to $\Delta = -1.3$.

It is important to note that, alongside the symmetric DW-cross configurations like the one shown in Fig. 3.107(d), *asymmetric* ones are possible too, with different numbers of atoms in the two species, $Q_1 \neq Q_2$. Indeed, the TF approximation (3.213) yields

$$Q_j = \left(\frac{\pi}{2\alpha_{jj}}\right)\left(\mu_j - \frac{1}{4}\Omega^2 R^2\right) R^2, \tag{3.218}$$

where R is the TF radius of the state; consequently, $\alpha_{11} \neq \alpha_{22}$ and/or $\mu_1 \neq \mu_2$ lead to $Q_1 \neq Q_2$, and, as a result, to asymmetric crosses. It was observed in Ref. [251] that the

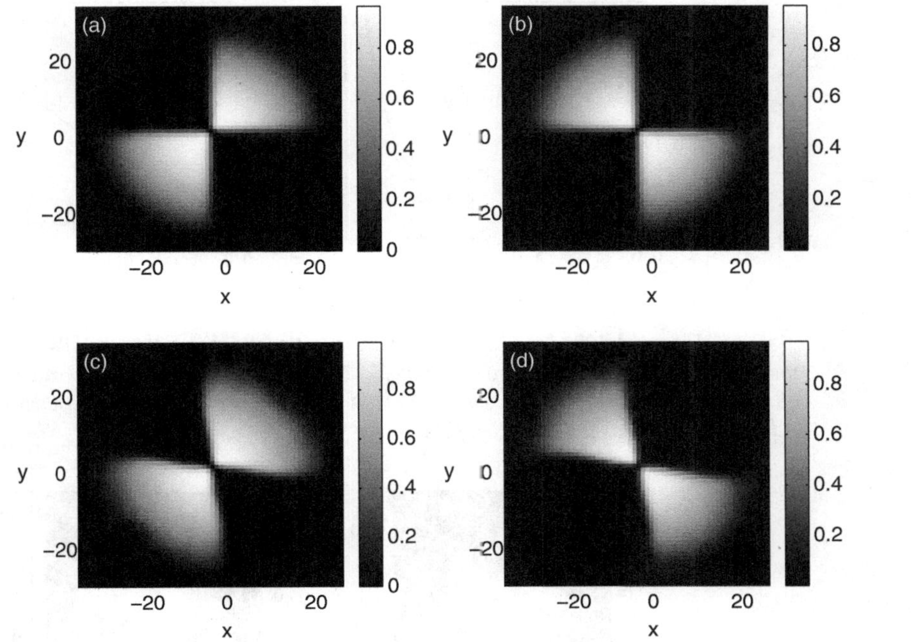

Figure 3.108. *Grayscale plots of the densities of the two species,* $|u_1|^2$ *and* $|u_2|^2$ *(left and right panels), in the symmetric [(a) and (b)] and asymmetric [(c) and (d)] DW-cross patterns. The parameters are* $\alpha_{11} = 1$, $\alpha_{22} = 1.01$, $\alpha_{12} = 1.52$, *and* $\Omega = 0.05$. *In the symmetric configuration,* $\mu_1 = \mu_2 = 1$, *and in the asymmetric one* $\mu_1 = 1.1025$, $\mu_2 = 1$, *and* $N_1/N_2 = 1.2$, $\theta_1 = 6\pi/11$, $\theta_2 = 5\pi/11$. *Adapted with permission from Ref. [251].*

numerical results always obey a natural relation, $N_1/N_2 \equiv Q_1/Q_2 = \theta_1/\theta_2$, where θ_j is the intrinsic angle of the jth sector, with $\theta_1 + \theta_2 \equiv \pi$. Examples of the symmetric and asymmetric DW crosses are shown, respectively, in the top and bottom panels of Fig. 3.108.

Stability and Rotation of the Domain-Wall Patterns

Static Domain-Wall Crosses. Dynamical stability of the symmetric and asymmetric DW crosses is an issue of importance in its own right. For a fixed trap's strength Ω, the stability strongly depends on the stoichiometry ratio N_1/N_2, more symmetric configurations being more robust. To show this, the nonlinearity coefficients α_{jj} were fixed, and $\mu_2 = 1$ was set and, then, for a given value of Ω in the interval $(0, 0.25)$, μ_1 was varied in Ref. [251] to induce variation of the ratio $N_1/N_2 \equiv Q_1/Q_2$ as per Eq. (3.218). Finally, Eqs. (3.210) were simulated in real time (up to $t = 1000$), to test the stability of the configuration. The resulting stability domain in the $(N_1/N_2, \Omega)$ parametric plane is displayed in Fig. 3.109(a). For example, at $\Omega = 0.05$, which corresponds to the above mentioned values (3.215) and (3.216) of the physical parameters, Fig. 3.109(a) shows that stable DW crosses exist in the interval $0.65 \leq N_1/N_2 \leq 1.65$. We note that, alternatively, it is possible to represent the domain of stability of the static DW crosses in the (μ_1, μ_2) plane, for a fixed value of Ω (then, each point of the domain would correspond to a given value of N_1/N_2).

Evolution of unstable DW crosses is also an issue of interest. A typical example of the instability development was found in Ref. [251] (see Fig. 3.110) at an initial stage to quickly rearrange the cross into a quasi-1D object, which is actually a dark soliton in the

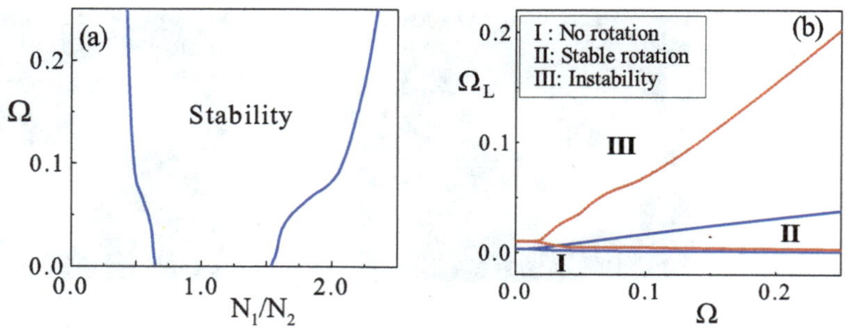

Figure 3.109. *Stability regions for the static* (a) *and rotating* (b) *DW crosses. The values of the nonlinear coefficients are fixed as in Eqs. (3.217). In* (b), *blue and red lines show, respectively, borders of the stable-rotation region for symmetric and asymmetric crosses (in the latter case, $N_1/N_2 = 1.5$). Adapted with permission from Ref. [251].*

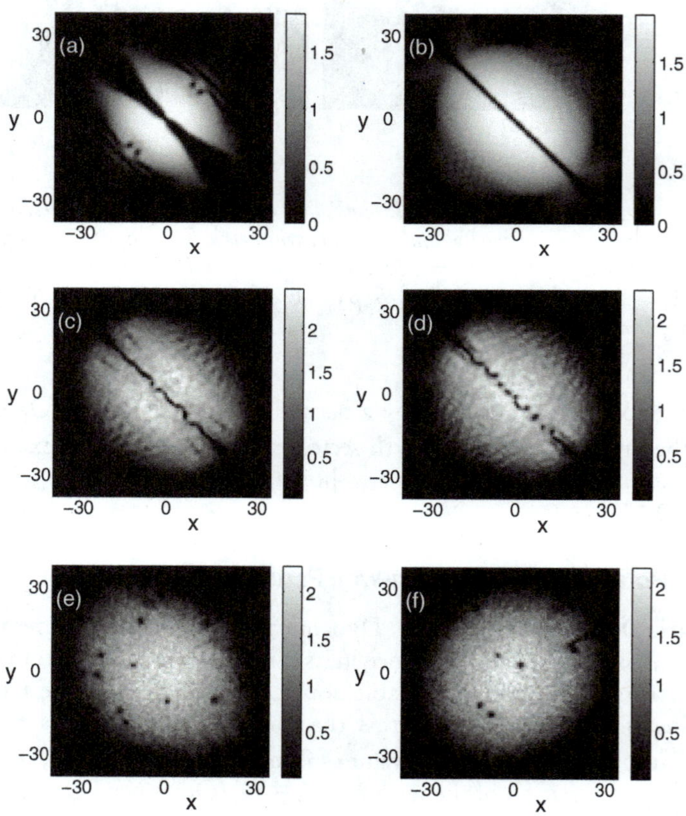

Figure 3.110. *Evolution of the density $|u_1|^2$ of the species with the larger number of atoms in an unstable DW cross, for the same parameters as in Fig. 3.108, but with $\mu_1 = 1.8025$. In this case, $N_1/N_2 = 2.5$ and $\theta_1 = 5\pi/7$, $\theta_2 = 2\pi/7$. The snapshots* (a)–(f) *correspond to $t = 0$, $t = 100$, $t = 310$, $t = 316$, $t = 1000$ and $t = 2200$, respectively (the physical time unit is 1.33 ms). The density $|u_2|^2$ is complementary to $|u_1|^2$, as in Figs. 3.108*(c) *and* (d)*. Adapted with permission from Ref. [251].*

species with the larger number of atoms, coupled to a bright soliton in the other species (this object resembles structures considered in Ref. [331]). The latter configuration is itself subject to a *snaking instability*, which is a general feature of quasi-1D dark solitons in

BECs filling 2D areas as discussed earlier in this chapter; it should be stressed that, unlike the dark solitons, the rectilinear DWs are not subject to this instability. The snaking instability initiates break-up of the dark stripe into four vortex-antivortex pairs in the first species, with the second species collecting itself into spots coupled to the pairs. Finally, the four vortex-antivortex pairs annihilate into two.

It should be added here that subsequent experimental work [308] was able to reach a range of values of Δ that varied roughly between -1 and 1, by exploiting interspecies scattering lengths in a mixture of ^{85}Rb and ^{87}Rb. In the latter case (and although the BEC was not 2D in nature), structures were observed that could be considered as tantamount to a rectilinear DW in the highly immiscible regime. On the other hand, the more general and quite interesting observation of Ref. [308] was concerning the formation of "droplet-like" states which are presumed to be long-lived metastable structures of one component surrounded by the second one; of course on the miscible side of the transition these authors observed overlapping clouds in accordance with the theory. The detailed exploration of the above droplet states is an interesting subject for further studies.

Domain-wall crosses in the rotating trap. We now turn to the case of rotating traps, with $\Omega_L \neq 0$ in Eqs. (3.210). In that setting, it was found in Ref. [251] that both the symmetric and asymmetric DW crosses revolve stably in a certain frequency interval,

$$(\Omega_L)_{\min} < \Omega_L < (\Omega_L)_{\max}. \tag{3.219}$$

If $\Omega_L < (\Omega_L)_{\min}$, the cross does not rotate at all, while at $\Omega_L > (\Omega_L)_{\max}$ it decays (see below). These results are summarized in Fig. 3.109(b). In region I, the DW cross remains quiescent, in region III it gets destroyed, while the stable rotation occurs in region II (see, for example, Fig. 3.111 that depicts a stably rotating propeller). Note that there are minimum values of the trap strength which are necessary for the rotation: $\Omega_{\min} = 0.012$ and 0.018 for the symmetric and asymmetric "double-vane propellers." Those values correspond to the edge points on the blue and red lines, respectively, in Fig. 3.109(b), the rotation frequencies at these points being $\Omega_L = 0.0034$ and 0.010. Note that for illustration of the relevant phenomenology the horizontal axis in Fig. 3.109(b) is extended up to a rather large value, $\Omega = 0.25$ although it was assumed above that the magnetic-trap's strength Ω is a small parameter, in order to derive the 2D GPE from the underlying 3D system.

As it was mentioned above, a dynamical property of the DW crosses which is crucially important for their relevance to the currently available experimental settings (where the immiscibility parameter $|\Delta|$ is small), is the fact that the crosses remain stable for all the negative values of Δ, up to $\Delta = 0$. This robustness carries over to the rotating crosses, as illustrated by Fig. 3.112, that shows a stable rotating cross in the case of $\Delta = 0$.

More general cross and propeller patterns. All the patterns produced by the intersection of more than two DWs in Ref. [251] were found to be unstable, rearranging themselves into the more fundamental patterns considered above. A typical example of that, with four intersecting DWs and $N_1/N_2 = 1$, is displayed in Fig. 3.113. The same instability of the higher-order crosses takes place in the rotating trap.

Rotating rectilinear domain walls. As was explained above, stable rectilinear DWs are less relevant for the currently (experimentally) tractable small values of the immiscibility parameter $|\Delta|$. Nevertheless, for completeness of the analysis, and also for the sake of possible future experiments in BEC mixtures of different atomic species, where $|\Delta|$ may be much larger than in the presently available mixtures of different spin states of the same isotope, it makes sense to briefly consider a possibility of stable rotation of the DWs of this type too. Numerical simulations show that generally, the rectilinear DW withstands rotation, within a certain interval of the angular velocities [cf. Eq. (3.219)], if it is stable in the static trap. Here, we do not aim to produce comprehensive results for the stability

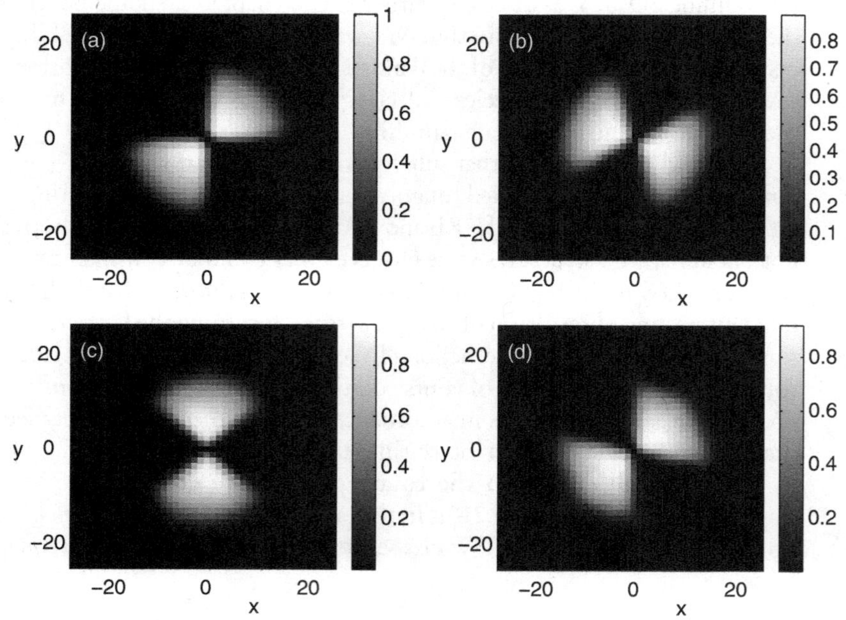

Figure 3.111. *A stable rotating asymmetric domain-wall cross ("double-vane propeller"). The four snapshots* (a), (b), (c), *and* (d) *show the distribution of the density* $|u_1|^2$ *at consecutive time moments—respectively,* $t = 0$, $T/6$, $T/3$, *and* $T/2$, *which cover a half of the rotation period,* $T/2 = \pi/\Omega_L \approx 52.3$ ($\simeq$ *70 ms, in physical units). The parameters are as in Figs. 3.108*(c) *and* (d) *but with* $\Omega = 0.1$ *and* $\Omega_L = 0.06$. *Adapted with permission from Ref. [251].*

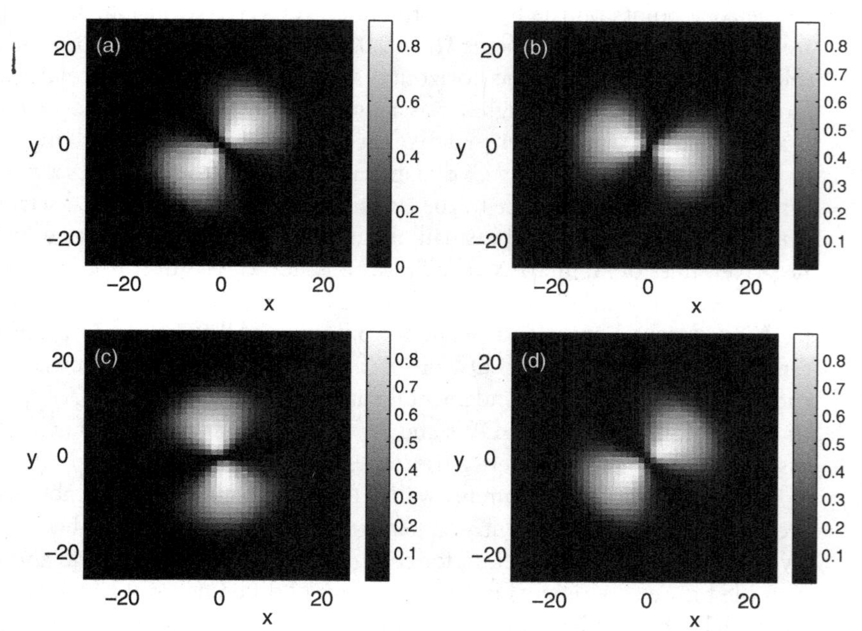

Figure 3.112. *A stable rotating symmetric domain-wall cross found exactly at the immiscibility border,* $\Delta = 0$ *(corresponding to* $\alpha_{11} = \alpha_{22} = \alpha_{12} = 1$*) and* $\Omega = 0.1$. *The arrangement of the figure is the same as in Fig. 3.111. Adapted with permission from Ref. [251].*

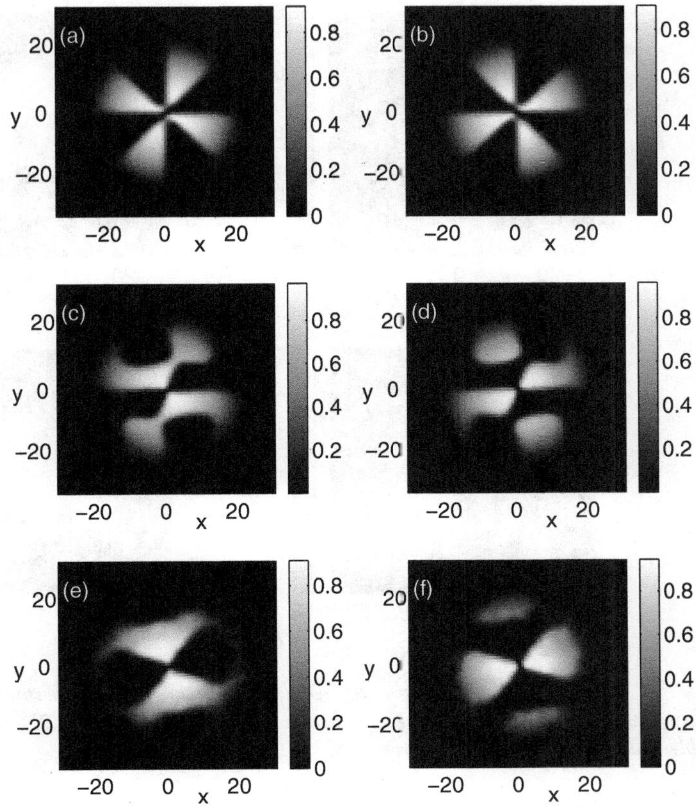

Figure 3.113. *Instability of a higher-order domain-wall cross (left and right panels show distributions of the densities $|u_1|^2$ and $|u_2|^2$). The frames* (a) *and* (b) *display the initial configuration at $t = 0$. Further frames show the evolving configurations:* (c), (d) *at $t = 50$, and* (e), (f) *at $t = 600$. The parameters are as in Figs. 3.108*(a) *and* (b). *Adapted with permission from Ref. [251].*

domain of the rotating rectilinear DWs. Instead, in Fig. 3.114 we display an example of a stably rotating DW of this type. This example has a special purport, as it shows the rotating rectilinear DW found at the minimum value of the immiscibility parameters at which it may be stable, $|\Delta| = 0.061$. Stable rotating rectilinear DWs found in the generic case are quite similar to the one shown in Fig. 3.114.

It is interesting to point out that in addition to the rectilinear DW states that were examined above, a more conventional example of DW states that was originally established in the work of Ref. [293] has a ball-shell structure. This implies that both components share the (radial) symmetry of the trap but due to their preference to phase separate, in order to minimize the intercomponent energy term, one is forming a ball-type domain inside, close to the center of the trap, while the other is forming a shell-like structure outside, being hollow inside. The proposal of Ref. [293] was further elaborated in the work of Ref. [292] which showed that the Thomas–Fermi approximation can also be used to predict many of the qualitative features of the two-component BECs and were able to identify also their collective excitations [634]. However, what was shown by Ref. [294] was that, in the immiscible regime of such two-component condensates, it may be energetically favorable (for an appropriate interplay of the number of atoms and the interspecies interaction) for the two-component phase separated state to break the radial symmetry and rather assume a form which is more reminiscent of the rectilinear DW states presented above.

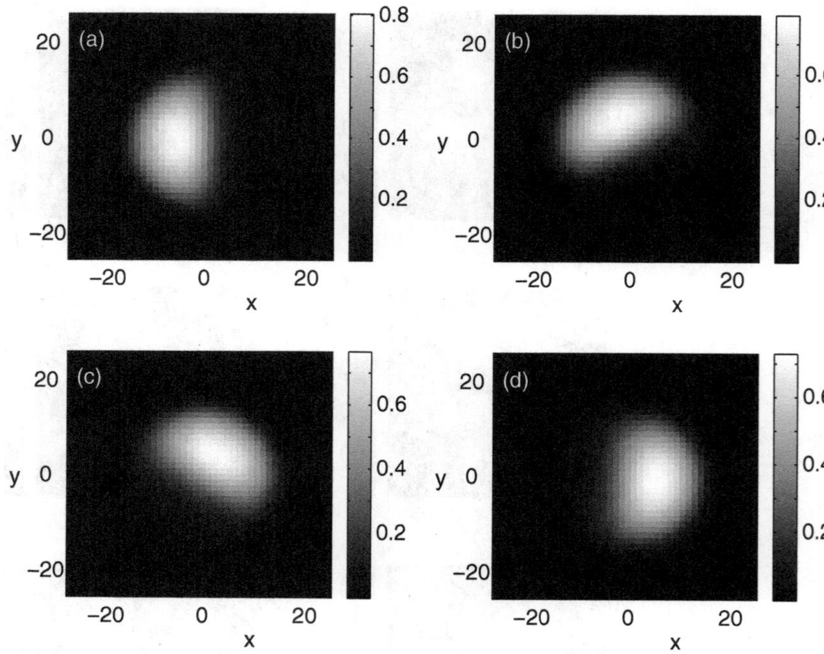

Figure 3.114. *Rotation of the rectilinear domain wall at the smallest value of the immiscibility parameter,* $|\Delta| = 0.061$ *(corresponding to* $\alpha_{11} = \alpha_{22} = 1$, $\alpha_{12} = 1.03$*), and* $\Omega = 0.1$ *at which it is stable. Adapted with permission from Ref. [251].*

3.11.5 ▪ The Role of Linear Coupling

In the work of Ref. [304] another example of the use of the SU(2) rotational symmetry, which commutes with the nonlinearity in the limit of $g_{ij} = 1$, was presented in order to produce novel states in a situation where the two components also feature linear (so-called Rabi) coupling between them.

In particular, using the matrix

$$\mathscr{U}(t) = e^{-i\alpha\mathscr{P}t} = \begin{pmatrix} \cos(\alpha t) & -i\sin(\alpha t) \\ -i\sin(\alpha t) & \cos(\alpha t) \end{pmatrix}, \tag{3.220}$$

where the matrix $\mathscr{P}$ is defined as

$$\mathscr{P} = \begin{pmatrix} 0 & 1 \\ 1 & 0 \end{pmatrix}, \tag{3.221}$$

it is straightforward to see that the time derivative of the 2D GPE produces a linear coupling of strength α between the two components. Hence, the matrix $\mathscr{U}(t)$ can be employed in order to translate from solutions of the problem without the linear coupling to (time-dependent) ones of the problem with linear coupling. If we denote the two components as ψ_1 and ψ_2, the density of the different components $n_i = |\psi_i|^2$ ($i = 1,2$) is given by

$$n_1 = \cos^2(\alpha t)|\psi_{01}|^2 + \sin^2(\alpha t)|\psi_{02}|^2 + |\psi_{01}||\psi_{02}|\sin(\theta_2 - \theta_1)\sin(2\alpha t), \tag{3.222}$$

$$n_2 = \cos^2(\alpha t)|\psi_{02}|^2 + \sin^2(\alpha t)|\psi_{01}|^2 + |\psi_{01}||\psi_{02}|\sin(\theta_1 - \theta_2)\sin(2\alpha t), \tag{3.223}$$

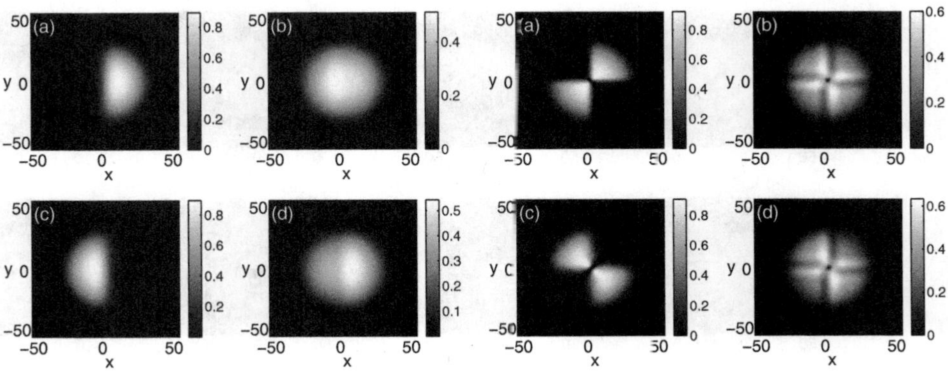

Figure 3.115. *Left panels: Contour plots of the density* $|\psi_1|^2$ *for a dipolar DW (the density* $|\psi_2|^2$ *of the other species is complementary to* $|\psi_1|^2$*) for* $t = 0$ (a), $T/4$ (b), $T/2$ (c), *and* $3T/4$ (d), *with* $T = \pi/\alpha \approx 15.7$ $(\alpha = 0.2)$; $\Omega = 0.045$, $\Delta = -0.061$. *The pattern persists for long times with the species "interchanging places." Right panels: Same as the top but for the quadrupolar DW with* $\Delta = -9 \times 10^{-4}$ *(the respective nonlinearity coefficients pertain to* ^{87}Rb*). The Rabi oscillation of this sloshing DW gives the impression of a rotating propeller. Adapted with permission from Ref. [304].*

where $\psi_{01} = |\psi_{01}|e^{i\theta_1}$, $\psi_{02} = |\psi_{02}|e^{i\theta_2}$. Note that if ψ_{01} and ψ_{02} are time independent, then $n_1 + n_2$ does not depend on time. Thus it is a *local* constant of the motion (as opposed to the global constant $\int (n_1 + n_2)\,dx$ which is conserved for *any* value of the nonlinearity parameters).

This idea is independent of the spatial dimension of the problem, and in the work of Ref. [304] it was used also in two spatial dimensions. The principle was illustrated in the case of DWs and vortices (relevant results, but for a rotating trap, were reported in Ref. [635]). As concerns the DW solutions, instead of using the more "standard" circular DWs between a less repulsive component in the middle of the trap and a more repulsive one forming an outer shell [293, 294], the above mentioned dipolar (i.e., rectilinear) and quadrupolar (i.e., cross-like) DWs of Ref. [251] were used. Once such a stationary state is reached for the 2D GPE (via imaginary time integration), using it as initial condition in the equation with the linear coupling produces a time periodic solution of the latter. Such examples are shown in Fig. 3.115 for the different types of DWs in the case of the usual harmonic trap $V(r) = (1/2)\Omega^2 r^2$. As indicated in the previous subsection, the quadrupolar DWs can persist even for $\Delta = 0$ and hence the resulting "rotating propeller" solutions are exact time periodic solutions of the linearly coupled equations. In the right set of panels of Fig. 3.115, a DW cross is shown for the realistic case of $a_{11} = 1.03$, $a_{12} = 1$, and $a_{22} = 0.97$. On the other hand, since the dipolar DWs only persist if $\Delta \leq -0.061$, these states cannot immediately arise for the ^{87}Rb parameters. In the left set of panels of Fig. 3.115 such a dipolar DW is shown, in the limiting case $\Delta = -0.061$; apparently, the time-periodic, dipolar DWs are only approximate solutions. Similarly to the 1D results, if $\Delta \neq 0$, we have found that there is a maximum (minimum) critical Δ_c up to which the Rabi oscillations persist (for the bulk of each component): $\Delta_c = -1.78$ for dipolar DWs and $\Delta_c = -0.64$ for the quadrupolar DWs.

Finally, one can use a similar construction for a pair of vortex structures with two components (see, e.g., Fig. 3.116). In this case, we initialize the imaginary time integration, in the absence of the linear coupling, with one component having a vortex centered at $(5, 0)$, while the other has a vortex at $(-5, 0)$. After the configuration relaxes to the stationary vortex pair solution of the two-component nonlinearly coupled GPEs, we again

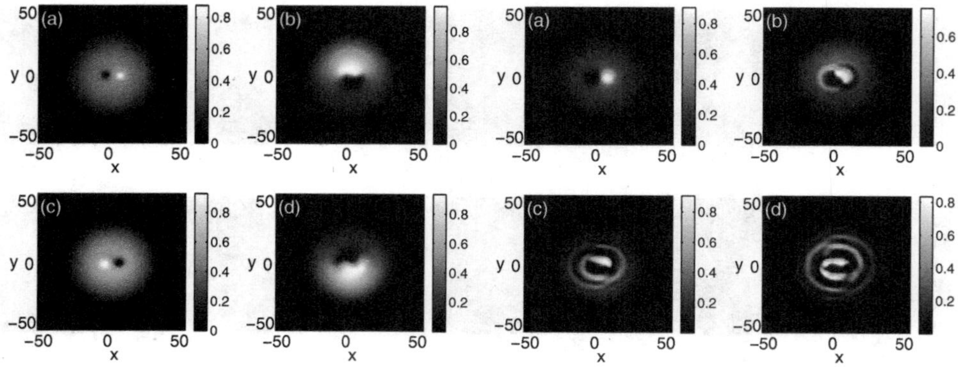

Figure 3.116. *Left panels: Contour plots of* $|\psi_1|^2$ *for two coupled vortices, initially placed at* $x = \pm 5$, *for* $t = 0$ (a), $T/4$ (b), $T/2$ (c), *and* $3T/4$ (d), *with* $T = \pi/\alpha \approx 15.7$ *(*$\alpha = 0.2$*);* $\Omega = 0.045$, $\Delta = -9 \times 10^{-4}$ *(*^{87}Rb*). The vortices "interchange locations" (in a structure resembling a spiral wave). Right panels: Same as the top but for* $\Delta = -3$ *(*$g = 1$, $h = 2$*). The configuration breaks up forming spiral patterns. Adapted with permission from Ref. [304].*

turn on the linear coupling and obtain a spiral rotation between the vortices resembling a spiral wave. In this case also, there is a critical $\Delta_c = -0.74$ beyond which the regularity of Rabi oscillations is destroyed. In this case, the breakup leads to the formation of spiral patterns in the condensate. It should be highlighted that similar vortex pair structures were also obtained in the work of Ref. [601]. In that setting, it was also explored how the distance of *stationary* (nonrotating) pairs of such vortices within the molecule decreases as the linear coupling between the components is increased. Finally, lattices were built from this fundamental building block of vortex molecules (or *merons* as they were referred to therein).

It should be added here that, in addition to using the Rabi coupling as a means of periodic alternation of wavefunctions between the components, it is also possible to use it as a switch in the spirit of Ref. [305]. There, a time-dependent Rabi frequency $\alpha(t)$ was considered to perform the following operation: at a time t_0, one has all the particles in $|1\rangle$ in the wavefunction $\psi_1(\vec{r}, t_0)$, and no particles in $|2\rangle$ ($\psi_2(\vec{r}, t_0) = 0$). At a time t_1 we wish to have all the particles in $|2\rangle$ in the same wavefunction $\psi_2(\vec{r}, t_1) = \psi_1(\vec{r}, t_0)$ (apart from a phase factor). To be more specific, we assume that $\alpha(t)$ depends on time as

$$\alpha(t) = \begin{cases} 0, & 0 \leq t < t_0, \\ \gamma, & t_0 \leq t \leq t_1 = t_0 + \delta, \\ 0, & t > t_1, \end{cases} \tag{3.224}$$

where t_0 is the switch-on time and δ denotes the duration of the Rabi pulse. Defining $\mathcal{I}(t) = \int_0^t \alpha(t')dt'$ and using the matrix $\mathcal{U}(t) = \exp[-i\mathcal{P}\mathcal{I}(t)]$, it was shown in Ref. [305] that the transformation still goes through in this general time-dependent case, enabling the coherent transfer (switch) of the wavefunction from one component to the other. In that light, for the appropriate duration of the pulse $\delta = \frac{\pi}{2\gamma}$, it was observed that at the end of the pulse

$$\begin{aligned} \psi_1(\vec{r}, t_1) &= 0, \\ \psi_2(\vec{r}, t_1) &= -ie^{-i\mu\delta}\psi_1(\vec{r}, t_0) \end{aligned} \tag{3.225}$$

which implies the direct exchange (up to a phase) of the component wavefunctions. This was successfully applied in the work of Ref. [305] not only for solitary waves and Thomas–Fermi clouds, but also for 2D vortices and even for vortex lattices. While the transfer was found to be exact in the limit of equal interactions $g_{ij} = 1$, g_{12} was also varied and fidelity of the transfer was evaluated with the conclusion that the transfer remains nearly exact for a wide range of parameters within the self-repulsive interaction regime (it most significantly deviated from unity when the interspecies interactions became attractive).

3.12 ▪ Exciton-Polaritons: An Emerging (Open System) Variant of Two-Dimensional BECs

In our explorations up to now, we have focused chiefly on Hamiltonian models of solitary waves and vortices, with the sole exception of the consideration of the effects of the coupling of the condensate with thermal atoms (see Secs. 2.6, 3.2, and 3.8.4). However, in recent years, one of the most rapidly developing branches of studies in the physics of Bose–Einstein condensation is that of exciton-polariton condensates in semiconductor micro-cavities. Although their experimental realization came about only a few years ago [636, 637, 638, 639], exciton-polariton BECs have already become a prototypical system for studies at the interface of nonequilibrium physics and nonlinear dynamics in the short time interval since then.

In addition to their similarities to prototypical Hamiltonian BECs which are the focus of this volume, exciton-polariton BECs have a number of distinctive characteristics. More specifically, the radiative lifetime of the polaritons provides a short relaxation time scale in the system of the order of 1–10 ps [640]. At the same time, the light mass of these quasi-particles provides them with a considerably higher condensation temperature. Moreover, the photonic component of the system only allows for a short lifetime and no thermalization. Instead, the exciton-polariton system produces a genuinely out of equilibrium condensate, requiring the external pumping of an excitonic reservoir, which in turn balances the polariton loss [640, 641]. This "open" nature of the system, featuring gain and loss, is then responsible for its rich pattern forming capabilities that have been recently summarized, e.g., in Refs [642, 643, 644].

Despite the short time since their experimental realization, numerous fundamental studies have already been conducted experimentally in such polariton condensates revealing, among others, a wide variety of 2D features such as flow without scattering (analogue of the flow without friction) [645], the existence of vortices [646] (see also Ref. [647] for vortex dipole dynamics and Ref. [648] for observations thereof), persistent currents, as well as higher-charge vortices [649] and collective dynamics [650].

The approach that has been used most commonly in theoretical studies of exciton-polariton BECs relies on the analysis of two coupled evolution equations for the polaritons and the exciton reservoir which enables their production. The corresponding model assumes the form of two coupled complex Ginzburg–Landau (cGL) equations describing the evolution of exciton and photon wavefunctions [651, 652, 653]. However, an alternative that has been proposed [654, 655, 656, 644] in the case of incoherent and/or far blue-detuned laser pumping (see, e.g., Ref. [644] and references therein), suggests that a single cGL equation for the macroscopically occupied polariton state may be used instead; such a model yields results consistent with experimental observations [657] (see also Refs. [643, 644]).

In the brief discussion of 2D exciton-polariton BECs herein, we will follow the recent exposition of Ref. [658]. More specifically, we will consider the case of incoherent pumping and use in our study a single cGL equation. Our aim is to briefly touch upon the existence and stability of fundamental states such as the nodeless cloud (i.e., the analogue of the atomic BEC Thomas–Fermi ground state) and the single vortex state. In addition to these more standard states, we will also consider states that such as the ring dark soliton (RDS) which, remarkably, although never stable in the context of atomic BECs (cf. the beginning of this chapter), can in fact be shown to be stable in suitable (gain) parametric regimes here. This is, effectively, a potentially stable radial form of a Nozaki–Bekki hole [659] that was previously explored in cGL contexts [660], yet was not found to be stable in these settings; instead, it was found there to potentially initiate a form of spiral wave turbulence. Importantly, this ring soliton structure was observed in polariton superfluids in a very recent experiment [661]. Interestingly, there also exist symmetry-breaking instabilities of this ring soliton structure which unveil a series of solutions without radial symmetry that may spontaneously emerge as a result of such instabilities. Among them, we highlight the potential for states with triangular or square/rectangular symmetry.

The relevant model in this case reads [654]

$$i\partial_t \psi = \{-\nabla_\perp^2 + r^2 + |\psi|^2 + i[(\chi(r) - \sigma|\psi|^2)]\}\,\psi, \tag{3.226}$$

where ψ denotes the polariton wavefunction trapped inside a 2D harmonic potential, $\nabla_\perp^2 \equiv \partial_x^2 + \partial_y^2$ is the transverse (2D) Laplacian and $r^2 \equiv x^2 + y^2$. The differences of Eq. (3.226) from the traditional form of the GPE can be traced in the presence of (i) the spatially dependent gain term with

$$\chi(r) = \alpha\Theta(r_m - |r|), \tag{3.227}$$

where Θ is the step function generating a symmetric spot of radius r_m and strength α for the gain, and (ii) the nonlinear saturation loss term, of strength σ. Estimates of the relevant physical time and space scales, as well as physically relevant parameter values, are given, e.g., in Ref. [654].

We can now identify stationary states using $\psi(r,t) = \psi_0(r)\exp(-i\mu t)$ and imposing the additional population balance constraint (balancing gain and loss in the 2D domain) as $dN/dt = 0$, which, in turn, leads to

$$\iint (\chi(r) - \sigma|\psi_0|^2)|\psi_0|^2\, dx\, dy = 0. \tag{3.228}$$

The fundamental difference with Hamiltonian BECs is that in this case, once external parameters are fixed such as the strength α and radius r_m of the gain and the strength of the saturable losses σ, the chemical potential of the solutions, μ, is fully determined. Hence, contrary to the monoparametric families of Hamiltonian BEC solutions, here we have *isolated* solutions [with specific μ chosen by Eq. (3.228)] due to the particle constraint. Once these solutions are identified, it is straightforward to perform a linearization around them, to examine through the BdG analysis their linear/spectral stability.

A typical two-parameter (α, r_m) plane of the nodeless cloud (NC) and single (central) vortex (CV) solution are shown in Fig. 3.117. An intriguing feature of this diagram is that, contrary to their Hamiltonian siblings, both the NC and the CV solutions may become dynamically unstable. Even more surprising is the fact that not only may the CV solution be dynamically unstable when the NC is stable (as one might naturally expect), but, conversely, the CV solution may be stable when the NC one is unstable. Hence, here the stability is critically affected not only by the trap and the nonlinearity, but by the interplay (and radius) of the gain and loss regions.

The profiles of the corresponding solutions do not appear offhand to significantly differ from their Hamiltonian counterparts (although subtle differences regarding their

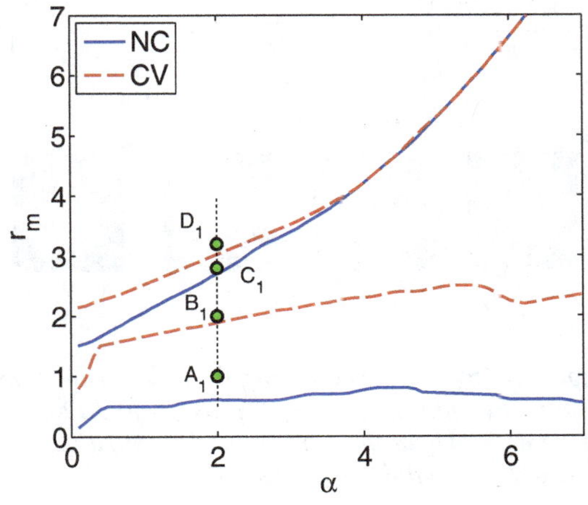

Figure 3.117. *Stability domains of nodeless cloud (NC) and central vortex (CV) solutions for* $\sigma = 0.35$*. The stable domains correspond to the regions between the curves. In a wide range of values within the 2D parameter space, these solutions coexist and may even be concurrently stable. Circles indicate points shown in following figures for NC and CV solutions. Adapted with permission from [658].*

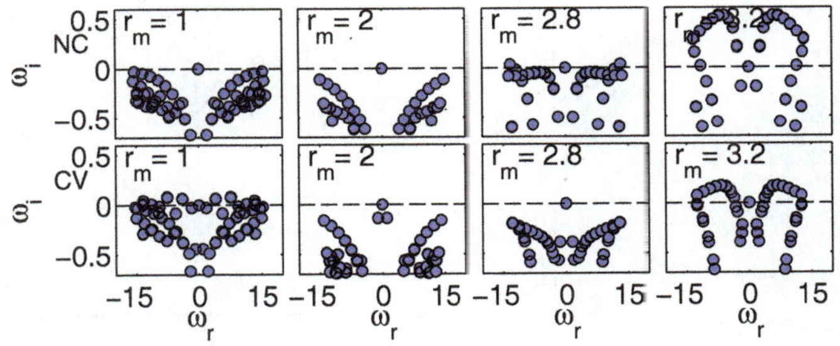

Figure 3.118. *Eigenfrequencies* ($\omega = \omega_r + i\,\omega_i$) *associated with the spectral analysis of the NC (top) and CV (bottom) waveforms. Notice how the instability of both the NC and the CV in these cases emerges from a band of continuous spectrum crossing into the unstable half-plane, as* r_m *is increased. Adapted with permission from Ref. [658].*

localization length may be relevant to aspects such as their stability). Nevertheless, let us proceed to visualize their stability, as the spectra of such nonlinear waveforms are critically different, given their dissipative behavior, from those of the Hamiltonian ones. This can be seen, e.g., in Fig. 3.118, where instabilities can be identified due to the crossing of whole bands of the spectrum to the unstable half-plane. This predisposes us towards a "dramatic" instability evolution that will significantly modify the background state and which, in fact, was observed in Refs. [654, 658] to result in the formation of vortex lattice configurations.

In addition to these more standard/fundamental configurations, the numerical exploration of Ref. [658] gave rise to a variety of additional configurations. Perhaps the most remarkable among them is the gray ring (GR) soliton, some typical density profiles of which are shown in Fig. 3.119. The depicted profiles correspond to the points shown as (green) dots in the 2D existence/stability (r_m, α) plane (i.e., width and amplitude of the parametric gain) shown in Fig. 3.120. The plane itself reveals some of the interesting features of such solutions, including the possibility of symmetry-breaking bifurcations giving rise to a new triangular form of solutions (denoted as TS; see panel F in Fig. 3.119).

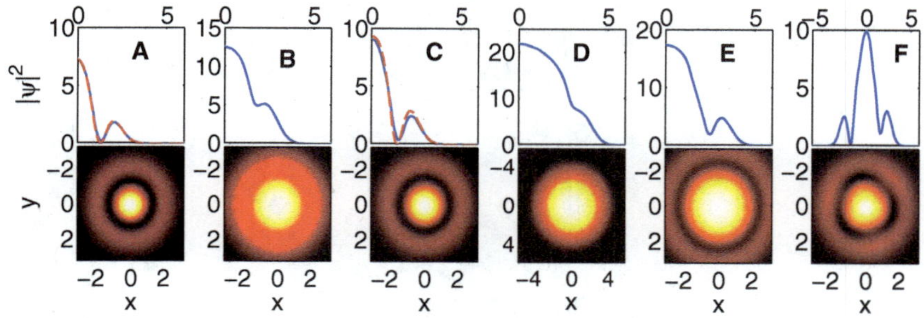

Figure 3.119. *Profiles of gray ring (GR) solitons and triangular solutions (TS) for the system parameters marked by (green) dots in the 2D plane of Fig. 3.120; $\sigma = 0.35$. Dashed (red) lines represent the Hamiltonian dark ring for the same chemical potential (panels A and C). Panel F represents a TS solution. Adapted with permission from Ref. [658].*

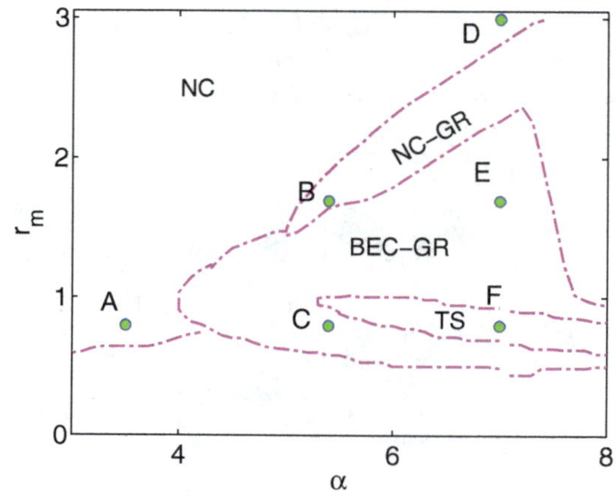

Figure 3.120. *Existence domains of gray ring (GR) solitons and triangular solutions (TS) for $\sigma = 0.35$. Solutions are stable except that the GR is unstable where the TS exists. For the distinction between GR solitons arising from the Hamiltonian BEC limit (BEC-GR) and those emerging from the NC (NC-GR), see details in Ref. [658]. Adapted with permission from Ref. [658].*

An additional feature is that while such ring dark solitons were proposed in both atomic condensates (as illustrated in the beginning of this Chapter) and in cGL gain/loss systems (as radial Nozaki–Bekki holes) in both contexts they were found to be unstable and thus break up into more prototypical coherent structures including vortices and spiral waves, respectively. In the context of polaritons, however, such structures were found to be potentially robust and, in fact, a recent experimental study has provided strong indications for their observation [661].

Contrary to what may be expected from the BEC siblings of such rings featuring a phase shift of π when stationary (and associated with a finite velocity when they are "gray"), in the polaritonic case, the rings are generically found to be gray even when they are stationary. This is, in fact, reminiscent of what was recently found also in the context of complex $\mathscr{P}\mathscr{T}$-symmetric potentials, e.g., in the work of Refs. [662, 663]. In both cases, the origin of the phenomenon is the same: in particular, the complex nature (of the potential in the $\mathscr{P}\mathscr{T}$-symmetric case and of the gain/loss structure in the cGL setting herein) of the terms in the equation produce a genuinely complex solution with a nontrivial phase structure and an associated "particle flux" along the stationary spatial profile. These features are absent in the BEC case, where the stationary ring dark soliton solution is genuinely real.

The spectral properties of the GR soliton state have been found to be significantly different than those of the previous two fundamental states (NC and CV). In particular, the instability of this state as r_m is increased stems from the fact that eigenfrequencies cross the origin of the spectral plane. This predisposes us towards a fundamentally different instability, arising through a symmetry breaking pitchfork bifurcation. This symmetry breaking bifurcation is substantiated in the parametric plane of Fig. 3.120 where a region has been denoted under TS (triangular solutions). As the curve outlining this region is crossed, the BEC-GR solutions undergo the above mentioned pitchfork bifurcation and spontaneously give rise to such TS. The TS states are generically found to be stable, a feature that renders them natural attractors for the dynamics. In addition to these triangular solutions, more complex waveforms were also identified in Ref. [658], including quadrupolar ones and were also found to be stable attractors of the dynamics in suitable regions of the parameter space of the problem.

Although these findings suggest a considerable wealth of solutions in the realm of semiconductor microcavities and exciton-polariton BECs, our current understanding of these states and especially their stability properties is, at present, rather limited. Naturally, this is a very promising field for further investigations and one that is bound to offer numerous additional insights in the near future.

Having completed our examination of 2D condensates both in the Hamiltonian and briefly in the above mentioned open system realm, we now turn our attention to 3D excitations and their corresponding study.

Chapter 4

The Three-Dimensional Case

4.1 ▪ Introduction

In this chapter, we give a brief survey of some of the results and literature on structures and phenomenology supported in three-dimensional (3D) NLS-type settings. Admittedly, this chapter is slanted towards the qualitative study of 3D structures rather than analytical results and approximations. This is mainly due to the difficulty in analytically treating such 3D settings which are, for the most part, studied using numerical explorations. Nonetheless, we do present some analytical results, most prominently ones coming from the application of classical fluid dynamics "technology" to the NLS equation when considering a nonviscous fluid (superfluid).

The material presented below is a blend of results in the defocusing NLS and, in particular, Bose–Einstein condensation, including experimental results. We start with the basic 3D structures in the form of vortex lines and review their stability, vibrational modes, internal (Kelvin) excitations and their reconnections. We then turn our attention to vortex rings (a vortex line looped into itself) as the basic localized excited state supported in 3D. For vortex rings, we cover their formation/nucleation, intrinsic velocity, internal Kelvin excitations, interactions (including leapfrogging behavior), mergers/splits, and (almost) inelastic scattering. We also give several examples whereby apparently unusual lower-dimensional behavior is uncovered and understood by the actual underlying 3D nature of the structures involved. A few examples of vortex ring nucleation in multi-component (two-component and spinor) BECs are also presented, together with a brief discussion of Skyrmion-like topological structures. Finally, we give an overview of a state-of-the-art efficient numerical integrator for NLS-type equations based on the massively parallel capabilities of graphical processor units (GPUs). The GPU code that we present (including its stability and boundary conditions) has sizable speedup factors when compared to its serial counterparts and thus greatly accelerates the numerics for the efficient investigation of NLS structures and phenomenology and, in particular, the highly computationally challenging 3D case.

4.2 ▪ Vortex Lines

4.2.1 ▪ Steady States, Stability, and Vibrational Modes

A vortex line (VL), or solitonic vortex, is the 3D extension of a 2D vortex by, infinitely and homogeneously, extending the solution into the axis perpendicular to the vortex plane. For physically relevant setups, where the background field is necessarily bounded due to

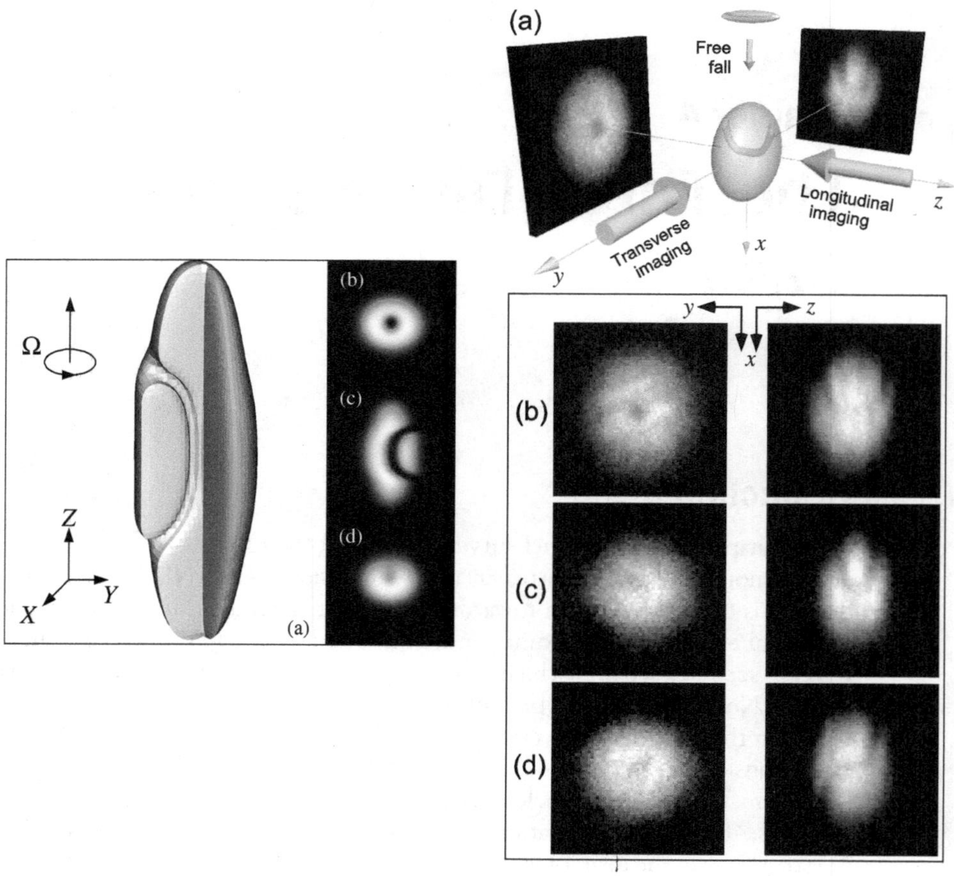

Figure 4.1. *Left panel: Example of a bent vortex line corresponding to the ground state of the Hamiltonian for a BEC.* (a) *3D cut of the density.* (b) *Transverse section at* $z = 0$. (c) *Transverse section at* $x = 0$. (d) *Integrated density plot along the z direction. Adapted with permission from Ref. [664]. Right panel:* (a) *Schematic of the imaging system. The cigar shaped condensate (top center) is imaged after 25 ms of time of flight (bottom center) leading to the inversion of the ellipticity in the* (x, y) *plane. Two beams image the atom cloud simultaneously along the longitudinal (z) and transverse (y) directions of the initial cigar-shaped cloud.* (b)–(d) *Simultaneous longitudinal and transverse images of condensates after 4 s* (b), *7.5 s* (c), *and 5 s* (d). *Adapted with permission from Ref. [665].*

the presence of an externally confining potential, the VLs are rendered finite in length. In that case, a single VL corresponds to a vorticity "tube" that traverses through the background density.

The stability and shape of a VL depends on several factors. For example, a straight VL in a spherically symmetric cloud (isotropic trapping in all directions) is only stable for a range of total angular momentum (cf. rotation of the cloud) [666]. In fact, when the straight VL becomes unstable, a bent VL becomes the minimizer of the energy and is thus stable [667]. The left panel in Fig. 4.1 shows an example of a bent VL and its corresponding cuts and z projection.

Vortex lines may be observed in BEC experiments. However, due to the lack of a full 3D imaging of the atom density, VLs can only be viewed from their 2D projection(s) corresponding to integrated densities. Such an example is depicted in the right panel of Fig. 4.1. In this experiment [665] a VL was obtained by introducing angular momentum through stirring the BEC, producing a small number of VLs (up to seven) and then al-

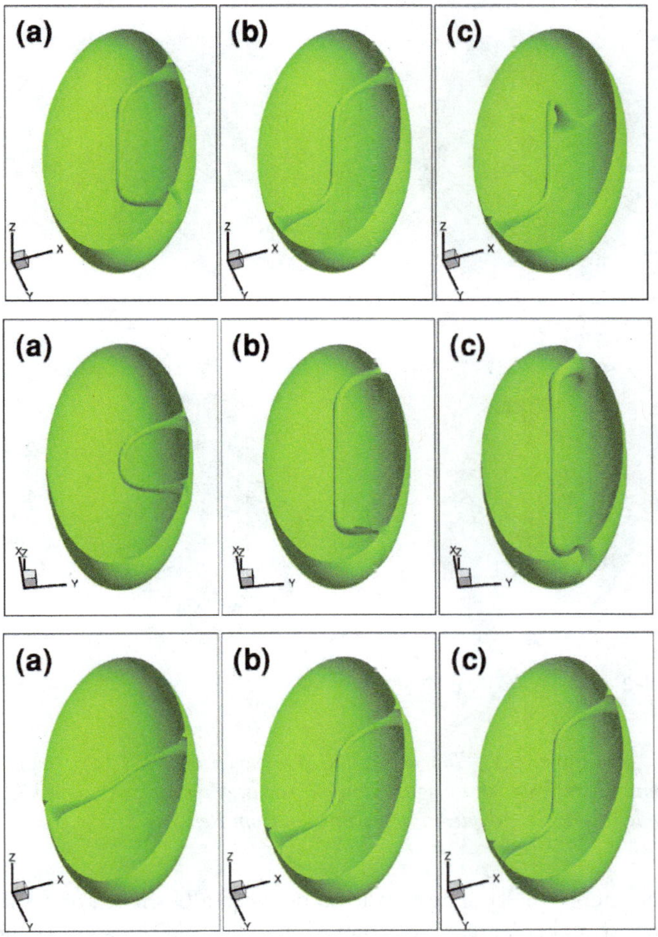

Figure 4.2. *Top panels:* (a) *U-shaped VL,* (b) *S-shaped VL, and* (c) *nonplanar S-shaped VL. Middle panels: Different U-shaped VL configurations for increasing angular momentum from left to right. Bottom panels: Different S-shaped VL configurations for increasing angular momentum (in the z component) from left to right. Adapted with permission from Ref. [668].*

lowing the condensate to relax to a single VL. Then, the BEC was allowed to expand (by switching off the magnetic trapping) and then photographed simultaneously along two perpendicular directions [cf. subpanel (a) on the right panel].

The specific shape of the bent vortex depends on the total angular momentum and, more importantly, on the aspect ratio of the background in the case of anisotropic traps [664, 667, 668, 670, 528, 671, 672, 673]. There are two main classes of bent VLs, namely U-shaped VLs and S-shaped VLs; see, respectively, subpanels (a) and (b) in the top panel of Fig. 4.2. The S-shaped VLs are not a global minimizer of the energy and only correspond to local minimizers [668, 672]. However, the difference in energy and angular momentum between U- and S-shaped VLs is very small since an S-shaped VL configuration is very close to a U-shape VL with a half part rotated by π. Although S-shaped VLs are not global minimizers, they have been clearly observed in experiments [665, 674] [see for example subpanel (d) in the right panel of Fig. 4.1]. Also, nonplanar S-shaped VLs are possible albeit only for right angles between the branches [see subpanel (c) in the top panel of Fig. 4.2] and, as in the case of coplanar S-shaped VLs, they are only local minimizers

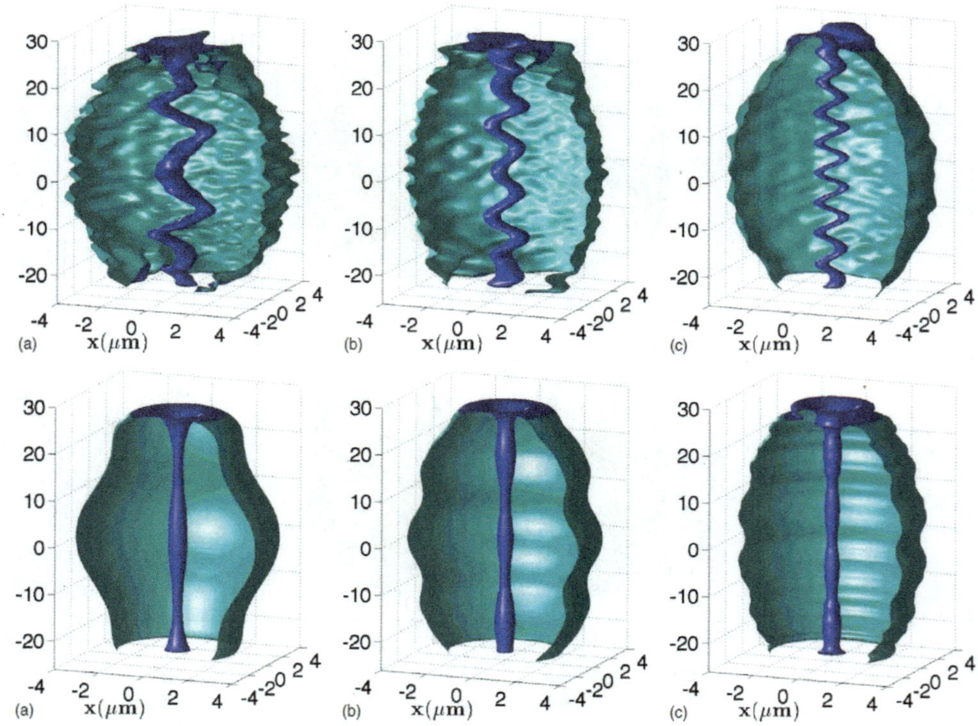

Figure 4.3. *Top panels: Kelvin modes of a VL in a trapped condensate for increasing wavenumbers (from left to right). Bottom panels: Varicose waves of a VL for increasing wavenumbers (from left to right). Adapted with permission from Ref. [669].*

of the energy [668]. The transition between a U-shaped or S-shaped VL to a straight (or almost straight) VL can be achieved by adjusting the angular momentum of the system [671, 667, 668]. Examples of this continuous transition between a bent VL and a straight VL are depicted in the middle and bottom panels of Fig. 4.2 for a U-shaped VL and an S-shaped VL, respectively.

A homogeneous condensate (i.e., in the absence of external potential) supports symmetric, straight VLs, when no perturbations are present. However, it is possible to excite helical traveling waves along the axis (z) of the VL, of the form $e^{i(kz-\omega_k t)}$, with wavenumber k and a frequency ω_k satisfying the dispersion relation [675]

$$\omega_k(k) \propto k^2 \ln\left(\frac{1}{|k|\xi}\right),$$

where ξ is the vortex core size (healing length). These helical waves, referred to as *Kelvin modes* [676] (or Kelvons), have been studied in great detail in classical fluids where their dispersion relation, for a hollow VL, involves Bessel functions [676, 677]. Kelvin modes have been observed in BEC experiments [674] by exciting the transverse quadrupolar modes in an elongated condensate (see also relevant theoretical results in Ref. [678]), and also in superfluid ^{4}He, from the reconnection of two VLs [679]. In nonuniform BECs (under the presence of a confining potential), Kelvin modes are also present and their corresponding dispersion relation needs to be adjusted by the presence of the inhomogeneity of the background cloud [675]. In fact, Kelvin modes can be effectively suppressed by the anisotropy of the trapping potential [616]. In the top row of Fig. 4.3 we depict a series

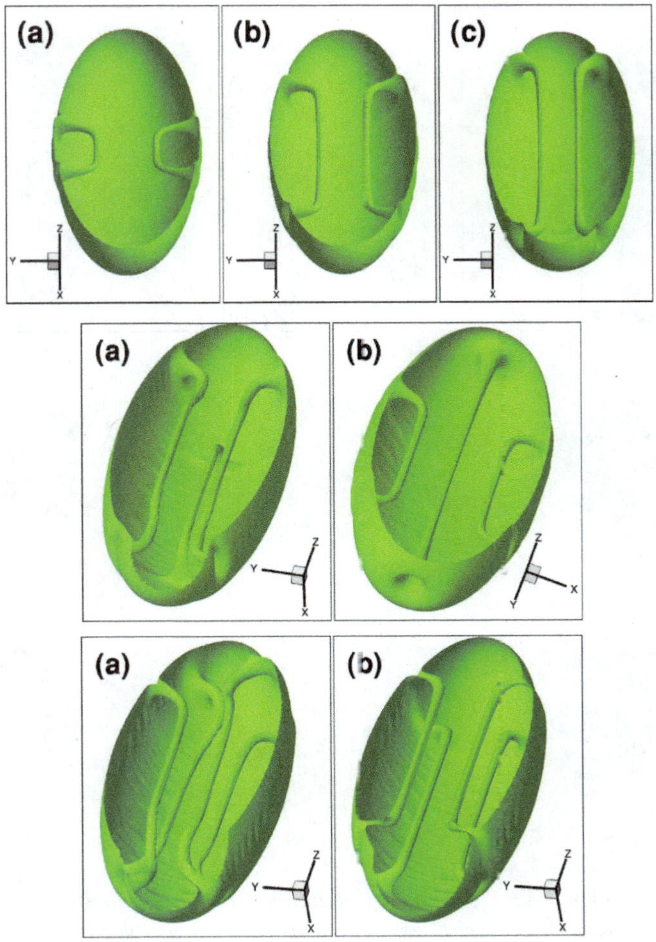

Figure 4.4. *Top panels: Two symmetric VLs for increasing angular momentum from left to right. Middle panels: Three VLs obtained from different initial conditions. Bottom panels: Four VLs obtained from different initial conditions. Adapted with permission from Ref. [668].*

of examples of Kelvin modes for different wavenumbers in a trapped BEC [669]. We will turn our attention to Kelvin modes when we study vortex rings, where the presence of a Kelvin mode excitation can have very strong effects on the inherent transverse velocity of vortex rings (see Sec. 4.3.2 below).

Another type of excitation that a VL might support is the so-called *varicose wave* [669, 680]. These excitations, corresponding to the periodic modulation of the vortex core size along the VL length, also possess their own dispersion relation [669]. Examples of such varicose waves are depicted in the bottom row of panels in Fig. 4.3.

When the angular momentum of the system is increased, progressively higher number of VLs are admissible within the condensate. For example, Fig. 4.4 depicts configurations with two, three, and four VLs configurations. These multi-VL configurations give rise to the 3D equivalents of the vortex lattices discussed in the previous chapter. The top two panels of Fig. 4.5 show these types of configurations. As it can be observed from the figure, as the total angular momentum of the system is increased, the system allows for a larger number of VLs. An interesting aspect of these crystalline VL lattices is their normal modes of vibration [681]. By linearizing about the neutrally stable equilibrium VL

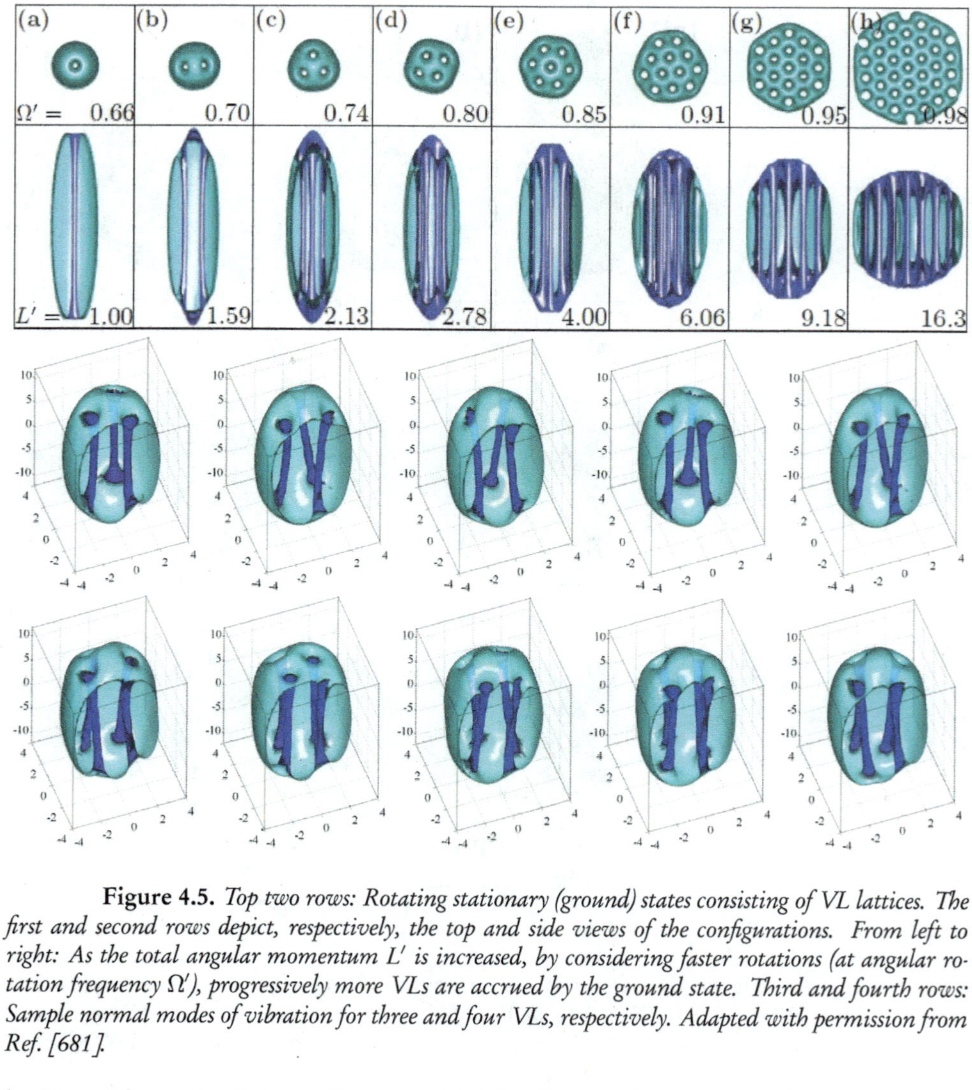

Figure 4.5. *Top two rows: Rotating stationary (ground) states consisting of VL lattices. The first and second rows depict, respectively, the top and side views of the configurations. From left to right: As the total angular momentum L' is increased, by considering faster rotations (at angular rotation frequency Ω'), progressively more VLs are accrued by the ground state. Third and fourth rows: Sample normal modes of vibration for three and four VLs, respectively. Adapted with permission from Ref. [681].*

lattice configuration it is possible to extract the eigenmodes corresponding to collective oscillations of the VL lattice. These collective modes, for large number of VLs in a VL lattice, are often referred to as *Tkachenko modes* and have been observed in BEC experiments [562, 682] and studied in detail in single [683, 567, 684] and two-component [685, 686] BECs. Figure 4.6 depicts a few experimental and theoretical examples of Tkachenko modes in quasi-2D BECs. It is interesting to contrast the standard 2D Tkachenko modes reported in quasi-2D BECs with the corresponding modes when the condensate is not in the quasi-2D regime. In particular, for 3D geometries, some of these collective modes correspond to periodic oscillations where the VLs perform nutation-type motion due to their 3D character. Two examples of these periodic solutions for three and four VLs are depicted in the bottom two rows of panels of Fig. 4.5.

4.2.2 ▪ Reconnections

Reconnections of vortex filaments [688] play an essential role in the description of fluid dynamics (cf. Ref. [689] and references therein) and particularly the description of turbu-

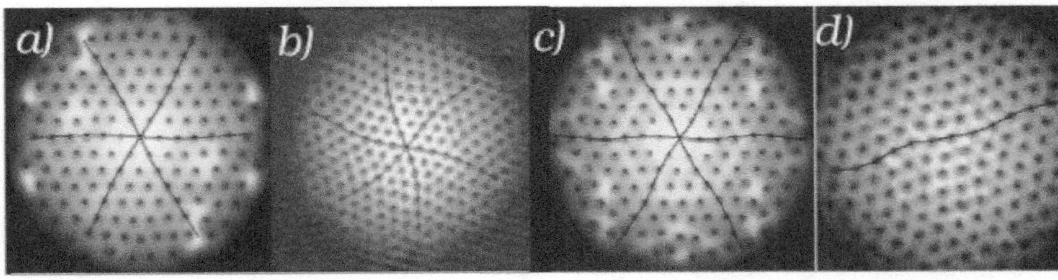

Figure 4.6. *Numerical and experimental Tkachenko modes in quasi-2D trapped BECs.* (a)–(b) *First Tkachenko mode:* (a) *theory and* (b) *experimental results. The curve is a fitted sine wave with wavelength* $1.33R_{\mathrm{TF}}$. (c)–(d) *Second Tkachenko mode:* (c) *theory and* (d) *experimental results. The curve is a fitted sine wave with wavelength* $1.33R_{\mathrm{TF}}$. *The curves are drawn to guide the eye. Adapted with permission from Ref. [567].*

lent regimes and its manifestation on energy cascading [690, 687]. The reconnection of vortex filaments is also important since it involves changes in the topology of the vorticity [691]. For a detailed account of the history and progress on vortex reconnections in classical and quantum fluids the interested reader is encouraged to refer to Refs. [689, 687] and references therein.

In this section, we summarize some of the numerical results on the GPE obtained in Ref. [687] illustrating the dynamical outcomes of different VL reconnections, depending on the relative angles between the VLs. For instance, when two VLs are placed next to each other, with their axes parallel but with opposite charges (namely, antiparallel VLs), they experience the so-called *Crow instability* [692, 693]. This instability induces the VLs to bend towards each other until they reconnect forming two U-shaped VLs that move apart from each other (see top row of panels in Fig. 4.7). In the reconnection process, a sizable rarefaction pulse is emitted from the location of the reconnection perpendicular to the VLs [687]. This rarefaction pulse corresponds to the emission of sound waves induced by the effective loss of energy carried by the vortex filaments after reconnection [694]. Reconnections for VLs initially oriented at different angles lead to similar outcomes as for the antiparallel case as it is depicted in the bottom two rows of panels in Fig. 4.7. Interestingly, there seems to be a universal power law behavior for the time evolution of the minimum distance $\delta(t)$ between the vortex filaments before and after reconnection, in the form [687]

$$\delta(t)=\begin{cases} A_b|t-t_0|^{\alpha_b} & \text{if} \quad t<t_0, \\ A_a|t-t_0|^{\alpha_a} & \text{if} \quad t>t_0, \end{cases} \tag{4.1}$$

where t_0 is the time of the reconnection, A_a and A_b are constants, while α_b and α_a are the exponents before and after reconnection, respectively. The exponents α_b and α_a depend slightly on the initial relative angles between the VLs. However, the most remarkable result is that the reconnection is not symmetric in that α_b is always smaller than α_a, indicating that the separation is faster after the reconnection. Typical (average) values for the power-law exponents for various initial angles correspond to $\alpha_b=0.39$ and $\alpha_a=0.68$ whose average is close to the predicted value $\alpha=0.5$ from scaling arguments [687].

4.2.3 ▪ Hydrodynamic Formulation: Connections with Fluid Mechanics

By employing a density $\rho(\vec{r},t)$ and phase $\phi(\vec{r},t)$ decomposition of the wavefunction, the so-called Madelung transformation, $\Psi(\vec{r},t)=\sqrt{\rho(\vec{r},t)}\,e^{i\phi(\vec{r},t)}$, it is possible to rewrite

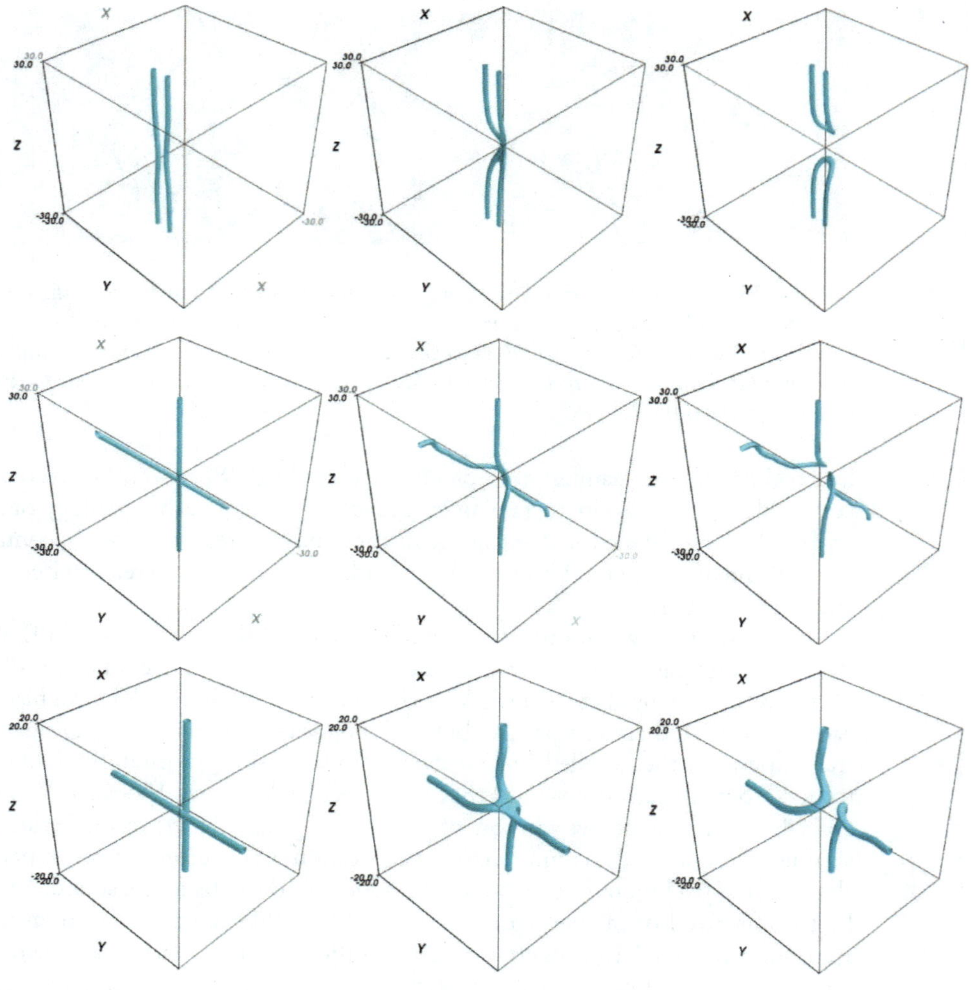

Figure 4.7. *Top row: Evolution of two initial parallel VLs of opposite vorticity which experience Crow instability and reconnect to form two U-shaped VLS. Middle row: Evolution two VLs initially at an angle $3\pi/4$. Bottom row: Evolution two VLs initially at an angle $\pi/2$. Adapted with permission from Ref. [687].*

the following hydrodynamic formulation of the GPE:

$$\frac{\partial \rho}{\partial t} + \nabla \cdot (\rho \vec{v}) = 0, \tag{4.2}$$

$$m\frac{\partial \vec{v}}{\partial t} + \nabla\left(\frac{1}{2}m|\vec{v}|^2 + g\rho\right) = \nabla(q - V(\vec{r})), \tag{4.3}$$

where the fluid velocity is obtained from the gradient of the BEC phase,

$$\vec{v}(\vec{r},t) = \frac{\hbar}{m}\nabla\phi(\vec{r},t), \tag{4.4}$$

and the so-called quantum pressure term corresponds to

$$q(\vec{r},t) = \frac{\hbar}{2m}\frac{\nabla^2\sqrt{\rho}}{\sqrt{\rho}}. \tag{4.5}$$

This hydrodynamic formulation allows for the interpretation of the BEC as a fluid with no viscosity (a superfluid) described by Euler's fluid equations (4.2) and (4.3) that include the quantum pressure term (4.5) and the effects of the external potential $V(\vec{r})$. The quantum pressure term becomes important near the core of vortices (and dark solitons, in the lower-dimensional setting). Therefore, if one is interested in the dynamics (a few healing lengths) away from any vortex core and in the absence of external potential, it is possible to approximate the BEC hydrodynamic description as a standard fluid without viscosity. This allows us to borrow tools developed from classical fluids and apply them to the dynamics of BECs. In particular, one may approach the study of the dynamics of VLs by using the Biot–Savart law [695, 696]. In a classical fluid, a vorticity field $\vec{\omega}$ induces a velocity field according to [695, 696]

$$\mathbf{u}_v(\vec{r}) = \frac{1}{4\pi}\int \frac{\vec{\omega}(\vec{r}\,')\times(\vec{r}-\vec{r}\,')}{|\vec{r}-\vec{r}\,'|^3}d\vec{r}\,', \tag{4.6}$$

where the integral is computed over all of space. In the context of the hydrodynamic formulation of BECs, the vorticity is computed in the usual manner, by taking the curl of the fluid velocity:

$$\vec{\omega} = \nabla \times \vec{v}.$$

Equation (4.6) can be used to account for the effect of a section of a VL onto a section of another VL. In fact, one can use this approach by considering a vorticity distribution of infinite strength but having a constant circulation $\varkappa$ along a vorticity curve. This curve might be a straight line as in a straight VL or, in general, a vortex filament described by an arbitrary curve. In that case, the fluid velocity induced at any given point by the entire vortex filament curve $\mathbf{R}$ parametrized by ℓ is given by

$$\mathbf{u}_v(\vec{r}) = \frac{\varkappa}{4\pi}\int \frac{\mathbf{s}\times(\vec{r}-\mathbf{R}(\ell))}{|\vec{r}-\mathbf{R}(\ell)|^3}\,d\ell, \tag{4.7}$$

where $\mathbf{s}$ is the unit tangent vector. In the case where one is interested in the effects of a closed vortex filament, as is the case of a vortex ring (see below), the integral in Eq. (4.7) is naturally replaced by a closed contour integral along the closed filament.

It is important to mention that the integral in Eq. (4.7) becomes singular when computing the fluid velocity induced on the filament itself [696], namely the self-induced velocity. This singularity can be circumvented by using the localized induction approximation [698, 699, 700, 701, 702, 703], which assumes a finite size for the vortex core. This regularization of the self-interacting term yields a powerful method to follow the dynamics of vortex filaments: the vortex filament method [704]. Within this approach, instead of following the 3D fluid equations of motion, the dynamics is reduced to following a 1D partial differential equation for each filament in the fluid. This technique allows for an efficient method to follow the dynamics of very large number of vortex filaments. For example, Fig. 4.8 showcases an example where the vortex filament method was employed to follow the decay of turbulence (see below) in superfluid ^{4}He [697]. Interestingly, the method can be adjusted to take into account the effects of VL reconnections [705] and the ensuing Kelvin modes along the filament [679].

4.2.4 ▪ Quantum Turbulence

The theme of superfluid turbulence has attracted considerable attention in the past few decades and, in particular, its manifestation in superfluid ^{4}He (cf. Refs. [677, 706, 707, 708, 607] and references therein). There is an intimate relation between superfluidity and the

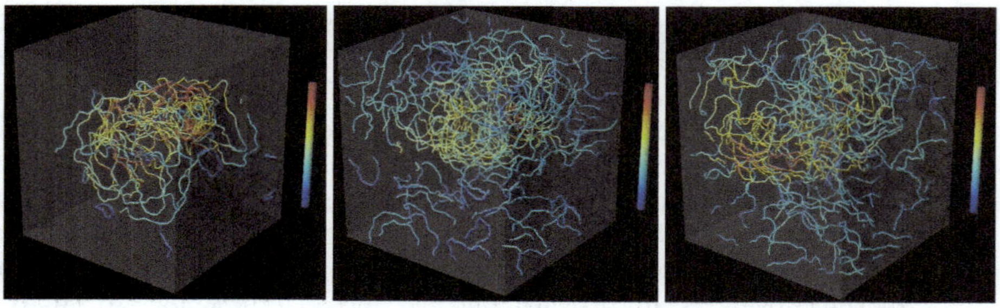

Figure 4.8. *Simulation of the decay of semiclassical turbulence in superfluid ^4He in a periodic domain. The panels show the evolution (left to right) of an initial distribution of vortex filaments (left). The color indicates the normalized vorticity. Adapted with permission from Ref. [697].*

GPE in that superfluid ^{4}He at zero temperature can be considered as a weakly interacting Bose gas (see the review [709] and references therein). In the GPE model at zero temperature, the effects of the noncondensed atoms (a.k.a. the thermal cloud) are completely negligible. Therefore, the natural mechanism whereby energy present at high wavenumbers is dissipated in a standard, viscous, fluid is not present. Nonetheless, it is possible to consider the total energy in the GPE, which is conserved, as having three contributions [619]: (a) the kinetic energy, (b) the internal energy, and (c) the "quantum" energy [corresponding to the quantum pressure term (4.5) and responsible for the emission of sound waves]. Using this separation of the energy, it is possible to obtain energy dissipation away from the kinetic energy part that is analogous to the Kolmogorov's power law decay [619, 602]. This energy cascade provides an analogue of fluid turbulence in the quantum superfluid described by the GPE, namely quantum turbulence (QT) [607]. Details on 2D QT can be found in Sec. 3.10.

On the other hand, experimental [710] and numerical [711] studies have confirmed that the velocity distributions in turbulent superfluids do not follow the statistics of classical fluids in three dimensions. Specifically, ordinary fluids present velocity components obeying Gaussian statistics, while in superfluid (quantum) counterparts power-law tails. This difference is attributed to the unbounded rotational motion present in the quantum superfluid, where the fluid velocity comes from the gradient of the phase of the wavefunction and thus has a $1/r$ blowup at the vortex cores [711].

4.3 ▪ Vortex Rings

If a VL is bent enough to close on itself, or if two VLs are close enough to each other, they can produce a vortex ring (VR) [692]. VRs are 3D structures whose core is a closed loop with vorticity around it [677] (i.e., a vortex filament that loops back into itself). Below we discuss several aspects of the generation, dynamics, and interactions of VRs.

4.3.1 ▪ Vortex Ring Generation

VRs can be produced by means of a variety of methods, including ones that have been discussed in the previous chapters for the generation of dark solitons and vortices. In particular, VRs can be formed

(i) by an impurity traveling faster than the speed of sound of the background [410] (see Sec. 4.4.1);

(ii) by nonlinear interference between colliding blobs of the background material [156, 159] (see Fig. 4.9);

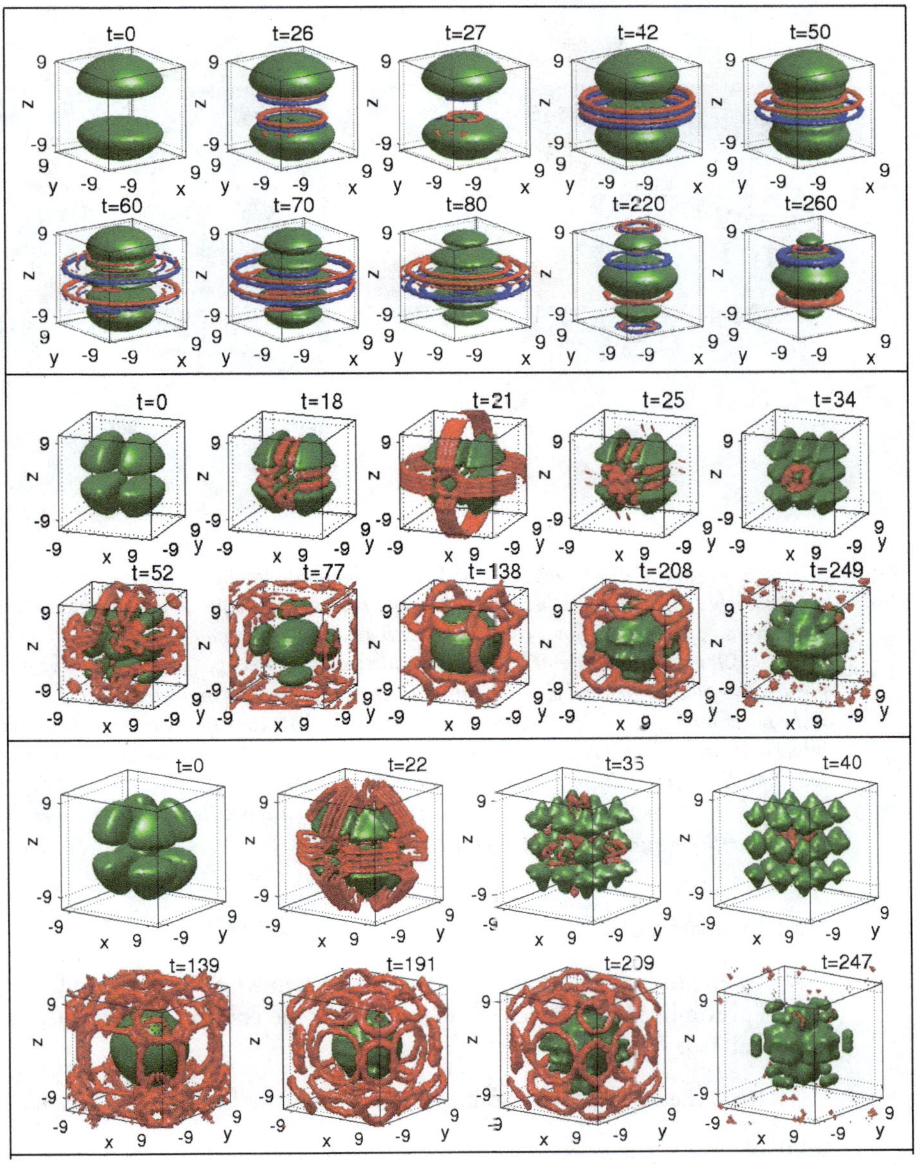

Figure 4.9. *Generation of VRs by nonlinear interference. A trapped condensate is split into several fragments by several laser sheets corresponding to repulsive potential barriers. Starting from the stationary fragmented condensate, the laser sheets are removed and the different fragments run towards the center of the harmonic (magnetic) trapping potential inducing their collision, producing an effective nonlinear interference pattern that contains VRs. Top panels: Two symmetric fragments collide and create a series of concentric VRs that impinge upon the condensate's density. Middle panels: Four colliding symmetric fragments. Bottom panels: Eight colliding symmetric fragments producing a VR cage around the bulk of the condensate. In all panels the 3D contour plot of atomic density* $|\Psi(x,y,z,t)|^2$ *at* $\max(|\Psi|^2)/2.5$ *is shown in green and the contour of the vorticity* $\omega(x,y,z,t)$ *at* $\max(\omega)/15$ *is shown in red (and blue) as a function of* (x,y,z) *at different times indicated. For the top series of panels, positively (negatively) charged VRs are depicted in blue (red). For the middle and bottom series of panels all VRs are depicted in red. Adapted with permission from Ref. [156].*

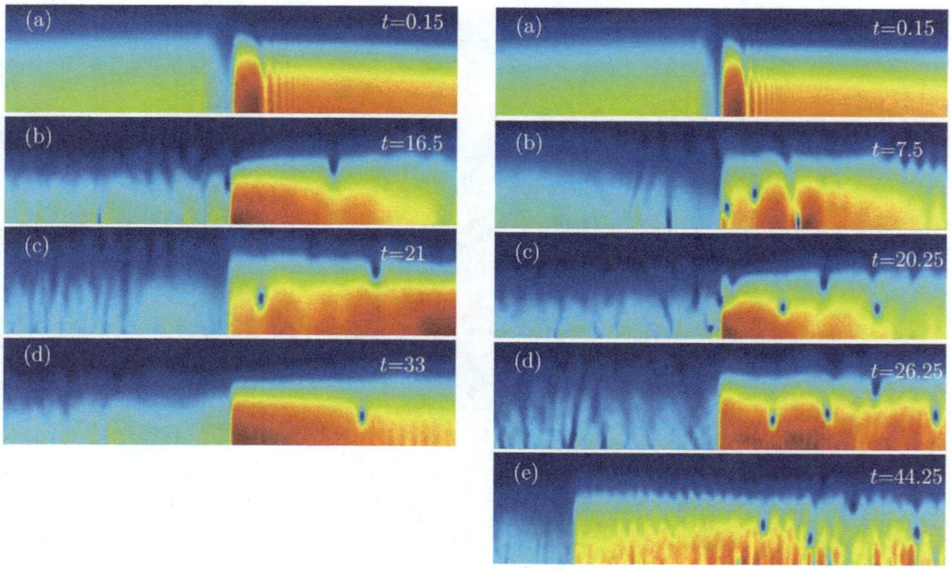

Figure 4.10. *Generation of VRs by the piston-like action of half of the condensate being pushed (right half) into the other half (left half). The piston-like procedure is achieved by starting with the stationary steady state for a condensate with a homogeneous nonlinearity g and then modifying (at time $t = 0$) the nonlinearity of half of the condensate such that $g_{\text{right}}/g_{\text{left}} > 1$. The panels present the pseudocolor plots of an axial cut of the atomic density at the times indicated in a cylindrical geometry (z is the horizontal axis while r is the vertical axis). The VRs correspond to the small circular depletions (blue) of density. The left and right series of panels correspond, respectively, to a relatively slower and relatively faster piston action where the ratio of nonlinearities is $g_{\text{right}}/g_{\text{left}} = 2.2$ for the left panels and $g_{\text{right}}/g_{\text{left}} = 2.4$ for the right panels. A faster piston action is able to produce a larger quantity of VRs. Adapted with permission from Ref. [712].*

(iii) by phase and density engineering techniques [713, 714];

(iv) by engineering the nonlinearity to have a step-wise function producing an effective "piston-like" phenomenon, where half of the condensate is pushed into the other half (see Fig. 4.10);

(v) from the decay of dark solitons or domain walls [49, 71] (see Sec. 4.3.5);

(vi) or even by introducing "bubbles" of one component in the other component in two-component NLS systems [715] (see Sec. 4.4.1).

Vortex rings have been observed experimentally in superfluid ^{4}He [716, 717], as well as in the context of BECs. In the latter case, VRs have been observed in the decay of dark solitons in two-component BECs [49], by direct density engineering [71] (see also the complementary theoretical proposals of Refs. [713, 718]), and in the evolution of colliding symmetric defects [66]. VRs have also been argued to be responsible for unusual experimental collisional outcomes of structures that may appear as dark solitons in cigar-shaped traps [73] (see Sec. 4.3.5). Numerical studies of experimentally feasible vortex ring generation in BECs have also been explored in the context of flow past an obstacle [719, 410], Bloch oscillations in an optical trap [720], collapse of bubbles [721], instability of 2D rarefaction pulses [722], flow past a positive ion [723, 724] or an electron bubble [725], Crow instability of two vortex pairs [693], and collisions of multiple BECs [159, 156].

4.3.2 ▪ Intrinsic Velocity for Vortex Rings

In contrast to single VLs which are stationary in the homogeneous background (i.e., in the absence of external potentials), a VR possesses an intrinsic velocity perpendicular to the ring plane due to Helmholtz's law [726]. The origin for the VR intrinsic velocity can be understood from following any cross-sectional cut along the VR axis: in that cut, the VR manifests itself as a pair of 2D vortices with *opposite* charge which thus travel parallel to each other (see previous chapter). Let us consider the following dimensionless form of the GPE in the absence of any trapping potential:

$$i\frac{\partial\Psi}{\partial t}+\frac{1}{2}\nabla^2\Psi-|\Psi|^2\Psi=0. \tag{4.8}$$

The transverse velocity of a single VR in the context of the NLS equation (4.8) has been studied analytically [727, 728, 729] and the results were shown to be consistent with numerical simulations of the NLS equation [730, 731]. For the NLS equation (4.8), an asymptotically approximate velocity for the VR is given by [729]

$$v_0 \approx -\frac{m}{2R}\left(\ln\frac{8R}{r_c}+L_0(m)-1\right), \tag{4.9}$$

where R is the ring's radius, m is its charge, and r_c is the vortex ring's core radius defined as

$$r_c=|m|\xi, \tag{4.10}$$

where ξ is the healing length given by

$$\xi=(2\mu)^{-1/2},$$

and μ is the system's chemical potential. The value $L_0(m)$ is referred to as the vortex core parameter and is defined as the convergent part of the energy per unit length of a VL in the NLS equation [729]. While $L_0(m)$ depends on the vortex ring charge m, it is independent of the NLS equation parameters. Since there are various values of $L_0(m)$ reported in the literature (cf. values given in Refs. [732, 729, 709]), while cases for $m>3$ have not been reported, in the table of Fig. 4.11 we provide the numerically obtained values for $L_0(m)$ for $m\in[1,10]$ by using refined numerics [733]. It is intriguing to note that one can produce a very close power-law fit yielding an approximation of $L_0(m)$, given by

$$L_0(m)\approx L_0(1)\,m^{-3/2}, \tag{4.11}$$

which may indicate the possibility that a closed-form expression for $L_0(m)$ might be achievable. In Fig. 4.11 we show a comparison of the values of L_0 given in the table of Fig. 4.11 to those given by Eq. (4.11).

It is important to mention that the transverse velocity discussed above corresponds to that of an *unperturbed* VR on a *homogeneous* background. As for VLs, it is possible

m	$L_0(m)$
1	0.380868
2	0.133837
3	0.070755
4	0.044567
5	0.030981
6	0.022954
7	0.017785
8	0.014247
9	0.011713
10	0.009833

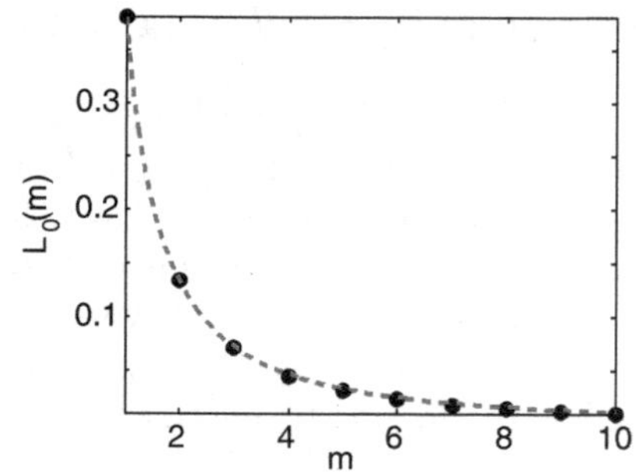

Figure 4.11. *Left: Values of the vortex core parameter L_0 for charges $m \in [1, 10]$. Right: Plot of L_0 vs. m. The dots correspond to the numerical values in the table while the dashed line is the fitted curve of Eq. (4.11).*

to perturb Kelvin modes along the VR. These Kelvin modes not only have their own dynamics, interactions and decay properties [670, 735], but they can also self-interact within a VR, resulting in a reduction or even reversal of its transverse velocity [736, 734, 731]. For example, we depict in Fig. 4.12 a sample of the results obtained in Ref. [734]. The left and middle panels of the figure show the evolution, over one period, for planar (left) and helical (middle) Kelvin modes. As depicted in the right panels, these modes are responsible for the slowdown of the VRs when compared to the unperturbed case. In fact, certain modes are able to completely stop or even reverse the velocity of the VRs (see velocity curves corresponding to higher-order $[m = 6, 8, 10]$ modes for sufficiently large relative mode amplitudes).

It is worth noting that a VR can also have more complicated geometries when it knots into itself. The vortex knotting may drastically modify the dynamics of the VRs by altering their motions and even modifying their stability [737, 738]. Finally, we mention that the velocity of a VR may be also modified by the inhomogeneity of the ensuing background density, induced by presence of an external trapping. Interestingly, a VR might be rendered stationary (stopped) by the presence of an external trap [739, 740, 741].

4.3.3 ▪ Vortex Ring Interactions: Leapfrogging

In addition to the internal dynamics (cf. Kelvin modes), VRs give rise to interesting interaction dynamics. For instance, if two rings of the same charge are initially placed in a coaxial configuration they will undergo *leapfrogging* dynamics. An example of this evolution is depicted in Fig. 4.13, where two identical VRs are placed next to each other and allowed to interact. The leapfrogging dynamics can be qualitatively understood by following the dynamics of the 2D vortices ensuing from any 2D cut containing the VRs' axis. In that cut, the VRs induce two vortex dipoles, each dipole containing a positively and a negatively charged vortex. In Fig. 4.14, we depict the evolution of two 2D vortex dipoles initialized precisely as the corresponding 2D cut of the initial configuration of the leapfrogging VRs of Fig. 4.13. As it is evident from Figs. 4.13 and 4.14, the 3D leapfrogging behaves very similarly to the 2D leapfrogging and, in fact, the leapfrogging periods and the average velocity of the leapfrogging for these two cases is similar.

Let us now formulate a reduced model for the leapfrogging of coaxial VRs by borrowing the ideas presented in Sec. 4.2.3. In particular, by employing the Biot–Savart law to the highly symmetric case for coaxial VRs, it is possible to obtain a reduced ODE model

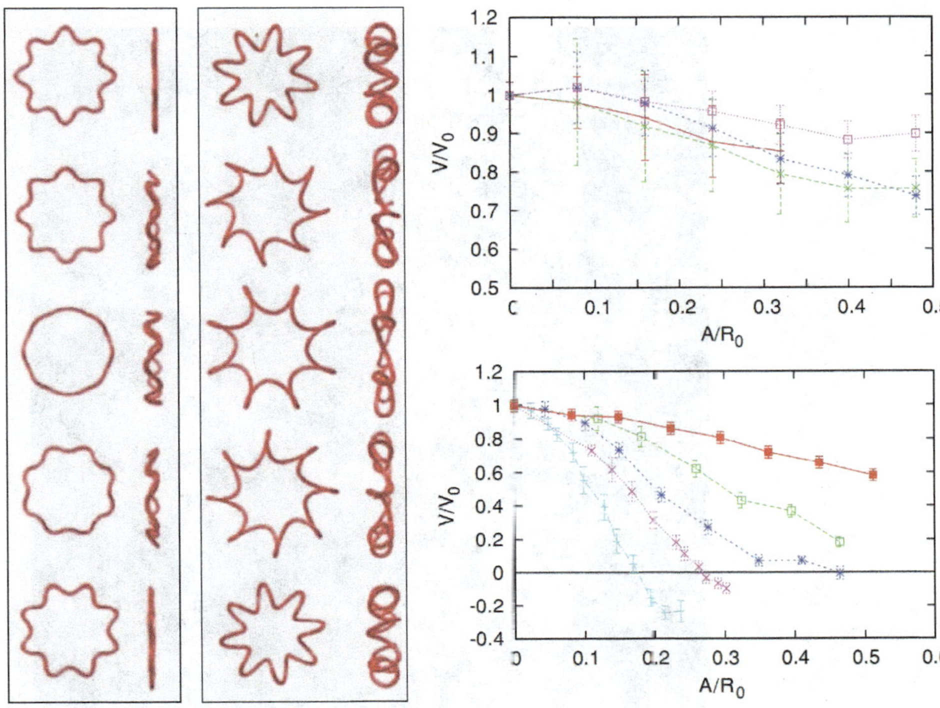

Figure 4.12. *Left: Time evolution of a VR of radius $R_0 = 25$ perturbed by a planar Kelvin wave of wavenumber $m = 8$ and relative amplitude $A/R_0 = 0.233$ during approximately one period of oscillation. Depicted are the isosurfaces $|\Psi|^2 = 0.4$ from frontal (left subpanels) and lateral (right subpanels) views. Middle: Same as in the left panel but for a helical Kelvin mode of large relative amplitude $A/R_0 = 0.302$. Top right: Relative translational velocity v/v_0 (v_0 is the velocity for an unperturbed VR [i.e., $A = 0$]) vs. relative Kelvin wave amplitude A/R_0 for VRs of radius $R_0 = 25$ (corresponding to $v_0 = 0.0857$) with planar Kelvin wave of wavenumbers $m = 2$ (□), 4 (∗), 6 (×), and 8 (+). Bottom right: Same as the top right subpanel but for helical Kelvin modes with wavenumbers (from top to bottom) $m = 2, 4, 6, 8$, and 10. Adapted with permission from Ref. [734].*

for their mutual interactions. Following Refs. [743, 744], the Biot–Savart law is evaluated along the VR configuration depicted in Fig. 4.15 to yield [742]

$$\dot{Z}_i = \frac{\varkappa_i}{4\pi R_i}\left(\ln\frac{8R_i}{r_{c,i}} - C\right) + \frac{1}{\varkappa_i R_i}\frac{\partial U}{\partial R_i}, \tag{4.12}$$

$$\dot{R}_i = -\frac{1}{\varkappa_i R_i}\frac{\partial U}{\partial Z_i}, \tag{4.13}$$

where the VR-VR interaction is captured by

$$U = \frac{1}{2\pi}\sum_{i=1}^{N}\sum_{j>i}^{N}\varkappa_i\varkappa_j I_{ij},$$

Figure 4.13. *Two vortex rings undergoing leapfrogging dynamics. The rings started with an initial radius of $R = 8$ and an initial vertical separation $H = 8$. Depicted are a series of snapshots at times $t = 0, 16, 27, 44, 58, 68, 85$ corresponding to approximately one period of oscillation of the leapfrogging. The snapshots depict the inverse transparency volumetric rendering of the density $|\Psi|^2$ such that $|\Psi|^2 = 1$ is fully transparent and $|\Psi|^2 = 0$ is fully opaque. Adapted with permission from Ref. [742].*

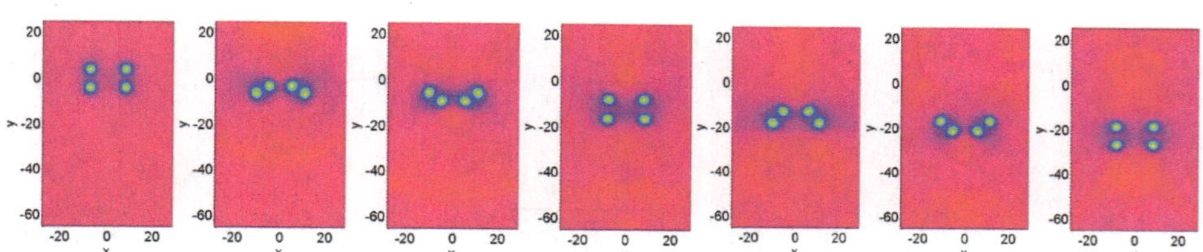

Figure 4.14. *Two 2D vortex dipoles emulating the leapfrogging dynamics of two VRs as depicted in Fig. 4.13. The vortex dipoles are initialized in the same manner as the VRs in Fig. 4.13, namely each vortex dipole has an internal horizontal separation of $2R = 16$ and a vertical separation of $H = 8$ between the two dipoles. Depicted are a series of snapshots at times $t = 0, 19, 29, 49, 53, 79, 86$ corresponding to approximately one period of oscillation of the leapfrogging. The snapshots depict the density $|\Psi|^2$.*

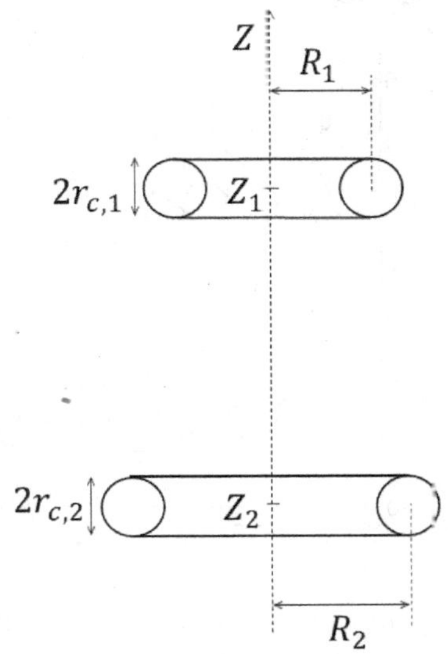

Figure 4.15. *Schematic of the variables to describe the reduced dynamics for the leapfrogging between two coaxial VRs located at Z_1 and Z_2 with respective radii R_1 and R_2. Adapted with permission from Ref. [742].*

through the integral

$$I_{ij} = \int_0^{\pi} \frac{R_i R_j \cos\theta \, d\theta}{\sqrt{\left(Z_i - Z_j\right)^2 + R_i^2 + R_j^2 - 2R_i R_j \cos\theta}}.$$

In the above description, R_i and Z_i are the radius and z coordinate of the ith ring, $\varkappa_i$ is its circulation or "vortex strength," which in our case is $\varkappa_i = 2\pi m$, $r_{c,i}$ is the VR core radius given by Eq. (4.10), and C is a constant. Let us now discuss the origin of the first term in Eq. (4.12) and the value of the constant C. This term, which only depends on the internal dynamics of a single VR, is precisely the self-induced velocity of the VR described in Eq. (4.9). In a standard fluid, in the absence of the quantum pressure term described in 4.2.3, the constant in Eq. (4.12) takes the value $C = 1/4$ [743]. However, in our case, the quantum pressure term modifies the self-induced velocity of a VR according to Eq. (4.9), and thus the constant C takes the value $C = L_0(m) - 1$, where $L_0(m)$ is the vortex core parameter discussed in the previous section. The above hybridization of the classical fluid case through the Biot–Savart law describing the interaction dynamics [743], together with the corrected self-induced velocity (4.9) [729], allow for a reduced ODE model that describes accurately the dynamics of leapfrogging VRs [742]. For example, Fig. 4.16 depicts leapfrogging orbits in the (R, Z) plane for several initial VR radii and separations in a comoving reference frame. It is clear from the figure that the reduced ODE model given by Eqs. (4.12) and (4.13) gives a very good approximation of the actual dynamics of the leapfrogging VRs in the original, fully 3D case.

It is interesting to note that Fig. 4.16 suggests the existence of a stable, comoving, equilibrium at the center of the periodic orbits. This point indeed corresponds to two overlapping ($Z_1 = Z_2$) VRs of the same radius, namely a vortex ring of charge two ($|m| = 2$). However, VRs of higher ($|m| > 1$) charge are unstable in a manner akin to 2D double-charged vortices in free space. The instability for higher charge vortices has been studied

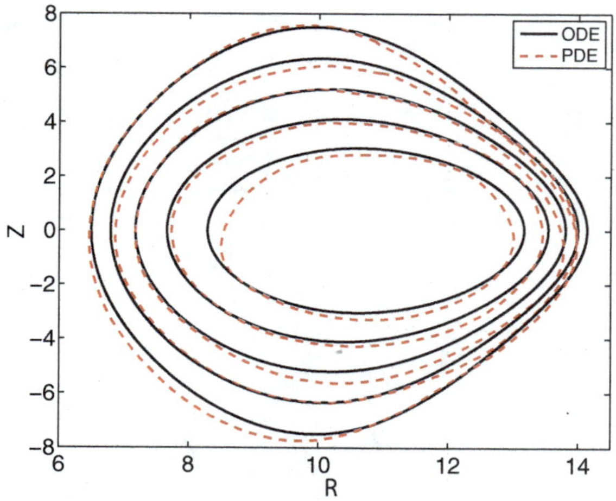

Figure 4.16. *Leapfrogging orbits for two VRs in the co-moving (R, Z) plane. The solid curves depict the results from the effective, reduced, equations of motion (4.12) and (4.13), while the dashed curves depict the corresponding results from those obtained from full 3D integrations of Eq. (4.8). The rings are initialized with radii $R = 11$ and initial separation distances of $Z_2(0) - Z_1(0) = 6, 8, 0, 12, 14$ (from inner to outer curves). Only one period of the leapfrogging motion is shown in a reference frame moving at the average velocity of the VR pair. Adapted with permission from Ref. [742].*

in detail [488, 745, 746, 621]; however, appropriate tuning of the external harmonic trapping potential might render the higher-charge vortex at the center of the trap to be stable for narrow parameter ranges. Furthermore, a localized confining potential can also render higher-order charge vortices to be stable at the center of the trap [747], or even at a certain distance away from the center of the trap [748]. The instability of multiply charged vortices splits the configuration into single-charge vortices while conserving the total winding number. Similarly, a doubly charged VR will dynamically break up into two individual singly charged VRs. An example of this instability is depicted in Fig. 4.17: there, an initial doubly charged VR of radius $R = 10$ evolves for some time before succumbing to the instability and splitting into two VRs leapfrogging into close proximity which undergo deformation and, eventually, merge together into a long, highly perturbed, vortex loop.

It is worth mentioning that there have been recent efforts in describing the leapfrogging dynamics of vortex ring bundles in classical fluids using the vortex filament method [749]. Examples of leapfrogging VR bundles with three and seven VRs are depicted in the top and bottom panels of Fig. 4.18. In particular, it is found that the leapfrogging dynamics is weakly unstable for a small number of vortex rings. An example of this instability for two VRs is depicted in Fig. 4.19. However, depending on the separation and radii of the vortex rings, the onset for this instability is not visible until tens of leapfrogging periods. Therefore, although weakly unstable, the time required for the instability to manifest itself might be long enough to be able to observe leapfrogging during the typical lifetime of current BEC experiments. Finally, it is also important to mention that the 2D equivalent of leapfrogging has been studied in the case of point vortices [750, 751]. This 2D leapfrogging corresponds to two vortex dipoles (vortex and antivortex pairs) and it is found to display three distinct regions depending on the distance between pairs (relative to the internal distance between vortices in each pair) as follows. (a) The small separation region does not yield leapfrogging. (b) Intermediate separations yield *unstable* leapfrogging. (c) Large separation distances yield *stable* leapfrogging. The 2D vortex dipole leapfrogging

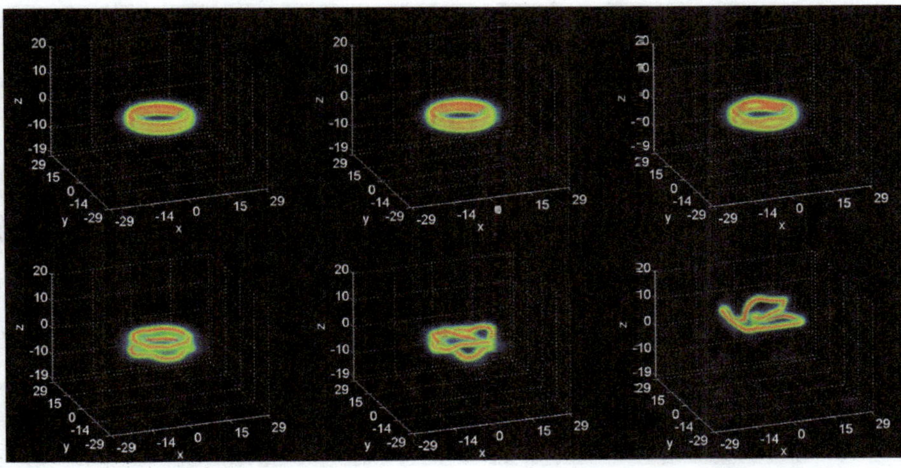

Figure 4.17. *Break-up of an optimized VR of charge $m = 2$ (a "double ring") and radius $R = 10$ in a comoving reference frame. The double-ring structure is shown from left to right, top to bottom, at times $t = 0, 140, 350, 1530, 1830, 2000$. The double ring evolves stably for a while, then exhibits oscillations leading to leapfrogging, and finally to deformation and merging.*

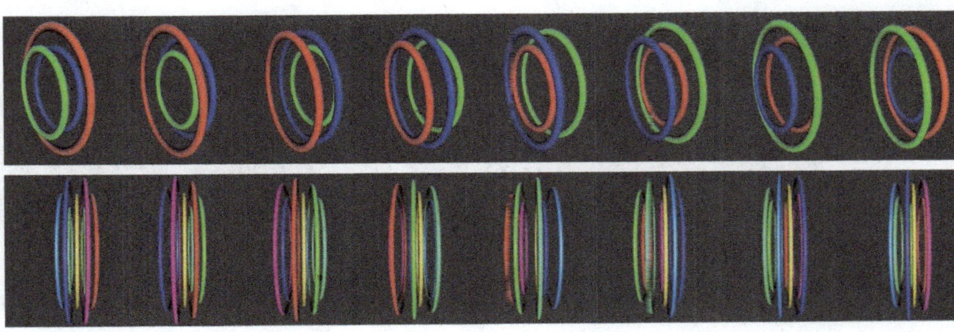

Figure 4.18. *Leapfrogging of VR bundles for three (top) and seven VRs (bottom). The series of snapshots represent approximately a full period for the leapfrogging. Adapted with permission from Ref. [749].*

is strongly connected to 3D vortex ring leapfrogging since any 2D cut passing through the axes of the VRs will produce a 2D pair of vortex dipoles. Therefore, 2D vortex dipole dynamics is the basic ingredient responsible for VR leapfrogging. Therefore, it would be interesting to test if the VR leapfrogging also displays the three regions described above and whether the weak instability that we have observed (cf. Fig. 4.19) can in fact be eliminated by using larger VR radii. We close this section by mentioning that the more general problem of reducing the dynamics of interacting VR for arbitrary orientations (i.e., not necessarily coaxial) is still an open problem.

4.3.4 ▪ Vortex Ring Scattering, Mergers, and Splits

The model described in the previous section assumes that the VRs are well separated, which precludes describing close encounters or collisions. The collision between VRs has a very rich dynamics [730] including mergers, annihilations, and, in suitable regimes, apparently near "elastic" scattering. An example displaying a VR merger is depicted in

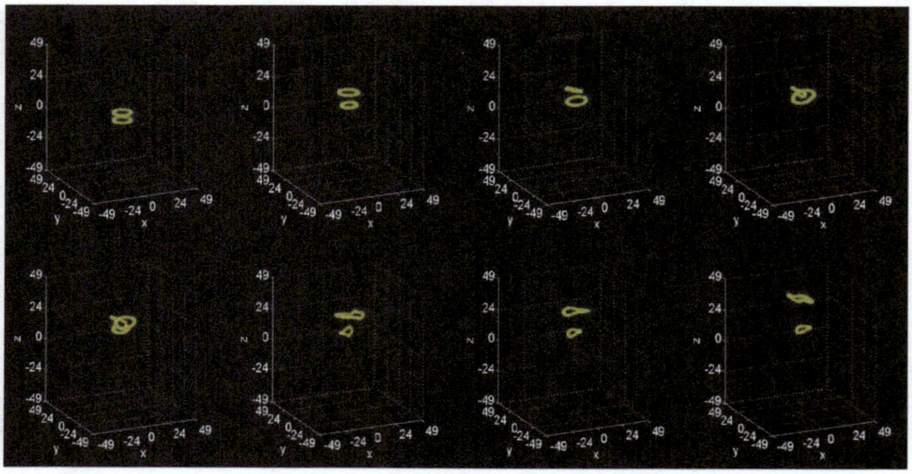

Figure 4.19. *Long-time simulation of leapfrogging VRs of radius $R = 8$ initially separated by a distance of $H = 6$ in a comoving reference frame. The rings are shown at times (left to right, top to bottom) $t = 0, 687, 715, 731, 739, 755, 771, 811$. Around $t = 650$ the configuration develops an asymmetry that is ultimately responsible for the collision of the VRs and their eventual separation.*

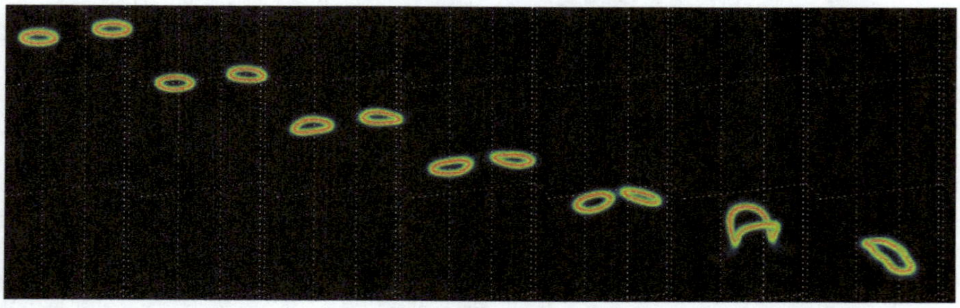

Figure 4.20. *Two initially parallel VRs eventually merge due to their effective mutual attraction (see Fig. 4.21) and leave behind a larger VR with a Kelvin mode oscillation.*

Fig. 4.20, where two initially parallel (parallel axes) VRs merge leaving behind a single VR with a highly excited quadrupolar (Kelvin) mode.

The effective mutual interaction for parallel VRs may be easily understood by following the dynamics of their corresponding 2D vortices in the cross-cut plane containing the two (parallel) VR axes. In that plane, the configuration corresponding to two identical VRs induces two aligned vortex dipoles. The dipoles have the same intrinsic linear velocity due to the interactions between the vortices of the dipole. However, the vortices of each pair closer to the other pair also interact creating a velocity *opposite* to the intrinsic vortex pair velocity. This extra velocity, that cannot be compensated by the interaction between the farther-away vortices, is responsible for the pairs to rotate towards each other as shown in the schematics of Fig. 4.21.

Another possible scenario for the interaction of VRs arises when two VRs of opposite charge undergo a head-on collision [730]. In that case, the two colliding VRs annihilate each other's topological charge resulting in an axisymmetric decaying rarefaction pulse. An example of such a case is shown in Fig. 4.22, where two VRs of radius $R = 6$ (one with charge $m = 1$ and one with $m = -1$) are positioned a distance of $6R$ apart from each other

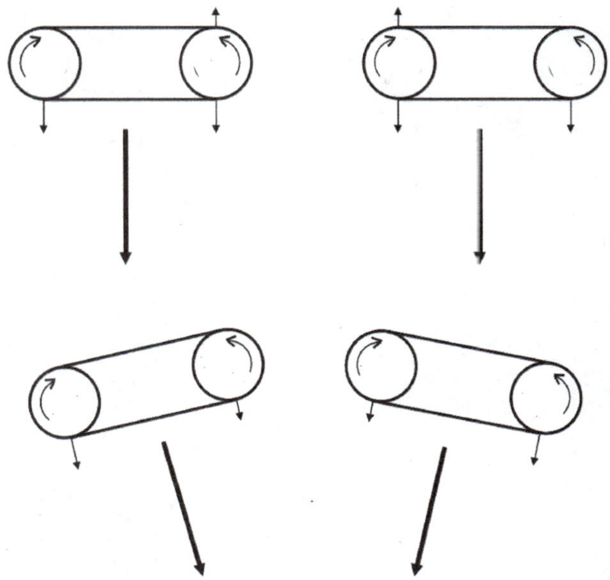

Figure 4.21. *Understanding the effective attraction between two parallel same-charge VRs leading to merger as shown in Fig. 4.20 by considering the 2D cut containing the VR axes. Top: Two initial parallel 2D vortices have their own intrinsic linear velocity (see long thick arrows pointing down). However, the interaction between the right vortex of the left VR and the left vortex of the right VR induces a small upward velocity (see small arrows pointing up). Bottom: The sum of all the velocities results in an effective rotation of the vortex pairs towards each other.*

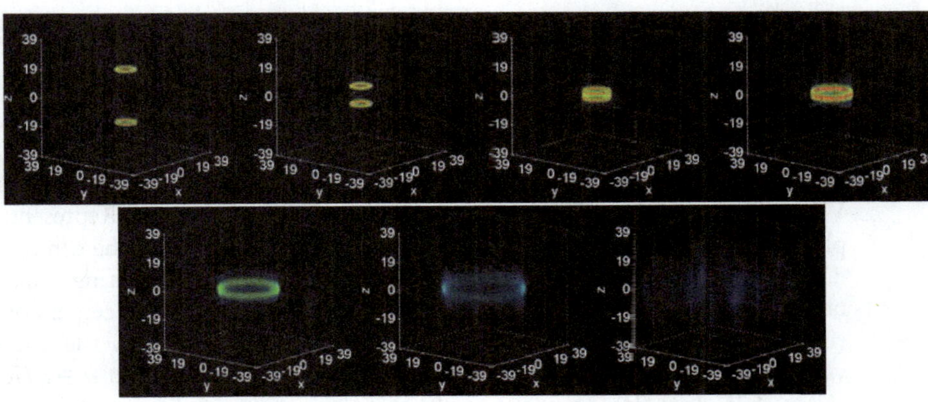

Figure 4.22. *Example of a head-on collision between VRs of charge* $m = 1$ *and* $m = -1$*. The rings have a radius of* $R = 6$ *and are shown at times* $t = 0, 22, 33, 39, 45, 51, 59, 68$*. Adapted with permission from Ref. [742].*

and allowed to collide and annihilate each other. This head-on collision example naturally raises the question of what happens if two colliding opposite charge VR do not start in a coaxial configuration but with an offset from each other. Numerical experiments [742] show that, typically, the two VRs merge and then separate into two new VRs moving away from each other at an angled trajectory, each exhibiting large Kelvin oscillations.

To quantify the relationship between offset distance and scattering angle, the VRs are positioned at a distance of $6R$ apart in the z direction, and in the y direction they are

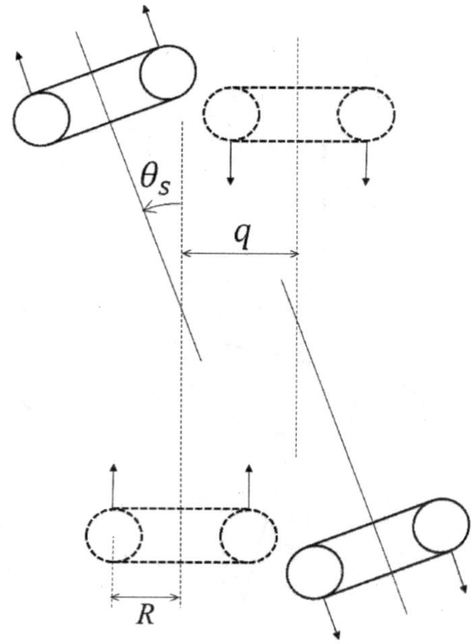

Figure 4.23. *Schematic of a scattering scenario between two coplanar vortex rings of initial radius R offset by a distance q (impact parameter). The initial, prescattering configuration is depicted with dashed lines while the post-scattering configuration is depicted with solid lines. The scattering angle θ_s is defined as the signed angle between incoming and outgoing axes as shown in the schematic. Adapted with permission from Ref. [742].*

placed with a planar offset distance (impact parameter) denoted as q (see Fig. 4.23). The scattering angle θ_s is defined as the angle away from vertical, so that if the VRs travel in a straight trajectory, the scattering angle is $\theta_s = 0$, while if they would scatter at right angles from each other, the angle would be $\theta_s = \pi/2$. After performing multiple numerical experiments for colliding VRs with offsets ranging from $q/R = 0$ to $q/R = 3$ for various VR radii, three different qualitative scenarios are observed [742]. Representative examples for these three types of scenarios are depicted in Fig. 4.24. The three qualitatively different scenarios correspond to (a) annihilation, (b) merging/splitting collision, and (c) nonmerging interactive flyby. Along the scattering dynamics, the centers of the vortex rings are tracked and the scattering angle recorded. The results for the scattering angle vs. the normalized (with respect to the VR radius) impact parameter $\tilde{q} \equiv q/R$ are shown in Fig. 4.25. It is clear from the results that there are three distinct sections in the relationship between scattering angle and offset, which correspond to qualitatively distinct topological outcomes as follows:

(a) For offsets with $q/R \in [0, 1/2)$ the two rings annihilate each other's topological charge, sending out a rarefaction pulse [see Fig. 4.24(a)], similar to the cases of head-on collision shown in Fig. 4.22. However, in this case, as the rings approach each other, they rotate causing the collisions to occur at the scattering angle. Another distinction of these collisions is that the resulting rarefaction pulse is not axisymmetric as in the head-on collision case, but more concentrated in the scattering direction.

(b) When the offset has $q/R \in (1/2, 2)$, the two rings undergo merging immediately followed by splitting into internally excited VRs with quadrupole oscillations [see

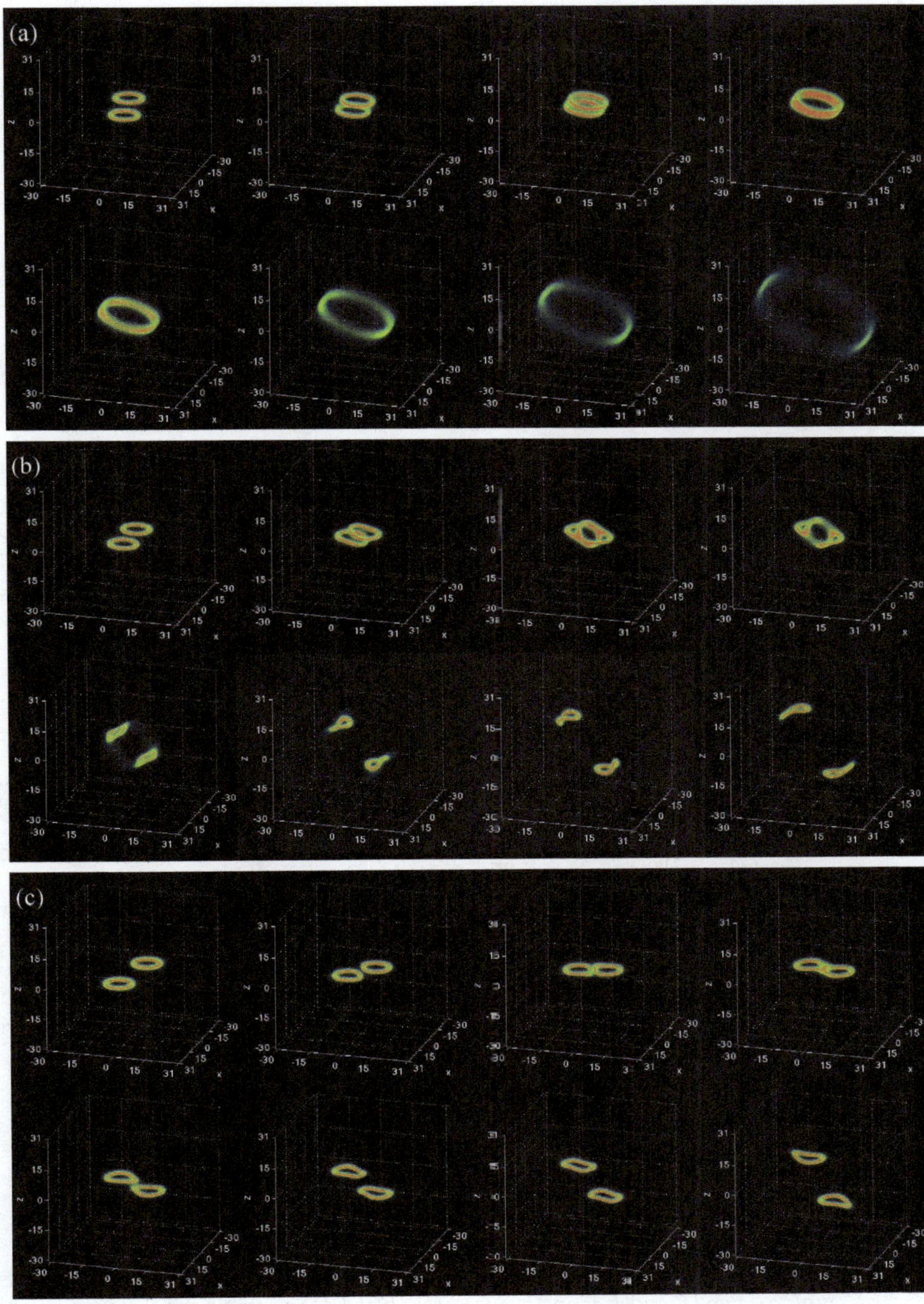

Figure 4.24. *Qualitatively distinct scattering scenarios for two colliding, initially offset, coplanar VRs corresponding to the three regions depicted in Fig. 4.25.* (a) *Annihilation scenario (with initial offset of $q/R = 0.35$) at times $t \approx 4, 8, 12, 16, 19, 23, 27, 32$.* (b) *Merging/splitting collision (with initial offset of $q/R = 1.0$) at times $t \approx 4, 10, 15, 16, 20, 26, 32, 37$.* (c) *Nonmerging interactive flyby (with initial offset of $q/R = 2.2$) at times $t \approx 1, 6, 9, 12, 16, 20, 24, 31$. Adapted with permission from Ref. [742].*

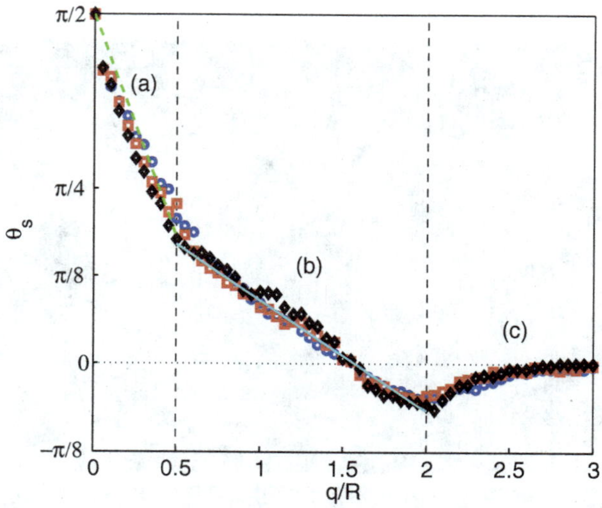

Figure 4.25. *Scattering angle versus the normalized offset distance q/R for two colliding coplanar unitary-charge VRs. The different regions correspond to the three qualitatively different scenarios depicted in Fig. 4.24. The (blue) circles are the results for VRs of radius $R = 6$, the (red) squares for $R = 8$, and the (black) diamonds for $R = 10$. The (green) dashed and (cyan) solid lines are given by the approximation of Eq. (4.14). Adapted with permission from Ref. [742].*

Fig. 4.24(b)]. Interestingly, when the offset is set to $q/R = 3/2$ or larger, the scattering angle becomes negative, but the qualitative behavior remains the same.

(c) For larger offset such that $q/R > 2$, the two VRs no longer merge and separate into new VRs, but are merely perturbed by each other in a flyby scenario, causing their trajectories to travel at a small negative angle from incidence, i.e., get weakly deflected "inward" [see Fig. 4.24(c)]. As the offset increases, this interaction decreases until approximately $q/R = 3$, where the VRs essentially ignore each other's presence and travel unperturbed.

It is interesting to note that the scattering angle as a function of offset in Fig. 4.25 appears linear in the first two qualitatively distinct intervals. In fact, the following simple phenomenological approximation for the relationship between the scattering angle and the normalized impact parameter can be formulated as [742]

$$\theta_s(\tilde{q}) \approx \begin{cases} \pi/2 - 2\tilde{q} & \text{if} \quad 0 < \tilde{q} < 1/2, \\ \pi/4 - \tilde{q}/2 & \text{if} \quad 1/2 < \tilde{q} < 2, \\ -f_{\rm d}(\tilde{q}) & \text{if} \quad \tilde{q} > 2, \end{cases} \tag{4.14}$$

where $\tilde{q} = q/R$ and $f_{\rm d}$ is an appropriately fitted decaying function. This approximation is shown along with the numerical results in Fig. 4.25. It is interesting to note that, despite the complex dynamics involved in the scattering of VRs, we are able to formulate a very simple, phenomenological scattering rule for coplanar collisions. This unexpectedly clean result tends to suggest that a reduced quasiparticle description may be possible even in some complex collision/scattering cases.

It is interesting to contrast the 3D scenario of coplanar VR scattering to its 2D counterpart. If one focuses on the dynamics restricted on the plane defined by the VR axes, each

VR induces a vortex pair in 2D. The case of scattering between 2D (point) vortex pairs has been studied in some detail [752, 753, 754]. The most notable differences between the ensuing dynamics for 2D vortex pairs and that of 3D VRs scattering is that the dynamics in 2D is apparently richer due to the fact that truly 2D vortices are more "free" to move (respecting the vortex-vortex interactions), while the vortex pairs produced by the cut of the VR in the plane are more tightly linked to each other through the ring. This extra freedom for 2D vortices allows for a rich variety of dynamical evolution scenarios. For instance, depending of the initial configuration (initial positions, orientations and vortex pair internal distances) the collision dynamics of two vortex pairs can display

(a) transient bound vortex pairs where the two vortex pairs interact by orbiting around each other and exchange partners for a few periods until scattering away from each other as two distinct vortex pairs,

(b) transient interactions between the four vortices that are chaotic and lead to the eventual expulsion of two distinct vortex pairs resulting in chaotic scattering, or

(c) transient bound states where three of the four vortices lock into a translational and precessional motion, while the fourth one moves separately, before the two pairs again scatter away, separating indefinitely.

4.3.5 ▪ The Role of Dimensionality

In this section we discuss some of the effects induced by the effective dimensionality on the dynamics of VRs and VLs. In particular, we focus on anisotropic external trappings with two trapping directions stronger than the third inducing an elongated background. The stability and collisions of dark soliton structures and VRs in a cylindrical geometry have been studied in some detail [170, 274, 273, 755, 756, 78]. In particular, it was shown in Ref. [170] that the decay of the dark soliton can be avoided if the transverse direction is sufficiently confined. Specifically, as discussed in Sec. 2.3.4, the parameter $\gamma = n_0 g_{3D}/(\hbar\omega_\rho)$, given by the ratio of the mean-field interaction energy at maximum density n_0 to the harmonic oscillator energy in the transverse direction $[\omega_\rho = (\omega_x, \omega_y)]$, was suggested as controlling the stability of the dark soliton stripe: in a cylindrical trap $(\omega_\rho = \omega_x = \omega_y)$, if $\gamma > \gamma_c$ where $\gamma_c \approx 2.4$ for $\omega_\rho \gg \omega_z$, the dark soliton becomes unstable and decays to a solitonic vortex (VL) [170].

An interesting regime ensues when the parameter γ, which measures the nonlinearity present in the system relative to the transverse direction, is slightly above critical. This regime can be achieved by appropriately tuning the trapping ratios along the different directions. For instance, in the experiment of Ref. [71] it was observed that the dark soliton produced by phase engineering oscillated between that original state and a VR. A similar evolution was observed in the experiments of Ref. [73], where a couple of dark solitons (again created by phase engineering) oscillated between their original profile, VRs and VLs. Figure 4.26 depicts a typical experimental evolution for this dynamics: the first row of snapshots corresponds to two initial dark soliton stripes that start bending [see the dark lines inside the ellipses in panel (a)] and decay into VRs [panel (b)]. As the VRs approach the opposite edge of the trap, they recombine into a dark soliton stripe due to the strong confinement of the trap's edge [see the dark soliton stripe very close to the right edge inside the right-most ellipse in panel (c)]. Subsequently, as the VRs return from the relevant turning point of their trajectory, a merger event occurs. After this merger between the two waves a single VL remains in the cloud performing oscillations back and forth along the longitudinal direction [see panels (d)–(f)].

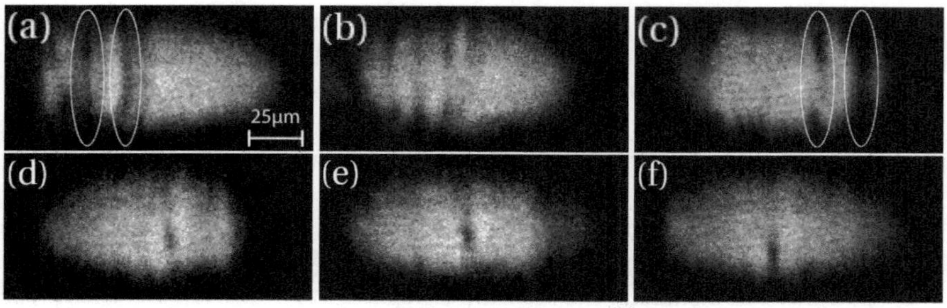

Figure 4.26. *Experimental snapshots for the dynamics of two initial dark soliton lines [panel (a)]. The different panels correspond to times* $t = (0, 12, 57, 119, 136, 155)$ *ms.* (a) *Initial state containing two dark soliton stripes. The ruler corresponds to the in-trap size of the condensate prior to time of flight.* (b) *The dark soliton stripes decay into VRs.* (c) *The resulting waveforms reach the opposite edge of the cloud and bounce back.* (d)–(f) *After the strong interaction a single, stable VL is left in the system. Adapted with permission from Ref. [73].*

A further interesting effect of the interplay between the dimensionalities is the counter-intuitive evolution of apparent dark solitons. For example, the top panel in Fig. 4.27 depicts the apparent "merger" of two dark solitons. Under closer inspection, see bottom panels of Fig. 4.27, the true 3D structure for this evolution reveals that the apparent dark solitons are in fact VRs that oscillate between VR and VL pairs. As it is clear from the figure, the two VLs pairs interact very strongly because their internal separation (i.e., the ring diameter) is almost identical. This strong interaction is responsible for one of the solitonic vortex pairs being sling shot away at a faster speed, leaving behind what appears ($t = 112$ ms) to clearly be a single vortex ring. Also, at the same time, this ejected pair is relegated to the periphery of the cloud in the x and y directions where the density is much weaker and so is its imprint in the integrated density plot depicted in the top panel of Fig. 4.27, resulting in a longitudinal observation of an apparent merger of two solitonic vortex pairs.

Yet another instance where the hidden 3D structure of apparent dark solitons plays an important role is in the recurrent collisions between dark soliton stripes. For instance, in quasi-1D geometries two dark solitons symmetrically displaced from the center of a parabolic trap undergo periodic oscillations that are well understood (see Sec. 2.2), and for which the theory and physical experiments match extremely well, as shown in Chap. 2. Nonetheless, using the same setup as the previous examples, it is possible to obtain collision dynamics with apparently inelastic collision between dark soliton stripes. For example, the top panels of Fig. 4.28 depict the apparent inelastic collisions between dark soliton stripes leading to an expanding array of "bubbles" (each spatio-temporal bubble amounting to a pair of collisions). Again, resorting to the three-dimensional nature of the original problem (see bottom panels of Fig. 4.28), it is clear that the we are in the presence of two VLs of the same charge that interact with each other (mutual rotation) and with the anisotropic background density induced by the external potential. The lack of periodicity in the collision "bubbles" is explained by the fact that the rotation between the two VLs remains "incomplete" due to its confined [anisotropic in (x, y)] nature. As the initial potential energy of interaction translates itself into rotational energy and partly gets emitted through phonon radiation, it draws the VLs gradually closer to the edge of the background cloud, hence renders them more mobile within the least confined z direction. The result of this gradually increased mobility is the more and more delayed repetition of their mutual interaction, which, in turn, is mirrored in the expanding bubbles in the (z, t) plane shown in the top panels of Fig. 4.28.

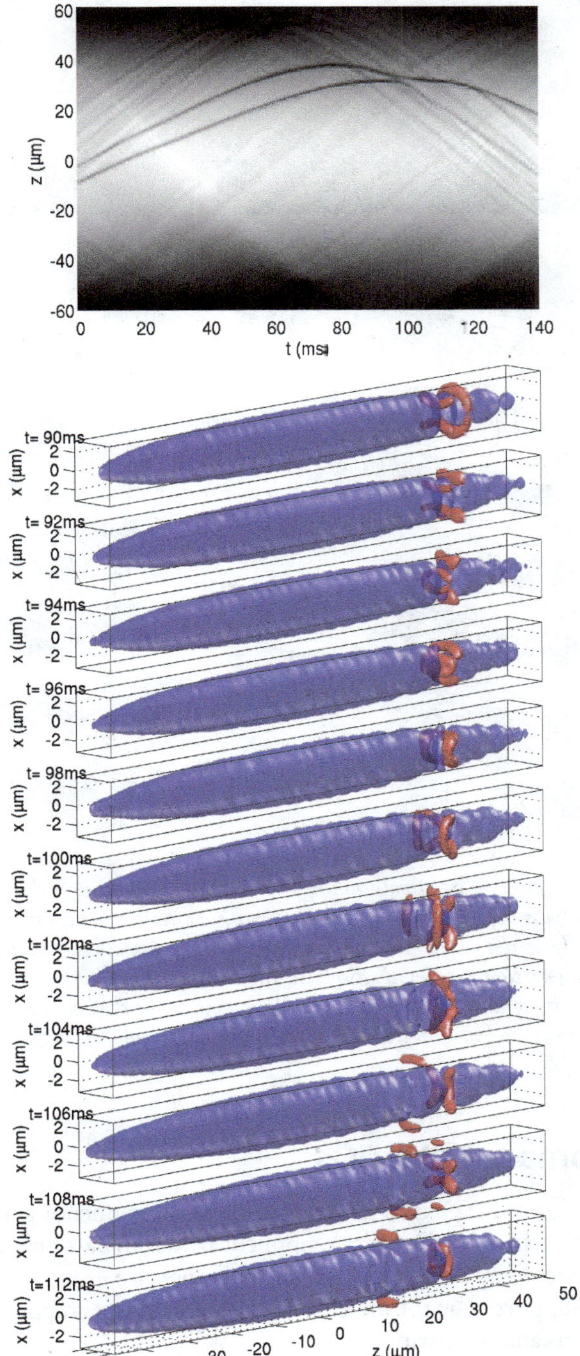

Figure 4.27. *Numerical evolution of two chasing dark soliton stripes. Top: Depicted is the time evolution of the density integrated about the x and y directions. Bottom: Corresponding 3D renderings depicting the sling shot effect suffered by one of the solitonic vortex pairs during collision. Density and vorticity isocontours at 40% and 85%, respectively. Adapted with permission from Ref. [73].*

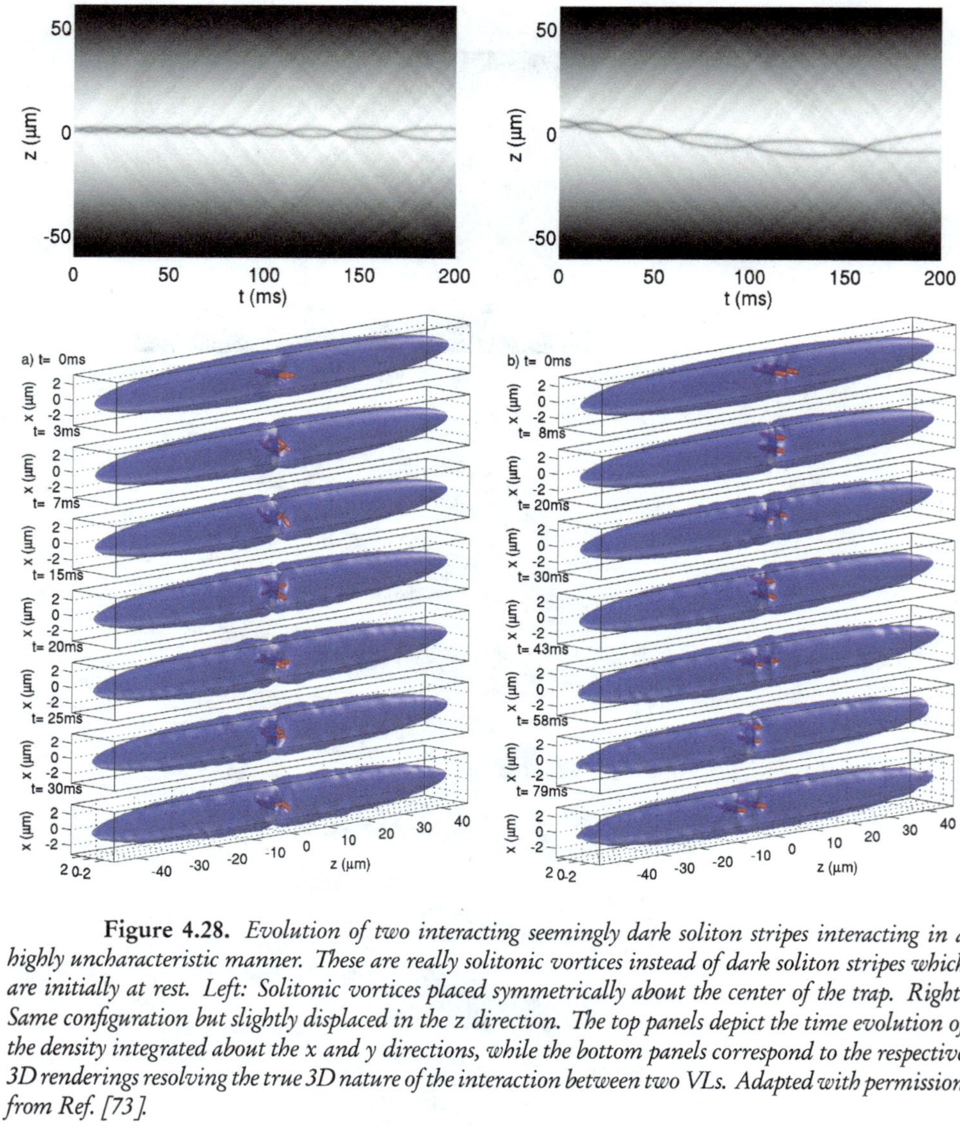

Figure 4.28. *Evolution of two interacting seemingly dark soliton stripes interacting in a highly uncharacteristic manner. These are really solitonic vortices instead of dark soliton stripes which are initially at rest. Left: Solitonic vortices placed symmetrically about the center of the trap. Right: Same configuration but slightly displaced in the z direction. The top panels depict the time evolution of the density integrated about the x and y directions, while the bottom panels correspond to the respective 3D renderings resolving the true 3D nature of the interaction between two VLs. Adapted with permission from Ref. [73].*

4.4 ▪ Multicomponent BECs

This section is devoted to the discussion of some phenomenology pertaining nonlinear structures arising in 3D multicomponent BECs. In particular, we cover the nucleation of VRs in two-component BECs. We also discuss two examples of spatio-temporal patterns that are supported by the interplay between the two components and the dimensionality: Skyrmions and oscillating ring excitations.

4.4.1 ▪ Vortex Ring Nucleation in Two-Component BECs

In this section we briefly discuss some case examples pertaining to the nucleation of VRs in multicomponent NLS equations. In the context of BECs, multicomponent systems arise in various different situations, as explained in Sec. 2.4. In the case of a binary BEC, the corresponding dimensionless version of the two coupled GPEs may be written as

$$i\frac{\partial \Psi_1}{\partial t} = \left[-\frac{1}{2}\nabla^2 + V_1 + g_{11}|\Psi_1|^2 + g_{12}|\Psi_2|^2\right]\Psi_1, \tag{4.15}$$

$$i\frac{\partial \Psi_2}{\partial t} = \left[-\frac{1}{2}\nabla^2 + V_2 + g_{22}|\Psi_2|^2 + g_{12}|\Psi_1|^2\right]\Psi_2, \tag{4.16}$$

where Ψ_i is the wavefunction corresponding to the ith ($i = 1,2$) component, V_i its corresponding trapping potential, and the interaction parameters g_{ij} define the intraspecies (g_{ii}) and interspecies (g_{12}) interactions.

In Ref. [715] a "bubble," namely a localized spherical mass, in the second component was engineered to nucleate a VR in the first component. Specifically, the bubble of component Ψ_2 was seeded in a homogeneous background of the Ψ_1 component ($V_1 = 0$). Then a gradient potential $V_2(x,y,z) = Fx$, where F is the strength of the accelerating potential, was applied to the Ψ_2 component thus accelerating the bubble that deformed into an oblate ellipsoid. Once the bubble reached a desired velocity, the accelerating V_2 potential is turned off. Provided that the final velocity of the bubble is larger than the speed of sound of the Ψ_1 component, the bubble induced the nucleation of VR in the Ψ_1 component. In fact, if the velocity of the bubble was large enough several VRs can be produced and may result in leapfrogging behavior [715].

On the other hand, let us consider the case of a spinor $F = 1$ BEC, and recall (see Sec. 2.4.7) that it can be described by the following system of three coupled NLS-type equations [137, 368]:

$$i\frac{\partial \Psi_{\pm1}}{\partial t} = H_0\Psi_{\pm1} + \delta\left[(|\Psi_{\pm1}|^2 + |\Psi_0|^2 - |\Psi_{\mp1}|^2)\Psi_{\pm1}\right] + \delta\Psi_0^2\Psi_{\mp1}^*, \tag{4.17}$$

$$i\frac{\partial \Psi_0}{\partial t} = H_0\Psi_0 + \delta\left[(|\Psi_{-1}|^2 + |\Psi_{-1}|^2)\Psi_0\right] + 2\delta\Psi_{-1}\Psi_0^*\Psi_{+1}, \tag{4.18}$$

where $H_0 \equiv -(1/2)\nabla^2 + V(\vec{r};t) + n_{\text{tot}}$, $n_{\text{tot}} = |\psi_{-1}|^2 + |\psi_0|^2 + |\psi_{+1}|^2$ is the total density, $V(\vec{r};t)$ is the external potential, and the parameter δ expresses the normalized spin-dependent interaction strength. As was discussed in Sec. 2.5.1, the spinor BEC system (4.17) and (4.18) admits two families of homogeneous solutions when $V(\vec{r};t) = 0$, corresponding to the "anti-phase-matched" and the "phase-matched" solutions [757, 715]. Then, analyzing the stability of "anti-phase-matched" stationary states $\Psi_i(\vec{r},t) = \psi_i \exp(-i\mu t)$ (where μ is the chemical potential) boosted into a comoving reference frame with velocity s, it is possible to obtain *two* critical speeds for the onset of instability:

$$c_1 = \sqrt{\delta}\, c_2, \quad c_2 = \sqrt{\mu}. \tag{4.19}$$

Similarly, when using the "phase-matched" homogeneous solutions, the following critical speeds can be obtained:

$$c_1 = \sqrt{\frac{-\delta}{1+\delta}}\, c_2, \quad c_2 = \sqrt{\mu}. \tag{4.20}$$

In this case, it is clear that the first critical speed will exist only in the ferromagnetic spinor BECs (such as ^{87}Rb) with $\delta < 0$, but not in anti-ferromagnetic ones (such as ^{23}Na) with $\delta > 0$.

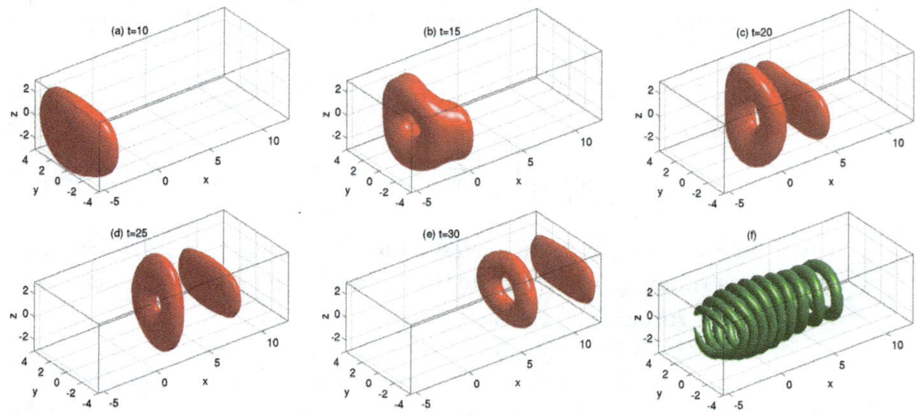

Figure 4.29. *3-spinor BEC VR nucleation by an impurity traveling faster than the speed of sound of the background (i.e., supercritical regime). Panels* (a)–(e) *depict isodensity contours corresponding to* $|\Psi_0|^2 = 0.3$ *at the indicated times. Note that the extent of the moving defect is clearly visible in these panels (it corresponds to the rightmost flat oval shape that is created by the atomic density depletion due to its presence). Panel* (f) *shows the isocontours of the norm of the vorticity of* Ψ_0 *at times* $t = 12, 14, \ldots, 30$ *(left to right). The running impurity has parameters* $w_y = 8$, $w_z = 4$, $a = 2$, $V_0 = 0.9$, *and speed* $s = 0.8 > c_2$. *Adapted with permission from Ref. [410].*

Figures 4.29 and 4.30 depict the nucleation of vortex rings when a repulsive impurity of the form

$$V = \frac{V_0}{16}\exp\left(-a(x-st)^2\right)\left[\tanh\left(y+\frac{w_y}{2}\right)+1\right]\left[\tanh\left(-y+\frac{w_y}{2}\right)+1\right]$$
$$\times\left[\tanh\left(z+\frac{w_z}{2}\right)+1\right]\left[\tanh\left(-z+\frac{w_z}{2}\right)+1\right]$$

is made to traverse the condensate at speed s. As it can be seen from Fig. 4.29, a VR is formed in the 3D spinor condensate, as a result of the supercritical nature of the chosen speed $s = 0.8 > c_2$. The isodensity contours of $|\psi_0|^2$ clearly show a depletion of atoms around the VR that is nucleated in the wake of the defect-induced region of density minima. It is interesting to note that the dynamics of the different components seems to be locked and, therefore, we only depict results for Ψ_0 [410]. Figure 4.30 depicts similar results to Fig. 4.29 but for a slightly larger velocity $s = 1$. In this figure, we first observe the nucleation of a VR in the wake of the defect at about $t = 12$. Around $t = 17$ a second VR is nucleated while the first vortex ring starts to shrink until it eventually disappears around $t = 23$, and shortly after this, the second ring deforms and splits into two separate VRs around $t = 30$.

4.4.2 ▪ Skyrmions

Vortex lines or rings, or more generally vortex filaments, are not the only topologically charged solutions that one may find in 3D BECs. In fact, another type of solution that carries topological charge, but that has no equivalent in single-component BECs, is the so-called *Skyrmion*. The term for Skyrmion is due to T.H.R. Skyrme, who introduced the pertinent model (the "Skyrme model") in the 1960s [759, 760] as an idea for incorporating baryons in the description of low-energy interactions of pions in the framework of the sigma model. Three-dimensional Skyrmions, namely solutions of the Skyrme model,

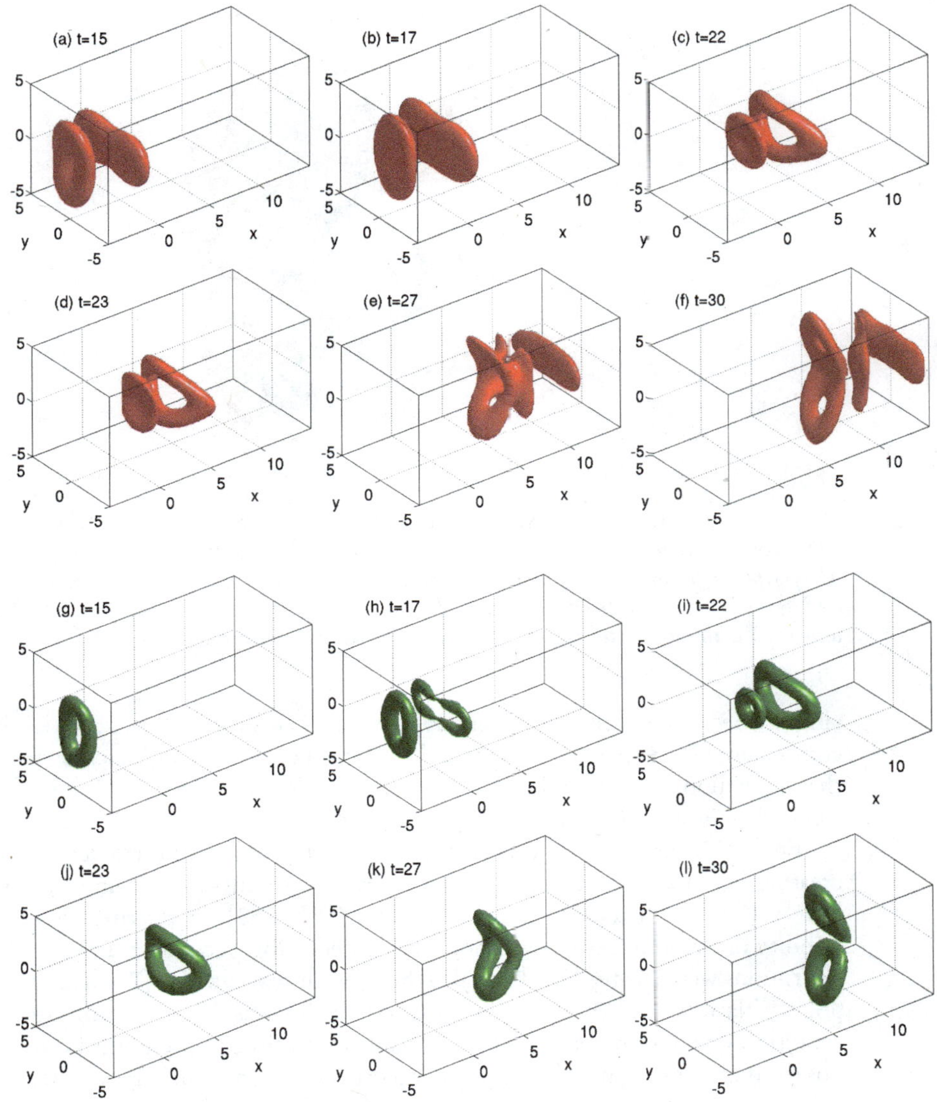

Figure 4.30. *Similar to Fig. 4.29 but for a slightly larger defect speed. Panels* (a)–(f) *depict the iso-density contours of* Ψ_0 *while panels* (g)–(l) *depict the respective vorticity isocontours. Note the successive nucleation of two vortex rings. The first one shrinks and collapses into itself between* $t = 22$ *and* $t = 23$, *while the second ring deforms and eventually splits into two separate vortex rings between* $t = 29$ *and* $t = 30$. *Same parameters as in Fig. 4.29 with a slightly large speed* $s = 1$. *Adapted with permission from Ref. [410].*

may thus represent baryons in field theory [761], while their 2D counterparts, namely *baby Skyrmions* in the sigma model, may account for the disappearance of antiferromagnetism and the onset of high-T_c superconductivity [762], as well as the formation of the ground state in quantum-Hall ferromagnets [763].

Skyrmion-like objects have also been predicted in BECs [718, 764, 765]. In particular, a two-component BEC may support structures that emulate the topology of a Skyrmion—but also a plethora of other topological structures (cf. the reviews [281, 282]). In the most basic setup, namely a two-component 3D BEC, a Skyrmion can be conceived as a vortex

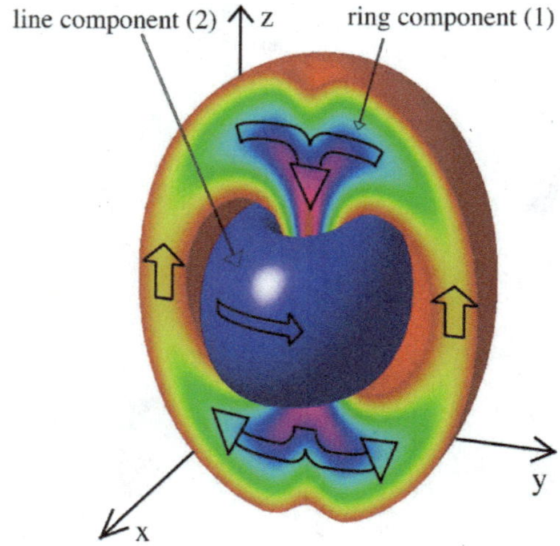

Figure 4.31. *Three-dimensional density profiles for a Skyrmion in a trapped, two-component BEC. The central (blue) torus is an isosurface of the vortex line in one of the components (component 2). The peripheral (colored) isosurface corresponds to the vortex ring in the other component (component 1) and its (y, z) cut depicts the density by a color map from red (lowest) to purple (highest). The circulation associated with each vortex is also indicated. Adapted with permission from Ref. [758].*

ring (see Sec. 4.3) in one component coupled to a 2D vortex line (see Sec. 4.2) in the other component [758]—see Fig. 4.31 for an illustration of such a topological structure. Notice that Skyrmions can be rendered stable in harmonically trapped BECs [766] and induced using phase imprinting techniques [767, 718, 713].

From the viewpoint of experiments, the first experimental efforts were focused on creating Skyrmion spin textures in spinor BECs by means of phase imprinting techniques in $F = 1$ spinor ^{23}Na [351] and $F = 2$ spinor ^{87}Rb [768] condensates. Nevertheless, the stability of the spin textures was not investigated. More recently, topologically protected Skyrmions were created [769] (see also the relevant work in Ref. [770]) in an $F = 1$ spinor polar ^{23}Na BEC and their stability was tested; this was done upon measuring the time evolution of the spin texture with a variable dwell time in the optical trap. It was thus observed that it dynamically deforms and decays to a uniform spin texture.

It is interesting to note that structures bearing the topology of Skyrmions can also be emulated in 2D (baby Skyrmions) and 3D, two component, NLS-type lattices [771].

4.4.3 ▪ Superfluid Ring Excitations in Binary BECs

Finally, let us describe some complex phenomenology that originates from the interaction between the components of a binary BEC. We follow closely the presentation given in Ref. [279] that describes the dynamics of superfluid ring excitations that are generated by the interatomic interactions between components in a binary mixture of *same* species (i.e., a homonuclear mixture) ^{87}Rb atomic BEC. These homonuclear systems allow for the exchange of mass between two (or more) hyperfine states of the atomic cloud. In Ref. [279] the two hyperfine states $|F = 1, m_f = -1\rangle \equiv |1\rangle$ and $|F = 2, m_f = +1\rangle \equiv |2\rangle$ of a ^{87}Rb condensate are considered. The condensate is initially prepared in the $|1\rangle$ state and, subsequently, a two-photon $\pi/2$ pulse is applied to transfer a fraction of the atoms from the $|1\rangle$ state to the $|2\rangle$ state. This coupling between the components, usually referred to as Rabi coupling (see also Sec. 3.11), is responsible for a periodic exchange between the two states. Therefore, it is possible, by directly controlling the strength and, more importantly, the duration of this coupling, to control the relative populations between the two components. In this manner, in the experiment described in Ref. [279], any desired relative population between the two components was achieved and its effects were monitored in the subsequent evolution after the Rabi coupling was turned off. Although the system starts (before the transfer) in its approximate ground state for a pure $|1\rangle$ condensate, after the transfer of a controlled fraction of the condensate to the $|2\rangle$ state, the system is no longer in its ground state. This excited initial condition is then responsible for interesting 3D patterns that can be observed in the experiment and further analyzed using its corresponding GPE model as we now describe.

After the transfer between the $|1\rangle$ and $|2\rangle$ components is performed, the evolution for their corresponding wavefunctions ψ_1 and ψ_2 can be described by a pair of coupled GPEs [772, 773, 774, 775, 634]:

$$i\hbar\frac{\partial\psi_1}{\partial t}=\left[-\frac{\hbar^2\nabla^2}{2m}+V_1+g_{11}|\psi_1|^2+g_{12}|\psi_2|^2-i\Gamma_1\right]\psi_1, \tag{4.21}$$

$$i\hbar\frac{\partial\psi_2}{\partial t}=\left[-\frac{\hbar^2\nabla^2}{2m}+V_2+g_{22}|\psi_2|^2+g_{12}|\psi_1|^2-i\Gamma_2\right]\psi_2, \tag{4.22}$$

where g_{ij} are the nonlinear interaction constants between atoms in states $|i\rangle$ and $|j\rangle$ and V_1 and V_2 represent the trapping potentials for species $|1\rangle$ and $|2\rangle$, respectively (having the two potentials being different turns out to be very important for the timing of the dynamical evolution, as we will see below). Furthermore, we allow for the system to include loss terms with rates Γ_1 and Γ_2. Specifically, a quantitative characterization of the loss rates can be given by $\Gamma_1=\frac{\hbar}{2}(\gamma_{111}|\psi_1|^4+\gamma_{12}|\psi_2|^2)$ and $\Gamma_2=\frac{\hbar}{2}(\gamma_{12}|\psi_1|^2+\gamma_{22}|\psi_2|^2)$. These loss rates can be accounted for by noting that three-body recombination losses dominate in a pure $|1\rangle$ condensate (with decay constant $\gamma_{111}=5.4(11)\times10^{-30}$ cm^6/s [539]), but two-body inelastic collisions dominate in mixed and pure $|2\rangle$ condensate (decay constants $\gamma_{12}=0.780(19)\times10^{-13}$ cm^3/s and $\gamma_{22}=1.194(19)\times10^{-13}$ cm^3/s [279]).

Figure 4.32 depicts typical experimental results and their corresponding numerical simulations for a 50:50 mixture of $N=3.50(5)\times10^5$ ^{87}Rb atoms. As mentioned above, since the initial prepared mixed state is out of equilibrium, we expect some nontrivial dynamical evolution as the two components interact (repel each other) through the nonlinear coupling terms. In fact, due to the mutual repulsion between the components a series of concentric rings with alternating hyperfine states are observed in the evolution of Fig. 4.32. The observed phase-separated states are reminiscent of the equivalent ones in 1D reported in Ref. [300]. There it was shown that the different alternating states bifurcate, as the interaction between the components is increased, from the mixed state where the two components are on top of each other. Careful modeling of the system using the GPEs (4.21) and (4.22) reveals the importance of considering the right potential for each of the components. In fact, if the two components are considered to experience indistinguishable effects from the trapping potential, namely $V_1=V_2$, the model tends to have a slower evolution for the ring excitation dynamics (compare the third row of panels which uses $V_1=V_2$ to the actual experimental results of the first row of panels in Fig. 4.32). However, proper accounting for the difference between the potentials for each component ($V_1\neq V_2$) allows one to obtain numerical results that compare very favorably to the experimental results [279]. This can be in fact observed by comparing the second row of panels in Fig. 4.32, where the correct difference between the two potentials has been used, with the experimental results of the first row of panels.

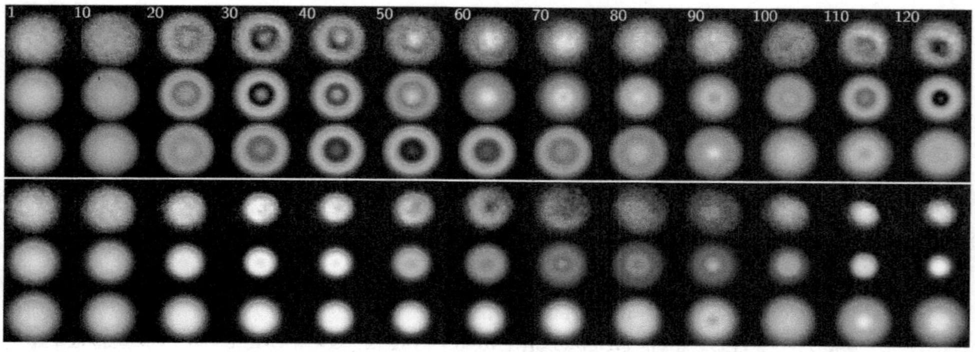

Figure 4.32. *Dynamical evolution of experimental and numerical density profiles for a binary ^{87}Rb prepared with equal populations in the $|1\rangle$ and $|2\rangle$ states. The first row depicts the experimental density profiles for the $|1\rangle$ atoms. The second and third rows depict the corresponding numerical results including losses and different trap frequencies (second row) and without these two features (third row). A similar arrangement is given for the $|2\rangle$ atoms in the bottom three rows of panels. The field of view in all pictures is approximately* 100 μm *on a side and the evolution time (in ms) for each column is indicated in the top row. Adapted with permission from Ref. [279].*

Furthermore, the results of Ref. [279] highlight the importance of appropriately taking into regard the fact that the system is indeed three-dimensional. Specifically, as depicted in the top panels of Fig. 4.33, there are strong excitations in the vertical (z) direction. In fact, shortly after the preparation of the initial state with the appropriate fraction of atoms in each component, the repulsion between the components induces the appearance of alternating cylindrically symmetric rings (see top-left panel of Fig. 4.33 taken at $t = 30$ ms) with no excitations about the z axis. However, letting the system to evolve further, it is evident that strong z-dependent excitations start occurring. See for example the top-right panel of Fig. 4.33 where the isocontour densities are depicted at $t = 80$ ms.[23] Interestingly, these z excitations were not detected in the actual physical experiments since no images projecting the density about any axis perpendicular to the z axis were taken. Nonetheless, it is worth pointing out that even if the, let us say, x-integrated densities were available from the experiments, it is not obvious that these projections would be able to give much information about these excitations. For example, see the side (x) projection on the top-right panel of Fig. 4.33 where the fine structure of the z excitation is not clearly visible due to the integrated projection present in any experimental setting.

Another very relevant, and important, fact of the modeling done via the GPEs (4.21) and (4.22) is the level of accuracy in extracting the density (modulus squared of the wavefunctions) along the evolution of the interacting components. For instance, in the bottom two panels of Fig. 4.33 we depict the experimental and numerical radial densities corresponding to the cases shown in the top panels. It is remarkable that the model (solid lines) is able to accurately reproduce the densities obtained from the experiments (symbols with error bars).

Finally, the effects of varying the overall initial atomic population and the population fraction between the components were also investigated in Ref. [279]. The results are summarized in Fig. 4.34. The top panel corresponds to snapshots taken during the evolution starting from an initial state with a 50:50 fraction between the two components for different number of atoms (increasing from left to right in the panels). As it is clear from

[23] Movies corresponding to the dynamical evolution of the ring excitations are available at http://nonlinear.sdsu.edu/~carreter/SuperfluidRingExcitations.html.

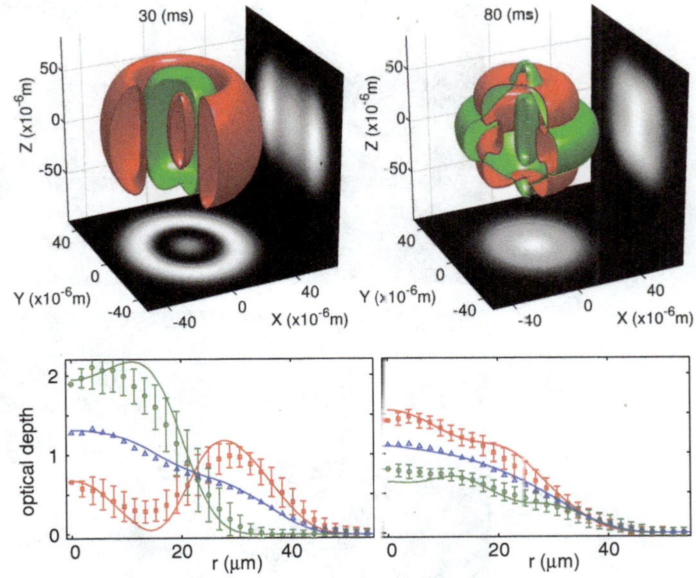

Figure 4.33. *(color) Top row: 3D renderings of the density distributions in a binary BEC computed from the model at 30 ms (left) and 80 ms (right). Each component is depicted by a isodensity contour slice at half of its corresponding maximal density. Red and green surfaces correspond to components* $|1\rangle$ *and* $|2\rangle$*, respectively. The bottom (side) projection corresponds to the z (x) integrated density for the* $|1\rangle$ *component in our model. Please visit http://nonlinear.sdsu.edu/~carreter/SuperfluidRingExcitations.html for movies showing the evolution of these density contours. Bottom row of panels: Radial profiles of the density distributions at 30 ms (left) and 80 ms (right). Symbols (squares for* $|1\rangle$ *and circles for* $|2\rangle$*) correspond to the experimental data, which are averaged over the azimuthal angle. Solid lines depict the model results. The middle curve depicts half of the total density, which is nearly preserved during the time evolution [277]. Adapted with permission from Ref. [279].*

the figure, larger number of atoms induces more interesting patterns (i.e., patterns with more concentric rings). This is simply due to the fact that for larger atom numbers the condensate is larger and thus there is more "space" to support a larger number of phase separated rings. Similarly, the bottom panel of Fig. 4.34 depicts snapshots when starting with a different fraction between components. It is evident that the increase in the number of atoms in the $|2\rangle$ state results in a larger number of rings. This behavior might be understood by noting that the excess energy after transfer places the system far from its ground state equilibrium and consequently induces a long oscillatory journey through the system's configuration space. Stronger interactions, larger transfer ratios, and greater differences in the trapping potentials should therefore produce more dramatic dynamics since they leave the initial configuration after transfer further away from the ground state.

We conclude this section by pointing out that, as speculated in Ref. [279], since the phase separation dynamics for the ring excitations is sensitive to the parameters of the system, optimization procedures to match experimental and numerical results could provide the means to reverse engineer some of the parameters and indirectly find reasonably accurate physical values for them. In particular, it is in general difficult to obtain precise (direct) measurement of the interatomic interaction coefficients g_{ij}. Nevertheless, in more recent experiments such as those of Ref. [776], the fitting of numerical simulations of the coupled GPEs to experimental results was used in order to obtain best fit

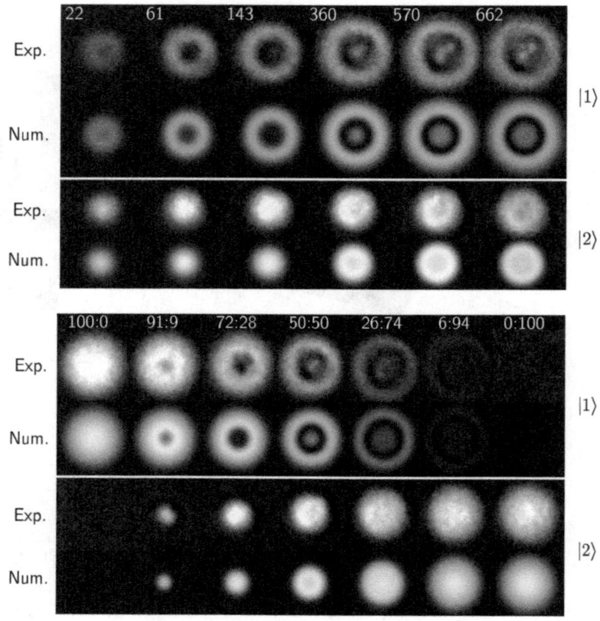

Figure 4.34. *Top set of panels: Experimental and numerical (respective top and bottom sub-rows) density profiles of a 50:50 mix of* $|1\rangle$ *and* $|2\rangle$ *atoms at* 30 *ms for the different populations indicated in thousands of atoms. Bottom set of panels: Same as the top row of panels for different relative numbers of atoms as indicated by the* $|1\rangle : |2\rangle$ *ratio. The atom number is fixed at* $N = 3.50(5) \times 10^5$*, as in Fig. 4.32. Adapted with permission from Ref. [279].*

predictions on the scattering lengths a_{12} and a_{22} (as well as loss coefficients γ_{12} and γ_{22}), improving on the earlier estimates of Ref. [279] discussed above.

4.5 ▪ New Computational Tools

The vast majority of possible dynamics, and even stationary solutions, of coherent structures in 2D and 3D NLS-type equations are not directly analytically tractable. Furthermore, approximate analytical (asymptotic, perturbational, variational, etc.) techniques are often limited and require a high degree of mathematical sophistication. On the other hand, numerical investigations have been instrumental in gaining understanding on the dynamics and interactions of coherent structures. Therefore, state-of-the-art numerical techniques and advances in the production of inexpensive high-level computational systems are enabling a deeper look into increasingly complex scenarios, particularly in numerically expensive 3D setups.

In this section, we describe a very efficient method to numerically integrate NLS-type equations. The efficiency of the proposed method (in terms of running time and hardware costs) is based on the processing capabilities of graphical processing units (GPUs), where the bulk of the computations are performed in a GPU card rather than on traditional CPUs. The GPU computational model is based on utilizing the many (several hundreds to potentially thousands) basic GPU processors of a dedicated graphics card as a massive parallel floating-point computational platform. The technology to numerically solve NLS-type equations is based on the numerical/computational building blocks described in the following sections. The full code to integrate the NLS in one, two, and three dimensions using the efficiency of GPU computing is described in detail in Ref. [777] and is available for free download at http://www.nlsemagic.com/.[24]

4.5.1 ▪ Two-Step High-Order Explicit Compact Scheme

One of the most widespread methods to integrate nonlinear PDEs is based on discretizing the solution in space by using finite-difference approximations. Let us describe a *high-order* scheme (one that exhibits greater than second-order accuracy) that, at the same time, is *compact* to minimize the amount of information to be shared between subdomains toward a more straightforward parallelization of the code. The use of a high-order scheme is crucial for higher-dimensional applications because it allows to simulate a solution with much fewer grid points. In this case, the desired grid point size is often bounded by the ability to resolve the structure of the solution, and not necessarily by the accuracy of the computation.

Higher-order finite-difference schemes are typically achieved by computing derivatives with a wider scheme stencil. This, however, complicates evaluations at the boundary and causes the parallel implementation to be more difficult to realize, as different nodes must share/transfer more boundary values to their neighboring nodes of the individual subdomains. Besides the added complexity of the codes, the extra communication and global-memory access (which is crucial in the case of GPUs) reduces the overall parallel performance [778]. Therefore, there is great interest in high-order compact (HOC) schemes that exhibit high-order accuracy but still only rely on the *closest* neighboring points for computations. HOC schemes have been developed for various multidimensional steady-state PDEs [779, 780, 781], as well as time-dependent PDEs [782, 783, 784, 785, 786] including the 1D NLS equation [787, 788].

Some major drawbacks of standard time-dependent HOC schemes is that their formulation is problem dependent [781] and they are intrinsically implicit, requiring

(i) solving a linear system in each time-step, as well as iterative processes when simulating nonlinear PDEs such as the NLS equation,

(ii) computational and storage requirements that can be prohibitive in large multidimensional settings, and

(iii) implementations that are, in general, difficult to optimally parallelize, especially in higher dimensions.

Therefore, it is crucial to formulate time-dependent schemes for NLS-type equations that are *high-order*, *compact*, and *fully explicit* [789]. The method is based on formulating a two-step (2S) procedure to compute the spatial derivatives, where each individual step is a compact computation. The first step of the 2SHOC scheme computes the standard second-order finite-difference approximation to the derivatives of the Laplacian, while the second step uses these computed derivative values to compactly approximate the Laplacian to fourth-order accuracy. Since the 2SHOC scheme computes the Laplacian independently of the governing PDE, it can be used for *any* time-dependent and time-independent PDEs that contain the Laplacian operator. Thus, an explicit 2SHOC scheme for the NLS equation can be formed by utilizing standard *explicit* ordinary differential equation (ODE) solvers for the set of equations ensuing from the discretized system. In particular, for our purpose, we will use a fourth-order Runge–Kutta (RK4) scheme [790].

---

[24] We are grateful to R.M. Caplan for sharing the code and assistance with some of the numerical results presented in this chapter.

Let us now briefly describe the 2SHOC in 2D (1D and 3D are very similar; for more details, see Ref. [789]). The Laplacian operator applied to Ψ can be approximated by

$$\nabla^2\Psi_{i,j} = D^{xx}\Psi_{i,j} + D^{yy}\Psi_{i,j} - \frac{h^2}{12}\left(\partial_x^4\Psi\Big|_{i,j} + \partial_y^4\Psi\Big|_{i,j}\right) + O(h^4), \tag{4.23}$$

where

$$D^{xx} = \frac{1}{h^2}(\Psi_{i+1,j} - 2\Psi_{i,j} + \Psi_{i-1,j})$$

is the 1D central-difference operator (a similar definition is used for D^{yy}). It is important to stress that the resulting 2SHOC formulation requires two storage matrices (D^{xx} and D^{yy}) in addition to that for Ψ — the 3D formulation requires three storage matrices (D^{xx}, D^{yy} and D^{zz}). In large-scale computations, memory use can become a bottleneck, so it is useful to limit the amount of extra storage required. Therefore, we devise a second scheme that only requires one extra storage matrix, at the cost of requiring more computations (a trade-off of this balance is presented below).

To formulate the lower-storage version of the 2SHOC, the standard five-point second-order finite-difference approximation to the 2D Laplacian is used for the first stage (stored in $D^{(2)}$), and the result is combined with a second-order cross-derivative stencil to yield the second-order approximations to the fourth derivative truncation terms in Eq. (4.23). First, note that the fourth derivative truncation terms in Eq. (4.23) can be written as

$$\Psi_{xxxx} + \Psi_{yyyy} = (\nabla^2\Psi)_{xx} + (\nabla^2\Psi)_{yy} - 2\Psi_{xxyy},$$

for which the cross-derivative term is known to have a nine-point second-order compact stencil [791]. Part of this stencil can be written in terms of the five-point second-order approximation to the Laplacian stored in $D^{(2)}$, in which case the single-storage 2SHOC scheme (after simplification) can be described by [789]

$$\nabla^2\Psi_{i,j} \approx -\frac{1}{12}\begin{array}{|c|c|c|}\hline & 1 & \\ \hline 1 & -12 & 1 \\ \hline & 1 & \\ \hline\end{array}\, D^{(2)}_{i,j} + \frac{1}{6}\begin{array}{|c|c|c|}\hline 1 & & 1 \\ \hline & -4 & \\ \hline 1 & & 1 \\ \hline\end{array}\,\Psi_{i,j}, \tag{4.24}$$

where only the single storage matrix $D^{(2)} = D^{xx} + D^{yy}$ is now required. A very similar formulation can be done in the 3D case but it is omitted for brevity of exposition. It can be easily shown that both (double and single storage) formulations of the proposed 2SHOC schemes (in any dimension) are computationally equivalent to the standard noncompact fourth-order wide stencil; for more details, including the 3D implementation, please refer to Ref. [789]. On top of the storage required by the HOC scheme, the RK4 time-stepping requires three storage matrices in addition to the storage for Ψ and external potential V_{ext}. Lower-storage RK4 schemes with comparable accuracy have been developed (a $3N$ in Ref. [792] and a five-stage $2N$ in Ref. [793] which requires an additional function evaluation, and whose coefficients are numerically derived), but for simplicity let us use the standard RK4. Table 4.1 presents the results on the comparison of the computational costs (operations and storage) for the various 3D components of the codes. We notice that the single-storage 2SHOC scheme ($1\times$) in the RK4 step for the NLS equation takes 70% more computations than the noncompact scheme, and for the Laplacian operator alone, takes a little more than twice the computations required for the noncompact scheme. However, since the main advantage of compact schemes is their parallel implementation, this increase in computation is easily offset by the parallelization.

Table 4.1. *3D computational cost analysis for 2SHOC schemes compared to noncompact schemes [789]. N is the total number of grid points and 3× and 1× stand for the triple storage and single storage versions.*

Scheme	Method	Operations	Storage	Operations ratio	Storage ratio
Laplacian:	Noncompact	14	N	–	–
	2SHOC (3×)	22	$4N$	1.57	4
	2SHOC (1×)	31	$2N$	2.21	2
RK4 Step:	Noncompact	97	$5N$	–	–
	2SHOC (3×)	129	$8N$	1.33	1.6
	2SHOC (1×)	165	$6N$	1.70	1.2

4.5.2 ▪ Modulus-Squared Dirichlet Boundary Conditions

When numerically solving time-dependent PDEs, proper handling of boundary conditions (BCs) may be challenging. Sometimes, an otherwise stable numerical scheme may become unstable through the BCs [794]. In addition, high-order schemes can degrade in accuracy to lower order when using BCs that are not compatible with the high-order accuracy [795]. A commonly used type of BCs consists of the Dirichlet BCs, where the function value at the boundary of the domain is set to a constant value—commonly this value is zero especially for structures decaying to zero at $\pm\infty$ (such as bright solitons). Problems involving PDEs whose function values are complex (like the NLS equation) and whose solutions do not necessarily decay to zero (as, e.g., the self-defocusing variant thereof) cannot straightforwardly make use of standard Dirichlet BCs; this is due to the constant oscillation of the real and imaginary parts of the function, caused by the intrinsic frequency of the underlying phase of the solutions. Thus, when dealing with the NLS with repulsive nonlinearity, that is the central theme of this book, the basic coherent structures consist of density dips embedded in a constant density background. There, it is desired that the *modulus squared* (density) of solution be constant at the boundary. Therefore, a simple yet efficient way to implement modulus-squared Dirichlet (MSD) BCs for time-dependent complex-valued PDEs has been proposed in Ref. [796]. This method eliminates the need for overly large grids or expensive and complicated BCs.

Let us briefly describe the MSD BC (cf. Ref. [796] for more details). The goal is to design appropriate BCs enforcing, for all times, that the modulus squared of the function Ψ at the boundaries (Ψ_b) is equal to some constant, positive, real value B:

$$|\Psi_b|^2 = B, \tag{4.25}$$

where the subindex b stands for the solution at the boundary. This is equivalent to stating that the solution at the boundary has the form

$$\Psi_b = \sqrt{B}\exp[-i\,\Omega_b(\vec{r},t)], \tag{4.26}$$

where $\Omega(\vec{r},t) \in \mathfrak{R}$ is the phase which can be a function of space and time. Taking the natural logarithm of Eq. (4.26) yields

$$\ln(\Psi_b) = \ln\left(\sqrt{B}\right) - i\,\Omega_b(\vec{r},t),$$

which, after taking a temporal derivative, yields

$$\frac{\partial}{\partial t}\ln(\Psi_b) = -i\frac{\partial}{\partial t}\Omega_b(\vec{r},t).$$

We now apply a spatial derivative in the normal direction to the boundary (denoted here by x) on each side yielding the continuous representation of the MSD boundary condition:

$$\frac{\partial}{\partial x}\left(\frac{\Psi_{t,b}}{\Psi_b}\right) = -i\frac{\partial}{\partial x}\left(\Omega_{t,b}(\vec{r},t)\right), \tag{4.27}$$

where the subindex t denotes $\partial/\partial t$. Discretizing Eq. (4.27) using first-order one-sided differencing, and effectively approximating the phase of the solution to vary linearly at the boundary, yields the discrete version of MSD BC [796]:

$$\Psi_{t,b} = i\,\mathrm{Im}\left[\frac{\Psi_{t,b-1}}{\Psi_{b-1}}\right]\Psi_b, \tag{4.28}$$

which can be directly evaluated, since the value of $\Psi_{t,b-1}$ is computed using the interior scheme being implemented and where the subscript $b-1$ stands for the nodes one cell inward from the boundary in the normal direction. The MSD BC (4.28) is given for the temporal derivative of the boundary point, which is ideally suited for Runge–Kutta type solvers, as the right-hand side of the PDEs (Ψ_t) is evaluated and used for the time-stepping [790]. For other methodologies, or in situations where the boundary value of the spatial derivatives is needed, the MSD BC can be inserted into the governing equation to formulate the required BCs.

It is worth listing at this stage some key features of the MSD BCs:

(a) they do not depend on the specific PDE;

(b) they can be used in multidimensional settings without modification since each boundary point only uses one interior point;

(c) they do not depend on the size of the grid spacing, and therefore do not need to be altered when using unequal grid spacing (cf., in adaptive mesh refinement);

(d) they are compact, and

(e) they are extremely easy to implement, making them an attractive alternative in parallel environments.

Many sophisticated BCs have been developed for both the linear and nonlinear Schrödinger equations that simulate transparent or artificial boundaries (see the review of Ref. [797] and references therein). However, most of these BCs focus on studying solutions which tend to zero at infinity, which preclude their use in our current setting with nonzero constant background at the boundaries.

To demonstrate the usefulness of the MSD BCs, we present some results for the dynamics of 2D NLS vortices (see Chap. 3). Let us start with a single unit charge vortex, which is known to be a stable, steady-state solution of the NLS equation. Then, we compare the MSD BC to the Laplacian-zero (L0) BC $\nabla^2\Psi_b = 0$. L0 BCs are tantamount to time-dependent Dirichlet BCs incorporating the (linear) time variation of the phase of the solution at the boundary, since $\nabla^2\Psi_b(t) = 0$ implies (locally) that $\Psi_b(t) \approx \Psi_b(0)\exp(i|\Psi_b(t)|^2 t)$. The L0 BC is chosen for comparison because it is an easy-to-implement BC that approximately keeps, for constant background density simulations, the

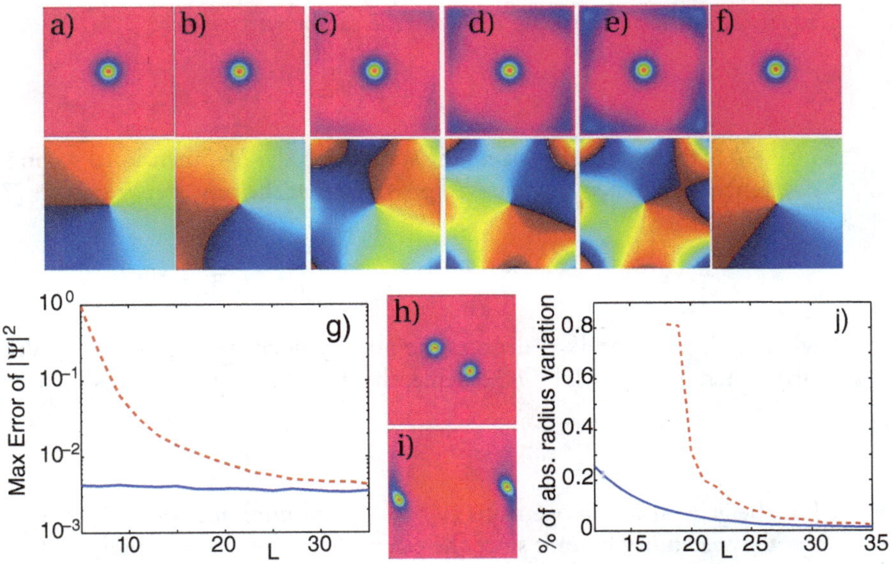

Figure 4.35. (a)–(f) *Vortex evolution (top shows density, bottom shows phase) for a* 120×120 *grid.* (a)–(e) *Laplacian-zero (L0) BCs at* $t = 0, 600, 1500, 2400, 3000$. (f) *MSD BC at* $t = 5\times10^4$. (g) *Modulus-squared error for the vortex using L0 (red dashed line) and MSD (solid blue line) BCs for varying domain size L (i.e., domain* $[-L,L]\times[-L,L]$*) at* $t = 300$. (h)–(i) *Two interacting vortices for, respectively, MSD* ($t = 1\times10^4$) *and L0* ($t = 1100$) *BCs.* (j) *Percentage of absolute radius (half distance between vortices) variation for two interacting vortices vs. domain size for* $t = 480$ *using L0 (red dashed line) and MSD (blue solid line) BCs. Adapted with permission from Ref. [796].*

phase structure of the solution at the boundary. We integrate the NLS equation without any external potential ($V_{\text{ext}} = 0$) for considerable time (up to $t = 5\times10^4$) using both the MSD and L0 BCs. The results are shown in panels (a)–(f) of Fig. 4.35. It is clear that the L0 BC is only useful for short times, as it quickly suffers from distortions that worsen as time progresses, to the point where the solution becomes unusable. In contrast, the MSD BCs create *no distortions* in the phase nor in the modulus squared of the solution for very long simulations even when the vortex gets very close to the boundary [see panel (g)]. In contrast, using L0 BCs would require a much larger domain (and thus many more mesh points) to reach the same level of accuracy as the MSD BCs. A similar conclusion is drawn when following the interaction dynamics between two vortices of equal charge [panels (h)–(j)]. As it is shown in the figure, MSD BCs allow for smaller domain sizes to capture the true dynamics consisting of pure rotation between vortices at constant separation and angular velocity when compared to L0 BCs, where the vortices tend to spiral out and collide with the boundaries. Allowing for the coherent structures to get closer to the boundaries without being distorted is the main advantage of using MSD BCs in large scale simulations (and particularly in higher dimensions) in order to "waste" as few mesh points as possible.

4.5.3 ▪ Numerical Stability

We now briefly describe the stability analysis for the methods described above; for more details please refer to Ref. [798]. For the linear ($g = 0$) and homogeneous ($V_{\text{ext}} = 0$) NLS equation, the corresponding RK4 scheme can be written as

$$\Psi_t = \mathscr{A}\Psi = \frac{i\,a}{h^2} A\Psi,$$

where h is the spatial step size. The stability of the D-dimensional second-order central difference (CD) scheme is achieved provided that the time step k satisfies [799, 800, 801]

$$k < k_{\text{CD}} = \frac{\sqrt{8}}{||\vec{\lambda}_A||_\infty} \frac{h^2}{a} = \frac{h^2}{\sqrt{2}aD}, \tag{4.29}$$

where $||\vec{x}||_\infty = \max\{|x_i|\}$ denotes the infinity norm and λ_A the eigenvalues of A. Similarly, when using the 2SHOC scheme, the stability bound gets modified as [798]

$$k < k_{\text{2SHOC}} = \frac{3}{4} k_{\text{CD}}.$$

For the general inhomogeneous ($V_{\text{ext}} \neq 0$) and nonlinear ($g \neq 0$) case, one cannot use the above stability bounds since the eigenvalues of A are not purely real. Nonetheless, it is possible to extract bounds for the eigenvalues of the corresponding matrices by using Gershgorin disks [802]. After considering the effects of the (a) nonlinear and (b) external potential terms, and (c) the BCs, it is possible to obtain a stability criterion for our full scheme. For more details please refer to Ref. [798]. It is important to mention that the precise stability bound depends on the nonlinearity and the external potential *evaluated* at the solution. Therefore, it is not, in general, possible to *a priori* give a closed estimate for k_{2SHOC}. However, it is crucial to note that we are mainly interested in coherent structures that are typically structurally stable and thus, in general, do not dramatically change their shapes. Therefore, one can compute, using the initial conditions, the initial bound $k_{\text{2SHOC}}(t = 0)$ and use it as a good approximation for later times. In practice, we use a more modest (and safe) bound by taking, for example, 80% of the initial bound such that $k_{\text{2SHOC}}(t > 0) = 0.8\, k_{\text{2SHOC}}(t = 0)$. This "global" bound has been found to work well for a wide range of numerical simulations in 1D, 2D, and 3D [798].

4.5.4 ▪ GPU Computing, CUDA Implementation, and Speedup Results

Having described the computational building blocks, we now illustrate their implementation in the context of GPU computing. GPUs, originally developed to parallelize massive computations to speedup video games and image processing, are now being adapted to expedite scientific codes. NVIDIA corporation is the leading company in this new form of computing, having developed the Compute Unified Device Architecture (CUDA), which can be compiled to run scientific codes on NVIDIA's line of GPU video cards, as well as nonvideo stand-alone GPUs designed specifically for parallel computation [803]. As of this writing, typical GPU can have up to about 3000 processing cores and up to 12GB of RAM, which allows a maximum throughput of over a teraFLOP on a single card! The price of the GPUs is another major factor, as one can purchase, as of this writing, an off-the-shelf GPU with over 1500 cores and 4GB of RAM for under \$450. The reasonable cost allows supercomputing capabilities to be realized on a single desktop PC. CUDA codes have recently been used for various finite-difference schemes including 2D electromagnetic scattering [804], light scattering and Maxwell's equations [805, 806], wave propagation in 3D [807, 808], and the complex Ginzburg–Landau equation [809].

Let us briefly outline the single-GPU CUDA implementation and its speedup results. For a more detailed exposition the interested reader is encouraged to consult Ref. [777].

To parallelize the operations for the NLS numerical scheme, each CUDA block must be able to run independently of all others; therefore, each step in our numerical method must be executed in a separate CUDA kernel. The kernel describes what each thread should compute. In our code, each thread computes on one grid cell. The input vector for each kernel is stored in global memory. Since access to this memory is very slow, we utilize the per-block shared memory. Each thread copies the needed value from global memory into shared memory, and after synchronizing the threads in the block, each thread has access to a block-sized chunk of the input vector. The code runs a defined number of time steps, transfers the current vector of the solution Ψ to the host CPU for plotting and analysis, and then returns to the GPU to continue integrating the solution. For each dimensionality (1D, 2D, and 3D), the implementation is slightly different since the domain has to be partitioned differently. We now summarize the key steps in the implementation for the different dimensionalities (for more details the interested reader is referred to Ref. [777]).

1D Implementation

For the 1D NLS case, the compute grid is set up to be 1D with a block size of 512 (therefore there are 512 compute threads in each block). In addition to the boundaries of the entire simulation grid size, a new boundary is now considered, which is the boundary of the compute block. For global-memory accesses to the Ψ vectors, this boundary poses no problem since all threads in all blocks have access to all of the global memory. However, to ensure performance, the per-block fast-shared memory is used in the computations. Each thread first transfers its position's grid point of Ψ to the relative position in the block's shared memory. Then, the shared-memory size is allocated to be 2 cells greater (514) to account for common boundaries. In the stage of the kernel where the threads transfer Ψ from global memory into shared memory, the boundary threads also transfer the neighboring points into the extra cell space at the boundary of the shared memory. Then, in the kernel computations, all threads can use shared memory.

2D Implementation

The code for the 2D NLS is similar to the 1D except that it uses a special 2D memory-copy command in CUDA to transfer the vectors to the GPU. This special 2D memory-copy command aligns the data with the card hardware, often padding the rows of the matrix to be a certain length. Therefore, the function returns a pitch value, which is used to compute the thread indices. The compute grid is set to be 2D with 2D blocks for easy indexing. As in the 1D case, the shared memory space is allocated to be 2 cells larger in width *and* height than the block size.

3D Implementation

For the 3D case, the CUDA provides a hardware-optimized memory transfer function similar to the 2D memory copy. However, tests on the 2D codes versus a naive 2D implementation (without the specialized memory copy) did not result in an appreciable performance boost when using the hardware-optimized routine [777]. Therefore, for the 3D implementation, the same approach as in lower dimensions is followed by splitting the 3D problem into a 3D grid with 3D blocks and then translate this to be a 2D grid of 3D blocks. However, since fully 3D block structures are now standard in GPU cards with compute capability 2.0 and higher [810], the full use of the hardware-optimized memory transfer functions is recommended.

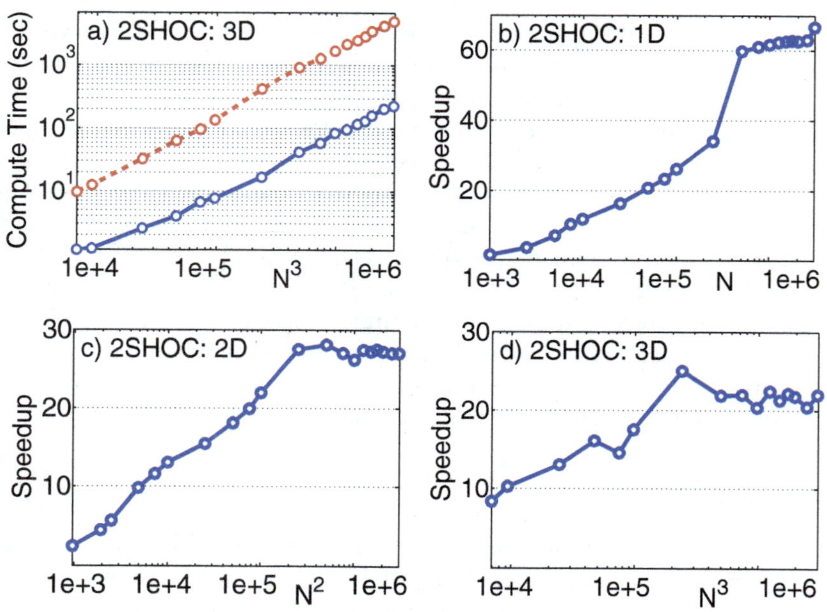

Figure 4.36. (a) *Results on compute time vs. number of grid points for the CPU (dashed) and GPU (solid) codes.* (b)–(d) *Computed speedup (ratio of computation time excluding graphics and overhead) between the GPU and CPU codes for the 2SHOC scheme in 1D, 2D, and 3D. Data adapted from Ref. [777].*

Speedup Results

Let us now present some speedup results for the GPU-accelerated codes for solving the NLS in 1D, 2D, and 3D PDE codes (for further details please refer to Ref. [777]). The speedup results were obtained in a very modest $450 GeForce GTX 580 card (512 CUDA cores and 1.5 MB) installed in a CPU with Intel Core i3 540 (dual core, 3.07 GHz and 4GB RAM) for a total price of about $1,500! All speedup tests use the homogeneous ($V_{\text{ext}} = 0$) and repulsive ($g = 1$) NLS. Specifically, for the 1D tests we used the gray soliton solution described in Chapter 2, for the 2D tests we used the vortex solutions described in Chapter 3 and, finally, for the 3D tests we used the VRs described in this chapter. Figure 4.36 summarizes typical speedups obtained for these 1D, 2D, and 3D cases. As it is clear from the figure, the speedup results are very encouraging (more than 60 in 1D and more than 20 in 3D). In all cases, the speedup increases until it saturates for large values of the grid size. The saturation results from the GPU achieving the maximum occupancy and efficiency both computationally and in terms of memory management versus the CPU. For higher dimensions the speedup is lower. This effect is due to (a) saturation of the L1 and L2 cache memories of the CPU and (b) the need for more shared memory to store the common edges between the increasing number of subdomains in higher dimensions.

Bibliography

[1] P. EMPLIT, J.P. HAMAIDE, F. REYNAUD, C. FROEHLY, AND A. BARTHELEMY, *Picosecond steps and dark pulses through nonlinear single-mode fibers.* Opt. Commun., **62** (1987) 374–379.

[2] D. KRÖKEL, N.J. HALAS, G. GIULIANI, AND D. GRISCHKOWSKY, *Dark-pulse propagation in optical fibers.* Phys. Rev. Lett., **60** (1988) 29–32.

[3] G.A. SWARTZLANDER, JR., D.R. ANDERSEN, J.J. REGAN, H. YIN, AND A.E. KAPLAN, *Spatial dark-soliton stripes and grids in self-defocusing materials.* Phys. Rev. Lett., **66** (1991) 1583–1586.

[4] G.R. ALLAN, S.R. SKINNER, D.R. ANDERSEN, AND A.L. SMIRL, *Observation of fundamental dark spatial solitons in semiconductors using picosecond pulses.* Opt. Lett., **16** (1991) 156–158.

[5] B. DENARDO, W. WRIGHT, S. PUTTERMAN, AND A. LARRAZA, *Observation of a kink soliton on the surface of a liquid.* Phys. Rev. Lett., **64** (1990) 1518–1521.

[6] B. DENARDO, B. GALVIN, A. GREENFIELD, A. LARRAZA, S. PUTTERMAN, AND W. WRIGHT, *Observations of localized structures in nonlinear lattices: Domain walls and kinks.* Phys. Rev. Lett., **68** (1992) 1730–1733.

[7] M. CHEN, M.A. TSANKOV, J.M. NASH, AND C.E. PATTON, *Microwave magnetic-envelope dark solitons in yttrium iron garnet thin films.* Phys. Rev. Lett., **70** (1993) 1707–1710.

[8] R. HEIDEMANN, S. ZHDANOV, R. SÜTTERLIN, H.M. THOMAS, AND G.E. MORFILL, *Dissipative dark soliton in a complex plasma.* Phys. Rev. Lett., **102** (2009) 135002.

[9] A. CHABCHOUB, O. KIMMOUN, H. BRANGER, N. HOFFMANN, D. PROMENT, M. ONORATO, AND N. AKHMEDIEV, *Experimental observation of dark solitons on the surface of water.* Phys. Rev. Lett., **110** (2013) 124101; A. CHABCHOUB, O. KIMMOUN, H. BRANGER, C. KHARIF, N. HOFFMANN, M. ONORATO, AND N. AKHMEDIEV, *Gray solitons on the surface of water.* Phys. Rev. E, **89** (2014) 011002(R).

[10] T. TSUZUKI, *Nonlinear waves in the Pitaevskii-Gross equation.* J. Low Temp. Phys., **4** (1971) 441–457.

[11] L.P. PITAEVSKII, *Vortex lines in an imperfect Bose gas.* Zh. Eksp. Teor. Fiz., **40** (1961) 646–649 [Sov. Phys. JETP, **13** (1961) 451–454]; E.P. GROSS, *Structure of a quantized vortex in boson systems.* Nuovo Cimento, **20** (1961) 454–457.

[12] V.E. ZAKHAROV AND A.B. SHABAT, *Exact theory of two-dimensional self-focusing and one-dimensional self-modulation of waves in nonlinear media.* Zh. Eksp. Teor. Fiz., **61** (1971) 118–134 [Sov. Phys. JETP, **34** (1971) 62–69].

[13] V.E. ZAKHAROV AND A.B. SHABAT, *Interaction between solitons in a stable medium.* Zh. Eksp. Teor. Fiz., **64** (1973) 1627–1639 [Sov. Phys. JETP, **37** (1973) 823–828].

[14] M.J. ABLOWITZ AND H. SEGUR, *Solitons and the Inverse Scattering Transform.* SIAM, Philadelphia, 1981.

[15] S.A. GREDESKUL AND YU.S. KIVSHAR, *Generation of dark solitons in optical fibers.* Phys. Rev. Lett., **62** (1989) 977–977.

[16] S.A. GREDESKUL, YU.S. KIVSHAR, AND M.V. YANOVSKAYA, *Dark-pulse solitons in nonlinear-optical fibers.* Phys. Rev. A, **41** (1990) 3994–4008.

[17] W. ZHAO AND E. BOURKOFF, *Propagation properties of dark solitons.* Opt. Lett., **14** (1989) 703–705.

[18] W. ZHAO AND E. BOURKOFF, *Interactions between dark solitons.* Opt. Lett., **14** (1989) 1371–1373.

[19] V.V. KONOTOP AND V.E. VEKSLERCHIK, *Randomly modulated dark soliton.* J. Phys. A: Math. Gen., **24** (1991) 767–785.

[20] A.N. SLAVIN, YU.S. KIVSHAR, E.A. OSTROVSKAYA, AND H. BENNER, *Generation of spin-wave envelope dark solitons.* Phys. Rev. Lett., **82** (1999) 2583–2586.

[21] K.J. BLOW AND N.J. DORAN, *Multiple dark soliton solutions of the nonlinear Schrödinger equation.* Phys. Lett. A, **107** (1985) 55–58.

[22] N. AKHMEDIEV AND A. ANKIEWICH, *First-order exact solutions of the nonlinear Schrödinger equation in the normal-dispersion regime.* Phys. Rev. A, **47** (1993) 3213–3221.

[23] L. GAGNON, *Solitons on a continuous-wave background and collision between two dark pulses: Some analytical results.* J. Opt. Soc. Am. B, 10 (1993) 469–474.

[24] R.N. THURSTON AND A.M. WEINER, *Collisions of dark solitons in optical fibers.* J. Opt. Soc. Am. B, **8** (1991) 471–477.

[25] L. LERNER, D.J. MITCHELL, AND A.W. SNYDER, *Interacting gray solitons.* Opt. Lett., **19** (1994) 1302–1304.

[26] YU.S. KIVSHAR AND W. KRÓLIKOWSKI, *Lagrangian approach for dark solitons.* Opt. Commun., **114** (1995) 353–362.

[27] D. FOURSA AND P. EMPLIT, *Investigation of black-gray soliton interaction.* Phys. Rev. Lett., **77** (1996) 4011–4014.

[28] I.M. UZUNOV AND V.S. GERDJIKOV, *Self-frequency shift of dark solitons in optical fibers.* Phys. Rev. A, **47** (1993) 1582–1585.

[29] YU.S. KIVSHAR AND X. YANG, *Perturbation-induced dynamics of dark solitons.* Phys. Rev. E, **49** (1994) 1657–1670.

[30] V.V. KONOTOP AND V.E. VEKSLERCHIK, *Direct perturbation theory for dark solitons.* Phys. Rev. E, **49** (1994) 2397–2407.

[31] X.J. CHEN, Z.D. CHEN, AND N.N. HUANG, *A direct perturbation theory for dark solitons based on a complete set of the squared Jost solutions.* J. Phys. A: Math. Gen., **31** (1998) 6929–6947.

[32] N.N. HUANG, S. CHI, AND X.J. CHEN, *Foundation of direct perturbation method for dark solitons.* J. Phys. A: Math. Gen., **32** (1999) 3939–3945.

[33] V.M. LASHKIN, *Perturbation theory for dark solitons: Inverse scattering transform approach and radiative effects.* Phys. Rev. E, **70** (2004) 066620.

[34] A. HASEGAWA AND F. TAPPERT, *Transmission of stationary nonlinear optical pulses in dispersive dielectric fibers. II. Normal dispersion.* Appl. Phys. Lett., **23** (1973) 171–172.

[35] YU.S. KIVSHAR AND B. LUTHER-DAVIES, *Dark optical solitons: Physics and applications.* Phys. Rep., **298** (1998) 81–197.

[36] YU.S. KIVSHAR AND G.P. AGRAWAL, *Optical Solitons: From Fibers to Photonic Crystals.* Academic Press, New York, 2003.

[37] M.H.J. ANDERSON, J.R. ENSHER, M.R. MATTHEWS, C.E. WIEMAN, AND E.A. CORNELL, *Observation of Bose-Einstein condensation in a dilute atomic vapor.* Science, **269** (1995) 198–201.

[38] K.B. DAVIS, M.-O. MEWES, M.R. ANDREWS, N.J. VAN DRUTEN, D.S. DURFEE, D.M. KURN, AND W. KETTERLE, *Bose-Einstein condensation in a gas of sodium atoms.* Phys. Rev. Lett., **75** (1995) 3969–3973.

[39] C.C. BRADLEY, C.A. SACKETT, J.J. TOLLETT, AND R.G. HULET, *Evidence of Bose-Einstein condensation in an atomic gas with attractive interactions.* Phys. Rev. Lett., **75** (1995) 1687–1690.

[40] E.A. CORNELL AND C.E. WIEMAN, *Nobel Lecture: Bose-Einstein condensation in a dilute gas, the first 70 years and some recent experiments.* Rev. Mod. Phys., **74** (2002) 875–893.

[41] W. KETTERLE, *Nobel Lecture: When atoms behave as waves: Bose-Einstein condensation and the atom laser.* Rev. Mod. Phys., **74** (2002) 1131–1151.

[42] C.J. PETHICK AND H. SMITH, *Bose-Einstein Condensation in Dilute Gases.* Cambridge University Press, Cambridge, 2001.

[43] L.P. PITAEVSKII AND S. STRINGARI, *Bose-Einstein Condensation.* Oxford University Press, Oxford, 2003.

[44] P.G. KEVREKIDIS, D.J. FRANTZESKAKIS, AND R. CARRETERO-GONZÁLEZ (EDS.), *Emergent Nonlinear Phenomena in Bose-Einstein Condensates: Theory and Experiment.* Springer, Heidelberg, 2008.

[45] R. CARRETERO-GONZÁLEZ, D.J. FRANTZESKAKIS, AND P.G. KEVREKIDIS, *Nonlinear waves in Bose-Einstein condensates: Physical relevance and mathematical techniques.* Nonlinearity, **21** (2008) R139–R202.

[46] S. BURGER, K. BONGS, S. DETTMER, W. ERTMER, K. SENGSTOCK, A. SANPERA, G.V. SHLYAPNIKOV, AND M. LEWENSTEIN, *Dark solitons in Bose-Einstein condensates.* Phys. Rev. Lett., **83** (1999) 5198–5201.

[47] J. DENSCHLAG, J.E. SIMSARIAN, D.L. FEDER, C.W. CLARK, L.A. COLLINS, J. CUBIZOLLES, L. DENG, E.W. HAGLEY, K. HELMERSON, W.P. REINHARDT, S.L. ROLSTON, B.I. SCHNEIDER, AND W.D. PHILLIPS, *Generating solitons by phase engineering of a Bose-Einstein condensate.* Science, **287** (2000) 97–101.

[48] Z. DUTTON, M. BUDDE, C. SLOWE, AND L.V. HAU, *Observation of quantum shock waves created with ultra-compressed slow light pulses in a Bose-Einstein condensate.* Science, **293** (2001) 663–668.

[49] B.P. ANDERSON, P.C. HALJAN, C.A. REGAL, D.L. FEDER, L.A. COLLINS, C.W. CLARK, AND E.A. CORNELL, *Watching dark solitons decay into vortex rings in a Bose-Einstein condensate.* Phys. Rev. Lett., **86** (2001) 2926–2929.

[50] K. BONGS, S. BURGER, S. DETTMER, D. HELLWEG, J. ARLT, W. ERTMER, AND K. SENGSTOCK, *Coherent manipulation and guiding of Bose-Einstein condensates by optical dipole potentials.* C. R. Acad. Sci. Paris, **2** (2001) 671–680.

[51] YU.S. KIVSHAR, T.J. ALEXANDER, AND S.K. TURITSYN, *Nonlinear modes of a macroscopic quantum oscillator.* Phys. Lett. A, **278** (2001) 225–230.

[52] P.G. KEVREKIDIS, V.V. KONOTOP, A. RODRIGUES, AND D.J. FRANTZESKAKIS, *Dynamic generation of matter solitons from linear states via time-dependent scattering lengths.* J. Phys. B: At. Mol. Opt. Phys., **38** (2005) 1173–1188.

[53] L.D. LANDAU AND E.M. LIFSHITZ, *Quantum Mechanics.* Pergamon Press, Oxford, 1987.

[54] L.M. PISMEN, *Vortices in Nonlinear Fields.* Clarendon Press, Oxford, 1999.

[55] T.W.B. KIBBLE, *Topology of cosmic domains and strings.* J. Phys. A: Math. Gen., **9** (1976) 1387–1398.

[56] W.H. ZUREK, *Cosmological experiments in superfluid helium?* Nature, **317** (1985) 505–508.

[57] C.N. WEILER, T.W. NEELY, D.R. SCHERER, A.S. BRADLEY, M.J. DAVIS, AND B.P. ANDERSON, *Spontaneous vortices in the formation of Bose-Einstein condensates.* Nature, **455** (2008) 948–951.

[58] W.H. ZUREK, *Causality in condensates: Gray solitons as relics of BEC formation.* Phys. Rev. Lett., **102** (2009) 105702.

[59] B. DAMSKI AND W.H. ZUREK, *Soliton creation during a Bose-Einstein condensation.* Phys. Rev. Lett., **104** (2010) 160404.

[60] J. ANGLIN, *Atomic dark solitons: Quantum canaries learn to fly.* Nature Phys., **4** (2008) 437–438.

[61] A. NEGRETTI AND C. HENKEL, *Enhanced phase sensitivity and soliton formation in an integrated BEC interferometer.* J. Phys. B: At. Mol. Opt. Phys., **37** (2004) L385–L390.

[62] A. NEGRETTI, C. HENKEL, AND K. MØLMER, *Quantum fluctuations in the image of a Bose gas.* Phys. Rev. A, **78** (2008) 023630.

[63] G.-B. JO, J.-H. CHOI, C.A. CHRISTENSEN, T.A. PASQUINI, Y.-R. LEE, W. KETTERLE, AND D.E. PRITCHARD, *Phase-sensitive recombination of two Bose-Einstein condensates on an atom chip.* Phys. Rev. Lett., **98** (2007) 180401.

[64] R.G. SCOTT, T.E. JUDD, AND T.M. FROMHOLD, *Exploiting soliton decay and phase fluctuations in atom chip interferometry of Bose-Einstein condensates.* Phys. Rev. Lett., **100** (2008) 100402.

[65] R.G. SCOTT, A.M. MARTIN, T.M. FROMHOLD, S. BUJKIEWICZ, F.W. SHEARD, AND M. LEADBEATER, *Creation of solitons and vortices by Bragg reflection of Bose-Einstein condensates in an optical lattice.* Phys. Rev. Lett., **90** (2003) 110404.

[66] N.S. GINSBERG, J. BRAND, AND L.V. HAU, *Observation of hybrid soliton vortex-ring structures in Bose-Einstein condensates.* Phys. Rev. Lett., **94** (2005) 040403.

[67] P. ENGELS AND C. ATHERTON, *Stationary and nonstationary fluid flow of a Bose-Einstein condensate through a penetrable barrier.* Phys. Rev. Lett., **99** (2007) 160405.

[68] C. BECKER, S. STELLMER, P. SOLTAN-PANAHI, S. DÖRSCHER, M. BAUMERT, E.-M. RICHTER, J. KRONJÄGER, K. BONGS, AND K. SENGSTOCK, *Oscillations and interactions of dark and dark-bright solitons in Bose-Einstein condensates.* Nature Phys., **4** (2008) 496–501.

[69] S. STELLMER, C. BECKER, P. SOLTAN-PANAHI, E.-M. RICHTER, S. DÖRSCHER, M. BAUMERT, J. KRONJÄGER, K. BONGS, AND K. SENGSTOCK, *Collisions of dark solitons in elongated Bose-Einstein condensates.* Phys. Rev. Lett., **101** (2008) 120406.

[70] A. WELLER, J.P. RONZHEIMER, C. GROSS, J. ESTEVE, M.K. OBERTHALER, D.J. FRANTZESKAKIS, G. THEOCHARIS, AND P.G. KEVREKIDIS, *Experimental observation of oscillating and interacting matter wave dark solitons.* Phys. Rev. Lett., **101** (2008) 130401.

[71] I. SHOMRONI, E. LAHOUD, S. LEVY, AND J. STEINHAUER, *Evidence for an oscillating soliton/vortex ring by density engineering of a Bose-Einstein condensate.* Nature Phys., **5** (2009) 193–197.

[72] G. THEOCHARIS, A. WELLER, J.P. RONZHEIMER, C. GROSS, M.K. OBERTHALER, P.G. KEVREKIDIS, AND D.J. FRANTZESKAKIS, *Multiple atomic dark solitons in cigar-shaped Bose-Einstein condensates.* Phys. Rev. A, **81** (2010) 063604.

[73] C. BECKER, K. SENGSTOCK, P. SCHMELCHER, P.G. KEVREKIDIS, AND R. CARRETERO-GONZÁLEZ, *Inelastic collisions of solitary waves in anisotropic Bose-Einstein condensates: Sling-shot events and expanding collision bubbles.* New J. Phys., **15** (2013) 113028.

[74] J.J. CHANG, P. ENGELS, AND M.A. HOEFER, *Formation of dispersive shock waves by merging and splitting Bose-Einstein condensates.* Phys. Rev. Lett., **101** (2008) 170404.

[75] D.J. FRANTZESKAKIS, *Dark solitons in atomic Bose-Einstein condensates: from theory to experiments.* J. Phys. A: Math. Theor., **43** (2010) 213001.

[76] A.L. FETTER AND A.A. SVIDZINKSY, *Vortices in a trapped dilute Bose-Einstein condensate.* J. Phys.: Cond. Mat., **13** (2001) R135–R194.

[77] P.G. KEVREKIDIS, R. CARRETERO-GONZÁLEZ, D.J. FRANTZESKAKIS, AND I.G. KEVREKIDIS, *Vortices in Bose-Einstein condensates: Some recent developments.* Mod. Phys. Lett. B, **18** (2004) 1481–1505.

[78] S. KOMINEAS, *Vortex rings and solitary waves in trapped Bose-Einstein condensates.* Eur. Phys. J. Spec. Topics, **147** (2007) 133–152.

[79] C.F. BARENGHI AND R.J. DONNELLY, *Vortex rings in classical and quantum systems.* Fluid Dyn. Res., **41** (2009) 051401.

[80] N.P. PROUKAKIS, N.G. PARKER, D.J. FRANTZESKAKIS, AND C.S. ADAMS, *Analogies between dark solitons in atomic Bose-Einstein condensates and optical systems.* J. Opt. B: Quantum Semiclass. Opt., **6** (2004) S380–S391.

[81] I. BLOCH, J. DALIBARD, AND W. ZWERGER, *Many-body physics with ultracold gases.* Rev. Mod. Phys., **80** (2008) 885–964.

[82] N.N. BOGOLIUBOV, *On the theory of superfluidity.* J. Phys. (Moscow), **11** (1947) 23–32.

[83] E.H. LIEB, R. SEIRINGER, AND J. YNGVASON, *Bosons in a trap: A rigorous derivation of the Gross-Pitaevskii energy functional.* Phys. Rev. A, **61** (2000) 043602.

[84] E.H. LIEB, R. SEIRINGER, J.P. SOLOVEJ, AND J. YNGVASON, *The Mathematics of the Bose Gas and its Condensation.* Oberwolfach Seminar Series, Birkhäuser, Basel, 2005.

[85] A. ELGART, L. ERDŐS, B. SCHLEIN, AND H.-T. YAU, *Gross-Pitaevskii equation as the mean field limit of weakly coupled bosons.* Arch. Ration. Mech. Anal., **179** (2006) 265–283.

[86] L. ERDŐS, B. SCHLEIN, AND H.-T. YAU, *Rigorous derivation of the Gross-Pitaevskii equation.* Phys. Rev. Lett., **98** (2007) 040404.

[87] L. ERDŐS, B. SCHLEIN, AND H.-T. YAU, *Derivation of the cubic non-linear Schrödinger equation from quantum dynamics of many-body systems.* Invent. Math., **167** (2007) 515–614.

[88] L. ERDŐS, B. SCHLEIN, AND H.-T. YAU, *Rigorous derivation of the Gross-Pitaevskii equation with a large interaction potential.* J. Amer. Math. Soc., **22** (2009) 1099–1156.

[89] L. ERDŐS, B. SCHLEIN, AND H.-T. YAU, *Derivation of the Gross-Pitaevskii equation for the dynamics of Bose-Einstein condensate.* Ann. of Math., **172** (2010) 291–370.

[90] E.H. LIEB AND R. SEIRINGER, *Derivation of the Gross-Pitaevskii equation for rotating Bose gases.* Commun. Math. Phys., **264** (2006) 505–537.

[91] E.H. LIEB, R. SEIRINGER, AND J. YNGVASON, *Superfluidity in dilute trapped Bose gases.* Phys. Rev. B, **66** (2002) 134529.

[92] F.J. DYSON, *Ground-state energy of a hard-sphere gas.* Phys. Rev., **106** (1957) 20–26.

[93] E.H. LIEB AND J. YNGVASON, *Ground state energy of the low density Bose gas.* Phys. Rev. Lett., **80** (1998) 2504–2507.

[94] P.G. KEVREKIDIS AND D.J. FRANTZESKAKIS, *Pattern-forming dynamical instabilities of Bose-Einstein condensates.* Mod. Phys. Lett. B, **18** (2004) 173–202.

[95] K.E. STRECKER, G.B. PARTRIDGE, A.G. TRUSCOTT, AND R.G. HULET, *Formation and propagation of matter-wave soliton trains.* Nature, **417** (2002) 150–153.

[96] L. KHAYKOVICH, F. SCHRECK, G. FERRARI, T. BOURDEL, J. CUBIZOLLES, L.D. CARR, Y. CASTIN, AND C. SALOMON, *Formation of a matter-wave bright soliton.* Science, **296** (2002) 1290–1293.

[97] S.L. CORNISH, S.T. THOMPSON, AND C.E. WIEMAN, *Formation of bright matter-wave solitons during the collapse of attractive Bose-Einstein condensates.* Phys. Rev. Lett., **96** (2006) 170401.

[98] F. KH. ABDULLAEV, A. GAMMAL, A.M. KAMCHATNOV, AND L. TOMIO, *Dynamics of bright matter wave solitons in a Bose-Einstein condensate.* Int. J. Mod. Phys. B, **19** (2005) 3415–3473.

[99] C. MENOTTI AND S. STRINGARI, *Collective oscillations of a one-dimensional trapped Bose-Einstein gas.* Phys. Rev. A, **66** (2002) 043610.

[100] V.M. PÉREZ-GARCÍA, H. MICHINEL, AND H. HERRERO, *Bose-Einstein solitons in highly asymmetric traps.* Phys. Rev. A, **57** (1998) 3837–3842.

[101] A.D. JACKSON, G.M. KAVOULAKIS, AND C.J. PETHICK, *Solitary waves in clouds of Bose-Einstein condensed atoms.* Phys. Rev. A, **58** (1998) 2417–2422.

[102] A.E. MURYSHEV, G.V. SHLYAPNIKOV, W. ERTMER, K. SENGSTOCK, AND M. LEWENSTEIN, *Dynamics of dark solitons in elongated Bose-Einstein condensates.* Phys. Rev. Lett., **89** (2002) 110401.

[103] L. SALASNICH, A. PAROLA, AND L. REATTO, *Effective wave equations for the dynamics of cigar-shaped and disk-shaped Bose condensates.* Phys. Rev. A, **65** (2002) 043614.

[104] F. GERBIER, *Quasi-1D Bose-Einstein condensates in the dimensional crossover regime.* Europhys. Lett., **66** (2004) 771–777.

[105] A. Muñoz Mateo and V. Delgado, *Ground-state properties of trapped Bose-Einstein condensates: Extension of the Thomas-Fermi approximation.* Phys. Rev. A, **75** (2007) 063610; A. Muñoz Mateo and V. Delgado, *Effective mean-field equations for cigar-shaped and disk-shaped Bose-Einstein condensates.* Phys. Rev. A, **77** (2008) 013617; A. Muñoz Mateo and V. Delgado, *Effective one-dimensional dynamics of elongated Bose-Einstein condensates.* Ann. Phys., **324** (2009) 709–724.

[106] G. Theocharis, P.G. Kevrekidis, M.K. Oberthaler, and D.J. Frantzeskakis, *Dark matter-wave solitons in the dimensionality crossover.* Phys. Rev. A, **76** (2007) 045601.

[107] P. Engels, C. Atherton and M.A. Hoefer, *Observation of Faraday waves in a Bose-Einstein condensate.* Phys. Rev. Lett., **98** (2007) 095301.

[108] A.I. Nicolin, R. Carretero-González, and P.G. Kevrekidis, *Faraday waves in Bose-Einstein condensates.* Phys. Rev. A, **76** (2007) 063609.

[109] A.J. Moerdijk, H.M.J.M. Boesten, and B.J. Verhaar, *Decay of trapped ultracold alkali atoms by recombination.* Phys. Rev. A, **53** (1996) 916–920.

[110] T. Köhler, *Three-body problem in a dilute Bose-Einstein condensate.* Phys. Rev. Lett., **89** (2002) 210404.

[111] L. Tonks, *The complete equation of state of one, two and three-dimensional gases of hard elastic spheres.* Phys. Rev., **50** (1936) 955–963; M. Girardeau, *Relationship between systems of impenetrable bosons and fermions in one dimension.* J. Math. Phys., **1** (1960) 516–523.

[112] B. Paredes, A. Widera, V. Murg, O. Mandel, S. Folling, I. Cirac, G.V. Shlyapnikov, T.W. Häansch, and I. Bloch, *Tonks-Girardeau gas of ultracold atoms in an optical lattice.* Nature, **429** (2004) 277–281.

[113] T. Kinoshita, T. Wenger, and D.S. Weiss, *Observation of a one-dimensional Tonks-Girardeau gas.* Science, **305** (2004) 1125–1128.

[114] E.B. Kolomeisky, and J.P. Straley, *Renormalization-group analysis of the ground-state properties of dilute Bose systems in d spatial dimensions.* Phys. Rev. B, **46** (1992) 11749–11756; E.B. Kolomeisky, T.J. Newman, J.P. Straley, and X. Qi, *Low-dimensional Bose liquids: Beyond the Gross-Pitaevskii approximation.* Phys. Rev. Lett., **85** (2000) 1146–1149.

[115] M.D. Girardeau and E.M. Wright, *Dark solitons in a one-dimensional condensate of hard core bosons.* Phys. Rev. Lett., **84** (2000) 5691–5694.

[116] V. Dunjko, V. Lorent, and M. Olshanii, *Bosons in cigar-shaped traps: Thomas-Fermi regime, Tonks-Girardeau regime, and in between.* Phys. Rev. Lett., **86** (2001) 5413–5416; P. Öhberg and L. Santos, *Dynamical transition from a quasi-one-dimensional Bose-Einstein condensate to a Tonks-Girardeau gas.* Phys. Rev. Lett., **89** (2002) 240402.

[117] E.H. Lieb, R. Seiringer, and J. Yngvason, *One-dimensional Bosons in three-dimensional traps.* Phys. Rev. Lett., **91** (2003) 150401.

[118] B.A. Malomed, *Variational methods in nonlinear fiber optics and related fields.* Prog. Opt., **43** (2002) 71–193.

[119] F. Kh. Abdullaev, N.K. Nurmanov, and E.N. Tsoy, *Variational approach to the problem of dark-soliton generation.* Phys. Rev. E, **56** (1997) 3638–3644.

[120] A.M. Kamchatnov, R.A. Kraenkel, and B.A. Umarov, *Asymptotic soliton train solutions of the defocusing nonlinear Schrödinger equation.* Phys. Rev. E, **66** (2002) 036609.

[121] V.A. BRAZHNYI AND A.M. KAMCHATNOV, *Creation and evolution of trains of dark solitons in a trapped one-dimensional Bose-Einstein condensate.* Phys. Rev. A, **68** (2003) 043614.

[122] YU.S. KIVSHAR, *On the soliton generation in optical fibers.* J. Phys. A: Math. Gen., **22** (1989) 337–340.

[123] A.M. WEINER, J.P. HERITAGE, R.J. HAWKINS, R.N. THURSTON, E.M. KIRSCHNER, D.E. LEARID, AND W.J. TOMLINSON, *Experimental observation of the fundamental dark soliton in optical fibers.* Phys. Rev. Lett., **61** (1988) 2445–2448.

[124] I.V. BARASHENKOV AND E.Y. PANOVA, *Stability and evolution of the quiescent and travelling solitonic bubbles.* Physica D, **69** (1993) 114–134.

[125] I.V. BARASHENKOV AND A.O. HARIN, *Nonrelativistic Cherns-Simons theory for the repulsive Bose gas.* Phys. Rev. Lett., **72** (1994) 1575–1579.

[126] P.O. FEDICHEV, A.E. MURYSHEV, AND G.V. SHLYAPNIKOV, *Dissipative dynamics of a kink state in a Bose-condensed gas.* Phys. Rev. A, **60** (1999) 3220–3224.

[127] A. JEFFREY AND T. KAWAHARA, *Asymptotic Methods in Nonlinear Wave Theory.* Pitman, Boston, 1982.

[128] E. INFELD AND G. ROWLANDS, *Nonlinear Waves, Solitons, and Chaos.* Cambridge University Press, Cambridge, 1990.

[129] V.E. ZAKHAROV AND E.A. KUZNETSOV, *Multi-scale expansions in the theory of systems integrable by the inverse scattering transform.* Physica D, **18** (1986) 445–463.

[130] F.G. BASS, V.V. KONOTOP, AND S.A. PUZENKO, *Dark solitary waves in a generalized version of the nonlinear Schrödinger equation.* Phys. Rev. A, **46** (1992) 4185–4191.

[131] YU.S. KIVSHAR AND V.V. AFANASJEV, *Dark optical solitons with reverse-sign amplitude.* Phys. Rev. A, **44** (1991) R1446–R1449.

[132] D.J. FRANTZESKAKIS, *Small-amplitude solitary structures for an extended nonlinear Schrödinger equation.* J. Phys. A: Math. Gen., **29** (1996) 3631–3639.

[133] H. LEBLOND, *The reductive perturbation method and some of its applications.* J. Phys. B: At. Mol. Opt. Phys., **41** (2008) 043001.

[134] G.X. HUANG, V.A. MAKAROV, AND M.G. VELARDE, *Two-dimensional solitons in Bose-Einstein condensates with a disk-shaped trap.* Phys. Rev. A, **67** (2003) 023604.

[135] G.X. HUANG, L. DENG, AND C. HANG, *Davey-Stewartson description of two-dimensional nonlinear excitations in Bose-Einstein condensates.* Phys. Rev. E, **72** (2005) 036621.

[136] V.A. BRAZHNYI AND V.V. KONOTOP, *Stable and unstable vector dark solitons of coupled nonlinear Schrödinger equations: Application to two-component Bose-Einstein condensates.* Phys. Rev. E, **72** (2005) 026616.

[137] H.E. NISTAZAKIS, D.J. FRANTZESKAKIS, P.G. KEVREKIDIS, B.A. MALOMED, AND R. CARRETERO-GONZÁLEZ, *Bright-dark soliton complexes in spinor Bose-Einstein condensates.* Phys. Rev. A, **77** (2008) 033612.

[138] M. AGUERO, D.J. FRANTZESKAKIS, AND P.G. KEVREKIDIS, *Asymptotic reductions of two coupled (2+1)-dimensional nonlinear Schrödinger equations: Application to Bose-Einstein condensates.* J. Phys. A: Math. Gen., **39** (2006) 7705–7718.

[139] L. Dobrek, M. Gajda, M. Lewenstein, K. Sengstock, G. Birkl, and W. Ertmer, *Optical generation of vortices in trapped Bose-Einstein condensates.* Phys. Rev. A, **60** (1999) R3381–R3384.

[140] B. Wu, J. Liu, and Q. Niu, *Controlled generation of dark solitons with phase imprinting.* Phys. Rev. Lett., **88** (2002) 034101.

[141] A.B. Shvartsburg, L. Stenflo, and P.K. Shukla, *Phase modulated dark solitons.* Phys. Scripta, **65** (2002) 164–166.

[142] R. Grimm, M. Weidemüller, and Y.B. Ovchinnikov, *Optical dipole traps for neutral atoms.* Adv. At. Mol. Opt. Phys., **42** (2000) 95–170.

[143] S. Burger, L.D. Carr, P. Öhberg, K. Sengstock, and A. Sanpera, *Generation and interaction of solitons in Bose-Einstein condensates.* Phys. Rev. A, **65** (2002) 043611.

[144] L.D. Carr, J. Brand, S. Burger, and A. Sanpera, *Dark-soliton creation in Bose-Einstein condensates.* Phys. Rev. A, **63** (2001) 051601(R).

[145] Th. Busch and J.R. Anglin, *Dark-bright solitons in inhomogeneous Bose-Einstein condensates.* Phys. Rev. Lett., **87** (2001) 010401.

[146] M.R. Matthews, B.P. Anderson, P.C. Haljan, D.S. Hall, C.E. Wieman, and E.A. Cornell, *Vortices in a Bose-Einstein condensate.* Phys. Rev. Lett., **83** (1999) 2498–2501.

[147] M.R. Matthews, B.P. Anderson, P.C. Haljan, D.S. Hall, M.J. Holland, J.E. Williams, C.E. Wieman, and E.A. Cornell, *Watching a superfluid untwist itself: Recurrence of Rabi oscillations in a Bose-Einstein condensate.* Phys. Rev. Lett., **83** (1999) 3358–3361.

[148] M.R. Andrews, C.G. Townsend, H.-J. Miesner, D.S. Durfee, D.M. Kurn, and W. Ketterle, *Observation of interference between two Bose condensates.* Science, **275** (1997) 637–641.

[149] Y. Castin and J. Dalibard, *Relative phase of two Bose-Einstein condensates.* Phys. Rev. A, **55** (1997) 4330–4337.

[150] A. Röhrl, M. Naraschewski, A. Schenzle, and H. Wallis, *Transition from phase locking to the interference of independent Bose condensates: Theory versus experiment.* Phys. Rev. Lett., **78** (1997) 4143–4146.

[151] W.P. Reinhardt and C.W. Clark, *Soliton dynamics in the collisions of Bose-Einstein condensates: An analogue of the Josephson effect.* J. Phys. B: At. Mol. Opt. Phys., **30** (1997) L785–L789.

[152] T.F. Scott, R.J. Ballagh, and K. Burnett, *Formation of fundamental structures in Bose-Einstein condensates.* J. Phys. B: At. Mol. Opt. Phys., **31** (1998) L329–L335.

[153] J. Ruostekoski, B. Kneer, W.P. Schleich, and G. Rempe, *Interference of a Bose-Einstein condensate in a hard-wall trap: From the nonlinear Talbot effect to the formation of vorticity.* Phys. Rev. A, **63** (2001) 043613.

[154] S.-J. Yang, Q.-S. Wu, S.-N. Zhang, S. Feng, W. Guo, Y.-C. Wen, and Y. Yu, *Generating ring dark solitons in an evolving Bose-Einstein condensate.* Phys. Rev. A, **76** (2007) 063606.

[155] S.-J. Yang, Q.-S. Wu, S. Feng, Y.-C. Wen, and Y. Yu, *Solitons and vortices in an evolving Bose-Einstein condensate.* Phys. Rev. A, **77** (2008) 035602.

[156] R. Carretero-González, N. Whitaker, P.G. Kevrekidis, and D.J. Frantzeskakis, *Vortex structures formed by the interference of sliced condensates.* Phys. Rev. A, **77** (2008) 023605.

[157] C. LEE, E.A. OSTROVSKAYA, AND YU.S. KIVSHAR, *Nonlinearity-assisted quantum tunnelling in a matter-wave interferometer.* J. Phys. B: At. Mol. Opt. Phys., 40 (2007) 4235–4244.

[158] D.R. SCHERER, C.N. WEILER, T.W. NEELY, AND B.P. ANDERSON, *Vortex formation by merging of multiple trapped Bose-Einstein condensates.* Phys. Rev. Lett., **98** (2007) 110402.

[159] R. CARRETERO-GONZÁLEZ, B.P. ANDERSON, P.G. KEVREKIDIS, D.J. FRANTZESKAKIS, AND C.N. WEILER, *Dynamics of vortex formation in merging Bose-Einstein condensate fragments.* Phys. Rev. A, 77 (2008) 033625.

[160] G. RUBEN, D.M. PAGANIN, AND M.J. MORGAN, *Vortex-lattice formation and melting in a nonrotating Bose-Einstein condensate.* Phys. Rev. A, **78** (2008) 013631.

[161] G. HUANG, M.G. VELARDE, AND V.A. MAKAROV, *Dark solitons and their head-on collisions in Bose-Einstein condensates.* Phys. Rev. A, **64** (2001) 013617.

[162] G.L. ALFIMOV AND D.A. ZEZYULIN, *Nonlinear modes for the Gross-Pitaevskii equation—a demonstrative computation approach.* Nonlinearity, **20** (2007) 2075–2092.

[163] YU.S. KIVSHAR AND B.A. MALOMED, *Dynamics of solitons in nearly integrable systems.* Rev. Mod. Phys., **61** (1989) 763–915.

[164] N.G. PARKER, N.P. PROUKAKIS, M. LEADBEATER, AND C.S. ADAMS, *Soliton-sound interactions in quasi-one-dimensional Bose-Einstein condensates.* Phys. Rev. Lett., **90** (2003) 220401.

[165] N.G. PARKER, N.P. PROUKAKIS, M. LEADBEATER, AND C.S. ADAMS, *Deformation of dark solitons in inhomogeneous Bose-Einstein condensates.* J. Phys. B: At. Mol. Opt. Phys., **36** (2003) 2891–2910.

[166] S.A. MORGAN, R.J. BALLAGH, AND K. BURNETT, *Solitary-wave solutions to nonlinear Schrödinger equations.* Phys. Rev. A, **55** (1997) 4338–4345.

[167] T. HONG, Y.Z. WANG, AND Y.S. HUO, *Temporal dark solitons in nonuniform Bose-Einstein condensates.* Phys. Rev. A, **58** (1998) 3128–3133.

[168] X.-J. CHEN, J.-Q. ZHANG, AND H.-C. WONG, *Dark solitons in Bose-Einstein condensates under a slowly varying trap.* Phys. Lett. A, **268** (2000) 306–308.

[169] TH. BUSCH AND J.R. ANGLIN, *Motion of dark solitons in trapped Bose-Einstein condensates.* Phys. Rev. Lett., **84** (2000) 2298–2301.

[170] A.E. MURYSHEV, H.B. VAN LINDEN VAN DEN HEUVELL, AND G.V. SHLYAPNIKOV, *Stability of standing matter waves in a trap.* Phys. Rev. A, **60** (1999) R2665–R2668.

[171] D.J. FRANTZESKAKIS, G. THEOCHARIS, F.K. DIAKONOS, P. SCHMELCHER, AND YU.S. KIVSHAR, *Interaction of dark solitons with localized impurities in Bose-Einstein condensates.* Phys. Rev. A, **66** (2002) 053608.

[172] G. HUANG, J. SZEFTEL, AND S. ZHU, *Dynamics of dark solitons in quasi-one-dimensional Bose-Einstein condensates.* Phys. Rev. A, **65** (2002) 053605.

[173] V.A. BRAZHNYI AND V.V. KONOTOP, *Evolution of a dark soliton in a parabolic potential: Application to Bose-Einstein condensates.* Phys. Rev. A, **68** (2003) 043613.

[174] V.V. KONOTOP AND L. PITAEVSKII, *Landau dynamics of a grey soliton in a trapped condensate.* Phys. Rev. Lett., **93** (2004) 240403.

[175] V.A. BRAZHNYI, V.V. KONOTOP, AND L.P. PITAEVSKII, *Dark solitons as quasiparticles in trapped condensates.* Phys. Rev. A, **73** (2006) 053601.

[176] D.J. FRANTZESKAKIS, N.P. PROUKAKIS, AND P.G. KEVREKIDIS, *Dynamics of shallow dark solitons in a trapped gas of impenetrable bosons.* Phys. Rev. A, 70 (2004) 015601.

[177] D.J. FRANTZESKAKIS, P.G. KEVREKIDIS, AND N.P. PROUKAKIS, *Crossover dark soliton dynamics in ultracold one-dimensional Bose gases.* Phys. Lett. A, **364** (2007) 129–134.

[178] M.D. GIRARDEAU, *Dynamics of Lieb-Liniger gases.* Phys. Rev. Lett., **91** (2003) 040401.

[179] M. ÖGREN, G.M. KAVOULAKIS, AND A.D. JACKSON, *Solitary waves in elongated clouds of strongly interacting bosons.* Phys. Rev. A, **72** (2005) 021603(R).

[180] B.B. BAIZAKOV, F. KH. ABDULLAEV, B.A. MALOMED, AND M. SALERNO, *Solitons in the Tonks-Girardeau gas with dipolar interactions.* J. Phys. B: At. Mol. Opt. Phys., **42** (2009) 175302.

[181] M.D. GIRARDEAU AND A. MINGUZZI, *Motion of an impurity particle in an ultracold quasi-one-dimensional gas of hard-core bosons.* Phys. Rev. A, **79** (2009) 033610.

[182] G. THEOCHARIS, P. SCHMELCHER, M.K. OBERTHALER, P.G. KEVREKIDIS, AND D.J. FRANTZESKAKIS, *Lagrangian approach to the dynamics of dark matter-wave solitons.* Phys. Rev. A, **72** (2005) 023609.

[183] N. BILAS AND N. PAVLOFF, *Dark soliton past a finite-size obstacle.* Phys. Rev. A, **72** (2005) 033618.

[184] D.E. PELINOVSKY, D.J. FRANTZESKAKIS, AND P.G. KEVREKIDIS, *Oscillations of dark solitons in trapped Bose-Einstein condensates.* Phys. Rev. E, **72** (2005) 016615.

[185] D.E. PELINOVSKY AND P.G. KEVREKIDIS, *Dark solitons in external potentials.* Z. Angew. Math. Phys., **59** (2008) 559–599.

[186] A.M. KAMCHATNOV AND M. SALERNO, *Dark soliton oscillations in Bose-Einstein condensates with multi-body interactions.* J. Phys. B: At. Mol. Opt. Phys., **42** (2009) 185303.

[187] N. ASANO, *Wave propagations in non-uniform media.* Prog. Theor. Phys. Suppl., **55** (1974) 52–79.

[188] K. KO AND H.H. KUEHL, *Korteweg–de Vries soliton in a slowly varying medium.* Phys. Rev. Lett., **40** (1978) 233–236.

[189] V.I. KARPMAN AND E.M. MASLOV, *Comments on "Soliton propagation in slowly varying media."* Phys. Fluids, **25** (1982) 1686–1687.

[190] TH. BUSCH AND G. HUYET, *Low-density, one-dimensional quantum gases in a split trap.* J. Phys. B: At. Mol. Opt. Phys., **36** (2003) 2553–2562.

[191] D.L. FEDER, M.S. PINDZOLA, L.A. COLLINS B.I. SCHNEIDER, AND C.W. CLARK, *Dark-soliton states of Bose-Einstein condensates in anisotropic traps.* Phys. Rev. A, **62** (2000) 053606.

[192] J. BRAND AND W.P. REINHARDT, *Solitonic vortices and the fundamental modes of the "snake instability": Possibility of observation in the gaseous Bose-Einstein condensate.* Phys. Rev. A, **65** (2002) 043612.

[193] J. DZIARMAGA AND K. SACHA, *Depletion of the dark soliton: The anomalous mode of the Bogoliubov theory.* Phys. Rev. A, **66** (2002) 043620.

[194] W. KOHN, *Cyclotron resonance and de Haas–van Alphen oscillations of an interacting electron gas.* Phys. Rev., **123** (1961) 1242–1244; J.F. DOBSON, *Harmonic-potential theorem: Implications for approximate many-body theories.* Phys. Rev. Lett., **73** (1994) 2244–2247.

[195] R.S. MacKay, *Stability of equilibria of Hamiltonian systems.* In R.S. MacKay and J.D. Meiss (eds.), *Hamiltonian Dynamical Systems.* A. Hilger, Bristol, 1987.

[196] T. Kapitula, P.G. Kevrekidis, and B. Sandstede, *Counting eigenvalues via the Krein signature in infinite-dimensional Hamiltonian systems.* Physica D, **195** (2004) 263–282.

[197] P.O. Fedichev, G.V. Shlyapnikov, and J.T.M. Walraven, *Damping of low-energy excitations of a trapped Bose-Einstein condensate at finite temperatures.* Phys. Rev. Lett., **80** (1998) 2269–2272.

[198] B. Wu and Q. Niu, *Superfluidity of Bose-Einstein condensate in an optical lattice: Landau-Zener tunnelling and dynamical instability.* New J. Phys., **5** (2003) 104.

[199] C.K. Law, P.T. Leung, and M.-C. Chu, *Quantum fluctuations of coupled dark solitons in a trapped Bose-Einstein condensate.* J. Phys. B: At. Mol. Opt. Phys., **35** (2002) 3583–3590.

[200] T. Kapitula and P.G. Kevrekidis, *Bose-Einstein condensates in the presence of a magnetic trap and optical lattice.* Chaos, **15** (2005) 037114.

[201] L.D. Carr, J.N. Kutz, and W.P. Reinhardt, *Stability of stationary states in the cubic nonlinear Schrödinger equation: Applications to the Bose-Einstein condensate.* Phys. Rev. E, **63** (2001) 066604.

[202] D.A. Zezyulin, G.L. Alfimov, V.V. Konotop, and V.M. Pérez-Garcia, *Stability of excited states of a Bose-Einstein condensate in an anharmonic trap.* Phys. Rev. A, **78** (2008) 013606.

[203] A. Radouani, *Soliton and phonon production by an oscillating obstacle in a quasi-one-dimensional trapped repulsive Bose-Einstein condensate.* Phys. Rev. A, **70** (2004) 013602.

[204] N. Bilas and N. Pavloff, *Propagation of a dark soliton in a disordered Bose-Einstein condensate.* Phys. Rev. Lett., **95** (2005) 130403.

[205] K. Sacha, C.A. Müller, D. Delande, and J. Zakrzewski, *Anderson localization of solitons.* Phys. Rev. Lett., **103** (2009) 210402.

[206] N.G. Parker, N.P. Proukakis, and C.S. Adams, *Dark soliton decay due to trap anharmonicity in atomic Bose-Einstein condensates.* Phys. Rev. A, **81** (2010) 033606.

[207] N.G. Parker, N.P. Proukakis, C.F. Barenghi, and C.S. Adams, *Dynamical instability of a dark soliton in a quasi-one-dimensional Bose-Einstein condensate perturbed by an optical lattice.* J. Phys. B: At. Mol. Opt. Phys., **37** (2004) S175–S185.

[208] G. Theocharis, D.J. Frantzeskakis, P.G. Kevrekidis, R. Carretero-González, and B.A. Malomed, *Dark soliton dynamics in spatially inhomogeneous media: Application to Bose-Einstein condensates.* Math. Comput. Simul., **69** (2005) 537–552.

[209] N.P. Proukakis, N.G. Parker, C.F. Barenghi, and C.S. Adams, *Parametric driving of dark solitons in atomic Bose-Einstein condensates.* Phys. Rev. Lett., **93** (2004) 130408.

[210] V.A. Brazhnyi and V.V. Konotop, *Theory of nonlinear matter waves in optical lattices.* Mod. Phys. Lett. B, **18** (2004) 627–651.

[211] I. Bloch, *Exploring quantum matter with ultracold atoms in optical lattices.* J. Phys. B: At. Mol. Opt. Phys., **38** (2005) S629–S643; I. Bloch, *Ultracold quantum gases in optical lattices.* Nature Phys., **1** (2005) 23–30.

[212] O. Morsch and M.K. Oberthaler, *Dynamics of Bose-Einstein condensates in optical lattices.* Rev. Mod. Phys., **78** (2006) 179–215.

[213] M. LEWENSTEIN, A. SANPERA, V. AHUFINGER, B. DAMSKI, A. SEN(DE), AND U. SEN, *Ultracold atomic gases in optical lattices: Mimicking condensed matter physics and beyond.* Adv. Phys., **56** (2007) 243–379.

[214] B. EIERMANN, TH. ANKER, M. ALBIEZ, M. TAGLIEBER, P. TREUTLEIN, K.P. MARZLIN, AND M.K. OBERTHALER, *Bright Bose-Einstein gap solitons of atoms with repulsive interaction.* Phys. Rev. Lett., **92** (2004) 230401.

[215] TH. ANKER, M. ALBIEZ, B. EIERMANN, M. TAGLIEBER, AND M.K. OBERTHALER, *Linear and nonlinear dynamics of matter wave packets in periodic potentials.* Opt. Express, **12** (2004) 11–18.

[216] D.W. JORDAN AND P. SMITH, *Nonlinear Ordinary Differential Equations.* Clarendon Press, Oxford, 1977.

[217] D. DIAKONOV, L.M. JENSEN, C.J. PETHICK, AND H. SMITH, *Loop structure of the lowest Bloch band for a Bose-Einstein condensate.* Phys. Rev. A, **66** (2002) 013604.

[218] G.L. ALFIMOV, P.G. KEVREKIDIS, V.V. KONOTOP, AND M. SALERNO, *Wannier functions analysis of the nonlinear Schrödinger equation with a periodic potential.* Phys. Rev. E, **66** (2002) 046608.

[219] P.G. KEVREKIDIS, K.Ø. RASMUSSEN, AND A.R. BISHOP, *The discrete nonlinear Schrödinger equation: A survey of recent results.* Int. J. Mod. Phys. B, **15** (2001) 2833–2900.

[220] P.G. KEVREKIDIS, *The Discrete Nonlinear Schrödinger Equation.* Springer, Heidelberg, 2009.

[221] C.M. DE STERKE AND J.E. SIPE, *Envelope-function approach for the electrodynamics of nonlinear periodic structures.* Phys. Rev. A, **38** (1988) 5149–5165.

[222] H. PU, L.O. BAKSMATY, W. ZHANG, N.P. BIGELOW, AND P. MEYSTRE, *Effective-mass analysis of Bose-Einstein condensates in optical lattices: Stabilization and levitation.* Phys. Rev. A, **67** (2003) 043605.

[223] E.A. OSTROVSKAYA AND YU.S. KIVSHAR, *Matter-wave gap vortices in optical lattices.* Phys. Rev. Lett., **93** (2004) 160405.

[224] P.G. KEVREKIDIS, R. CARRETERO-GONZÁLEZ, G. THEOCHARIS, D.J. FRANTZESKAKIS, AND B.A. MALOMED, *Stability of dark solitons in a Bose-Einstein condensate trapped in an optical lattice.* Phys. Rev. A, **68** (2003) 035602 [see also erratum: Phys. Rev. A, **72** (2005) 069908(E)].

[225] G. THEOCHARIS, D.J. FRANTZESKAKIS, R. CARRETERO-GONZÁLEZ, P.G. KEVREKIDIS, AND B.A. MALOMED, *Controlling the motion of dark solitons by means of periodic potentials: Application to Bose-Einstein condensates in optical lattices.* Phys. Rev. E, **71** (2005) 017602.

[226] F. KH. ABDULLAEV, B.B. BAIZAKOV, S.A. DARMANYAN, V.V. KONOTOP, AND M. SALERNO, *Nonlinear excitations in arrays of Bose-Einstein condensates.* Phys. Rev. A, **64** (2001) 043606.

[227] A.V. YULIN AND D.V. SKRYABIN, *Out-of-gap Bose-Einstein solitons in optical lattices.* Phys. Rev. A, **67** (2003) 023611.

[228] G.L. ALFIMOV, V.V. KONOTOP, AND M. SALERNO, *Matter solitons in Bose-Einstein condensates with optical lattices.* Europhys. Lett., **58** (2002) 7–13.

[229] V.V. KONOTOP AND M. SALERNO, *Modulational instability in Bose-Einstein condensates in optical lattices.* Phys. Rev. A, **65** (2002) 021602(R).

[230] P.J.Y. Louis, E.A. Ostrovskaya, and Yu.S. Kivshar, *Matter-wave dark solitons in optical lattices.* J. Opt. B: Quantum Semiclass. Opt., **6** (2004) S309–S317.

[231] M.A. Porter, P.G. Kevrekidis, R. Carretero-González, and D.J. Frantzeskakis, *Dynamics and manipulation of matter-wave solitons in optical superlattices.* Phys. Lett. A, **352** (2006) 210–215.

[232] S. Peil, J.V. Porto, B. Laburthe Tolra, J.M. Obrecht, B.E. King, M. Subbotin, S.L. Rolston, and W.D. Phillips, *Patterned loading of a Bose-Einstein condensate into an optical lattice.* Phys. Rev. A, **67** (2003) 051603(R).

[233] J.H. Feng and F.K. Kneubühl, *Solitons in a periodic structure with Kerr nonlinearity.* IEEE J. Quantum Electron., **29** (1993) 590–597.

[234] Yu.S. Kivshar, W. Królikowski, and O.A. Chubykalo, *Dark solitons in discrete lattices.* Phys. Rev. E, **50** (1994) 5020–5032.

[235] M. Johansson and Yu.S. Kivshar, *Discreteness-induced oscillatory instabilities of dark solitons.* Phys. Rev. Lett., **82** (1999) 85–88.

[236] A.A. Sukhorukov and Yu.S. Kivshar, *Spatial optical solitons in nonlinear photonic crystals.* Phys. Rev. E, **65** (2002) 036609.

[237] H. Susanto and M. Johansson, *Discrete dark solitons with multiple holes.* Phys. Rev. E, **72** (2005) 016605.

[238] E.P. Fitrakis, P.G. Kevrekidis, H. Susanto, and D.J. Frantzeskakis, *Dark solitons in discrete lattices: Saturable versus cubic nonlinearities.* Phys. Rev. E, **75** (2007) 066608.

[239] V.V. Konotop and S. Takeno, *Stationary dark localized modes: Discrete nonlinear Schrödinger equations.* Phys. Rev. E, **60** (1999) 1001–1008.

[240] D.E. Pelinovsky and P.G. Kevrekidis, *Stability of discrete dark solitons in nonlinear Schrödinger lattices.* J. Phys. A: Math. Theor., **41** (2008) 185206.

[241] O.M. Braun and Yu.S. Kivshar, *Nonlinear dynamics of the Frenkel-Kontorova model.* Phys. Rep., **306** (1998) 2–108.

[242] M. Greiner, O. Mandel, T. Esslinger, T.W. Hansch, and I. Bloch, *Quantum phase transition from a superfluid to a Mott insulator in a gas of ultracold atoms.* Nature, **415** (2002) 39–44.

[243] D. Jaksch, C. Bruder, J.I. Cirac, C.W. Gardiner, and P. Zoller, *Cold bosonic atoms in optical lattices.* Phys. Rev. Lett., **81** (1998) 3108–3111.

[244] R.V. Mishmash and L.D. Carr, *Quantum entangled dark solitons formed by ultracold atoms in optical lattices.* Phys. Rev. Lett., **103** (2009) 140403.

[245] R.V. Mishmash, I. Danshita, C.W. Clark, and L.D. Carr, *Quantum many-body dynamics of dark solitons in optical lattices.* Phys. Rev. A, **80** (2009) 053612.

[246] K.V. Krutitsky, J. Larson, and M. Lewenstein, *Dark solitons near the Mott-insulator-superfluid phase transition.* Phys. Rev. A, **82** (2010) 033618.

[247] L. Di Menza and C. Gallo, *The black solitons of one-dimensional NLS equations.* Nonlinearity, **20** (2007) 461–496.

[248] R. Onofrio, C. Raman, J.M. Vogels, J.R. Abo-Shaeer, A.P. Chikkatur, and W. Ketterle, *Observation of superfluid flow in a Bose-Einstein condensed gas.* Phys. Rev. Lett., **85** (2000) 2228–2231.

[249] A.S. RODRIGUES, P.G. KEVREKIDIS, M.A. PORTER, D.J. FRANTZESKAKIS, P. SCHMELCHER, AND A.R. BISHOP, *Matter-wave solitons for a periodic piecewise-constant nonlinearity.* Phys. Rev. A, **78** (2008) 013611.

[250] G. THEOCHARIS, P. SCHMELCHER, P.G. KEVREKIDIS, AND D.J. FRANTZESKAKIS, *Matter-wave solitons of collisionally inhomogeneous condensates.* Phys. Rev. A, **72** (2005) 033614.

[251] B.A. MALOMED, H.E. NISTAZAKIS, D.J. FRANTZESKAKIS, AND P.G. KEVREKIDIS, *Static and rotating domain-wall cross patterns in Bose-Einstein condensates.* Phys. Rev. A, **70** (2004) 043616.

[252] B.A. MALOMED, H.E. NISTAZAKIS, D.J. FRANTZESKAKIS, AND P.G. KEVREKIDIS, *Domain-wall crosses and propellers in binary Bose-Einstein condensates.* Math. Comput. Simul., **69** (2005) 400–412.

[253] E.A. KUZNETSOV AND S.K. TURITSYN, *Instability and collapse of solitons in media with a defocusing nonlinearity.* Zh. Eksp. Teor. Fiz., **94** (1988) 119–124 [Sov. Phys. JEPT, **67** (1988) 1583–1588].

[254] E.A. KUZNETSOV AND J.J. RASMUSSEN, *Instability of two-dimensional solitons and vortices in defocusing media.* Phys. Rev. E, **51** (1995) 4479–4484.

[255] D.E. PELINOVSKY, YU.A. STEPANYANTS, AND YU.S. KIVSHAR, *Self-focusing of plane dark solitons in nonlinear defocusing media.* Phys. Rev. E, **51** (1995) 5016–5026.

[256] YU.S. KIVSHAR AND D.E. PELINOVSKY, *Self-focusing and transverse instabilities of solitary waves.* Phys. Rep., **331** (2000) 117–195.

[257] V. TIKHONENKO, J. CHRISTOU, B. LUTHER-DAVIES, AND YU.S. KIVSHAR, *Observation of vortex solitons created by the instability of dark soliton stripes.* Opt. Lett., **21** (1996) 1129–1131.

[258] A.V. MAMAEV, M. SAFFMAN, AND A.A. ZOZULYA, *Propagation of dark stripe beams in nonlinear media: Snake instability and creation of optical vortices.* Phys. Rev. Lett., **76** (1996) 2262–2265.

[259] YU.S. KIVSHAR AND X. YANG, *Ring dark solitons.* Phys. Rev. E, **50** (1994) R40–R43.

[260] D.J. FRANTZESKAKIS AND B.A. MALOMED, *Multiscale expansions for a generalized cylindrical nonlinear Schrödinger equation.* Phys. Lett. A, **264** (2000) 179–185.

[261] H.E. NISTAZAKIS, D.J. FRANTZESKAKIS, B.A. MALOMED, AND P.G. KEVREKIDIS, *Head-on collisions of ring dark solitons.* Phys. Lett. A, **285** (2001) 157–164.

[262] W. DREISCHUH, A. FLIESSER, I. VELCHEV, S. DINEV, AND L. WINDHOLZ, *Phase measurements of ring dark solitons.* Appl. Phys. B, **62** (1996) 139–142.

[263] D. NESHEV, A. DREISCHUH, V. KAMENOV, I. STEFANOV, S. DINEV, W. FLIESSER, AND L. WINDHOLZ, *Generation and intrinsic dynamics of ring dark solitary waves.* Appl. Phys. B, **64** (1997) 429–433.

[264] A. DREISCHUH, D. NESHEV, G.G. PAULUS, F. GRASBON, AND H. WALTHER, *Ring dark solitary waves: Experiment versus theory.* Phys. Rev. E, **66** (2002) 066611.

[265] G. THEOCHARIS, D.J. FRANTZESKAKIS, P.G. KEVREKIDIS, B.A. MALOMED, AND YU.S. KIVSHAR, *Ring dark solitons and vortex necklaces in Bose-Einstein condensates.* Phys. Rev. Lett., **90** (2003) 120403.

[266] L.D. CARR AND C.W. CLARK, *Vortices in attractive Bose-Einstein condensates in two dimensions.* Phys. Rev. Lett., **97** (2006) 010403; L.D. CARR AND C.W. CLARK, *Vortices and ring solitons in Bose-Einstein condensates.* Phys. Rev. A, **74** (2006) 043613.

[267] J.-K. XUE, *The nonlinear evolution of ring dark solitons in Bose-Einstein condensates.* J. Phys. A: Math. Gen., **37** (2004) 11223–11228.

[268] J.-K. XUE AND P. PENG, *Head-on collision of ring dark solitons in Bose-Einstein condensates.* Chin. Phys., **15** (2006) 1149–1153.

[269] G. HERRING, L.D. CARR, R. CARRETERO-GONZÁLEZ, P.G. KEVREKIDIS, AND D.J. FRANTZESKAKIS, *Radially symmetric nonlinear states of harmonically trapped Bose-Einstein condensates.* Phys. Rev. A, **77** (2008) 023625.

[270] X.-H. HU, X.-F. ZHANG, D. ZHAO, H.-G. LUO, AND W.M. LIU, *Dynamics and modulation of ring dark solitons in two-dimensional Bose-Einstein condensates with tunable interaction.* Phys. Rev. A, **79** (2009) 023619.

[271] P.G. KEVREKIDIS, G. THEOCHARIS, D.J. FRANTZESKAKIS, AND B.A. MALOMED, *Feshbach resonance management for Bose-Einstein condensates.* Phys. Rev. Lett., **90** (2003) 230401.

[272] S. KOMINEAS AND J. BRAND, *Collisions of solitons and vortex rings in cylindrical Bose-Einstein condensates.* Phys. Rev. Lett., **95** (2005) 110401.

[273] S. KOMINEAS AND N. PAPANICOLAOU, *Solitons, solitonic vortices, and vortex rings in a confined Bose-Einstein condensate.* Phys. Rev. A, **68** (2003) 043617.

[274] S. KOMINEAS AND N. PAPANICOLAOU, *Vortex rings and Lieb modes in a cylindrical Bose-Einstein condensate.* Phys. Rev. Lett., **89** (2002) 070402.

[275] X.Y. TANG AND P.K. SHUKLA, *Solution of the one-dimensional spatially inhomogeneous cubic-quintic nonlinear Schrödinger equation with an external potential.* Phys. Rev. A, **76** (2007) 013612.

[276] C.J. MYATT, E.A. BURT, R.W. GHRIST, E.A. CORNELL, AND C.E. WIEMAN, *Production of two overlapping Bose-Einstein condensates by sympathetic cooling.* Phys. Rev. Lett., **78** (1997) 586–589.

[277] D.S. HALL, M.R. MATTHEWS, J.R. ENSHER, C.E. WIEMAN, AND E.A. CORNELL, *Dynamics of component separation in a binary mixture of Bose-Einstein condensates.* Phys. Rev. Lett., **81** (1998) 1539–1542.

[278] D.M. STAMPER-KURN, M.R. ANDREWS, A.P. CHIKKATUR, S. INOUYE, H.-J. MIESNER, J. STENGER, AND W. KETTERLE, *Optical confinement of a Bose-Einstein condensate.* Phys. Rev. Lett., **80** (1998) 2027–2030.

[279] K.M. MERTES, J.W. MERRILL, R. CARRETERO-GONZÁLEZ, D.J. FRANTZESKAKIS, P.G. KEVREKIDIS, AND D.S. HALL, *Nonequilibrium dynamics and superfluid ring excitations in binary Bose-Einstein condensates.* Phys. Rev. Lett., **99** (2007) 190402.

[280] J. STENGER, S. INOUYE, D.M. STAMPER-KURN, H.-J. MIESNER, A.P. CHIKKATUR, AND W. KETTERLE, *Spin domains in ground-state Bose-Einstein condensates.* Nature, **396** (1998) 345–348.

[281] Y. KAWAGUCHI AND M. UEDA, *Spinor Bose-Einstein condensates.* Phys. Rep., **520** (2012) 253–381.

[282] D.M. STAMPER-KURN AND M. UEDA, *Spinor Bose gases: Symmetries, magnetism, and quantum dynamics.* Rev. Mod. Phys., **85** (2013) 1191–1244.

[283] G. MODUGNO, G. FERRARI, G. ROATI, R.J. BRECHA, A. SIMONI, AND M. INGUSCIO, *Bose-Einstein condensation of potassium atoms by sympathetic cooling.* Science, **294** (2001) 1320–1322.

[284] B. DEMARCO AND D.S. JIN, *Onset of Fermi degeneracy in a trapped atomic gas.* Science, **285** (1999) 1703–1706.

[285] K.M. O'HARA, S.L. HEMMER, M.E. GEHM, S.R. GRANADE, AND J.E. THOMAS, *Observation of a strongly interacting degenerate Fermi gas of atoms.* Science, **298** (2002) 2179–2182.

[286] R.J. BALLAGH, K. BURNETT, AND T.F. SCOTT, *Theory of an output coupler for Bose-Einstein condensed atoms.* Phys. Rev. Lett., **78** (1997) 1607–1611.

[287] A. AMORUSO, I. MECCOLI, A. MINGUZZI, AND M.P. TOSI, *Density profiles and collective excitations of a trapped two-component Fermi vapour.* Eur. Phys. J. D, **8** (2000) 361–369.

[288] R. ROTH AND H. FELDMEIER, *Phase diagram of trapped degenerate Fermi gases including effective s- and p-wave interactions.* J. Phys. B: At. Mol. Opt. Phys., **34** (2001) 4629–4644.

[289] P. CAPUZZI, A. MINGUZZI, AND M.P. TOSI, *Collective excitations of a trapped boson-fermion mixture across demixing.* Phys. Rev. A, **67** (2003) 053605.

[290] V.P. MINEEV, *The theory of the solution of two near-ideal Bose gases.* Zh. Eksp. Teor. Fiz., **67** (1974) 263–267 [Sov. Phys. JETP, **40** (1974) 132–136].

[291] P. AO AND S.T. CHUI, *Binary Bose-Einstein condensate mixtures in weakly and strongly segregated phases.* Phys. Rev. A, **58** (1998) 4836–4840.

[292] H. PU AND N.P. BIGELOW, *Properties of two-species Bose condensates.* Phys. Rev. Lett., **80** (1998) 1130–1133.

[293] T.-L. HO AND V.B. SHENOY, *Binary mixtures of Bose condensates of alkali atoms.* Phys. Rev. Lett., **77** (1996) 3276–3279.

[294] B.D. ESRY, C.H. GREENE, J.P. BURKE, JR., AND J.L. BOHN, *Hartree-Fock theory for double condensates.* Phys. Rev. Lett., **78** (1997) 3594–3597.

[295] E. TIMMERMANS, *Phase separation of Bose-Einstein condensates.* Phys. Rev. Lett., **81** (1998) 5718–5721.

[296] M. TRIPPENBACH, K. GORAL, K. RZAZEWSKI, B.A. MALOMED, AND Y.B. BAND, *Structure of binary Bose-Einstein condensates.* J. Phys. B: At. Mol. Opt. Phys., **33** (2000) 4017–4031.

[297] S. COEN AND M. HAELTERMAN, *Domain wall solitons in binary mixtures of Bose-Einstein condensates.* Phys. Rev. Lett., **87** (2001) 140401.

[298] M.I. MERHASIN, B.A. MALOMED, AND R. DRIBEN, *Transition to miscibility in a binary Bose-Einstein condensate induced by linear coupling.* J. Phys. B: At. Mol. Opt. Phys., **38** (2005) 877–892.

[299] K. KASAMATSU AND M. TSUBOTA, *Modulation instability and solitary-wave formation in two-component Bose-Einstein condensates.* Phys. Rev. A, **74** (2006) 013617.

[300] R. NAVARRO, R. CARRETERO-GONZÁLEZ, AND P.G. KEVREKIDIS, *Phase separation and dynamics of two-component Bose-Einstein condensates.* Phys. Rev. A, **80** (2009) 023613.

[301] B. SUN AND M.S. PINDZOLA, *Bogoliubov modes and the static structure factor for a two-species Bose-Einstein condensate.* J. Phys. B: At. Mol. Opt. Phys., **43** (2010) 055301.

[302] S. Gautam and D. Angom, *Phase separation of binary condensates in harmonic and lattice potentials.* J. Phys. B: At. Mol. Opt. Phys., **44** (2011) 025302.

[303] J. Sabbatini, W.H. Zurek, and M.J. Davis, *Phase separation and pattern formation in a binary Bose-Einstein condensate.* Phys. Rev. Lett., **107** (2011) 230402.

[304] B. Deconinck, P.G. Kevrekidis, H.E. Nistazakis, and D.J. Frantzeskakis, *Linearly coupled Bose-Einstein condensates: From Rabi oscillations and quasiperiodic solutions to oscillating domain walls and spiral waves.* Phys. Rev. A, **70** (2004) 063605.

[305] H.E. Nistazakis, Z. Rapti, D.J. Frantzeskakis, P.G. Kevrekidis, P. Sodano, and A. Trombettoni, *Rabi switch of condensate wave functions in a multicomponent Bose gas.* Phys. Rev. A, **78** (2008) 023635.

[306] S.G. Bhongale and E. Timmermans, *Phase separated BEC for high-sensitivity force measurement.* Phys. Rev. Lett., **100** (2008) 185301.

[307] G. Thalhammer, G. Barontini, L. De Sarlo, J. Catani, F. Minardi, and M. Inguscio, *Double species Bose-Einstein condensate with tunable interspecies interactions.* Phys. Rev. Lett., **100** (2008) 210402.

[308] S.B. Papp, J.M. Pino, and C.E. Wieman, *Tunable miscibility in a dual-species Bose-Einstein condensate.* Phys. Rev. Lett., **101** (2008) 040402.

[309] C.P. Search and P.R. Berman, *Manipulating the speed of sound in a two-component Bose-Einstein condensate.* Phys. Rev. A, **63** (2001) 043612.

[310] S.D. Jenkins and T.A.B. Kennedy, *Dynamic stability of dressed condensate mixtures.* Phys. Rev. A, **68** (2003) 053607.

[311] E. Nicklas, H. Strobel, T. Zibold, C. Gross, B.A. Malomed, P.G. Kevrekidis, and M.K. Oberthaler, *Rabi flopping induces spatial demixing dynamics.* Phys. Rev. Lett., **107** (2011) 193001.

[312] J. Dalibard, F. Gerbier, G. Juzeliūnas, and P. Öhberg, *Colloquium: Artificial gauge potentials for neutral atoms.* Rev. Mod. Phys., **83** (2011) 1523–1543.

[313] Y.-J. Lin, R.L. Compton, K. Jimenez-Garcia, J.V. Porto, and I.B. Spielman, *Synthetic magnetic fields for ultracold neutral atoms.* Nature, **462** (2009) 628–632.

[314] Y.-J. Lin, K. Jimenez-Garcia, and I.B. Spielman, *Spin-orbit-coupled Bose-Einstein condensates.* Nature, **471** (2011) 83–86.

[315] T.L. Ho and S. Zhang, *Bose-Einstein condensates with spin-orbit interaction.* Phys. Rev. Lett., **107** (2011) 150403.

[316] Y. Li, L.P. Pitaevskii, and S. Stringari, *Quantum tricriticality and phase transitions in spin-orbit coupled Bose-Einstein condensates.* Phys. Rev. Lett., **108** (2012) 225301.

[317] X.X. Liu, H. Pu, B. Xiong, W.M. Liu, and J.B. Gong, *Formation and transformation of vector solitons in two-species Bose-Einstein condensates with a tunable interaction.* Phys. Rev. A, **79** (2009) 013423.

[318] D. Schumayer and B. Apagyi, *Stability of static solitonic excitations of two-component Bose-Einstein condensates in finite range of interspecies scattering length a_{12}.* Phys. Rev. A, **69** (2004) 043620.

[319] M. Vijayajayanthi, T. Kanna, and M. Lakshmanan, *Bright-dark solitons and their collisions in mixed N-coupled nonlinear Schrödinger equations.* Phys. Rev. A, **77** (2008) 013820.

[320] H. Li, D.N. Wang, and Y.S. Cheng, *Dynamics of dark-bright vector solitons in a two-component Bose-Einstein condensate.* Chaos Solitons Fract., **39** (2009) 1988–1993.

[321] S. Rajendran, P. Muruganandam, and M. Lakshmanan, *Interaction of dark-bright solitons in two-component Bose-Einstein condensates.* J. Phys. B: At. Mol. Opt. Phys., **42** (2009) 145307.

[322] V.A. Brazhnyi and V.M. Pérez-García, *Stable multidimensional soliton stripes in two-component Bose-Einstein condensates.* Chaos Solitons Fract., **44** (2011) 381–389.

[323] C. Yin, N.G. Berloff, V.M. Pérez-García, D. Novoa, A.V. Carpentier, and H. Michinel, *Coherent atomic soliton molecules for matter-wave switching.* Phys. Rev. A, **83** (2011) 051605.

[324] V. Achilleos, P.G. Kevrekidis, V.M. Rothos, and D.J. Frantzeskakis, *Statics and dynamics of atomic dark-bright solitons in the presence of impurities.* Phys. Rev. A, **84** (2011) 053626.

[325] V. Achilleos, D. Yan, P.G. Kevrekidis, and D.J. Frantzeskakis, *Dark-bright solitons in Bose-Einstein condensates at finite temperatures.* New J. Phys., **14** (2012) 055006.

[326] H. Susanto, P.G. Kevrekidis, R. Carretero-González, B.A. Malomed, D.J. Frantzeskakis, and A.R. Bishop, *Čerenkov-like radiation in a binary superfluid flow past an obstacle.* Phys. Rev. A, **75** (2007) 055601.

[327] Yu.G. Gladush, A.M. Kamchatnov, Z. Shi, P.G. Kevrekidis, D.J. Frantzeskakis, and B.A. Malomed, *Wave patterns generated by a supersonic moving body in a binary Bose-Einstein condensate.* Phys. Rev. A, **79** (2009) 033623.

[328] P. Öhberg and L. Santos, *Dark solitons in a two-component Bose-Einstein condensate.* Phys. Rev. Lett., **86** (2001) 2918–2921.

[329] P. Öhberg and L. Santos, *Solitons in two-component Bose-Einstein condensates.* J. Phys. B: At. Mol. Opt. Phys., **34** (2001) 4721–4735.

[330] Yu.S. Kivshar and S.K. Turitsyn, *Vector dark solitons.* Opt. Lett., **18** (1993) 337–339.

[331] P.G. Kevrekidis, H.E. Nistazakis, D.J. Frantzeskakis, B.A. Malomed, and R. Carretero-González, *Families of matter-waves in two-component Bose-Einstein condensates.* Eur. Phys. J. D, **28** (2004) 181–185.

[332] X.-F. Zhang, X.-H. Hu, X.-X. Liu, and W.M. Liu, *Vector solitons in two-component Bose-Einstein condensates with tunable interactions and harmonic potential.* Phys. Rev. A, **79** (2009) 033630.

[333] C. Hamner, J.J. Chang, P. Engels, and M.A. Hoefer, *Generation of dark-bright soliton trains in superfluid-superfluid counterflow.* Phys. Rev. Lett., **106** (2011) 065302.

[334] S. Middelkamp, J.J. Chang, C. Hamner, R. Carretero-González, P.G. Kevrekidis, V. Achilleos, D.J. Frantzeskakis, P. Schmelcher, and P. Engels, *Dynamics of dark-bright solitons in cigar-shaped Bose-Einstein condensates.* Phys. Lett. A, **375** (2011) 642–646.

[335] D. Yan, J.J. Chang, C. Hamner, P.G. Kevrekidis, P. Engels, V. Achilleos, D.J. Frantzeskakis, R. Carretero-González, and P. Schmelcher, *Multiple dark-bright solitons in atomic Bose-Einstein condensates.* Phys. Rev. A, **84** (2011) 053630.

[336] M.A. Hoefer, J.J. Chang, C. Hamner, and P. Engels, *Dark-dark solitons and modulational instability in miscible two-component Bose-Einstein condensates.* Phys. Rev. A, **84** (2011) 041605(R).

[337] D. Yan, J.J. Chang, C. Hamner, M. Hoefer, P.G. Kevrekidis, P. Engels, V. Achilleos, D.J. Frantzeskakis, and J. Cuevas, *Beating dark-dark solitons in Bose-Einstein condensates.* J. Phys. B: At. Mol. Opt. Phys., **45** (2012) 115301.

[338] S.V. Manakov, *On the theory of two-dimensional stationary self-focusing of electromagnetic waves.* Zh. Eksp. Teor. Fiz., **65** (1973) 505–516 [Sov. Phys. JETP, **38** (1973) 248–253].

[339] V.E. Zakharov and S.V. Manakov, *The theory of resonance spectra of wave packets in nonlinear media.* Zh. Eksp. Teor. Fiz., **71** (1976) 203–211 [Sov. Phys. JETP, **42** (1976) 842–850].

[340] V.E. Zakharov and E.I. Schulman, *To the integrability of the system of two coupled nonlinear Schrödinger equations.* Physica D, **4** (1982) 270–274.

[341] V.G. Makhan'kov and O.K. Pashaev, *Nonlinear Schrödinger Equation with noncompact isogroup.* Theor. Math. Phys., **53** (1982) 979–987.

[342] R. Radhakrishnan and M. Lakshmanan, *Bright and dark soliton solutions to coupled nonlinear Schrödinger equations.* J. Phys. A: Math. Gen., **28** (1995) 2683–2692.

[343] A.P. Sheppard and Yu.S. Kivshar, *Polarized dark solitons in isotropic Kerr media.* Phys. Rev. E, **55** (1997) 4773–4782.

[344] Q.-H. Park and H.J. Shin, *Systematic construction of multicomponent optical solitons.* Phys. Rev. E, **61** (2000) 3093–3106.

[345] N.S. Manton, *An effective Lagrangian for solitons.* Nucl. Phys. B, **150** (1979) 397–412; J.P. Gordon, *Interaction forces among solitons in optical fibers.* Opt. Lett., **8** (1983) 596–598; P.G. Kevrekidis, A. Khare, and A. Saxena, *Solitary wave interactions in dispersive equations using Manton's approach.* Phys. Rev. E, **70** (2004) 057603.

[346] E.A. Ostrovskaya, Yu.S. Kivshar, Z. Chen, and M. Segev, *Interaction between vector solitons and solitonic gluons.* Opt. Lett., **24** (1999) 327–329.

[347] M. Coles, D.E. Pelinovsky, and P.G. Kevrekidis, *Excited states in the large density limit: a variational approach.* Nonlinearity, **23** (2010) 1753–1770.

[348] J.-C. van der Meer, *Hamiltonian Hopf bifurcation with symmetry.* Nonlinearity, **3** (1990) 1041–1056.

[349] H. Susanto, J. Cuevas, and P. Krüger, *Josephson tunnelling of dark solitons in a double-well potential.* J. Phys. B: At. Mol. Opt. Phys., **44** (2011) 095003.

[350] T.R.O. Melvin, A.R. Champneys, P.G. Kevrekidis, and J. Cuevas, *Travelling solitary waves in the discrete Schrödinger equation with saturable nonlinearity: Existence, stability and dynamics.* Physica D, **237** (2008) 551–567.

[351] A.E. Leanhardt, Y. Shin, D. Kielpinski, D.E. Pritchard, and W. Ketterle, *Coreless vortex formation in a spinor Bose-Einstein condensate.* Phys. Rev. Lett., **90** (2003) 140403.

[352] L.E. Sadler, J.M. Higbie, S.R. Leslie, M. Vengalattore, and D.M. Stamper-Kurn, *Spontaneous symmetry breaking in a quenched ferromagnetic spinor Bose-Einstein condensate.* Nature, **443** (2006) 312–315.

[353] M.S. Chang, Q.S. Qin, W.X. Zhang, L. You, and M.S. Chapman, *Coherent spinor dynamics in a spin-1 Bose condensate.* Nature Phys., **1** (2005) 111–116.

[354] J. Ieda, T. Miyakawa, and M. Wadati, *Exact analysis of soliton dynamics in spinor Bose-Einstein condensates.* Phys. Rev. Lett., **93** (2004) 194102.

[355] J. IEDA, T. MIYAKAWA, AND M. WADATI, *Matter-wave solitons in an $F = 1$ spinor Bose-Einstein condensate.* J. Phys. Soc. Jpn., **73** (2004) 2996–3007.

[356] L. LI, Z. LI, B.A. MALOMED, D. MIHALACHE, AND W.M. LIU, *Exact soliton solutions and nonlinear modulation instability in spinor Bose-Einstein condensates.* Phys. Rev. A, **72** (2005) 033611.

[357] M. UCHIYAMA, J. IEDA, AND M. WADATI, *Dark solitons in $F = 1$ spinor Bose-Einstein condensate.* J. Phys. Soc. Jpn., **75** (2006) 064002.

[358] W. ZHANG, Ö.E. MÜSTECAPLIOGLU, AND L. YOU, *Solitons in a trapped spin-1 atomic condensate.* Phys. Rev. A, **75** (2007) 043601.

[359] B.J. DABROWSKA-WÜSTER, E.A. OSTROVSKAYA, T.J. ALEXANDER, AND YU.S. KIVSHAR, *Multicomponent gap solitons in spinor Bose-Einstein condensates.* Phys. Rev. A, **75** (2007) 023617.

[360] T.-L. HO, *Spinor Bose condensates in optical traps.* Phys. Rev. Lett., **81** (1998) 742–745.

[361] T. OHMI AND K. MACHIDA, *Bose-Einstein condensation with internal degrees of freedom in alkali atom gases.* J. Phys. Soc. Jpn., **67** (1998) 1822–1825.

[362] E.G.M. VAN KEMPEN, S.J.J.M.F. KOKKELMANS, D.J. HEINZEN, AND B.J. VERHAAR, *Interisotope determination of ultracold rubidium interactions from three high-precision experiments.* Phys. Rev. Lett., **88** (2002) 093201.

[363] N.N. KLAUSEN, J.L. BOHN, AND C.H. GREENE, *Nature of spinor Bose-Einstein condensates in rubidium.* Phys. Rev. A, **64** (2001) 053602.

[364] T. BERGEMAN, M.G. MOORE, AND M. OLSHANII, *Atom-atom scattering under cylindrical harmonic confinement: Numerical and analytic studies of the confinement induced resonance.* Phys. Rev. Lett., **91** (2003) 163201.

[365] H. PU, C.K. LAW, S. RAGHAVAN, J.H. EBERLY, AND N.P. BIGELOW, *Spin-mixing dynamics of a spinor Bose-Einstein condensate.* Phys. Rev. A, **60** (1999) 1463–1470; H. PU, S. RAGHAVAN, AND N.P. BIGELOW, *Manipulating spinor condensates with magnetic fields: Stochastization, metastability, and dynamical spin localization.* Phys. Rev. A, **61** (2000) 023602.

[366] J. IEDA AND M. WADATI, *Nonlinear dynamics of spin structure in confined Bose-Einstein condensates.* J. Low Temp. Phys., **148** (2007) 405–410.

[367] N.P. ROBINS, W. ZHANG, E.A. OSTROVSKAYA, AND YU.S. KIVSHAR, *Modulational instability of spinor condensates.* Phys. Rev. A, **64** (2001) 021601(R).

[368] H.E. NISTAZAKIS, D.J. FRANTZESKAKIS, P.G. KEVREKIDIS, B.A. MALOMED, R. CARRETERO-GONZÁLEZ, AND A.R. BISHOP, *Polarized states and domain walls in spinor Bose-Einstein condensates.* Phys. Rev. A, **76** (2007) 063603.

[369] P.G. KEVREKIDIS, D.J. FRANTZESKAKIS, B.A. MALOMED, A.R. BISHOP, AND I.G. KEVREKIDIS, *Dark-in-bright solitons in Bose-Einstein condensates with attractive interactions.* New J. Phys., **5** (2003) 64.1–64.17.

[370] J. IEDA, M. UCHIYAMA, AND M. WADATI, *Inverse scattering method for square matrix nonlinear Schrödinger equation under nonvanishing boundary conditions.* J. Math. Phys., **48** (2007) 013507.

[371] Z.D. LI, P.B. HE, L. LI, J.Q. LIANG, AND W.M. LIU, *Magnetic soliton and soliton collisions of spinor Bose-Einstein condensates in an optical lattice.* Phys. Rev. A, **71** (2005) 053611.

[372] Z.-D. LI AND Q.-Y. LI, *Dark soliton interaction of spinor Bose-Einstein condensates in an optical lattice.* Ann. Phys., **322** (2007) 1961–1971.

[373] H.Q. CAI, S. R YANG, AND J.K. XUE, *Weak nonlinear matter waves in a trapped spin-1 condensates.* Commun. Theor. Phys., **55** (2011) 583–588.

[374] H. PU, S. RAGHAVAN, AND N.P. BIGELOW, *Creation of topological states in spinor condensates.* Phys. Rev. A, **63** (2001) 063603.

[375] J.-P. MARTIKAINEN, A. COLLIN, AND K.-A. SUOMINEN, *Creation of a monopole in a spinor condensate.* Phys. Rev. Lett., **88** (2002) 090404; D.E. CHANG, *Method for monopole creation in spinor Bose-Einstein condensates.* Phys. Rev. A, **66** (2002) 025601.

[376] H. SAITO, Y. KAWAGUCHI, AND M. UEDA, *Topological defect formation in a quenched ferromagnetic Bose-Einstein condensates.* Phys. Rev. A, **75** (2007) 013621.

[377] H. TERCAS, D.D. SOLNYSHKOV, AND G. MALPUECH, *Topological Wigner crystal of half-solitons in a spinor Bose-Einstein condensate.* Phys. Rev. Lett., **110** (2013) 035303.

[378] R.S. TASGAL AND Y.B. BAND, *Continuous-wave solutions in spinor Bose-Einstein condensate.* Phys. Rev. A, **87** (2013) 023626.

[379] Z. YAN, K.W. CHOW, AND B.A. MALOMED, *Exact stationary wave patterns in three coupled nonlinear Schrödinger/Gross-Pitaevskii equations.* Chaos Solitons Fract., **42** (2009) 3013–3019.

[380] T. KUROSAKI AND M. WADATI, *Matter-wave bright solitons with a finite background in spinor Bose-Einstein condensates.* J. Phys. Soc. Jpn., **76** (2007) 084002.

[381] E.V. DOKTOROV, J. WANG, AND J. YANG, *Perturbation theory for bright spinor Bose-Einstein condensate solitons.* Phys. Rev. A, **77** (2008) 043617.

[382] P. SZAŃKOWSKI, M. TRIPPENBACH, E. INFELD, AND G. ROWLANDS, *Class of compact entities in three-component Bose-Einstein condensates.* Phys. Rev. A, **83** (2011) 013626.

[383] P. SZANKOWSKI, M. TRIPPENBACH, AND E. INFELD, *An extended representation of three-spin-component Bose-Einstein condensate solitons.* Physica D, **241** (2012) 1811–1814.

[384] V.S. GERDJIKOV, N.A. KOSTOV, AND T.I. VALCHEV, *Solutions of multi-component NLS models and Spinor Bose-Einstein condensates.* Physica D, **238** (2009) 1306–1310.

[385] M. UCHIYAMA, J. IEDA, AND M. WADATI, *Multicomponent bright solitons in $F = 2$ spinor Bose-Einstein condensates.* J. Phys. Soc. Jpn., **76** (2007) 074005.

[386] A.A. ZABOLOTSKII, *Inverse scattering transform for the Yajima-Oikawa equations with non-vanishing boundary condition.* Phys. Rev. A, **80** (2009) 063616.

[387] V.E. ZAKHAROV, *Collapse of Langmuir waves.* Zh. Eksp. Teor. Fiz., **62** (1972) 1745–1751 [Sov. Phys. JETP, **35** (1972) 908–914].

[388] N. YAJIMA AND M. OIKAWA, *Formation and interaction of sonic-Langmuir solitons inverse scattering method.* Prog. Theor. Phys., **56** (1976) 1719–1739.

[389] L.D. LANDAU, *The theory of superfluidity of helium II.* J. Phys. (Moscow), **5** (1941) 71.

[390] C. RAMAN, M. KÖHL, R. ONOFRIO, D.S. DURFEE, C.E. KUKLEWICZ, Z. HADZIBABIC, AND W. KETTERLE, *Evidence for a critical velocity in a Bose-Einstein condensed gas.* Phys. Rev. Lett., **83** (1999) 2502–2505.

[391] T. FRISCH, Y. POMEAU, AND S. RICA, *Transition to dissipation in a model of superflow.* Phys. Rev. Lett., **69** (1992) 1644–1647.

[392] V. HAKIM, *Nonlinear Schrödinger flow past an obstacle in one dimension.* Phys. Rev. E, **55** (1997) 2835–2845.

[393] N. PAVLOFF, *Breakdown of superfluidity of an atom laser past an obstacle.* Phys. Rev. A, **66** (2002) 013610.

[394] P. LEBOEUF AND N. PAVLOFF, *Bose-Einstein beams: Coherent propagation through a guide.* Phys. Rev. A, **64** (2001) 033602.

[395] A. RADOUANI, *Gray solitons and phonons in repulsive Bose-Einstein condensates confined in one-dimensional nonharmonic traps.* Phys. Rev. A, **68** (2003) 043620.

[396] G. THEOCHARIS, P.G. KEVREKIDIS, H.E. NISTAZAKIS, D.J. FRANTZESKAKIS, AND A.R. BISHOP, *Generation of dark solitons in oscillating Bose-Einstein condensates.* Phys. Lett. A, **337** (2005) 441–448.

[397] R. CARRETERO-GONZÁLEZ, P.G. KEVREKIDIS, D.J. FRANTZESKAKIS, B.A. MALOMED, S. NANDI, AND A.R. BISHOP, *Soliton trains and vortex streets as a form of Cerenkov radiation in trapped Bose-Einstein condensates.* Math. Comput. Simul., **74** (2007) 361–369.

[398] A.M. LESZCZYSZYN, G.A. EL, YU.G. GLADUSH, AND A.M. KAMCHATNOV, *Transcritical flow of a Bose-Einstein condensate through a penetrable barrier.* Phys. Rev. A, **79** (2009) 063608.

[399] G.E. ASTRAKHARCHIK AND L.P. PITAEVSKII, *Motion of a heavy impurity through a Bose-Einstein condensate.* Phys. Rev. A, **70** (2004) 013608.

[400] I. CARUSOTTO, S.X. HU, L.A. COLLINS, AND A. SMERZI, *Bogoliubov-Čerenkov radiation in a Bose-Einstein condensate flowing against an obstacle.* Phys. Rev. Lett., **97** (2006) 260403.

[401] YU.G. GLADUSH, G.A. EL, A. GAMMAL, AND A.M. KAMCHATNOV, *Radiation of linear waves in the stationary flow of a Bose-Einstein condensate past an obstacle.* Phys. Rev. A, **75** (2007) 033619.

[402] YU.G. GLADUSH AND A.M. KAMCHATNOV, *Generation of linear waves in Bose-Einstein condensate flow past an obstacle.* Zh. Eksp. Teor. Fiz., **132** (2007) 589–595 [JETP, **105** (2007) 520–525].

[403] YU.G. GLADUSH, L.A. SMIRNOV, AND A.M. KAMCHATNOV, *Generation of Cherenkov waves in the flow of a Bose-Einstein condensate past an obstacle.* J. Phys. B: At. Mol. Opt. Phys., **41** (2008) 165301.

[404] G.A. EL AND A.M. KAMCHATNOV, *Spatial dispersive shock waves generated in supersonic flow of Bose-Einstein condensate past slender body.* Phys. Lett. A, **350** (2006) 192–196 [see also erratum: *Spatial dispersive shock waves generated in supersonic flow of Bose-Einstein condensate past slender body.* Phys. Lett. A, **352** (2006) 554–555].

[405] G.A. EL, A. GAMMAL, AND A.M. KAMCHATNOV, *Oblique dark solitons in supersonic flow of a Bose-Einstein condensate.* Phys. Rev. Lett., **97** (2006) 180405.

[406] G.A. EL, YU.G. GLADUSH, AND A.M. KAMCHATNOV, *Two-dimensional periodic waves in supersonic flow of a Bose-Einstein condensate.* J. Phys. A: Math. Theor., **40** (2007) 611–619.

[407] E.A. CORNELL, report at *Conference on Nonlinear Waves, Integrable Systems and their Applications*, Colorado Springs, 2005: http://jilawww.colorado.edu/bec/papers.html.

[408] A.M. KAMCHATNOV AND L.P. PITAEVSKII, *Stabilization of solitons generated by a supersonic flow of Bose-Einstein condensate past an obstacle.* Phys. Rev. Lett., **100** (2008) 160402.

[409] L.Y. Kravchenko and D.V. Fil, *Stationary waves in a supersonic flow of a two-component Bose gas.* J. Low Temp. Phys., **155** (2009) 219–234.

[410] A.S. Rodrigues, P.G. Kevrekidis, R. Carretero-González, D.J. Frantzeskakis, P. Schmelcher, T.J. Alexander, and Yu.S. Kivshar, *Spinor Bose-Einstein condensate past an obstacle.* Phys. Rev. A, **79** (2009) 043603.

[411] M. Albiez, R. Gati, J. Fölling, S. Hunsmann, M. Cristiani, and M.K. Oberthaler, *Direct observation of tunneling and nonlinear self-trapping in a single bosonic Josephson junction.* Phys. Rev. Lett., **95** (2005) 010402.

[412] T. Zibold, E. Nicklas, C. Gross, and M.K. Oberthaler, *Classical bifurcation at the transition from Rabi to Josephson dynamics.* Phys. Rev. Lett., **105** (2010) 204101.

[413] S. Raghavan, A. Smerzi, S. Fantoni, and S.R. Shenoy, *Coherent oscillations between two weakly coupled Bose-Einstein condensates: Josephson effects, p oscillations, and macroscopic quantum self-trapping.* Phys. Rev. A, **59** (1999) 620–633; S. Raghavan, A. Smerzi, and V.M. Kenkre, *Transitions in coherent oscillations between two trapped Bose-Einstein condensates.* Phys. Rev. A, **60** (1999) R1787–R1790; A. Smerzi and S. Raghavan, *Macroscopic quantum fluctuations in the Josephson dynamics of two weakly linked Bose-Einstein condensates.* Phys. Rev. A, **61** (2000) 063601.

[414] E.A. Ostrovskaya, Y.S. Kivshar, M. Lisak, B. Hall, F. Cattani, and D. Anderson, *Coupled-mode theory for Bose-Einstein condensates.* Phys. Rev. A, **61** (2000) 031601(R).

[415] K.W. Mahmud, J.N. Kutz, and W.P. Reinhardt, *Bose-Einstein condensates in a one-dimensional double square well: Analytical solutions of the nonlinear Schrödinger equation.* Phys. Rev. A, **66** (2002) 063607.

[416] V.S. Shchesnovich, B.A. Malomed, and R.A. Kraenkel, *Solitons in Bose-Einstein condensates trapped in a double-well potential.* Physica D, **188** (2004) 213–240.

[417] D. Ananikian and T. Bergeman, *Gross-Pitaevskii equation for Bose particles in a double-well potential: Two-mode models and beyond.* Phys. Rev. A, **73** (2006) 013604.

[418] P. Zin, E. Infeld, M. Matuszewski, G. Rowlands, and M. Trippenbach, *Method for obtaining exact solutions of the nonlinear Schrödinger equation for a double-square-well potential.* Phys. Rev. A, **73** (2006) 022105.

[419] T. Kapitula and P.G. Kevrekidis, *Bose-Einstein condensates in the presence of a magnetic trap and optical lattice: two-mode approximation.* Nonlinearity, **18** (2005) 2491–2512.

[420] G. Theocharis, P.G. Kevrekidis, D.J. Frantzeskakis, and P. Schmelcher, *Symmetry breaking in symmetric and asymmetric double-well potentials.* Phys. Rev. E, **74** (2006) 056608.

[421] D.R. Dounas-Frazer, A.M. Hermundstad, and L.D. Carr, *Ultracold bosons in a tilted multilevel double-well potential.* Phys. Rev. Lett., **99** (2007) 200402.

[422] C. Cambournac, T. Sylvestre, H. Maillotte, B. Vanderlinden, P. Kockaert, Ph. Emplit, and M. Haelterman, *Symmetry-breaking instability of multimode vector solitons.* Phys. Rev. Lett., **89** (2002) 083901.

[423] P.G. Kevrekidis, Z. Chen, B.A. Malomed, D.J. Frantzeskakis, and M.I. Weinstein, *Spontaneous symmetry breaking in photonic lattices: Theory and experiment.* Phys. Lett. A, **340** (2005) 275–280.

[424] M. Ornigotti, G. Della Valle, D. Gatti, and S. Longhi, *Topological suppression of optical tunneling in a twisted annular fiber.* Phys. Rev. A, **76** (2007) 023833.

[425] N. Tsukada, M. Gotoda, Y. Nomura, and T. Isu, *Laser-assisted coherent atomic tunneling between two trapped Bose-Einstein condensates.* Phys. Rev. A, **59** (1999) 3862–3867; M. Holthaus and S. Stenholm, *Coherent control of the self-trapping transition.* Eur. Phys. J. B, **20** (2001) 451–467; R. Khomeriki, S. Ruffo, and S. Wimberger, *Driven collective quantum tunneling of ultracold atoms in engineered optical lattices.* Europhys. Lett., **77** (2007) 40005; Q. Xie and W. Hai, *Nonlinear Floquet states and quasienergies of a Bose-Einstein condensate in a driven double-well potential.* Phys. Rev. A, **80** (2009) 053603; H.E. Nistazakis, B.A. Malomed, P.G. Kevrekidis, and D.J. Frantzeskakis, *Control of the symmetry breaking in double-well potentials by the resonant nonlinearity management.* Chaos, **21** (2011) 013114.

[426] B.A. Malomed (ed.), *Spontaneous Symmetry Breaking, Self-Trapping, and Josephson Oscillations in Nonlinear Systems*. Springer, Berlin, 2013.

[427] J.A. Stickney and A.A. Zozulya, *Wave-function recombination instability in cold-atom interferometers.* Phys. Rev. A, **66** (2002) 053601.

[428] B.A. Malomed, *Soliton Management in Periodic Systems.* Springer, New York, 2006.

[429] Y.V. Kartashov, B.A. Malomed, and L. Torner, *Solitons in nonlinear lattices.* Rev. Mod. Phys., **83** (2011) 247–305.

[430] S. Inouye M.R. Andrews, J. Stenger, H.J. Miesner, D.M. Stamper-Kurn, and W. Ketterle, *Observation of Feshbach resonances in a Bose-Einstein condensate.* Nature, **392** (1998) 151–154.

[431] M. Theis, G. Thalhammer, K. Winkler, M. Hellwig, G. Ruff, R. Grimm, and J.H. Denschlag, *Tuning the scattering length with an optically induced Feshbach resonance.* Phys. Rev. Lett., **93** (2004) 123001.

[432] M. Marinescu and L. You, *Controlling atom-atom interaction at ultralow temperatures by dc electric fields.* Phys. Rev. Lett., **81** (1998) 4596–4599.

[433] J.L. Roberts, N.R. Claussen, J.P. Burke, Jr., C.H. Greene, E.A. Cornell, and C.E. Wieman, *Resonant magnetic field control of elastic scattering in cold ^{85}Rb.* Phys. Rev. Lett., **81** (1998) 5109–5112.

[434] J. Herbig, T. Kraemer, M. Mark, T. Weber, C. Chin, H.C. Nagerl, and R. Grimm, *Preparation of a pure molecular quantum gas.* Science, **301** (2003) 1510–1513; C.A. Regal, C. Ticknor, J.L. Bohn, and D.S. Jin, *Creation of ultracold molecules from a Fermi gas of atoms.* Nature, **424** (2003) 47–50; M. Greiner, C.A. Regal, and D.S. Jin, *Emergence of a molecular Bose-Einstein condensate from a Fermi gas.* Nature, **426** (2003) 537–540.

[435] M. Bartenstein, A. Altmeyer, S. Riedl, S. Jochim, C. Chin, J.H. Denschlang, and R. Grimm, *Collective excitations of a degenerate gas at the BEC-BCS crossover.* Phys. Rev. Lett., **92** (2004) 203201; T. Bourdel, L. Khaykovich, J. Cubizolles, J. Zhang, F. Chevy, M. Teichmann, L. Tarruell, S.J.J.M.F. Kokkelmans, and C. Salomon, *Experimental study of the BEC-BCS crossover region in lithium 6.* Phys. Rev. Lett., **93** (2004) 050401.

[436] H. Saito and M. Ueda, *Dynamically stabilized bright solitons in a two-dimensional Bose-Einstein condensate.* Phys. Rev. Lett., **90** (2003) 040403.

[437] F. Kh. Abdullaev, J.G. Caputo, R.A. Kraenkel, and B.A. Malomed, *Controlling collapse in Bose-Einstein condensates by temporal modulation of the scattering length.* Phys. Rev. A, **67** (2003) 013605.

[438] M. MATUSZEWSKI, E. INFELD, B.A. MALOMED, AND M. TRIPPENBACH, *Fully three dimensional breather solitons can be created using Feshbach resonances.* Phys. Rev. Lett., **95** (2005) 050403.

[439] D.E. PELINOVSKY, P.G. KEVREKIDIS, AND D.J. FRANTZESKAKIS, *Averaging for solitons with nonlinearity management.* Phys. Rev. Lett., **91** (2003) 240201.

[440] D.E. PELINOVSKY, P.G. KEVREKIDIS, D.J. FRANTZESKAKIS, AND V. ZHARNITSKY, *Hamiltonian averaging for solitons with nonlinearity management.* Phys. Rev. E, **70** (2004) 047604.

[441] F. KH. ABDULLAEV, A.M. KAMCHATNOV, V.V. KONOTOP, AND V.A. BRAZHNYI, *Adiabatic dynamics of periodic waves in Bose-Einstein condensates with time dependent atomic scattering length.* Phys. Rev. Lett., **90** (2003) 230402.

[442] V.M. PÉREZ-GARCÍA, V.V. KONOTOP, AND V.A. BRAZHNYI, *Feshbach resonance induced shock waves in Bose-Einstein condensates.* Phys. Rev. Lett., **92** (2004) 220403.

[443] F.KH. ABDULLAEV AND M. SALERNO, *Adiabatic compression of soliton matter waves.* J. Phys. B, **36** (2003) 2851–2859.

[444] M.I. RODAS-VERDE, H. MICHINEL, AND V.M. PÉREZ-GARCÍA, *Controllable soliton emission from a Bose-Einstein condensate.* Phys. Rev. Lett., **95** (2005) 153903; A.V. CARPENTIER, H. MICHINEL, M.I. RODAS-VERDE, AND V.M. PÉREZ-GARCÍA, *Analysis of an atom laser based on the spatial control of the scattering length.* Phys. Rev. A, **74** (2006) 013619.

[445] G. THEOCHARIS, P. SCHMELCHER, P.G. KEVREKIDIS, AND D.J. FRANTZESKAKIS, *Dynamical trapping and transmission of matter-wave solitons in a collisionally inhomogeneous environment.* Phys. Rev. A, **74** (2006) 053614.

[446] J. GARNIER AND F.KH. ABDULLAEV, *Transmission of matter-wave solitons through nonlinear traps and barriers.* Phys. Rev. A, **74** (2006) 013604.

[447] F.KH. ABDULLAEV AND J. GARNIER, *Propagation of matter-wave solitons in periodic and random nonlinear potentials.* Phys. Rev. A, **72** (2005) 061605(R).

[448] G. DONG, B. HU, AND W. LU, *Ground-state properties of a Bose-Einstein condensate tuned by a far-off-resonant optical field.* Phys. Rev. A, **74** (2006) 063601.

[449] H. SAKAGUCHI AND B.A. MALOMED, *Two-dimensional solitons in the Gross-Pitaevskii equation with spatially modulated nonlinearity.* Phys. Rev. E, **73** (2006) 026601.

[450] G. FIBICH, Y. SIVAN, AND M.I. WEINSTEIN, *Bound states of nonlinear Schrödinger equations with a periodic nonlinear microstructure.* Physica D, **217** (2006) 31–57; Y. SIVAN, G. FIBICH, AND M.I. WEINSTEIN, *Waves in nonlinear lattices: Ultrashort optical pulses and Bose-Einstein condensates.* Phys. Rev. Lett., **97** (2006) 193902.

[451] S.K. TURITSYN, B.G. BALE, AND M.P. FEDORUK, *Dispersion-managed solitons in fibre systems and lasers.* Phys. Rep., **521** (2012) 135.

[452] N.J. SMITH AND N.J. DORAN, *Modulational instabilities in fibers with periodic dispersion management.* Opt. Lett., **21** (1996) 570–572.

[453] M. CENTURION, M.A. PORTER, Y. PU, P.G. KEVREKIDIS, D.J. FRANTZESKAKIS, AND D. PSALTIS, *Modulational instability in a layered Kerr medium: Theory and experiment.* Phys. Rev. Lett., **97** (2006) 234101.

[454] Z. RAPTI, G. THEOCHARIS, P.G. KEVREKIDIS, D.J. FRANTZESKAKIS, AND B.A. MALOMED, *Modulational instabilitity in Bose-Einstein condensates under Feshbach resonance management.* Phys. Scripta, **T107** (2004) 27.

[455] C. KITTEL, *Introduction to Solid State Physics*. Wiley, New York, 1986.

[456] W. MAGNUS AND S. WINKLER. *Hill's Equation*, Wiley, New York, 1966.

[457] I.I. GOLD'MAN AND V.D. KRIVCHENKOV, *Problems in Quantum Mechanics*. Dover, New York, 1993.

[458] C. PARÉ AND P.-A. BÉLANGER, *Antisymmetric soliton in a dispersion-managed system*. Opt. Commun., **168** (1999) 103–109.

[459] M. REED AND B. SAIMON, *Methods of Modern Mathematical Physics IV: Analysis of Operators*. Academic Press, New York, 1978.

[460] R. D'AGOSTA, B.A. MALOMED, AND C. PRESILLA, *Stationary solutions of the Gross-Pitaevskii equation with linear counterpart*. Phys. Lett. A, **275** (2000) 424–434.

[461] N. PROUKAKIS, S. GARDINER, M. DAVIS, AND M. SZYMAŃSKA (EDS.), *Quantum gases. Finite temperature and non-equilibrium dynamics*. Imperial College Press, London, 2013.

[462] L.P. PITAEVSKII, *Phenomenological theory of superfluidity near the λ point*. Zh. Eksp. Teor. Fiz., **35** (1958) 408–415 [Sov. Phys. JETP, **35** (1959) 282–287].

[463] B. JACKSON AND N.P. PROUKAKIS, *Finite-temperature models of Bose-Einstein condensation*. J. Phys. B: At. Mol. Opt. Phys., **41** (2008) 203002.

[464] B. JACKSON, N.P. PROUKAKIS, AND C.F. BARENGHI, *Dark-soliton dynamics in Bose-Einstein condensates at finite temperature*. Phys. Rev. A, **75** (2007) 051601.

[465] B. JACKSON, C.F. BARENGHI, AND N.P. PROUKAKIS, *Matter wave solitons at finite temperatures*. J. Low Temp. Phys., **148** (2007) 387–391.

[466] E. ZAREMBA, T. NIKUNI, AND A. GRIFFIN, *Dynamics of trapped Bose gases at finite temperatures*. J. Low Temp. Phys., **116** (1999) 277–345; B. JACKSON AND E. ZAREMBA, *Modeling Bose-Einstein condensed gases at finite temperatures with N-body simulations*. Phys. Rev. A, **66** (2002) 033606.

[467] A. GRIFFIN, T. NIKUNI, AND E. ZAREMBA, *Bose-Condensed Gases at Finite Temperatures*. Cambridge University Press, Cambridge, 2009.

[468] S.P. COCKBURN, H.E. NISTAZAKIS, T.P. HORIKIS, P.G. KEVREKIDIS, N.P. PROUKAKIS, AND D.J. FRANTZESKAKIS, *Matter-wave dark solitons: Stochastic versus analytical results*. Phys. Rev. Lett., **104** (2010) 174101.

[469] S.P. COCKBURN, H.E. NISTAZAKIS, T.P. HORIKIS, P.G.KEVREKIDIS, N.P. PROUKAKIS, AND D.J. FRANTZESKAKIS, *Fluctuating and dissipative dynamics of dark solitons in quasi-condensates*. Phys. Rev. A, **84** (2011) 043640.

[470] H.T.C. STOOF, *Coherent versus incoherent dynamics during Bose-Einstein condensation in atomic gases*. J. Low Temp. Phys., **114** (1999) 11–108.

[471] H.T.C. STOOF AND M.J. BIJLSMA, *Dynamics of fluctuating Bose-Einstein condensates*. J. Low Temp. Phys., **124** (2001) 431–442.

[472] C.W. GARDINER AND M.J. DAVIS, *The stochastic Gross-Pitaevskii equation: II*. J. Phys. B, **36** (2003) 4731–4753.

[473] A.D. MARTIN AND J. RUOSTEKOSKI, *Quantum and thermal effects of dark solitons in a one-dimensional Bose gas*. Phys. Rev. Lett., **104** (2010) 194102; A.D. MARTIN AND J. RUOSTEKOSKI, *Nonequilibrium quantum dynamics of atomic dark solitons*. New J. Phys., **12** (2010) 055018.

[474] M.J. STEEL, M.K. OLSEN, L.I. PLIMAK, P.D. DRUMMOND, S.M. TAN, M.J. COLLETT, D.F. WALLS, AND R. GRAHAM, *Dynamical quantum noise in trapped Bose-Einstein condensates.* Phys. Rev. A, **58** (1998) 4824–4835.

[475] A. SINATRA, C. LOBO, AND Y. CASTIN, *The truncated Wigner method for Bose-condensed gases: Limits of validity and applications.* J. Phys. B: At. Mol. Opt. Phys., **35** (2002) 3599–3631.

[476] D.M. GANGARDT AND A. KAMENEV, *Quantum decay of dark solitons.* Phys. Rev. Lett., **104** (2010) 190402.

[477] S. CHOI, S.A. MORGAN, AND K. BURNETT, *Phenomenological damping in trapped atomic Bose-Einstein condensates.* Phys. Rev. A, **57** (1998) 4057–4060.

[478] M. TSUBOTA, K. KASAMATSU, AND M. UEDA, *Vortex lattice formation in a rotating Bose-Einstein condensate.* Phys. Rev. A, **65** (2002) 023603.

[479] R.A. DUINE AND H.T.C. STOOF, *Stochastic dynamics of a trapped Bose-Einstein condensate.* Phys. Rev. A, **65** (2001) 013603.

[480] P.G. KEVREKIDIS AND D.J. FRANTZESKAKIS, *Multiple dark solitons in Bose-Einstein condensates at finite temperatures.* Discrete Contin. Dyn. Syst. S, **4** (2011) 1199–1212.

[481] H. TAKEUCHI, N. SUZUKI, K. KASAMATSU, H. SAITO, AND M. TSUBOTA, *Quantum Kelvin-Helmholtz instability in phase-separated two-component Bose-Einstein condensates.* Phys. Rev. B, **81** (2010) 094517.

[482] Y.B. BAND, I. TOWERS, AND B.A. MALOMED, *Unified semiclassical approximation for Bose-Einstein condensates: Application to a BEC in an optocal potential.* Phys. Rev. A, **67** (2003) 023602.

[483] P.A. RUPRECHT, M.J. HOLLAND, K. BURNETT, AND M. EDWARDS, *Time-dependent solution of the nonlinear Schrödinger equation for Bose-condensed trapped neutral atoms.* Phys. Rev. A, **51** (1995) 4704–4711.

[484] E L. ALLGOWER AND K. GEORG, *Numerical path following.* In P.G. CIARLET AND J.L. LIONS (EDS.), *Handbook of Numerical Analysis*, Vol. 5, pp. 3–207. North-Holland, Amsterdam, 1997.

[485] H.B. KELLER, *Numerical solution of bifurcation and nonlinear eigenvalue problems.* In P.H. RABINOWITZ (ED.), *Applications of Bifurcation Theory*, pp. 359–384. Academic Press, New York, 1977.

[486] R.L. PEGO AND H. WARCHALL, *Spectrally stable encapsulated vortices for nonlinear Schrödinger equations.* J. Nonlin. Sci., **12** (2002) 347–394.

[487] YU.S. KIVSHAR AND T.J. ALEXANDER, *Trapped Bose-Einstein Condensates: Role of Dimensionality.* in *Proceedings of the APCTP-Nankai Symposium on Yang-Baxter Systems, Nonlinear Models and Their Applications,* B.K. CHUNG, Q-HAN. PARK, AND C. RIM (EDS.) World Scientific, Singapore, 1999.

[488] H. PU, C.K. LAW, J.H. EBERLY, AND N.P. BIGELOW, *Coherent disintegration and stability of vortices in trapped Bose condensates.* Phys. Rev. A, **59** (1999) 1533–1537.

[489] C. SULEM AND P.L. SULEM, *The Nonlinear Schrödinger Equation.* Springer, New York, 1999.

[490] A.S. DESYATNIKOV, A.A. SUKHORUKOV, AND YU.S. KIVSHAR, *Azimuthons: Spatially modulated vortex solitons.* Phys. Rev. Lett., **95** (2005) 203904.

[491] YU.S. KIVSHAR, J. CHRISTOU, V. TIKHONENKO, B. LUTHER-DAVIES, AND L.M. PISMEN, *Dynamics of optical vortex solitons.* Opt. Commun., **152** (1998) 198–206.

[492] A. Dreischuh, S. Chevrenkov, D. Neshev, G.G. Paulus, and H. Walther, *Generation of lattice structures of optical vortices.* J. Opt. Soc. Am. B, **19** (2002) 550–556.

[493] A.S. Desyatnikov, Yu.S. Kivshar, and L. Torner, *Optical vortices and vortex solitons.* Prog. Optics, **47** (2005) 291–391.

[494] J.E. Williams and M.J. Holland, *Preparing topological states of a Bose-Einstein condensate.* Nature, **401** (1999) 568–572.

[495] K.W. Madison, F. Chevy, W. Wohlleben, and J. Dalibard, *Vortex formation in a stirred Bose-Einstein condensate.* Phys. Rev. Lett., **84** (2000) 806–809.

[496] A. Recati, F. Zambelli, and S. Stringari, *Overcritical rotation of a trapped Bose-Einstein condensate.* Phys. Rev. Lett., **86** (2001) 377–380.

[497] S. Sinha and Y. Castin, *Dynamic instability of a rotating Bose-Einstein condensate.* Phys. Rev. Lett., **87** (2001) 190402.

[498] K.W. Madison, F. Chevy, V. Bretin, and J. Dalibard, *Stationary states of a rotating Bose-Einstein condensate: Routes to vortex nucleation.* Phys. Rev. Lett., **86** (2001) 4443–4446.

[499] C. Raman, J.R. Abo-Shaeer, J.M. Vogels, K. Xu, and W. Ketterle, *Vortex nucleation in a stirred Bose-Einstein condensate.* Phys. Rev. Lett., **87** (2001) 210402.

[500] A.E. Leanhardt, A. Görlitz, A.P. Chikkatur, D. Kielpinski, Y. Shin, D.E. Pritchard, and W. Ketterle, *Imprinting vortices in a Bose-Einstein condensate using topological phases.* Phys. Rev. Lett., **89** (2002) 190403.

[501] Y. Shin, M. Saba, M. Vengalattore, T.A. Pasquini, C. Sanner, A.E. Leanhardt, M. Prentiss, D.E. Pritchard, and W. Ketterle, *Dynamical instability of a doubly quantized vortex in a Bose-Einstein condensate.* Phys. Rev. Lett., **93** (2004) 160406.

[502] B.Y. Rubinstein and L.M. Pismen, *Vortex motion in the spatially inhomogeneous conservative Ginzburg-Landau model original.* Physica D, **78** (1994) 1–10.

[503] S. Middelkamp, P.G. Kevrekidis, D.J. Frantzeskakis, R. Carretero-González, and P. Schmelcher, *Stability and dynamics of matter-wave vortices in the presence of collisional inhomogeneities and dissipative perturbations.* J. Phys. B: At. Mol. Opt. Phys., **43** (2010) 155303.

[504] A.L. Fetter, *Vortex dynamics in spin-orbit-coupled Bose-Einstein condensates.* Phys. Rev. A, **89** (2014) 023629.

[505] D.E. Pelinovsky and P.G. Kevrekidis, *Variational approximations of trapped vortices in the large-density limit.* Nonlinearity, **24** (2011) 1271–1289.

[506] V.I. Arnold, *Mathematical Methods of Classical Mechanics.* Springer, Berlin, 1989.

[507] D.V. Skryabin, *Instabilities of vortices in a binary mixture of trapped Bose-Einstein condensates: Role of collective excitations with positive and negative energies.* Phys. Rev. A, **63** (2000) 013602; D.V. Skryabin, *Energy of internal modes of nonlinear waves and complex frequencies due to symmetry breaking.* Phys. Rev. E, **64** (2001) 055601(R).

[508] S.A. McGee and M.J. Holland, *Rotational dynamics of vortices in confined Bose-Einstein condensates.* Phys. Rev. A, **63** (2001) 043608.

[509] A.L. Fetter and J.-K. Kim, *Vortex precession in a rotating nonaxisymmetric trapped Bose-Einstein condensate.* J. Low Temp. Phys., **125** (2001) 239–248.

[510] J. Dziarmaga and J. Meisner, *Quantum fluctuations of a vortex in a dilute Bose-Einstein condensate.* J. Phys. B: At. Mol. Opt. Phys., **38** (2008) 4211–4219.

[511] T. Kapitula, P.G. Kevrekidis, and D.J. Frantzeskakis, *Disk-shaped Bose-Einstein condensates in the presence of an harmonic trap and an optical lattice.* Chaos, **18** (2008) 023101.

[512] B. Jackson, J.F. McCann, and C.S. Adams, *Vortex formation in dilute inhomogeneous Bose-Einstein condensates.* Phys. Rev. Lett., **80** (1998) 3903–3906.

[513] B. Jackson, N.P. Proukakis, C.F. Barenghi, and E. Zaremba, *Finite-temperature vortex dynamics in Bose-Einstein condensates.* Phys. Rev. A, **79** (2009) 053615.

[514] R.A. Duine, B.W.A. Leurs, and H.T.C. Stoof, *Noisy dynamics of a vortex in a partially Bose-Einstein condensed gas.* Phys. Rev. A, **69** (2004) 053623.

[515] D. Yan, R. Carretero-González, D.J. Frantzeskakis, P.G. Kevrekidis, N.P. Proukakis, and D. Spirn, *Exploring vortex dynamics in the presence of dissipation: Analytical and numerical results.* Phys. Rev. A, **89** (2014) 043613.

[516] E. Lundh and P. Ao, *Hydrodynamic approach to vortex lifetimes in trapped Bose condensates.* Phys. Rev. A, **61** (2000) 063612.

[517] P.G. Kevrekidis, G. Theocharis, D.J. Frantzeskakis, and A. Trombettoni, *Avoiding infrared catastrophes in trapped Bose-Einstein condensates.* Phys. Rev. A, **70** (2004) 023602.

[518] S. Middelkamp, P.G. Kevrekidis, D.J. Frantzeskakis, R. Carretero-González, and P. Schmelcher, *Bifurcations, stability and dynamics of multiple matter-wave vortex states.* Phys. Rev. A, **82** (2010) 013646.

[519] E.W. Kirr, P.G. Kevrekidis, E. Shlizerman, and M.I. Weinstein, *Symmetry-breaking bifurcation in nonlinear Schrödinger/Gross–Pitaevskii equations.* SIAM J. Math. Anal., **40** (2008) 566–604.

[520] M. Möttönen, S.M.M. Virtanen, T. Isoshima, and M.M. Salomaa, *Stationary vortex clusters in nonrotating Bose-Einstein condensates.* Phys. Rev. A, **71** (2005) 033626.

[521] V. Pietilä, M. Möttönen, T. Isoshima, J.A.M. Huhtamäki, and S.M.M. Virtanen, *Stability and dynamics of vortex clusters in nonrotated Bose-Einstein condensates.* Phys. Rev. A, **74** (2006) 023603.

[522] T. Kapitula, P.G. Kevrekidis, and R. Carretero-González, *Rotating matter waves in Bose-Einstein condensates.* Physica D, **233** (2007) 112–137.

[523] S. Middelkamp, P.G. Kevrekidis, D.J. Frantzeskakis, R. Carretero-González, and P. Schmelcher, *Emergence and stability of vortex clusters in Bose-Einstein condensates: A bifurcation approach near the linear limit.* Physica D, **240** (2011) 1449–1459.

[524] A.M. Kamchatnov and S.V. Korneev, *Dynamics of ring dark solitons in Bose-Einstein condensates and nonlinear optics.* Phys. Lett. A, **374** (2010) 4625–4628.

[525] P.G. Kevrekidis and D.E. Pelinovsky, *Distribution of eigenfrequencies for oscillations of the ground state in the Thomas-Fermi limit.* Phys. Rev. A, **81** (2010) 023627.

[526] Manjun Ma, R. Carretero-González, P.G. Kevrekidis, D.J. Frantzeskakis, and B.A. Malomed, *Controlling the transverse instability of dark solitons and nucleation of vortices by a potential barrier.* Phys. Rev. A, **82** (2010) 023621.

[527] P.J. Torres, P.G. Kevrekidis, D.J. Frantzeskakis, R. Carretero-González, P. Schmelcher, and D.S. Hall, *Dynamics of vortex dipoles in confined Bose-Einstein condensates.* Phys. Lett. A, **375** (2011) 3044–3050.

[528] A. Aftalion and R.L. Jerrard, *Shape of vortices for a rotating Bose-Einstein condensate.* Phys. Rev. A, **66** (2002) 023611.

[529] A.L. Fetter, *Rotating trapped Bose-Einstein condensates.* Rev. Mod. Phys., **81** (2009) 647–691.

[530] P. Noziéres and D. Pines, *The Theory of Quantum Fluids, Vol. 2: Superfluid Bose Liquids.* Perseus Books, Cambridge, MA, 1999.

[531] D.V. Freilich, D.M. Bianchi, A.M. Kaufman, T.K. Langin, and D.S. Hall, *Real-time dynamics of single vortex lines and vortex dipoles in a Bose-Einstein condensate.* Science, **329** (2010) 1182–1185.

[532] S. McEndoo and Th. Busch, *Small numbers of vortices in anisotropic traps.* Phys. Rev. A, **79** (2009) 053616.

[533] P.K. Newton and G. Chamoun, *Vortex lattice theory: A particle interaction perspective.* SIAM Rev., **51** (2009) 501–542.

[534] C.D. Meyer, *Matrix Analysis and Applied Linear Algebra.* SIAM, Philadelphia, 2000.

[535] W. Li, M. Haque, and S. Komineas, *Vortex dipole in a trapped two-dimensional Bose-Einstein condensate.* Phys. Rev. A, **77** (2008) 053610.

[536] T.W. Neely, E.C. Samson, A.S. Bradley, M.J. Davis, and B.P. Anderson, *Observation of vortex dipoles in an oblate Bose-Einstein condensate.* Phys. Rev. Lett, **104** (2010) 160401.

[537] S. Middelkamp, P.J. Torres, P.G. Kevrekidis, D.J. Frantzeskakis, R. Carretero-González, P. Schmelcher, D.V. Freilich, and D.S. Hall, *Guiding-Center Dynamics of Vortex Dipoles in Bose-Einstein Condensates.* Phys. Rev. A, **84** (2011) 011605.

[538] W. Petrich, M.H. Anderson, J.R. Ensher, and E.A. Cornell, *Stable, tightly confining magnetic trap for evaporative cooling of neutral atoms.* Phys. Rev. Lett., **74** (1995) 3352–3355.

[539] E.A. Burt, R.W. Ghrist, C.J. Myatt, M.J. Holland, E.A. Cornell, and C.E. Wieman, *Coherence, correlations, and collisions: What one learns about Bose-Einstein condensates from their decay.* Phys. Rev. Lett., **79** (1997) 337–340.

[540] D.J. McCarron, H.W. Cho, D.L. Jenkin, M.P. Köppinger, and S.L. Cornish, *Dual-species Bose-Einstein condensate of ^{87}Rb and ^{133}Cs.* Phys. Rev. A, **84** (2011) 011603.

[541] A.M. Barry and P.G. Kevrekidis, *Few-particle vortex cluster equilibria in Bose-Einstein condensates: existence and stability.* J. Phys. A: Math. Theor., **46** (2013) 445001.

[542] J.A. Seman, E.A.L. Henn, M. Haque, R.F. Shiozaki, E.R.F. Ramos, M. Caracanhas, O. Castilho, C. Castelo Branco, P.E.S. Tavares, F.J. Poveda-Cuevas, G. Roati, K.M.F. Magalhaes, and V.S. Bagnato, *Three-vortex configurations in trapped Bose-Einstein condensates.* Phys. Rev. A, **82** (2010) 033616.

[543] T. Kolokolnikov, P.G. Kevrekidis, and R. Carretero-González, *A tale of two distributions: from few to many vortices in quasi-two-dimensional Bose-Einstein condensates.* Proc. R. Soc. A, **470** (2014) 20140048.

[544] T. HAVELOCK, *The stability of motion of rectilinear vortices in ring formation.* Philos. Mag., **11** (1931) 617.

[545] L.J. CAMPBELL AND R. ZIFF, *A Catalog of Two-Dimensional Vortex Patterns.* Los Alamos Scientific Laboratory Report No. LA-7384-MS (1978); see also L.J. CAMPBELL AND R.M. ZIFF, *Vortex patterns and energies in a rotating superfluid.* Phys. Rev. B, **20** (1979) 1886–1902.

[546] T. KOLOKOLNIKOV, H. SUN, D. UMINSKY, AND A.L. BERTOZZI, *Stability of ring patterns arising from two-dimensional particle interactions.* Phys. Rev. E, **84** (2011) 015203(R).

[547] A.L. BERTOZZI, J. VON BRECHT, H. SUN, T. KOLOKOLNIKOV, AND D. UMINSKY, *Ring patterns and their bifurcations in a nonlocal model of biological swarms.* Commun. Math. Sci., **13** (2015) 955–985.

[548] R. NAVARRO, R. CARRETERO-GONZÁLEZ, P.J. TORRES, P.G. KEVREKIDIS, D.J. FRANTZESKAKIS, M.W. RAY, E. ALTUNTAŞ, AND D.S. HALL, *Dynamics of few co-rotating vortices in Bose-Einstein condensates.* Phys. Rev. Lett., **110** (2013) 225301.

[549] A.V. ZAMPETAKI, R. CARRETERO-GONZÁLEZ, P.G. KEVREKIDIS, F.K. DIAKONOS, AND D.J. FRANTZESKAKIS, *Exploring rigidly rotating vortex configurations and their bifurcations in atomic Bose-Einstein condensate.* Phys. Rev. E, **88** (2013) 042914.

[550] E. DOEDEL, H.B. KELLER, AND J.P. KERNÉVEZ, *Numerical analysis and control of bifurcation problems, Part II: Bifurcation in infinite dimensions.* Int. J. Bif. Chaos, **1** (1991) 745–772.

[551] J.R. ABO-SHAEER, C. RAMAN, J.M. VOGELS, AND W. KETTERLE, *Observation of vortex lattices in Bose-Einstein condensates.* Science, **292** (2001) 476–479.

[552] H. AREF AND D. VAINCHTEIN, *Point vortices exhibit asymmetric equilibria.* Nature, **392** (1998) 769–770.

[553] H.E. CABRAL AND D.S. SCHMIDT, *Stability of relative equilibria in the problem of $N+1$ vortices.* SIAM J. Math. Anal., **31** (2000) 231–250.

[554] A. BARRY, G. HALL, AND C. WAYNE, *Relative equilibria of the $(1+N)$-vortex problem.* J. Nonlin. Sci., **22** (2012) 63–83.

[555] H. AREF, *Point vortex dynamics: A classical mathematics playground.* J. Math. Phys., **48** (2007) 065401.

[556] H. AREF, *Vortices and polynomials.* Fluid Dyn. Res., **39** (2007) 5–23.

[557] T.J. STIELTJES, *Sur certains polynôme qui vérifient une équation différrentielle linéaire du second ordre et sur la theorie des fonctions de Lamé.* Act. Math., **6** (1885) 321–326.

[558] M. MARDEN, *Geometry of Polynomials.* American Mathematical Society, Providence, RI, 1949.

[559] G. SZEGÖ, *Orthogonal Polynomials.* American Mathematical Society, Providence, RI, 1959.

[560] A.M. BARRY, F. HAJIR, AND P.G. KEVREKIDIS, *Generating functions, polynomials and vortices with alternating signs in Bose-Einstein condensates.* J. Phys. A: Math. Theor., **48** (2015) 155205.

[561] M. ADLER AND J. MOSER, *On a class of polynomials connected with the Korteweg–de Vries equation.* Commun. Math. Phys., **61** (1978) 1–30.

[562] I. CODDINGTON, P. ENGELS, V. SCHWEIKHARD, AND E.A. CORNELL, *Observation of Tkachenko oscillations in rapidly rotating Bose-Einstein condensates.* Phys. Rev. Lett., **91** (2003) 100402.

[563] V. SCHWEIKHARD, I. CODDINGTON, P. ENGELS, S. TUNG, AND E.A. CORNELL, *Vortex-lattice dynamics in rotating spinor Bose-Einstein condensates.* Phys. Rev. Lett., **93** (2004) 210403.

[564] A.A. ABRIKOSOV, *On the magnetic properties of superconductors of the second group.* Sov. Phys. JETP, **5** (1957) 1174–1182.

[565] R.C. FETECAU, Y. HUANG, AND T. KOLOKOLNIKOV, *Swarm dynamics and equilibria for a nonlocal aggregation model.* Nonlinearity, **24** (2011) 2681.

[566] A.J. BERNOFF AND C.M. TOPAZ, *A primer of swarm equilibria.* SIAM J. Appl. Dyn. Syst., 10 (2011) 212–250.

[567] L.O. BAKSMATY, S.J. WOO, S. CHOI, AND N.P. BIGELOW, *Tkachenko waves in rapidly rotating Bose-Einstein condensates.* Phys. Rev. Lett., **92** (2004) 160405.

[568] L.J. CAMPBELL, *Transverse normal modes of finite vortex arrays.* Phys. Rev. A, **24** (1981) 514–534.

[569] T. MIZUSHIMA, Y. KAWAGUCHI, K. MACHIDA, T. OHMI, T. ISOSHIMA, AND M.M. SALOMAA, *Collective oscillations of vortex lattices in rotating Bose-Einstein condensates.* Phys. Rev. Lett., **92** (2004) 060407.

[570] J. STOCKHOFE, S. MIDDELKAMP, P.G. KEVREKIDIS, AND P. SCHMELCHER, *Impact of anisotropy on vortex clusters and their dynamics.* Europhys. Lett. **93** (2011) 20008.

[571] S. MCENDOO AND T. BUSCH, *Vortex dynamics in anisotropic traps.* Phys. Rev. A, **82** (2010) 013628.

[572] N. LO GULLO, T. BUSCH, AND M. PATERNOSTRO, *Structural change of vortex patterns in anisotropic Bose-Einstein condensates.* Phys. Rev. A, **83** (2011) 053612.

[573] C.T. KELLEY, *Solving Nonlinear Equations with Newton's Method* Fundamentals of Algorithms. SIAM, Philadelphia, 2003.

[574] R.H. GOODMAN, P.G. KEVREKIDIS, AND R. CARRETERO-GONZÁLEZ, *Dynamics of vortex dipoles in anisotropic Bose-Einstein condensates.* SIAM J. Appl. Dyn. Sys. **14** (2015) 699–729.

[575] C. WANG, G. THEOCHARIS, P.G. KEVREKIDIS, N. WHITAKER, D.J. FRANTZESKAKIS, AND B.A. MALOMED, *Nonlinear Schrödinger equations with a four-well potential in two dimensions: bifurcations and stability analysis.* In J.A. TENREIRO MACHADO, A.C.J. LUO, R.S. BARBOSA, M.F. SILVA, AND L.B. FIGUEIREDO (EDS.), *Nonlinear Science and Complexity*, pp. 173–180. Springer, Dordrecht, 2011.

[576] C. WANG, G. THEOCHARIS, P.G. KEVREKIDIS, N. WHITAKER, K.J.H. LAW, D.J. FRANTZESKAKIS, AND B.A. MALOMED, *A two-dimensional paradigm for symmetry-breaking: The nonlinear Schrödinger equation with a four-well potential.* Phys. Rev. E, **80** (2009) 046611.

[577] D.E. PELINOVSKY, P.G. KEVREKIDIS, AND D.J. FRANTZESKAKIS, *Persistence and stability of discrete vortices in nonlinear Schrödinger lattices.* Physica D, **212** (2005) 20–53.

[578] K.J.H. LAW, L. QIAO, P.G. KEVREKIDIS, AND I. G . KEVREKIDIS, *Stability of quantized vortices in a Bose-Einstein condensate confined in an optical lattice.* Phys. Rev. A, **77** (2008) 053612.

[579] P.G. KEVREKIDIS, H. SUSANTO, AND Z. CHEN, *High-order-mode soliton structures in two-dimensional lattices with defocusing nonlinearity.* Phys. Rev. E, **74** (2006) 066606.

[580] K.J.H. LAW, P.G. KEVREKIDIS, B.P. ANDERSON, R. CARRETERO-GONZÁLEZ, AND D.J. FRANTZESKAKIS, *Structure and stability of two-dimensional Bose-Einstein condensates under both harmonic and lattice confinement.* J. Phys. B: At. Mol. Opt. Phys., **41** (2008) 195303.

[581] P.G. KEVREKIDIS, R. CARRETERO-GONZÁLEZ, G. THEOCHARIS, D.J. FRANTZESKAKIS, AND B.A. MALOMED, *Vortices in a Bose-Einstein condensate confined by an optical lattice.* J. Phys. B: At. Mol. Opt. Phys., **70** (2003) 3467–3476.

[582] N.G. PARKER, N.P. PROUKAKIS, C.F. BARENGHI, AND C.S. ADAMS, *Controlled vortex-sound interactions in atomic Bose-Einstein condensates.* Phys. Rev. Lett., **92** (2004) 160403.

[583] N.G. BERLOFF, P.G. KEVREKIDIS, V.V. KONOTOP, M. LEWENSTEIN, AND B.A. MALOMED, *Nonlinear phenomena in degenerate quantum gases.* Physica D, **238** (2009) 1289–1298.

[584] S. THANVANTHRI, K.T. KAPALE, AND J.P. DOWLING, *Arbitrary coherent superpositions of quantized vortices in Bose-Einstein condensates via orbital angular momentum of light.* Phys. Rev. A, **77** (2008) 053825.

[585] Y.V. KARTASHOV, V.A. VYSLOUKH, AND L. TORNER, *Power-dependent shaping of vortex solitons in optical lattices with spatially modulated nonlinear refractive index.* Opt. Lett., **33** (2008) 2173–2175.

[586] L. WU, L. LI, J.-F. ZHANG, D. MIHALACHE, B.A. MALOMED, AND W.M. LIU, *Exact solutions of the Gross-Pitaevskii equation for stable vortex modes in two-dimensional Bose-Einstein condensates.* Phys. Rev. A, **81** (2010) 061805(R).

[587] M.C. DAVIS, R. CARRETERO-GONZÁLEZ, Z. SHI, K.J.H. LAW, P.G. KEVREKIDIS, AND B.P. ANDERSON, *Manipulation of vortices by localized impurities in Bose-Einstein condensates.* Phys. Rev. A, **80** (2009) 023604.

[588] K.J.H. LAW, T.W. NEELY, P.G. KEVREKIDIS, B.P. ANDERSON, A.S. BRADLEY, AND R. CARRETERO-GONZÁLEZ, *Dynamic and energetic stabilization of persistent currents in Bose-Einstein condensates.* Phys. Rev. A, **89** (2014) 053606.

[589] G. HERRING, P.G. KEVREKIDIS, R. CARRETERO-GONZÁLEZ, B.A. MALOMED, D.J. FRANTZESKAKIS, AND A.R. BISHOP, *Trapped bright matter-wave solitons in the presence of localized inhomogeneities.* Phys. Lett. A, **345** (2005) 144–153.

[590] P.G. KEVREKIDIS, D.J. FRANTZESKAKIS, R. CARRETERO-GONZÁLEZ, B.A. MALOMED, G. HERRING, AND A.R. BISHOP, *Statics, dynamics and manipulation of bright matter-wave solitons in optical lattices.* Phys. Rev. A, **71** (2005) 023614.

[591] N.G. PARKER AND C.S. ADAMS, *Emergence and decay of turbulence in stirred atomic Bose-Einstein condensates.* Phys. Rev. Lett., **95** (2005) 145301.

[592] B.A. MALOMED AND P.G. KEVREKIDIS, *Discrete vortex solitons.* Phys. Rev. E, **64** (2001) 026601.

[593] T.J. ALEXANDER, A.A. SUKHORUKOV, AND YU.S. KIVSHAR, *Asymmetric vortex solitons in nonlinear periodic lattices.* Phys. Rev. Lett., **93** (2004) 063901.

[594] V. KOUKOULOYANNIS AND R.S. MACKAY, *Existence and stability of 3-site breathers in a triangular lattice.* J. Phys. A: Math. Theor., **38** (2005) 1021–1030.

[595] V. KOUKOULOYANNIS, P.G. KEVREKIDIS, K.J.H. LAW, I. KOURAKIS, AND D.J. FRANTZESKAKIS, *Existence and stability of multisite breathers in honeycomb and hexagonal lattices.* J. Phys. A: Math. Theor., **43** (2010) 235101.

[596] M.J. DAVIS, R.J. BALLAGH, AND K. BURNETT, *Dynamics of thermal Bose fields in the classical limit.* J. Phys. B, **34** (2001) 4487–4512; P.B. BLAKIE AND M.J. DAVIS, *Projected Gross-Pitaevskii equation for harmonically confined Bose gases at finite temperature.* Phys. Rev. A, **72** (2005) 063608.

[597] E.A.L. HENN, J.A. SEMAN, E.R.F. RAMOS, M. CARACANHAS, P. CASTILHO, E.P. OLIMPIO, G. ROATI, D.V. MAGALHAES, K.M.F. MAGALHAES, AND V.S. BAGNATO, *Observation of vortex formation in an oscillating trapped Bose-Einstein condensate.* Phys. Rev. A, **79** (2009) 043618.

[598] E.A.L. HENN, J.A. SEMAN, G. ROATI, K.M.F. MAGALHAES, AND V.S. BAGNATO, *Emergence of turbulence in an oscillating Bose-Einstein condensate.* Phys. Rev. Lett., **103** (2009) 045301.

[599] T. WINIECKI, J.F. MCCANN, AND C.S. ADAMS, *Pressure drag in linear and nonlinear quantum fluids.* Phys. Rev. Lett., **82** (1999) 5186–5189.

[600] C.-T. PHAM, C. NORE, AND M.-E. BRACHET, *Boundary layers and emitted excitations in nonlinear Schrödinger superflow past a disk.* Physica D, **210** (2005) 203–226.

[601] K. KASAMATSU, M. TSUBOTA, AND M. UEDA, *Vortex molecules in coherently coupled two-component Bose-Einstein condensates.* Phys. Rev. Lett., **93** (2004) 250406.

[602] T.W. NEELY, A.S. BRADLEY, E.C. SAMSON, S.J. ROONEY, E.M. WRIGHT, K.J.H. LAW, R. CARRETERO-GONZÁLEZ, P.G. KEVREKIDIS, M.J. DAVIS, AND B.P. ANDERSON, *Characteristics of two-dimensional quantum turbulence in a compressible superfluid.* Phys. Rev. Lett., **111** (2013) 235301.

[603] A.S. BRADLEY AND B.P. ANDERSON, *Energy spectra of vortex distributions in two-dimensional quantum turbulence.* Phy. Rev. X, **2** (2012) 041001.

[604] R.H. KRAICHNAN, *Inertial ranges in two-dimensional turbulence.* Phys. Fluids, **10** (1967) 1417–1423; C. LEITH, *Diffusion approximation for two-dimensional turbulence.* Phys. Fluids, **11** (1968) 671–673; G.K. BATCHELOR, *Computation of the energy spectrum in homogeneous two-dimensional turbulence.* Phys. Fluids, **12** (1969) 233–239; R.H. KRAICHNAN AND D. MONTGOMERY, *Two-dimensional turbulence.* Rep. Prog. Phys., **43** (1980) 547–619.

[605] J. SOMMERIA, *Two-dimensional turbulence.* Les Houches Summer School 2001: New Trends in Turbulence **74** (2001) 385–447; P. TABELING, *Two-dimensional turbulence: A physicist approach.* Phys. Rep., **362** (2002) 1–62; G. BOFFETTA AND R.E. ECKE, *Two-dimensional turbulence.* Annu. Rev. Fluid Mech., **44** (2012) 427–451.

[606] C.F. BARENGHI, R.J. DONNELLY, AND W.F. VINEN (EDS.), *Quantized Vortex Dynamics and Superfluid Turbulence.* Springer, Berlin, New York, 2001.

[607] W.F. VINEN AND J.J. NIEMELA, *Quantum turbulence.* J. Low Temp. Phys., **128** (2002) 167–231.

[608] L. ONSAGER, *Statistical hydrodynamics.* Il Nuovo Cimento, **6** (1949) 279.

[609] G. EYINK AND K. SREENIVASAN, *Onsager and the theory of hydrodynamic turbulence.* Rev. Mod. Phys., **78** (2006) 87–135.

[610] S. NAZARENKO AND M. ONORATO, *Freely decaying turbulence and Bose-Einstein condensation in Gross-Pitaevski model.* J. Low Temp. Phys., **146** (2006) 31–46.

[611] T.-L. HORNG, C.-H. HSUEH, S.-W. SU, Y.-M. KAO, AND S.-C. GOU, *Two-dimensional quantum turbulence in a nonuniform Bose-Einstein condensate.* Phys. Rev. A, **80** (2009) 023618.

[612] R. NUMASATO AND M. TSUBOTA, *Possibility of inverse energy cascade in two-dimensional quantum turbulence.* J. Low Temp. Phys., **158** (2010) 415–421.

[613] R. NUMASATO, M. TSUBOTA, AND V.S. LVOV, *Direct energy cascade in two-dimensional compressible quantum turbulence.* Phys. Rev. A, **81** (2010) 063630.

[614] B. NOWAK, D. SEXTY, AND T. GASENZER, *Superfluid turbulence: Nonthermal fixed point in an ultracold Bose gas.* Phys. Rev. B, **84** (2011) 020506R.

[615] B. NOWAK, J. SCHOLE, D. SEXTY, AND T. GASENZER, *Nonthermal fixed points, vortex statistics, and superfluid turbulence in an ultracold Bose gas.* Phys. Rev. A, **85** (2012) 043627.

[616] S.J. ROONEY, P.B. BLAKIE, B.P. ANDERSON, AND A.S. BRADLEY, *Suppression of Kelvon-induced decay of quantized vortices in oblate Bose-Einstein condensates.* Phys. Rev. A, **84** (2011) 023637.

[617] P.B. BLAKIE, A.S. BRADLEY, M.J. DAVIS, R.J. BALLAGH, AND C.W. GARDINER, *Dynamics and statistical mechanics of ultra-cold Bose gases using c-field techniques.* Adv. Phys., **57** (2008) 363–455.

[618] S.J. ROONEY, A.S. BRADLEY, AND P.B. BLAKIE, *Decay of a quantum vortex: Test of nonequilibrium theories for warm Bose-Einstein condensates.* Phys. Rev. A, **81** (2010) 023630.

[619] C. NORE, M. ABID, AND M.E. BRACHET, *Decaying Kolmogorov turbulence in a model of superflow.* Phys. Fluids, **9** (1997) 2644–2669.

[620] K.J.H. LAW, P.G. KEVREKIDIS, AND L.S. TUCKERMAN, *Stable vortex-bright-soliton structures in two-component Bose-Einstein condensates.* Phys. Rev. Lett., **105** (2010) 160405.

[621] R. KOLLÁR AND R.L. PEGO, *Spectral stability of vortices in two-dimensional Bose-Einstein condensates via the Evans function and Krein signature.* Appl. Math. Res. Express, **2012** (2012) 1–46.

[622] C. HUEPE, L.S. TUCKERMAN, S. METENS, AND M.E. BRACHET, *Stability and decay rates of nonisotropic attractive Bose-Einstein condensates.* Phys. Rev. A, **68** (2003) 023609.

[623] T. KAPITULA, K.J.H. LAW, AND P.G. KEVREKIDIS, *Interaction of excited states in two-species Bose–Einstein condensates: A case study.* SIAM J. Appl. Dyn. Syst., **9** (2010) 34–61.

[624] L.N. TREFETHEN, *Spectral Methods in MATLAB.* SIAM, Philadelphia, 2000.

[625] P. SONNEVELD AND M.B. VAN GIJZEN, *IDR(s): A Family of Simple and Fast Algorithms for Solving Large Nonsymmetric Systems of Linear Equations.* SIAM J. Sci. Comput., **31** (2008) 1035–1062.

[626] *The Induced Dimension Reduction Method.* http://ta.twi.tudelft.nl/NW/users/gijzen/IDR.html.

[627] ARPACK homepage. http://www.caam.rice.edu/software/ARPACK/.

[628] H.M. NILSEN AND E. LUNDH, *Splitting dynamics of doubly quantized vortices in Bose-Einstein condensates.* Phys. Rev. A, **77** (2008) 013604.

[629] J. STOCKHOFE, P.G. KEVREKIDIS, D.J. FRANTZESKAKIS, AND P. SCHMELCHER, *Dark-bright ring solitons in Bose-Einstein condensates.* J. Phys. B: At. Mol. Opt. Phys., **44** (2011) 191003.

[630] S. GAUTAM, P. MURUGANANDAM, AND D. ANGOM, *Coreless vortex dipoles and trapped droplets in phase-separated binary condensates.* J. Phys. B, **45** (2012) 055303.

[631] M. POLA, J. STOCKHOFE, P. SCHMELCHER, AND P.G. KEVREKIDIS, *Vortex-bright soliton dipoles: Bifurcations, symmetry breaking and soliton tunneling in a vortex-induced double well.* Phys. Rev. A, **86** (2012) 053601.

[632] Y. CASTIN AND R. DUM, *Bose-Einstein condensates with vortices in rotating traps.* Eur. Phys. J. D, **7** (1999) 399–412.

[633] B.P. ANDERSON, P.C. HALJAN, C.E. WIEMAN, AND E.A. CORNELL, *Vortex precession in Bose-Einstein condensates: Observations with filled and empty cores.* Phys. Rev. Lett., **85** (2000) 2857–2860.

[634] H. PU AND N.P. BIGELOW, *Collective excitations, metastability, and nonlinear response of a trapped two-species Bose-Einstein condensate.* Phys. Rev. Lett., **80** (1998) 1134–1137.

[635] J.J. GARCÍA-RIPOLL, V.M. PÉREZ-GARCÍA, AND F. SOLS, *Split vortices in optically coupled Bose-Einstein condensates.* Phys. Rev. A, **66** (2002) 021602(R).

[636] J. KASPRZAK, M. RICHARD, S. KUNDERMANN, A. BAAS, P. JEAMBRUN, J. KEELING, F.M. MARCHETTI, M.H. SZYMAŃSKA, R. ANDRÉ, J.L. STAEHLI, V. SAVONA, P.B. LITTLEWOOD, B. DEVEAUD, AND L.S. DANG, *Bose-Einstein condensation of exciton polaritons.* Nature, **443** (2006) 409–414.

[637] R. BALILI, V. HARTWELL, D. SNOKE, L. PFEIFFER, AND K. WEST, *Bose-Einstein condensation of microcavity polaritons in a trap.* Science, **316** (2007) 1007–1010;

[638] W. LAI, N.Y. KIM, S. UTSUNOMIYA, G. ROUMPOS, H. DENG, M.D. FRASER, T. BYRNES, P. RECHER, N. KUMADA, T. FUJISAWA, AND Y. YAMAMOTO, *Coherent zero-state and π-state in an exciton-polariton condensate array.* Nature, **450** (2007) 529–532.

[639] H. DENG, G.S. SOLOMON, R. HEY, K.H. PLOOG, AND Y. YAMAMOTO, *Spatial coherence of a polariton condensate.* Phys. Rev. Lett., **99** (2007) 126403.

[640] B. DEVEAUD (ED.), *The Physics of Semiconductor Microcavities.* Wiley-VCH, Weinheim, 2007.

[641] D. SANVITTO AND V. TIMOFEEV (EDS.), *Exciton Polaritons in Microcavities, New Frontiers,* Springer Series in Solid-State Sciences, Vol. 172, Springer, New York, 2012.

[642] H. DENG, H. HAUG, AND Y. YAMAMOTO, *Exciton-polariton Bose-Einstein condensation.* Rev. Mod. Phys., **82** (2010) 1489–1537.

[643] J. KEELING AND N.G. BERLOFF, *Exciton-polariton condensation.* Contemp. Phys., **52** (2011) 131–151.

[644] I. CARUSOTTO AND C. CIUTI, *Quantum fluids of light.* Rev. Mod. Phys., **85** (2013) 299–366.

[645] A. AMO, J. LEFRÈRE, S. PIGEON, C. ABRADOS, C. CIUTI, I. CARUSOTTO, R. HOUDRÉ, E. GIACOBINO, AND A. BRAMATI, *Superfluidity of polaritons in semiconductor microcavities.* Nature Phys., **5** (2009) 805–809.

[646] K.G. LAGOUDAKIS, M. WOUTERS, M. RICHARD, A. BAAS, I. CARUSOTTO, R. ANDRÉ, L.S. DANG, AND B. DEVEAUD-PLÉDRAN, *Quantized vortices in an exciton-polariton condensate.* Nature Phys., **4** (2008) 706–710; see also for half-quantum vortices the work of K.G. LAGOUDAKIS, T. OSTATNICKÝ, A.V. KAVOKIN, Y.G. RUBO, R. ANDRÉ, AND B. DEVEAUD-PLÉDRAN, *Observation of half-quantum vortices in an exciton-polariton condensate.* Science, **326** (2009) 974–976.

[647] M.D. FRASER, G. ROUMPOS, AND Y. YAMAMOTO, *Vortex-antivortex pair dynamics in an exciton-polariton condensate.* New J. Phys., **11** (2009) 113048.

[648] G. ROUMPOS, M.D. FRASER, A. LOFFLER, S. HÖFFLING, A. FORCHEL, AND Y. YAMAMOTO, *Single vortex-antivortex pair in an exciton-polariton condensate.* Nature Phys., **7** (2011) 129–133; G. NARDIN, G. GROSSO, Y. LÉGER, B. PIETKA, F. MORIER-GENOUD, AND B. DEVEAUD-PLÉDRAN, *Hydrodynamic nucleation of quantized vortex pairs in a polariton quantum fluid.* Nature, **7** (2011) 635–641; G. TOSI, F.M. MARCHETTI, D. SANVITTO, C. ANTÓN, M.H. SZYMAŃSKA, A. BERCEANU, C. TEJEDOR, L. MARRUCCI, A. LEMAÎTRE, J. BLOCH, AND L. VIÑA, *Onset and dynamics of vortex-antivortex pairs in polariton optical parametric oscillator superfluids.* Phys. Rev. Lett., **107** (2011) 036401.

[649] D. SANVITTO, F.M. MARCHETTI, M.H. SZYMAŃSKA, G. TOSI, M. BAUDISCH, F.P. LAUSSY, D.N. KRIZHANOVSKII, M.S. SKOLNICK, L. MARRUCCI, A. LEMAÎTRE, J. BLOCH, C. TEJEDOR, AND L. VIÑA, *Persistent currents and quantized vortices in a polariton superfluid.* Nature Phys., **6** (2010) 527–533.

[650] A. AMO, D. SANVITTO, F.P. LAUSSY, D. BALLARINI, E. DEL VALLE, M.D. MARTIN, A. LEMAÎTRE, J. BLOCH, D.N. KRIZHANOVSKII, M.S. SKOLNICK, C. TEJEDOR, AND L VIÑA, *Collective fluid dynamics of a polariton condensate in a semiconductor microcavity.* Nature, **457** (2009) 291–296.

[651] M. WOUTERS AND I. CARUSOTTO, *Excitations in a nonequilibrium Bose-Einstein condensate of exciton polaritons.* Phys. Rev. Lett., **99** (2007) 140402.

[652] M. WOUTERS, I. CARUSOTTO, AND C. CIUTI, *Spatial and spectral shape of inhomogeneous nonequilibrium exciton-polariton condensates.* Phys. Rev. B, **77** (2008) 115340.

[653] C. CIUTI AND I. CARUSOTTO, *Quantum fluid effects and parametric instabilities in microcavities.* Phys. Status Solidi B **242** (2005) 2224–2245.

[654] J. KEELING AND N.G. BERLOFF, *Spontaneous rotating vortex lattices in a pumped decaying condensate.* Phys. Rev. Lett., 100 (2008) 250401.

[655] M.O. BORGH, J. KEELING, AND N.G. BERLOFF, *Spatial pattern formation and polarization dynamics of a nonequilibrium spinor polariton condensate.* Phys. Rev. B, **81** (2010) 235302.

[656] M.O. BORGH, G. FRANCHETTI, J. KEELING, AND N.G. BERLOFF, *Robustness and observability of rotating vortex lattices in an exciton-polariton condensate.* Phys. Rev. B, **86** (2012) 035307.

[657] G. TOSI G. CHRISTMANN, N.G. BERLOFF, P. TSOTSIS, T. GAO, Z. HATZOPOULOS, P.G. SAVVIDIS, AND J.J. BAUMBERG, *Sculpting oscillators with light within a nonlinear quantum fluid.* Nature Phys., **8** (2012) 190–194.

[658] A.S. RODRIGUES, P.G. KEVREKIDIS, R. CARRETERO-GONZÁLEZ, J. CUEVAS, D.J. FRANTZESKAKIS, AND F. PALMERO, *From nodeless clouds and vortices to gray ring solitons and symmetry-broken states in two-dimensional polariton condensates.* J. Phys.: Cond. Mat., **26** (2014) 155801.

[659] K. NOZAKI AND N. BEKKI, *Exact solutions of the generalized Ginzburg-Landau equation.* J. Phys. Soc. Jpn., **53** (1984) 1581–1582; K. NOZAKI AND N. BEKKI, *Formations of spatial patterns and holes in the generalized Ginzburg-Landau equation.* Phys. Lett. A, **110** (1985) 133–135.

[660] M. BAZHENOV AND M. RABINOVICH, *The screening of a spiral field in a 2D complex Ginzburg-Landau equation.* Phys. Lett. A, **179** (1993) 191–197; M. BAZHENOV AND M. RABINOVICH, *Synchronized disorder in a 2D complex Ginzburg-Landau equation.* Physica D, **73** (1994) 318–334.

[661] D. BALLARINI, M. DE GIORGI, E. CANCELLIERI, B. SILVA FERNÁNDEZ, A. BRAMATI, G. GIGLI, F. LAUSSY, AND D. SANVITTO, *Backjet, shock waves and ring solitons in the quantum pond of a polariton superfluid.* Preprint, arXiv:1309.3083.

[662] V. ACHILLEOS, P.G. KEVREKIDIS, D.J. FRANTZESKAKIS, AND R. CARRETERO-GONZÁLEZ, *Dark solitons and vortices in PT-symmetric nonlinear media: From spontaneous symmetry breaking to nonlinear phase transitions.* Phys. Rev. A, **86** (2012) 013808.

[663] D.A. ZEZYULIN AND V.V. KONOTOP, *Nonlinear modes in the harmonic PT-symmetric potential.* Phys. Rev. A, **85** (2012) 043840.

[664] J.J. GARCÍA-RIPOLL AND V.M. PÉREZ-GARCÍA, *Vortex nucleation and hysteresis phenomena in rotating Bose-Einstein condensates.* Phys. Rev. A, **63** (2001) 041603R.

[665] P. ROSENBUSCH, V. BRETIN, AND J. DALIBARD, *Dynamics of a single vortex line in a Bose-Einstein condensate.* Phys. Rev. Lett., **89** (2002) 200403.

[666] J.J. GARCÍA-RIPOLL AND V.M. PÉREZ-GARCÍA, *Stability of vortices in inhomogeneous Bose condensates subject to rotation: A three-dimensional analysis.* Phys. Rev. A, **60** (1999) 4864–4874.

[667] A. AFTALION AND T. RIVIERE, *Vortex energy and vortex bending for a rotating Bose-Einstein condensate.* Phys. Rev. A, **64** (2001) 043611.

[668] A. AFTALION AND I. DANAILA, *Three-dimensional vortex configurations in a rotating Bose-Einstein condensate.* Phys. Rev. A, **68** (2003) 023603.

[669] T.P. SIMULA, T. MIZUSHIMA, AND K. MACHIDA, *Vortex waves in trapped Bose-Einstein condensates.* Phys. Rev. A, **78** (2008) 053604.

[670] J.J. GARCÍA-RIPOLL AND V.M. PÉREZ-GARCÍA, *Vortex bending and tightly packed vortex lattices in Bose-Einstein condensates.* Phys. Rev. A, **64** (2001) 053611.

[671] M. MODUGNO, L. PRICOUPENKO, AND Y. CASTIN, *Bose-Einstein condensates with a bent vortex in rotating traps.* Eur. Phys. J. D **22** (2003) 235–257.

[672] S. KOMINEAS, N.R. COOPER, AND N. PAPANICOLAOU, *Single vortex states in a confined Bose-Einstein condensate.* Phys. Rev. A, **72** (2005) 053624.

[673] A. AFTALION, *Vortices in Bose-Einstein Condensates*, Progress in Nonlinear Differential Equations and Their Applications, Vol. **67**, Birkhäuser, Basel, 2006.

[674] V. BRETIN, P. ROSENBUSCH, F. CHEVY, G.V. SHLYAPNIKOV, AND J. DALIBARD, *Quadrupole oscillation of a single-vortex condensate evidence for Kelvin modes ranges where they exist.* Phys. Rev. Lett., **90** (2003) 100403.

[675] A.L. FETTER, *Kelvin mode of a vortex in a nonuniform Bose-Einstein condensate.* Phys. Rev. A, **69** (2004) 043617.

[676] W. THOMSON (LORD KELVIN), *Vibrations of a columnar vortex.* Philos. Mag., **10** (1880) 155–168.

[677] R.J. DONNELLY, *Quantized Vortices in Helium II.* Cambridge University Press, New York, 1991.

[678] T. MIZUSHIMA, M. ICHIOKA, AND K. MACHIDA, *Beliaev damping and Kelvin mode spectroscopy of a Bose-Einstein condensate in the presence of a vortex line.* Phys. Rev. Lett., **90** (2003) 180401.

[679] E. FONDA, D.P. MEICHLE, N.T. OUELLETTE, S. HORMOZ, AND D.P. LATHROP, *Direct observation of Kelvin waves excited by quantized vortex reconnection.* Proc. Natl. Acad. Sci., **111** (2014) 4707–4710.

[680] K. SASAKI, N. SUZUKI, AND H. SAITO, *Capillary instability in a two-component Bose-Einstein condensate.* Phys. Rev. A, **83** (2011) 053606.

[681] T.P. SIMULA AND K. MACHIDA, *Kelvin-Tkachenko waves of few-vortex arrays in trapped Bose-Einstein condensates.* Phys. Rev. A, **82** (2010) 063627.

[682] V. SCHWEIKHARD, I. CODDINGTON, P. ENGELS, V.P. MOGENDORFF, AND E.A. CORNELL, *Rapidly rotating Bose-Einstein condensates in and near the lowest Landau level.* Phys. Rev. Lett., **92** (2004) 040404.

[683] G. BAYM, *Tkachenko modes of vortex lattices in rapidly rotating Bose-Einstein condensates.* Phys. Rev. Lett., **91** (2003) 110402.

[684] E.B. SONIN, *Ground state and Tkachenko modes of a rapidly rotating Bose-Einstein condensate in the lowest-Landau-level state.* Phys. Rev. A, **72** (2005) 021606(R).

[685] M. KE CELI AND M. Ö. OKTEL, *Tkachenko modes and structural phase transitions of the vortex lattice of a two-component Bose-Einstein condensate.* Phys. Rev. A, **73** (2006) 023611.

[686] S.J. WOO, S. CHOI, L.O. BAKSMATY, AND N.P. BIGELOW, *Dynamics of vortex matter in rotating two-species Bose-Einstein condensates.* Phys. Rev. A, **75** (2007) 031604(R).

[687] S. ZUCCHER, M. CALIARI, A.W. BAGGALEY, AND C.F. BARENGHI, *Quantum vortex reconnections.* Phys. Fluids, **24** (2012) 125108.

[688] J. KOPLIK AND H. LEVINE, *Vortex reconnection in superfluid helium.* Phys. Rev. Lett., **71** (1993) 1375–1378.

[689] S. KIDA AND M. TAKAOKA, *Vortex reconnection.* Annu. Rev. Fluid Mech., **26** (1994) 169–177.

[690] L. SKREBK AND K.R. SREENIVASAN, *Developed quantum turbulence and its decay.* Phys. Fluids, **24** (2012) 011301.

[691] D.R. POOLE, H. SCOFFIELD, C.F. BARENGHI, AND D.C. SAMUELS, *Geometry and topology of superfluid turbulence.* J. Low Temp. Phys., **132** (2003) 97–117.

[692] T.P. SIMULA, *Crow instability in trapped Bose-Einstein condensates.* Phys. Rev. A, **84** (2011) 021603(R).

[693] N.G. BERLOFF AND P.H. ROBERTS, *Motion in a Bose condensate: IX. Crow instability of antiparallel vortex pairs.* J. Phys. A: Math. Gen., **34** (2001) 10057–10066.

[694] M. LEADBEATER, T. WINIECKI, D.C. SAMUELS, C.F. BARENGHI, AND C.S. ADAMS, *Sound emission due to superfluid vortex reconnections.* Phys. Rev. Lett., **86** (2001) 1410–1413.

[695] G.K. BATCHELOR, *An Introduction to Fluid Dynamics.* Cambridge University Press, Cambridge, 1967.

[696] P.G. SAFFMAN, *Vortex Dynamics.* Cambridge University Press, Cambridge, 1992.

[697] R. HÄNNINENA AND A.W. BAGGALEYB, *Vortex filament method as a tool for computational visualization of quantum turbulence.* Proc. Natl. Acad. Sci., **111** (2014) 4667–4674.

[698] L.S. DA RIOS, *On the motion of an unbounded fluid with a vortex filament of any shape.* Rend. Circ. Mat. Palermo, **22** (1906) 117–135.

[699] R. BETCHOV, *On the curvature and torsion of an isolated vortex filament.* J. Fluid Mech., **22** (1965) 471–479.

[700] E.D. SIGGIA, *Collapse and amplification of a vortex filament.* Phys. Fluids, **28** (1985) 794–805

[701] C.F. BARENGHI, R.J. DONNELLY, AND W.F. VINEN, *Quantized Vortex Dynamics and Superfluid Turbulence.* Springer, Berlin, 2001.

[702] R.J. ARMS AND F.R. HAMA, *Localized-induction concept on a curved vortex and motion of an elliptic vortex ring.* Phys. Fluids, **8** (1965) 553–559.

[703] S.A. STRONG AND L.D. CARR, *Generalized local induction equation, elliptic asymptotics, and simulating superfluid turbulence.* J. Math. Phys., **53** (2012) 033102.

[704] A. LEONARD, *Vortex methods for flow simulation.* J. Comput. Phys., **37** (1980) 289–335.

[705] K.W. SCHWARZ, *Three-dimensional vortex dynamics in superfluid ^{4}He: Homogeneous superfluid turbulence.* Phys. Rev. B, **38** (1988) 2398–2417.

[706] J.T. TOUGH, *Superfluid turbulence.* In D.F. BREWER (ED.), *Progress in Low Temperature Physics, VIII*, p. 133. North-Holland, Amsterdam, 1982.

[707] C.F. BARENGHI, C.J. SWANSON, AND R.J. DONNELLY, *Emerging issues in helium turbulence.* J. Low Temp. Phys., **100** (1995) 385–413.

[708] C.F. BARENGHI, G.G. BAUER, D.C. SAMUELS, AND R.J. DONNELLY, *Superfluid vortex lines in a model of turbulence.* Phys. Fluids, **9** (1997) 2631–2643.

[709] P.H. ROBERTS AND N.G. BERLOFF, *Nonlinear Schrödinger equation as a model of superfluid helium.* In C.F. BARENGHI, R.J. DONNELLY, AND W.F. VINEN (EDS.), *Quantized Vortex Dynamics and Superfluid Turbulence*, Lecture Notes in Physics, Vol. **571**, Springer, Berlin, 2001.

[710] M.S. PAOLETTI, M.E. FISHER, K.R. SREENIVASAN, AND D.P. LATHROP, *Velocity statistics distinguish quantum turbulence from classical turbulence.* Phys. Rev. Lett., **101** (2008) 154501.

[711] A.C. WHITE, C.F. BARENGHI, N.P. PROUKAKIS, A.J. YOUD, AND D.H. WACKS, *Nonclassical velocity statistics in a turbulent atomic Bose-Einstein condensate.* Phys. Rev. Lett., **104** (2010) 075301.

[712] F. PINSKER, N.G. BERLOFF, AND V.M. PÉREZ-GARCÍA, *Nonlinear quantum piston for the controlled generation of vortex rings and soliton trains.* Phys. Rev. A, **87** (2013) 053624.

[713] J. RUOSTEKOSKI AND Z. DUTTON, *Engineering vortex rings and systems for controlled studies of vortex interactions in Bose-Einstein condensates.* Phys. Rev. A, **72** (2005) 063626.

[714] M.D. REICHL AND E.J. MUELLER, *Vortex ring dynamics in trapped Bose-Einstein condensates.* Phys. Rev. A, **88** (2013) 053626.

[715] K. SASAKI, N. SUZUKI, AND H. SAITO, *Dynamics of bubbles in a two-component Bose-Einstein condensate.* Phys. Rev. A, **83** (2011) 033602.

[716] G.W. RAYFIELD AND F. REIF, *Quantized vortex rings in superfluid helium.* Phys. Rev., **136** (1964) A1194–A1208.

[717] G. GAMOTA, *Creation of quantized vortex rings in superfluid helium.* Phys. Rev. Lett., **31** (1973) 517–520.

[718] J. RUOSTEKOSKI AND J.R. ANGLIN, *Creating vortex rings and three-dimensional Skyrmions in Bose-Einstein condensates.* Phys. Rev. Lett., **86** (2001) 3934–3937.

[719] B. JACKSON, J.F. MCCANN, AND C.S. ADAMS, *Vortex rings and mutual drag in trapped Bose-Einstein condensates.* Phys. Rev. A, **60** (1999) 4882–4885.

[720] R.G. SCOTT, A.M. MARTIN, S. BUJKIEWICZ, T.M. FROMHOLD, N. MALOSSI, O. MORSCH, M. CRISTIANI, AND E. ARIMONDO, *Transport and disruption of Bose-Einstein condensates in optical lattices.* Phys. Rev. A, **69** (2004) 033605.

[721] N.G. BERLOFF AND C.F. BARENGHI, *Vortex nucleation by collapsing bubbles in Bose-Einstein condensates.* Phys. Rev. Lett., **93** (2004) 090401.

[722] N.G. BERLOFF, *Evolution of rarefaction pulses into vortex rings.* Phys. Rev. B, **65** (2002) 174518.

[723] N.G. BERLOFF, *Vortex nucleation by a moving ion in a Bose condensate.* Phys. Lett. A, **277** (2000) 240–244.

[724] N.G. BERLOFF AND P.H. ROBERTS, *Motions in a Bose condensate: VII. Boundary-layer separation.* J. Phys. A: Math. Gen., **33** (2000) 4025–4038.

[725] N.G. BERLOFF AND P.H. ROBERTS, *Motion in a Bose condensate: VIII. The electron bubble.* J. Phys. A: Math. Gen., **34** (2001) 81–91.

[726] P. ENGELS, *Viewpoint: Observing the dance of a vortex–antivortex pair, step by step.* Physics, **3** (2010) 33.

[727] D. AMIT AND E.P. GROSS, *Vortex rings in a Bose fluid.* Phys. Rev., **145** (1966) 130–136.

[728] A.L. FETTER, *Vortices in an imperfect Bose gas. IV. Translational velocity.* Phys. Rev., **151** (1966) 100–104.

[729] P.H. ROBERTS AND J. GRANT, *Motions in a Bose condensate I. The structure of the large circular vortex.* J. Phys. A: Gen. Phys., **4** (1971) 55–72.

[730] J. KOPLIK AND H. LEVINE, *Scattering of superfluid vortex rings.* Phys. Rev. Lett., **76** (1996) 4745–4748.

[731] J.L. HELM, *What Is the Speed of a Perturbed Vortex Ring?* Technical report, Newcastle University, http://research.ncl.ac.uk/quantum-fluids/files/helm-mmath.pdf, April 2010.

[732] V.L. GINZBURG AND L.P. PITAEVSKII, *On the theory of superfluidity.* Sov. Phys. JETP, **7** (1958) 858–861.

[733] R.M. CAPLAN, *Study of Vortex Ring Dynamics in the Nonlinear Schrödinger Equation Utilizing GPU-Accelerated High-Order Compact Numerical Integrators.* PhD thesis, Claremont Graduate University and San Diego State University, 2012.

[734] J.L. HELM, C.F. BARENGHI, AND A.J. YOUD, *Slowing down of vortex rings in Bose-Einstein condensates.* Phys. Rev. A, **83** (2011) 045601.

[735] D. PROMENT, C.F. BARENGHI, AND M. ONORATO, *Interaction and decay of Kelvin waves in the Gross-Pitaevskii model.* Preprint, arXiv:1308.0852.

[736] F. MAGGIONI, S. ALAMRI, C.F. BARENGHI, AND R.L. RICCA, *Velocity, energy, and helicity of vortex knots and unknots.* Phys. Rev. E, **82** (2010) 026309.

[737] R.L. RICCA, D.C. SAMUELAS, AND C.F. BARENGHI, *Evolution in vortex knots.* J. Fluid Mech., **391** (1999) 29–44.

[738] D. PROMENT, M. ONORATO, AND C.F. BARENGHI, *Vortex knots in a Bose-Einstein condensate.* Phys. Rev. E, **85** (2012) 036306.

[739] M. ABAD, M. GUILLEUMAS, R. MAYOL, AND M. PI, *Vortex rings in toroidal Bose-Einstein condensates.* Laser Phys., **18** (2008) 648–652.

[740] M. GUILLEUMAS, D.M. JEZEK, R. MAYOL, M. PI, AND M. BARRANCO, *Generating vortex rings in Bose-Einstein condensates in the line-source approximation.* Phys. Rev. A, **65** (2002) 053609.

[741] C.H. HSUEH, S.C. GOU, T.L. HORNG, AND Y.M. KAO, *Vortex-ring solutions of the Gross-Pitaevskii equation for an axisymmetrically trapped Bose-Einstein condensate.* J. Phys. B, **40** (2007) 4561–4571.

[742] R.M. CAPLAN, J.D. TALLEY, R. CARRETERO-GONZÁLEZ, AND P.G. KEVREKIDIS, *Scattering and leapfrogging of vortex rings in a superfluid.* Phys. Fluids, **26** (2014) 097101.

[743] M. KONSTANTINOV, *Chaotic phenomena in the interaction of vortex rings.* Phys. Fluids, **6** (1994) 1752–1767.

[744] B.N. SHASHIKANTH AND J.E. MARSDEN, *Leapfrogging vortex rings: Hamiltonian structure, geometric phases and discrete reduction.* Fluid Dyn. Res., **33** (2003) 333–356.

[745] P. KUOPANPORTTI AND M. MÖTTÖNEN, *Splitting dynamics of giant vortices in dilute Bose-Einstein condensates.* Phys. Rev. A, **81** (2010) 033627.

[746] P. KUOPANPORTTI, E. LUNDH, J.A.M. HUHTAMÄKI, V. PIETILÄ, AND M. MÖTTÖNEN, *Core sizes and dynamical instabilities of giant vortices in dilute Bose-Einstein condensates.* Phys. Rev. A, **81** (2010) 023603.

[747] P. KUOPANPORTTI AND M. MÖTTÖNEN, *Stabilization and pumping of giant vortices in dilute Bose-Einstein condensates.* J. Low Temp. Phys., **161** (2010) 561–573.

[748] K.J.H. LAW, T.W. NEELY, P.G. KEVREKIDIS, B.P. ANDERSON, A.S. BRADLEY, AND R. CARRETERO-GONZÁLEZ, *Dynamic and energetic stabilization of persistent currents in Bose-Einstein condensates.* Phys. Rev. A, **89** (2014) 053606.

[749] D.H. WACKS, A.W. BAGGALEY, AND C.F. BARENGHI, *Coherent laminar and turbulent motion of toroidal vortex bundles.* Phys. Fluids, **26** (2014) 027102.

[750] D.J. ACHESON, *Instability of vortex leapfrogging.* Eur. J. Phys., **21** (2000) 269–273.

[751] L. TOPHØJ AND H. AREF, *Instability of vortex pair leapfrogging.* Phys. Fluids, **25** (2013) 014107.

[752] B. ECKHARDT AND H. AREF, *Integrable and chaotic motions of four vortices II. Collision dynamics of vortex pairs.* Philos. Trans. R. Soc. Lond. A, **326** (1988) 655–696.

[753] T. PRICE, *Chaotic scattering of two identical point vortex pairs.* Phys. Fluids, **5** (1993) 2479–2483.

[754] L. TOPHØJ AND H. AREF, *Chaotic scattering of two identical point vortex pairs revisited.* Phys. Fluids, **20** (2008) 093605.

[755] S. KOMINEAS AND N. PAPANICOLAOU, *Nonlinear waves in a cylindrical Bose-Einstein condensate.* Phys. Rev. A, **67** (2003) 023615.

[756] S. KOMINEAS AND J. BRAND, *Collisions of solitons and vortex rings in cylindrical Bose-Einstein condensates.* Phys. Rev. Lett., **95** (2005) 110401.

[757] M. MATUSZEWSKI, T.J. ALEXANDER, AND YU.S. KIVSHAR, *Spin-domain formation in antiferromagnetic condensates.* Phys. Rev. A, **78** (2008) 023632.

[758] S. WÜSTER, T.E. ARGUE, AND C.M. SAVAGE, *Numerical study of the stability of Skyrmions in Bose-Einstein condensates.* Phys. Rev. A, **72** (2005) 043616.

[759] T.H.R. SKYRME, *A non-linear field theory.* Proc. R. Soc. Lond. A, **260** (1961) 127–138.

[760] T.H.R. SKYRME, *A unified field theory of mesons and baryons.* Nucl. Phys., **31** (1962) 556–569.

[761] S. WONG, *What Exactly Is a Skyrmion?* Preprint, arXiv:hep-ph/0202250.

[762] F. WILCZEK, *Fractional Statistics and Anyon Superconductivity.* World Scientific, Singapore, 1990.

[763] D.H. LEE AND C.L. KANE, *Boson-vortex-Skyrmion duality, spin-singlet fractional quantum Hall effect, and spin-1/2 anyon superconductivity.* Phys. Rev. Lett., **64** (1990) 1313–1317.

[764] U. AL KHAWAJA AND H. STOOF, *Skyrmions in a ferromagnetic Bose-Einstein condensate.* Nature, **411** (2001) 918–920.

[765] R.A. BATTYE, N.R. COOPER, AND PAUL M. SUTCLIFFE, *Stable Skyrmions in two-component Bose-Einstein condensates.* Phys. Rev. Lett., **88** (2002) 080401.

[766] C.M. SAVAGE AND J. RUOSTEKOSKI, *Energetically stable particlelike Skyrmions in a trapped Bose-Einstein condensate.* Phys. Rev. Lett., **91** (2003) 010403.

[767] J. RUOSTEKOSKI, *Topological phase preparation in a pair of atomic Bose-Einstein condensates.* Phys. Rev. A, **61** (2000) 041603.

[768] L.S. LESLIE, A. HANSEN, K.C. WRIGHT, B.M. DEUTSCH, AND N.P. BIGELOW, *Creation and detection of Skyrmions in a Bose-Einstein condensate.* Phys. Rev. Lett., **103** (2009) 250401.

[769] J.-Y. CHOI, W. J. KWON, AND Y.-I. SHIN, *Observation of topologically stable 2D Skyrmions in an antiferromagnetic spinor Bose-Einstein condensate.* Phys. Rev. Lett., **108** (2012) 035301.

[770] J.-Y. CHOI, S. KANG, S.W. SEO, W.J. KWON, AND Y.-I. SHIN, *Observation of a geometric Hall effect in a spinor Bose-Einstein condensate with a Skyrmion spin texture.* Phys. Rev. Lett., **111** (2013) 245301.

[771] P.G. KEVREKIDIS, R. CARRETERO-GONZÁLEZ, D.J. FRANTZESKAKIS, B.A. MALOMED, AND F.K. DIAKONOS, *Skyrmion-like states in two- and three-dimensional dynamical lattices.* Phys. Rev. E, **75** (2007) 026603.

[772] E.V. GOLDSTEIN AND P. MEYSTRE, *Quasiparticle instabilities in multicomponent atomic condensates.* Phys. Rev. A, **55** (1997) 2935–2940.

[773] T. BUSCH, J.I. CIRAC, V.M. PÉREZ-GARCÍA, AND P. ZOLLER, *Stability and collective excitations of a two-component Bose-Einstein condensed gas: A moment approach.* Phys. Rev. A, **56** (1997) 2978–2983

[774] R. GRAHAM AND D. WALLS, *Collective excitations of trapped binary mixtures of Bose-Einstein condensed gases.* Phys. Rev. A, **57** (1998) 484-487.

[775] B.D. ESRY AND C.H. GREENE, *Low-lying excitations of double Bose-Einstein condensates.* Phys. Rev. A, **57** (1998) 1265–1271.

[776] M. EGOROV, B. OPANCHUK, P. DRUMMOND, B.V. HALL, P. HANNAFORD, AND A.I. SIDOROV, *Measurement of s-wave scattering lengths in a two-component Bose-Einstein condensate*, Phys. Rev. A **87** (2013) 053614.

[777] R.M. CAPLAN, *NLSEmagic: Nonlinear Schrödinger equation multi-dimensional Matlab-based GPU-accelerated integrators using compact high-order schemes.* Comput. Phys. Commun., **184** (2013) 1250–1271.

[778] W.F. Spotz and G.F. Carey, *Iterative and parallel performance of high-order compact systems.* SIAM J. Sci. Comput., **19** (1998) 1–14.

[779] J. Zhang, *An explicit fourth-order compact finite difference scheme for three-dimensional convection-diffusion equation.* Commun. Numer. Methods Eng., **14** (1998) 209–218.

[780] W.F. Spotz and G.F. Carey, *A high-order compact formulation for the 3D Poisson equation.* Numer. Methods Partial Diff. Eq., **12** (1996) 235–243.

[781] L. Ge and J. Zhang, *Symbolic computation of high-order compact difference schemes for three dimensional linear elliptic partial differential equations with variable coefficients.* J. Comput. Appl. Math., **143** (2002) 9–27.

[782] W. Liao, *An implicit fourth-order compact finite difference scheme for one-dimensional Burgers' equation.* Appl. Math. Comput., **206** (2008) 755–764.

[783] J.C. Kalita, P. Chhabra, and S. Kumar, *A semi-discrete higher order compact scheme for the unsteady two-dimensional Shrödinger equation.* J. Comput. Appl. Math., **197** (2006) 141–149.

[784] A. Mohebbi and M. Dehghan, *The use of compact boundary value method for the solution of two-dimensional Schrödinger equation.* J. Comput. Appl. Math., **225** (2009) 124–134.

[785] M. Ciment and S.H. Leventhal, *Higher order compact implicit schemes for the wave equation.* Math. Comput., **29** (1975) 985–994.

[786] W.F. Spotz and G.F. Carey, *Extension of high-order compact schemes to time-dependent problems.* Numer. Methods Partial Diff. Eq., **17** (2001) 657–672.

[787] S. Xie, G. Li, and S. Yi, *Compact finite-difference schemes with high accuracy for one-dimensional nonlinear Schrödinger equation.* Comput. Methods Appl. Math. Eng., **198** (2009) 1052–1060.

[788] M. Dehghan and A. Taleei, *A compact split-step finite difference method for solving the nonlinear Schrödinger equations with constant and variable coefficients.* Comput. Phys. Commun., **181** (2010) 43–51.

[789] R.M. Caplan and R. Carretero-González, *A two-step high order compact scheme for the Laplacian operator and its implementation in a fully explicit method for integrating the nonlinear Schrödinger equation.* J. Comput. Appl. Math., **251** (2013) 33–46.

[790] G.H. Golub and J.M. Ortega, *Scientific Computing and Differential Equations. An Introduction to Numerical Methods,* 2nd ed. Academic Press, San Diego, California, 1992.

[791] W.F. Spotz, *High-order Compact Finite Difference Schemes for Computational Mechanics.* PhD thesis, University of Texas at Austin, 1995.

[792] D.J. Fyfe, *Economical evaluation of Runge-Kutta formulae.* Math. Comput., **20** (1966) 392–398.

[793] M.H. Carpenter and C.A. Kennedy, *Fourth-Order 2N-Storage Runge-Kutta Schemes.* NASA Technical Memorandum 109112 NASA, Hampton, VA, 1994.

[794] J.C. Strikwerda, *Finite Difference Schemes and Partial Differential Equations,* 2nd ed. SIAM, Philadelphia, PA, 2004.

[795] B. Fornberg, *A Practical Guide to Pseudospectral Methods.* Cambridge University Press, Cambridge, 1996.

[796] R.M. CAPLAN AND R. CARRETERO-GONZÁLEZ, *A modulus-squared Dirichlet boundary condition for time-dependent complex partial differential equations and its application to the nonlinear Schrödinger equation.* SIAM J. Sci. Comput., **36** (2014) A1–A19.

[797] X. ANTOINE, A. ARNOLD, C. BESSE, M. EHRHARDT, AND A. SCHÄDLE, *A review of transparent and artificial boundary conditions techniques for linear and nonlinear Schrödinger equations.* Commun. Comput. Phys., **4** (2008) 729–796.

[798] R.M. CAPLAN AND R. CARRETERO-GONZÁLEZ, *Numerical stability of explicit Runge-Kutta finite-difference schemes for the nonlinear Schrödinger equation.* Appl. Numer. Math., **71** (2013) 24–40.

[799] J.D. LAMBERT, *Numerical Methods for Ordinary Differential Systems.* Wiley, New York, 1991.

[800] M.M. CERIMELE, M.L. CHIOFALO, F. PISTELLA, S. SUCCI, AND M.P. TOSI, *Numerical solution of the Gross-Pitaevskii equation using an explicit finite-difference scheme: An application to trapped Bose-Einstein condensates.* Phys. Rev. E, **62** (2000) 1382–1389.

[801] M.S. ISMAIL, *A fourth-order explicit scheme for the coupled nonlinear Schrödinger equation.* Appl. Math. Comput., **196** (2008) 273–284.

[802] G.H. GOLUB, *Matrix Computations,* 3rd ed. Johns Hopkins University Press, Baltimore, 1996.

[803] NVIDIA, *NVIDIA's next generation CUDA compute architecture: Fermi.* White paper, http://www.nvidia.com/content/pdf/fermi_white_papers/nvidia_fermi_compute_architecture_whitepaper.pdf, 2009.

[804] S.E. KRAKIWSKY, L.E. TURNER, AND M.M. OKONIEWSKI, *Graphics processor unit (GPU) acceleration of finite-difference time-domain (FDTD) algorithm.* In *Circuits and Systems 2004, ISCAS '04, Proceedings of the 2004 International Symposium*, Vol. 5, pp. 265–268, 2004.

[805] A. BALEVIC, L. ROCKSTROH, A. TAUSENDFREUND, S. PATZELT, G. GOCH, AND S. SIMON, *Accelerating simulations of light scattering based on finite-difference time-domain method with general purpose GPUs.* In *Computational Science and Engineering 2008, CSE'08*, pp. 327–334, 2008.

[806] S. ADAMS, J. PAYNE, AND R. BOPPANA, *Finite difference time domain (FDTD) simulations using graphics processors.* HPCMP Users Group Conference, pp. 334–338, 2007.

[807] P. MICIKEVICIUS, *3D finite difference computation on GPUs using CUDA*, in *Proceedings of the 2nd Workshop on General Purpose Processing on Graphics Processing Units, GPGPU-2, New York*, pp. 79–84, 2009.

[808] D. MICHÉA AND D. KOMATITSCH, *Accelerating a three-dimensional finite-difference wave propagation code using GPU graphics cards.* Geophys. J. Int., **182** (2010) 389–402.

[809] K.A. HAWICK AND D.P. PLAYNE, *Numerical simulation of the complex Ginzburg-Landau equation on GPUs with CUDA*, CSTN-070, 2010. http://complexity.massey.ac.nz/cstn/070/cstn-070.html.

[810] *CUDA Toolkit v7.0 Programming Guide.* http://docs.nvidia.com/cuda/cuda-c-programming-guide/index.html#compute-capabilities

Index